MW00845477

INTRODUCTION TO COSMETIC FORMULATION AND TECHNOLOGY

GABRIELLA BAKI, Ph.D. AND KENNETH S. ALEXANDER, Ph.D.

The University of Toledo, College of Pharmacy and Pharmaceutical Sciences

WILEY

Copyright © 2015 by John Wiley & Sons, Inc. All rights reserved

Published by John Wiley & Sons, Inc., Hoboken, New Jersey
Published simultaneously in Canada

No part of this publication may be reproduced, stored in a retrieval system, or transmitted in any form or by any means, electronic, mechanical, photocopying, recording, scanning, or otherwise, except as permitted under Section 107 or 108 of the 1976 United States Copyright Act, without either the prior written permission of the Publisher, or authorization through payment of the appropriate per-copy fee to the Copyright Clearance Center, Inc., 222 Rosewood Drive, Danvers, MA 01923, (978) 750-8400, fax (978) 750-4470, or on the web at www.copyright.com. Requests to the Publisher for permission should be addressed to the Permissions Department, John Wiley & Sons, Inc., 111 River Street, Hoboken, NJ 07030, (201) 748-6011, fax (201) 748-6008, or online at http://www.wiley.com/go/permissions.

Limit of Liability/Disclaimer of Warranty: While the publisher and author have used their best efforts in preparing this book, they make no representations or warranties with respect to the accuracy or completeness of the contents of this book and specifically disclaim any implied warranties of merchantability or fitness for a particular purpose. No warranty may be created or extended by sales representatives or written sales materials. The advice and strategies contained herein may not be suitable for your situation. You should consult with a professional where appropriate. Neither the publisher nor author shall be liable for any loss of profit or any other commercial damages, including but not limited to special, incidental, consequential, or other damages.

For general information on our other products and services or for technical support, please contact our Customer Care Department within the United States at (800) 762-2974, outside the United States at (317) 572-3993 or fax (317) 572-4002.

Wiley also publishes its books in a variety of electronic formats. Some content that appears in print may not be available in electronic formats. For more information about Wiley products, visit our web site at www.wiley.com.

Library of Congress Cataloging-in-Publication Data:

Baki, Gabriella.
Introduction to cosmetic formulation and technology / Gabriella Baki, Ph.D., and Kenneth S. Alexander, Ph.D.
pages cm
Includes bibliographical references and index.
ISBN 978-1-118-76378-0 (cloth)
1. Cosmetics. 2. Cosmetics industry. 3. Toilet preparations. I. Alexander, Kenneth S., 1942- II. Title.
TP983.B27 2015
646.7′2–dc23

2014041090

Cover image courtesy of Gabe Balazs, President and Founder of Gabe Balazs Media, LLC

Typeset in 10/12 pt TimesLTStd by Laserwords Private Limited, Chennai, India

Printed in the United States of America
SKY10020862_082920

1 2015

CONTENTS

PREFACE

Cosmetics and personal care products have played an essential role in our lives for thousands of years. Today, an increasing number of products are aimed at special target audiences and gain importance in groups that have not received as much attention in the past. Innovation, creative thinking, and problem solving are essential skills needed in the cosmetic industry. However, the most important skill of all is a strong foundation and basic understanding of the main principles. Cosmetic science is a delicate blend of a variety of knowledge, including chemistry, biology, formulation science, pharmacology, marketing, and law. Until now, these knowledge bases were presented in separate books written for specific professional audiences. Non-professionals were often left with no options to find a single, comprehensive source containing all the information they need.

This book seeks to bridge the gap between basic science books and the highly advanced published literature and is intended to serve as a comprehensive reference for the field of cosmetic science. It covers all the major aspects of cosmetic formulation and technology at both fundamental and practical levels. In this regard, this book aims at broadening the readers' knowledge and introduces them to the intriguing world of cosmetics and personal care products. It is designed to provide a basic understanding of cosmetics and personal care products, including their regulation, main characteristics, ingredients, formulation, testing, and packaging, with the primary focus on the products available in the United States.

This book is divided into seven chapters. The first two chapters are concerned with general concepts and basic definitions and include an overview of the current legislation covering cosmetics and drugs in the United States. The next five chapters discuss the various types of color cosmetics and personal care products. The most beneficial way to discuss products used in our everyday lives is by application surface and function. This book is organized in such a manner. Chapter 3 focuses on the various types of skin care products, including skin cleansing products, skin moisturizing products, products for special skin concerns, such as aging and acne, sunscreens, as well deodorants and antiperspirants. Chapter 4 relates to makeup products for the lips, eyes, face, and nails. Chapter 5 discusses various products applied to the hair, including hair cleansing products and conditioners, hair styling products, and hair coloring products. Chapter 6 provides an overview of dental and oral care products, including toothpaste and mouthwash. Finally, Chapter 7 discusses smaller, but significant, product categories, including hair removal products, sunless tanners, baby products, and feminine hygiene products.

Each of these five chapters focusing on various products follows the same structure and includes an overview of the structure and function of the relevant application surfaces; basic principles relevant to the chapter; types and definition of various products related to the chapter; history of product use; how the products may affect the application surfaces; required qualities and characteristics of products and consumer needs; an overview of the main ingredients, characteristics, and formulation of the products; typical quality problems related to the formulation and/or use of products; product testing and, finally, packaging.

Each chapter of this textbook features a unique structure to enhance the readers' experience and facilitate the learning process. At the beginning of each section is a list of learning objectives telling the readers what important information they will be expected to know after reading and studying the chapter. The terms they will need to know in each section are also provided at the beginning of the section in a box. Throughout the text, brief "Did you know?" paragraphs draw attention to interesting information that will enhance the understanding of the material in the text. FYI (For Your Information) boxes direct readers to a particular website or other resource for further information. Each section ends with questions designed to test the readers' understanding of the section's information. The answers to these review questions can be found at the end of the book. All key terms and their definitions are included in a glossary of terms at the end of each section.

The wide variety of topics found in this book should benefit a large audience interested in cosmetic science, including undergraduate students, graduate students, researchers, students in beauty schools, and those concerned with regulation, sales, and marketing of cosmetics and personal care products. The up-to-date and comprehensive content of this book may also be of interest to the non-specialist reader. Finally, it is our hope that this book would serve as a standard textbook for cosmetic science education.

ACKNOWLEDGMENTS

The authors would like to express their gratitude to graphic designer Gabe Balazs from Gabe Balazs Media, LLC, for his excellence in preparing the graphic artwork. We are also grateful to our publisher, John Wiley & Sons, Inc., for all the guidance and assistance they provided for this book. We would also like to thank the University of Toledo for the continued support. Finally, we wish to thank our family members for their encouragement and unstinted support during the preparation of this book.

BIOGRAPHY OF GABRIELLA BAKI

Dr. Gabriella Baki is an Assistant Professor of Pharmaceutics at the University of Toledo College of Pharmacy and Pharmaceutical Sciences (UT CPPS). She serves as an instructor for various lectures and laboratories in UT CPPS's new and unique undergraduate program, the BS in Pharmaceutical Sciences Cosmetic Science and Formulation Design. Dr. Baki is a pharmacist, graduated in 2008 from the University of Szeged, Hungary. She also has a Ph.D. in Pharmaceutics from the same university.

Dr. Baki has been a faculty member at UT CPPS since January 2012. Her research area focuses on cosmetics and topical pharmaceutical products as well as solid oral dosage forms with the ultimate goal of increasing consumer experience, patient adherence, and product performance. Dr. Baki has made several technical presentations and over a dozen poster presentations. She has also written more than a dozen publications and contributed to several book chapters.

BIOGRAPHY OF KENNETH S. ALEXANDER

Dr. Alexander is a native Philadelphian who graduated from the Philadelphia College of Pharmacy & Science and the University of Rhode Island. He is currently a Professor of Pharmacy at The University of Toledo and has been there since 1972. Dr. Alexander developed the Industrial Pharmacy graduate program at this institution and is currently its Coordinator. He has pioneered a number of new theories for the hindered settling of suspensions, sublimation/evaporation of liquids and solids, stability of drugs with their excipients in solid dosage forms as well as the effect of grinding and compaction on the stability of solids. In 2003, he developed the BSPS Pharmaceutics Major and in 2011, the Cosmetic Science Major, and he is currently the Coordinator for both undergraduate programs.

1

GENERAL CONCEPTS

SECTION 1: BASIC DEFINITIONS

LEARNING OBJECTIVES

Upon completion of this section, the reader will be able to

1. **define the following terms:**

Claim	**Personal care product**	**Cosmetic science**	**Drugs**
Cosmeceutical	**FD&C Act**	**Dietary supplement**	**Soap**
Cruelty-free	**USDA**	**Cosmetics**	**Organic**
Intended use	**Hypoallergenic**	**FDA**	**Toiletries**

2. **discuss whether cosmetic science is really a science or not;**
3. **discuss what knowledge and background education are necessary if one wants to work in the cosmetic industry;**
4. **differentiate between a cosmetic and a drug;**
5. **explain what the main factor is that legally differentiates cosmetics and drugs in the United States;**

Introduction to Cosmetic Formulation and Technology, First Edition. Gabriella Baki and Kenneth S. Alexander.
© 2015 John Wiley & Sons, Inc. Published 2015 by John Wiley & Sons, Inc.

6. explain how certain products can be both drugs and cosmetics;
7. explain how a cosmetic product's intended use is established in the United States;
8. explain why cosmeceuticals represent a gray zone between cosmetics and over-the-counter (OTC) drug–cosmetic products in the United States;
9. discuss if dietary supplements are cosmetics or not in the United States;
10. explain what the following terms mean and how their use is regulated in the United States: organic, hypoallergenic, cruelty-free, preservative-free, dermatologist recommended, clinically proven, patented formula, and pH balanced;
11. explain how soaps are regulated in the United States;

KEY CONCEPTS

1. Cosmetic science is a real science, and it is a multidisciplinary field since it includes basic knowledge and a wide range of information from a number of different scientific fields.
2. Cosmetics are articles intended to be rubbed, poured, sprinkled, or sprayed on, introduced into, or otherwise applied to the human body or any part thereof for cleansing, beautifying, promoting attractiveness, or altering the appearance without affecting structure or function.
3. Drugs are articles intended for use in the diagnosis, cure, mitigation, treatment, or prevention of disease and articles (other than food) intended to affect the structure or any function of the body of man or other animals.
4. The legal difference between a cosmetic and a drug in the US is determined by a product's intended use, i.e., what the product is used for.
5. In the US, certain products can be cosmetics and drugs at the same time since they meet the definitions of both cosmetics and drugs.
6. A cosmetic product's intended use can be established in a number of ways, including claims, consumer perception of the product, and the history of an ingredient.
7. Many claims commonly used today for cosmetic products are not recognized or used by the Food and Drug Administration (FDA) in any sense and have only limited scientific evidence behind them. However, the use of these words is not prohibited, which is why they are used.
8. True soaps are regulated by the Consumer Product Safety Commission and not by the FDA.

What is Cosmetic Science?

Generally speaking, science is the organized body of knowledge that is derived from a systematic observation of natural events and conditions and that can be verified or tested by further investigations.[1] Examples for science include chemistry, biology, and physics. Is cosmetic science considered a real science? When trying to answer this question, we tend to think of information we saw, heard, or read in the news or on TV. In addition, public opinion and religion may also influence the perception of what constitutes a real science. Today, there are a number of doubtful consumers and even professionals who are wondering what cosmetic science is all about. Depending on the source of the information, cosmetic science has been identified as follows:

- **Commercial science** that tries to find reasons for selling a product.
- **Comparative science** based on the fact that many manufacturers compare their own products to other manufacturers' products and try to convince consumers why to buy their products instead of other companies' products.
- **Traditional science**, such as chemistry or physics, where there are hypotheses and scientists try to justify or deny them by performing a number of tests and reactions.
- **Borderline science** as it is a transition among a number of different scientific fields, including pharmacy, chemistry, dermatology, and marketing, among others.
- Some consumers believe that it is really **not a science.**[2]

Let us review what knowledge and background education is needed if someone wants to work in the cosmetic industry as a scientist.

- Basic knowledge of **anatomy and physiology** is needed to understand the structure and function of the skin, hair, lips, teeth, and so on, to where products are usually applied.
- To be able to formulate effective, stable, and safe products that have appealing aesthetics, appropriate performance, and compatibility with the application surfaces, it is necessary to understand the basic physical, chemical, and physicochemical properties of the raw ingredients that are typically used. Therefore, a **chemical** background, including organic, inorganic, colloid, and polymer chemistry, is also required.
- To be able to choose appropriate ingredients, the basic properties and therapeutic effects of the raw materials on the target surfaces have to be known. Therefore, a basic **pharmacological** education is also inevitable.
- Future formulators also need to be aware of and understand the different dosage forms from which they can choose to incorporate the ingredients. Additionally, they have to know the various manufacturing techniques that are used to produce the dosage forms. Therefore, they need to be taught **formulation technology**.
- It goes without saying that basic knowledge and understanding of the current **guidelines, rules, and regulations** relevant for cosmetics and OTC (over-the-counter) drug–cosmetic products are essential. As part of the regulations, one needs to be aware of and understand the rules that regulate **labeling and packaging** of a final cosmetic product.
- Education in **analytical sciences** as well as **microbiology** is also important in order to understand the different types of tests and testing methods that are

performed for cosmetics and OTC drug–cosmetic products to evaluate their performance, efficacy, safety, and stability.

- Additionally, understanding what consumers expect from products and what their needs are is also required in order to be able to target those needs and satisfy consumers (**consumer needs**).
- Finally, basic understanding of **marketing** and **business** is essential to understand how a business, such as the cosmetic industry, works.

Based on all of the above, we can conclude that ❶ **cosmetic science is a real science, and it is a multidisciplinary field since it includes basic knowledge and a wide range of information from a number of different scientific fields.** It is involved with developing, formulating, and producing cosmetics and personal care products. See a summary of this information in Figure 1.1.

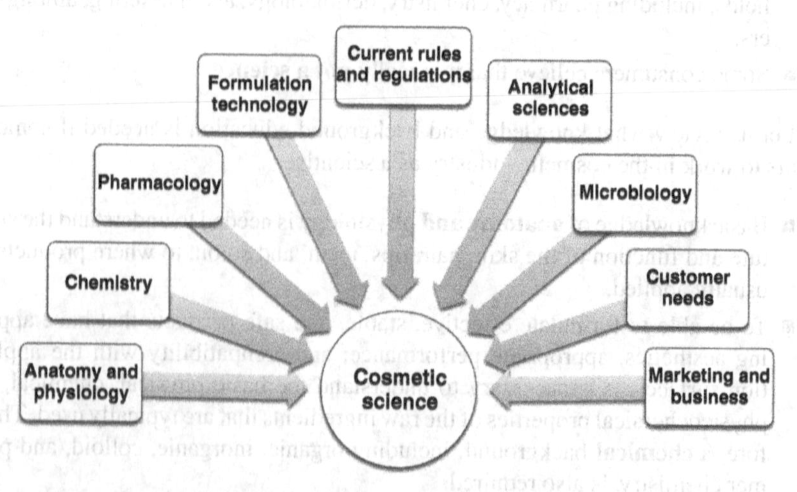

Figure 1.1 Scientific areas contributing to cosmetic science.

Basic Definitions

Is It a Drug or a Cosmetic? Today, we can find cosmetics and personal care products almost everywhere, including grocery stores, pharmacies, beauty salons, or even gas stations. But, what are cosmetics? Is there a definition for them? When people are asked the question "What do you think cosmetics are?", the answer is often "The hundreds of makeup products that my wife puts on her face," "The products that an aesthetician uses," "I guess my shaving cream," or "My antiwrinkle cream." These answers do not completely cover the spectrum of cosmetics available, and some of them are even not true. You as a health science professional have to be aware of the basic definitions and able to apply them and act in accordance with them.

United States In the United States, the Food and Drug Administration (FDA) has authority over cosmetic products, drugs, and foods. Within the FDA, the Office of Cosmetics and Colors, which is within the Center for Food Safety and Applied Nutrition (CFSAN), regulates cosmetics (see more information in Section 2 of Chapter 2).[3] The legal responsibilities and requirements are laid down in the Food, Drug and Cosmetic (FD&C) Act, which was introduced in 1938 as a revision of the Food and Drugs Act of 1906. The FD&C Act defines two main categories of products, namely cosmetics and drugs.

- **Cosmetics** The FD&C Act defines ● **cosmetics** by their intended use as follows:[4] "Articles intended to be rubbed, poured, sprinkled, or sprayed on, introduced into, or otherwise applied to the human body or any part thereof for cleansing, beautifying, promoting attractiveness, or altering the appearance without affecting structure or function." Among the products included in this definition are skin moisturizers, lipsticks, nail polishes, eye and facial makeup products, shampoos, permanent waves, hair coloring products, and deodorants as well as any material intended for use as a component of a cosmetic product.
- **Drugs** Drugs are also regulated under the FD&C Act by a different office, namely the Office of Nonprescription Drugs, within the Center for Drug Evaluation and Research (CDER) (see more information in Section 2 of Chapter 2). According to the FD&C Act, ● **drugs** are[5]: "Articles intended for use in the diagnosis, cure, mitigation, treatment, or prevention of disease" and "articles (other than food) intended to affect the structure or any function of the body of man or other animals."

As regulated by the FDA, there are two categories of drugs: OTC drugs and prescription-only drugs. OTC drugs can be purchased without a prescription, as their name states. These products are considered safe and effective for use by the general public without a prescriber's authorization. Examples include products we can buy for headache, sore throat, and allergy. On the other hand, prescription medications require a prescription written by a licensed doctor before they can be purchased by patients. These drugs are not safe for self-treatment for several reasons; this is why they cannot be used without a doctor's supervision. Examples include antibiotics, contraceptives, and drugs taken for high blood pressure.

The two definitions (i.e., that of cosmetics and drugs) provided by the FD&C Act legally determine whether a formulation is a drug or a cosmetic. It is important to note that ● **the legal difference between a cosmetic and a drug in the US is determined by a product's intended use,[6] i.e., what the product is used for,** and not the ingredients in the product. Therefore, if the intended use relates to the prevention and treatment of a disease, the substance is a drug; if its intended use is described in advertisements as promoting attractiveness, the product is a cosmetic.

You may wonder why you have to know the definition of both cosmetics and drugs if you want to work for the cosmetic industry. At this point, you have to understand that, ● **in the US, certain products can be cosmetics and drugs at the same time if they meet the definitions of both cosmetics and drugs.** These products will be referred to as OTC drug–cosmetic products in this textbook. As described previously, the main factor that legally differentiates a drug from a cosmetic in the US is the intended use of the product. Double function may happen when a product has two intended uses. For example, a shampoo is a cosmetic because its intended use is to cleanse the hair. An antidandruff ingredient is considered a drug because it is intended to be used to treat dandruff. Consequently, an antidandruff shampoo is both a cosmetic and a drug (see Figure 1.2).

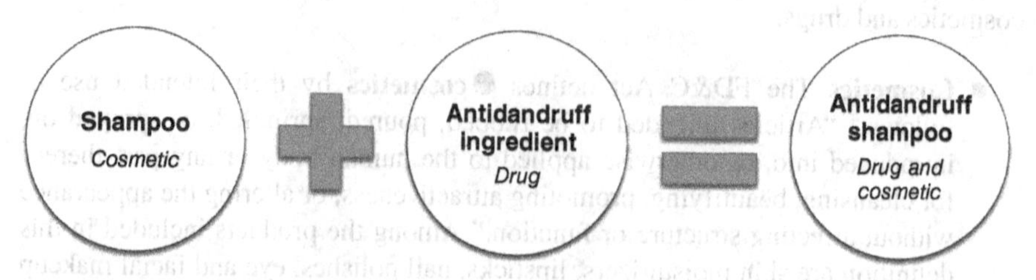

Figure 1.2 An antidandruff can be considered an OTC drug–cosmetic product.

Additional examples for products that have both drug and cosmetic functions include the following:

- **Toothpaste** that contains fluoride to prevent tooth decay. Its cosmetic function is to clean and refresh the teeth and oral cavity. The presence of fluoride (e.g., sodium fluoride), however, makes this product to be considered an OTC drug product, which, in addition to cleaning and freshening, also prevents a disease, i.e., tooth decay.
- **Deodorants** that not only mask bad body odor but also alter the normal process of perspiration, i.e., **antiperspirants**.
- **Mouthwash** that contains ingredients to prevent and/or treat gingivitis (i.e., inflammation of the gums).
- **Facial foundations** that also contain sunscreens to protect the skin from the harmful radiation of the sun.
- **Facial cleansers** that contain antiacne active ingredients to prevent and/or treat acne vulgaris.
- **Hand soaps** that contain antibacterial agents to kill germs.[7]

It has to be emphasized that neither the term "OTC drug-cosmetic product" is recognized by the FDA, nor there is an official definition for this terminology. However, these products are subject to the FDA's regulations for both drugs and cosmetics (not just for drugs as allergy medications). Therefore, this term will be used in this textbook. The purpose of using this terminology is to make readers understand the uniqueness of this category from a regulatory perspective. Simple OTC products, such as painkiller tablets, have only one intended use and do not have any cosmetic function. However, products called OTC drug–cosmetic combination products have both drug and cosmetic functions, which makes them unique among OTC products.

In the US, most cosmetics that are also drugs are considered OTC drugs. These products are regulated by both CDER and CFSAN.[3,8] There are only a small number of products that are prescription-only drugs and still offer cosmetic benefits for the users.

DID YOU KNOW?

While the FDA only defines cosmetics and drugs, consumers and companies often use the terms "personal care product," "decorative care product," "makeup," "color cosmetic," and "toiletries." But what are these terms? The term "toiletries" is often used for products that are used to clean the body, hair, and teeth, for example, a bodywash, shampoo, and toothpaste, respectively. This term is quite often used interchangeably with personal care products. The terms "color cosmetics," "makeup," and "decorative care products" are generally used for products primarily applied by women to make themselves more attractive, for example, a lipstick, mascara, and nail polish. However, it should be kept in mind that these terms do not reflect the legal state of the products, i.e., whether they are cosmetics or drugs. Even a lipstick can be a drug if it contains sunscreens.

Definition of Cosmetics in Other Markets In addition to the US, the major markets for cosmetics are considered to be in Europe, Canada, and Japan. The definitions and regulations of cosmetics in these markets are slightly or significantly different from those in the US and also from one another. It means that the same product types may be categorized in a different way in different markets. It is especially important for companies that export their products from one country to another, since in this case, they have to meet the other countries' definitions and regulations. To provide readers with a general picture, the basic definitions in these markets are reviewed here.

European Union In the European Union (EU), cosmetics are regulated under the Cosmetic Products Regulation (EU Regulation 1223/2009), which replaced the Cosmetics Directive (76/768/EEC of 1976). The provisions of the EU Cosmetic Products Regulation aim at ensuring that consumers' health is protected and that they are well informed by monitoring the composition and labeling of products. According to the Regulation, a cosmetic is[9]: "Any substance or mixture intended to be placed in contact with the external parts of the human body (epidermis, hair system, nails, lips and external genital organs) or with the teeth and the mucous membranes of the oral cavity with a view exclusively or mainly to cleaning them, perfuming them, changing their appearance, protecting them, keeping them in good condition or correcting body odors."

The requirements and procedures for the marketing authorization of medicinal products for human use, as well as the rules for the constant supervision of products after they have been authorized, are primarily laid down in different regulations, namely Directive 2001/83/EC (which has been amended in 2004 by Directive 2004/27/EC) and Regulation (EC) No. 726/2004. A medicinal product (i.e., drug) is defined as follows:[10,11] "(a) Any substance or combination of substances presented for treating or preventing disease in human beings or animals. (b) Any substance or combination of substances which may be used in or administered to human beings or animals with a view to making a medicinal diagnosis or to restoring, correcting or modifying physiological functions by exerting a pharmacological, immunological or metabolic action."

The definition of a medicinal product covers substances that have an effect on the human body. However, as it is agreed in the EU, this definition is only relevant to ingredients that significantly affect the metabolism of the human body. Products such as toothpastes that simply help protect against certain diseases, for example, dental caries, do not qualify them as a medicinal product in the EU member states.[12] Now you understand that a cosmetic in the EU can have a mild activity and possess pharmaceutical activity. It results in the fact that sunscreens and antidandruff shampoos, for instance, are considered cosmetics in the EU; however, they are considered drugs in the US. An important difference is that in the EU, a product can be either a cosmetic or a drug but not the combination of both.

CANADA In Canada, cosmetics and drugs are regulated by Health Canada under the Food and Drugs Act and its Cosmetic Regulations.[13] Let us review how they define these product categories. Legislation in Canada identifies two main categories of products: cosmetics and drugs, including nonprescription (OTC) drugs.

According to Health Canada, a cosmetic product is:[14] "Any substance used to clean, improve or change the complexion, skin, hair, nails or teeth. Cosmetics include beauty preparations (makeup, perfume, skin cream, nail polish) and grooming aids (soap, shampoo, shaving cream, deodorant)."

The Food and Drug Act defines drugs as follows:[14] "Any substance or mixture of substances manufactured, sold or represented for use in (1) the diagnosis, treatment, mitigation or prevention of a disease, disorder, abnormal physical state, or the symptoms thereof in man or animal, (2) restoring, correcting or modifying organic functions in man or animal, or (3) disinfection in premises in which food is manufactured, prepared or kept."

According to Health Canada, a personal care product can be defined as a substance or mixture of substances that is generally recognized by the public for use in daily cleansing or grooming. Depending on the ingredients and the claims of a product (i.e., statements referring to the expected effect of products found on product labels, on the Internet, in advertisements, and in any promotional materials), a personal care product can be regulated as a cosmetic or a drug. As you can see, Canada also differentiates between drugs and cosmetics; however, unlike in the US, a product can be included only within a single category.

JAPAN In Japan, cosmetics and drugs are regulated by the Ministry of Health, Labour and Welfare (MHLW) under the Pharmaceuticals Affairs Law (PAL). The PAL was first adopted in 1943 and was amended several times until 2001. The current PAL was implemented in 2001, and it is often referred to as the "deregulation" since a number of mandatory requirements were ablated by the new regulation. The PAL defines three relevant categories of products: cosmetics, quasi-drugs, and drugs. According to the PAL of Japan, the term "cosmetic" refers to:[15] "Items (other than quasi-drugs) with mild action on the human body and which are intended to be applied to the human body by means of rubbing, sprinkling and the like for the purpose of cleaning, beautifying, adding to the attractiveness, altering the appearance, or keeping the skin or hair in good condition."

Under the same act, the term "drug" is defined as follows:[15] "(1) Items recognized in the Japanese Pharmacopoeia; (2) items (other than quasi-drugs) which are intended for use in the diagnosis, cure or prevention of disease in humans or animals, and which are not equipment or instruments (including dental materials, medical supplies and sanitary materials), and (3) items (other than quasi-drugs and cosmetics), which are intended to affect the structure or functions of the body of humans or animals, and which are not equipment or instruments."

Unlike the US, Canada, and EU, Japan has an additional category, the so-called quasi-drugs, and includes products that fall in between the two abovementioned categories. According to the PAL, the term "quasi-drug" applies to:[15] "Items that have mild action on the human body but are not intended for the uses in the diagnosis, cure or prevention of disease or to affect the structure or function of the body. The purposes of use of quasi-drugs are specified in the PAL as: prevention of nausea or other discomfort, foul breath or body odor; prevention of prickly heat, sores and the like; prevention of hair loss, to promote hair growth, or for hair removal; and eradication of or repellence of rats, flies, mosquitoes, fleas, etc. for the health of man or other animals."

A quasi-drug is defined in Japan as a product that has minimal to moderate pharmacologic activity but is restricted in use to specific indications. Products in this class include bath preparations, skin whitening products, acne products, antidandruff shampoos, fluorinated toothpaste, hair dyes, and many others. These products are considered borderline medicinal products, which are categorized differently in various markets.

Summary Now that we reviewed the four major markets' definitions of cosmetics and drugs, you understand that these definitions are not the same and do not cover the same types of products. It results in the fact that a simple product, such as a sunscreen, is defined and regulated in a different way in the US, EU, Canada, and Japan. The most important consequence of this is that it makes import quite difficult. After reading Chapter 2, which is about the rules and regulations of cosmetics and drugs in the US, you will understand that the deviations previously make big differences from a regulatory aspect.

How is a Product's Intended Use Established in the United States?

You saw that the intended use of the product has an essential role in categorizing a product in the US. But how is the intended use of a product determined? ❻ **A cosmetic product's intended use can be established in a number of ways, including claims, consumer perception of the product, and the history of an ingredient.**[6]

- **Claims** Claims reflect the expected effects of a particular product. It should be noted that certain claims may cause a product to be considered a drug, even if the product is marketed as a cosmetic. Examples for cosmetic claims include "moisturizes skin," "cleans hair," and "freshens breath," while drug claims include "reduces wrinkles," "helps prevent chapped lips," and "restructures the deepest epidermal layers." Cosmetics cannot be marketed with drug claims; therefore, products sold as cosmetics but advertised with drug claims are subject to regulatory action. In the majority of the cases, the FDA issues a warning letter to the companies making drug claims to their cosmetic products and recommends them to change the wording of their claims in order to avoid regulatory difficulties.
- **Consumer Perception** Consumer perception is established through the product's reputation. This means asking consumers to fill out questionnaires about why they buy a particular product and what they expect the product to do. It provides a basic understanding of the claims by the consumer. In addition, it reflects whether a claim that was intended to be a cosmetic claim is understood as a cosmetic claim and is not misinterpreted.
- **The History of an Ingredient's Use** The presence of a pharmacologically active ingredient in a therapeutically active concentration can make a product a drug, even in the absence of explicit drug claims. When an ingredient is known to have drug-like effects (such as sodium fluoride can prevent cavities), and it is incorporated into a product, it will automatically make the product to be considered a drug as the intended use will be the prevention and/or treatment of a disease. Although the explicitly stated intended use is the primary factor in determining the cosmetic versus drug product category, the type and amount of the ingredient(s) present in a product must be considered in determining its regulatory status, even if a product does not make explicit drug claims.

Popular Cosmetic Claims

Claims are statements found on product labels, in TV and radio ads, and in magazines, indicating the expected positive effects of a product (e.g., softens and smoothens skin, visibly tightens pores), the look the product provides (e.g., vibrant color, flawless finish), or the absence of ingredients that may cause safety concerns (e.g., paraben-free, noncomedogenic). This part reviews some of the most popular cosmetic claims in the US, such as cosmeceutical, organic, hypoallergenic, and cruelty-free. What do these terms mean? Do they have an official definition for them? Are these recognized by the FDA? The truth is that many of these terms were created by marketing people to make products sound catchy, innovative, trustworthy and appealing and raise consumer's interest. **Many claims commonly used today for cosmetic products are not recognized or used by the FDA in any sense and have only limited scientific evidence behind them. However, the use of these words is not prohibited, which is why they are used.** Since these words are used in TV commercials, in advertisements, and on product labels, it is worth discussing them.

Cosmeceuticals When the term "cosmeceutical" was introduced in the 20th century, it was used for prescription-only products that addressed appearance issues, such as acne.[16] Today, this term is mainly used for multifunctional products that can be purchased as cosmetics and that are advertised to offer additional skin benefits over simple cosmetics. The term itself sounds like the combination of the terms "cosmetics" and "pharmaceuticals" (i.e., drugs), which catches the consumers' interest.

Cosmeceuticals are generally advertised to contain bioactive ingredients that, although are not drugs, have visible and measurable short-term and long-term effects on the skin, such as improvement of fine lines. Examples for bioactive ingredients include vitamins, antioxidants, proteins, anti-inflammatory agents, and many others.

Although it is a frequently used word by skin care professionals and physicians, the term is not recognized by the FDA. The FDA states "a product can be a drug, a cosmetic, or a combination of both; therefore, the term cosmeceutical has no meaning under the law." As discussed previously, Japan has a specific category of products that are in between cosmetics and drugs (called quasi-drugs). In the US, however, there is currently no such category.

These products represent a gray zone between cosmetics and drugs (see Figure 1.3) as many of them are sold as cosmetics; however, they may have drug-like

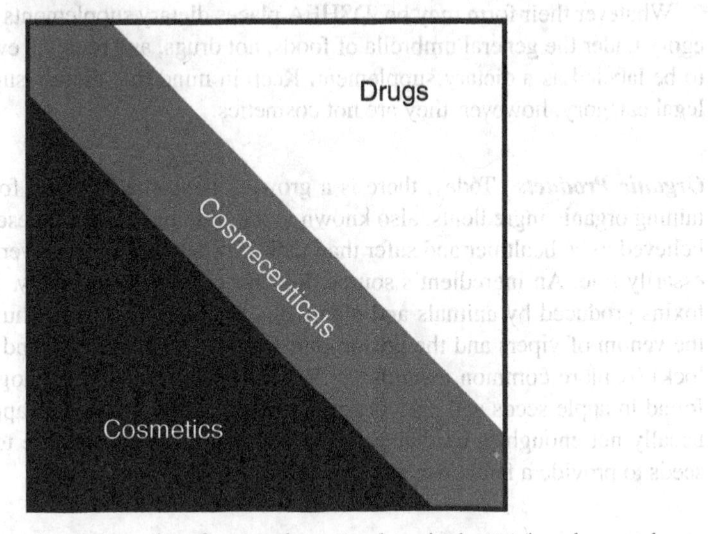

Figure 1.3 The two legal categories of personal care products in the US, i.e., drugs and cosmetics, are shown as black and white, while cosmeceuticals represent a gray zone between these two categories.

effects on the applied surfaces. For example, a number of today's anti-wrinkle formulations are advertised as cosmeceuticals and sold as cosmetics. As we do not have a long history of use for many bioactive ingredients, we lack the established pharmacological profile for them. Cosmeceuticals are currently defined by the claims made about their intended use and ingredients they contain. For example, a product that "eliminates wrinkles" is a drug, while a product that "minimizes the appearance of wrinkles" is a cosmetic, even though they both may contain the same ingredients.[16]

Nutraceuticals There is a distinct category of products, known as dietary supplements or nutraceuticals, that are often thought to be cosmetics by consumers based on the claims heard on TV and seen on the Internet and printed media (i.e., "beauty from the inside out" and "beauty from within"). These products often claim to make the hair, skin, and nails look healthier, shinier, and stronger.

Although they may be believed to be cosmetics, dietary supplements represent a specific category separate from foods, drugs, and cosmetics. The word "nutraceutical" refers to the combination of natural ingredients and pharmaceuticals. Dietary supplements are also regulated by the FDA; however, they are defined in the Dietary Supplement Health and Education Act (DSHEA) of 1994; it is not the FD&C Act that regulates cosmetics.[17] These products contain dietary ingredients such as vitamins, minerals, herbs or other botanicals, amino acids, and substances such as enzymes, organ tissues, glandulars, and metabolites. Dietary supplements can also be extracts or concentrates and may be found in many forms such as tablets, capsules, softgels, gelcaps, liquids, or powders. According to the FDA, a dietary supplement is: "A product taken by mouth that contains a dietary ingredient intended to supplement the diet."

Whatever their form may be, DSHEA places dietary supplements in a special category under the general umbrella of foods, not drugs, and requires every supplement to be labeled as a dietary supplement. Keep in mind that dietary supplements are a legal category; however, they are not cosmetics.

Organic Products Today, there is a growing consumer demand for products containing organic ingredients, also known as natural, ingredients. These ingredients are believed to be healthier and safer than their synthetic pairs. However, this is not necessarily true. An ingredient's source does not determine its safety. There are many toxins produced by animals and plants that are poisonous to the human body (e.g., the venom of vipers and the poison-dart frog or the alkaloids found in poison hemlock). A more common example is the "poison" (known as cyanogenic glycoside) found in apple seeds and apricot seeds. Their amount in a single apple or apricot is usually not enough to be dangerous to humans, but it is possible to ingest enough seeds to provide a fatal dose and eventually die.

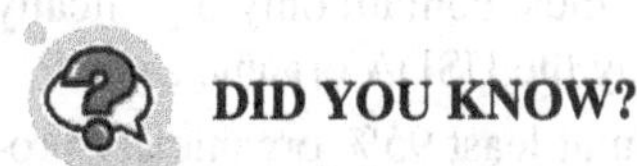

DID YOU KNOW?

Socrates, a Greek philosopher, was killed by the poison of poison hemlock in 399 BC. Its main alkaloid is a neurotoxin; ingestion in any quantity could result in respiratory collapse and death. The alkaloid causes death by blocking the neuromuscular junction. It results in an ascending muscular paralysis with eventual paralysis of the respiratory muscles, which results in death due to lack of oxygen to the heart and brain.

To many consumers, naturally sounding ingredients, such as aloe extract, chamomile extract, and lemon seed extract, mean natural ingredients and equal safety. However, it should be known that these ingredients can be synthetized in laboratories, which results in the exact same chemical structure and physical characteristics. Therefore, the name of an ingredient does not mean it has a natural origin. A general concern with natural ingredients is that they often contain a mixture of various ingredients, depending on the amount of light, humidity, temperature, and nutrients the plants received during cultivation. In the case of synthetic ingredients, their exact composition is always known and is easy to control; however, this is not true for natural products. Therefore, natural ingredients may have a higher chance of interactions with other ingredients in products. Additionally, we do not have an established safety profile for a number of natural ingredients, and we do not know whether they can cause allergic reactions, and if yes, to what extent. Moreover, the actual prevalence of adverse effects when using organic ingredients is often unrecognized or underreported.[18] Therefore, consumers should be careful with natural ingredients as adverse effects, such as skin irritation, sensitizations, phototoxicity, and allergy, have been reported.[19,20]

The term "organic" is not defined in any of the FDA's laws or regulations. However, the term is regulated by the Agricultural Marketing Service of the US Department of Agriculture (USDA) as it applies to agricultural products through its National Organic Program (NOP) regulation.[21] In addition to agricultural products (organic foods and beverages), the USDA also certifies cosmetics and personal care products if they contain or are made of agricultural ingredients and can meet the USDA/NOP organic production, handling, processing, and labeling standards. The USDA, however, has not created specific organic standards for formulating and labeling personal care products that contain organic ingredients; the standards used are the same for foods, beverages, and personal care products. The USDA has four categories of ingredients/products based on the amount of organic ingredients in a product and other factors. The categories are the following: 100% organic, organic, made with more than 70% organic ingredients, and made with less than 70% organic ingredients.

- Products in order to be labeled "100% organic" must contain only organically produced ingredients. These products can display the USDA organic seal.
- Products in the "organic" category must contain at least 95% organically produced ingredients. The regulation has requirements for the remaining 5% as well. These products can also display the USDA organic seal.
- Products in the "made with more than 70% organic ingredients" contain at least 70% organic ingredients. They can be labeled "made with organic ingredients." These products can list up to three of the organic ingredients on the principal display panel (see more information on this type of panel in Section 1 of Chapter 2). The products may not display the USDA organic seal.
- Products in the last category cannot use the term "organic" anywhere on the principal display panel. However, they may identify specific ingredients that are USDA certified on the information panel (see more information on this type of panel in Section 1 of Chapter 2). These products may not display the USDA organic seal.[21]

In addition to the USDA, cosmetics and personal care products may be certified by other certification programs, including the ANSI 305 established by the National Sanitation Foundation (NSF) International[22] and the OASIS (Organic and Sustainable Industry Standards) program established by the cosmetic industry.[23]

DID YOU KNOW?

Cosmetic products labeled with organic claims must comply with both USDA regulations (or other certifier's regulations) for the organic claim and FDA regulations for labeling and safety requirements for cosmetics.

When purchasing organic cosmetics and personal care products, consumers should be looking for the certifying agency's name and/or logo as well as the ingredients being identified as organic. As discussed previously, the name of an ingredient does not refer to its source.

DID YOU KNOW?

The NOP is the federal regulatory framework governing organic food. It is administered by USDA. The NOP covers in detail all aspects of food production, processing, delivery, and retail sale.[24]

Hypoallergenic Products Hypoallergenic cosmetics are products claimed to produce fewer allergic reactions than other non-hypoallergenic cosmetic products. Consumers with hypersensitive skin and even those with normal skin may be led to believe that these products will be gentler to their skin than non-hypoallergenic cosmetics. However, it should be noted that there is no federal standard or definition for the use of the term "hypoallergenic." The term means whatever the particular company or consumers want it to mean. Manufacturers of hypoallergenic cosmetics are not required to submit data and test results to the FDA to substantiate their hypoallergenicity claims.[25] The term usually refers to products that do not contain ingredients known to cause allergic reactions, such as fragrances. However, as the use of the term is not regulated, it is recommended that consumers with sensitive skin check the list of ingredients on cosmetic labels and see whether there are any ingredients in the product that may cause problems to them.

Cruelty-Free Products "Cruelty-free" or "Not tested on animals" claims can often be found on labels or advertisements. Animal testing is a hot topic all over the world. Various animal protection organizations have protested against testing cosmetic products on animals, and in certain markets, testing has already been prohibited. In the EU, testing of finished products and ingredients on animals has already been prohibited (it is referred to as the testing ban). In addition, marketing of finished cosmetics or cosmetic products that have been tested on animals or that contain ingredients that have been tested on animals is being prohibited after March 11, 2013 (it is referred to as the marketing ban).[26]

In the US, the FD&C Act does not specifically require the use of animals in testing cosmetics for safety. However, the agency advises cosmetic manufacturers to employ whatever testing is appropriate and effective for substantiating the safety of their products. Therefore, it remains the responsibility of the manufacturer to substantiate the safety of both ingredients and finished cosmetic products prior to marketing. It means that animal testing may be used to establish product safety. Alternative methods to replace animal experiments, such as *ex vivo* studies (i.e., studies using tissues from an organism in an external artificial environment), have been developed in the past decade. However, they are still not accepted officially by the regulatory agencies in the US due to various reasons. One of the biggest challenges is to try mimicking the complexity of the human tissues as they act together in an artificial environment.

With respect to the "cruelty-free" and "not tested on animal" claims, there is no legal definition for them. Therefore, companies can use these phrases as they like. An important note is that, even if they use these terms for products that were not tested on animals, it does not mean that the raw materials were not tested on animals years ago when they were first introduced. The FDA says that "a cosmetic manufacturer might only use those raw materials and base their cruelty-free claims on the fact that the materials or products are not currently tested on animals."[27]

Preservative-Free Products Preservatives protect cosmetic formulations from microbiological contamination, for example, overgrowth of molds, yeast, and bacteria in lotions (see more information in Section 2 of this chapter). As the majority of cosmetic formulations contain water, protection against bacteria, molds, and yeast is essential. All products containing water should contain some types of preservative to provide appropriate beyond use date for their products and safety for consumers. Therefore, a "preservative-free" claim is questionable most of the time. We can rarely find products that have an acceptable shelf life without any preservatives. As the major problem is water, which necessitates the use of preservatives, products not containing water (the so-called anhydrous formulations) do not have to contain preservatives and can still have acceptable shelf life. Other product types that can claim to be "preservative-free" include formulations containing a higher percentage of ethanol, which is widely known to have an antimicrobial activity. In addition, certain products have a specific pH value that does not favor the growth of microorganisms. There are also special types of packaging materials, such as airtight packaging, which ensures the absence of organisms.

DID YOU KNOW?

Water-based cosmetic products provide a perfect environment for microbial growth, and the products' additional components can serve as nutrients for these microorganisms. It should be kept in mind that a contaminated product (which may show no visible signs of contamination) is much more dangerous for users than preservatives.

It should be noted that one of the most widely and most frequently used preservatives are called the parabens. These ingredients are very effective even in very low concentration; however, they can cause allergic reactions in sensitive consumers. Additionally, concerns arose with regard to the safe use of parabens (see more information in Section 2 of Chapter 3). They were linked to breast cancer and endocrine disruption.[28,29] Although no study has confirmed the potential risks of using parabens on human health, the claims that they can cause breast cancer and endocrine disruption have been widely spread. As many consumers are afraid of using products containing parabens, many formulators have substituted parabens with other types of preservatives to ensure product longevity. They usually also claim their products to be "paraben-free." If you see this claim on a product, it refers to the fact that the product does not contain parabens. However, it does not mean that it does not contain any sort of preservatives.

Another frequent claim with regard to preservatives is "no added preservatives." It means that the product formulators did not add any ingredients to the formulation whose primary function would be preservation. However, there are a number of cosmetic ingredients that have a primary effect, such as skin conditioning, and have a limited antimicrobial property as well. In such cases, the ingredient added for its conditioning activity will also prevent microbial contamination in the product to a certain extent. These types of preservatives are usually called "nonpreservative preservatives" as their primary function is not the prevention of contamination. Their efficacy, however, may be not as good as that of parabens; therefore, formulations often have to use a higher amount of these ingredients.

"Dermatologist Recommended" Products The claim "dermatologist recommended" is commonly used on cosmetic products. It may lead consumers to believe that a medical panel of dermatologists has evaluated the product thoroughly and recommends it based on proven results. The truth is that there is no governing body in the US requiring cosmetic companies to show data on whether a dermatologist, a few, or a large number of them tested and recommend a cosmetic product. A "dermatologist recommended" claim is probably based on a product survey, and one or many dermatologists could have endorsed the product. It would also be valuable to know how many of them were neutral, disliked, or hated the product. In addition, it is also worth mentioning that even ingredients considered safe can initiate allergic reactions in sensitive patients. Therefore, unless the products were tested on a wide majority of users over an extended period of time, such a claim has only a little scientific value.

"Clinically Proven" Claims The claim "clinically proven" is scientific and powerful in the consumers' mind. Since claims made about cosmetic products must be truthful and proven (as required by the Federal Trade Commission, see more information on the FTC in Section 3 of Chapter 2), usually there is a science behind such claims, and companies perform tests to back up their claims. It should be noted that clinical testing is not required for cosmetic products that do not have drug claims. The "clinically proven" claim refers to the fact that a product was tested in a clinical environment on humans; however, the details of the clinical testing are usually not provided. Important factors to consider in the case of clinical studies include the number of participants, whether they truly represent a wider group of usual consumers, skin condition of the participants (whether they had any skin sensitivity, allergy, skin disease, etc.), the use and type of a reference product, length of the study, frequency of application, use of other products, the type of data analysis used in the study, and many other factors. Therefore, the claim "clinically proven" without being aware of the study details is still not very informative.

"Patented Formula" Claims Another commonly used claim is the "patented formula." Consumers believe that a product that has been patented must be more serious and scientific; therefore, it works better than other products. The truth is that patenting a product is often related to the technology of how the product is manufactured and not the actual effect of the product. Therefore, it does not necessarily mean that the product is more effective or has a longer performance.

"pH Balanced" Claims Companies that make "pH balanced" claims try to imply some level of superiority over products that do not make this claim. They want consumers to believe that the products will be less irritating and will work better. However, any decently formulated product is formulated in a pH range that is compatible with the skin, hair, underarms, or other application surfaces. A consumer will never notice a difference between a product that is "pH balanced" and one that is just normally formulated.

DID YOU KNOW?

In Canada, the regulatory agency for cosmetics, called Health Canada, has an entire list of acceptable and unacceptable claims for cosmetic products, which provides guidance for manufacturers. In the US, currently there is no such list.

A Special Category: Soap

Soap is a category that needs special explanation. This is because the regulatory definition of a soap is different from the way in which people commonly use the word. People usually associate the word "soap" with a cleaning aid used for washing our hands and body. In a strictly chemical sense—and this is the definition that the FDA takes into account—soap is a salt produced from fatty acids (e.g., stearic acid) and alkalis (e.g., sodium hydroxide). It is referred to as ordinary or true soap. **True soaps are regulated by the Consumer Product Safety Commission, and not by the FDA.** The reason for this is that soap was not included in the first regulation of cosmetics, which is the FD&C Act of 1938, and it has not been included since then. Based on the FDA's definition for cosmetics, a soap would be clearly a cosmetic, as it will clean our hands and body; however, the regulation excludes it from the definition of a cosmetic. In the following box, you can find the official, legal definition of a true soap.[30] According to the FDA, a true soap is: "(1) The bulk of the nonvolatile matter in the product consists of an alkali salt of fatty acids and the product's detergent properties are due to the alkali-fatty acid compounds; (2) and the product is labeled, sold, and represented solely as soap."

Today, there are only a small number of true soaps in the traditional sense on the market. Most hand and body cleansers on the market are actually synthetic surfactant-based products and go under the jurisdiction of the FDA. These surfactant cleansers are popular because they easily make suds in water and do not form deposits. Some of these synthetic surfactant products are actually marketed as soaps; however, they are not true soaps by the legal definition of the word.

With regard to the claims made by soap manufacturers, a soap can be considered a simple noncosmetic product, a cosmetic, or a drug. If no cosmetic claims are made for a soap, other than that it cleanses our hands and body, and no drug claims are made, a soap is considered a noncosmetic, nondrug product, similar to a dishwashing detergent. If a cosmetic claim is made on the label of a true soap, such as moisturizing or deodorizing, it is considered a cosmetic and regulated by the FDA. If a drug claim is made for a soap, such as antibacterial, antiperspirant, or anti-acne, the product is considered an OTC drug–cosmetic product and regulated by the FDA.

GLOSSARY OF TERMS FOR SECTION 1

Claim: A statement referring to the expected effect of a product found on product labels, on the Internet, in advertisements, and in any promotional materials.

Cosmeceutical: This term is used for multifunctional products that can be purchased as cosmetics and that are advertised to offer additional skin benefits over simple cosmetics.

Cosmetic science: An interdisciplinary science involved with developing, formulating, and producing cosmetics and personal care products.

Cosmetics: The FD&C Act defines *cosmetics* by their intended use as follows"Articles intended to be rubbed, poured, sprinkled, or sprayed on, introduced into, or otherwise applied to the human body or any part thereof for cleansing, beautifying, promoting attractiveness, or altering the appearance without affecting structure or function."

Cruelty-free: A term used on cosmetic labels to indicate that no animals were involved in product testing. This term does not have an official FDA definition, and its use is not regulated by the FDA.

Dietary supplement: A product that is intended to supplement the body with vitamins, minerals, herbs, and other ingredients, which may not be consumed in a sufficient amount.

Drugs: The FD&C Act defines drugs by their intended use as follows "Articles intended for use in the diagnosis, cure, mitigation, treatment, or prevention of disease" and "articles (other than food) intended to affect the structure or any function of the body of man or other animals."

FDA: Food and Drug Administration; in the United States, cosmetics and drugs are regulated by the FDA, which is an agency of the US Department of Health and Human Services.

FD&C Act: Food, Drug and Cosmetic Act; the FDA regulates cosmetics, foods, and drugs under the authority of the Food, Drug and Cosmetic Act.

Hypoallergenic: A term used on cosmetic labels to indicate that the product is unlikely to cause allergic reactions. This term does not have an official FDA definition, and its use is not regulated by the FDA.

Intended use: The purpose for what a product is used.

Organic: A term often used on cosmetic labels to indicate the presence of naturally derived ingredients. This term does not have an official FDA definition, and its use is not regulated by the FDA.

Personal care product: A term often used for products that are used to clean the body, hair, and teeth, for example, a bodywash, shampoo, and toothpaste, respectively.

Soap: In a strictly chemical sense, soap is a salt produced from fatty acids (e.g., stearic acid) and alkalis (e.g., sodium hydroxide).

Toiletries: A synonym for personal care products.

USDA: The United States Department of Agriculture, which certifies ingredients and products as organic if they meet the USDA standards and requirements.

REVIEW QUESTIONS FOR SECTION 1

Multiple Choice Questions

1. Which of the following is necessary if you want to work in the cosmetic industry?
   a) Pharmacology
   b) Chemistry
   c) Marketing
   d) All of the above
2. Which of the following regulates cosmetics in the United States?
   a) Consumer Product Safety Admission
   b) European Commission
   c) Food and Drug Administration
   d) Product manufacturers
3. In the United States, cosmetics are regulated under the ___.
   a) DSHEA Act
   b) FD&C Act
   c) IUPAC Act
   d) CDC Act
4. In the United States, OTC drug–cosmetic products are ___.
   a) Regulated as drugs
   b) Regulated as cosmetics
   c) Not regulated
   d) Regulated as dietary supplements
5. Which of the following is true for OTC drug–cosmetic products?
   a) They are subject to the FDA's regulations for both drugs and cosmetics
   b) They are cosmetics that also have drug functions
   c) They are regulated by the FDA
   d) All of the above
6. Which of the following is an example for an OTC drug–cosmetics product?
   a) A dry hair shampoo
   b) A deodorant
   c) An anticavity toothpaste
   d) A lipstick

7. Cosmeceuticals represent a "gray" zone between drugs and cosmetics in the US because ___.

a) They are considered dietary supplements but should be considered drugs
b) They are sold as drugs but are actually cosmetics
c) They are sold as cosmetics but may have drug-like effects
d) They are packaged into gray boxes

8. Which of the following is NOT true with respect to dietary supplements in the United States?

a) They are regulated as cosmetics
b) They are a legal category of products
c) They contain a dietary ingredient intended to supplement the diet
d) All of the above

9. Which of the following can certify a cosmetic product as organic in the United States?

a) FDA
b) USDA
c) Manufacturer
d) Customer

10. What is the legal difference between a drug and a cosmetic product in the United States?

a) Color
b) Intended use
c) Application surface
d) All of the above

11. Which of the following is used to determine the intended use of a product in the United States?

a) Questionnaire
b) Claims
c) Ingredients
d) All of the above

12. What happens if you add an active ingredient to a cosmetic product?

a) It will be considered a drug
b) It will be considered a cosmetic
c) I do not know, but I will not use it
d) It will be hypoallergenic

13. Which of the following is TRUE for organic cosmetics in the United States?

a) They are considered drugs
b) They are a subcategory of cosmetics, which are recognized by the FDA
c) They are always safer than cosmetics containing synthetic ingredients
d) They are cosmetics that contain naturally derived ingredients

14. What is the FDA's statement on hypoallergenic products?

a) They are less allergenic than non-hypoallergenic products
b) There is no such category
c) Every cosmetic product should be hypoallergenic
d) They should be used by hypersensitive consumers only

15. Which of the following is NOT true for "true soaps" in the United States?

a) They can only claim to clean the skin
b) They are salts of fatty acids and alkalis
c) They are regulated by the FDA
d) They are regulated by the Consumer Product Safety Commission

16. In the United States, cosmetic companies are _____ test their products on animals.

a) Not allowed to
b) Allowed to
c) Forced by the government to
d) None of the above

Fact or Fiction?

_____ a) Natural ingredients are always safer than synthetic ingredients.
_____ b) Even if a product is labeled hypoallergenic, it may contain substances that can cause allergic reactions for some people.
_____ c) Animal testing is prohibited in the United States.
_____ d) The commonly used term "cosmeceutical" is not recognized by the FDA.
_____ e) Dietary supplements are cosmetics.

You can see two products in the below. Which one is a cosmetic and which one is an OTC drug–cosmetic product? Justify your answer.

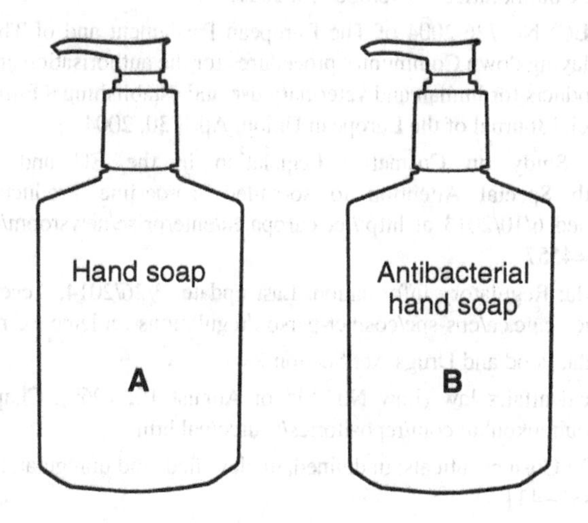

REFERENCES

1. Morris, C. G.: Academic Press Dictionary of Science and Technology, Houston: Gulf Professional Publishing, 1992.
2. Wiechers, J. W.: Is Cosmetic Science Really "Bad"? Part II: Detecting Baloney Science, *Cosmet Toiletries*, 2009. Accessed 5/15/2012 at http://www.cosmeticsandtoiletries.com/research/chemistry/54804732.html
3. FDA: CFSAN—What We Do, Last update: 4/11/2012, Accessed on 5/15/2013 at http://www.fda.gov/AboutFDA/CentersOffices/OfficeofFoods/CFSAN/WhatWeDo/default.htm
4. FD&C Act Section 201(i)
5. FD&C Act Section 201(g)(1)
6. FDA: Is It a Cosmetic, a Drug, or Both? (Or Is It Soap?), Last update: 3/30/2012, Accessed 1/17/2014 at http://www.fda.gov/cosmetics/guidancecomplianceregulatoryinformation/ucm074201.htm
7. Hutt, P. B.: Legal Distinction in USA between Cosmetic and Drug, In: Elsner, P., Maibach, H., eds: Cosmeceuticals and Active Cosmetics: Drugs versus Cosmetics, 2nd Edition, Boca Raton: CRC Press, 2005.
8. FDA: FAQs About CDER, Last update: 10/20/2010, Accessed on 5/10/2013 at http://www.fda.gov/AboutFDA/CentersOffices/OfficeofMedicalProductsandTobacco/CDER/FAQsaboutCDER/default.htm
9. Regulation (EC) No 1223/2009 of The European Parliament and of The Council of 30 November 2009 on cosmetic products, Official Journal of the European Union, December 22, 2009.

10. Directive 2001/83/EC of The European Parliament and of the Council of 6 November 2001 on the Community code relating to medicinal products for human use, Official Journal of the European Communities, November 28, 2001.
11. Regulation (EC) No 726/2004 of The European Parliament and of The Council of 31 March 2004 laying down Community procedures for the authorisation and supervision of medicinal products for human and veterinary use and establishing a European Medicines Agency, Official Journal of the European Union, April 30, 2004.
12. Comparative Study on Cosmetics Legislation in the EU and Other Principal Markets with Special Attention to so-called Borderline Products, RPA August 2004, Accessed 6/10/2013 at http://ec.europa.eu/enterprise/newsroom/cf/_getdocument.cfm?doc_id=4557
13. Health Canada: Regulatory Information, Last update: 3/26/2014, Accessed 4/2/2014 at http://www.hc-sc.gc.ca/cps-spc/cosmet-person/regulations-reglements/index-eng.php
14. Health Canada, Food and Drugs Act Section 2
15. Pharmaceutical affairs law (Law No. 145 of August 10, 1960), Chapter 1, Article 2 http://www.jouhoukoukai.com/repositories/source/pal.htm
16. Draelos, Z. D.: Cosmeceuticals: undefined, unclassified, and unregulated. *Clin Dermatol.* 2009;27(5):431–434.
17. Dietary Supplement Health and Education Act of 1994
18. Walji, R., Boon, H., Barnes, J., et al.: Consumers of natural health products: natural-born pharmacovigilantes? *BMC Complim Altern Med.* 2010;10(8):1–10.
19. Antignac, E, Nohynek, G. J., Re, T., et al.: Safety of botanical ingredients in personal care products/cosmetics. *Food Chem Toxicol.* 2011;49(2):324–341.
20. Bakkali, F., Averbeck, S., Averbeck, D., et al.: Biological effects of essential oils. *Food Chem Toxicol.* 2008;46(2):446–475.
21. CRF Title 7 Part 205
22. NSF International: Cosmetics, Accessed 4/30/2014 at http://www.nsf.org/services/by-industry/consumer-products/cosmetics
23. OASIS, Accessed 4/30/14 at http://www.oasisseal.org/
24. National Organic Program, Last update: 10/17/2012, Accessed 4/10/2014 at http://www.ams.usda.gov/AMSv1.0/ams.fetchTemplateData.do?template=TemplateC&navID=NationalOrganicProgram&leftNav=NationalOrganicProgram&page=NOPConsumers&description=Consumers&acct=nopgeninfo
25. FDA: Hypoallergenic, Last update: 3/23/2014, Accessed 4/10/2014 at http://www.fda.gov/Cosmetics/Labeling/Claims/ucm2005203.htm
26. Directive 2010/63/EU of The European Parliament and of The Council of 22 September 2010 on the protection of animals used for scientific purposes, Official Journal of the European Union, October 20, 2010.
27. FDA: Cruelty Free/Not Tested on Animals, Last update: 3/19/2014, Accessed 4/10/2014 at http://www.fda.gov/Cosmetics/Labeling/Claims/ucm2005202.htm
28. Darbre, P. D., Aljarrah, A., Miller, W. R., et al.: Concentrations of parabens in human breast tumours. *J Appl Toxicol.* 2004;24(1):5–13.
29. Routledge, E. J., Parker, J., Odum, J., et al.: Some alkyl hydroxy benzoate preservatives (parabens) are estrogenic. *Toxicol Appl Pharmacol.* 1998;153(1):12–19.
30. CFR Title 21 Part 701.20

SECTION 2: CLASSIFICATION OF COSMETICS AND OTC DRUG–COSMETIC PRODUCTS. COSMETIC INGREDIENTS AND ACTIVE INGREDIENTS USED IN COSMETICS AND OTC DRUG–COSMETIC PRODUCTS

LEARNING OBJECTIVES

Upon completion of this section, the reader will be able to

1. Define the following terms:

Target group	Dosage form	Application surface	Color additive
Active ingredient	Antioxidant	Chelating agent	Anticaries ingredient
Flavoring agent	Preservative	Moisturizer	Antidandruff ingredient
Plasticizer	Solvent	Propellant	Anti-acne ingredient
Surfactant	Astringent	Thickener	Lip protectant
pH buffer	Sunscreen	Antiperspirant	Cosmetic ingredient
Abrasive	Sweetener		

2. list various factors based on which cosmetics and OTC drug–cosmetic products can be classified;
3. list some examples for target groups of cosmetic products;
4. list some examples for application surfaces for cosmetics;
5. list some examples for functions of cosmetic products;
6. differentiate between cosmetic ingredients and active ingredients;
7. explain why cosmetics cannot contain active ingredients;
8. discuss how color additives are regulated in the United States, including premarket approval and batch certification;
9. explain the function of the following types of cosmetic ingredients and provide some examples for each group: abrasives, antioxidants, chelating agents, color additives, flavoring agents, fragrances, moisturizers, pH buffers, plasticizers, preservatives, propellants, solvents, surfactants, sweeteners, and thickeners;
10. explain where a formulator can find a list of active ingredients approved for OTC drugs;
11. explain the function of the following types of active ingredients and provide some examples for each group: antiacne ingredients, anticaries ingredients, antidandruff ingredients, antiperspirants, skin protectants, and sunscreens.

KEY CONCEPTS

1. Cosmetics and OTC drug–cosmetic products can be classified based on the target groups, dosage forms, legal status, application surfaces, function, and numerous other ways.
2. Since cosmetics do not treat any conditions or affect the structure of the skin, they do not contain active ingredients. Cosmetic products contain cosmetic ingredients only.
3. Since OTC drug–cosmetic products are used for the prevention, diagnosis, cure, or treatment of a disease, they contain active ingredients, in addition to the cosmetic ingredients. In these products, cosmetic ingredients are referred to as inactive ingredients.
4. Cosmetic ingredients are used in products to provide them with appropriate aesthetics, texture, pH, color, and smell as well as to fulfill the cosmetic claims for products.
5. Although color additives are not classified as active ingredients, they are still subject to a stricter regulation in the US.
6. Active ingredients deliver the claimed therapeutic action and have an effect on the human body, i.e., to prevent and/or treat a disease.
7. Active ingredients for cosmetic products are either listed in OTC monographs or are new ingredients.

Classification of Cosmetics and OTC Drug–Cosmetic Products

❶ **Cosmetics and OTC drug–cosmetic products can be classified** in many ways. They can be classified **based on the target groups, dosage forms, legal status, application surfaces, function, and numerous other ways**. This section provides an overview of these classification systems (see Figure 1.4).

- *Target Groups*: A target group or target audience can be defined as a specific group of customers at whom a cosmetic product or OTC drug–cosmetic product is aimed. Target groups for cosmetics can include women, men, teenagers, and babies. The groups, such as women, can be further categorized into subcategories, such as pregnant women, African American women, women with sensitive skin, young women, and aging women.
- *Dosage Forms*: A dosage form is the final physical form of a mixture of ingredients that consumers can take in their hands, purchase, and use. The various dosage forms available for cosmetics and OTC drug–cosmetic products are discussed in Section 3 of this chapter.
- *Legal Status*: As discussed in Section 1, a product can be a cosmetic, a drug, or a combination of both, depending on the intended use. As a reminder, other product categories, such as cosmeceuticals, are not recognized by the FDA.

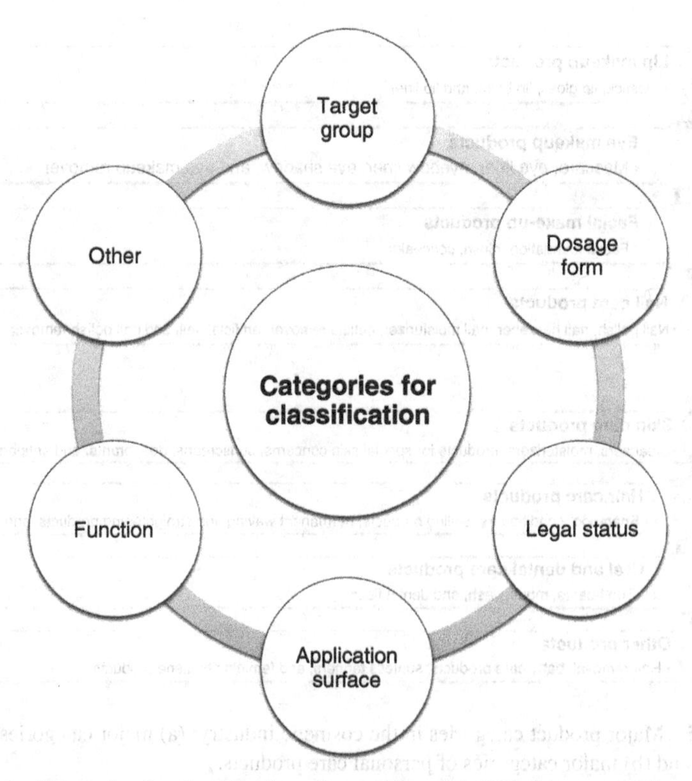

Figure 1.4 Categories for classification of cosmetics and OTC drug–cosmetic products in the United States.

- *Application Surfaces*: An application surface or area of application can be defined as a body surface to which cosmetics or OTC drug–cosmetic products are applied. These can include the skin of the face, lips, eyelids, body, and hands; hair on the scalp and body and eyelashes; teeth and oral cavity; and nails, among others.
- *Function*: Cosmetics and OTC drug–cosmetic products can have a variety of functions. Examples include cleansing, which is the function of shampoos, bodywashes, hand soaps, and eye makeup removers; moisturizing, which is the function of facial creams, hand creams, body lotions, aftershave balms, and cuticle softeners; coloring, which is the function of many color cosmetics, such as lipstick, mascara, and blush; and protection, which is the function of sunscreens, diaper rash creams, and lip balms. Additionally, there are many other functions cosmetics and OTC drug–cosmetic products can fulfill. They will be further reviewed in the relevant chapters under the various product types.
- *Others*: The cosmetic industry often classifies products into 5–6 different groups based on the combination of their function and application surface.

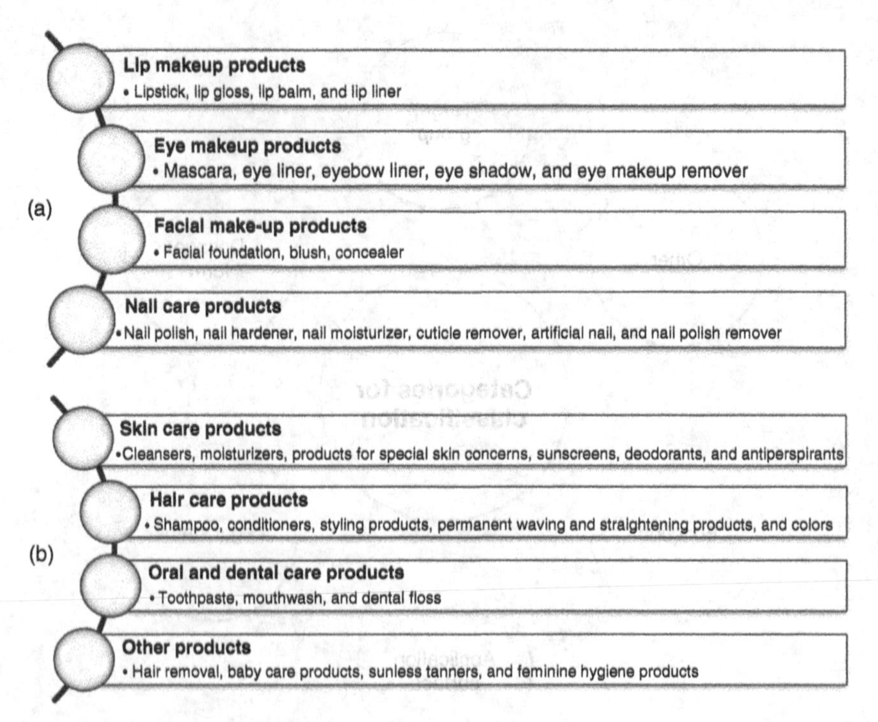

Figure 1.5 Major product categories in the cosmetic industry: (a) major categories of color cosmetics and (b) major categories of personal care products.

Popular categories include color cosmetics, skin care products, hair care products, oral care products, perfumes, and other products.

This textbook reviews color cosmetics in one chapter with the ultimate function of coloring. The various product types to be covered are shown in Figure 1.5a. Personal care products are broken down into various chapters based on the application surfaces and main functions (see Figure 1.5b).

Major Ingredient Types in Cosmetics and OTC Drug–Cosmetic Products and Their Functions

Cosmetic Ingredients ● **Since cosmetics do not treat any conditions or affect the structure of the skin, they do not contain active ingredients. Cosmetic products contain cosmetic ingredients only. ● Since OTC drug–cosmetic products are used for the prevention, diagnosis, cure, or treatment of a disease; they contain active ingredients, in addition to the cosmetic ingredients. In these products, cosmetic ingredients are referred to as inactive ingredients.**

● **Cosmetic ingredients are used in products to provide them with appropriate aesthetics, texture, pH, color, and smell as well as to fulfill the cosmetic**

claims for products. They have a variety of functions in cosmetics and OTC drug–cosmetic products, including helping mix immiscible ingredients, stabilizing formulations, moisturizing the skin, adding luster to the hair, providing color to products, helping expel the content of aerosol cans, removing stain from the teeth, just to mention a few. A commonly used and practical way to classify ingredients is based on the functions of the ingredients. As many ingredients have multiple functions, they can belong to more than one group. This section is a part that reviews the most commonly used types of cosmetic ingredients (without completeness) and their main functions, including examples for each type of ingredient.

Abrasives The term "abrasive" refers to an ingredient that is capable of polishing or cleaning a harder surface by rubbing or grinding. Abrasives are solid particles and are generally used in toothpastes and skin care products, such as face, hand, foot, and body scrubs. Although the effect these ingredients provide is the same, the types used in oral and skin care products are different.

- In skin care formulations, abrasives provide an exfoliating effect, which means that they help rub off and peel the outer layer of the skin, known as the stratum corneum (SC). The skin has its own normal peeling process known as desquamation. However, in certain cases, it can be beneficial to help the skin peel off the dead cells from its surface.
  - Examples for abrasives used in facial and body formulations include seeds of fruits, such as peach, apple, and apricot; nutshells, such as almond and walnut; grains, such as oats and wheat; synthetic components, such as polyethylene and polypropylene beads; nylon powder; synthetic waxes; and natural waxes, e.g., rice bran wax.
- Abrasives also contribute to the physical cleaning effect and stain removal of toothpastes as well as increase the gloss of the teeth. Abrasives in toothpastes are usually finely ground particles that will not hurt and wear away the enamel but are able to clean the teeth and remove discoloration to a certain extent.
  - Examples for abrasives used in toothpastes include mineral powders, such as hydrated alumina, dehydrated silica, magnesium and calcium carbonate, dicalcium phosphate, and sodium bicarbonate.

Antioxidants Antioxidants, as their name implies, provide protection against oxidative reactions. This property is generally utilized to provide stability to cosmetic formulations and can also be used to slow down skin aging caused by various oxidative mechanisms. Based on these, antioxidants can be used to fulfill two different functions in cosmetics and OTC drug–cosmetic products.

- Antioxidants can prevent undesirable chemical changes (such as decomposition, rancidity, color change, and odor formation) within a formulation triggered by oxygen in the presence of light, heat, or metal ions. Therefore, they contribute to the stability of cosmetic products. These ingredients are found in the majority of cosmetics and OTC drug–cosmetic products, especially in those containing oils, fats, butters, and waxes since these ingredients are generally more sensitive to oxidation reactions than other ingredients.

- Examples for antioxidants used as stabilizers in cosmetic products include mainly synthetic compounds, such as butylated hydroxytoluene (generally referred to as "BHT"), butylated hydroxyanisole (generally referred to as "BHA"), and propyl gallate.

■ In addition, they are also beneficial for the users' skin since they fight against free radicals in the skin. It is known that oxidative stress initiated by free radicals accelerate skin aging and contribute to the formation of lines and wrinkles, pigmentation, or even malignant processes. We have natural antioxidants and defense mechanisms in our skin to neutralize free radicals; however, applying antioxidants helps inactivate these reactive molecules and prevent symptoms related to sunlight-induced skin aging.[1] Antioxidants for this purpose are often found in skin moisturizing products.

- Examples for skin antioxidants include vitamins, such as vitamin A, vitamin C (ascorbic acid), and vitamin E (tocopherol), as well as natural extracts, isoflavones and polyphenols.

Chelating Agents Chelating agents are molecules with a specific three-dimensional structure that are able to complex with metal ions. Metallic impurities can come from many different sources, including cosmetic ingredients, water system, metallic equipment, and storage containers. If not deactivated, they can deteriorate cosmetic products by reducing clarity, compromising fragrance integrity, and causing rancidity. Chelating agents can help stabilize cosmetics and prevent their deterioration by catching (sequestering) metal ions.

■ Examples for chelating agents used in cosmetics and OTC drug–cosmetic products include ethylenediaminetetraacetic acid (generally referred to as "EDTA"), its derivatives, including disodium and tetrasodium EDTA; phosphoric and phosphonic acid derivatives; as well as citric acid and its derivatives.

DID YOU KNOW?

The word "chelate" comes from the Greek word for crab's claw. It refers to a three-dimensional, pincer-like structure of the chelating molecule and the metal ion. The chelating agent seizes the metal ion as if with a claw and keeps it from reacting with other substances.

Color Additives Color additives add color to cosmetic and OTC drug–cosmetic products, making them attractive, appealing, appetizing, and informative. ❺ **Although color additives are not classified as active ingredients, they are still subject to a stricter regulation in the US.** Therefore, they are discussed here in more detail.

DID YOU KNOW?

The terms "colorant," "coloring agent," and "color additive" can be used interchangeably; however, the official term used by the FDA is "color additive."

According to the FD&C Act, a color additive is:[2] "Any material that is a dye, pigment, or other substance […] when added or applied to a food, drug, or cosmetic or to the human body or any part thereof, is capable of imparting a color thereto." Based on the properties and composition of these ingredients, several subclasses can be distinguished.

- A **dye** is a chemical compound that is soluble in the particular solvent in which it is dispersed (e.g., oil or water). Examples include indigo and Green 3.
- A **pigment** is a component that is insoluble in the particular solvent in which it is dispersed. An example is black iron oxide.
- A **lake** is a water-insoluble pigment formed by chemically reacting dyes with a substratum, for example, aluminum, calcium, or barium. These pigments are widely used in lipsticks and nail lacquers. An example is Yellow 5 Al Lake.

As mentioned previously, color additives are subject to a strict system of approval under US law. Color additives (except for coal-tar hair dyes) must go through a premarket approval process before they may be used in cosmetics (see more information on coal-tar hair dyes in Chapter 5).[3] There is a list available for approved color additives, otherwise known as the positive list. The list specifies areas to which color additives can be applied, as well as limitations, restrictions, and comments on their use. It may happen that a color additive is approved for cosmetic use in general, but not for the eye area. Product manufacturers can only use color additives that have been approved by the FDA for intended uses stated in the regulations that pertain to them. The reason for this approval is that many color additives can cause skin irritation and allergic-type reactions; therefore, not all of them can be used in cosmetics, and there are some restrictions even for approved ingredients.

DID YOU KNOW?

The FDA has not approved any tattoo pigments for injection into the skin. This applies to all tattoo pigments, including those used for ultraviolet (UV) and glow-in-the-dark tattoos. Many pigments used in tattoo inks are industrial-grade colors suitable for printers' ink or automobile paint.[4]

In addition to premarket approval, a number of color additives must be batch certified in the US. Based on whether a color additive has to be certified or not, the FD&C Act classifies color additives into two main categories: those subject to certification (called certifiable color additives) and those exempt from certification.[5,6] In general, batch certification is required when the composition of a color additive needs to be controlled to protect public health. Some color additives may contain impurities of toxicological concern. For ingredients that are certifiable, the law requires each and every batch of these color additives to be tested and certified by the FDA before it can be used in cosmetics.

- **Colors subject to certification** include synthetic organic dyes, lakes, or pigments. These ingredients have either three-part names, including a prefix, a color, and a number (such as FD&C Yellow No. 5), or a shorter name without a prefix (such as Yellow 5).
- **Colors exempt from certification** are generally obtained from mineral, plant, or animal sources. They usually have simple or chemical names (such as bismuth citrate, caramel, dihydroxyacetone, lead acetate, and titanium dioxide).

Flavoring Agents A flavor is generally described as the sensory impression of food or other substance and is determined by the texture, taste, and smell of a product. Flavoring agents provide a characteristic taste and/or smell to products. In addition, they can also mask bad taste. Overall, they contribute to product acceptance. Flavoring ingredients are primarily used in products that come into contact with the taste buds, including lip care formulations, such as lipstick, as well as dental and oral care products, such as toothpaste.

- Examples for flavoring agents used in cosmetics and OTC drug–cosmetic products include ingredients providing natural flavors, such as peppermint, wintergreen, menthol, eucalyptol, strawberry, and banana, and those providing artificial flavor, such as chocolate, bubble gum, and punch.

Fragrances Fragrances are natural or synthetic compounds with a characteristic smell that are added to cosmetics and OTC drug–cosmetic products to create an aesthetic impression for consumers and make them feel more attractive due to the nice smell. Fragrances can also be added to cosmetics and OTC drug–cosmetic products to mask the undesirable odor of one or more of the raw ingredients. Unlike perfumes, which are hydroalcoholic solutions with a high fragrance content sprayed on the skin to transit a pleasant redolence to the user, personal and cosmetic care fragrances have a lower perfume content and are generally used in makeup products and skin and hair care formulations to increase product acceptance and mask the natural smell of the ingredients.

- Examples for fragrances used in cosmetics and OTC drug–cosmetic products include natural components, such as essential oils obtained from different parts of flowers, fruits, roots, leaves, and seeds. They may also be obtained from animal glands and organs. Another class includes synthetic fragrance components such as linalool and citronellol. Nowadays, mainly synthetic fragrances are used to ensure reproducibility. In addition, synthetic fragrances can be stronger, longer lasting, more complex, easier to manufacture and sophisticated, more reproducible from lot to lot, and less expensive than natural fragrances.[7]

Moisturizers The term "moisturizer" is an umbrella term used to describe ingredients that add moisture to the skin and help retain moisture in the skin. They make the skin feel softer and smoother and reduce roughness, cracking, and irritation. These ingredients are used in many of today's formulations either as a main component of a formulation, for example, in a daily facial moisturizer, or as ingredients that provide additional benefits, for example, in a nail polish remover. Currently, four subclasses of moisturizers are distinguished (see more detail on the various types of moisturizers in Section 3 of Chapter 3).

- **Humectants** are hygroscopic ingredients. In general, they can serve two functions in cosmetics and OTC drug–cosmetic products, which are the following:
  - They can contribute to skin hydration by drawing water from the deeper layers of the epidermis and dermis to the outer layer of the skin (SC). Examples for humectants used for their moisturization properties include glycerin, sorbitol, urea, and propylene glycol.
  - In addition, they inhibit water evaporation from cosmetic products, i.e., provide protection against drying out. Typically, sorbitol and glycerin are used for this purpose.
- **Emollients** replenish oils and lipids in the skin. They soften and smooth the skin by filling void spaces on the skin surface and replacing lost lipids in the SC. They also provide protection and lubrication on the skin surface, minimize chafing, and enhance the skin's aesthetic properties.
  - Examples for emollients used in cosmetics and OTC drug–cosmetic products include vegetable oils; seed and nut oils; fruit butters; lanolin; synthetic esters of fatty alcohols and fatty acids, such as isopropyl palmitate and glyceryl stearate; polymers, such as polyquaterniums; hydrocarbons, such as mineral oil and paraffin; siloxanes, such as dimethicone and cyclopentasiloxane; and many others.

- **Occlusives** are hydrophobic in nature and form a water-repellent layer over the skin. It physically blocks, or at least retards, water loss through the skin.
  - Examples for occlusives used in cosmetics include hydrocarbon oils and waxes, such as petrolatum, mineral oil, paraffin, carnauba, and candelilla wax; silicone oils, such as dimethicone; vegetable oils and animal fats; fatty acids, such as stearic acid; fatty alcohols, such as cetyl alcohol; and many other ingredients.
- **Enhancers of the skin barrier** (otherwise known as skin rejuvenators) help restore, protect, and enhance the skin's barrier function. Additionally, they create a film over the skin surface that aesthetically smoothens the skin and stretches out fine lines.[8]
  - Examples for enhancers of the skin barrier used in cosmetics include proteins, such as collagen, keratin, and elastin.

As the mechanism of action is different for the various moisturizers, they are generally used in combination with each other to provide tailored benefits for consumers.

pH Buffers The pH buffers can change the pH of cosmetics and OTC drug–cosmetic formulations. pH adjustment may be necessary in formulations for many reasons. Examples include matching the formulation's pH with that of the application surface, stabilizing formulations since certain ingredients are stable at specific pH values only, and thickening formulations as certain thickeners must be neutralized in order to achieve optimum viscosity. An example for these types of thickeners is the carbomers.

- Examples for pH buffers used in cosmetics and OTC drug–cosmetic products include citric acid and lactic acid as acidic ingredients as well as sodium hydroxide and the commonly used triethanolamine as alkaline ingredients.

Plasticizers Plasticizers are ingredients that can soften films and impart flexibility to films, such as nail polish film or hair spray film, as the film dries. Often, the developing film is very rigid and brittle, which makes nail polish sensitive to chipping and cracking or hair spray film stiff and brittle. Plasticizers can help prevent these undesired effects. Plasticizers are often used in nail polishes, nail hardeners, sunscreens, and film-former-based hair styling products.

- Examples for plasticizers used in nail polishes include camphor, castor oil, glyceryl tribenzoate, triphenyl phosphate, citrate esters, and acetyl tributyl citrate, among others. In film-forming ingredient-based hair styling products, mainly mineral oil, dimethicone, and castor oil are used.

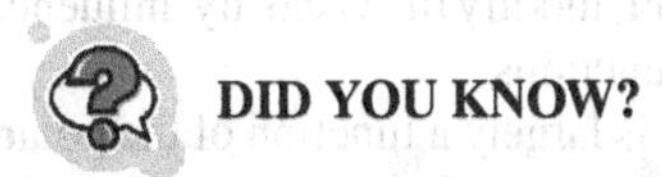

Plasticizers are added to hard, brittle plastics to make them more flexible. Many cosmetic packaging materials, e.g., shower gel containers, liquid soap containers, and toothpaste tubes, also contain plasticizers, so that we are able to squeeze the bottle and get a small amount of it out of the container. If no plasticizers were added to these plastics, they would be too hard and easy to break.

Preservatives Preservatives are used to prevent undesirable growth of molds, yeast, and bacteria in liquid, semisolid, and powder products. Their use is especially important in water-based products since water provides an ideal environment for microbial growth. The mechanism of action and range of efficacy usually vary among the different types of preservatives. Therefore, they are generally used in combination with each other to provide protection against a wide variety of microorganisms.

- Examples for preservatives used in cosmetics and OTC drug–cosmetic products include parabens, such as methylparaben and propylparaben; formaldehyde donors, such as DMDM hydantoin, imidazolidinyl urea, and glutaraldehyde; cationic surfactants (generally referred to as “quats”), such as benzalkonium chloride and benzethonium chloride; alcohols, such as ethanol and benzyl alcohol; phenol derivatives, such as phenoxyethanol; isothiazolones, such as methylchloroisothiazolinone; and other components, such as sorbic acid.

Propellants Propellants are added to aerosol formulations to maintain a suitable pressure within the aerosol can and expel the content of the container when the valve is open. They are usually compressed or liquefied gases. Propellants are used in aerosol products, such as shaving creams, hair sprays, antiperspirants, and sunscreen products, among others.

- Examples for propellants used in cosmetics and OTC drug–cosmetic products include isopentane (liquefied gas), butane, isobutane, and propane (compressed gases).

Solvents Solvents are important parts of most cosmetics and OTC drug–cosmetic formulations. They are typically liquids used to dissolve solid ingredients, mix with liquids, provide a vehicle for formulations, and contribute to the texture of products. Solvents can aid in the development of body lotions, facials creams, foundations, nail polish, hair spray, sunscreens, hand sanitizer, and many other products. They can contribute to the stability of formulations, regulate evaporation rate, provide a cooling effect, aid in product application, modify the skin feel, modify the viscosity, influence the film-forming properties, and have many other functions.

The solubility of an ingredient in a given solvent is largely a function of the polarity of the solvent. Solvents work under the general rule of like dissolves like. The term "like" refers to the overall polarity of the solvent molecule (whether polar or nonpolar) and the overall polarity of the solute (i.e., ingredient to be dissolved). Based on the dielectric constant, which refers to polarity, solvents are typically broadly classified into three categories: polar, semi-polar, and non-polar solvents.

- **Polar** solvents contain strong dipolar molecules and exhibit hydrogen-bonding properties. Therefore, their dielectric constant is high. Polar solvents generally dissolve polar solutes.
  - Typical polar solvents used in cosmetics and OTC drug–cosmetic products include water and glycols, such as glycerin and propylene glycol.
- **Semi-polar solvents** are also made up of strong dipolar molecules; however, they do not form hydrogen bonds. Semipolar solvents are capable of dissolving both polar and non-polar substances. Therefore, they can serve as a medium for a multicomponent homogeneous system containing polar and non-polar solvents.
  - Examples for semipolar solvents used in cosmetics and OTC drug–cosmetic products include alcohols, such as ethanol and isopropyl alcohol; ketones, such as acetone; and esters, such as ethyl acetate.
- **Non-polar solvents** contain molecules that have only a small or no dipolar character (i.e., their dielectric constant is low). They dissolve non-polar molecules.
  - Examples for non-polar solvents used in cosmetics and OTC drug–cosmetic products include oils, such as mineral oil, petrolatum, sunflower oil, almond oil; silicone oils; hexane; toluene; and dimethyl ether.

This widely used classification for solvents is not exclusive. There are many solvents that can fit into more than one of these broad categories. For example, glycerin can be considered both polar and semi-polar although it can form hydrogen bonds.

Solvent selection for cosmetics and OTC drug–cosmetic products typically depends on the types of ingredients in the formulation, types of dosage form, and compatibility with the application surface. For example, nail polish formulas usually contain semipolar solvents, such as ethyl acetate and isopropyl alcohol, since the resins used as film-formers are only soluble in these solvents. However, these solvents are not used for sunscreens as sunscreen ingredients are not soluble in these solvents, and they could also irritate the skin. Many organic UV filters are oil-soluble crystalline solids, which need to be dissolved in oils. If the solvent properties of the oil phase are poor, the filters can recrystallize during storage, dramatically reducing the efficacy. See more examples for solvents in the individual chapters for various products.

Surfactants Surfactants, also known as surface active ingredients, are the most widely used ingredients in cosmetics and OTC drug–cosmetic products. They have a very unique chemical structure, including both a hydrophilic (i.e., water loving) and a hydrophobic (i.e., oil loving) portion (depicted in Figure 1.6), which enables them to be dissolved both in water and in oil. Surfactants can lower the surface tension between two liquids or between a liquid and a solid, making them suitable for many applications. They contribute to the formulation, stability, and applicability of personal care and cosmetic products. They can fulfill a variety of functions in cosmetics and OTC drug–cosmetic products, including emulsification (they help two immiscible phases, e.g., water and oil, mix with each other to form an emulsion), solubilization, cleansing, foaming, foam boosting, wetting, antifoaming, conditioning, preserving, stabilizing, controlling viscosity, and many others.

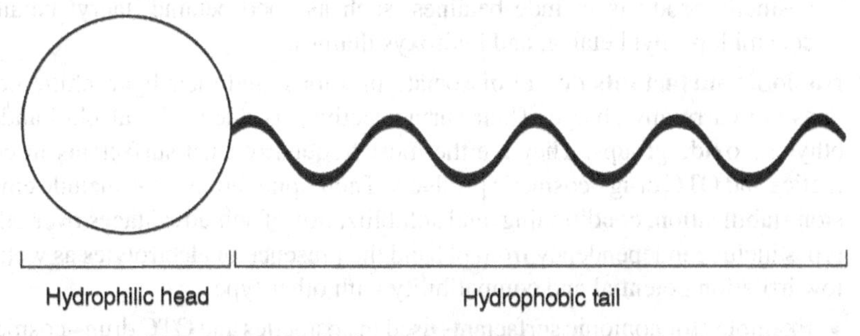

Figure 1.6 The basic structure of a surfactant molecule.

Surfactants can be classified based on their hydrophobic chain into hydrocarbon and silicone surfactants. A more general classification is based on the charge of their hydrophilic part, and the four subtypes are anionic, cationic, amphoteric, and non-ionic surfactants.

- **Anionic surfactants** contain a negative charge in their hydrophilic head. Anionic surfactants are typically utilized for their foaming and excellent cleaning properties. Their major drawback is the irritating potential, especially for sulfates, which may cause concerns for many consumers.
  - Examples for anionic surfactants used in cosmetics and OTC drug–cosmetic products include carboxylic acid compounds, such as stearic acid; soaps, such as triethanolamine stearate and potassium laurate; sulfates, such as sodium lauryl sulfate, ammonium lauryl sulfate, and sodium laureth sulfate; sulfonates, including taurates, isothionates, and olefin sulfonates; and sulfosuccinates, such as disodium laureth sulfosuccinate.
- **Cationic surfactants** contain a positive charge in their hydrophilic head. They represent the most powerful conditioning agents for the skin and hair. As the

overall surface charge of the skin and hair is negative, cationic surfactants are electrostatically attracted to these negative sites.

- Examples for cationic surfactants used in cosmetics and OTC drug–cosmetic products include amines and their derivatives and the most frequently used quaternized ammonium compounds (often referred to as "quats"), such as cetrimonium chloride, stearalkonium chloride and benzalkonium chloride, and quaternium and polyquaternium molecules.

■ **Amphoteric surfactants** have both a negative and a positive charge in their hydrophilic head. They have good cleansing, bactericidal, bacteriostatic, lathering, and softening properties and are able to stabilize and induce foam formation. Therefore, they are used in shampoos, baby products, and aerosols.

- Examples for amphoteric surfactants used in cosmetics and OTC drug–cosmetic products include betaines, such as coco betaine, lauryl betaine, cocamidopropyl betaine, and hydroxysultaines.

■ **Nonionic surfactants** do not dissociate into ions, and their hydrophilic head does not carry any charge. Their surface activity is due to the alcohol and/or ethylene oxide groups. They are the most frequently used surfactants in cosmetics and OTC drug–cosmetic products. Their application areas include emulsion stabilization, conditioning, and solubilization. Their advantages over other types include independency from pH and the presence of electrolytes as well as low irritation potential and compatibility with other types.

- Examples for nonionic surfactants used in cosmetics and OTC drug–cosmetic products include glycol and glycerol esters, such as glyceryl monostearate; sorbitan esters, such as sorbitan stearate and sorbitan palmitate; polysorbates (otherwise known as ethoxylated sorbitan esters), such as polysorbate 20; fatty alcohols, such as cetyl alcohol and stearyl alcohol; polyoxyethylene–polyoxypropylene block copolymers (otherwise known as poloxamers), such as Poloxamer 407; amine oxides, such as cocamine oxide and cocamidopropylamine oxide; alkanolamides, such as cocamide monoethanolamine (MEA) and diethanolamine (DEA); alkylglucosides, such as lauryl glucoside; and many others.

Surfactants are typically characterized by their HLB numbers. The abbreviation "HLB" stands for hydrophile–lipophile balance. HLB is an empirical expression for the relationship of the hydrophilic and hydrophobic groups of a surfactant. The HLB system uses a scale of 1–20 based on the affinity of the surfactant to oil and water; the higher the HLB value, the more water soluble the surfactant. In general, emulsifiers with HLB values of 1–3 are antifoaming agents, those with values of 4–6 are water-in-oil (W/O) emulsifiers, those with values of 7–9 are wetting agents, those with values of 8–18 are oil-in-water (O/W) emulsifiers, those with values of 13–15 are cleansing agents, and those with values of 10–18 are solubilizing agents.[9]

DID YOU KNOW?

Based on the cleansing and foaming properties, surfactants can be divided into primary and secondary surfactants (the latter are also known as co-surfactants). The main function of primary surfactants is cleansing and foaming. Co-surfactants are used to reduce irritation and drying caused by primary surfactants. Additionally, some co-surfactants also have conditioning effects. Therefore, many skin and hair cleansing products contain a mixture of primary and secondary surfactants.

Sweeteners Sweeteners provide a sweet flavor and contribute to product acceptance in cosmetics and OTC drug–cosmetic products that come into direct contact with the taste buds. These products include toothpastes, mouthwashes, lipsticks, lip balms, and lip glosses. Sweeteners are generally categorized into two major groups: true sweeteners (also known as no-calorie sweeteners) that do not provide any calories, and low-calorie sweeteners that add some calories to products. They are usually used in combination with flavors as the combination of flavors and sweeteners provides an acceptable and attractive taste for the formulations.

- Examples for no-calorie sweeteners used in cosmetics and OTC drug–cosmetic products include sodium saccharin, acesulfame, aspartame, sucralose, and stevia, and examples for low-calorie sweeteners include xylitol, mannitol, and sorbitol.

DID YOU KNOW?

Most artificial sweeteners are not sugars; only sorbitol and xylitol are sugar derivatives, i.e., sugar alcohols. However, these molecules are still able to provide an intensive sweet taste ranging from 10 to 100 times sweeter than table sugar. For example, sodium saccharine is 700 times sweeter than table sugar. We can find these components in gums and diet beverages that have a sweet taste but provide a low- or no-calorie intake. An important property of these ingredients is that they do not promote the growth of bacteria that causes cavities (i.e., *Streptococcus mutans*).

Thickeners Thickeners are ingredients that can increase the viscosity of cosmetics and OTC drug–cosmetic products. They also improve stability, modify appearance and products aesthetics, improve applicability, and modify the rheology of a product. Thickeners can also be used to build viscosity in suspensions and act as suspending agents, for example, in nail polish formulations.

We generally differentiate between viscosity-increasing agents for aqueous systems, which increase the viscosity of the aqueous (water) phase, such as the water phase of an O/W emulsion, and for nonaqueous systems, which increase the thickness of the oil phase of cosmetic products, such as the oil phase of a W/O emulsion.

- Examples for water-based thickeners include gums, such as xanthan gum and guar gum; cellulose and its derivatives, such as hydroxyethyl cellulose; hydrophilic clays, such as hectorites, bentonites, and magnesium aluminum silicates; polyethylene glycols (PEGs), such as PEG 200; and synthetic polymers, such as carbomers; and also sodium chloride. Examples for non-water-based systems include waxes, such as carnauba wax; long-chain alcohols, such as cetyl alcohol; organoclays; fumed silica; synthetic polymers; and polyethylenes, among others.

For the selection of thickeners, a variety of factors should be taken into consideration, including the products' use, application surface, compatibility with other ingredients in the formula, pH (certain thickeners, e.g., carbomers, are alkali swellable, and they need an alkaline pH to reach optimum viscosity; therefore, they cannot be used in an acidic environment), clarity, the presence of electrolytes, temperature during processing (waxes have to be melted in order to be mixed with oils; if a product is made without heating, waxes cannot be used), and shear during processing (some ingredients, such as carbomer, require shear in order to be activated and gain optimum viscosity, while others may be sensitive for shearing, such as fumed silica).

DID YOU KNOW?

Separation of emulsions and sedimentation of suspensions can be described with the following equation, called Stokes' law:

$$V = \frac{2r^2 g(D_1 - D_2)}{9\mu}$$

where V is the particle's settling velocity (m/s), r is the radius of the particle, g is the gravitational acceleration (m/s^2), D_1 is the density of the particles (kg/m^3), D_2 is the density of the fluid (kg/m^3), and μ is the viscosity of the fluid (Pa·s). It is obvious from the equation that the higher the viscosity, the lower the settling velocity. Therefore, an emulsion or suspension will be more stable if it has a higher viscosity. An essential role in the formulation of emulsions and suspensions is to have a slow settling velocity. This is one of the main reasons for using thickeners in the formulations.

Active Ingredients As discussed previously, OTC drug–cosmetic combination products also contain active ingredients. **Active ingredients deliver the claimed**

therapeutic action and have an effect on the human body, i.e., to prevent and/or treat a disease.

● **Active ingredients for cosmetic products are either listed in OTC monographs or are new ingredients.** OTC monographs can be described as a "recipe book" that tells formulators what ingredients and in what concentration can be used. Active ingredients are broken down into categories and characterized together in the monographs. The monographs also describe what types of formulations an ingredient can be used for and also provide information on claims and labeling. For some product types, such as sunscreens, testing is described as well. OTC monographs define the safety, effectiveness, and labeling of all OTC active ingredients. They are continually updated by adding, changing, and removing ingredients, labeling, and other pertinent information, as needed.

The OTC monograph system was set up in 1972 with the FDA's OTC Drug Review. The Drug Review is an ongoing process by which the safety and efficacy of OTC ingredients are assessed. An expert advisory panel reviews the data relating to claims and active ingredients for different therapeutic classes. The panel recommendations become mandated in an official OTC monograph through a three-step public rule-making process (see Figure 1.7).[10]

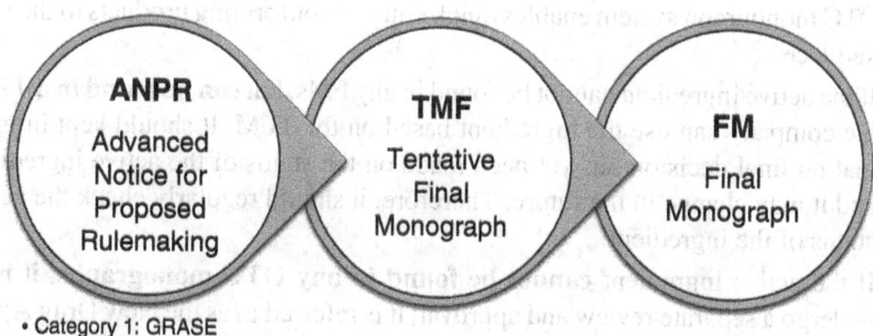

Figure 1.7 The three-step OTC monograph process.

- **Step 1:** In this step, the panel reviews active ingredients to determine whether they could be generally recognized as safe and effective for use in self-treatment. Panel reports are published in the Federal Register as an advanced notice of proposed rulemaking (ANPR). The ingredients in this phase are classified into one of these three categories:
  - **Category I:** Generally recognized as safe and effective for the claimed therapeutic indication (GRASE).
  - **Category II:** Not generally recognized as safe and effective or unacceptable indications (not GRASE).
  - **Category III:** Insufficient data available to permit final classification.

- **Step 2:** Following FDA review and public comment, a tentative final monograph (TFM) is issued by the FDA and published in the Federal Register, proposing approved ingredients, uses, doses, required warning, and appropriate claims.
- **Step 3:** The final step is the implementation of the final monograph (FM). The monographs establish conditions under which certain OTC drug products are generally recognized as safe and effective. FMs are also published in the Federal Register.

If an OTC active ingredient does not have an FM, it means that there is not enough data on its safety and effectiveness to make a final decision about the ingredient. Since TMFs are in place for all categories of OTC drugs, in such cases, formulators should follow the guidelines of the TMFs. This is the case with many antimicrobial ingredients used in hand sanitizers.

If a company wishes to formulate an OTC drug–cosmetic product and use an active ingredient, it has several options to choose from, including the following:

- If the active ingredient **can be found in an FM**, the company can make and market the OTC product using the ingredient in its approved concentration for its approved indication. No FDA preapproval is necessary for this to occur. The OTC monograph system enables quick entry of conforming products to the marketplace.
- If the active ingredient cannot be found in any FMs, but **can be found in a TFM**, the company can use the ingredient based on the TFM. It should kept in mind that no final decision has yet been made on the status of the active ingredient, and it may change in the future. Therefore, it should regularly check the actual status of the ingredient.
- If the active ingredient **cannot be found in any OTC monographs**, it must undergo a separate review and approval; it is referred to as the New Drug Application (NDA). The company has to submit information on the safety and efficacy of the active ingredient with regard to the particular OTC drug–cosmetic product.
- If the active ingredient **can be found in an OTC monograph** (either a TFM or an FM); however, the indication for which the company wants to use the active ingredient **cannot be found**, and it must also undergo the NDA process (similar to the previous situation).
- A company may also **petition to change** an FM to include additional ingredients or to modify current labeling practice. In this case, it has to submit information on the safety and efficacy of the active ingredient. It cannot market the product with the new condition (e.g., higher dose) until the FM is amended.

Now, let us review the major active ingredient types (without completeness) that can be found in various types of OTC drug–cosmetic products.

FYI

The electronic Code of Federal Regulations (eCFR) and OTC monographs are freely accessible for everyone. For more information, visit www.ecfr.gov and select Title 21 (Foods and Drugs).

Anti-Acne Ingredients Acne is a complex skin disorder that develops when the hair follicles become plugged with oil and dead skin cells (see more information in Section 4 of Chapter 3). Acne most commonly occurs on the face, neck, back, and chest. Bacteria can grow in this plug and cause inflammation. Acne symptoms may vary from invisible clogged pores, through small, hardly noticeable clogged pores with a small white or black head, to red, enlarged, inflamed, and painful lesions, including papules, pustules, nodules, or cysts (depending on the severity).

Anti-acne ingredients are listed in the OTC monograph[11] for topical antimicrobial drugs. The listed ingredients have abrasive, exfoliating (peeling), and antiseptic effects to remove the excess amount of oil and peel the skin. These ingredients can be found in skin care products, such as facial washes, toners, and moisturizers, as well as in makeup products, such as concealers.

- Examples for FDA-approved anti-acne ingredients used in the topical treatment of mild acne include benzoyl peroxide (in 2.5–10%), salicylic acid (in 0.5–2%), sulfur (in 3–10%), and resorcinol (in 2% when combined with sulfur).

Anticaries Ingredients Caries (otherwise known as tooth decay or cavities) is characterized by brown spots or holes on the chewing surface of the teeth. Cavities usually do not hurt, unless they grow very large and affect the nerves or cause a tooth fracture. However, untreated tooth decay can destroy the inside of the tooth (pulp), which can lead to tooth loss (more information on cavities can be found in Chapter 6).

Anticaries ingredients are listed in the OTC monograph[12] for anticavities drugs. These ingredients are used to prevent caries; strengthen the tooth enamel and slow down the formation of an invisible, sticky film (called plaque) that leads to caries formation; as well as restore and harden the teeth. These ingredients are available in all types of oral care products, including toothpaste, toothpaste gel, and mouthwash.

- Examples for FDA-approved anticaries ingredients used in OTC drug–cosmetic products include fluoride components, such as sodium monofluorophosphate, sodium fluoride, and stannous fluoride. Concentrations are specified for each ingredient for each product type, such as gel, paste, and mouthwash.

Antidandruff Ingredients Dandruff is a common non-contagious condition of the scalp, characterized by an increased rate of shedding of dead cells from the scalp.

Although shedding of dead cells is continuous and imperceptible for most people, the rate of shedding is greatly accelerated in dandruff. The usual symptoms include white, oily-looking flakes on the hair and shoulders and itching. The causes of dandruff are thought to be the combination of an overproduction of skin oil and irritation from yeast called *Malassezia*.

Antidandruff ingredients are listed in the OTC monograph[13] for miscellaneous external drug products. These ingredients have antimicrobial properties, and some of them also have exfoliating properties. Generally, they are formulated into shampoos.

- Examples for FDA-approved antidandruff ingredients used in OTC drug–cosmetic combination products include coal tar (in 0.5–5%), zinc pyrithione (in 0.3–2% when washed off), salicylic acid (in 1.8–3%), selenium sulfide (in 1%), and sulfur (in 2–5%).

Antiperspirant Ingredients Antiperspirants affect the function of the body by reducing the amount of sweat that reaches the skin surface and not just masking bad body odor as do deodorants do. Underarm odor is caused by the bacterial breakdown of sweat (see more detail in Section 6 of Chapter 3).

Antiperspirants are listed in the OTC monograph[14] for antiperspirant drug products.

- Examples for FDA-approved antiperspirant used in OTC drug–cosmetic products include aluminum chloride (up to 25%), aluminum chlorohydrate (up to 25%), aluminum sesquichlorohydrate (up to 25%), aluminum dichlorohydrate (up to 25%), and PEG and zirconium complexes, such as aluminum chlorohydrex PEG (up to 25%) and aluminum zirconium octachlorohydrate (up to 20%).

Skin Protectant Ingredients Skin protectant ingredients are listed in the OTC monograph[15] for skin protectant drug products. This category includes several subcategories, such as astringents, lip protectant ingredients, and skin protectant ingredients.

Astringents are applied to the skin or mucous membranes for a local and limited protein coagulant effect. They are commonly used in facial toners and aftershave solutions to tighten pores.

- Examples for FDA-approved astringents used in OTC drug–cosmetic products include aluminum acetate (in 0.13–0.5%), aluminum sulfate (in 46–63%), and witch hazel.

Lip protectants temporarily prevent dryness and help relieve chapping of the exposed surfaces of the lips. They are traditionally called lip balms. Drying of the lips is a typical sign of dehydration. Ingredients used as lip protectants moisturize the lips as well as provide protection against further water loss.

- Examples for FDA-approved lip protectants used in OTC drug–cosmetic products include allantoin (in 0.5–2%), cocoa butter (in 50–100%), dimethicone (in 1–30%), and petrolatum (in 30–100%), among many others.

Skin protectants temporarily protect injured or exposed skin or mucous membrane surfaces from harmful or annoying stimuli. In addition, they may provide relief to such surfaces. Skin protectants are usually formulated as lotions, creams, and ointments. Examples for skin protectants are the same as those for lip protectants.

Sunscreens Sunscreens (also known as UV filters) protect the skin from the harmful radiation of the sun. Sunscreens are listed in the OTC monograph[16] for sunscreen drug products. They can be formulated into a variety of products forms, including aerosol sprays, lotions, sticks, and gels (see more detail on this in Section 5 of Chapter 3).

UV filters protect the skin via two separate mechanisms: either reflecting sunlight (physical filters) or absorbing sunlight and converting it to heat (chemical filters). Currently, there are 16 UV filters listed in the OTC monograph.

- Examples for FDA-approved sunscreen ingredients used in OTC drug–cosmetic products include physical filters, such as titanium dioxide (up to 25%) and zinc oxide (up to 25%), as well as chemical filters, such as avobenzone (up to 3%), octocrylene (up to 10%), and padimate O (up to 8%).

Cosmetics and OTC drug–cosmetic products can contain additional types of ingredients. The purpose of this section was to provide the readers with a general understanding of the major and commonly used minor ingredients types before moving on to discuss the various product categories. More information is provided on the ingredient types, their function, and characteristics in the individual chapters for the various products.

GLOSSARY OF TERMS FOR SECTION 2

Abrasive: An ingredient that is capable of polishing or cleaning a harder surface by rubbing or grinding.

Active ingredient: An ingredient in OTC drugs or prescription-only drugs that delivers the claimed therapeutic action.

Antiacne ingredient: An ingredient that has abrasive, exfoliating (peeling), and/or antiseptic effects and can treat acne skin.

Anticaries ingredient: An ingredient that can help prevent the formation of brown holes in the teeth (i.e., tooth decay), strengthen the tooth enamel, slow down plaque formation, and harden the teeth.

Antidandruff ingredient: An ingredient that has antimicrobial and exfoliating effect and can treat dandruff.

Antioxidant: An ingredient that provides protection against oxidative reactions.

Antiperspirant: An ingredient that can reduce the amount of sweat reaching the skin surface.

Application surface: A body surface to which cosmetics or OTC drug–cosmetic products are applied.

Astringent: An ingredient that can provide the mucous membranes with a local and limited protein coagulant effect.

Chelating agent: An ingredient that is able to complex with metal ions.

Color additive: An ingredient that adds color to a product, making it attractive, appealing, appetizing, and informative.

Cosmetic ingredient: An ingredient that is used in cosmetic products to provide them with appropriate aesthetics, texture, pH, color, and smell, as well as to fulfill cosmetic claims for products. OTC drug–cosmetic products also contain cosmetic ingredients that are called inactive ingredients in their case.

Dosage form: The final physical form of a mixture of ingredients that consumers can take in their hands, purchase, and use.

Flavoring agent: An ingredient that provides a characteristic taste and/or smell to a product.

Lip protectant: An ingredient that can temporarily help prevent lip dryness and help relieve chapped lips.

Moisturizer: An ingredient that adds moisture to the skin and help keep moisture in the skin.

pH buffer: An ingredient that can change or maintain the pH of a product.

Plasticizer: An ingredient that can soften films and impart flexibility to films.

Preservative: An ingredient that can prevent undesirable growth of molds, yeast, and bacteria in a product.

Propellant: An ingredient that is added to aerosol formulations to maintain a suitable pressure within the can and expel the content of the container.

Solvent: An ingredient that is used to dissolve solid ingredients, mix with liquids, provide a vehicle for formulations, and contribute to the texture of products.

Sunscreen: An ingredient that can provide protection to the skin against the harmful UV radiation.

Surfactant: A surface active ingredient that can lower the surface tension between two liquids or between a liquid and a solid.

Sweetener: An ingredient that provides a sweet flavor to a product.

Target group: A specific group of customers at whom a cosmetic product or OTC drug–cosmetic is aimed.

Thickener: An ingredient that can increase the viscosity of a product.

REVIEW QUESTIONS FOR SECTION 2

Multiple Choice Questions

1. Which of the following ingredients do OTC drug–cosmetic products contain?
   a) Cosmetic ingredients, which are called inactive ingredients
   b) Active ingredients
   c) A and B
   d) None of the above

2. The use of preservatives is extremely crucial in the case of ___-based formulations.
   a) Oil
   b) Water
   c) Preservative
   d) Silicone

3. Anionic surfactants are primarily used as ___ agents in cosmetic products.
   a) Cleansing
   b) Conditioning
   c) Flavoring
   d) Coloring

4. What does the abbreviation "HLB" refer to?
   a) An empirical expression for the relationship of hydrophilic and hydrophobic groups in surfactants.
   b) A type of certification that has to be received before selling color additives to customers.
   c) A type of antioxidant that is used in anti-aging products.
   d) An expression for the relationship of sedimentation rate of suspensions.

5. Who certifies certifiable color additives?
   a) WHO
   b) EPA
   c) FDA
   d) OTC

6. Why is batch certification necessary for certain color additives?
   a) To make sure that these color additives have a natural origin
   b) To certify that they can be used for tattoos
   c) To ensure that there are no impurities of toxicological concern
   d) To check whether they are soluble in water

7. Flavors and sweeteners are typically included in formulations that can come into contact with the ___.

a) Eyes
b) Hair
c) Sweat glands
d) Taste buds

8. OTC monographs contain ___ ingredients that can be used in OTC products ___ receiving an FDA premarket approval.

a) Active/with
b) Inactive/with
c) Active/without
d) Inactive/without

Matching

Match the terms in column A with their appropriate definition in column B.

	Column A	Column B
_____	1. Abrasives	A. Ingredients that are strictly regulated in the United States.
_____	2. Antioxidants	B. Ingredients that bind to metal ions.
_____	3. Antiperspirants	C. Ingredients that help adjust the pH of cosmetic formulations.
_____	4. Astringents	D. Ingredients that help expel the contents of aerosol containers.
_____	5. Chelating agents	E. Ingredients that make cleansing products, e.g., shampoo's bubble.
_____	6. Color additives	F. Ingredients that make films more flexible.
_____	7. Moisturizers	G. Ingredients that make the skin softer and smoother.
_____	8. pH buffers	H. Ingredients that prevent the growth of bacteria, molds, and yeast.
_____	9. Plasticizers	I. Ingredients that provide a local and limited protein coagulant effect.
_____	10. Preservatives	J. Ingredients that provide a sweet taste for formulations.
_____	11. Propellants	K. Ingredients that provide a vehicle for formulations.
_____	12. Solvents	L. Ingredients that provide protection against free radicals.

	Column A	Column B
______	13. Sunscreens	M. Ingredients that provide protection from the sun.
______	14. Surfactants	N. Ingredients that provide viscosity control for cosmetic formulations.
______	15. Sweeteners	O. Ingredients that reduce the amount of sweat.
______	16. Thickeners	P. Ingredients used in facial scrubs for their exfoliating effect.

Matching

Match the terms in column A with their appropriate definition in column B.

	Column A	Column B
______	1. Amphoteric	A. A type of solvent that have a high dielectric constant.
______	2. Anionic	B. A type of solvent that have a low dielectric constant.
______	3. Aqueous	C. A type of surfactant that do not have any charges in their head part.
______	4. Cationic	D. A type of surfactant that have a negative charge in their head part.
______	5. Dye	E. A type of surfactant that have a positive charge in their head part.
______	6. Hydrophilic	F. A type of surfactant that has both negative and positive charges in the head part.
______	7. Hydrophobic	G. Ingredients insoluble in the solvent in which they are dispersed.
______	8. Non-aqueous	H. Ingredients soluble in the solvent in which they are dispersed.
______	9. Nonionic	I. Non-water-based.
______	10. Non-polar	J. Water-based.
______	11. Pigments	K. Water hating.
______	12. Polar	L. Water loving.

REFERENCES

1. Masaki, H.: Role of antioxidants in the skin: anti-aging effects. *J Dermatol Sci.* 2010;58(2):85–90.
2. CFR Title 21 Part 70.3(f)
3. CFR Title 21 Part 71
4. FDA: Think before you Ink, Last update: 4/23/2014, Accessed 4/30/2014 at http://www.fda.gov/forconsumers/consumerupdates/ucm048919.htm
5. FD&C Act Part 721(c)

6. CFR Title 21 Parts 70 and 80
7. Salvador, A., Chisvert, A.: Perfumes in Cosmetics. Regulatory Aspects and Analytical Methods for Fragrance Ingredients and other Related Chemicals in Cosmetics, In: Salvador, A., Chisvert, A., eds: Analysis of Cosmetic Products, Amsterdam: Elsevier, 2007.
8. Nolan, K., Marmur, E.: Moisturizers: reality and the skin benefits. *Dermatol Ther.* 2012;25:229–233.
9. Swarbick, J., Rubino, J. T., Rubino, O. P.: Coarse Dispersions, In: Felton, L. E., ed.: Remington, London: Pharmaceutical Press, 2013.
10. Derbis, J., Evelyn, B., McMeekin, J.: FDA Aims to Remove Unapproved Drugs from Market-Risk-Based Enforcement Program Focuses on Removing Potentially Harmful Products. *Pharmacy Today* 2008, Accessed 4/22/2014 at http://www.fda.gov/downloads/Drugs/GuidanceComplianceRegulatoryInformation/EnforcementActivitiesbyFDA/SelectedEnforcementActionsonUnapprovedDrugs/ucm119899.pdf
11. CFR Title 21 Part 333.301–333.350
12. CFR Title 21 Part 355
13. CFR Title 21 Part 358.701–358.760
14. CFR Title 21 Part 350
15. CFR Title 21 Part 347
16. CFR Title 21 Part 352

SECTION 3: DOSAGE FORMS FOR COSMETICS AND OTC DRUG–COSMETIC PRODUCTS

LEARNING OBJECTIVES

Upon completion of this section, the reader will be able to

1. Define the following terms:

Anhydrous	Hydroalcoholic	Solution	Dosage form
Liquid dosage form	Solid dosage form	Semisolid dosage form	Emulsion
Lotion	Cream	Suspension	Gel
Paste	Ointment	Powder	Capsule
Stick	Aerosol	Sedimentation	Stokes' law
Aqueous			

2. explain the concept of a dosage form;
3. differentiate between a solution and an emulsion;
4. distinguish between an emulsion and a suspension;
5. explain the difference between a lotion and a cream;

6. explain the difference between regular pastes and toothpastes;
7. discuss the advantages of gels when used as cosmetic products;
8. discuss the advantages of using sticks;
9. briefly discuss the main advantages and disadvantages of creams and ointments;
10. briefly discuss the importance of thickeners as it applies to suspensions;
11. briefly discuss the potential use of capsules as cosmetic products;
12. explain the concept of an aerosol product;
13. provide a few cosmetic and/or OTC drug–cosmetic examples for the following dosage forms: solution, cream, lotion, ointment, suspension, loose powder, pressed powder, capsule, gel, stick, aerosol, foam, and paste.

KEY CONCEPTS

1. Cosmetics and OTC drug–cosmetic products are available as hundreds of types of formulations, which can be broken down into some basic groups, called dosage forms, based on their physical and pharmaceutical properties.
2. A dosage form is defined as the final physical form of a mixture of chemical ingredients that consumers can take in their hands, purchase, and use as a cosmetic product or an OTC drug–cosmetic product.
3. Dosage forms can be classified based on their physical form as liquids, semisolids, and solids.
4. Liquid dosage forms available as cosmetics and OTC drug–cosmetic products include solutions, lotions, and suspensions.
5. Solid dosage forms available as cosmetics and OTC drug–cosmetic products include loose powders, pressed powders, sticks, and capsules.
6. Semisolid dosage forms available as cosmetics and OTC drug–cosmetic products include creams, ointments, pastes, and gels.

---

❶ **Cosmetics and OTC drug–cosmetic products are available as hundreds of types of formulations**, such as a shower gel, baby powder, lipstick, bath bomb, liquid eyeliner, shaving foam, lip liner pencil, toothpaste, roll-on deodorant, and many others. However, all these different formulations **can be broken down into some basic groups, called dosage forms, based on their physical and pharmaceutical properties.** This section provides an overview of the most common dosage forms available for cosmetics and OTC drug–cosmetic products and their main characteristics. It also summarizes the main advantages and disadvantages of the different dosage forms and provides ideas on how and when the various dosage forms should be selected.

What Is a Dosage Form?

Before starting to discuss the different dosage forms, let us first understand what a dosage form is. ● **A dosage form is defined as the final physical form of a mixture of chemical ingredients** (cosmetic ingredients and/or active ingredients) **that consumers can take in their hands, purchase, and use as a cosmetic product or an OTC drug–cosmetic product.**

A relevant question would be: "Why are there dosage forms?" Cosmetic and/or active ingredients cannot be applied individually to the skin, hair, or mucous membranes; for example, consider a simple mouthwash. It is composed of at least 6–8 different ingredients. The consumer cannot just buy the individual ingredients and apply them one after another to mimic a mouthwash formulation. The ingredients have to be put into a form (a dosage form) that contains all ingredients in a specific ratio and can be applied to the oral mucous membrane.

● **Dosage forms can be classified based on their physical form as liquids, semisolids, and solids.** Figure 1.8 illustrates the major cosmetic dosage form subclasses.

- If a dosage form is pourable and consists of liquid ingredients, it is classified as a **liquid** dosage form. These dosage forms cannot be directly taken into the hands as they would flow off the hands. ● **Liquid dosage forms available as cosmetics and OTC drug–cosmetic products include solutions, lotions, and suspensions.**

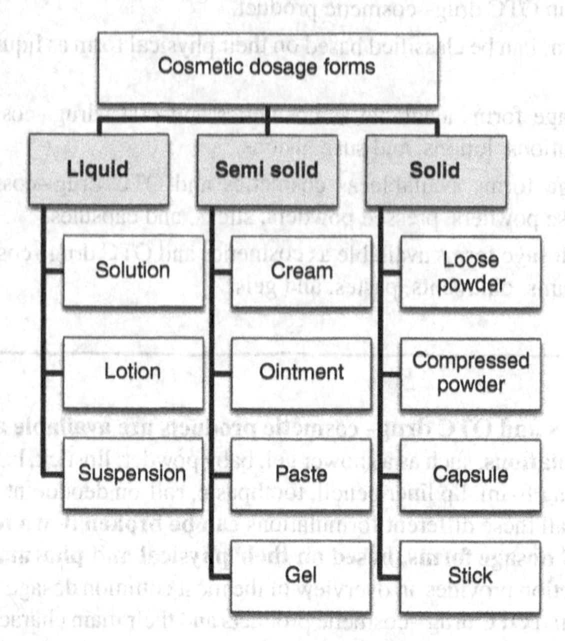

Figure 1.8 The major types of dosage forms and their subclasses typically used for cosmetic products.

- If a dosage form consists of dry solid particles, mixed and/or pressed together, or waxy materials that are present in a solidified form having a characteristic shape, the dosage form is classified as a **solid** dosage form. ● **Solid dosage forms available as cosmetics and OTC drug–cosmetic products include loose powders, pressed powders, sticks, and capsules.**
- If a dosage form has a consistency in between solid and liquid forms, it is classified as a **semisolid** dosage form. The difference is usually in the viscosity of these dosage forms. If a dosage form has a low viscosity, meaning that it can be easily poured and will quickly flow off the hands, it is a liquid. If it is more viscous, and more force is needed to dispense and apply it, it is a semisolid. ● **Semisolid dosage forms available as cosmetics and OTC drug–cosmetic products include creams, ointments, pastes, and gels.**

When selecting dosage forms for a product, a formulator has to take many factors into account. These include the properties of the ingredients used: whether they are water-soluble (hydrophilic) or oil-soluble (lipophilic), liquid or solid, and their compatibility with each other. Formulators must also consider the purpose of the product's application and the application surface. For example, for the treatment of hair, an easily applicable and washable dosage form should be chosen, e.g., a lotion instead of a greasy ointment. If a longer-lasting effect is desired, such as for diaper rash formulations, the product type can be greasier and more water resistant, containing a higher amount of oils.

Dosage Forms for Cosmetic Applications

The FDA has a standard for drug dosage forms, including their definitions.[1] This standard will be used as the basis of our discussion since the definition of a dosage form is the same regardless of the product's status, i.e., a cosmetic or an OTC drug–cosmetic product. In addition, all dosage forms can be used for both cosmetics and OTC drug–cosmetic products.

Solution Solution is the simplest type of formulation. Solutions are liquids, they are pourable; they flow and conform to their container at room temperature. According to the FDA's definition, "... a solution is a clear, homogeneous liquid dosage form that contains one or more chemical substances dissolved in a solvent or mixture of mutually miscible solvents."

Solutions can be classified based on the types of solvent used. Three major types are usually distinguished, which are the following:

- **Water-based** (i.e., aqueous) solutions contain water as the vehicle. Examples for water-based solutions include eye makeup remover, hand soap, and many shampoos.
- **Hydroalcoholic** solutions contain a mixture of water and alcohol as the vehicle. Examples for hydroalcoholic solutions include hair spray, mouthwash, after-shave cologne, and facial toner.

- **Anhydrous** (i.e., waterless) solutions contain ingredients other than water as the vehicle. The solvents can be organic solvent, such as for base coat or nail polish remover. Solvents can also be oily components, such as for bath oils or prewaxing oils.

A solution is considered a thermodynamically stable dosage form; it does not tend to change over time. The formulation of solutions is usually simple. Soluble solid ingredients are dissolved first, starting with those that need heating for the dissolution process. Smelly, easily evaporating ingredients are usually added at the end to prevent their loss from the product. Similarly, colored ingredients and color additives are also added at the end since they could make the completion of the dissolution process more difficult to detect.

DID YOU KNOW?

Cleansing wipes can also be classified as solutions since the dry, usually nonwoven clothes are soaked into a cleansing solution.

Emulsion The majority of cosmetic raw materials are not miscible or just partially mixable with each other; in these cases, formulators have to choose a dosage form that can contain immiscible ingredients together. According to the FDA's definition, "... an emulsion consists of a two-phase system comprised of at least two immiscible liquids, one of which is dispersed as droplets (this is usually referred to as the internal or dispersed phase) within the other liquid (otherwise known as the external or continuous phase), generally stabilized with one or more emulsifying agents."

Based on this definition, there are three essential ingredients for an emulsion, namely an oil phase, a water phase, and an emulsifier. There are two main types of emulsions: O/W and W/O. In an O/W emulsion, oil is dispersed in water; the oil is present in the form of droplets, while the water is present as the outer (continuous) phase. The emulsifier molecules coat the surface of the oil droplets, making them more miscible with water and preventing them from flowing to the surface (see an O/W emulsion depicted in Figure 1.9). In an O/W emulsion, the surfactant molecules have their hydrophilic head groups oriented towards the continuous water phase and their hydrophobic tails facing the oil droplets.

Emulsions are the most widely used dosage forms in the cosmetic industry due to their advantages over other dosage forms. They have a unique texture and provide a nice skin feel, and they are used as vehicles to deliver both hydrophilic and hydrophobic ingredients.

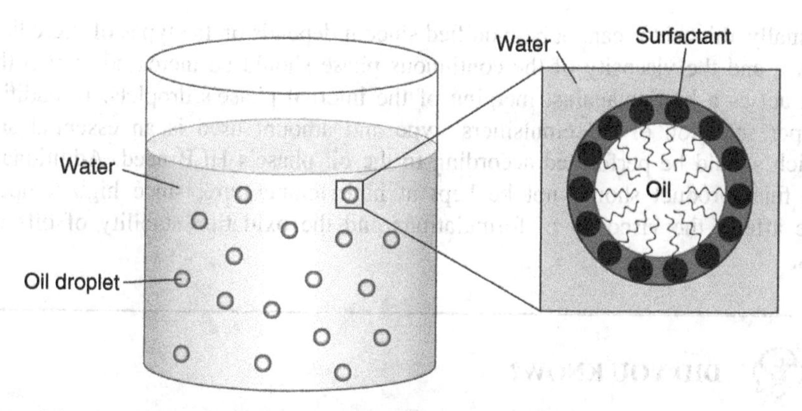

Figure 1.9 An oil-in-water (O/W) emulsion consisting of an oil phase, a water phase, and surfactant molecules surrounding the oil droplets.

- Generally, **O/W emulsions** are chosen for applications that require a relatively small amount of fatty materials, such as for hair conditioners, shaving creams, or facial moisturizing creams.
- On the other hand, W/O emulsions are preferred when a larger amount of oil is desired in the formulation. **W/O emulsions** are greasier, leave a longer-lasting residue, and are more water resistant (as they contain oils in the outer phase). Typical products for which W/O emulsions are preferred include diaper rash products, sunscreens, and barrier creams.
- From a cosmetic perspective, **water-in-silicon (W/Si) emulsions** are also important. These formulations provide a unique, nongreasy skin feel and quick drying effect, leaving the skin smooth. Examples for such formulations include facial foundations, cream eyeshadows, and certain sunscreens.
- In addition to the simple two-phase emulsions, more **complex** types may also be formed, such as water-in-oil-in-water (W/O/W) emulsions, which are basically an emulsion in an emulsion.

Emulsions are generally opaque formulations due to the size of the internal phase's droplets. This white, creamy appearance is generally appealing to consumers. They are thermodynamically instable, which means that they tend to separate over time. As we know, water and oil do not mix. Emulsions incorporate these two ingredients using emulsifiers, which help keep the phases mixed in each other for a longer period of time. However, even with emulsifiers, the phases try to reach the state of lowest energy[2] to be stable. According to Stokes' law, the stability of emulsions can be increased by decreasing the droplet size of the internal phase (which can be achieved by intensive mixing); the densities of the two phases should be as close as possible (usually, this factor cannot be modified since it depends on the types of ingredients used); and the viscosity of the continuous phase should be increased, which then can act as a barrier against merging of the internal phase's droplets. In addition, proper selection of the emulsifiers' type and amount used is an essential step, which should be performed according to the oil phase's HLB need. Additionally, the final product should not be kept at high temperature, since high temperature affects the viscosity of formulations and the oxidation stability of oils and fats.

DID YOU KNOW?

The proportion of the phases does not necessarily correlate to the type of the emulsion. For example, a formulation that consists of 65% water phase and 35% oil phase does not necessarily mean that it is an O/W emulsion. The main factor that determines the type of an emulsion is the solubility of the emulsifier agents. According to Bancroft's rule, "… the phase in which the emulsifier is most soluble becomes the continuous phase."[3] Therefore, water-soluble emulsifiers (with high HLB values) promote the formation of O/W emulsions, whereas oil-soluble emulsifiers (with low HLB values) promote the formation of W/O emulsions.

Based on their viscosity, usually two types of emulsions are distinguished for cosmetic applications, namely lotions and creams.

- **Lotions** are low-viscosity (thin) emulsions behaving as liquids; therefore, they can be poured from a bottle or pumped from a jar. They are designed to be applied without heavy rubbing. According to the FDA's definition, "… a lotion is an emulsion, liquid dosage form." Lotions contain a higher amount of water in the continuous phase than creams. Due to the higher amount of water phase, they are less greasy and easily washable. Lotions are often referred to as "milks" and "balms." Examples for lotions include facial cleansing milks, liquid foundations, aftershave balms, and non-aerosol sunscreen sprays.
- **Creams** are high-viscosity (thicker), semisolid emulsions. According to the FDA's definition, "… a cream is an emulsion usually containing > 20% water and volatiles and/or < 50% hydrocarbons, waxes, or polyols as the vehicle." Since creams contain a higher amount of oil phase, they are generally more greasy, even the O/W types. Creams do not flow readily; therefore, they can be packed into a jar or a tube for dispensing. Examples for creams include facial moisturizing creams, leave-in hair conditioners, sunscreens, cream eyeshadows, and depilatory creams.

The conventional method of preparing a cosmetic emulsion is the hot method. It includes heating the two phases separately and, when both of their temperatures are the same, mixing them together vigorously. Mixing should generally continue until the emulsion is cool to prevent immediate separation. Once cool, additional temperature- and shear-sensitive ingredients can be added and mixed into the emulsion.

Another method, known as the cold process, can also be used for certain emulsions. In this process, the phases can be mixed at room temperature without heating. The advantage of this technique is that it saves energy since no heating is needed, and also saves the processing time as all ingredients, including the temperature-sensitive ingredients, can be combined prior to mixing the two phases. However, it limits the types of thickeners that can be used for the formulations; for example, waxes cannot be used as they require heating.

DID YOU KNOW?

Foams can also be categorized as emulsions where the gas phase is dispersed in a liquid continuous medium.

Ointment Ointments are superthick semisolid products compared to lotions or creams. According to the FDA's definition, "... an ointment is a semisolid dosage form, usually containing <20% water and volatiles and >50% hydrocarbons, waxes, or polyols as the vehicle. They are usually used topically for protection or as medicated skin products."

Ointments have an occlusive nature and provide a seal over the skin. They can contain a small amount of water or can be anhydrous. In the case of anhydrous formulations, the chance for microbiological contamination is low, which is a distinct advantage. However, ointments have a less aesthetic appeal for skin care and dermatology products as they are oily, waxy, greasy, sticky, tacky, and heavy. They are advantageous for smaller skin areas that are extremely dry and need moisture retention, and for areas that are prone to friction from clothing and need protection. Ointments are often opaque and yellowish due to the high amount of oils. Due to the undesired skin feel, there are only a small number of cosmetic products formulated as ointments. Examples include some hair styling products, such as hair pomade; and diaper rash ointments.

The formulation steps for ointments depend on whether they contain water. If they contain water, they can be considered very thin W/O emulsions. These are processed with the hot method. If anhydrous, their formulation consists of mixing the oily ingredients until a homogeneous mixture develops. Heating may be needed.

Paste Pastes are very thick semisolid formulations that are difficult to apply and spread over the skin surface due to their high solid content. They are similar to ointments but contain more solids and, therefore, are stiffer. According to the FDA's definition, "... a paste is a semisolid dosage form that contains a large proportion (20–50%) of solids finely dispersed in a fatty vehicle. This dosage form is generally used for external application to the skin or mucous membranes."

From a cosmetic perspective, pastes as dosage forms can be used as diaper rash treatment products. Additionally, toothpastes are also pastes that are meant to clean and/or polish the teeth. According to the FDA's definition, "... a toothpaste is intended to clean and/or polish the teeth, and it may contain certain additional agents." However, there is a huge difference between "regular" paste and toothpaste from the vehicle's perspective. Regular pastes are anhydrous formulations, based on a fatty vehicle; therefore, they are highly adhesive to surfaces and hard to remove with water. Toothpastes, on the other hand, are water-based formulations that mix well with the saliva. If they were based on a fatty vehicle, their application would be uncomfortable.

Suspension A suspension is a dosage form for delivering insoluble solid ingredients in a liquid medium. According to the FDA's definition, "... a suspension is a liquid dosage form that contains solid particles dispersed in a liquid vehicle."

Based on the type of the liquid vehicle, usually three types of suspensions are distinguished, including the following:

- **Water-based** suspensions;
- **Hydroalcoholic** formulations, such as certain facial toners;
- **Anhydrous** formulations, such as silicone-based antiperspirant sprays, organic solvent-based nail polishes, and any liquid colored cosmetics containing pigments as color additives, such as mascara, liquid eyeliner, and lip gloss.

Suspensions, similar to emulsions, are thermodynamically unstable. The insoluble particles—due to gravitation—tend to sediment to the bottom of the container over time. The rate of sedimentation can be characterized by the Stokes' equation that was discussed under emulsions. According to Stokes' law, factors influencing the stability of suspensions include the viscosity of the liquid phase, difference in densities between the two phases, particle size of the insoluble particles, and gravitational acceleration. Of these factors, the viscosity of the liquid phase and particle size of the dispersed phase can be easily modified to increase suspension stability. Thickeners are essential parts of suspensions since they can increase viscosity and, therefore, slow down the rate of sedimentation of insoluble solid particles and increase the systems' overall stability (see Figure 1.10). The smaller the particle size of the insoluble ingredient(s), the slower the sedimentation. Size reduction can be achieved by grinding the powders using a mortar and a pestle on a laboratory-scale basis or using mills on a larger scale.

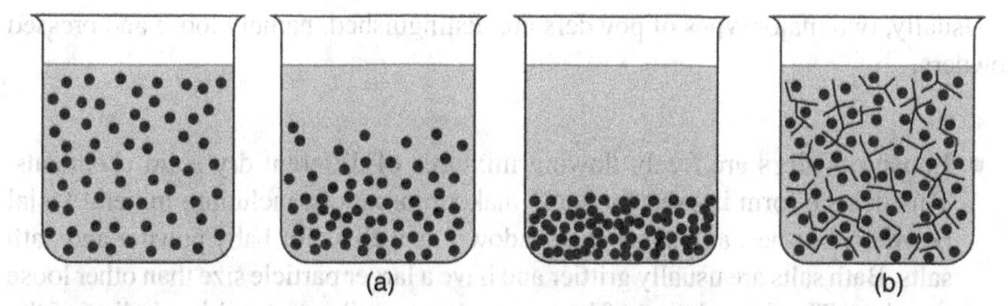

Figure 1.10 Sedimentation of solid particles in a suspension over time: (a) if no thickeners are used, the insoluble solid particles sediment at the bottom of the container in a short period of time and (b) when thickeners are used, they can increase the viscosity of the liquid phase and keep the solid particles dispersed in the liquid medium, slowing down the rate of sedimentation.

The formulation of suspensions usually starts with the size reduction step as well as preparation of the thickener solution, which can be a lengthy procedure. After proper hydration and swelling of the thickener, other liquid ingredients can be mixed into the liquid phase. Finally, the solid particles are wetted and a concentrated slurry is made using a small portion of the liquid phase to get a smooth, uniform preparation. Additional liquid phase can be added after the mixture becomes uniform and no clumps or non-wetted powder aggregates are present.

DID YOU KNOW?

The labeling of certain facial toners, body toners, and eyeshadows indicates that the product should be shaken before use to activate the main ingredients. Usually, this description designates suspensions since suspensions sediment over time; therefore, shaking redisperses ("activates") the insoluble powders.

DID YOU KNOW?

The major difference between a solution and a suspension is the type of solid ingredient dispersed in the liquid vehicle. Solutions contain soluble ingredients; therefore, they are clear formulations. Suspensions contain insoluble ingredients, which make the formulations cloudy.

Powder (Including Loose and Pressed Powder) Powders are solid dosage forms. According to the FDA's definition, " … a powder is an intimate mixture of dry, finely divided chemicals."

Usually, two major types of powders are distinguished, namely loose and pressed powders.

- **Loose powders** are freely flowing mixtures of different dry solid chemicals. This dosage form is used for some makeup products, including mineral facial powders, blushes, and some eye shadows, as well as for baby powder and bath salts. Bath salts are usually grittier and have a larger particle size than other loose powders. The formulation of loose powders usually starts with grinding of the raw materials to provide a similar, fine particle size for them (unless they are purchased as a fine powder). After grinding, the ingredients are mixed together and then sieved before filling them into the container. Geometric dilution is used when blending two or more powder ingredients of unequal quantities. In this method, ingredients are mixed in approximately equal volumes. Geometric dilution provides uniform distribution for powders, which contain various powder ingredients in varying quantities.
- **Pressed powders** are made of a blend of freely flowing powders via compression. Pressed powders are popular for eyeshadows, facial powders, finishing powders, and blushes as well. Bath bombs are also examples for pressed powders, which are available in various shapes.
  The first steps of their formulation are similar to those of loose powders (grinding and mixing); however, instead of filling them into the container, they are pressed into shape. Today, a wide variety of shapes are available on the market, including rectangular, triangle, round, flower, heart, and many others. In the case of cosmetic formulations, it is usual that the powders are directly compressed into their final container (called a godet, i.e., an aluminum plate). Pressed powders usually contain binders, which keep the particles together and prevent breaking and crumbling upon shaking. Examples for binders include solid ingredients, such as zinc stearate, and starches, as well as liquids, such as isopropyl isostearate, triglyceride, and dimethicone.

Capsule According to the FDA's definition, "… a capsule is a solid dosage form consisting of a shell and a filling. The shell is composed of a single sealed enclosure, or two halves that fit together and which are sometimes sealed with a band. Capsule shells may be made from gelatin, starch, or cellulose, or other suitable materials, may be soft or hard, and are filled with solid or liquid ingredients that can be poured or squeezed."

From a cosmetic perspective, capsules are always soft gelatin capsules that contain an oily liquid ingredient inside. Examples for such products include bath oil beads and anti-aging serum capsules. Bath oil beads are used in the bathtub. Due to the gelatin, the capsules easily dissolve when they come into contact with bathwater and release their contents into the bathtub. Anti-aging serum capsules usually have a small tab that can be gently twisted or cut to open the capsule. Since soft capsules are easy to squeeze, the product can be get released the shell by gently squeezing it.

Gel Gel is usually a transparent, semisolid dosage form. According to the FDA's definition, "... a gel is a semisolid dosage form that contains a gelling agent to provide stiffness to a solution or a colloidal dispersion."

Gelling agents are synonymous with thickeners, which increase viscosity and provide a complex internal structure (to find examples for thickeners, refer to Section 2 of this chapter). Gels may also contain fragrance beads and exfoliating beads, as some skin cleansing gels do. Additional examples for gels include hair styling gels, facial cleansing gels, shaving gels, aftershave gels, after-sun gels, extrudable deodorant and/or antiperspirant gels, and hand sanitizer gels, among others.

Based on the nature of the vehicle, two main types of gels are distinguished, including the following:

- **Water-based** formulations, such as facial cleansers; and
- **Hydroalcoholic** formulations, such as hair styling gels and hand sanitizers.

Gels contain a higher amount of water compared to other semisolid dosage forms. As water evaporates after application, it provides a cooling effect. This can be advantageous in the case of sunburn products when the cooling sensation can also be perceived as an analgesic (i.e., pain reliever) effect, and also in the case of aftershave gels where water and/or alcohol evaporation has a refreshing effect.

The formulation of gels usually starts with hydration of the thickener, which may take hours. As discussed in Section 2 of this chapter, certain thickeners are not stable at extreme pH values, and they may be sensitive to shearing. These characteristics should be taken into account when selecting the thickener. After the thickener is completely hydrated, the additional ingredients can be added and mixed until uniform. Smelly and colored ingredients are usually added at the end, similar to solutions.

DID YOU KNOW?

Today, toothpaste gels are also available on the market. They can be considered a combination of a toothpaste and a gel. Their purpose of use is identical to that of toothpastes, i.e., cleaning and polishing the teeth and stain removal.

Stick A stick is a solid dosage form that is made of waxes and a smaller amount of oils. According to the FDA's definition, "... a stick is a dosage form prepared in a relatively long and slender often cylindrical form."

Examples for sticks include color cosmetics, such as lipsticks, lip liners, eyeshadow sticks, eyeliners, blush pencils, and concealers, as well as personal care products, such as deodorant/antiperspirant sticks and sunscreen sticks. Sticks are advantageous when consumers do not want to touch and apply products with their fingers.

They deliver the colors or active ingredients, such as antiperspirant or sunscreen ingredients, by a rubbing action.

Sticks are made by a specific technique called molding. First, the waxy ingredients that are solid at room temperature are melted and mixed with the oils and additional ingredients. While still hot, the mix is poured either into the final containers, such as lip balm cases, or into metal or plastic molds and allowed to cool. As the sticks cool, they take the shape of the mold/package. For lipsticks, a final step, called flaming, is usually also included to provide the sticks with a shiny surface (see more detail on the molding technique in Section 1 of Chapter 4).

Aerosol Aerosols are more of a packaging choice than a specific product type. Many of the above-discussed dosage forms can be produced in the form of an aerosol, including lotions, creams, and suspensions, if a proper can, propellant, and nozzle setup is used. They are composed of a product concentrate and a liquefied or compressed gas propellant. According to the FDA's definition, "… an aerosol is a product that is packaged under pressure and contains various ingredients that are released upon activation of an appropriate valve system."

Aerosols are easy to use and provide a quick drying effect, which makes them popular for certain applications. Examples for aerosol products include hair sprays, hair mousse, shaving cream, deodorants/antiperspirants, sunscreens, and sunless tanners.

The product concentrate is made as if it were a regular product, such as a cream. Then, the concentrate and the propellant are filled into a can, usually using the pressure filling technique. During this technique, first the product concentrate is filled into the can and the valve is crimped in place. Then, the propellant is filled into the can under pressure through the valve. This procedure requires special caution; therefore, aerosols are usually packaged in specialized buildings by specialized personnel.

Aerosol cans are made up of four components: the container, the valve, the actuator, and the cap. The type of actuator and product concentrate has to be selected according to the required application. For shaving foam, an actuator with a larger nozzle and a good foaming emulsion are optimal. These together can provide a product that is able to stand on the palms/face. On the other hand, for an aerosol sunscreen, an actuator with a small nozzle and a light, easily dispersible, and quickly evaporating product is recommended.

DID YOU KNOW?

Foams can be generated from aerosol cans and non-aerosol containers as well. In the case of foams generated from aerosol cans, the product concentrate is an emulsion, which is filled into the can with propellants. An example for this type of foam is shaving foam.

GLOSSARY OF TERMS FOR SECTION 3

Aerosol: A product that is packaged under pressure and contains various ingredients that are released upon activation of an appropriate valve system.

Anhydrous: Waterless, water free.

Aqueous: Water-based.

Capsule: A solid dosage form consisting of a shell and a powder or liquid filling.

Cream: A semisolid emulsion with medium viscosity. It is more viscous than a lotion, but less viscous than an ointment.

Dosage form: The final physical form of a mixture of ingredients that consumers can take in their hands, purchase, and use.

Emulsion: Usually a white, opaque system that consists of at least two immiscible liquids, one of which is dispersed as droplets (internal phase) in the other (external phase). The system is generally stabilized with emulsifiers.

Gel: A clear semisolid dosage form that contains a gelling agent, which provides stiffness to the product.

Hydroalcoholic: Containing a mixture of water and alcohol.

Liquid dosage form: A dosage form that has a liquid consistency and is freely flowing. It cannot be directly taken into the hands since it would flow off the hand.

Loose powder: A solid dosage form containing a freely flowing mixture of different dry solid ingredients.

Lotion: A low-viscosity liquid emulsion.

Ointment: A highly viscous, usually greasy, semisolid dosage form. It is more viscous than a cream.

Paste: A very thick semisolid dosage form containing a high amount of solids finely dispersed in the vehicle.

Pressed powder: A solid dosage form that contains a freely flowing mixture of different dry solid ingredients in a compressed form.

Sedimentation: A process during which suspended particles in a suspension slowly settle down at the bottom of the container.

Semisolid dosage form: A dosage form that is highly viscous; it is thinner than solids but thicker than liquids.

Solid dosage form: A dosage form that consists of primarily dry solid particles mixed and/or pressed together, or waxy ingredients molded into a specific shape.

Solution: A clear, homogeneous liquid dosage form that contains one or more chemical substances dissolved in a solvent or mixture of mutually miscible solvents.

Stick: A solid dosage form that is made of waxes and a smaller amount of oils and is prepared in a relatively long cylindrical form.

Stokes' law: A mathematical equation that describes the separation of emulsions and sedimentation of suspensions.

Suspension: An opaque liquid dosage form that contains solid particles dispersed in a liquid vehicle.

REVIEW QUESTIONS FOR SECTION 3

Multiple Choice Questions

1. Which of the following dosage forms is made of a gelatin shell?
a) Paste
b) Gel
c) Capsule
d) Lotion

2. Which of the following is a liquid dosage form?
a) Solution
b) Paste
c) Gel
d) Powder

3. What is the internal phase in an oil-in-water (O/W) emulsion?
a) Oil
b) Water
c) Both
d) None of them

4. Which of the following contains the highest amount of water typically?
a) Oil-in-water emulsion
b) Water-in-oil emulsion
c) Ointment
d) Gel

5. Which of the following dosage forms typically contain thickeners?
a) Gels and suspensions
b) Solutions and gels
c) Ointments and gels
d) Sticks and gels

6. What is the major difference between a solution and a suspension?
   a) The type of the propellant
   b) The type of the vehicle
   c) The solubility of solid ingredients dispersed in the liquid vehicle
   d) All of the above
7. What is the major difference between a cream and a lotion?
   a) Color
   b) Viscosity
   c) pH
   d) The type of the vehicle
8. Which of the following increases an emulsion's instability according to Stokes' law?
   a) Viscosity of the continuous phase
   b) Internal phase's droplet size
   c) Difference in the phases' densities
   d) All of the above
9. The term "dosage form" can be defined as ___.
   a) The physical form of the ingredients (cosmetic ingredients and/or active ingredients)
   b) The amount of the cosmetic ingredients and/or active ingredients used in a product
   c) The final physical form of a mixture of ingredients (cosmetic ingredients and/or active ingredients)
   d) The amount of a product that is recommended to be used daily
10. When mixing oils with water in the presence of an emulsifier, what does the type of the forming emulsion depend on?
   a) The amount of the water phase
   b) The amount of the oil phase
   c) The solubility of the emulsifier, if it is water-soluble, O/W emulsion forms
   d) The solubility of the emulsifier, if it is water-soluble, W/O emulsion forms

Fact or Fiction?

_____ a) Both lotions and creams are emulsions.
_____ b) All foams are aerosols.
_____ c) Suspensions are thermodynamically instable formulations.
_____ d) Creams are more viscous than lotions.

Matching

Match the dosage forms in column A with their appropriate definition in column B.

	Column A	Column B
______	1. Aerosol foam	A. A long, slender solid dosage form based on waxes and oils
______	2. Capsule	B. Clear, homogeneous liquid dosage form
______	3. Cream	C. Compressed solid dosage form made of a blend of powders
______	4. Gel	D. Emulsion containing a propellant
______	5. Loose powder	E. Liquid dosage form containing insoluble solid particles
______	6. Lotion	F. Opaque, thick semisolid dosage form
______	7. Ointment	G. Opaque, thin liquid dosage form
______	8. Paste	H. Solid dosage form consisting of a shell and a filling
______	9. Pressed powder	I. Solid dosage form, a mixture of freely flowing powders
______	10. Solution	J. Usually transparent, viscous semisolid dosage form
______	11. Stick	K. Very thick semisolid dosage form with a high amount of solids
______	12. Suspension	L. Yellowish, opaque, greasy, and sticky semisolid dosage form

Matching

Match the examples in column A with their appropriate dosage form in column B.

	Column A	Column B
______	1. Aftershave balm	A. Aerosol
______	2. Baby powder	B. Capsule
______	3. Bath bead	C. Cream
______	4. Compact blush	D. Gel
______	5. Hair pomade	E. Loose powder
______	6. Hair spray	F. Lotion
______	7. Hand sanitizer gel	G. Ointment
______	8. Leave-in hair conditioner	H. Pressed powder
______	9. Lip balm	I. Solution
______	10. Mouthwash	J. Stick
______	11. Nail polish	K. Suspension

REFERENCES

1. FDA: Dosage Form, Last update: 4/30/2009, Accessed 4/13/2014 at http://www.fda.gov/drugs/developmentapprovalprocess/formssubmissionrequirements/electronicsubmissions/datastandardsmanualmonographs/ucm071666.htm
2. Gibbs, J. W.: On the equilibrium of heterogeneous substances. *Trans Connect Acad.* 1878;3:108–248.
3. Bancroft, W. D.: The theory of emulsification. *J Phys Chem.* 1913;17:501–519.

2

LEGISLATION FOR COSMETICS AND OTC DRUG–COSMETIC PRODUCTS

SECTION 1: CURRENT RULES AND REGULATIONS FOR COSMETICS AND OTC DRUG–COSMETIC PRODUCTS IN THE UNITED STATES AND EUROPEAN UNION

LEARNING OBJECTIVES

Upon completion of this section, the reader will be able to

1. define the following terms:

Adulterated	Import	Inspection	Misbranded
Premarket approval	Prohibited ingredient	Recall	Responsible person
Restricted ingredient	VCRP		

2. name the authority that regulates cosmetics in the US;

Introduction to Cosmetic Formulation and Technology, First Edition. Gabriella Baki and Kenneth S. Alexander.
© 2015 John Wiley & Sons, Inc. Published 2015 by John Wiley & Sons, Inc.

3. name the authority that regulates OTC cosmetic–drug products in the US;
4. name the authority that regulates cosmetics in the EU;
5. explain what options cosmetic manufacturers and distributors have for registering their facilities and products;
6. explain what the Voluntary Cosmetic Registration Program (VCRP) is and how it works;
7. define the term "premarket approval" and discuss whether it is necessary for cosmetic products in the US and the EU, respectively;
8. explain how a manufacturer can assure and maintain high quality of its products;
9. discuss how the use of cosmetic ingredients is restricted in the US and the EU, respectively;
10. explain who is responsible for the safety of cosmetic products in the US and the EU, respectively;
11. discuss how the FDA can monitor if the cosmetics are safe;
12. explain what makes a cosmetic product adulterated under the US law;
13. explain how the safety of OTC drug–cosmetic products is established in the US;
14. discuss how animal testing of cosmetic products is regulated in the US and the EU, respectively;
15. explain what makes a cosmetic product misbranded under the US law;
16. discuss what rules cosmetics imported to the US and the EU must meet, respectively;
17. define the term "recall" and explain who can initiate a recall in the US and the EU, respectively;
18. discuss what the FDA can do with an unsafe cosmetic product.

KEY CONCEPTS

1. From the consumers' perspective, it does not make a difference if a mouthwash is considered a drug or a cosmetic. However, the regulatory consequences of classifying a product as a drug rather than as a cosmetic are substantial.
2. In the US, cosmetic manufacturers are not required to register either their cosmetic manufacturing establishments or their cosmetic products and ingredients with the FDA.
3. In the US, cosmetic products are not subject to FDA premarket approval.
4. The FDA does not require all manufacturers to use the same quality standards and systems.
5. In the United States, cosmetic manufacturers can use almost any raw materials they want as cosmetic ingredients without any approval or limitation. Restrictions

only apply for the use of color additives, a small number of prohibited and restricted ingredients, as well as active ingredients.

6. In the US, cosmetic products have to be safe, and cosmetic manufacturers are responsible for substantiating safety of their products and ingredients before marketing.
7. All cosmetic products imported into the US are subject to the same laws and regulations as those produced in the US.
8. The FDA has no authority to order a mandatory recall of a cosmetic product; this process is based on a voluntary action. However, the FDA can take authoritative legal actions against manufacturers that persist in marketing a defective product.

Introduction

After reading Chapter 1, you understand how a cosmetic and an OTC drug are defined in the US. ❶ **From the consumers' perspective, it does not make a difference if a mouthwash is considered a drug or a cosmetic. However, the regulatory consequences of classifying a product as a drug rather than as a cosmetic are substantial.** It is due to the fact that the requirements for drugs are more extensive than the requirements applicable to cosmetics. When consumers understand that the FDA regulates cosmetic products, the majority of them assume that their regulation is very similar to that of drugs. However, this is not true. To tell the truth, the FDA has only limited regulatory authority over cosmetic products (unlike over drugs). The FDA can and does monitor companies on a randomized basis; however, many actions with respect to cosmetics are voluntary, and it is the manufacturers' responsibility to meet the current requirements. A reason for this limited authority is that in the US, the basic legislation regulating cosmetics has remained largely unchanged for a significant period of time (the FD&C Act was introduced in 1938). The definition of a cosmetic was developed when the range of cosmetic products and ingredients used in them was quite limited.

By contrast, the legislation in the European Union (EU) was subject to numerous amendments, corrections, and adaptations since it was introduced due to technical progress. It enabled definitions and controls to keep pace with product developments. As a result, many new products, including sunscreens and toothpaste, were included within the category "cosmetics" instead of being categorized as "drugs." Due to continuous updates, the EU Cosmetics Directive, and then the EU Cosmetic Product Regulation, became a model of modern cosmetic regulations worldwide. Countries around the world are trying to harmonize their regulations covering cosmetics and tend to get their own regulation closer to the one of the EU. Therefore, this section reviews both the US and the EU regulations.

Regulation in the United States

Registration of Manufacturing Facilities and Products

Cosmetics ● **In the US, cosmetic manufacturers are not required to register either their cosmetic manufacturing establishments or their cosmetic products and ingredients with the FDA.** However, manufacturers and distributors of cosmetic products are encouraged to register their facilities and submit information on their cosmetic products through the FDA's voluntary reporting system, the VCRP.[1,2] As its name implies, it is a voluntary step that might be taken if the cosmetic company desires to do so. An important fact to emphasize is that it does not apply to cosmetic products for professional use only, such as products used in beauty salons, and products that are not for sale, such as hotel samples, free gifts, or cosmetic products you make at your home to give them to your friends.[3] The purpose of this program is to provide a way for the cosmetic industry to report information to the FDA on their manufacturing facilities as well as cosmetics products after they are on the market. Initially, the VCRP was divided into three parts: establishment registration, product notifications, and adverse event reporting. Today, adverse event reporting is no longer part of the VCRP.[4] The VCRP helps the FDA assure the safety and proper labeling of cosmetic products sold in the US. In addition, information reported is also used by the Cosmetic Ingredient Review (CIR) panel (see more detail in Section 3 of this chapter) in assessing ingredient safety and determining priorities for ingredient safety reviews.

DID YOU KNOW?

Since registration of cosmetic products is not required in the US, the use of an "FDA-registered" logo is not allowed. It could mislead consumers to believe that a product bearing the logo is better than another not bearing the logo.

OTC Drugs Companies manufacturing OTC products, including OTC cosmetic–drug products, such as a sunscreen, are required to register their companies every year with the FDA and update their list of all manufactured drugs twice annually.[5,6]

DID YOU KNOW?

Participation in the VCRP does not mean that a cosmetic manufacturing establishment or cosmetic product is approved. Uninformed consumers might assume that a company's product participating in the VCRP is better than another company's product that does not participate in the VCRP. However, it is not true since taking part in this program only means that a company voluntarily provides information to the FDA. As a result, using the VCRP participation as a promotional tool is prohibited by the FDA. Companies using the fact of participation in their product advertisements, for instance, are misleading the consumers, and it is against the law.[7,8]

Premarket Approval

Cosmetics Premarket approval is a process in which a product is approved by a regulatory authority before entering the market and being sold to consumers to ensure its safety and efficacy for its intended use(s). ● **In the US, cosmetic products are not subject to FDA premarket approval.**

OTC Drugs In the case of OTC medications, including drug–cosmetic products, premarket approval may be necessary. If a manufacturer uses ingredients listed in the OTC monograph and complies with the monograph, premarket approval is not necessary since the conditions stated in a monograph ensure safety and efficacy. In contrast, new active ingredients are subject to the FDA's premarket approval. In the premarket approval, companies have to prove that their product is effective and safe for its intended use(s). This process requires companies to perform clinical trials. Clinical trials are experiments that use human subjects to see whether a drug is effective and what side effects it may cause.[9]

Manufacturing

Cosmetics Manufacture of cosmetic products should be performed in such a way that products are safe and effective and their quality is constant throughout all batches. ● **The FDA does not require all manufacturers to use the same quality standards and systems.** However, it encourages them to follow the Good Manufacturing Practices (GMPs). It is worth emphasizing that, although following the GMPs is not a requirement in the case of cosmetics, the FDA has a cosmetic GMP guidance for the industry. This document provides guidance for the industry in identifying the standards and potential issues that can affect the quality of cosmetics (see more information on the cosmetic GMPs in Section 4 of this chapter). In addition, quality assurance guidelines are also published by the Personal Care Product Council (PCPC) to help cosmetic manufacturers establish their GMPs and quality assurance programs.

OTC Drugs The law requires strict adherence to GMP requirements for drugs, and there are regulations specifying the minimum current GMP requirements.[10,11]

Use of Ingredients

Cosmetics ● **In the US, cosmetic manufacturers can use almost any raw materials they want as cosmetic ingredients without any approval or limitation.**

Restrictions only apply for the use of color additives, a small number of prohibited and restricted ingredients, as well as active ingredients.

- As discussed in Chapter 1, there is a positive list of color additives. It includes ingredients that are specifically approved for cosmetics. All color additives must be tested for safety and approved for their intended use by the FDA before they can be marketed in the US.
- In the US, there is only a short list of prohibited ingredients, including bithionol, chlorofluorocarbon propellants, chloroform, halogenated salicylanilides, methylene chloride, vinyl chloride, zirconium-containing complexes, and some prohibited cattle materials.
- In addition, there are some restrictions regarding the use of hexachlorophene and mercury compounds in cosmetics as well as the use of the term "sunscreens."[12,13]
- As discussed previously, cosmetics cannot contain any active ingredients.

OTC Drugs OTC drug-active ingredients are strictly regulated. In general, they must either receive premarket approval by FDA through the New Drug Application (NDA) process or conform to an OTC monograph for a particular drug category (as discussed previously in Section 2 of Chapter 1).[9]

DID YOU KNOW?

An NDA is an application submitted to the FDA if a company would like to introduce a new drug product (i.e., active ingredient) into the US market.

As for the inactive ingredients, there is no list of "approved" ingredients that manufacturers can select their ingredients from. For OTC drug–cosmetic products subject to FDA premarket approval (new drugs), inactives are only approved as components of new drugs. For OTC drug–cosmetic products that are regulated under the FDA's monograph system, there is no required approval for the excipients. Inactive ingredients that are incorporated into OTC drug–cosmetic products must only be determined to be safe and suitable by the manufacturer before use.[14]

Testing and Product Safety

Cosmetics ❶ **In the US, cosmetic products have to be safe and cosmetic manufacturers are responsible for substantiating safety of their products and ingredients before marketing.** However, there is no specific protocol in the relevant regulations on how to perform safety testing. It is each manufacturer's task to determine how to test a product and ensure that it is safe for human use. They usually follow guidelines developed by the industry.

As for the safety of cosmetic ingredients, the US has an independent expert panel, usually referred to as the "CIR," that reviews relevant data on cosmetic ingredients and decides whether they are safe under their current conditions of use. It helps manufacturers choose ingredients for their products and continuously be informed about the safety of ingredients. In addition, in 2008, the PCPC (see more information on PCPC in Section 3 of this chapter) initiated a program called the Consumer Commitment Code (CCC) on voluntary basis for all companies in the cosmetic industry. The code goes beyond the existing law, and its purpose is to help companies ensure safety of their products. Under this code, participating companies agree to follow the recommendation of the CIR, use ingredients that are substantiated for safety, participate in the VCRP, report serious and unexpected adverse reactions with cosmetic products, produce their products in a manner consistent with good manufacturing procedures, and substantiate the safety of their raw materials and finished goods.[15]

In the US, manufacturers also have an option of not performing safety testing. In such cases, they have to put the following warning statement on the principal display panel (PDP) of the product's label: "Warning – The safety of this product has not been determined."[16] It should be noted that this action is very rare.

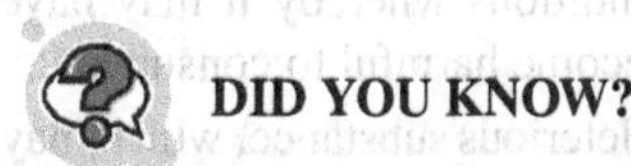

In the US, cosmetic companies can still test their cosmetic products on animals. The Food, Drug, and Cosmetic (FD&C) Act does not specifically require the use of animals in testing cosmetics for safety. However, the agency has consistently advised cosmetic manufacturers to employ any testing appropriate and effective for substantiating the safety of their products. It remains the responsibility of the manufacturer to substantiate the safety of both ingredients and finished cosmetic products prior to marketing.

As in-market surveillance, the FDA monitors the safety of cosmetic products that are being marketed. The FD&C Act authorizes the FDA to conduct inspections of cosmetic firms (on the basis of complaints or suspicion of violation of law) without prior notice in order to assure compliance with the regulations. If they find proof for the harmfulness of a product to consumers when used as intended, the FDA takes regulatory action. In addition to inspections, the FDA has a number of ways to monitor the safety of cosmetic products. A short list of the possible ways is provided as follows:[17]

- FDA encourages cosmetic firms to report product information through the VCRP since it uses the information provided in the database.
- In addition to the on-site inspections, FDA works with US Customs and Border Protection to examine imported cosmetics and make sure that they meet the US regulations.

- FDA periodically buys cosmetics and analyzes them, especially if it is aware of a potential problem. The information obtained can be used to alert consumers, support regulatory actions, or issue guidance for industry.
- FDA takes the results from the CIR into consideration when evaluating safety.
- FDA also monitors peer-reviewed literature and takes data provided by other governmental agencies, such as the National Toxicology Program (NTP) into consideration.
- Bad reactions to cosmetics can be reported by consumers and health-care providers to the FDA to notify it about unsafe cosmetics. This information is also being used by FDA in monitoring the safety of cosmetics.

To ensure safety of marketed cosmetic products, the FD&C Act prohibits the distribution of **adulterated** cosmetics. Under this law, a cosmetic can become adulterated in a number of ways, such as if[18]

- it contains any poisonous or deleterious substance, which may render the product harmful to consumers under customary conditions of use;
- it consists of any filthy, putrid, or decomposed substance;
- it is manufactured or held under insanitary conditions whereby it may have become contaminated with filth, or may have become harmful to consumers;
- its container is composed of any poisonous or deleterious substance, which may render the contents injurious to health;
- it contains a color additive that is not approved for the intended use.

DID YOU KNOW?

A coal-tar hair dye is not considered adulterated in the US as long as it bears the required caution statement on its label and gives instructions for performing a patch test, even if it causes irritation to the skin or is otherwise harmful to the human body.

As part of the safety requirements for cosmetic products, there are a number of mandatory warning and caution statements for specific products that have to appear on the label of a cosmetic (see more detail on labeling statements in Section 2 of this chapter).

OTC Drugs In the case of drugs, evidence of efficacy and safety is essential in order to receive premarket approval for a product from the FDA. If a manufacturer uses OTC-monographed active ingredients, safety and effectiveness are ensured since monographs ensure these characteristics. In such cases, clinical studies are generally not required to prove safety and efficacy; however, there may be exceptions. Products that confirm to a final monograph may be marketed without further FDA review.

Those that do not conform, or are new ingredients, must be reviewed by the NDA process, including clinical trials, to gain information on the drug. In addition, to further ensure the safe use of OTC products, the FDA conducts studies to learn if potential consumers understand the label, can properly choose the medicine for their needs, and use it according to the directions. An additional requirement for OTC drug–cosmetic product – which is only voluntary in the case of cosmetics – is the mandatory reporting of serious adverse events.

Packaging and Labeling

Cosmetics Products must be packaged and labeled before they can go on the market to inform consumers about the intended uses, ingredients, safety issues, and other important facts required by the law. There are certain rules for labeling that manufacturers are required to meet. In general, labels of cosmetic products and OTC drugs have to be legible, easy to understand, and truthful (i.e., not misleading). Labeling and packaging of cosmetic products are regulated under the FD&C Act and the Fair Packaging and Labeling (FP&L) Act of 1967, while that of OTC drugs are laid down in the Code of Federal Regulations (CFR).[19] The regulation on labeling for both types of products is very detailed and requires all consumer products in interstate commerce to be honestly and informatively labeled. The FDA has a Labeling Manual on its web site providing information on the labeling requirements for cosmetics. As part of the manual, it has a Labeling Guide that helps manufacturers and distributors answer their labeling questions through examples.

All cosmetic products are subject to the FP&L Act excluding products for professional use only (e.g., preparations used by professionals on customers at their place of work) or which are not distributed for retail sale (e.g., free product samples). This is why only limited information can be found on free products received in hotels. (See details of labeling rules in Section 2 of this chapter.)

As part of ensuring compliance with the relevant labeling regulations, the FDA prohibits the marketing of **misbranded** cosmetics. The term "misbranding" applies to labeling and packaging. Under the FD&C Act, a cosmetic is considered misbranded if[20]

- its labeling is false or misleading;
- its label does not include all required information;
- the required information is not adequately prominent and conspicuous;
- its container is formed or filled as to be misleading;
- it is a color additive and its packaging and labeling are improper;
- its packaging or labeling does not comply with the Poison Prevention Packaging Act of 1970, which requires child-resistant packaging (for some products only; e.g., primers and glue removers of artificial nails).

When sold for retail, certain products, such as liquid oral hygiene products (e.g., mouthwashes) and all cosmetic vaginal products (e.g., douches), must be packaged in tamper-resistant packages, which if breached or missing, alert a consumer that tampering has occurred.[21]

Since OTC medications are used directly by consumers without consultation with a physician or pharmacist, the labeling of these products has to ensure that patients have enough information in an easily understandable form so they can appropriately self-select the products. The information panel has to bear a so-called Drug Facts box, which contains information in a standardized format (see more information on this in Section 2 of this chapter).

Import of Products

Cosmetics Companies importing products considered to be solely cosmetics in the US are not required to register with the FDA, and no registration number is required for importing cosmetics into the US. However, ❼ **all cosmetic products imported into the US are subject to the same laws and regulations as those produced in the US.** It means that they must be safe for their intended uses and must comply with all labeling and packaging rules.[22]

As discussed previously, there are certain products that are considered drugs in the US, but cosmetics in another country. Examples include antidandruff shampoos, cosmetics with ultraviolet (UV) filters, or antiacne medications. The importer of such products has to pay attention to the US classification of the product and make sure that it complies with all relevant rules. When cosmetics enter the US border, they are subject to examination by the US Customs. Products that do not comply with FDA laws and regulations are subject to refusal of admission into the US. The most common reasons for import refusals of cosmetics are labeling violations, the illegal use of color additives, and the presence of poisonous or deleterious substances such as pathogenic microorganisms.

As not all cosmetics are inspected or sampled on entry, the FDA issues import alerts to help inspectors target their efforts. Products listed in import alerts are subject to detention without physical examination. In general, import alerts on cosmetics focus on therapeutic claims, use of unapproved color additives and poisonous or deleterious substances, microbial contaminations, and labeling failures.

OTC Drugs All foreign drug establishments whose products are imported or offered for import into the US are required to register their establishment with the FDA and renew it annually. In addition, similar to US manufacturers, they are required to list all of their drug products in commercial distribution in the US and update it twice annually. Imported products are subject to inspection at the time of entry by the US Customs, and all shipments that do not comply with the US are subject to retention. In addition, to ensure that the FDA is notified of all drugs imported into the US, the importer has to file an entry notice and an entry bond with the US Customs, pending a decision regarding the admissibility of the product.[23]

Recalls

Cosmetics Recall is a voluntary action taken by manufacturers and distributors as their responsibility to protect the public health and well-being from products that present a risk of injury or gross deception or are otherwise defective. Recalls of cosmetics and OTC drug–cosmetic products may be conducted on a firm's own initiative or on a FDA request.[24] ● **Under the FD&C Act, the FDA has no authority to order a mandatory recall of a cosmetic product; this process is based on a voluntary action. However, the FDA has roles in facilitating recalls and can take authoritative legal actions against manufacturers that persist in marketing a defective product.** A summary of all actions the FDA can take is provided as follows:[17,25]

- It can request that a firm recall a product by issuing a warning letter. In a warning letter, the FDA informs firms that it is aware of violations and a corrective action is needed. Warning letters are usually posted on the FDA's web site.
- It can have targeted establishment inspections to check whether compliance with the regulations.
- It can monitor the progress of a recall.
- It can conduct an audit at the wholesale or retail level for customers to verify the recall's effectiveness.
- It can evaluate the health hazard represented by the product and assign it a classification.
- If notification of the public is necessary, it assures that the company performs the notification or if the company fails in this duty then the FDA itself notifies the consumers.
- It can ask a federal court to issue an injunction (the court issues an order requiring a company to do or cease doing a specific action, e.g., production or distribution of a defective product).
- It can request that US marshals seize the products (it means that the products are taken in possession by a district court until the problem is resolved).
- It can initiate criminal action.
- It can also refuse entry of an imported product.

Manufacturers and/or distributors may initiate a recall at any time to fulfill their responsibility to protect the public health from products that present a risk of injury or gross deception or are otherwise defective. Firms may also initiate a recall following notification of a problem by FDA or a state agency, in response to a formal request by the FDA, or as ordered by the FDA.[26]

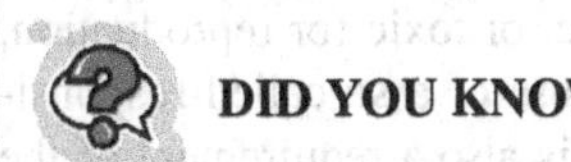

DID YOU KNOW?

You can often hear the terms "withdrawal" and "recall." They both refer to removal or correction of a marketed product; however, they are not the same. Recall is initiated when a product, according to the FDA, is in violation of the laws and against which the FDA would initiate legal action. Therefore, manufacturers typically initiate voluntary recalls when a defect is found to avoid a potentially more significant enforcement action by FDA. On the other hand, a withdrawal is performed when there is only a minor violation that would not be subject to legal action by the FDA or there is no violation at all, e.g., routine equipment repairs.

OTC Drugs Similar to cosmetics, recall is a voluntary action for OTC drugs. The FDA does not have authority to recall OTC drugs; however, it can initiate the same actions for OTC drug–cosmetic combination products as for cosmetics.

Summary In conclusion, the regulations are stricter for OTC drug–cosmetic products and their manufacturers and distributors have more obligations as well. Table 2.1 summarizes the rules and regulations for cosmetics versus OTC drug–cosmetic products in the US.

Regulation in the European Union

The provisions of the EU Cosmetic Products Regulation (EU Regulation 1223/2009) aim at ensuring that the consumers' health is protected and they are well informed by monitoring the composition and labeling of products. The Regulation also provides information on the assessment of product safety and the prohibition of animal testing. The European Commission (EC) has overall responsibility for cosmetics legislation within the EU. Each member state designates a competent authority that enforces the legislation. This part reviews the EU cosmetic regulations. It should be kept in mind that certain EU cosmetics, e.g., sunscreens and antiacne products, are considered drugs in the US.

Registration of Manufacturing Facilities and Products The EU regulation does not require registration of cosmetic products in the EU. However, completion of a premarket notification of each cosmetic product is mandatory as per Article 13 of the Regulation. Notification is an administrative process performed electronically requiring submission of some essential data regarding cosmetic products. The data include the following among others: category of cosmetic product, member state in which the cosmetic product is to be placed on the market, identity information of the responsible person, the presence of substances in the form of nanomaterials, information on substances classified as carcinogenic, mutagenic, or toxic for reproduction, and relevant product specifications. Providing the address of a so-called responsible person, which is the manufacturer in certain cases, is also a requirement of the regulation.[27]

TABLE 2.1 Summary of the Rules and Regulations for Cosmetics and OTC Drug–Cosmetic Products in the United States

	Cosmetics	OTC Drug–Cosmetic Products
Registration of companies	• Voluntary through VCRP	• Required • Has to be done every year
Registration of products	• Voluntary through VCRP	• Required • Has to be done twice annually
Premarket approval	• Not a requirement	• Required for new drugs • Not required for ingredients listed in the OTC monograph
Product manufacturing	• GMP is recommended	• GMP is required
Use of ingredients	• Strict regulation for color additives • Eight prohibited ingredients • Three restricted ingredients • No active ingredients can be used	• Active ingredients: either new drugs or are listed in the OTC monograph • Inactive ingredients are approved as part of the final product
Testing and product safety	• Products have to be safe • Testing is manufacturers' responsibility • No official protocol • CIRs ingredient safety	• Safety and efficacy have to be proven
Packaging and labeling	• Strict regulation • Labeling manual available	• Strict regulation • Drug Facts table is required
Import	• Registration is not required • Imported products are subject to the same rules as products manufactured in the United States	• Companies importing products to the United States have to register with the FDA
Recalls	• Voluntary by manufacturer/distributor • FDA does not have authority to order a mandatory recall	• Voluntary by manufacturer/distributor • FDA does not have authority to order a mandatory recall

Premarket Approval In the EU, cosmetics are not subject to premarket approval. However, the regulation requires the designation of a responsible person for each cosmetic product in order to place a product on the market. As its name implies, this person will be responsible for the cosmetic product. Without a responsible person, cosmetic products cannot be placed on the EU market.[28]

Manufacturing The EU Cosmetic Product Regulation states that the manufacture of cosmetic products shall comply with the GMPs. EN ISO 22716:2007 Cosmetics, GMPs, Guidelines on GMPs provides a basis for the cosmetic GMP, and it has already become universally accepted across the EU.[28]

Use of Ingredients The EU also has restrictions on the use of cosmetic ingredients. It has lists of permitted color additives, preservatives, UV filters and prohibited and restricted ingredients in Annexes II–VI of the Cosmetic Product Regulation. However, it has to be emphasized that these lists are more exhaustive and include many more ingredients than the US lists.

- Annex II lists over 1300 substances that are prohibited from use in cosmetic products.
- Annex III lists over 250 substances that cosmetic products must not contain except those subject to the restrictions laid down (this is the restricted ingredient list).
- There is a separate positive list of color additives allowed in cosmetic products (Annex IV).
- The EU has a positive list of preservatives permitted in cosmetics products (Annex V) as well as UV filters permitted in cosmetic products (Annex VI).

In addition, the Regulation prohibits the use of substances recognized as carcinogenic, mutagenic, or toxic for reproduction, excluding certain exceptions.[29–33] EU manufacturers and EU importers of cosmetic ingredients or finished cosmetics products also have to comply with the so-called REACH regulation (EC No. 1907/2006) before they place their cosmetic products on the EU market. REACH is the EU regulation concerning the Registration, Evaluation, Authorization and Restriction of Chemicals. Under this regulation, chemical ingredients must be (pre)registered with the European Chemical Agency if their annual tonnage exceeds 1 ton.[34]

Testing and Product Safety Similar to the US, the safety of cosmetic products placed on the EU market is not the responsibility of the competent authority. In the EU, it is the responsibility of an assigned responsible person. Safety assessment has to be performed before placement of a product on the market. Although the regulation requires safety assessment, the types of safety testing are not included in the Regulation itself. Guidelines for safety testing have been prepared by the Scientific Committee on Consumer Safety (SCCS). Under the EU regulation, the responsible person has to prepare a product information file (PIF), including data on safety and efficacy of a product. The PIF also serves as a basis for premarket notification of cosmetic products. Safety of products is assured by an in-market surveillance system controlled by each member state.[35]

The Regulation provides (and the previous regulation, the Cosmetic Directive also provided) regulatory framework for phasing out of animal testing for cosmetic purposes. Today, performing animal tests to evaluate cosmetic products is completely prohibited in the EU. Animal testing of finished cosmetic products and cosmetic ingredients has been banned since 2004 and 2009, respectively (these two bans on testing are referred to as the "testing ban"). The marketing ban, i.e., ban on the marketing of cosmetic products tested on animals and products containing ingredients tested on animals, has applied since 2009 for all human health effects with the exception of repeated-dose toxicity, reproductive toxicity, and toxicokinetics. For these specific health effects, the marketing ban has applied since March 11, 2013, regardless of the availability of alternative nonanimal tests.[28]

Packaging and Labeling General labeling requirements are listed in Article 19 of the Regulation. Labeling requirements are similar to those of the FDA (see more information on the US label information in Section 2 of this chapter). The EU regulation requires the following information to be present on a cosmetic label:[36]

- Name and address of the responsible person
- Country of origin for imported cosmetic products
- Nominal content
- Expiry date
- Precautions
- Batch number reference for identifying cosmetic products
- Function of the cosmetic product, unless it is clear from its presentation
- List of ingredients in descending order down to 1%, under 1% in any order.

A main difference is that in the EU, expiration date (shelf life) of cosmetic products must be indicated.[37]

- For products having a minimum durability of more than 30 months, the period after opening (PAO) should be indicated by an open jar symbol. The open jar is accompanied by a number indicating the durability of the product after it has been opened and the letter "M" referring to months (see Figure 2.1a).

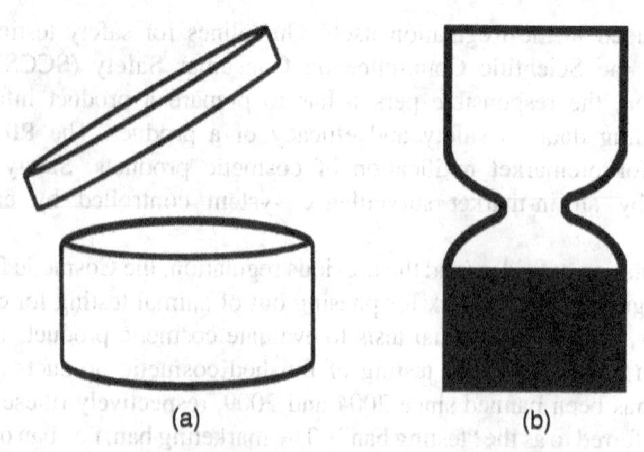

Figure 2.1 Expiration date of cosmetic products in the European Union. (a) Period after opening. (b) Hourglass symbol. Adapted from the Regulation (EC) 1223/2009 of the European Parliament and of the Council of 30 November 2009 on cosmetic products.

- For products having a minimum durability of less than 30 months, the actual expiration date must be indicated using an hourglass symbol or the following words: "best used before the end of" and the date (see Figure 2.1b).

Another difference is that in the EU, nanomaterials have to be listed among the ingredients using the name "nano."

The Commission has a glossary of common ingredient names that must be used for labeling of cosmetics, and it continuously updates this list. For the glossary, the Commission takes into account the internationally recognized nomenclatures, including the International Nomenclature of Cosmetic Ingredients (INCI).[38] The glossary is published in the *Official Journal of the European Union*.

Import of Products

Companies importing products to the EU have to ensure that they comply with all EU regulations. A PIF should be prepared for each cosmetic product – similar to cosmetics manufactured in the EU – and made available to the authorities. Rules relevant for products manufactured in the EU are also relevant for cosmetics imported. Examples for these include the prohibition of use of banned ingredients, competence with labeling information (including the date of minimum durability), notification of competent authority about the product, and prohibition of animal testing.

Recalls

In the EU, product recall is the responsibility of the responsible person assigned by the manufacturer. However, the Regulation states that the competent authority can also withdraw or recall a product from the market if immediate action is necessary or if the responsible person does not take all appropriate measures within the set time limit.[39–41]

GLOSSARY OF TERMS FOR SECTION 1

Adulterated: A product that is generally impure or unsafe due to its composition, manufacturing process, storage location, or container.

Import: The process of bringing products into a country from abroad for sale.

Inspection: A detailed examination performed by the FDA at establishments that manufacture, process, pack, or hold FDA-regulated products to determine the establishment's compliance with laws determined by the FDA.

Misbranded: A product that is false or misleading due to its labeling, packaging, or container.

Premarket approval: A process in which a product is approved by a regulatory authority before entering the market and being sold to consumers to ensure its safety and efficacy for its intended use(s).

Prohibited ingredient: An ingredient that cannot be used in cosmetic products in the US.

Recall: A voluntary action taken by manufacturers and distributors to protect the public health and well-being from products that present a risk of injury or gross deception or are otherwise defective.

Responsible person: A person who is liable for the regulatory compliance of cosmetic products on the EU market. According to the EU Cosmetics Regulation, each cosmetic product has to be guaranteed by a responsible person.

Restricted ingredient: An ingredient that can be used in cosmetic products in the US, but only under the restrictions stated in the regulations.

VCRP: The Voluntary Cosmetic Registration Program is a program established by the FDA to provide a way for the cosmetic industry to report information on their manufacturing facilities and cosmetic products after they are on the market.

REVIEW QUESTIONS FOR SECTION 1

Multiple Choice Questions

1. What is a cosmetic product recall?
   a) A voluntary removal of the product from the market
   b) The approval of the product by the FDA before selling it to consumers
   c) Removal of the product from the market by the FDA
   d) A type of advertisement for cosmetic products that are approved by the FDA

2. What is the FDA's statement on animal testing with respect to cosmetics?
   a) The FDA does not specifically require the use of animals in testing cosmetics for safety
   b) Manufacturers may perform animal tests to establish product safety
   c) The FDA prohibits animal tests in safety evaluation of cosmetics
   d) a and b are true

3. Which of the following cosmetic products has to meet the FDA's cosmetic regulations?
   a) Products marketed in the US
   b) Products made in the US
   c) Products imported to the US
   d) All of the above

4. Which of the following is responsible for the safety of cosmetic products in the US and the EU, respectively?
   a) FDA/EC
   b) Product manufacturer/responsible person
   c) Personnel performing safety testing/personnel performing safety testing
   d) Responsible person/product manufacturer

5. Premarket notification of cosmetic products in the EU is:
   a) A voluntary action
   b) A mandatory action
   c) There is no premarket notification in the EU
   d) None of the above

6. The use of the following ingredients in cosmetics is strictly regulated in the US:
   a) Color additives and restricted and prohibited ingredients
   b) Color additives, UV filters, and preservatives
   c) Prohibited and restricted ingredients
   d) UV filters, fragrances, and preservatives

7. What is the current regulation for the registration of cosmetic companies in the US?
   a) Registration is required for each and every new company
   b) Registration is required every year after the initial registration
   c) Registration is required twice annually after the initial registration
   d) Registration is not required; it is voluntary

8. How can the FDA monitor cosmetic product safety?
 a) By performing an on-site inspection
 b) By buying and analyzing cosmetics
 c) Using consumer reports on bad reactions to cosmetics
 d) All of the above

9. Which of the following is true for cosmetic products imported to the US?
 a) The products have to be registered in the US
 b) The company importing products has to register with the FDA
 c) Product labels have to be in accordance with the US rules
 d) Color additives not approved in the US can be used in products that are imported to the US

10. Which of the following is true for active ingredients used in OTC drug–cosmetic combination products in the US?
 a) Any ingredients listed in the OTC monographs can be used in the concentration listed and under the conditions detailed
 b) Any active ingredients can be used in any concentration
 c) New active ingredients can be used in less than 1% without any approval
 d) Only ingredients that have been used for at least 50 years can be used

11. The "FDA-registered" logo can be used on ___.
 a) Any cosmetic products
 b) Skin care products only
 c) Cosmetics containing preservatives
 d) No products

12. Premarket approval is:
 a) A voluntary action to remove a cosmetic product from the market
 b) A legal action that can be taken by the FDA against cosmetic companies that do not manufacture safe products
 c) A process through which a regulatory authority (e.g., the FDA) approves the marketing of a product in the US before it is actually sold to consumers
 d) A type of an inspection the FDA regularly performs at shampoo manufacturing companies

Fact or Fiction?

_________ a) Cosmetics must be safe in the US.
_________ b) The FDA can inspect cosmetic companies without prior notice.
_________ c) The FDA can initiate recalls for cosmetics.
_________ d) The VCRP is a cosmetic approval program.
_________ e) The FDA can buy and analyze cosmetic products to evaluate their safety.

Misbranded or Adulterated?

1. Here is a list of problems that may occur in and/or related to cosmetics. Categorize the problems based on which makes a product adulterated (A) or misbranded (M).

Misbranded or adulterated	
_______	1. Contains any poisonous substance
_______	2. Labeling is false
_______	3. Contains a nonpermitted ingredient
_______	4. Label does not include all required information
_______	5. Contains a noncertified color additive
_______	6. Information is not legible on the labeling
_______	7. Product is prepared under insanitary conditions
_______	8. Misleading container presentation or fill
_______	9. Container is composed of poisonous substance
_______	10. Labeling is misleading
_______	11. Contains filth
_______	12. Product is stored under insanitary conditions

REFERENCES

1. CFR Title 21 Part 710.1
2. FDA: Voluntary Cosmetic Registration Program (VCRP), Last update: 11/15/2012, Accessed 5/11/2013 at http://www.fda.gov/Cosmetics/GuidanceComplianceRegulatoryInformation/VoluntaryCosmeticsRegistrationProgramVCRP/
3. CFR Title 21 Part 710.9
4. S. F. Wright, D. C. Havery: The voluntary cosmetic registration program – how it works, *Cosmetiscope.* 2010;16(5):1–6.
5. Food Drug and Cosmetic Act, Section 510
6. CFR Title 21 Part 207
7. CFR Title 21 Part 710.8
8. CFR Title 21 Part 720.9
9. FDA: Is it a Cosmetic, a Drug, or Both? (or is it a Soap?), Last update: 3/6/2013. Accessed 4/23/2013 at http://www.fda.gov/cosmetics/guidancecomplianceregulatoryinformation/ucm074201.htm
10. CFR Title 21 Part 210
11. CFR Title 21 Part 211
12. CFR Title 21 Part 250.250
13. CFR Title 21 Parts 700.11 through 700.35
14. Katdare, A., Chaubal, M.: *Excipient Development for Pharmaceutical, Biotechnology, and Drug Delivery Systems*, Boca Raton: CRC Press, 2006.

15. PCPC: Questions and Answers Consumer Commitment Code Personal Care Products Council (the Council), Accessed 5/07/2013 at http://www.personalcarecouncil.org/questions-and-answers-consumer-commitment-code
16. CFR Title 21 Part 740.10
17. FDA: How FDA Evaluates Regulated Products: Cosmetics, Last update: 4/4/2012, Accessed 5/11/2013 at http://www.fda.gov/aboutfda/transparency/basics/ucm262353.htm
18. Food Drug and Cosmetic Act, Section 601 (21 USC 361)
19. CFR Title 21 Part 201
20. Food Drug and Cosmetic Act, Section 602 (21 USC 362)
21. Food Drug and Cosmetic Act, Section 700.25
22. FDA: Information for Cosmetics Importers, Last update: 4/15/2013, Accessed 5/21/2013 at http://www.fda.gov/Cosmetics/InternationalActivities/ImportsExports/CosmeticImports
23. FDA: Small Business Assistance: Import and Export of Human Drugs and Biologics, Last update: 11/01/2011, Accessed 5/23/2013 at http://www.fda.gov/Drugs/DevelopmentApprovalProcess/SmallBusinessAssistance/ucm052787.htm
24. CFR Title 21 Part 7.40–7.50
25. FDA: Questions and Answers on Current Good Manufacturing Practices, Good Guidance Practices, Level 2 Guidance – Holding and Distribution, Human Drug Recalls, Last update: 2/25/2011, Accessed 5/9/2013 at http://www.fda.gov/Drugs/GuidanceComplianceRegulatoryInformation/Guidances/ucm221671.htm
26. FDA: Regulatory Precedes Manual, Chapter 7, Recall Procedures, Accessed 11/10/13 at http://www.fda.gov/downloads/iceci/compliancemanuals/regulatoryProceduresManual/UCM074312.pdf
27. EU: Article 13 of Regulation (EC) No 1223/2009 of the European Parliament and of the Council of 30 November 2009 on cosmetic products
28. EU: Regulation (EC) No 1223/2009 of the European Parliament and of the Council of 30 November 2009 on cosmetic products
29. EU: Annex II of Regulation (EC) No 1223/2009 of the European Parliament and of the Council of 30 November 2009 on cosmetic products
30. EU: Annex III of Regulation (EC) No 1223/2009 of the European Parliament and of the Council of 30 November 2009 on cosmetic products
31. EU: Annex IV of Regulation (EC) No 1223/2009 of the European Parliament and of the Council of 30 November 2009 on cosmetic products
32. EU: Annex V of Regulation (EC) No 1223/2009 of the European Parliament and of the Council of 30 November 2009 on cosmetic products
33. EU: Annex VI of Regulation (EC) No 1223/2009 of the European Parliament and of the Council of 30 November 2009 on cosmetic products
34. CIRS: REACH and Cosmetics, Accessed 5/28/13 at http://www.cirs-reach.com/pdf/REACH_and_Cosmetics.pdf
35. EU: Article 22 of Regulation (EC) No 1223/2009 of the European Parliament and of the Council of 30 November 2009 on cosmetic products
36. EU: Article 19 of Regulation (EC) No 1223/2009 of the European Parliament and of the Council of 30 November 2009 on cosmetic products

37. EU: Annex VII of Regulation (EC) No 1223/2009 of the European Parliament and of the Council of 30 November 2009 on cosmetic products
38. EU: Article 33 of Regulation (EC) No 1223/2009 of the European Parliament and of the Council of 30 November 2009 on cosmetic products
39. EU: Article 5 of Regulation (EC) No 1223/2009 of the European Parliament and of the Council of 30 November 2009 on cosmetic products
40. EU: Article 2 of Regulation (EC) No 1223/2009 of the European Parliament and of the Council of 30 November 2009 on cosmetic products
41. EU: Article 26 of Regulation (EC) No 1223/2009 of the European Parliament and of the Council of 30 November 2009 on cosmetic products

SECTION 2: LABELING TUTORIAL FOR COSMETICS AND OTC DRUG–COSMETIC PRODUCTS MARKETED IN THE UNITED STATES

LEARNING OBJECTIVES

Upon completion of this section, the reader will be able to

1. define the following terms:

Alphabetical order	Country of origin	Descending order of predominance	
Directions for safe use	Drug Facts box	Expiration date	FP&L Act
Identity statement	Immediate container	INCI	INCI names
Information panel	Ingredient declaration	Label	Labeling
Net quantity of content	Outer container	Principal display panel	
Retail sale	Warning statement		

2. name the regulations that regulate the labeling and packaging of cosmetic products in the US;
3. differentiate between a package and a label;
4. differentiate between an outer and an inner container;
5. differentiate between a label and labeling;
6. differentiate between the PDP and the information panel;
7. list what information has to appear on the PDP of a cosmetic product in the US;

8. list what information has to appear on the information panel of a cosmetic product in the US;
9. explain how flavors and fragrances can be indicated on the label of a cosmetic product;
10. list what information has to appear on the PDP of an OTC drug–cosmetic product in the US;
11. list what information has to appear on the information panel of an OTC drug–cosmetic product in the US;
12. name the official language for labeling a cosmetic product marketed in the US;
13. briefly discuss the requirements for ingredient declaration in the case of a cosmetic product and an OTC drug–cosmetic product, respectively;
14. explain what the official rule is for ingredient declaration on a cosmetic product that is distributed free of charge, e.g., a free shampoo sample in a hair salon;
15. explain what labeling rules a cosmetic product imported to the US has to meet;
16. explain what the abbreviation "INCI" stands for and how INCI names are established;
17. briefly discuss whether consumers know the exact composition of a cosmetic product or an OTC drug–cosmetic product in the US.

KEY CONCEPTS

1. Cosmetics distributed in the US must comply with the labeling regulations published by the FDA under the authority of the FD&C Act and the FP&L Act. OTC drug–cosmetic products must comply with the cosmetic regulations and OTC drug regulations as well.
2. Labeling has a significant role in the classification of a product in the US.
3. Cosmetic products must have both a PDP and information panels.
4. According to the current US regulations, the following information must be indicated on the PDP of cosmetics marketed in the US: an identity statement, net quantity of contents, and, in specific cases, a warning statement referring to the safety of the product
5. According to the current US regulations, the following information has to be included on the information panels of cosmetics marketed in the US: name and place of business, country of origin, directions for safe use, warning and caution statements, and ingredient declaration.
6. OTC drug–cosmetic products must have a combination of cosmetic and OTC labeling.
7. According to the current US regulations, the following information must be indicated on the PDP of OTC drug–cosmetic products marketed in the US: an identity statement and net quantity of contents.

8. According to the US regulations, the following information must be indicated on information panels of OTC drug–cosmetic products marketed in the US: a Drug Facts box, the name and place of business, expiration date, lot number, and country of origin.
9. In the US, cosmetic ingredients must be listed by their established or adopted names, which in general are the INCI names.
10. You never know the exact composition of a cosmetic product or OTC drug–cosmetic product if you are not the product manufacturer.

Introduction

● **Cosmetics distributed in the US must comply with the labeling regulations published by the FDA under the authority of the FD&C Act and the FP&L Act. OTC drug–cosmetic products must comply with the cosmetic regulations and OTC drug regulations as well.** Requirements for the labeling of OTC drugs are regulated under the CFR.[1] This section reviews the FDA's labeling requirements with regard to cosmetics and OTC drug–cosmetic products.

As mentioned previously, neither the FD&C Act nor the FP&L Act requires cosmetic products, including their labeling to undergo premarket approval by the FDA. It is the manufacturer's and/or distributor's responsibility to ensure that products are labeled properly. Failure to comply with labeling requirements may result in a misbranded product. As discussed in Section 1 of this chapter, the FDA prohibits the marketing of misbranded cosmetics.

It is important to note that these regulations are relevant for **all** cosmetics marketed in the US, whether they were manufactured in the US or were imported from a foreign county. It has to be emphasized that ● **labeling has a significant role in the classification of a product in the US** since the major differentiating factor between a drug and a cosmetic, i.e., the intended use, is determined from the statements on a product's labeling. The courts, in deciding whether a product is a cosmetic, a drug, or a combination of a drug and a cosmetic, have relied principally on the consumer's perception of the meaning of a label statement and less so on the interpretation of the meaning of a label statement by the labeler or a regulatory agency.

Definitions

Before discussing the particulars of a label, there are some definitions we should understand.

- The **package** is the **outer** container of a product, such as a box or folding carton. However, the package can also be the **immediate** container, for example, bottle, jar, or aerosol container that holds the product if the immediate container is not displayed in a box or folding carton. Figure 2.2 shows the difference between an outer and an immediate container.

- The term "label" refers to a written, printed, or graphic display of information on the container of a cosmetic or affixed to or appearing on a package containing a product.[2–4]
- **Labeling** refers to all labels and other written, printed, or graphic matter on or accompanying a product. This includes labels, inserts, risers, display packs, leaflets, promotional literature, or any other written or printed information distributed with a product.[5,6]
- The **principal display panel** (known for short as PDP) is the panel of a package that is most likely to be shown or examined under customary conditions of display for retail sale.[7,8]

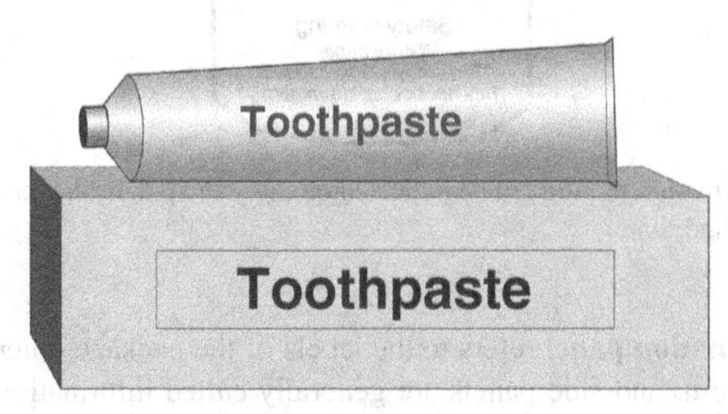

Figure 2.2 A cosmetic package, including an outer container (i.e., a folding carton) and an immediate container (i.e., a plastic tube).

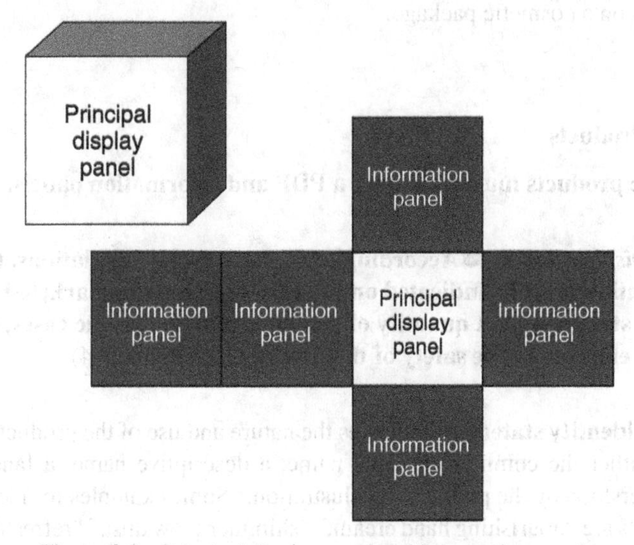

Figure 2.3 Principal display panel and information panels.

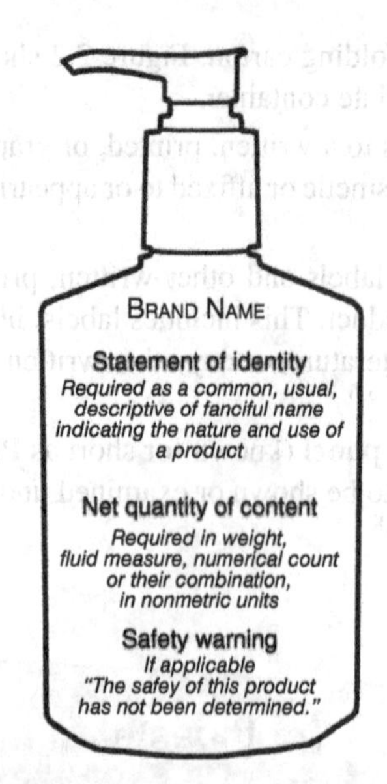

BRAND NAME
Statement of identity
Required as a common, usual, descriptive of fanciful name indicating the nature and use of a product
Net quantity of content
Required in weight, fluid measure, numerical count or their combination, in nonmetric units
Safety warning
If applicable
"The safey of this product has not been determined."

Figure 2.4 Information required to be indicated on the principal display panel of a cosmetic product in the United States.

- The **information panel** refers to the labels of the package other than the PDP. Back, bottom, and side panels are generally called information panels. Since the information must be legible, the bottom of the package is generally not acceptable for placement of required information such as the cosmetic ingredient declaration. Figure 2.3 shows a general layout of the PDP and information panels on a cosmetic package.

Cosmetic Products

❸ Cosmetic products must have both a PDP and information panels.

Principal Display Panel **❹ According to the current US regulations, the following information must be indicated on the PDP of cosmetics marketed in the US: an identity statement, net quantity of contents, and, in specific cases, a warning statement referring to the safety of the product** (see Figure 2.4).

1. The **identity statement** indicates the nature and use of the product, by means of either the common or usual name, a descriptive name, a fanciful name understood by the public or an illustration.[9] Some examples for identity statements are "nourishing hand cream," "shimmer glow dust," "refreshing shower gel," "matte-wear liquid foundation," and "shave gel."

DID YOU KNOW?

You can find the brand name on the front panel of the majority of products. However, it is not a requirement of the current regulations.

2. The **net quantity of content** indicates the amount of product in the container in terms of weight, fluid measure, numerical count, or a combination of these. In the case of an aerosol product, the net content statement must express the net quantity of the contents expelled. In the US, the net content must be in terms of nonmetric (US) system, e.g., ounce or fluid ounce, and it may additionally be stated in terms of the metric system, e.g., grams or milliliters. For cosmetic containers smaller than 1/4 oz. or 1/8 fl. oz. in capacity, declaration of the net quantity of contents may appear on a tear-away tag, tape, or card affixed to the container.[10] Examples for the net quantity of contents include 2 fl. oz./59 ml, Net Wt. 0.21 oz./6 g, and 25 wipes.
3. If the safety of the cosmetic product has not been evaluated, the required **warning statement** ("Warning – The safety of this product has not been determined") must be stated on the PDP.[11]

Note: If there is an outer and an inner package, the identity statement, net quantity, and warning statement always go on the outer package, while on the inner package, only the identity statement (i.e., name of the product) is required.

Information Panels ❺ **According to the current US regulations, the following information has to be included on the information panels of cosmetics marketed in the US: name and place of business, country of origin, directions for safe use, warning and caution statements, and ingredient declaration** (see Figure 2.5).

1. **Name and place of business,** which may be of the manufacturer, packer, or distributor.[12] If the distributor is not the manufacturer, the label must say "Manufactured for … " or "Distributed by … " or similar.
2. **Country of origin** has to be indicated only if the product is imported to the US.[13,14]
3. **Directions for safe use** are necessary to be indicated if misuse of a cosmetic product may be hazardous to consumers. It is a type of material facts that if missing can result in misleading labeling and therefore renders a product misbranded.[15] An example can be a shampoo where a direction for safe use can be given as "Massage into wet hair and rinse off with warm water. Avoid eye contact." For a facial foundation, it may be provided as follows: "Blend on face with brush or fingertips."

Directions for safe use
Required if misuse may be hazardous to consumers
Warning and caution statements
Required for the safe use of specific products
Ingredients
Required in descending order of predominance (down to 1%)
Name and place of business
Required
Country of origin
Required if imported to the United States

Figure 2.5 Information required to be indicated on the information panels of a cosmetic product in the United States.

4. The law says that cosmetic labels must bear **warning and caution statements**. Warning statements can be general warning statements issued whenever necessary to prevent health hazard. They may include statements on where the product should be used, how it should be used safely, when it should be discontinued, or whether it should be stored in a place where children cannot have an access to it. In addition to these general statements, some cosmetics must bear specific warnings or cautions required and worded by regulation.[16] Cosmetics in self-pressurized containers (aerosol products), suntanning preparations, feminine deodorant sprays, foaming detergent bath products, and children's bubble bath products are examples of products requiring such statements. For example, the labeling of suntanning preparations that do not contain a sunscreen ingredient must display the following warning: "Warning – This product does not contain a sunscreen and does not protect against sunburn. Repeated exposure of unprotected skin while tanning may increase the risk of skin aging, skin cancer, and other harmful effects to the skin even if you do not burn."[17]
5. **Ingredient Declaration**: Ingredients have to be indicated on cosmetic products that are produced or distributed for retail sale to consumers.[18] However, cosmetics not customarily distributed for retail sale, for example, hair products used by professionals on customers at their establishments, hotel samples, or samples distributed free of charge, are exempt from this requirement. Since these products are not sold to consumers, they are not required to bear an ingredient label.

DID YOU KNOW?

If you see the term "professional" on a product, it does not mean that it is used by professionals only. Many products labeled as "For professional use only" are produced for retail sale; therefore, they have to bear an ingredient label.

An important requirement is that the ingredient declaration must be legible. It may appear on any information panel of a package and may also appear on a firmly affixed tag, tape, or card, i.e., in case of a mascara or eyeliner pencil. Cosmetic ingredients must be declared in descending order of predominance if present in greater than 1%. Ingredients other than color additives present at 1% or less, i.e., fragrances and preservatives, may be listed in any order after all the ingredients listed present at more than 1%. This 1% limit is not visible on the label, only the manufacturer is aware of the actual concentrations. Color additives of any concentration may be listed in any order after listing all other ingredients.[18]

Ingredients must be identified by names established or adopted by regulation. There are some names that have been established by the FDA by regulation, such as chlorofluorocarbon 11 for trichlorofluoromethane.[18] If the ingredient is not listed in this group, i.e., the majority of ingredients, the next source of information is the INCI dictionary (see more information on this in Section 3 of this chapter). If there is no INCI name, the name on the label may be listed in the US Pharmacopoeia (USP), National Formulary (NF), US Homeopathic Pharmacopoeia, and Merck Index. If none lists a name for an ingredient, the common or usual name must be used, which is understood by consumers. As a last choice, a chemical or technical name or description can be used; however, trade names are not acceptable.

DID YOU KNOW?

Some color additives are added to a cosmetic for color matching to provide the same shade for each batch, i.e., in blushes. These ingredients are not necessarily present in each batch; however, they still have to be labeled. These ingredients can be found at the end of the ingredient list after the phrase "May Contain."

Fragrances and flavors may be declared in descending order of predominance as "fragrance" and "flavor" or can be declared individually by their appropriate names. If a fragrance also serves as a flavor, it must be declared as flavor and fragrance.[18]

Some ingredients do not even have to be labeled. The FDA regulations do not require the disclosure of incidental ingredients that are present at an insignificant level and have no technical or functional effect in the cosmetic product. An example is a preservative of a raw ingredient. When the raw material preserved with preservative X is added to a cosmetic product, the concentration of preservative X will reduce to a level at which it will not be effective as a preservative any longer. So, in the final formulation, it will only be an incidental ingredient. In addition, the name of an ingredient accepted by the FDA as a trade secret need not be disclosed on the label. It may be listed as "and other ingredients" at the end of the ingredient declaration.

An additional requirement of the regulations is that all label and labeling statements must be in English. The only exception to this rule is for products distributed solely in a US territory where a different language is predominant, such as Puerto Rico.[19] If a second language is also used for labeling, all information present in English must be present in the other language as well.

OTC Drug–Cosmetic Products

❻ **OTC drug–cosmetic products must have a combination of cosmetic and OTC labeling.** They also bear a PDP and one or more information panels; however, the content and format of these labels are different from those of cosmetic products.

Principal Display Panel ❼ **According to the current US regulations, the following information must be indicated on the PDP of OTC drug–cosmetic products marketed in the US: an identity statement and net quantity of contents** (see Figure 2.6).

1. The **identity statement** indicates the established name of the active ingredient (if there is any) and its general pharmacological category or primary intended action.[20] Some examples for identity statements are "antibacterial hand soap," "hand sanitizer," "foundation primer sunscreen," "acne cleanser salicylic acid acne medication," and "antiperspirant and deodorant."
2. The **net quantity of contents** indicates the amount of product in the container in terms of weight, fluid measure, numerical count, or a combination of these. In the US, for example, for cosmetics, the net content must be in terms of nonmetric system, and it may additionally be stated in terms of the metric system.[21]

Information Panels ❽ **According to the US regulations, the following information must be indicated on information panels of OTC drug–cosmetic products marketed in the US: a Drug Facts box, the name and place of business, expiration date, lot number, and country of origin** (see Figure 2.7).

1. **Drug Facts box** includes information about the product in a standardized form. This type of labeling is intended to make it easier for consumers to read

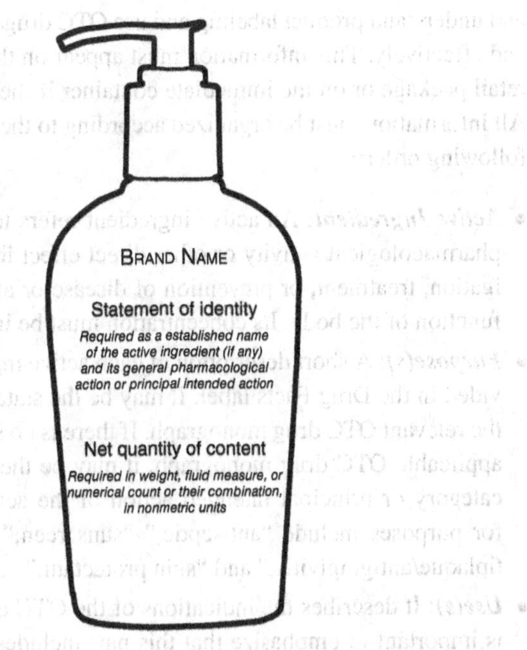

Figure 2.6 Information required to be indicated on the principal display panel of an OTC drug product in the United States.

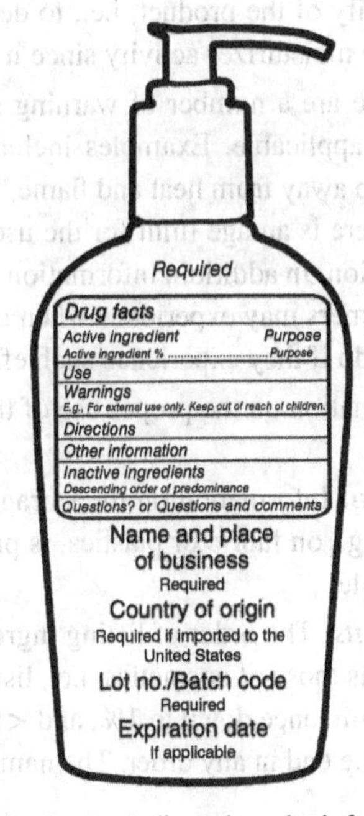

Figure 2.7 Information required to be indicated on the information panels of an OTC drug product in the United States.

and understand product labeling and use OTC drug–cosmetic products safely and effectively. This information must appear on the outside container of the retail package or on the immediate container if there is no outside container. All information must be organized according to the following headings in the following order:[22]

- ***Active Ingredient***: An active ingredient refers to any component that has pharmacological activity or other direct effect in the diagnosis, cure, mitigation, treatment, or prevention of disease or affects the structure or any function of the body. Its concentration must be indicated in percentage.
- ***Purpose(s)***: A short description of each active ingredient must also be provided in the Drug Facts label. It may be the statement of identity found in the relevant OTC drug monograph. If there is no statement of identity or no applicable OTC drug monograph, it may be the general pharmacological category or principal intended action of the active ingredient. Examples for purposes include "antiseptic," "sunscreen," "acne medication," "antiplaque/antigingivitis," and "skin protectant."
- ***Use(s)***: It describes the indications of the OTC drug–cosmetic product. It is important to emphasize that this part includes information only on the active ingredients. An example would be an antibacterial moisturizing hand lotion. **Use in the Drug** Facts label would include information only on the antibacterial activity of the product, i.e., to decrease bacteria on the skin, but nothing on the moisturizer activity since it is a cosmetic function.
- ***Warning(s)***: There are a number of warning statements that may appear in this section if applicable. Examples include "For external use only," "Flammable. Keep away from heat and flame," and "Keep out of the reach of children." If there is an age limit for the use of the product, it will also appear in this section. In addition, information should be provided on what side effects consumers may experience when using the product, as well as what they should do if they experience such effects.
- ***Directions***: Information on the proper use of the product must be listed in this section.
- ***Other Information***: Information on the storage of a product, its potential harmful effects, e.g., on fabric or plastics, is provided here. However, this section is applicable.
- ***Inactive Ingredients***: The order of listing ingredients and their nomenclature are the same as those of cosmetics, i.e., listing is provided in descending order of predominance down to 1%, and <1% in any order, color additives are listed at the end in any order. The names used should be primarily INCI names.

- ***Questions? Or Questions or Comments?***: This section is optional; a phone number may be indicated here, which may be called if consumers would like to obtain more information on the product.

In addition to the Drug Facts box, the following information must also be provided on the information panels:

2. **Name and place of business** of manufacturer, packer, or distributor. If the distributor is named on the label, its name must be provided using phrases such as "Manufactured for ...," "Distributed by ...," or similar. If the packer is identified, the name must be identified by the phrase "Packed by ..." or "Packaged by ...."[23]
3. **Expiration date** of a product if required by the regulations.[24]
4. **Lot number** or **batch code** is generally required to be indicated on OTC drug–cosmetic products. It is a specific number that helps identify a product for manufacturers.[25]
5. **Country of origin** is required if a product is imported to the US.[13]

INCI Naming

As mentioned earlier, ● **in the US, cosmetic ingredients must be listed by their established or adopted names, which in general are the INCI names**. The abbreviation "INCI" stands for the International Nomenclature of Cosmetic Ingredients. INCI names were created by the PCPC (formerly known as CTFA, Cosmetic, Toiletry, and Fragrance Association) and can be found in the International Cosmetic Ingredient Dictionary and Handbook (otherwise known as INCI dictionary). The first edition of the dictionary was published in 1973; since then, many changes have taken place. In 2013, its 14th edition is used. A committee of the PCPC, called the International Nomenclature Committee (INC), has responsibility to maintain the integrity and validity of the data incorporated into the dictionary. The main purpose of the INCI names as well as dictionary is to standardize ingredient names, minimize language barriers, and help international trade. The same cosmetic ingredient may be manufactured and distributed by multiple suppliers, resulting in different trade names. Using trade names would make understanding ingredient labels more difficult and export/import almost impossible. However, with a standardized nomenclature, it is easier for concerned consumers to make educated decisions by checking ingredient lists and finding ingredients they might be allergic to. In addition, it helps manufacturers use standardized names on their products, which are mutually accepted by many countries.

INCI names and the dictionary are used in a number of markets, including the US, China, Japan, countries in Europe, and many others for listing ingredients on cosmetic product labels. It should be noted that despite the efforts at harmonization of the INCI nomenclature, there are still some differences between the INCI system in different countries in terms of the nomenclature of colors, botanicals, and trivial ingredients. For example, the INCI name for H_2O in Europe is aqua, while in the US, it is water. Or the US requires the use of US color names, such as Red 4, while the EU requires the use of color index numbers, such as CI 14700. In such cases, the alternative name can be provided brackets, for example, Red 4 (CI 14700). They seem to be minor differences; however, they can cause serious difficulties for importers and exporters. Additionally, in some countries, INCI names are being translated into the local language, which also results in differences.

INCI names are assigned on the basis of the chemical composition and structure of an ingredient; however, they are generally different from the official chemical names (i.e., IUPAC names; IUPAC stands for the International Union of Pure and Applied Chemists) of a molecule that can be used to write a structural formula for any ingredient.[26] For example, the IUPAC name of a frequently used preservative is methyl 4-hydroxybenzoate, while its INCI name is methylparaben. Based on the IUPAC names, one can define the proper structure of the molecule; however, the INCI name will not provide a basis for that. INCI names are specifically used to identify ingredients in cosmetic and personal care products and OTC drug–cosmetic products, but not in OTC drugs, prescription-only drugs, or foods.

New ingredients that are not listed in the dictionary are continuously named by the INC and added to the dictionary. In the INCI naming process, the INC works from the written information provided by manufacturers; therefore, it is the suppliers' responsibility to ensure that the information submitted is complete and accurate. The dictionary provides additional information and references for cosmetic ingredients provided by manufacturers.[27]

How Do You Know the Exact Composition of a Cosmetic Product or OTC Drug–Cosmetic Product?

The true answer is that ⓾ **you never know the exact composition of a cosmetic product or OTC drug–cosmetic product if you are not the product manufacturer** (more specifically the formulator working for the manufacturer). Manufacturers are required to appropriately label their products according to the relevant labeling regulations; however, there is certain information that you as a consumer will never be aware of.

It is a fact that you can see the list of ingredients on product labels. However, as it discussed earlier, there are some ingredients that can be listed as umbrella terms, and you will not know what ingredients those terms include. This is the case with fragrances and flavors. The term "fragrance" may include over 50 ingredients, so you are not able to identify the individual components. In addition, incidental ingredients are not listed since they do not have any technical or functional effect in the formulation. The list of ingredients may be helpful for customers with sensitive skin in making decisions about a product. However, consumers still cannot see the quantity of ingredients that remain as confidential information, and only a limited number of people are privy to it.

The bottom line is that you never know the exact composition of a cosmetic product or OTC drug–cosmetic product. However, it is worth reading the product labels before using a product to understand how it should be used, whether there is any warning or caution with regard to its use, or to check the ingredients if you are sensitive or allergic.

GLOSSARY OF TERMS FOR SECTION 2

Alphabetical order: An indexing method in which names are arranged in the same sequence as the letters of the alphabet.

Country of origin: The country where the product was manufactured and imported from.

Descending order of predominance: An indexing method in which names are arranged based on their amount in the product. The ingredient presenting in the highest amount goes first and the ingredient presenting in the lowest amount is the last item on the list.

Directions for safe use: Statements on a product label that indicate how the product should be used safely.

Drug Facts box: A box found on the information panel of OTC drugs, which contains information about the products' active ingredients, purpose, use, warnings, directions, and inactive ingredients, among other information. It has a standard format in the US.

Expiration date: A date that indicates until when a product should be used. After the expiration date, the product is no longer expected to be stable and provide the claimed benefits.

FP&L Act: Fair Packaging and Labeling Act regulates the labeling and packaging of cosmetic products.

Identity statement: A statement on the PDP that indicates the nature and use of a product.

Immediate container: A container that is in direct contact with the product. An example is an aerosol can.

INCI names: Standard names used to indicate cosmetic ingredients on cosmetic product labels in a number of countries, including US, China, Japan, and many countries in Europe. These are often different from the chemical names of the ingredients.

INCI: International Nomenclature of Cosmetic Ingredients.

Information panel: Labels of a cosmetic package other than the PDP.

Ingredient declaration: A list of ingredients printed on the label of cosmetic products.

Label: A written, printed, or graphic display of information on the container of a cosmetic or affixed to or appearing on a package containing a product.

Labeling: All labels and other written, printed, or graphic matter on or accompanying a product.

Net quantity of content: It is a statement on a product label that indicates the amount of product in the container.

Outer container: A container that is in indirect contact with a product. An example is a carton box.

Principal display panel: PDP, the panel of a cosmetic package that is most likely to be shown or examined under customary conditions of display for retail sale.

Retail sale: The direct sale of products in small quantities to consumers.

Warning statement: A statement on a product label, which contains important information regarding the composition, safety, and storage of the product that consumers should know before using the product.

REVIEW QUESTIONS FOR SECTION 2

Multiple Choice Questions

1. Which of the following regulates cosmetic packaging and labeling in the US?
   a) FDA
   b) WHO
   c) ABC
   d) CDC
2. In an OTC drug–cosmetic product, cosmetic ingredients are listed in __ and active ingredients are listed in __.
   a) Alphabetical order/descending order
   b) Descending order/alphabetical order
   c) Descending order/descending order
   d) Increasing order/descending order
3. Cosmetic ingredients have to be identified on product labels using ____ names.
   a) INCI
   b) IUPAC
   c) Chemical
   d) Physical

4. Which of the following is true for labeling of cosmetic products?
   a) It is not regulated in the US
   b) It is strictly regulated in the US
   c) Rules are the same for cosmetics and drugs
   d) Rules are the same for each and every country in the world

5. Which of the following is true for expiration dates regarding cosmetics in the US?
   a) An open jar symbol is used to indicate the expiration date
   b) The FDA requires manufacturers to indicate the number of years for which the product can be used
   c) No expiration date has to be indicated on the labels
   d) Expiration date has to be indicated on skin care products only

6. What is the official language for labeling cosmetics and OTC drug–cosmetic products in the US?
   a) English
   b) Spanish
   c) The importing country's official language
   d) None of the above

7. The "Drug Facts" box can be found on ___.
   a) All cosmetic products
   b) All color cosmetics
   c) All OTC drug products, excluding OTC drug–cosmetic combination products
   d) All OTC drug products, including OTC drug–cosmetic combination products

8. Which of the following ingredients do not have to be indicated on cosmetic product labels?
   a) Colors used for color matching
   b) Incidental ingredients
   c) Active ingredients
   d) Preservatives

Matching

1. Match the numbered parts of the cosmetic label with the appropriate terms found in the following box. There are extra terms that you do not have to use. Use every term only once.

2. Indicate whether the items are mandatory or not in the US according to the current FDA regulations.

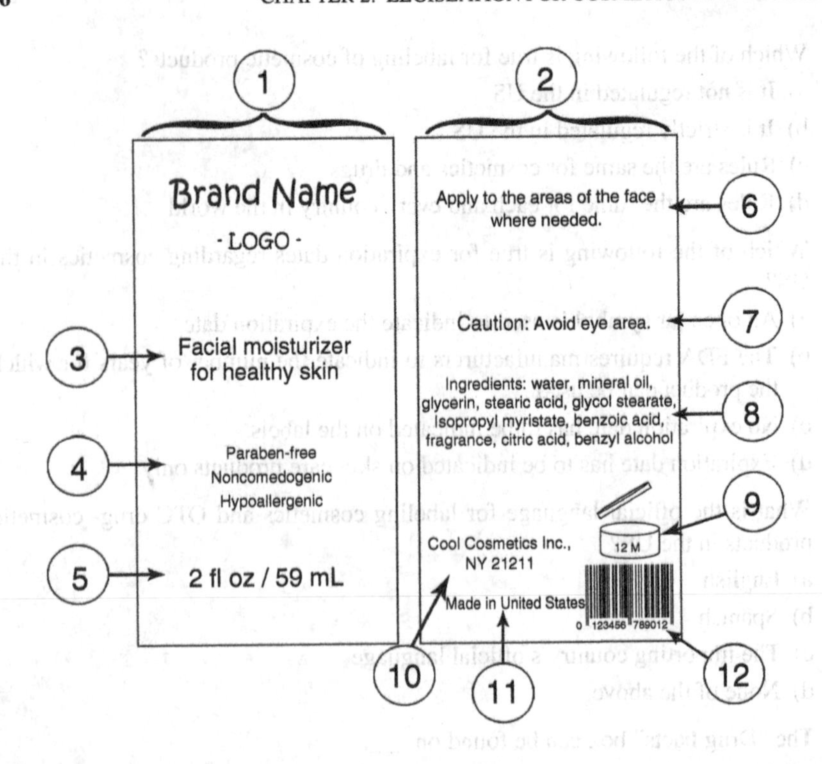

Company's logo	Identity statement	Principal display panel
Net weight	Ingredients causing irritation	Warning statement if a product's safety has not been tested
List of ingredients	Country of origin	Manufacturer's name and address
Expiration date	Date of manufacture	Directions for safe use
List of employees	Bar code	General warning and caution statement(s)
Product claims	Information panel	Packaging type, i.e., jar

Parts of a cosmetic label	Mandatory in the US?
1. ______	1. ______
2. ______	2. ______
3. ______	4. ______
4. ______	4. ______
5. ______	5. ______
6. ______	6. ______
7. ______	7. ______
8. ______	8. ______
9. ______	9. ______
10. ______	10. ______
11. ______	11. ______
12. ______	12. ______

Matching

Match the terms in column A with their appropriate description in column B.

	Column A	Column B
_____	1. "And other ingredients"	A. A product type that does not have to bear labeling that meets the current US rules and regulations
_____	2. Active ingredients	B. A product type that has to bear labeling that meets the current US rules and regulations
_____	3. Color additives	C. A written, printed, or graphic matter on or accompanying a cosmetic product
_____	4. Cosmetic ingredients in <1%	D. All written, printed, or graphic matter on or accompanying a cosmetic product
_____	5. Cosmetic ingredients in >1%	E. An act that regulates cosmetic packaging and labeling
_____	6. FD&C Act	F. An act that regulates cosmetics in general
_____	7. FP&L Act	G. Container in direct contact with the cosmetic product
_____	8. Free hotel sample	H. Container in indirect contact with the cosmetic product
_____	9. Immediate container	I. Ingredients that are listed in alphabetical order on cosmetic labels
_____	10. Incidental ingredients	J. Ingredients that are listed in descending order of predominance on cosmetic labels
_____	11. Label	K. Ingredients that are used as processing aids and have no functional effect in cosmetic products
_____	12. Labeling	L. Ingredients that are used to match the color of a colored cosmetic with the standard color
_____	13. "May contain" ingredients	M. Ingredients that can be listed in any order after all other ingredients on cosmetic labels
_____	14. Outer container	N. Ingredients that can be listed in any order after ingredients present in >1% on cosmetic labels
_____	15. Retail sale product	O. Trade secret ingredients accepted by the FDA

REFERENCES

1. CFR Title 21 Part 201
2. Food, Drug and Cosmetic Act Section 201(k)
3. Fair Packaging and Labeling Act, Section 10(c)
4. CFR Title 21 Part 1.3(b)
5. Food, Drug and Cosmetic Act Section 201(m)
6. CFR Title 21 Part 1.3(a)
7. Fair Packaging and Labeling Act, Section 10(t)
8. CFR Title 21 Part 701.10
9. CFR Title 21 Part 701.11
10. CFR Title 21 Part 701.13
11. CFR Title 21 Part 740.10
12. CFR Title 21 Part 701.12
13. Title 19 United States Code, Chapter 4 (Tariff Act of 1930), § 1304(b)
14. CFR Title 19 Part 134
15. CFR Title 21 Part 1.21
16. CFR Title 21 Part 740
17. CFR Title 21 Part 740.19
18. CFR Title 21 Part 701.3
19. CFR Title 21 Part 701.2(b)
20. CFR Title 21 Part 201.61
21. CFR Title 21 Part 201.62
22. CFR Title 21 Part 201.66
23. CFR Title 21 Part 201.10
24. CFR Title 21 Part 201.17
25. CFR Title 21 Part 201.18
26. Abrutyn, E. S.: The INCI process: a chairman's perspective. *Cosmet Toiletries.* 2007;122(9):22–30.
27. PCPC: INCI Name Process, Accessed 5/14/2013 at http://www.personalcarecouncil.org/science-safety/inci-name-process

SECTION 3: GOVERNMENT AND INDEPENDENT ORGANIZATIONS IN THE COSMETIC INDUSTRY

LEARNING OBJECTIVES

Upon completion of this section, the reader will be able to

1. define the following terms:

CDER	CFSAN	CIR	Cosmetics Europe
EC	FDA	FTC	Office of Cosmetics and Colors
PCPC	SCCS		

2. describe what types of products are regulated by the FDA;
3. name which center/office is responsible for cosmetics within the FDA;
4. name which center/office is responsible for OTC drug–cosmetic products within the FDA;
5. discuss the responsibilities of the PCPC;
6. discuss the responsibilities of the Federal Trade Commission (FTC);
7. discuss how the safety of cosmetic ingredients is evaluated in the US and the EU, respectively;
8. describe the responsibilities of the EC with respect to cosmetics;
9. describe what the function of Cosmetics Europe is in the EU and name some activities in which it is involved.

KEY CONCEPTS

1. The FDA is the main consumer protection agency of the US government.
2. The Center for Food Safety and Applied Nutrition (CFSAN) is responsible for foods and cosmetic products.
3. Within the FDA, the Center for Drug Evaluation and Research (CDER) is responsible for ensuring that drugs, including both prescription-only and OTC drugs, are safe and effective.
4. The PCPC is the leading US personal care trade association that represents the cosmetic and personal care industry at the federal, state, and local levels.
5. The CIR panel is an independent, voluntary panel that assesses the safety of cosmetic ingredients under normal conditions of use in the US.
6. The FTC has principal authority over advertising for foods, dietary supplements, OTC drugs, medical devices, and cosmetics.

7. The EC has overall responsibility for cosmetic legislation within the EU.
8. The SCCS is a scientific advisory board of the EC. It evaluates health and safety risks of nonfood consumer products.
9. The Cosmetics Europe is a representation of European cosmetic companies.

Introduction

In addition to the FDA, there are a number of organizations that have an impact on the cosmetic industry and take part in developing various guidelines. In this section, you can find a summary of the most important organizations related to cosmetic and OTC drug–cosmetic products.

Food and Drug Administration

❶ **The FDA is the main consumer protection agency of the US government.** Its origins can be traced back to about the 1850s. Its modern regulatory function began in 1906 with the Pure Food and Drugs Act, which was the predecessor of the FD&C Act.

The FDA has a number of responsibilities in protecting the public health in the US. It assures safety, efficacy, and security of human and veterinary drugs, biological products, medical devices, food supply, cosmetics, and products that emit radiation. The FDA also has responsibility for regulating the manufacturing, marketing, and distribution of tobacco products to protect public health and to reduce tobacco use by minors. In addition, the FDA plays a significant role in counterterrorism capability.[1] The FDA is also responsible for advancing public health by helping to speed up product innovations. It provides accurate, science-based information for professionals and consumers on its web site to educate people and continuously inform them about the news related to drugs, cosmetics, medical devices, foods, and other topics.

With respect to cosmetics, the FDA is responsible for assuring that cosmetics are safe and properly labeled. If it finds noncompliance with the relevant regulations, it can take various regulatory actions. The FDA is working together with a number of national and international governmental and nongovernmental groups and organizations to ensure safety of cosmetic products and harmonize the rules and regulations of these products. The FDA's goals with respect to international harmonization include facilitation of international trade and promote mutual understanding; facilitation of exchange of scientific and regulatory knowledge with foreign governments; and acceptance of equivalent standards, compliance activities, and enforcement programs of other countries. In the case of OTC drug–cosmetic products, it is responsible for safety and labeling as well as efficacy of product. The FDA inspects cosmetic and OTC drug–cosmetic manufacturing facilities to verify that all procedures performed meet the regulations. It also works with the US Customs and inspects imported products at the border to ensure that they meet US regulations.

Center for Food Safety and Applied Nutrition

● **The CFSAN is part of the FDA. CFSAN is responsible for foods and cosmetic products.** It helps promote and protect public health by ensuring safety and honest labeling of these products. The CFSAN plays an important role in consumer education and international harmonization of standards as well. Within CFSAN, the Office of Cosmetics and Colors regulates and monitors the safety and labeling of cosmetic products. They provide in-market surveillance for marketed cosmetic products and are responsible for the related compliance activities. They are responsible for the VCRP and its daily operation. In addition, they deal with batch certification of certifiable color additives as well.[2]

Center for Drug Evaluation and Research

● **The CDER is also part of the FDA. CDER is responsible for ensuring that drugs, including both prescription-only and OTC drugs, are safe and effective.** This center regulates all cosmetics that are also drugs, such as sunscreens, antiacne medications, antidandruff shampoos, and antiperspirants. The CDER evaluates new drugs before they are sold and checks drugs on the market to ensure that they meet the highest standards. In addition, it oversees the research, development, and manufacture of drugs and monitors their marketing to make sure they are truthful. They also play an important role in informing health professionals and consumers about drugs and their safe use.[3]

Personal Care Product Council

● **The PCPC is the leading US personal care trade association that represents the cosmetic and personal care industry at the federal, state, and local levels.** It was founded in 1894 as the CTFA. Currently (as of June 2014), it has more than 600 member companies who represent manufacturers and distributors of finished products and suppliers of raw materials, packaging, and other services used in the production and marketing of finished products, as well as consumer and trade publications. The council promotes safety of cosmetic products by establishing product safety assessments (through its review panel, see CIR) and keeps consumers informed through its dedicated web site (CosmeticsInfo.org). In addition, PCPC is actively engaged in international efforts to align global regulatory standards for consumer products and eliminate trade barriers worldwide.[4]

Cosmetic Ingredient Review Panel

● **The CIR panel is an independent, voluntary panel** that was established in 1976 by the US industry trade association (CTFA, now known as the PCPC), with the support of the FDA and the Consumer Federation of America. Although funded by PCPC, the CIR and the review process are distinctly separate and independent from PCPC and the cosmetic industry. The CIR **assesses the safety of cosmetic ingredients under normal conditions of use in the US.** It typically evaluates cosmetic ingredients whose safety is questionable or related to which some issues occurred in the past.[5] The CIR is a nonprofit panel of scientific and medical experts, including toxicologists, dermatologists, and chemists, which serve as voting members. Nonvoting members of the panel include representatives from the FDA, PCPC, and consumer groups. CIR assesses ingredient safety in an unbiased, expert manner.

When selecting ingredients, CIR monitors data in the published literature, safety reports of cosmetic products in the case of adverse reactions, and reports made by manufacturers and distributors through the VCRP. Referring back to the VCRP, it means that by participating in the VCRP, companies support this safety review process. Before CIR prepares its final safety assessment, comments from the public can also be made. CIR's findings have established a public record of the safety of cosmetic ingredients. CIR safety assessments are made available as monographs and are also accessible in the *International Journal of Toxicology*. Each year, CIR publishes the CIR Compendium, a comprehensive collection of findings from all CIR reports.

Although findings provide information for the FDA, they have no legal authority and the FDA is not obliged to act on the findings. However, the FDA usually takes into account the CIR's conclusion on ingredients in its decision-making process.

Federal Trade Commission

● **The FTC** was established in 1914 with the purpose of preventing unfair methods of competition in commerce. Since then, the FTC has been directed to administer a wide variety of consumer protection laws. The FTC works to prevent fraudulent, deceptive, and unfair business practices in the marketplace. It is an independent executive-branch agency whose members are appointed by the President and confirmed by the Senate. The FTC **has principal authority over advertising for foods, dietary supplements, OTC drugs, medical devices, and cosmetics.** As discussed in the previous section, labeling for these products is regulated by the FDA.[6] Under the FTC Act, advertising must be truthful and nondeceptive, advertisers must have evidence to back up their claims, and advertisements cannot be unfair. The FTC investigates advertisements and can take action against an ad if it does not meet the FTC Act's requirements. The aims of the FTC include the following: to identify illegal practices; stop and take action against illegal practices through law enforcement; prevent consumer injury through education of consumers and businesses; enhance consumer benefit through research, reports, and advocacy; and protect American consumers globally.[6]

European Commission

The EC is the EU's executive body and represents the interests of Europe as a whole (as opposed to the interests of individual countries). It is one of three main EU institutions; the other two are the European Parliament and the Council (of the EU). The EC prepares legislation for adoption by the Council (representing the member countries) and the Parliament (representing the citizens).[7] With respect to cosmetics, ● **the EC has overall responsibility for cosmetic legislation within the EU.** Individual responsibility over cosmetics remains in each member state, which designates its own competent authority that enforces the legislation and monitors if companies comply with it or not.

Scientific Committee on Consumer Safety

❽ **The SCCS is a scientific advisory board of the EC. It evaluates health and safety risks of nonfood consumer products.** Examples of the types of issues within the scope of the SCCS include cosmetic products and personal care products, toys, textiles and clothing, and etc., as well as services, such as tattooing and artificial suntanning. It reviews a particular ingredient in question and issues an opinion prior to any regulation changes by the EC. This committee is similar to the US CIR. It usually evaluates ingredients on request and publishes the findings of their risk assessment on the EC's web site. Similar to the CIR, the SCCS also provides an opportunity for interested parties to comment on the findings. At the end of the risk assessment process, the Committee adopts SCCS opinions.[8]

Cosmetics Europe – The Personal Care Association

❾ **Cosmetics Europe is a representation of European cosmetic companies**, including major international cosmetics manufacturers and small businesses, as well as associations involved in the cosmetic industry. It is a nonprofit association that was established in 1962. It was formerly known as the "Colipa" (an acronym initially standing for "Comité de liaison de la parfumerie") and renamed in 2012 to "Cosmetics Europe – the Personal Care Association." Its purpose is to promote the cosmetic industry in Europe.[9] It can be identified as a sister association of the US PCPC. Cosmetics Europe works with the EC and individual member states and many other international and national groups and organizations. It is committed to help EU member states implement the new cosmetic regulations. It is involved in international activities to align regulatory issues and help trade. It also supports the development of animal-free safety assessment methods for cosmetic ingredients and products. It has some of its own publications for the cosmetic industry, including guidelines for safety assessment of cosmetics, efficacy testing of cosmetics, *in vitro* determination of UVA protection, and others.

GLOSSARY OF TERMS FOR SECTION 3

CDER: Center for Food Safety and Applied Nutrition is part of the FDA it is responsible for ensuring that drugs are safe and effective.

CFSAN: Center for Food Safety and Applied Nutrition is part of the FDA. It is responsible for foods and cosmetic products.

CIR: Cosmetic Ingredient Review is an independent, voluntary panel that assesses the safety of cosmetic ingredients under normal conditions of use in the US.

Cosmetics Europe: It is a representation of European cosmetic companies.

EC: European Commission has the overall responsibility for cosmetics legislation within the EU.

FDA: Food and Drug Administration is the main consumer protection agency of the US government.

FTC: Federal Trade Commission has principal authority over advertising for foods, dietary supplements, OTC drugs, medical devices, and cosmetics in the US.

Office of Cosmetics and Colors: Part of the CFSAN, it regulates and monitors the safety and labeling of cosmetic products.

PCPC: Personal Care Product Council represents the cosmetic and personal care industry at the federal, state, and local levels in the US.

SCCS: Scientific Committee on Consumer Safety is a scientific advisory board of the EC. It evaluates health and safety risks of nonfood consumer products.

REVIEW QUESTIONS FOR SECTION 3

Multiple Choice Questions

1. Which of the following is NOT true for the FDA?
   a) It regulates drugs, cosmetics, and foods among others
   b) It is the main consumer protection agency of the US government
   c) It has authority over advertising for drugs, cosmetics and foods among others
   d) It works with the US Customs to ensure imported products meet the US regulations

2. Who has authority over cosmetic advertising in the US?
   a) Federal Trade Commission
   b) Food and Drug Administration
   c) Cosmetic Ingredient Review
   d) Personal Care Product Council

3. What is the function of the Personal Care Product Council?
   a) It represents the cosmetic and personal care industry in the US
   b) It represents the cosmetic and personal care industry in the EU
   c) It regulates cosmetics in the US
   d) It represents the US personal care industry in the EU

4. What is the function of the Cosmetic Ingredient Review panel?
   a) To assess the quality of cosmetic ingredients in the US
   b) To assess the safety of cosmetic ingredients in the US
   c) To help cosmetic companies register in the Voluntary Cosmetic Registration Program
   d) To assess the FDA's regulatory activity

5. With who is the CIR affiliated?

a) No one; it is independent
b) Food and Drug Administration
c) European Commission
d) World Health Organization

6. Which of the following is TRUE for the European Commission?

a) It evaluates the safety of cosmetic ingredients in the US
b) It revaluates the safety of cosmetic ingredients in the EU
c) It regulates drugs in the US
d) It regulates cosmetics in the EU

Matching

Match the government and independent organizations in column A with their appropriate function in column B.

	Column A	Column B
_____	1. Center for Drug Evaluation and Research	A. Advisory board in the EU, evaluates the health and safety risks of nonfood products
_____	2. Center for Food Safety and Applied Nutrition	B. Independent panel, assesses the safety of cosmetic ingredients
_____	3. Cosmetic Ingredient Review Panel	C. Part of the FDA, ensures that drugs are safe and effective
_____	4. Cosmetics Europe	D. Part of the FDA, regulates the safety and labeling of cosmetics
_____	5. European Commission	E. Regulates cosmetic and OTC drug advertising in the US
_____	6. Federal Trade Commission	F. Regulatory authority for cosmetics and drugs in the US
_____	7. Food and Drug Administration	G. Regulatory authority for cosmetics in the EU
_____	8. Personal Care Product Council	H. Represents cosmetic companies in Europe
_____	9. Scientific Committee on Consumer Safety	I. Represents the cosmetic and personal care industry at the federal, state, and local levels in the US

REFERENCES

1. FDA: About the FDA, Last update: 6/19/2012, Accessed 5/15/2013 at http://www.fda.gov/AboutFDA/WhatWeDo/default.htm
2. FDA: CFSAN – What We Do, Last update: 4/11/2012, Accessed 5/15/2013 at http://www.fda.gov/AboutFDA/CentersOffices/OfficeofFoods/CFSAN/WhatWeDo/default.htm
3. FDA: FAQs About CDER, Last update: 10/20/2010, Accessed 5/10/2013 at http://www.fda.gov/AboutFDA/CentersOffices/OfficeofMedicalProductsandTobacco/CDER/FAQsaboutCDER/default.htm
4. PCPC: 2008 Annual Report, Accessed 5/16/2013 at http://www.personalcarecouncil.org/sites/default/files/2008CouncilAnnualReport.pdf
5. CIR: Cosmetic Ingredient Review Procedures, October 2010, Accessed 5/13/2013 at http://www.cir-safety.org/how-does-cir-work
6. FTC: Protecting America's Consumers, Accessed 5/19/2013 at http://www.ftc.gov/
7. EC: European Commission at Work, Accessed 5/20/2013 at http://ec.europa.eu/atwork/index_en.htm
8. EC: Health and Consumers: SCCS, Accessed 5/19/2013 at http://ec.europa.eu/health/scientific_committees/consumer_safety/index_en.htm
9. Cosmetics Europe: Colipa Annual Report 2010, Accessed 5/20/2013 at https://cosmeticseurope.eu/about-cosmetics-europe.html

SECTION 4: COSMETIC GOOD MANUFACTURING PRACTICES

LEARNING OBJECTIVES

Upon completion of this section, the reader will be able to

1. define the following terms:

ISO 22716	Cosmetic GMPs	Rule	Standard
ICCR	Risk minimization	Quality	Batch
ISO	GMP	cGMPs	Internal audit

2. define the term "manufacturing" and differentiate among products, raw ingredients, and packaging material manufacturers;
3. discuss how the use of cosmetic GMPs is regulated in the US and the EU, respectively;
4. differentiate between a standard and a rule;
5. define the term "quality";

6. name some areas for which GMPs contain guidelines regarding cosmetic products;
7. explain through some examples why a manufacturer should follow the GMPs during the manufacturing of cosmetic products.

KEY CONCEPTS

1. General requirements for cosmetic products, whether sold in the US or the EU, are that they must be safe, effective, stable, and have the same quality over batches.
2. In the US, there is no official mandatory quality assurance system for cosmetic manufacturing. However, the FDA highly recommends cosmetic manufacturers to follow the GMPs.
3. In the EU, all cosmetic products placed on the market must be manufactured according to the cosmetic GMPs.
4. The FDA's guideline can effectively help manufacturers self-inspect their operations. The guideline provides recommendations for the following areas: documentation, records, buildings and facilities, equipment, personnel, raw materials, production, laboratory controls, internal audit, complaints, adverse events, and recalls.
5. The cosmetic GMPs do not specify the type of technique or equipment manufacturers have to use; it just states minimum requirements.
6. The GMPs are designed to minimize the risks involved in any production that cannot be eliminated through testing the final product.
7. An additional important reason for using the GMPs is to ensure that all manufactured products, including those that are not tested, meet the quality specifications.

Cosmetic Good Manufacturing Practices

Manufacturing is a process during which a new cosmetic product is produced from raw materials based on a formula using specific techniques and equipment in a controlled environment. The person or entity responsible for the production of a cosmetic product is usually referred to as the product manufacturer (or finished good manufacturer). It should also be mentioned that manufacturing may refer to the production of raw materials. Companies responsible for the production of raw materials are usually referred to as the raw material manufacturers or suppliers. In addition, manufacturing may refer to the production of packaging materials. These companies are usually known as packaging material suppliers. This section focuses on the manufacture of cosmetic products.

❶ **General requirements for cosmetic products, whether sold in the US or the EU, are that they must be safe, effective, stable, and have the same quality over batches.** As discussed in Section 1 of this chapter, ❷ **in the US,**

there is no official mandatory quality assurance system for cosmetic manufacturing. However, the FDA highly recommends cosmetic manufacturers to follow the Good Manufacturing Practices (hereinafter referred to as "GMP"). The FDA issued a guidance for the industry regarding cosmetic GMP (as of June 2014, the draft version is available). This guidance reflects the FDA's current thinking. It is a recommendation that should be followed; however, its implementation is not required. The guidance is intended to help the industry in identifying the standards and potential issues that can affect the quality of cosmetic products.[1]

Unlike cosmetics, OTC drug–cosmetic products must be manufactured in accordance with the GMP requirements, or otherwise, they are considered adulterated in the US.[2] As discussed previously, ❸ **in the EU, all cosmetic products placed on the market must be manufactured according to the Cosmetic GMPs** described in the ISO 22716 standard. Therefore, all non-European markets importing to the EU are concerned since they have to meet the European standard to be able to enter their products into the market. To keep up with the EU regulations, many companies follow the GMP outside the EU as well, even if they do not want to import to the EU. Understanding the basics of GMP is essential for everyone who wants to be employed at any level of the cosmetic industry. This section provides an overview of the cosmetic GMPs.

DID YOU KNOW?

The abbreviation "ISO" refers to International Organization of Standardization. ISO is a nongovernmental organization, the world's largest developer of voluntary international standards. These standards are state-of-the-art and help the industry to be more efficient. The ISO standards are developed through global consensus, and their purpose is to help break down the barriers to international trade.[3]

DID YOU KNOW?

The terms "rule" and "standard" are not synonyms. Rules are mandatory, constraining, and rigid directions or instructions. If not met or applied, legal consequences may follow them. However, standards, also called guidelines, are more flexible, providing a greater range of choices. Standards are recommended to be followed in order to meet rules; however, they are not mandatory. An example for the rules and standards can be the GMP for cosmetics in the US and the EU. The FDA does not require manufacturers to follow GMPs, it only recommends it. The GMPs in the US are considered a guideline. Of course, the best way to provide safety to cosmetic products is to follow the GMPs, but manufacturers can follow other appropriate standards as well. However, application of the GMPs is a rule in the EU that must be followed in order to place a product on the European market.

The International Cooperation on Cosmetics Regulation (ICCR) is an international voluntary partnership of the health regulatory authorities from the United States, Japan, the EU, and Canada, with participation from the cosmetic industry associations. This group maintains the highest level of global consumer protection, while minimizing barriers to international trade. ICCR provides a forum for discussions on alignment of cosmetic regulations between the four members.[4] The ICCR recognized the importance of the GMPs and the need to work toward a common approach. In 2007, the regulators agreed to take the ISO Standard 22716 (cosmetic GMPs) into consideration when developing or updating their guidelines on the cosmetic GMPs. The FDA updated its guideline in 2013 and incorporated elements of ISO 22716 into the new guideline as appropriate.

FYI

The GMP guidance is freely accessible for every manufacturer through the FDA's website. For more information, visit the FDA's website: Home, Cosmetics, Guidance and Regulation, Guidance Documents, *Draft Guidance for Industry: Cosmetic Good Manufacturing Practices*.

The GMP is part of quality assurance, which ensures that products are consistently produced and controlled according to quality standards appropriate to their intended use and as required by the product specification.[5] Talking about quality assurance and quality standards, let us stop here for a minute and think about what the term "quality" means. Quality can be defined as the totality of characteristics of an entity that bears on its ability to satisfy stated and implied needs.[6] It is more a theoretical definition stating that quality is suitable for the purpose for which the product will be used. However, it might still differ depending on the product type, since different groups have different needs and may require different characteristics for a product. It is especially true for cosmetics. For example, a "high-quality mascara" might mean five different characteristics if you ask five women. Consumer satisfaction is a very important part of a product's quality; however, in itself it would be too subjective to outline the required characteristics. To make it more objective, we usually use requirements based on measurable characteristics, e.g., viscosity, color, and pH, just to mention a few. By meeting these requirements, manufacturers are able to provide the same quality for each batch of their products. This is the main purpose of following quality assurance systems such as the GMPs.

DID YOU KNOW?

The term "batch" refers to the total amount of products manufactured in one operation.

GMPs are often referred to as the current GMPs and are abbreviated as "cGMPs." The term "current" refers to techniques and equipment that are state of the art, innovative, and help manufacturers meet the minimum requirements of GMPs. However, what is considered current; is not defined, it is left to the manufacturers to decide that. Systems and equipment that may have been the best to prevent contamination, mix-ups, and errors a decade ago may be less than adequate by today's standards. Therefore, manufacturers should keep up with the technical advances and upgrade their equipment from time to time.

As discussed in the previous sections of this chapter, the FD&C Act prohibits the introduction of adulterated or misbranded cosmetics into the US market. Finished goods manufacturers are recommended to follow the cosmetic GMPs in order to reduce the risk of producing adulterated or misbranded cosmetics. ❶ **The FDA's guideline can effectively help manufacturers self-inspect their operations. The guideline provides recommendations for the following areas:**[1]

- **Documentation**, i.e., the process of documenting information in either electronic or paper-based format during all operations performed at a company, including manufacturing and testing for future reference. The term can also refer to the documents compiled.
- **Records**, i.e., documentation created.
- **Buildings and facilities**, i.e., all buildings and facilities used for manufacturing cosmetics at a company.
- **Equipment**, i.e., all equipment and utensils used in processing, holding, transferring, and packaging of cosmetics at a company.
- **Personnel**, i.e., everyone supervising or performing cosmetic manufacturing or quality control at a company.
- **Raw materials**, i.e., starting materials used for cosmetic manufacturing.
- **Production**; i.e., the process of manufacturing cosmetics.
- **Laboratory controls**, i.e., sample collection techniques, specification, test methods, laboratory equipment, and technician qualifications for testing newly manufactured and retained cosmetic samples.

- **Internal audit**, i.e., systematic and independent examination made by competent personnel inside the company, the aim of which is to determine whether activities covered by these guidelines and related results comply with planned arrangements and whether these arrangements are implemented effectively and are suitable for achieving objectives.[7]
- **Complaints, adverse events, and recalls.**

● **The cosmetic GMPs do not specify the type of technique or equipment manufacturers have to use, it just states minimum requirements**; e.g., equipment should be clean. The GMP requirements are general and flexible; each manufacturer can meet them on its own way. It is each manufacturer's responsibility to find out how it can best meet requirements. They do not necessarily have to implement the same cleaning measures and use the same cleaning aids. The most important aspect is the end result; i.e., cleanliness must be provided.

After reading the GMP guideline, you will have a general idea of how deeply it controls manufacturing. Let us now review the main reasons for the use of the cosmetic GMPs.

- ● **The GMPs are designed to minimize the risks involved in any production that cannot be eliminated through testing the final product.** The main risks may include unexpected contamination of products, causing health damage; incorrect labels on containers, which could mean that consumers receive the wrong product; insufficient or too much ingredients, resulting in undesired effects; etc.. It is more expensive to find mistakes after they have been made than preventing them in the first place. If a company does not follow any type of quality assurance system and after producing a batch, it finds out that the batch does not meet the quality standards, the manufacturer has to throw the entire batch away and start producing another one. We can admit that this process would be time and money consuming, so quality assurance should be part of each and every manufacturing procedure. GMPs prevent errors that cannot be eliminated through quality control of the finished product.
- Quality and maintaining quality are essential characteristics of cosmetic products and their production. Every consumer is looking for reliable, safe, and high-quality products. Quality is continuously checked during manufacturing; however, only a small percentage of finished products is actually tested. Without the implementation of a good in-place quality assurance system, the manufacturer cannot ensure that every item of a batch is of the same safety and quality. Therefore, ● **an additional important reason for using the GMPs is to ensure that all manufactured products, including those that are not tested, meet the quality specifications.**

A very important thought is that "good quality must be built in during the manufacturing process; it cannot be tested into the product afterwards."[8]

GLOSSARY OF TERMS FOR SECTION 4

Batch: The total amount of products manufactured in one operation.

cGMPs: Current GMPs refers to techniques and equipment that are state of the art, innovative, and help manufacturers meet the minimum requirements of GMP.

Cosmetic GMPs: Cosmetic good manufacturing Practices.

GMP: GMP is part of quality assurance which ensures that products are consistently produced and controlled according to quality standards appropriate to their intended use and as required by the product specification.

ICCR: International Cooperation on Cosmetics Regulation is an international voluntary partnership between the health regulatory authorities of the United States, Japan, the EU, and Canada. It provides a forum for discussions on alignment of cosmetic regulations between the four members.

Internal audit: Systematic and independent examination made by competent personnel inside the company, the aim of which is to determine whether activities covered by these guidelines and related results comply with planned arrangements and whether these arrangements are implemented effectively and are suitable for achieving objectives.

ISO 22716: It is a standard that describes the Cosmetics Good Manufacturing Practices.

ISO: International Organization of Standardization is a nongovernmental organization and the world's largest developer of voluntary international standards.

Quality: The totality of characteristics of an entity that bear on its ability to satisfy the stated and implied needs.

Risk minimization: GMPs are designed to minimize the risks involved in any production; e.g., contamination and incorrect labeling, which cannot be eliminated through testing the final product.

Rule: Mandatory, constraining, and rigid direction or instruction. If not met or applied, legal consequences may follow.

Standard: Flexible direction or instruction that is recommended to be followed; however, it is not mandatory. No legal consequences may follow if not met.

REVIEW QUESTIONS FOR SECTION 4

Multiple Choice Questions

1. Which of the following is TRUE with respect to the regulation of cosmetic product manufacturing in the US?
   a) The regulation in the US is the same as the regulation in the EU
   b) The FDA requires manufacturers to follow the cosmetic GMPs
   c) The FDA has a cosmetic GMP guidance and recommends that manufacturers follow this guidance
   d) The FDA prohibits manufacturers to follow the cosmetic GMPs
2. Which of the following is the main purpose of following the cosmetic GMPs?
   a) To ensure that all cosmetic companies participate in the VCRP
   b) To help minimize risks that could affect the quality, safety, and/or effectiveness of cosmetic products
   c) To help the FDA approve premarket applications for cosmetics
   d) To eliminate the need for quality and safety testing of cosmetics
3. Which of the following is NOT true for the FDA's GMP guidance?
   a) It provides specific directions on how to meet the requirements
   b) It does not provide specific directions on how to meet the requirements
   c) It includes general requirements that should be met
   d) It is freely accessible for every manufacturer
4. Which of the following are members of the International Cooperation on Cosmetic Regulation (ICCR)?
   a) Japan, Canada, China, European Union
   b) Japan, Canada, United States, European Union
   c) Canada, United Kingdom, Japan, European Union
   d) China, Canada, United States, European Union
5. What is the ISO 22716?
   a) A standard that provides guidance on the cosmetic Good Manufacturing Practices
   b) A chapter of the FDA's guidance on the cosmetic Good Manufacturing Practices
   c) A world meeting organized annually to discuss barriers to international trade of cosmetics
   d) None of the above

6. Rules are ___, while standards are ___.

a) Recommended/required (mandatory)
b) Recommended/recommended
c) Required (mandatory)/required (mandatory)
d) Required (mandatory)/recommended

Fact or Fiction

______ a) GMP is designed to ensure that every manufacturer has the same type of processing, transferring, and filling equipment.
______ b) Rigorous adherence to Good Manufacturing Practices minimizes risk of adulteration or mislabeling of cosmetics.
______ c) The FDA recommends manufacturers to follow the cosmetic GMPs.
______ d) During product testing, only a small percentage of a batch is tested.
______ e) Quality can be easily defined through questionnaires given to consumers.

REFERENCES

1. FDA: Guidance for Industry, Cosmetic Good Manufacturing Practices, Draft Guidance, Last revised 6/2013, Accessed 11/11/2013 at http://www.fda.gov/downloads/Cosmetics/GuidanceComplianceRegulatoryInformation/GuidanceDocuments/UCM358287.pdf
2. FD&C Act Section 501
3. ISO: About ISO, Accessed 11/112013 at http://www.iso.org/iso/home/about.htm
4. FDA: Cosmetics: International Cooperation on Cosmetic Regulation (ICCR), Last update: 5/9/2014, Accessed 5/24/2014 at http://www.fda.gov/Cosmetics/InternationalActivities/ICCR/default.htm
5. Sharp, J.: *Good Manufacturing Practices, Philosophy and Applications*, Buffalo Grove: Interpharm Press, 1991.
6. International Organization for Standardization; *Technical Committee ISO/TC 176. ISO 8402: Quality Management and Quality Assurance—Vocabulary*, 2nd Edition, Geneva: International Organization for Standardization, 1994.
7. Infernal Audit: *Cosmetics – Good Manufacturing Practices (GMPs) – Guidelines on Good Manufacturing Practices, ISO 22716:2007*, Geneva, Switzerland: ISO, 2007.
8. WHO: Training Workshops on Good Manufacturing Practices (GMP), Accessed 5/1/2013 at http://apps.who.int/prequal/trainingresources/Training_gmp_info.htm

3

SKIN CARE PRODUCTS

INTRODUCTION

Skin care products are the most widely used product types among all cosmetics and personal care products. They have been the largest sector of the cosmetics market for many years, and their market still continues to grow. In the past, skin care products were primarily used to clean and moisturize the skin. Over the years, with the introduction of new raw materials and advanced technologies, a variety of innovative products have been developed that perform better and have cosmetically more appealing properties. Today, we have a variety of products that offer multiple functions, including moisturization, protection from the sun, and prevention of aging among others, and they can even help prevent and/or treat some skin care problems.

This chapter provides a basic understanding of the structure of human skin and its main functions as well as various products that are applied to the skin. These include products used to remove dirt and makeup from the skin, improve the skin quality, maintain and/or restore the skin's youthful appearance, and aid in alleviating the symptoms of various skin diseases. It also reviews the main characteristics, ingredients, formulation technology, and testing methods and packaging materials for skin cleansing products, skin moisturizing products, products for special skin concerns, and sunscreens as well as deodorants and antiperspirants.

Introduction to Cosmetic Formulation and Technology, First Edition. Gabriella Baki and Kenneth S. Alexander.
© 2015 John Wiley & Sons, Inc. Published 2015 by John Wiley & Sons, Inc.

SECTION 1: SKIN ANATOMY AND PHYSIOLOGY

LEARNING OBJECTIVES

Upon completion of this section, the reader will be able to

1. define the following terms:

Acid mantle	Basal cell layer	Brick-and-mortar model	Collagen
Corneocyte	Corneodesmosomes	Cornified lipid envelope	Dermis
Desquamation	Elastin	Epidermis	Intercellular lamellar lipid
Keratinization	Keratinocyte	Langerhans cell	Melanin
Melanocyte	Merkel cell	NMF	Pilosebaceous unit
Resident flora	Stratum corneum	Subcutis	TEWL
Transient flora			

2. name the layers of the human skin;
3. explain the functions of the stratum corneum;
4. briefly describe the structure of the stratum corneum;
5. explain the importance of the natural moisturizing factor;
6. briefly discuss the consequences of losing the natural moisturizing factor;
7. differentiate between keratinization and desquamation;
8. briefly discuss the importance of the basal cell layer in the epidermis;
9. explain where melanocytes can be found and what their function is;
10. explain where vitamin D is produced in the human skin;
11. briefly discuss the main functions of the epidermis;
12. describe the main structure of the dermis;
13. differentiate between collagen and elastin;
14. briefly discuss the main functions of the dermis;
15. describe the main functions of the hypodermis;
16. explain the normal process of water loss through the skin;
17. explain what the term "transepidermal water loss" refers;
18. explain what keeps water loss from the skin at a minimum level;
19. differentiate between the resident and transient flora of the skin;
20. explain what the pH and surface charge of the human skin are;

21. explain what the practical advantages of the skin surface charge are;
22. briefly describe Fitzpatrick's classification system for the human skin;
23. differentiate between normal, dry, oily, combination, and sensitive skin;
24. explain how the hormonal differences between males and females influence the skin's structure;
25. explain how the differences in the skin's function and structure between males and females can be utilized in product formulation.

KEY CONCEPTS

1. The human skin consists of two main layers, namely the epidermis and dermis. Underneath the dermis, there is a third layer, called the hypodermis.
2. Epidermis is the outer layer of the skin that functions as a protective layer against external influences.
3. The skin's outermost layer, the SC, is a natural barrier, which has a very unique structure, often referred to as the "brick and mortar" structure.
4. The dermis is located under the epidermis, and it functions as a supporting frame to the epidermis, supplying it with nutrients and oxygen via the blood capillaries.
5. Hypodermis is a loose connective tissue that stores fat in the fat cells.
6. In normal skin, there is a continuous movement of water from the deeper layers toward the superficial layer where water eventually evaporates.
7. Human skin is continuously inhabited by many different bacteria and fungi, which under normal circumstances in a healthy individual are harmless and are even beneficial. Microbes on the skin are generally divided into two categories: resident flora and transient flora.
8. Normally, the skin surface is slightly acidic and ranges between pH 4.5 and 5.5.
9. Since the isoelectric point of the proteins found in the skin's upper layers is between 3.5 and 4.5, the skin has a negative charge at physiological pH.
10. Human skin can be categorized based on gender, its color, UV sensitivity, vulnerability, oiliness, healthiness, and special needs, among others.

Introduction

Skin is the number one target for most cosmetics and personal care products. Consumers apply products to their skin to cleanse, protect, moisturize, peel, or cover it. As mentioned earlier, cosmetics are not intended to change the structure of the skin. However, OTC drug–cosmetic products and prescription-only drugs and even some cosmeceutical products alter the structure of the skin. Therefore, understanding the structure and function of the skin is essential.

Structure and Function of Human Skin

Skin is the largest sensory and contact organ in the human body. Its surface area in adults is approximately 1.5–2 m^2.[1] ❶**The human skin consists of two main layers, namely the epidermis and dermis.**[2] **Underneath the dermis, there is a third layer, called the hypodermis**, which consists mainly of fat cells and is not considered a component of the skin. Figure 3.1 depicts the basic anatomical parts of the human skin. The skin is a complex organ made up of dead cells, epithelium, connective tissue, muscles, nerves, blood vessels, as well as the so-called appendages (i.e., accessory structures), including the nails, hair, and glands, such as sebaceous glands, eccrine and apocrine sweat glands.[3]

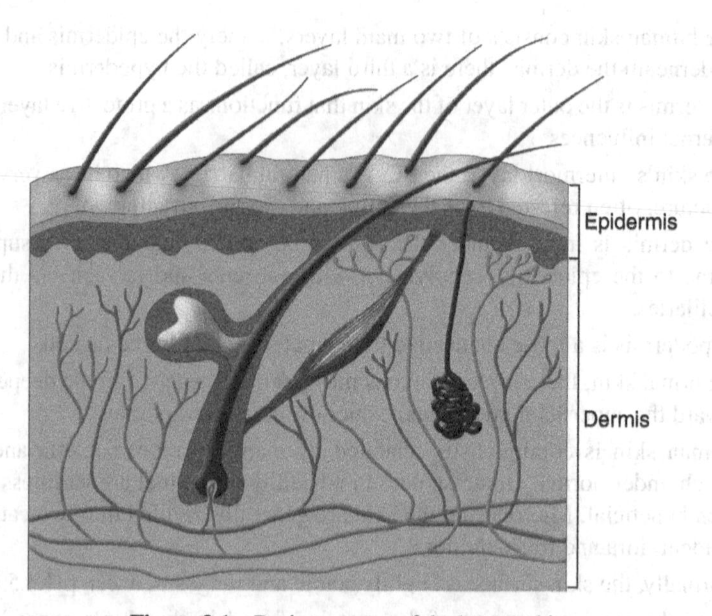

Figure 3.1 Basic structure of the human skin.

Epidermis

❷**Epidermis is the outer layer of the skin that functions as a protective layer against external influences.** The layers are depicted in Figure 3.2. It is composed of five main layers, which have Latin names, including the following[3,4]:

- Stratum corneum (hereinafter referred to as the "SC"), otherwise known as the horny layer, is made up of dead cells that continuously shed and are replaced by cells in the adjacent layer. This layer is very thick compared to the others; it contains 15–30 layers of dead cells.
- Stratum lucidum, the translucent or clear layer, contains 3–5 rows of densely packed flat dead cells.

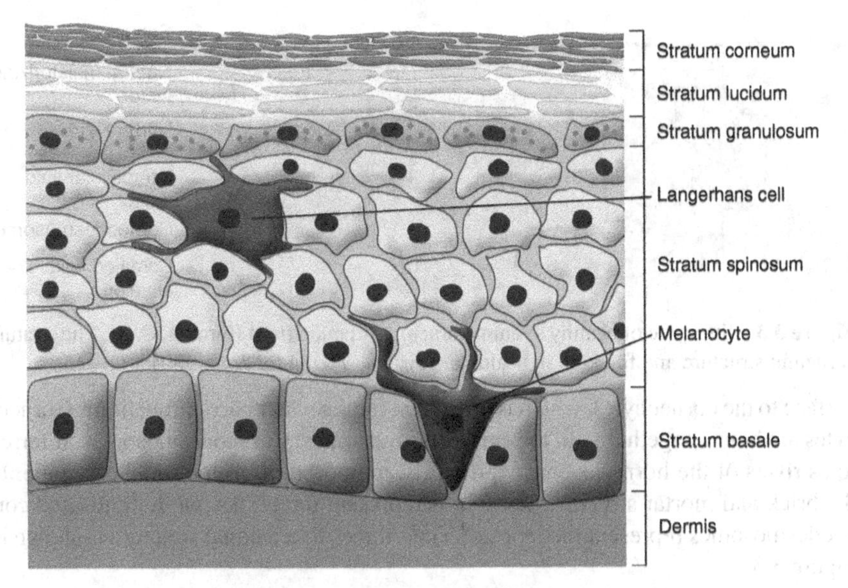

Figure 3.2 Main layers of the epidermis.

- Stratum granulosum, the granular layer, consists of 3–5 layers of flattened keratinocytes that begin to die. In this layer, granules can be observed in the cells, this is from where the name comes.
- Stratum spinosum, the prickle cell layer, contains 8–10 rows of cells. This layer is responsible for lipid and protein synthesis.
- Stratum basale (or stratum germinativum), the basal cell layer, is made up of a single layer of cells. This is the layer where cells divide continuously to form new keratinocytes. Melanocytes, Langerhans cells, and Merkel cells are also found in this layer.

The epidermis differs in thickness in thin and thick skin. Thick skin is found on the palms of the hands and soles of the feet and has all five layers. Thin skin, which covers the rest of the body, does not have a stratum lucidum and has a thinner SC than thick skin.

●The skin's outermost layer, the SC, is a natural barrier, which has a very unique structure, often referred to as the "brick and mortar" structure. The SC is made up of tightly packed, water-resistant dead cells (called **corneocytes**), which are embedded in a complex lipid material (called **intercellular lamellar lipids**). The intercellular lamellar lipid membrane is made up of layers of lipids (the term "lamellar" refers to this layered appearance), and it is mainly made up of ceramides, cholesterol, and fatty acids.[5–7] This lipid matrix waterproofs the epidermis, prevents dehydration, and provides necessary moisture permeability to the SC. Corneocytes are protected by a **cornified envelope**, which consist of two components: proteins and ceramide lipids. It is believed that the covalently bound lipids provide a hydrophobic

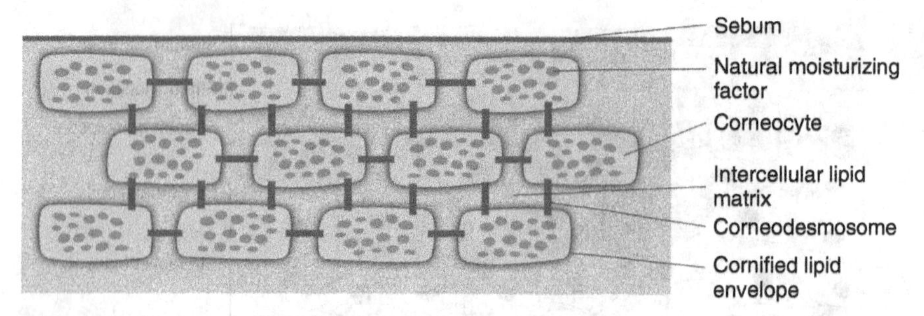

Figure 3.3 Structure of healthy stratum corneum. Adapted from Harding, C. R.: The stratum corneum: structure and function in health and disease. *Dermatol Ther.* 2004;17:6–15.

surface to the corneocytes, which is important for the water barrier function.[8] Corneocytes are bound together with the **corneodesmosomes**. Corneodesmosomes, referred to as rivets of the horny layer, are primarily composed of proteins and ceramide. In the brick and mortar structure, dead cells represent the bricks, while lipids and corneodesmosomes represent the mortar.[9] Their three-dimensional structure is shown in Figure 3.3.

Corneocytes contain a blend of hygroscopic compounds, collectively referred to as the natural moisturizing factor (hereinafter referred to as the "NMF"), which plays a vital role in maintaining the hydration of the SC.[10] NMF acts as a natural humectant, i.e., it keeps the skin hydrated. Its main ingredients include amino acids and their derivatives, such as pyrrolidone carboxylic acid, urocanic acid; lactic acid; urea; citrate; sugars and minerals, such as potassium, calcium, sodium, and magnesium.[11] The lipid matrix helps prevent the loss of NMF. Since the NMF components are water soluble, they are easily leached from the cells, especially when the skin is drier.[12] For this reason, NMF can be washed away, if not protected by the cornified lipid envelope. This is why skin cleansing with soaps can be detrimental to its barrier function. If NMF is lost from the corneocytes, the skin becomes dry, flaky, and uncomfortable (see Figure 3.4). In healthy skin, the SC is impenetrable to irritants and allergens, and water loss is minimized. However, when the skin dehydrates, the corneocytes shrink, causing cracks to appear between them, allowing penetration of external irritants, which trigger a local inflammatory response, causing discomfort, itching and scratching, and further skin damage.

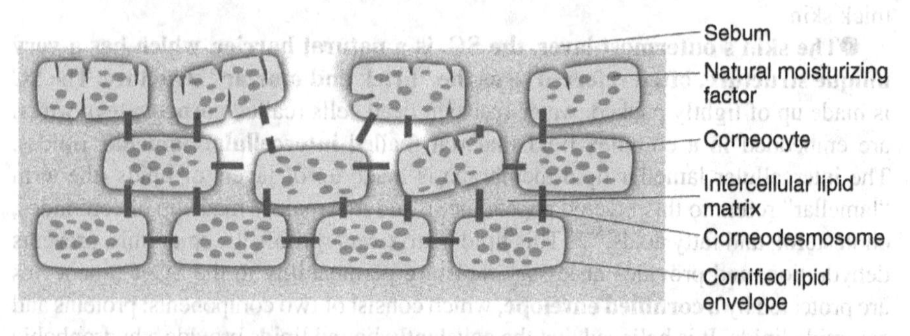

Figure 3.4 Structure of damaged stratum corneum.

Keratinization, also known as cornification, is the dynamic process of epidermal renewal. It begins in the basal layer where new skin cells, known as keratinocytes, are formed and are continually pushed upward. As keratinocytes are pushed upward through the different layers of the epidermis, they go through a "makeover" where their composition and shape change significantly. By the time the cells reach the third layer, stratum granulosum, most have stopped dividing and have started producing keratin. As keratin develops, keratinocytes becomes flattened cells their membrane thickens, and becomes less permeable. By the time they reach the SC, they are dead skin cells, which are eventually shed from the surface of the skin. This latter process is referred to as **desquamation**.[13] The turnover of the epidermis normally proceeds in about 4 weeks. During that time, the cells are part of the SC for about 2 weeks before being shed from the skin surface.[14] As the connection provided by the corneodesmosomes is very tight, keratinized cells are generally shed in large groups or sheets rather than individual cells.

DID YOU KNOW?

Keratinization and shedding are continuous processes, they happen every moment in our lives. Normally, these processes are invisible. When the physiological balance is disturbed, shedding becomes visible and may be accompanied by additional symptoms, such as dryness, itching, and redness.

In addition to the keratinocytes, the epidermis contains other types of cells as well, including melanocytes, Langerhans cells, and Merkel cells. **Melanocytes** are found in the basal cell layer and are responsible for melanin production (see Figure 3.2). Melanin is a pigment that gives the skin its color. Another pigment responsible for skin color is called carotene. It is also found in the epidermis and has an orange-yellowish color. Carotene is also found in a variety of orange vegetables, such as carrot. One eating of a lot of vegetables containing carotene may actually get an orange skin tone. Melanin also provides photoprotection to the skin against ultraviolet (UV) light.[15] **Langerhans cells** play an important role in the body's immune system.[16] **Merkel cells** tend to lie close to sensory nerve endings in the stratum basale. They function as touch receptors.[3]

DID YOU KNOW?

Albinism is a condition resulting from the skin's inability to synthesize melanin and is characterized by milky or translucent skin, pale or colorless hair, and pink or blue iris.[17]

Vitamin D production also occurs in the epidermis, which is known to be the main source of vitamin D in humans. UV light stimulates the conversion of the precursor 7-dehydrocholesterol (also known as provitamin D_3) to vitamin D_3.[18] Vitamin D is essential for skeletal development. An interesting fact is that since melanin acts as a photoprotective substrate, it affects the time required for the synthesis of vitamin D. African American skin absorbs more UVB light than Caucasian skin; therefore, it requires 3–5 times longer exposure to make the same amount of vitamin D as Caucasian skin does.[19]

DID YOU KNOW?

When the skin is immersed into water, osmotic forces may move water into or out of the cells. Sitting in a bathtub filled with tap water causes water to move into the epidermis since tap water is hypotonic (i.e., contains fewer dissolved molecules) than body fluids. Cells may swell four times their normal volume, a phenomenon particularly noticeable on the palms and soles. Swimming in the ocean reverses the direction of osmotic flow since the ocean is a hypertonic solution. It makes water leave the body. The process is slow but may lead to severe dehydration on a longer-term basis.

Epidermis contains no blood vessels; it is entirely dependent on the underlying dermis for nutrient delivery. As discussed earlier, this skin layer is made up of a large quantity of keratin, which is responsible for the skin's strength. In summary, the basic functions of the epidermis include the following:

- Maintaining an optimal water content for the skin.
- Limiting water loss through the skin.
- Sustaining optimal lipid content.
- Providing immune protection;
- Acting as an antioxidant barrier against reactive oxygen species.
- Synthetizing vitamin D.
- Providing photoprotection.
- Providing skin color.
- Allowing for desquamation of SC cells.

Dermis

●The dermis is located under the epidermis, and it functions as a supporting frame to the epidermis, supplying it with nutrients and oxygen via the blood capillaries. Its internal three-dimensional structure can be described as an amorphous (i.e., without shape or structure) substance, which acts like a mortar for all components of the dermis. This amorphous substance includes the following elements:

fibroblasts (which produce the intercellular substance and collagen fibers), nerves and sensory organs, blood vessels, sebaceous glands, sweat glands, and hair follicles as well as connective tissue containing collagen and elastin fibers.[3] **Collagen** fibers give the skin its strength, while **elastin** is responsible for the skin's elasticity, i.e., its ability to spring back after being stretched.[2] If these fibers are damaged, for example, as a result of aging, the skin becomes loose and looks thin and wrinkled. In addition, collagen plays an important role in wound healing.[20]

Sebaceous glands are found all over the body, except for the palms, soles, and dorsum of the feet. They are the largest and most concentrated in the face and scalp, where they are sites of origin of acne and other skin disorders. Sebaceous glands are part of the pilosebaceous unit, which also includes hair follicles and a small muscle. There are only a small percentage of sebaceous glands that are not attached to hair follicles. These are referred to as "free" glands and mainly occur in the vermilion border of the lips or in the eyelids (for more detail, see Chapter 4). The normal function of sebaceous glands is to produce and secrete sebum, a group of complex oils, consisting of triglycerides and fatty acid breakdown products, wax esters, squalene, cholesterol, and cholesterol esters.[21,22] Sebum gets to the skin surface through small openings, which are often referred to as pores. It lubricates the skin to protect against friction and provides a protective layer on the skin to reduce water loss from the skin. In addition, it transports antioxidants and has antibacterial activity and pre- and anti-inflammatory function.[23]

Nerve endings are responsible in transmitting sensory signals, such as touch, pressure, pain, and temperature from the skin. **Blood vessels** play an essential role in supplying the epidermis with oxygen and nutrients. In addition, they are particularly important in the regulation of body temperature, along with the sweat glands. As water from sweat evaporates from the skin, it has a cooling effect, which subsequently decreases the body temperature.

Hypodermis

As mentioned previously, there is a fat layer beneath the dermis, which is called the **hypodermis** (hypo means "below"), subcutaneous (sub means "under," cutaneous means "skin") layer, or subcutis. ●**Hypodermis is a loose connective tissue that stores fat in the fat cells.** It acts as a cushioning layer to protect the vital organs from trauma and provides protection against cold. In addition, the fat serves as an energy deposit for the body and defines the body's contours.

Moisture Content of Normal Skin

The water content of the skin, including the epidermis and dermis, is approximately 80%. The SC has a lower water content of approximately 10–30%.[24] The water level in the superficial layers of the human skin is of utmost importance in determining many of its properties. When the water content of the skin is normal, it appears smooth, soft, and glowing. When its water content is lower than normal, lines are more visible, the skin feels tight and dry, and itching and redness can be experienced.

●In normal skin, there is a continuous movement of water from the deeper layers toward the superficial layer where water eventually evaporates. A commonly used term with regard to the skin's hydration state is "transepidermal water loss" (TEWL). It describes the total amount of water lost from the skin (g of water/meter square per hour), a loss that occurs constantly by passive diffusion through the epidermis.[25] Keratinized skin cells, NMF, and intercellular lipids keep TEWL to an acceptable minimum.

DID YOU KNOW?

Water has two options for crossing the skin from the viable tissues toward the outer environment: active transport by sweating through the sweat glands and passive diffusion through the horny layer. Sweating is a mechanism for body temperature control. In addition, it can express psychological stress. In contrast, TEWL is not visible to the naked eye. With no air turbulence, the skin is covered by a transition layer where moisture is transferred from the skin surface to the atmosphere. The quantity of water leaving the SC is estimated at 300–400 ml per day under normal conditions, which is about 1/10–1/20 of sweating.[26]

TEWL and skin hydration have been widely used as indices in evaluating skin barrier function.[27] TEWL can be defined as the diffusion of water through the skin,[28] while skin hydration reflects the water content of the SC.[29] If the skin is damaged by either physical or chemical agents, the barrier function is somewhat compromised and an increase in TEWL can be observed.[30] This problem is further compounded by the fact that the resulting water loss will decrease the barrier's ability even more.

Water is necessary for the skin to maintain its flexibility. When the skin is overly dry, it loses its ability to stretch, causing it to crack and peel more easily. Faster peeling means the skin cells are being shed more rapidly, which triggers an increase in the rate of cell production in the basal layer. When the process of cell growth, migration, and shedding is accelerated, the barrier is significantly weakened since cells providing the barrier do not have time to fully mature.[31] As the new cells rise to the surface of the skin, they lose all their components, except their protein skeleton structure.[32] It is this layer of protein that provides the barrier against water loss. When the process of cell turnover is sped up, the cells reach the top of the skin before they have completely flattened and lost the rest of their components. These immature cells cannot prevent water from moving across them as well as the older ones can. Therefore, water is lost from the skin at a faster rate, and the skin is less able to defend against foreign invaders.

However, water alone will not increase the skin's moisture level. A protective lipid layer is necessary to prevent water from evaporating off the skin. There are various types of moisturizer ingredients acting in different ways to increase the moisture content of the skin. These ingredients are discussed in Section 2 of this chapter.

Skin Flora and Skin pH

❼Human skin is continuously inhabited by many different bacteria and fungi, which under normal circumstances in a healthy individual are harmless and are even beneficial. Microbes on the skin are generally divided into two categories: resident flora and transient flora.[33]

- The **resident** flora consists of relatively fixed types of microorganisms; however, species contained in the resident flora cannot be rigidly defined for all humans since the type and density of species vary from individual to individual, depending on the anatomic location, local humidity, amount of sebum and sweat production, physiological differences, diet, age, geographic location, hormonal status, medicines, and other factors.[34] The most common bacteria include *Corynebacterium*, *Streptococcus*, *Staphylococcus*, *Neisseria*, *Peptococcus*, *Acinetobacter*, and *Proprionibacterium* species, while the most common fungal genus is *Malassezia.*[35,36] The resident flora protects the host from pathogenic bacteria.[37] It cannot be easily removed from the skin by mechanical friction.
- The **transient** flora consists of nonpathogenic or potentially pathogenic (i.e., capable of causing a disease) microorganisms that inhibit the skin and mucous membranes for hours, days, or even weeks. The transient flora changes all the time depending on what we touch. Members of the transient flora are generally of little significance as long as the normal resident flora is intact. However, if the resident flora is disturbed, transient microorganisms can colonize, proliferate, and produce disease. Because of its constant exposure to and contact with the environment, the skin is particularly apt to contain transient microorganisms. The transient flora is easier to remove by mechanical friction than the resident flora.

Maintenance of the resident flora and their barrier function requires the skin pH to be maintained at the physiological level as well as the ability of the bacteria to adhere to the skin surface and rapidly readhere during the normal process of desquamation.[38] **❽Normally, the skin surface is slightly acidic and ranges between pH 4.5 and 5.5**, which may vary depending on the gender and body site.[39–41] This acidic environment is often referred to as the “acid mantle” of the skin. Skin pH influences a number of parameters, including barrier homeostasis, SC integrity, and cohesion as well as the bacterial defense mechanisms. Increase in pH leads to reduced desquamation, dry and scaly skin, and an optimal environment for the growth of pathogenic bacteria. It is essential, therefore, to maintain the skin’s acidic pH.[42]

Surface Charge of Normal Skin

Isoelectric point (pI) of a protein occurs at the pH where the positive and negative charges are balanced. At their isoelectric point, proteins behave like zwitterions (which have both negative and positive charges; however, their overall charge is neutral). At any pH below its pI, the protein has a positive charge, and at pH above its pI, it has a negative charge. **❾Since the isoelectric point of the proteins found in the skin's upper layers is between 3.5 and 4.5, the skin has a negative charge at physiological pH.**[43] Therefore, ingredients with positive charges (i.e., cationic compounds) are attracted to it. The interactions of positively charged molecules with skin surfaces contribute to surface improvements, such as smoothness, softness, and enhanced manageability. This process is usually referred to as "conditioning."

The negative surface charge has another practical advantage. Antiseptic hand washes and hand sanitizers may contain positively charged surfactant molecules that have antibacterial activity. The attraction of these cationic entities is very high to the negatively charged skin surface, which makes them stay on the skin for an extended period of time.

Skin Types

When it comes to skin types, there is a number of different classification systems published in the current literature. **❿Human skin can be categorized based on gender, its color, UV sensitivity, vulnerability, oiliness, healthiness, and special needs among others.** The most relevant classification systems for the use of cosmetics and personal care products are discussed here.

Skin Types Based on Sensitivity to Ultraviolet Light Skin color (otherwise known as skin phototype) and hair type account for the major differences in the skin of the various ethnic groups. Fitzpatrick developed a skin type classification in 1975 based on how the skin behaves to UV radiation exposure. It correlates the skin color (i.e., its melanin content) with its ability to tan or burn with UV light exposure. It has six categories, including the following[44]:

- **Type I** includes people with red and blond hair, blue eyes as well as freckles, and very fair skin. This is the type of skin that never tans and burns very easily. People with this skin type are extremely sun sensitive.
- **Type II** includes people with fair skin, red or blond hair, and blue, hazel, or green eyes. This skin is very sun sensitive, burns easily, and tans minimally.
- **Type III** includes people with cream white to olive skin, fair with mainly brown or sandy hair, and any eye color. This skin type is sun sensitive; however, it can gradually tan to a light brown and sometimes burns. It is a very common skin type.
- **Type IV** includes those who have dark brown skin, dark brown hair, and green, hazel, or brown eyes. This skin type always tans to a moderate brown with minimal burning and has minimal sun sensitivity.
- **Type V** includes people with dark brown skin who tan well, very rarely burn, and have sun-insensitive skin. They usually have dark black hair and dark brown eyes.

- **Type VI** describes deeply pigmented black skin. They have black hair and dark brown eyes. This skin type is sun insensitive, which always tans and never burns.

DID YOU KNOW?

The Fitzpatrick classification was developed to classify persons with white skin to select appropriate doses of UVA light for phototherapy patients. Although it was never designed for cosmetic purposes, it is frequently used as a criterion to determine the safety of a variety of cosmetic procedures in individuals with white, olive, brown, or black skin.[44]

It is important to keep in mind that everybody, regardless of race or ethnicity, is subject to potential adverse effects from overexposure to the sun. However, some skin types are more vulnerable, than others to the harmful effects of the sun.

The FDA has a classification system similar to the Fitzpatrick system; it takes into account the skin's reaction to sun exposure. It has six categories as shown in Table 3.1.[45]

TABLE 3.1 FDA's Classification of Skin Types

Skin Type	Skin Color	Reaction to Sun Exposure
I	Pale white	Always burns – never tans
II	White to light beige	Burns easily – tans minimally
III	Beige	Burns moderately – tans gradually to light brown
IV	Light brown	Burns minimally – tans well to moderately brown
V	Moderate brown	Rarely burns – tans profusely to dark brown
VI	Dark brown or black	Never burns – tans profusely

Skin Types Based on Hydration State and Lipid Content

As discussed earlier, it is essential to keep the skin hydrated in order to maintain the integrity of the skin barrier and prevent loss of water as well as penetration of physical and chemical substances.[46] Scientists differentiate among oily, dry, combination, and normal skin. These types have different characteristic features. It should be emphasized that the skin type of an individual is not constant; it may change over time depending on several internal factors, such as hydration state, lipid content, pH, moisture binding capacity, as well as some external factors, such as UV light, wind, temperature, and humidity content. This classification is often used when selecting cosmetics, such as cleansers and moisturizers.

- **Normal skin** has no exact definition; it is usually compared to other skin types as a reference. It is generally described as not too oily and not too dry. At a cosmetological level, normal skin is structurally and functionally balanced, and it has fine pores; it is smooth and well supplied with blood. In addition, it has no or only a few imperfections, no severe sensitivity, and a radiant complexion.

- **Dry skin** is relatively common; most people experience it from time to time due to various factors. It can be characterized as scaly, rough, and dull, which can lead to tautness and itchiness. In addition, it generally has red patches and can be characterized with less elasticity and a rough complexion. Dry skin tends more toward premature aging and is likely to have more wrinkles. Environmental factors, such as low relative humidity, cold weather, and sunlight, in addition to repeated contact with water, surfactants, and solvents, plus numerous skin diseases and dietary deficiencies, can produce dry skin.
- **Oily skin** has enlarged pores; therefore, it is very shiny as a result of overactivity of the sebaceous glands. Oiliness is most visible on the forehead, nose, and chin, and these parts are oily to the touch. Oily skin usually develops with the onset of puberty and affects a large percentage of young people. There are several factors that can cause and/or contribute to oily or greasy skin, including genetic inheritance, hormonal changes, diet, stress, and external agents (such as cosmetics, chemicals, UV light). Individuals with this skin type often tend to suffer from acne and dandruff as adolescents.
- **Combination skin**, as its name implies, is the combination of normal and oily skin, or of oily and dry skin. This type of skin has a tendency to be greasy in the central T-zone of the forehead, nose, and chin. The skin on the other areas (cheeks and hairline) is normal or dry.

DID YOU KNOW?

Skin types should be determined when the face is clean, i.e., without any moisturizers and makeup products. Since facial cleansers may have a drying effect, its type cannot be properly determined right after washing the face.

- **Sensitive skin** is a complex dermatological condition, defined by abnormal sensory symptoms, for example, tingling, chafing, burning, or prickling, and possibly pain or pruritus by various chemicals (e.g., cosmetics, soaps, water, pollution), physical factors (e.g., UV light, heat, cold and wind), microorganisms, psychological factors (e.g., stress), and hormones (e.g., menstrual cycles).[47] It is often thought to be a specific skin type, similar to oily or dry skin. However, it is more of a condition since normal, oily, dry, and combination skin can also be sensitive to various irritants. The term "sensitive skin" mainly refers to facial skin, but it can also concern other body areas, such as hands, scalp, or genital area.[48] It is a very common condition in the US, affecting a larger percentage of women.[49] Its pathophysiology includes the alteration of the skin barrier, allowing potential irritants and microorganisms to penetrate the skin and generate an inflammatory reaction.[50]

According to the American Academy of Dermatology (AAD), there are four distinct types of sensitive skin: acne, rosacea, burning and stinging, and contact dermatitis. They all have one characteristic in common: inflammation.[51]

Skin Types Based on Gender

The basic structure, function, and biochemical processes of male and female skin are similar; however, distinct differences exist.

- There are obvious differences driven by **hormonal** differences between the sexes. It is well known that testosterone plays a key role in facial and body hair growth, sebum production, and overall masculine features. Androgen stimulation causes an increase in the thickness of the skin; male skin is reported to be approximately 25% thicker than that of women.[52] Estrogen, however, negatively regulates body hair growth, affects body fat distribution, and positively impacts wound repair rates.
- Men appear to **age** slower than women, which may be related to the increased thickness of the skin, its higher collagen content [54], and the presence of facial hair that covers up the fine lines. Although both genders experience the same rate of collagen loss, women have lower collagen content at baseline. Therefore, this decrease results in more visible signs of aging.
- Regarding **skin thickness**, it has also been shown that male skin gradually thins with advancing age, whereas female skin stays more constant until menopause. Upon entering menopause, female skin progressively thins, suggesting a hormonal regulation of skin thickness in females versus males.[55]
- As for other skin features, it has been shown that the **sebum content** is higher in men than in women at all body locations (e.g., forearm, hands) and in all age groups. The sebaceous gland activity remains stable in men with aging, whereas it decreases over lifetime in women, especially from the age of 50–60.[56] The reduction in sebum is also accompanied by a reduction in the SC lipids in women, which may be attributed to a reduction in estrogen with advancing age.
- In addition to the biochemical differences, there are also differences in skin functionality. Studies have demonstrated that **TEWL** is lower in men than in women under the age of 50. It can be assumed that sebum lipids might have a potential occlusive effect on the skin surface; the higher sebum content in men might be a possible explanation for their lower TEWL compared with women. However, with aging, gender-related differences in TEWL assimilate.[57]
- Most recent studies have shown that young men show higher SC **hydration** in comparison with women. SC hydration is stable or even increases in women over lifetime, whereas the skin hydration in men progressively decreases, beginning at the age of 40.[56,58]

- An additional difference between male and female skin biochemistry is that men tend to sweat more than women. It creates an environment favorable for bacterial growth, which results in odor production. Male sweat also remains longer on the skin. In addition, more body hair in the case of men increases the body surface for bacterial colonization. The presence of sweat may contribute to differing skin pH between men and women. It has been shown that women's skin is significantly more alkaline than men's skin (average pH = 5.4 vs 4.4, respectively); however, the underarm pH is similar in both genders.[58]
- Male skin appears to be more sensitive to **UV radiation**, both from an acute exposure and from a longer time frame associated with more chronic exposure. Therefore, they have an increased risk for skin cancer.[55]

These differences can be utilized to formulate targeted cosmetics and personal care products for both genders to better meet their needs.

After discussing the basic characteristics of human skin and its general classifications, detailed discussion of products applied to the skin follows. Additional anatomical features, such as sweat glands as well as skin concerns, for example, acne vulgaris and aging, are reviewed in other sections.

GLOSSARY OF TERMS FOR SECTION 1

Acidic mantle: Normally, the skin has an acidic pH; usually, this is referred to as the "acidic mantle."

Basal cell layer: The innermost layer of the epidermis, which includes the melanocytes, Langerhans cells, and Merkel cells.

Brick and mortar model: The unique three-dimensional structure of the stratum corneum, in which dead cells represent the bricks, while lipids and corneodesmosomes represent the mortar.

Collagen: A structural protein found in the dermis, which provides the skin with strength.

Corneocyte: Water-resistant dead cells in the stratum corneum.

Corneodesmosomes: Units primarily composed of proteins and ceramide that bind corneocytes together.

Cornified lipid envelope: A unit consisting of proteins and lipids that protects the corneocytes.

Dermis: A skin layer located under the epidermis. It functions as a supporting frame to the epidermis, supplying it with nutrients and oxygen via the blood capillaries.

Desquamation: A process during which dead skin cells shed from the skin surface.

Elastin: A structural protein found in the dermis, which provides the skin with elasticity.

Epidermis: The outer layer of the skin, which functions as a protective layer against external influences.

Intercellular lamellar lipid: Lipids mainly made of ceramides, cholesterol, and fatty acids structured into a layered form. Corneocytes are embedded into this lipid matrix in the stratum corneum.

Keratinization: A process during which epidermal cells renew.

Keratinocyte: Newly formed epidermal cell.

Langerhans cell: A cell located in the basal cell layer of the epidermis, which plays an important role in the body's immune system.

Melanin: A pigment that gives the skin its color.

Melanocyte: Cells found in the basal cell layer of the epidermis that are responsible for melanin production.

Merkel cell: A cell located in the basal cell layer of the epidermis, which functions as a touch receptor.

NMF: Natural moisturizing factor, a blend of hygroscopic compounds located in the corneocytes, which acts as a natural humectant and keeps the skin hydrated.

Pilosebaceuos unit: A unit located in the dermis that includes the sebaceous gland, a hair follicle, and a small muscle.

Resident flora: Species of microorganisms that are always present on the skin and are not easily removed by mechanical friction.

Sebaceous gland: A gland located in the dermis, which produces sebum. It is part of the **pilosebaceuos** unit.

Stratum corneum: It is the horny layer that is present at the outermost layer of the epidermis.

Subcutis: A skin layer located under the dermis. It is a loose connective tissue that stores fat in fat cells.

TEWL: Transepidermal water loss, the amount of water lost (gram of water per meter square per hour) through the epidermis by passive diffusion.

Transient flora: Species of microorganisms that may be present on the skin under certain conditions for certain lengths of time. It changes all the time depending on what we touch. By applying mechanical friction it is easier to remove the transient flora than the resident flora.

REVIEW QUESTIONS FOR SECTION 1

Multiple Choice Questions

1. Which of the following is NOT true for the basal cell layer in the epidermis?
   a) Melanocytes are found here
   b) Cosmetics are applied to this layer
   c) New skin cells form here
   d) All of the above

2. Which of the following is responsible for the elasticity of the skin?
   a) Collagen
   b) Keratin
   c) Elastin
   d) Melanin
3. How long does it take on an average basis for a newly formed skin cell to travel to the top layer of the skin and be shed?
   a) About 1 month
   b) About 3 months
   c) About 1 year
   d) More than 1 year
4. What happens if the NMF is lost from the skin?
   a) Skin becomes oily
   b) Skin becomes dry and flaky
   c) Skin becomes hydrated
   d) Skin becomes wrinkled
5. Cosmetics and personal care products are applied to the following skin layer:
   a) Epidermis
   b) Dermis
   c) Fat cell layer
   d) Cuticle
6. Corneodesmosomes are ___.
   a) Dead cells in the skin
   b) Newly formed living cells in the skin
   c) Building units of the stratum corneum made of proteins and ceramide
   d) Building blocks of the dermis that bind dead cells together
7. In which layer of the human skin is melanin found?
   a) Dermis
   b) Fat cell layer
   c) Stratum corneum
   d) Stratum basale
8. Which of the following is NOT true for melanin?
   a) It is a pigment that provides skin color
   b) It provides photoprotection to the skin against UV radiation
   c) It plays a role in the body's immune system
   d) It is produced by melanocytes

9. What is the normal water content of the human skin, including the epidermis and dermis?

a) 0%
b) 100%
c) 10–30%
d) 80%

10. What is the normal pH of the human skin?

a) 1–2
b) 4.5–5.5
c) 7
d) 9–10

11. The skin's overall surface charge is ___.

a) Negative
b) Neutral
c) Positive

12. Which of the following is part of the pilosebaceuos unit?

a) Sebaceous gland
b) Hair follicle
c) Small muscle
d) All of the above

13. What does the term "transepidermal water loss" refer to?

a) The amount of water lost from the skin through the sweat glands
b) The amount of water lost from the skin through the sebaceous glands
c) The rate of water lost from the skin through the epidermis
d) The amount of water lost from the skin through the epidermis

14. What is Fitzpatrick's skin classification system based on?

a) The skin's sensitivity to detergents
b) The skin's sensitivity to UV exposure
c) The skin's sensitivity to water loss
d) The skin's sensitivity to fragrances

15. Which of the following is NOT true for the resident flora of the human skin?

a) It changes all the time depending on what we touch
b) It consists of relatively fixed types of microorganisms
c) It protects the host from pathogenic bacteria
d) If it is disturbed, transient microorganisms can colonize and produce disease

Fact or Fiction?

_________ a) Skin loses water continuously even without sweating.
_________ b) The outer surface of the skin is made up of dead cells.
_________ c) All germs found on the skin should be properly removed with skin cleansing products.
_________ d) Women's skin is more acidic than men's skin.
_________ e) NMF acts as a natural humectant in the epidermis.

REFERENCES

1. Agache, P.: The Human Skin: An Overview, In: Agache, P., Agache, P. G., Humbert, P., eds: *Measuring the Skin*, New York: Springer, 2004: 3.
2. Carlson, B. M.: Integumentary, Skeletal, and Muscular Systems, In: *Human Embryology and Developmental Biology*, St. Louis, MO: Mosby, 1994: 153–181.
3. Krause, J. W.: *Krause's Essential Human Histology for Medical Students*, Boca Raton: Universal-Publishers, 2005.
4. Allen, C., Harper, V.: *Laboratory Manual for Anatomy and Physiology*, New Jersey: John Wiley and Sons, 2011: 84.
5. Schaefer, H., Redelmeier, T. E.: *Skin Barrier: Principles of Percutaneous Absorption*, Basel: Karger, 1996: 310–336.
6. Wertz, P. W., van den Bergh, B.: The physical, chemical and functional properties of lipids in the skin and other biological barriers. *Chem Phys Lipids*. 1998:91:85–96.
7. Del Rosso, J. Q., Levin, J.: The clinical relevance of maintaining the functional integrity of the stratum corneum in both healthy and disease-affected skin. *J Clin Aesthet Dermatol*. 2011;4(9):22–42.
8. Harding, C. R., Long, S., Richardson, J., et al.: The cornified cell envelope: an important marker of the stratum corneum maturation in healthy and dry skin. *Int J Cosm Sci*. 2003;25:157–167.
9. Nemes, Z., Steinert, P. M.: Bricks and mortar of the epidermal barrier. *Exp Mol Med*. 1999;31(1):5–19.
10. Rawlings, A. V., Scott, I. R., Harding, C. R., et al.: Stratum corneum moisturization at the molecular level. *J Invest Dermatol*. 1994;103(5):731–740.
11. Rawlings, A. V., Harding, C. R.: Moisturization and skin barrier function. *Dermatol Ther*. 2004;17:43–48.
12. Robinson, M., Visscher, M., Laruffa, A., et al.: Natural moisturizing factors (NMF) in the stratum corneum (SC). II. Regeneration of NMF over time after soaking. *J Cosmet Sci*. 2010;61(1):23–29.
13. Marks, R.: The stratum corneum barrier: the final frontier. *J Nutr*. 2004;134(8 Suppl):2017S–2021S.
14. Matts, P. J.: Water, water everywhere. *IFSCC Magazine*. 2008;11(3):201–205.

15. Brenner, M., Hearing, V. J.: The protective role of melanin against uv damage in human skin. *Photochem Photobiol.* 2008;84(3):539–549.
16. Chomiczewska, D., Trznadel-Budźko, E., Kaczorowska, A., et al.: The role of Langerhans cells in the skin immune system. *Pol Merkur Lekarski.* 2009;26(153):173–177.
17. Summer, G. S.: Albinism: classification, clinical characteristics, and recent findings. *Optom Vis Sci.* 2009;86:659–662.
18. Segaert, S., De Haes, P., Bouillon, R.: The Epidermal Vitamin D System, In: Michael F., ed. *Holick: Biologic Effects of Light*, New York: Springer, 2002: 245–253.
19. Clemens, T. L., Henderson, S. L., Adams, J. S., et al.: Increased skin pigment reduces the capacity of skin to synthesise vitamin D3. *Lancet.* 1982;1:74–86.
20. Baranoski, S., Ayello, E. A.: *Wound Care Essentials: Practice Principles*, Philadelphia: Lippincott Williams & Wilkins, 2008: 149.
21. Downing, D. T., Stewart, M. E., Wertz, P. W., et al.: Skin lipids: an update. *J Invest Dermatol.* 1987;88:2–6.
22. Nikkari, T., Schreibman, P. H., Ahrens, E. H.: In vivo studies of sterol and squalene secretion by human skin. *J Lipid Res.* 1974;15:563–573.
23. Zouboulis, C. C.: Acne and sebaceous gland function. *Clin Dermatol.* 2004;22:360–366.
24. Agache, P.: Stratum Corneum Histopathology, In: Agache, P., Humbert, P., ed.: *Measuring the Skin*, Berlin: Springer-Verlag, 2004.
25. Watson, A., Clarke, T. S., Yates, D., et al.: Reliable use of the ServoMed Evaporimeter EP-2™ to assess transepidermal water loss in the canine. *J Nutr.* 2002;132(6):1661S–1664S.
26. Gabard, B., Treffel, P.: Transepidermal Water Loss, In: Agache, P., Agache, P. G., Humbert, P., eds: *Measuring the Skin*, New York: Springer. 2004.
27. Rogiers, V.: EEMCO guidance for the assessment of transepidermal water loss in cosmetic sciences. *Skin Pharmacol Appl Skin Physiol.* 2001;14:117–128.
28. Imhof, R. E., De Jesus, M. E., Xiao, P., et al.: Closed-chamber transepidermal water loss measurement: microclimate, calibration and performance. *Int J Cosmet Sci.* 2009;31:97–118.
29. Gabard, B., Clarys, P., Barel, A. O.: Comparison of Commercial Electrical Measurement Instruments for Assessing the Hydration State of the Stratum Corneum, In: Serup, J., Jemec, G. B. E., Grove, G. L., eds: *Handbook of Non-Invasive Methods and the Skin*, 2nd Edition, Boca Raton: CRC Press, 2006: 351–358.
30. Gioia, F., Celleno, L.: The dynamics of transepidermal water loss from hydrated skin. *Skin Res Technol.* 2002;8:178–186.
31. Maes, D. H., Marenus, K. D.: Main Finished Products: Moisturizing and Cleansing Creams, In: Baran, R., Maibach, H. I., eds: *Cosmetic Dermatology*, Baltimore: Martin Dunitz Ltd., 1994:77–88.
32. Courtenay, M.: Preparations for skin conditions. *Nursing Times.* 1998;7:54–55.
33. Price, P. B.: The bacteriology of normal skin: a new quantitative test applied to a study of the bacterial flora and the disinfectant action of mechanical cleansing. *J Infect Dis.* 1938;63:301–318.
34. Chiller, K., Selkin, B. A., Murakawa, G. J.: Skin microflora and bacterial infections of the skin, *J Invest Derm Symp Proc.* 2001;6:170–174.

35. Grice, E. A.: Topographical and temporal diversity of the human skin microbiome. *Science*. 2009;324:1190–1192.

36. Gao, Z., Perez-Perez, G. I., Chen, Y., et al.: Quantification of major human cutaneous bacterial and fungal populations, *J Clin Microbiol*. 2010,48(10):3575.

37. Lambers, H., Piessens, S., Bloem, A., et al.: Natural skin surface pH is on average below 5, which is beneficial for its resident flora. *Int J Cosmet Sci*. 2006;28(5):359–370.

38. Feingold, D. S.: Bacterial adherence, colonization, and pathogenicity. *Arch Dermatol*. 1986;122:161–163.

39. Fluhr, J. W., Dickel, H., Kuss, O., et al.: Impact of anatomical location on barrier recovery, surface pH and stratum corneum hydration after acute barrier disruption. *Brit J Dermatol*. 2002;146(5):770–776.

40. Chikakane, K., Takahashi, H.: Measurement of skin pH and its significance in cutaneous diseases. *Clin Dermatol*. 1995;13:299–306.

41. Thune, P., Nilsen, T., Hanstad, I. K., et al.: The water barrier function of the skin in relation to the water content of stratum corneum, pH and skin lipids. *Acta Derm Venereol*. 1988;68:277–283.

42. Saba, M., Yosipovitch, A., Yosipovitch, G.: Skin pH: from basic science to basic skin care. *Acta Derm Venereol*. 2013;93:261–267.

43. Wilkerson, V. J.: The chemistry of human epidermis. II. The isoelectric points of the stratum corneum, hair and nails as determined by electrophoresis. *J Biol Chem*. 1935;112:329–335.

44. Fitzpatrick, T. B.: The validity and practicality of sun-reactive skin types I through VI. *Arch Dermatol*. 1988;124:869–871.

45. FDA: Radiation-Emitting Products, Your Skin, Last updated 09/26/2013, Accessed 01/07/2014 at http://www.fda.gov/Radiation-EmittingProducts/RadiationEmittingProductsandProcedures/Tanning/ucm116428.htm

46. Zhai, H., Maibach, H. I.: Occlusion vs. skin barrier function. *Skin Res Technol*. 2002;8:1–6.

47. Primavera, G., Berardesca, E.: Sensitive skin: mechanisms and diagnosis. *Int J Cosmet Sci*. 2005;27:1–10.

48. Saint-Martory, C., Roguedas-Contios, A. M., Sibaud, V., et al.: Sensitive skin is not limited to the face. *Brit J Dermatol*. 2008;158:130–133.

49. Misery, L., Sibaud, V., Merial-Kieny, C., Taieb, C.: Sensitive skin in the American population: prevalence, clinical data, and role of the dermatologist. *Int J Dermatol*. 2011;50(8):961–967.

50. Luger, T. A.: Neuromediators: a crucial component of the skin immune system. *J Dermatol Sci*. 2002;30:87–93.

51. AAD: Sensitive Skin, Accessed 11/2/13 at http://www.aad.org/media-resources/stats-and-facts/prevention-and-care/sensitive-skin

52. Shuster, S., Black, M. M., Bottoms, E.: Skin collagen and thickness in women with hirsuties, *Br Med J*. 1970;4:772.

53. Phillips, T. J., Demircay, Z., Sahu, M.: Hormonal effects on skin aging. *Clin Geriatr Med*. 2001;17:661–672.

54. Oblong, J. E.: Comparison of the impact of environmental stress on male and female skin. *Brit J Dermatol*. 2012;166(2):41–44.

55. Luebberding, S., Krueger, N., Kerscher, M.: Age-related changes in skin barrier function – quantitative evaluation of 150 female subjects. *Int J Cosmet Sci*. 2013;35:183–190.
56. Jacobi, U., Gautier, J., Sterry, W., et al.: Gender-related differences in the physiology of the stratum corneum. *Dermatology*. 2005;211(4):312–317.
57. Luebberding, S., Krueger, N., Kerscher, M.: Skin physiology in men and women: in vivo evaluation of 300 people including TEWL, SC hydration, sebum content and skin surface pH. *Int J Cosmet Sci*. 2013;35:477–483.

SECTION 2: SKIN CLEANSING PRODUCTS

LEARNING OBJECTIVES

Upon completion of this section, the reader will be able to

1. define the following terms:

Actuation force	Antiseptic	Bath product	Chemical cleansing
CMC	Disinfectant	Extrudability	Facial cleanser
Firmness	Foamability	Foaming cleanser	Hand cleanser
Hand sanitizer	Hardness	*In vitro* test	*In vivo* test
Low-foaming cleanser	Non-foaming cleanser	pH	Physical cleansing
Rancidification	Separation of emulsions	Shower product	Soap
Spreadability	Syndet	Texture	Viscosity

2. explain the basic function of skin cleansing;
3. briefly discuss how skin cleansers can negatively affect the stratum corneum;
4. list the main concerns with regard to the use of alcohol-based hand sanitizers;
5. list the various required cosmetic qualities and characteristics that an ideal skin cleanser product should possess;
6. list the various required technical qualities and characteristics that an ideal skin cleanser product should possess;
7. differentiate between soaps and syndets;
8. differentiate between chemical and physical skin cleansing;
9. explain how surfactants remove dirt from the skin;
10. explain how solvents remove dirt from the skin;
11. differentiate between foaming and non-foaming cleansing products;
12. explain why nonfoaming products do not foam;

13. list the major ingredient types of facial cleansers, and provide some examples for each type;
14. list some problems that the pH of traditional soaps can cause;
15. differentiate between facial scrubs and facial toners;
16. differentiate between rinse-off and peel-off facial cleansers;
17. differentiate between dry and wet cleansing cloths;
18. discuss the advantages of liquid body cleansers over bar soaps;
19. explain what the effects of body cleansers containing abrasives are;
20. differentiate between regular, moisturizer, and specialty body washes;
21. briefly discuss how bath oils should be used;
22. differentiate between floating and soluble bath oils;
23. explain the function of surfactants in floating bath oils;
24. name a few commonly used salts in bath salts;
25. explain how effervescent bath bombs work when they are placed in water;
26. differentiate between antiseptics and disinfectants;
27. briefly discuss how alcohol kills germs on the hands;
28. name active ingredients used in alcohol-based hand sanitizers;
29. name active ingredients used in water-based hand sanitizers;
30. explain why alcohol-based hand sanitizers do not have persistent killing effect;
31. list some typical quality issues that may occur during the formulation and/or use of skin cleansing products, and explain why they may occur;
32. list the typical quality parameters that are regularly tested for skin cleanser products, and briefly describe their method of evaluation;
33. briefly discuss the potential safety issues with regard to the use of parabens, triclosan, sodium lauryl sulfate, and microbeads in skin cleansers;
34. name the efficacy parameter that is generally tested for topical antiseptics claiming antimicrobial effects, and describe the method of evaluation;
35. list the typical containers available for skin cleanser products.

KEY CONCEPTS

1. Skin cleansing products, including facial cleansers, bath and shower products, and hand cleansing products, can be classified as either cosmetics or drugs in the US, depending on the claims made and ingredients present.
2. Skin cleansing products contain surfactants that are capable of emulsifying water-insoluble ingredients into micelles, which can be easily washed away from the skin. Ideally, cleansers should not damage the skin's complex structure and lead to irritation, dryness, redness, and itching. Unfortunately, many skin

cleansers do cause changes in the skin's structure and barrier function, leading to various signs and symptoms.

3. Cleansing products can be categorized in several ways, including their cleansing mechanism, chemical nature, harshness (mildness), and dosage form.
4. In general, skin cleansers are used to remove dirt, makeup, environmental pollutants, germs, and other types of soilage from the skin.
5. In addition to simple cleansing, hand washing has an essential role in infection control as well.
6. Although the public belief is that hand sanitizers may replace hand washing, the CDC recommends that hand sanitizers should not be used in place of soap and water but only as an adjunct.
7. Typical quality-related issues of skin cleansing products include sticky bath salts and bath bombs, broken bath bombs, discoloration of bar soaps, poor foaming activity of foaming formulations, separation of emulsions, cloudy solution, microbiological contamination, clumping, and rancidification.
8. Parameters commonly tested to evaluate the quality of skin cleansing products include spreadability, extrudability, texture, and firmness of lotions, creams, and gels; actuation force; foaming property, foam stability, foam viscosity, foam density, and foam structure; hardness of batch bombs; disintegration time and dissolution time of bath bombs, bath salts, and bath beads; preservative efficacy; viscosity; and pH.
9. The most frequently tested efficacy parameter is the antimicrobial activity of hand sanitizers and antibacterial skin cleansers.
10. Ingredients causing safety concerns regarding skin cleansing products include parabens, triclosan, sodium lauryl sulfate, and microbeads.

Introduction

Human skin is in continuous contact with the environment, which can modify its normal flora, pH, and barrier properties, leading to unwanted conditions. Skin cleansing serves as a foundation for healthy skin with an intact barrier function, and it also contributes to the skin's aesthetic appearance. It includes the removal of dirt, oil, cosmetics, and dead skin cells. In addition, skin cleansing is the initial step in overall skin care and prepares the skin for the application of moisturizing, protective, and nourishing products. Furthermore, beyond skin care, it also plays an important role in psychological well-being.

In general, both genders are concerned about their skin and feel that they have to take care of it. Skin cleansing products include a number of different formulations. The characteristics of an ideal skin cleanser vary according to the needs of the consumers.

This section provides a general overview of the various cleansing mechanisms and cleansing ingredients. It also reviews the commonly used cleansing products for the face, body, and hands as well as their basic ingredients, formulation, and testing methods as well as packaging. In addition, it also provides a summary on how these products may affect the skin and what consumers' general requirements are.

Types and Definition of Skin Cleansing Products

❶Skin cleansing products, including facial cleansers, bath and shower products, and hand cleansing products, can be classified as either cosmetics or drugs in the US, depending on the claims made and ingredients present. The majority of skin cleansing products are intended to remove dirt and oil from the skin, which are cosmetic claims. Many skin cleansing products contain moisturizing ingredients today, which help avoid the skin from getting dry. These products are also considered cosmetics. However, products removing and/or killing bacteria that may cause infections, such as antibacterial hand soaps and hand sanitizers and facial washes proven to fight against bacteria causing acne, are considered drugs in the US.

- **Facial cleansers** are designed to clean the facial skin, remove dirt and makeup, provide exfoliation, and remove potentially harmful microorganisms. Available products include lathering and foamless emulsions, gels, scrubs, toners, masks, and cleansing wipes.
- **Bath and shower products** are designed to remove dirt, perspiration, and dead cells from the body skin as well as enhance the bathing experience, soften and moisten the skin, provide a relaxing experience, and leave the skin feeling clean and fresh. Bath salts can change the salinity of bath water to increase buoyancy and make the body feel lighter in the bath, mimicking the properties of natural mineral baths or hot springs. Product types may vary from bar soaps, bath salts, bath bombs, and bath oils to bubble bath products and shower gels.
- **Hand cleansing products** are designed to clean the hands. These products are identical to facial cleansers and body washes. They include bar soaps and syndet bars as well as liquid soaps. They are often enhanced with additional moisturizing ingredients. **Hand sanitizers** are used specifically to remove microorganisms from the hands with the intent of preventing infections and reducing the spread of infectious diseases. These product types are classified as OTC drug–cosmetic products in the US. The classification includes alcohol-based and non-alcohol-based hand sanitizers in the form of bar soaps, gels, lotions, creams, and cleaning wipes.

History of Using Skin Cleansing Products

Cleaning the skin has been part of humans' lives from prehistoric times. As early as the 2700 BC, the Chinese were already using salt for medicinal purposes. Egyptians and later Greeks also recognized the antiseptic qualities of salt.

Many different civilizations can be given credit for discovering soap. The first recorded evidence of the manufacture of soap-like materials dates back to around 2800 BC in ancient Babylon. Soap was made from fats boiled with ashes and water and was mostly for cleaning textile as well as treating skin diseases.[1] A big part of

Egyptian culture was dedicated to beauty and cleanliness through skin care. Cleopatra is well-known for her skin care regimen, including bathing in milk high in lactic acid, which turned out to be quite favorable for skin exfoliation.[2] The early Greeks bathed for aesthetic reasons. They used clay, sand, pumice, and ashes as well as a metal tool known as a "strigil" to scrape off the oil and dirt from their body. As the Romans dwelled in a water-rich area, they built aqueducts to bring clean water into towns and used soaps to clean themselves. The baths were luxurious, and bathing became very popular.[3]

DID YOU KNOW?

According to an ancient Roman legend, the name "soap" originates from Mount Sapo (Rome), where burnt animals were sacrificed. Rain washed a mixture of melted animal fat, or tallow, and wood ashes down into the clay soil along the Tiber River. Women found that this clay mixture made their wash cleaner with much less effort.[4]

After the fall of Rome in the Middle Ages, bathing declined. Poor sanitary habits lead to people suffering from a number of hygiene-related diseases and illnesses. One of the most well-known diseases caused by the lack of personal hygiene was the Black Death of the 14th century. Personal cleanliness started to become popular again in most of Europe only in the 17th century. Small soap manufacturing businesses were established all over Europe using animal and vegetable oils with ashes of plants to produce soaps. Italy, Spain, and France were early centers of soap manufacturing, due to their ready supply of raw materials such as oil from olive trees.

A major step toward large-scale commercial soapmaking occurred in the late 18th century when a French chemist, Nicolas Leblanc, patented a process for making soda ash (i.e., sodium carbonate) from common salt. Modern soapmaking (i.e., saponification) was born in the early 19th century. Another French chemist, Michel Eugène Chevreul, is credited with discovering the chemical nature and relationship between fats, glycerin, and fatty acids. In the 1860s, a Belgian chemist, Ernest Solvay, further improved the process by reducing the cost of soda and, at the same time, improving both the quality and the quantity of this material, which was vital to support the growth of the soapmaking industry. The chemistry of soap manufacturing stayed essentially the same until the early 20th century, when the first detergent was developed in Germany.[5] After that, there was a great improvement in the manufacturing of bubble bath products and liquid hand cleansers, and they started to become popular.[3]

The first concerns about the importance of hand sanitizers as part of the infection control arose in the mid-1800s. Professor Ignaz Semmelweis (called the savior of mothers) is credited with the first groundbreaking work in this field. He was a Hungarian obstetrician working in a hospital in Vienna. He disproved the belief that childbed fever (an infection of women after childbirth or miscarriage) was caused by "poison air." The hospital where he worked had two obstetric departments. The first was reserved for doctors and trainee doctors, while Ward 2 was where only midwives learned their profession. At the Vienna Hospital, it was very common for obstetricians to carry out autopsies (i.e., examination of a cadaver to determine or confirm the cause of death) in the morning and then carried on with their other work in Ward 1. Midwives did not do autopsies. The incidence of maternal death was high in Ward 1, but low in Ward 2. Semmelweis believed that there had to be a link between the work done in the postmortem room and the obstetricians coming into Ward 1. He concluded that the cause of high death rate is related to the cadavers and infection was transmitted by hand to women attended by medical students in the first department. He ordered hand washing in a chlorine solution before entering Ward 1 and examining women; after which the incidence of childbed fever significantly decreased.[6,7]

The evolution of skin cleansing products through the ages, cultures, and traditions did not stop its pace in our modern times. Synthetic detergents now form the basis of many present-day skin cleansing products. New ingredients, technologies, and products are continuously developed and introduced into the market. In addition, today, hand washing with soap and water and hand antisepsis using hand sanitizers are the cornerstone of many infection control programs.

How Skin Cleansing Products May Affect the Skin?

Skin cleansing is essential in maintaining the skin health and contributing to its aesthetic appearance. Dirt found on our skin consists of sweat, sebum and its breakdown products, dead skin cells, residues of cosmetics and personal care products applied to the skin, dust, and other environmental impurities carried in the air. Most of these compounds are not soluble in water, so washing the skin with simple water would not be sufficient to remove dirt. ●**Skin cleansing products contain surfactants that are capable of emulsifying water-insoluble ingredients into micelles, which can be easily washed away from the skin. Ideally, cleansers should not damage the skin's complex structure and lead to irritation, dryness, redness, and itching. Unfortunately, many skin cleansers do cause changes in the skin's structure and barrier function, leading to various signs and symptoms.** The major negative effects of skin cleansers and additional concerns are summarized as follows:

- Many skin cleansers **solubilize lipids** that are found on the skin surface to provide protection and may even **extract skin components**, such as the NMF during cleansing. Additionally, surfactants may also **remain in the SC** after even rinsing the product with water. Surfactants can disrupt the SC's structure, likely in the lipid lamellae,[8] and weaken its barrier function (see Figure 3.4 in Section 1). In addition, cleansers can lead to a reduction in the level of NMF in the skin.[9] The disruption of the SC lipid order by surfactants contributes to the barrier-damaging side effects of skin cleansing.[10] It has been shown that cleansing can increase TEWL, leading to dehydration of the skin.[11] In addition to damaging the skin barrier, surfactant penetration can cause irritation and inflammation [12] and alter barrier renewing processes by affecting keratinocyte differentiation and desquamation.[13]

- It has been shown that the tendency of surfactants to damage skin proteins is related to the **charge density** of the surfactant aggregates.[14] This explains the following well-known order for the irritation potential of surfactants, namely, anionic surfactants > amphoteric surfactants > nonionic surfactants. As cleansing products are primarily based on anionic surfactants, approaches have been developed to decrease the tendency of these surfactants to damage skin proteins. Common approaches include increasing the size of the head/polar group of the surfactant[15] and using a combination of anionic surfactants with amphoteric or nonionic surfactants.[16]
- Another factor that may contribute to SC damage is the cleanser's **pH**. Soap-based cleansers are alkaline in nature, while the pH of most syndets (synthetic surfactant-based cleansers) is close to neutral or slightly acidic. It has been shown that soap-based cleansers have a higher potential to irritate skin than cleansers with synthetic surfactants (syndets).[17] The balance between the soap damage and the recovery power of skin determines the skin's condition of soap users. In winter, with the additional stress of cold, drying weather, many soap users will experience dry, itchy skin. Soap is also drying in hot, dry climates, but usually causes minimal dryness when the weather is warm and humid.[18]
- A further concern with regard to the use of skin cleansing products is **irritation, itching, inflammatory responses,** and **allergies** to product ingredients, including antibacterial actives, preservatives, and perfumes. It is not necessary for the surfactant to penetrate into dermal layers to trigger such conditions. An increase in the production of cytokines can also elicit a response from the dermis.[19] If soaps damage the skin barrier, they can easily get into the deeper layers of the SC, leading to irritation, reddening, and itching.
- Hand washing has traditionally been identified as the most important infection control intervention to prevent the transmission of microorganisms and diseases[20] ever since Semmelweis observed its immense effect on the incidence of childbed fever.[21] In recent years, a number of hand hygiene products, including many with antimicrobial activity, have become available, and some are marketed to the general public. It has been demonstrated in clinical studies that antibacterial agents, such as ethanol, isopropanol, chlorhexidine, are effective in preventing diseases.[22,23] However, concerns have arisen with regard to the **long-term safety** of some ingredients. It is questionable whether some ingredients may increase the resistance of microorganisms to antimicrobials. If these agents increased the resistance of antibiotic resistance, it may lead to an increase in the number of multiresistant microorganisms, increasing financial burden and costs for patients and their families, and even the number of deaths in healthcare settings.

- There are some occupations where **frequent hand washing** is a must; examples include healthcare settings, such as pharmacies and hospitals; pharmaceutical manufacturing companies; as well as the food industry. In such cases, frequent handwashing may lead to long-term changes in the skin, such as chronic damage, irritant contact dermatitis and eczema, and concomitant changes in skin flora.
- Products containing alcohol are potentially **flammable**; therefore, it is an additional safety concern. Extra precautions have to be taken during manufacturing, shipping, and handling of such products.
- An additional concern regarding the use of alcohol-based hand sanitizers is that they may be accidentally (or intentionally) **consumed**.[24] Based on the National Poison Data System data,[25] intentional hand sanitizer exposures have greatly increased in the last several years. The term "alcohol" may sound inviting for people with alcoholic problems. In addition, it may be accidentally ingested by children. Acute ethanol intoxication can result in several serious, even life-threatening, clinical effects. These include decreased body temperature, central nervous system and respiratory depression, abnormal heart rhythm, low blood pressure, nausea and vomiting, liver injury, and many other severe symptoms.[26] This problem raises questions about the unrestricted use of such products. Hand sanitizers are a must in promoting and maintaining hand hygiene in healthcare settings; however, their use may be restricted in locations with at-risk patients.[27]
- Alcohol also removes dirt and oils from the skin, which may **dry out** the skin. It may be beneficial for patients with oily skin; however, it will lead to dryness if overused. Alcohol-based formulations, however, should not be used on dry or aging skin since it will worsen their condition.

Required Qualities and Characteristics and Consumer Needs

From a consumer perspective, a quality skin cleansing product should possess the following characteristics:

- Neutral or pleasant odor and color
- Easy to rub on with appropriate foaming property
- Easy to spread
- Pleasant feeling during application
- Non-oily/non-greasy feeling
- Leaves no residue
- Moisturizes the skin while cleaning
- Non-comedogenic
- Well tolerated and non-allergenic
- Hand sanitizers: do not dry the skin, but kill bacteria and viruses.

The technical qualities of skin cleansing products can be summarized as follows:

- Long-term stability
- Smooth texture
- No microbiological contamination and growth
- Appropriate rheological properties
- Appropriate foaming activity
- Appropriate performance
- Appropriate pH
- Dermatological safety.

Cleansing Products – Basic Concepts

●**Cleansing products can be categorized in several ways, including their cleansing mechanism, chemical nature, harshness (mildness), and dosage form.** Before moving on to discuss the various types of products and their major characteristics, it is worth to understand the difference between soaps and syndets.

Classification Based on Chemical Nature and Mildness

Based on chemistry, three basic types of compounds can be found in skin cleansing products, including soaps, synthetic surfactants, and solvents.

- **Soaps** are salts of fatty acids. If the alkali used contains sodium, potassium, or ammonium ions, water-soluble soaps are formed, whereas zinc and magnesium make insoluble so-called metallic soaps. Skin cleansing products contain water-soluble soaps. The pH of soaps is alkaline and is usually in the range of 9.5–10. This is one of the main reasons why soap-based cleansers can irritate the skin. Soaps are amphiphilic molecules having both hydrophilic and hydrophobic groups, i.e., they are soluble in both oil and water. They are anionic surfactants. Soaps are often referred to as natural surfactants since most oils and fats used for their production can be found in nature.
- **Synthetic surfactants**, which are often referred to as "soapless" soaps or syndets, are also amphiphilic compounds. The most frequently used surfactants are anionic in nature, similar to soaps; however, they are much milder to the skin and are, therefore, more popular. The difference in the chemical structure of the molecule and the pH of the final product (usually around pH 7) makes them much milder.
- **Solvents**, as reviewed earlier, can be classified broadly as polar, semipolar, and nonpolar type. Nonpolar solvent-based products, such as those containing mineral oil, may be potentially advantageous for dry skin consumers since they can deposit a thin oil layer on the skin surface; however, they may be disadvantageous for users with oily skin, for the same reason. Skin cleansing products

containing alcohol can dry the skin, which may be beneficial for oily skin but not for dry skin. It is very important, therefore, to select skin cleansing products in accordance with the skin type.

Classification Based on Cleaning Principle

With regard to the working principle, two basic mechanisms can be attributed to the cleaning effect, namely, chemical cleaning and physical cleaning.

- **Chemical cleaning** can be achieved through emulsifying and dissolving the dirt on the face. The skin feel and application profile are different for these products.
  - Surfactants work by reducing the interfacial tension between oil and water, i.e., **emulsifying** oily components on the surface of the skin with water. The process is depicted in Figure 3.5. The stronger the surfactant, the more hydrophobic material removed, the greater the potential skin damage from excessive removal of naturally occurring skin lipids and the greater the compromise of the skin barrier function. Therefore, correct and careful selection of surfactants is required to ensure proper mildness. Soap- and surfactant-based cleansers usually require water and generally include a rinsing step.

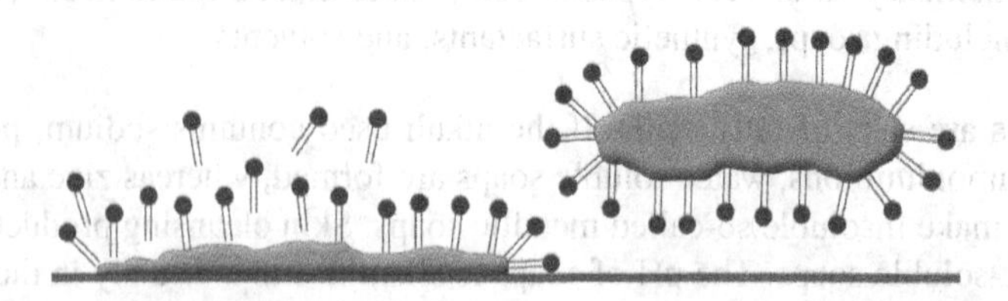

Figure 3.5 Removal of oily dirt from the skin surface by emulsifiers.

  - Solvent-based systems clean the skin by **dissolving** sebum and external oils present on the skin as residues of cosmetics and similar materials. Solvents work under the chemical premise that "like dissolves like." Solvent-based cleansers are usually not used in conjunction with water; rather, they are applied and then wiped off with a tissue or cotton ball. Solvents used in cleansing products include non-polar solvents, such as mineral oil; semi-polar solvents, such as alcohol; and polar solvents, such as water. There are some non-lathering cleansers that are emulsions; however, they do not foam. They also remove dirt by dissolving it on the face.
- **Physical cleaning** is an alternative to chemical cleaning; the working principle is **abrasion** (friction), which is generated primarily by the direct interaction of a washcloth, tissue, cotton ball, cleansing cloth, or abrasive particles and the surface of the skin. Friction works to help remove dirt and increase the interaction of chemical cleaning agents with oils.

Classification Based on Foamability

Segmenting by foaming activity, there are three major categories of skin cleansing products on the market: foaming products, low-foaming products, and non-foaming products.

- **Foaming** cleansers contain a significant amount of well-foaming surfactants. These products are the most popular today and often provide the most refreshing sensation afterward. Most body washes, hand soaps, and facial cleansing products belong to this category. These products are typically surfactant solutions, gels, scrubs, and O/W emulsions with a high cleaning power. Foaming products can contain the same basic ingredients for the body and the face. However, for the face, a different sensation and greater mildness are required. Therefore, generally, combinations of milder surfactants are used. Although these products contain mild surfactants, they can still significantly damage the skin barrier if left on the skin for a longer period of time.
- **Low-foaming** products contain a lower level of well-foaming surfactants compared to foaming products and are milder to the skin. Product forms available as low-foaming products include lotions, gels, scrubs, and creams. They still foam when mixed with water; however, mildness is often obtained at the expense of effective cleansing and lathering. These products primarily contain nonionic surfactants, often combined with amphoteric and polymeric types. Foam-booster secondary surfactants may also be added to increase the user experience. Low-foaming products are typically marketed for the face, but in some cases, for the body as well.
- **Non-foaming** cleansers include surfactant solutions, creams, lotions, bath oils, bath salts, and toners. Cleansers in this category tend to be the mildest due to their low well-foaming surfactant or soap content. Non-foaming cleansers can be solvent-based, such as facial toners and hand sanitizers, and emulsion-based, such as most facial cleansers, body washes, and hand soaps (which solubilize dirt). Emulsion-based non-foaming products contain the highest level of oils compared to the other two groups; therefore, they are ideal for depositing a thin layer of oil onto the skin, which remains on the skin even after rinsing. For this reason, these preparations are generally more effective for dry skin users and not recommended for oily and acne-prone skin. They are primarily formulated for the face and, in some cases, for the body. Emulsion-based products are often referred to as moisturizers or emollient body washes. Body washes with high emollient content have been shown to be beneficial for very dry skin users and for those whose only source of topical moisturization is their body cleanser.[28]

Classification Based on Product Types

Skin cleansing products classified based on their form include **solid** cleansing aids, such as soap bars, bath salts, bath beads, and cleansing wipes; **liquid** products, such as surfactant solutions, low-viscosity emulsions, toners, two-phase cleanser; as well as **semisolid** products, such as creams, pastes as facial masks, scrubs, and gels. It is important to understand that a cream may be foaming, low-foaming, or non-foaming, depending on the ingredients it is made of.

Typical Ingredients and Formulation of Skin Cleansing Products

❶In general, skin cleansers are used to remove dirt, makeup, environmental pollutants, germs, and other types of soilage from the skin. Since the basic needs of the skin on the face, body, and hands are slightly different, products are discussed according to these main areas of the body.

Face

Facial skin and its cleansing has always been a big concern for both men and women. According to studies, most women in the US wash their face twice a day (in the morning and at bedtime), which is twice as often as they typically wash their bodies.[29] As already discussed, the main ingredients in cleansing products are soaps and/or surfactants and solvents. A short summary of the typical ingredient types is provided as follows:

- **Surfactants** act as cleansing agents and emulsifiers. There are four main groups of surfactants with different characteristics. Usually, they are used in combination with each other to build appropriate properties into the formulations.
  - **Anionic surfactants** have good lathering and detergent properties, which are necessary to remove dirt. As mentioned earlier, natural soaps are anionic molecules as well. Examples for anionic surfactants include lauryl sulfates, such as sodium lauryl sulfate (SLS); laureth sulfates, such as sodium laureth sulfate; sarcosinates, such as triethanolamine lauroyl sarcosinate; isethionates, such as sodium cocoyl isothionate; taurates, such as sodium methyl cocoyl taurate; sulfosuccinates, such as sodium dioctyl sulfosuccinate; and more recently the monoalkyl phosphates, such as potassium lauryl phosphate.
  - **Cationic surfactants** have a positive charge which makes them attracted to the skin. Therefore, they can be employed as conditioning agents. Examples include amines, alkylimidazolines, alkoxylated amines, and quaternary ammonium.
  - **Amphoteric surfactants** are well tolerated and lather well and, therefore, are also often used in facial cleansers as secondary surfactants to help boost foam, improve conditioning, and reduce irritation. Examples include betaines, such as cocamidopropyl betaine; imidazolinium derivatives; amine oxides, such as cocamidopropylamine oxide; and alkylamino acids.
  - **Nonionic surfactants**, such as fatty alcohols; poloxamers; alkylene oxides; polyglucosides, such as lauryl glucoside; amides, such as cocamide diethanolamine (DEA), are very mild; thus, they are commonly used as emulsifiers, conditioning agents, and solubilizers. Their main drawback is that they do not lather particularly well. However, they form a perfect combination with anionics.

- **Solvents** act as cleansing agents as well as provide a vehicle for various formulations.
  - General examples include water (the most commonly used vehicle in emulsions), ethanol, isopropyl alcohol, or mineral oil as a non-polar solvent for cleaning purposes.
- **Thickeners** are structuring agents, primarily used for gels, lotions, and creams. They provide appropriate rheological properties for the systems as well as contribute to their stability.
  - Examples include hydrophilic ingredients, such as cellulose derivatives, gums, acrylates, and other types of polymers, as well as waxes for the oil phase.
- **Skin conditioning agents** (otherwise known as moisturizers) counteract the SC-disruptive properties of soaps and surfactants.
  - Examples include glycerin, olive oil, almond oil, mineral oil, silicone oils, waxes, panthenol, and allantoin.

DID YOU KNOW?

Cleansing agents may contain various types of moisturizers, although part of them is removed from the skin when rinsing with water. Therefore, significant amounts of moisturizers will not be left on the skin surface after cleansing. This is why proper moisturization is important after cleansing the skin for all skin types.

- **pH buffers** may be used for various reasons in facial cleansing products.
  - Alkaline solutions are used for saponification. Examples include potassium hydroxide, sodium hydroxide, and ammonium hydroxide.
  - Additional ingredients that may alter the formulation's pH include citric acid and lactic acid. They shift the pH into the acidic range, which is closer to the natural pH of the skin and are less irritant.
  - pH buffers, such as triethanolamine, may also be needed to thicken the formulation via neutralizing the thickeners.
- **Abrasives** Facial scrubs contain specific exfoliating components that are responsible for physical cleaning.
  - Examples include natural components, seeds of many fruits (such as peach, apple, apricot), nut shells (such as almond, walnut), and grains (such as oats, wheat). Synthetic scrub particles include polyethylene or polypropylene beads. In addition, aluminum oxide particles and sodium tetraborate decahydrate can also be used.

- **Colorants** may contribute to the marketing appeal of the product. Both natural and synthetic colorants can be used as facial cleansers. In certain products, titanium dioxide or glycol stearate is used as an opacifier.
- **Fragrances** are often added to facial cleansing preparations to mask the odor of the raw ingredients. It should be kept in mind, however, that they may be highly irritative, especially for users with sensitive skin.
- **Preservatives** provide protection against microbiological contamination.
  - Most systems contain preservatives, including parabens, phenoxyethanol, and benzoates.
- **Antibacterial agents** are widely used in today's formulations. They may be beneficial for controlling certain skin conditions, such as acne (it is discussed in detail in Section 4 of this chapter); superficial skin infections, such as folliculitis; and control infections after exposure to dirt or other potential sources of contamination. These are generally considered active ingredients in products.
  - The most commonly used compound is triclosan; however, its safety and efficacy are currently being investigated in cleansing products by the FDA. Additional examples include benzoyl peroxide and lactic acid (soaps containing a higher amount of lactic acid have an acidic pH, which is thought to be antibacterial).
- **Absorbents** are mainly used in facial masks to absorb sebum from the skin. These are water insoluble, mainly inorganic compounds.
  - Examples include zinc oxide, titanium dioxide, kaolin, calamine, clay, and natural mud.
- **Astringents** are the major ingredients in facial toners. They tighten pores and refresh the skin. Most of them are generally considered active ingredients.
  - Examples include alcohol and witch hazel.
- Certain soaps contain **other** ingredients, such as vitamins, and a variety of exotic natural ingredients (usually derived from fruits, other plants, etc.).

DID YOU KNOW?

Lathering formulations are generally more popular among consumers. Although it is possible to formulate cleaning products that do not foam or just slightly foam and are effective in removing dirt, most users expect skin cleansing agents to generate foam when cleaning. They feel that foaming is a sign of proper cleaning.

There is a wide variety of skin cleansing products available today, including bar soaps, syndet bars, gels, scrubs, creams and lotions, toners, masks, and cleansing wipes. Let us review these product types and their main properties:

- Traditionally, **bar soap** made with natural surfactants was used as a skin cleanser. Soap bars are available in several major types, differing in the types of surfactants used, lather provided, rate of lathering, color, smell, and skin compatibility. It is effective in removing grime and is relatively inexpensive; however, the washing solution formed by soap is extremely alkaline (pH 9.5–11.0)[29] and can cause irritation, dryness, and scaling. Although several subtypes of bar soap have been developed over the years to decrease its harshness, they are gradually losing their popularity.
  - Superfatted soaps contain more oil than required to a stoichiometric reaction. The excess oil may serve as a moisturizer and an emollient and improves the mildness of the product.[30]
  - Transparent soaps have a higher concentration of glycerin, which is a skin moisturizer. Although they contain natural soaps and therefore their pH is alkaline, they are still considered milder due to the presence of glycerin.
- **Syndet bars** are similar to soap bars, although they contain synthetic emulsifiers instead of natural soaps. Therefore, syndet bars have a better skin compatibility profile than traditional soaps. Their cleaning effect is very good, and the residue left on the skin is minimal. Due to their more gentle nature, they are becoming more important as skin cleansers. Nearly all common synthetic detergent bars are based on an anionic surfactant, acyl isethionate.

DID YOU KNOW?

Soap bars are not necessarily made of soap (i.e., natural surfactants). Today, a variety of soap bars containing synthetic surfactants are available on the market. However, many consumers relate the shape and name to the harsh effect regular products would cause. Most consumers are not aware that a soap in the form of a bar can be made of different types of surfactants and is, therefore, not necessarily harsh.

- Cleansing **gels** are water-based systems, containing various types of thickening agents. As discussed in Chapter 1, they typically have a transparent appearance. Gels are a popular cleansing form; they are used with water since they are foaming formulations. Inert particles (i.e., beads) are often incorporated into gels, which may provide an additional exfoliating effect.
- **Scrubs** are generally O/W emulsions or gels that contain small particles of natural or synthetic origin. Scrubs are intended to provide a deep cleansing effect, including skin exfoliation from abrasion with the particles, and they also polish the skin. It is important to keep in mind that the skin regularly and continuously sheds; therefore, additional exfoliators should not be used every day. They affect the skin barrier and can lead to damage if used too vigorously. Massaging and

rubbing these preparations onto the skin must be done with the utmost care and gentleness and in strict accordance with the manufacturer's instructions.

- **Cleansing creams and lotions** are typically O/W emulsions. As discussed in Chapter 1, the main difference between a cream and lotion is their viscosity; lotions have lower viscosity. They are usually made of relatively delicate surfactants, and their irritation potential is much lower than that of soaps. They can be lathering, low-foaming, or non-lathering formulations.
- **Toners** are clear solutions, generally based on various types of solvents and a low amount of surfactants. Waterless cleansers for the face are often based on alcohols and glycols since sebum is soluble in these solvents. Toners are usually applied with cotton balls, tissues, or washcloths and evaporate quickly after application. These products may be beneficial when there is no access to water and can be effective for very oily skin; however, long-term use may be harmful to the skin barrier. Toners are popular for younger users because of the perceived acne benefits and pore-tightening effect.
- **Facial masks** are a unique product type for cleaning the face. Certain products are rinsed off with water, while others can be peeled off as a film.
  - Masks that **rinse off** are removed from the skin with warm water 15–30 min after application. They usually contain insoluble, absorbent materials, such as clays, zinc oxide, kaolin, and others. These masks can effectively absorb sebum on the skin and, therefore, are primarily recommended for users with oily skin. Certain rinse-off masks are based on moisturizers and/or cleansing agents and do form an actual dried layer on the skin. They are more beneficial for users with dry skin.
  - Masks that **peel off** are made of rubbery substances, such as polyvinyl alcohol or rubber-based substances, such as latex. As these masks dry on the skin, they harden and form a thin, flexible, and usually transparent film on the skin. This film can be easily peeled off the face. Their major benefit is that they increase the skin's water content by inhibiting water evaporation. Therefore, these masks are recommended for persons with drier facial skin.
- **Cleansing wipes**, often referred to as cloths, represent one of the latest technologies among facial cleansing products. They are quite popular as they offer ease of use. Cleansing wipes are beneficial in cases when there is no access to water. They are usually designed to be used for one time only; therefore, they can be more hygienic than bar soaps. They consist of low levels of mild detergents with additional conditioning ingredients that may be deposited on the skin. The wipes can be made of natural fibers, such as cotton, synthetic fibers, such as polypropylene, or a blend of these. Wipes clean the facial skin by a combination of physical and chemical cleaning, which may provide much cleaner skin. Two popular product types are available today: dry wipes and wet wipes.
  - **Dry cleansing cloths** consist of lathering surfactants that are deposited onto a disposable cloth. In addition, moisturizing ingredients such as glycerin can also be deposited onto the wipes. The cloth is to be wetted before use and

rubbed to generate lather. A unique advantage of the dry cleansing cloth technology is that the product can be manufactured so that different ingredients can be placed in different "zones" on a cloth.[31] This approach enables formulators to use ingredients that are not compatible in a liquid cleanser.

- **Wet cleansing cloths** are pre-wetted by the manufacturer and packaged and distributed in a ready-to-use form. Wet clothes are used without additional water; they are generally non-foaming types. The major advantage of wet cloths is that small amounts of moisturizers can be deposited on the skin. This property makes these products highly favorable for users with dry skin.

DID YOU KNOW?

Although cleansing creams and lotions may deposit moisturizer ingredients on the skin, they should not be used as moisturizers and left on the skin. These products contain cleansing agents that are mild; however, if left on, it may lead to irritation. For the same reason, it is preferable to wash them off the skin than just wiping them off with a dry cloth.

The formulation methods of various products are attributed to the basic dosage forms.

- The formulation of facial cleanser gels usually start with hydration of the thickener. This process can be aimed by premixing the gelling agent with glycerin. After full hydration, all other components can be added and mixed thoroughly to prepare a uniform final product.
- Cleansing toners are usually solutions of water and alcohol with additional soluble and miscible ingredients added. The formulation typically starts with dissolving solid ingredients in the liquid phase taking into consideration their solubility (i.e., whether they are soluble in the water phase or alcohol phase). After complete dissolution, all liquid ingredients can be mixed to the proper liquid phase, which is followed by mixing the two phases together.
- The formulation of cleansing creams and lotions follows the general steps of emulsification.
- Scrubs are usually formulated as creams or gels, and the abrasive particles are mixed into them in a later stage of the formulation.
- Abrasive masks can be formulated as powders that will be mixed with the liquid phase right before use by the users. In such cases, the formulation follows the general steps of powder mixing (including grinding at the beginning and sieving at the end). When formulated as creams or gels, the formulation follows the general steps of emulsification or gel formulation. The formulation of rinse-off masks in the form of pastes requires special techniques to avoid lumping, that

is, formation of dry powder aggregates. Usually, the absorbents, such as kaolin, are slowly dispersed and wetted in water with constant, intensive mixing. After hydration, additional ingredients can be added to the phase. In order to avoid lumping, the final product, which has a high viscosity, is usually also mixed in high-shear mixers for 15–30 min and can be passed through a special mill.

- Soaps can be made by two processes, namely saponification and neutralization. Saponification involves heating fats and oils and reacting them with a liquid alkali to produce soap and water (neat soap) plus glycerin. In the case of neutralization, fats and oils are hydrolyzed to yield crude fatty acids and glycerin. The fatty acids are then purified and neutralized with an alkali to produce neat soap (water and soap).[3] The industrial production of soap involves a continuous process, including continuous addition of fat and removal of the product. Smaller scale-size production involves the traditional batch process. Both continuous and batch processes produce soap in liquid form, called neat soap, and a valuable by-product, glycerin. Glycerin is recovered by chemical treatment, followed by evaporation and refining. The next processing step after saponification or neutralization is drying, i.e., converting the neat soap into dry soap pellets. The moisture content of the pellets will vary depending on the desired properties of the soap bar. The dry soap pellets are then mixed with other ingredients and refined. In the final processing step, the mix is extruded and cut into bar-size units and shaped into the final shape.

Body

Cleansing products for the body include bath products that are used in bathtubs and shower products that are used in the shower. The most popular forms of body cleansers are liquid body washes, including bubble bath, shower gel, and shower cream. These products often include emollients and other moisturizing ingredients and, therefore, can be used to hydrate the skin in addition to cleaning it. By delivering greater emollient deposition and milder surfactants than simple soap bars, liquid body washes can actually improve the skin over time.[29]

Body skin, including the trunk, arms, legs, and genitalia, accounts for about 90% of the total body surface area. However, interestingly, facial cleansing products still receive more attention than body cleansing products. Many consumers focus on whether they have acne-prone or sensitive facial skin, but do not take into account their body skin type. As mentioned earlier, the type of the skin changes over time and may vary even among different body parts in individuals. One may have combination facial skin, normal body skin, and extremely dry hand skin. Therefore, the type of product used should be carefully selected according to the skin's needs.

Bath and Shower Products Bath and shower products include a wide variety of formulations, starting from solid dosage forms, such as bar soaps, bath salts, bath bombs, and bath beads (i.e., capsules), to semisolid forms, including shower gels and creams to liquids such as bubble bath products and bath oils.

- Similar to facial cleansing, **soap bar** was used as the traditional product to clean the body. It contained natural soaps, i.e., alkali salts of fatty acids. Due to the disadvantages, which were discussed in the previous part, and the appearance of syndet bars, the use of soap bars has significantly diminished in the past decades. An additional disadvantage of natural surfactants when used in bathtubs is that they tend to form scum and rings on the wall of the tub in the presence of hard water. Soap rings do not rinse away easily. In addition, they tend to remain behind and produce visible deposits on clothing and make fabrics feel stiff. Syndet bars for cleaning the body are similar to those for facial cleansing. They contain synthetic surfactants, which are known to be gentler to the skin.

DID YOU KNOW?

Hardness of water is caused by the presence of mineral salts, mostly calcium and magnesium salts, but sometimes also iron and manganese. These mineral salts can react with natural soap to form an insoluble precipitate known as scum.[3]

- Liquid cleaning products offer an improved skin feel and more convenient and hygienic dispensing than wash bars. In addition, most liquid body washes include more emollients and, therefore, can benefit the skin in more ways than just cleansing it. **Bubble bath** products, also known as foam bath products, as their name implies, are intended to fill the bath with a light, frothy lather. These liquid formulations are one of the most popular bath preparations today. **Shower gels** are transparent gel-like products. They are very popular as they are easy to apply and rinse. Inert particles are often incorporated into gels, providing an additional exfoliating and polishing property. **Shower creams** are offered as an alternative to shower gels. They have a milk- or cream-like appearance and are usually O/W emulsions. They have an opaque appearance and usually contain skin conditioning agents. Inert exfoliating particles can also be incorporated into this form of body cleaning products. In general, shower gels and shower creams do not foam as much as bubble bath products do, and their viscosity is higher than that of bubble bath products.

There are three main types of body washes currently available in the market:

- **Regular** body washes contain mild surfactants and other additives to modify the product feel. Their primary function is to provide skin cleansing.
- Body washes containing a higher level of moisturizers (such as glycerin and emollient oils) are usually referred to as **moisturizing** body washes. These products may provide additional benefits to dry skin in addition to performing the base skin cleansing function. It has been clearly demonstrated that incorporation of high levels of emollients into liquid cleansers improves the mildness and

moisturization of these cleansers.[32] It has also been shown that body washes rich in humectants tend to be less beneficial than the lipophilic-rich products. The reason for that is humectants are water-soluble ingredients and leave significantly lower levels of deposit on the skin than lipophilic materials and thus do not provide the same level of clinical benefit. The only way to tell whether a formula is humectant rich or emollient rich is to check the list of ingredients in the product label.

- There are products that fall into a broad category usually referred to as **specialty** body washes. They may be considered a subcategory of the aforementioned two groups since they contain cleansing and moisturizing ingredients as well as ingredients with special functions. Examples for such ingredients include exfoliating beads or other grit materials (e.g., pulverized fruit seeds) to provide exfoliation. These products are referred to as body scrubs. An additional example is ingredients providing a warming or cooling sensation to the skin after application. Menthol, spearmint oil, and peppermint oils provide a cooling effect, while pepper oil stimulates blood flow and provides a warming sensation.[33] In addition, body washes may contain antibacterial ingredients, most often triclosan, for additional benefits. These products are usually referred to as antibacterial body washes. The use of such products has decreased due to safety and efficacy concerns, which are discussed under the safety parameters of skin cleansing products.

DID YOU KNOW?

Solutions of anionic surfactants can be thickened with the addition of sodium chloride. The reason for this phenomenon is the change in micellar size. As discussed in Chapter 1, surfactants form micelles in solutions above their critical micelle concentration (CMC). Micelles with anionic heads repulse each other. Salts reduce the repulsive forces between the charged head groups on the micellar surface, and it decreases the charge density of micelles, which in turn decreases the CMC and increases micellar size. As viscosity depends on the size and packing structure of micelles, if they are getting larger, viscosity increases.[34] Viscosity can be increased only until a certain point. If too much salt is added, the formulation will get thinner. Plotting a salt curve is used in practice to determine the optimal salt concentration needed in such solutions.[35]

The types of ingredients used in bubble bath products, shower gels, and creams are generally the same; therefore, they are summarized together.

- **Surfactants** are the main ingredients as in most cleansing products. The principal foaming agents used in bubble baths are anionic surfactants, including alkyl ethoxylated sulfates, such as sodium laureth sulfate; alkyl sulfates, such as SLS and triethanolamine lauryl sulfate; isothionates, such as sodium cocoyl isethionate; as well as sulfosuccinates and sarcosinates. Anionic surfactants have excellent foaming and cleansing power; however, they are generally irritating. Therefore, they are often combined with nonionic and amphoteric surfactants, which act as secondary surfactants and decrease the irritating potential of anionic surfactants and leave the skin with a pleasant feel as well as contribute to foam stability.

- **Thickeners** provide the appropriate rheological properties for the systems as well as contribute to the foam stability.
    - Surfactant-based products can be thickened by increasing the surfactant concentration; using various hydrophilic thickening agents, such as cellulose derivatives, gums, and acrylic acid derivatives; and, in certain cases, adding sodium chloride. For example, sodium laureth sulfate can be easily thickened with sodium chloride.
- **Foam stabilizers** are generally surfactants, which do not have a good foaming property by themselves; however, they can improve the stability of foam generated by anionic surfactants.
    - The most frequently used ingredients include nonionic surfactants, such as cocamide DEA.
- **Water** is the main vehicle for the formulations.
- **Skin conditioning agents** are moisturizers, which can be deposited on the skin surface. They are especially beneficial for consumers with dry skin.
    - Examples include petrolatum, olive oil, almond oil, mineral oil, shea butter, silicone oils, waxes, vitamins, panthenol, and allantoin. Certain nonionic surfactants have emollient properties as well.
- **Abrasives** may be added to specialty body washes to provide an exfoliating effect. See examples under facial cleanser scrubs.
- **Preservatives** are essential ingredients in water-based formulations to prevent microbiological contamination.
    - Examples include parabens, phenoxyethanol, methylisothiazolinone, and benzoates.
- **Additional ingredients** may include FDA-approved colorants; fragrances; pearls to provide a special and unique appearance to the products; chelating agents, such as EDTA and its derivatives; as well as natural ingredients, such as herbal extracts, vitamins, and minerals.

These products are either solutions containing surfactants or O/W emulsions, if emollients are included as well. Therefore, the formulation steps should follow the general routine established for solution making or emulsification processes.

Bath Oils Bath oils and essences represent a unique category of bathing preparations today. These are oily products containing a high amount of emollients, often combined with fragrances. They are intended to be used in the bathtub to moisturize the skin. Bath oils and essences are typically non-foaming formulations.

The main ingredients of such formulations are the **emollients**. Originally, mineral oil was incorporated as the main emollient since it has excellent hydrating properties. It forms an occlusive layer on the skin and prevents water loss through the skin. Its major disadvantage is that it may feel heavy and greasy on the skin. Newer emollients used in bath oils include isopropyl myristate, isopropyl palmitate, and other isopropyl esters; polypropylene glycol (PPG) ethers; natural oils, such as grape-seed oil, olive oil, sweet cherry oil, and tea tree oil; as well as vegetable oil. **Surfactants** act as solubilizers in these formulations and help the oils spread on the surface of water instead of forming smaller concentrated patches (what simple oils would normally do) or even disperse with water. Generally, nonionic surfactants are used for this reason, many of which may actually act as emollients as well. Examples include polysorbates, other ethoxylated sorbitan esters, such as PEG-40 sorbitan peroleate, and ethoxylated carboxylic acids, such as PEG-12 laurate. Additional ingredients may include fragrances and antioxidants and colorants in some cases.

The major types of these products include floating oils, water-dispersible oils, and soluble oils.[36]

- **Floating oils** contain a higher percentage of oils and a smaller amount of surfactants, which help spread the oil. However, surfactants do not solubilize the oil in the bath water. The choice of surfactant is important and depends on the spreadability and HLB value of the oils to be spread. Ideal surfactants permit the oil to form a continuous film rather than exist as individual droplets. Ideal surfactants should demonstrate a reasonably high HLB (approximately 9) yet remain soluble in the oily composition.[37] It maximizes the surface of the film on the water and allows for better evaporation of the perfume as well as adherence to the skin. The main disadvantages of floating oils is that they may form an oily ring around the edge of the bathtub, and it is difficult to get the soap to lather in this environment since mineral oil, lanolin, and many other emollients likely to be present are excellent defoaming agents (i.e., decreasing the foaming activity of the surfactants). In addition, oils can make the bathtub or floor quite slippery and pose a serious hazard. As these are anhydrous formulations, they can be packed into gelatin capsules of spherical or some other shape. The capsules can be coated with a pearling agent, and these products are often referred to as bath pearls. When the gelatin capsules are added to the hot bath, the gelatin dissolves in the hot water and the contents disperse.

DID YOU KNOW?

Aromatherapy products are very popular today. They contain a high level of fragrances and essential oils, which can provide various experiences for users, such as relaxation, stimulation, and stress relief. Aromatherapy products are generally considered cosmetics in the United States; however, products claiming that they help a person to quit smoking, act as a sleeping aid, and treat depression or high blood pressure are regulated as drugs.[38]

- **Dispersible oils** contain high levels of fragrances, emollient oils, and surfactants, which are capable of dispersing these oils in the bath water to produce an *in situ* emulsion (i.e., instantly formed emulsion). When the emulsion forms, it may give a white cloud due to micelle formation. As the surfactant level is higher in this type of product, there is a lower chance for the formation of an oily ring around the edge of the bathtub.
- **Soluble oils** are solutions of nonionic surfactants in water containing a small amount of emollients in a solubilized form. Due to the smaller amount of oils, these products can produce some foam when mixed with the bath water. However, nonionic surfactants do not have excellent foaming properties; therefore, these are only slightly foaming products.

The formulation of floating oils and water-dispersible oils is a simple mixing process where all the components can be mixed together until a homogenous product is formed. The formulation of soluble bath oils is a solution preparation where emollients are captured by the surfactant molecules and transformed into a solubilized form.

Bath Salts and Bath Bombs Solid body cleansing aids, including bath salts and bath bombs, represent the oldest form of products to clean the skin and give it a pleasant odor. Originally, they were intended to simulate the salt content of natural spas. Most of these products contain sodium salts of weak acids and are therefore alkaline. When greatly diluted, just like in the bathtub, they are considered harmless. Salts often used in these products include sodium carbonate, sodium bicarbonate, sodium sesquicarbonate, disodium phosphate, sodium chloride, sodium borate (it has a mild bacteriostatic action and slight astringent properties), and sodium perborate. Additional minerals, such as magnesium sulfate, sodium sulfate, sodium thiosulfate, magnesium chloride, and potassium bitartarate, may also be incorporated to mimic spa water. Surfactants, colorants, and fragrances as well as various oils and botanical extracts may also be added to the formulations.

Effervescent bath bombs usually contain sodium bicarbonate and an acid, such as citric acid or tartaric acid. The chemical reaction between these ingredients releases carbon dioxide when the product is placed in water. The intention is to simulate the effect of natural carbonated spas. Such products must be carefully formulated and packaged to avoid moisture, which can start the carbonation reaction.

Colorants are typically sprayed onto the powder mixture as an aqueous, alcoholic, or hydroalcoholic solution. Another method, which is preferred for larger quantities of products, is to immerse salts in a tank containing the solution of colorant for a short period of time. After the coloring process, the salts are spread evenly on trays and allowed to dry. Perfumes and botanical extracts can also be sprayed onto the salt followed by quick drying. The final product should then be packed immediately to avoid loss of perfume.

Bath bombs can contain binders, i.e., ingredients that hold individual particles together and prevent breaking of the cubes/balls into smaller pieces during packaging and shipping. Bath bombs are usually pressed into final shape after slightly wetting the powder mixture with a mixture of water, oils, and fragrances. Water dissolves a small amount of the salts during mixing. When water evaporates during drying, the dissolved salts recrystallize and act as a bridge for the particles. Air humidity level should be taken into account during formulation of bath salts and bath bombs since many of these ingredients are highly hygroscopic and can readily absorb water from the air.

Hand

Hands represent the body parts that are the most frequently cleaned during a day. ●**In addition to simple cleansing, hand washing has an essential role in infection control.** There are numerous diseases that can be spread by not washing the hands; therefore, keeping the hands clean is one of the most important steps that can be taken to avoid getting sick and spreading germs to others. There is information posted at many communal places, which draws attention to the importance of hand washing. Often the proper technique to be used in order to remove all the necessary dirt and microorganisms is also depicted. Studies demonstrated that washing the hands with soap and water is more effective in removing pathogen bacteria than hand washing with water alone.[39] According to the US Centers for Disease Control and Prevention (hereinafter referred to as the "CDC"), if soap and water are not available, alcohol-based hand sanitizers should be used to clean the hands.[40] It should also be kept in mind, however, that cleansing the hands without moisturizing them afterward will damage the skin barrier and lead to dry skin symptoms.

Hand cleansing products include bar soaps, liquid soaps without and with antibacterial agents, hand cleansing wipes, as well as alcohol-based and non-alcohol-based hand sanitizers.

- Similar to facial and body cleansing, **bar soap** is mostly replaced by syndet bars due to their milder effect on the skin. These products clean the hands and remove bacteria by chemical and physical means. Although bars are popular forms of hand cleaning in most parts of the world, it should be kept in mind that there is a high incidence of soap contamination and, therefore, the spread of infections. The soaps are in contact with dirty hands, and if the soap bar itself is not completely rinsed off, microorganisms may remain and grow on its surface.[41]
- For this reason, at most communal restrooms, such as in restaurants, healthcare institutions, and educational institutions, soap bars have been generally replaced with **liquid soaps**, including gels and creams.

Most ingredient types used in hand cleansing products are the same as those employed in facial and body cleansing products. A brief summary of the common ingredients is provided here (for more detail, refer back to the previous parts of this section).

- **Water** is the vehicle for the formulations.
- **Surfactants** are typically used as a blend of various types to optimize foam and cleansing properties while minimizing negative effects. Types include mild **anionic** agents, such as fatty acid carboxylates, sarcosinates, isethionates, sulfosuccinates; **nonionic** surfactants, such as amine oxides; **amphoteric** agents, such as betaines; and even certain **cationic** components, such as cetrimonium chloride, which have an antiseptic effect as well.
- **Skin conditioning agents** include various types of emollients and humectants to provide the hands with a soft and smooth feeling.
- **Thickeners** adjust the product's viscosity to an optimal level. They are typically of the hydrophilic type. Certain hand soaps may be thickened with sodium chloride (for more detail, refer back to the previous part).
- **Preservatives** are usually added to liquid hand cleansers although they may contain other antibacterial agents. Antibacterial agents are designed to kill microorganisms on the skin and may not be adequate to protect the product from microbes that generally contaminate water-based products. Examples include DMDM hydantoin, parabens, methylchloroisothiazolinone, and methylisothiazolinone.
- **Natural components** include aloe extract, chamomile extract, lavender extract, cinnamon extract, and many others. These ingredients may have various effects, such as antiseptic, anti-inflammatory, and soothing.
- **Additional** ingredients may include chelating agents, fragrances, colorants, and pH buffers.

Topical Antiseptic (Antimicrobial) Products In recent years, numerous hand hygiene products, including many with antimicrobial activity, have become available for the general public. Antibacterial products combine the cleaning action of the physical removal of foreign materials with an antiseptic agent that kills microorganisms. These products are primarily targeted toward reducing the level of transient bacteria and viruses on the hands.

DID YOU KNOW?

The terms "antiseptics" and "disinfectants" may be used sometimes interchangeably; however, they are not exactly the same. Antiseptics are chemical agents used to prevent infections by killing or inhibiting the growth of microorganisms in or on living tissue (e.g., healthcare personnel hand washes). In contrast, disinfectants are used on inanimate surfaces or objects to destroy or irreversibly inactivate infectious microorganisms. Currently, antiseptics, such as antibacterial soaps and hand sanitizers, are regulated by the FDA, whereas disinfectants are regulated by the US Environmental Protection Agency (EPA).[42]

Currently available antiseptic products are diverse; they can be divided into three broad categories based on the proposed use: healthcare antiseptics, consumer antiseptics, and food handler antiseptics.

- **Healthcare antiseptics** are intended for use by healthcare professionals and consist of healthcare personnel hand washes, hand sanitizers, surgical hand scrubs, and patient preoperative skin preparations.
- **Consumer antiseptics**, also called antiseptic hand washes, are intended to be used by the general public in a variety of settings. They are largely marketed as antibacterial soaps, hand sanitizers, and antibacterial wipes.
- **Food handler antiseptics** are marketed for hand washing in a variety of food handling establishments. These include hand soaps and hand sanitizers.

Currently, the majority of the topical antiseptic products are marketed under the Tentative Final Monograph (TFM) for OTC Healthcare Antiseptic Drug Products, established in 1994 (for more information on what a TFM is, refer back to Section 2 of Chapter 1). Similar to final monographs, TFMs also specify the active ingredients that can be used for OTC drug products and for labeling, product testing, and other general requirements. The TFM makes no distinction between healthcare personnel hand washes and antiseptic hand washes for consumer use with regard to the testing criteria and effectiveness requirements.

In the US, antimicrobial actives that can be incorporated into these products are classified into three categories as follows:

- **Category I:** ingredients **G**enerally **R**ecognized **A**s **S**afe and **E**ffective (GRASE) for the claimed therapeutic indication
- **Category II:** ingredients generally not recognized as safe and effective or have unaccepted indications (not GRASE)
- **Category III:** insufficient data available to permit final classification; ingredients for which there is insufficient evidence; however, the FDA is not objecting to marketing or sale of these products.

The TMF currently lists only two active ingredients as Category I for antiseptic and healthcare hand washes: ethanol and povidone iodine. Ethanol is widely used in consumer products as well as in products for healthcare professional, while povidone iodine is primarily used in healthcare antiseptics. Other ingredients used in consumer antiseptics are currently classified as Category III. They include benzalkonium chloride, benzethonium chloride, *para*-chloro-*meta*-xylenol, triclocarban, and triclosan. As the monograph is currently in a tentative state, manufacturers are able to market hand sanitizers based on Category III ingredients. However, they are required by the FDA to submit further safety and/or efficacy data to prevent these ingredients from being excluded in the Final Monograph as GRASE active ingredients.

The majority of consumer liquid hand soaps labeled as "antibacterial" contain triclosan.[44] Although the FDA does not formally regulate the levels of triclosan used in consumer products, most liquid hand soaps contain between 0.1% and 0.45% weight/volume.[45] A chemically related compound, triclocarban, is used in antibacterial bar soap formulations. Currently, no natural active ingredients, such as thyme, are listed in the TFM. Thus, natural antimicrobials are precluded from legal use in the US, unless an approved New Drug Application (NDA) is obtained. Hand sanitizer wipes usually contain other antibacterial actives, such as benzalkonium chloride (in a concentration of approximately 0.1%) and benzethonium chloride (in a concentration of approximately 0.3%).

Antibacterial hand sanitizers are applied to dry hands followed by rubbing. These products dry fast and rinsing is not necessary. Hand sanitizer products are usually categorized according to their alcohol content, based on this fact we can distinguish between alcohol-based and water-based formulations.

- Most **alcohol-based** formulations contain either ethanol or isopropyl alcohol, or a combination of these two ingredients. The antimicrobial activity of alcohols results from their ability to denature proteins.[46] Alcohol solutions containing 60–95% alcohol are recognized as being the most effective, with higher concentrations being less potent.[47,48] This paradox results from the fact that proteins are not denatured easily in the absence of water.[46] Alcohol concentrations in antiseptic hand sanitizers are often expressed as a percentage by volume. It has been shown that products containing less than 60% alcohol, which may be marketed to consumers, are less effective in reducing the number of microorganisms on the hands.[49] It is important, therefore, that consumers check the alcohol concentration in hand sanitizers before purchasing any products. In addition to alcohol, alcohol-based hand sanitizers typically contain water; skin conditioning agents, such as glycerin, propylene glycol, isopropyl myristate, and vitamin E; thickeners; colorants; and fragrances. Preservatives are also added to most formulations since alcohol may not be effective against all microorganisms that may contaminate the product. Alcohol-based products' ability to kill bacteria ends once the product has dried on the skin where products with other antimicrobial ingredients continue to provide protection well after the solution has dried.
- **Water-based** formulations are generally based on water, surfactant, and antimicrobial ingredients to which emollients can be added. They are typically supplied as liquids, gels, and foams. Commonly applied antimicrobial ingredients include benzalkonium chloride and triclosan. These water-based formulations are often labeled as "alcohol-free" formulations. The main reason for their introduction to the market was to offer a hand sanitizer product without the negative, drying effect of alcohols. Water-based formulations are better for the skin, they pose much less of a threat in cases of accidental ingestion, and they are not

flammable. Another clear benefit is that they offer immediate and persistent killing activity.

DID YOU KNOW?

The FDA does not allow hand sanitizer manufacturers to make claims against specific types of pathogens (such as **Methicillin-resistant** Staphylococcus aureus, known as MRSA, as well as *Salmonella* and *Escherichia coli*) due to FDA concerns that consumers will presume that the particular illness caused by these germs will be completely prevented through use of the product, although they may claim that the products "help reduce bacteria that can potentially cause disease."[50]

The formulation of the products can be attributed to the classical dosage forms.

- The formulation of hand sanitizer gels usually starts with the hydration of the thickener. After the process is completed, all other components are added and mixed thoroughly to prepare a homogenous final product.
- The formulation of soap and syndet bars is identical to that of the facial cleansing soaps and syndets.
- Foaming hand sanitizers are usually supplied as liquid soaps, and the foam is generated by a special disperser. Liquid soaps are either simple solutions or O/W emulsions if emollients are added; therefore, their formulation is identical to the formulation of liquid facial and body cleansers.
- Hand cleansing wipes are produced by wetting the dry cloths, folding them together, or packing them in separate foils for single use. Formulation of the solution or O/W emulsion usually follows the general solution mixing or emulsification process.

CDC Recommendations on Proper Hand Hygiene Hand washing done properly and regularly is considered to be the gold standard for removing transient bacteria from the hands and preventing the spread of infection and illness. **●Although the public belief is that hand sanitizers may replace hand washing, the CDC recommends that hand sanitizers should not be used in place of soap and water but only as an adjunct.** In addition, it states that hand sanitizers may not be as effective when the hands are visibly dirty. Even in healthcare settings, CDC guidelines recommend soap and water on the hands that are visibly soiled or contaminated with proteins, rather than using the alcohol-based sanitizers. The reason for this is that alcohol cannot adequately reduce a number of important pathogens. Fats and proteins may not be visible on the hands, and these materials can coat and protect pathogens from the action of alcohol and may interfere with and neutralize alcohol efficacy.[51,52] In addition, alcohols have very poor activity against bacterial spores, protozoan oocysts, and certain nonenveloped viruses.[53,54] Soap, friction, and running water effectively remove the fatty materials and reduce pathogens of concern. Proper hand hygiene involves the use of soap and warm, running water and rubbing the hands vigorously for at least 20 s. The use of a nail brush is not necessary or desired, but close attention should be paid to the nail areas, as well as areas between the fingers. Wet hands have been known to transfer pathogens much more readily than dry hands or hands not washed at all. Careful hand drying is a critical factor for bacterial transfer to the skin, food, and environmental surfaces. Hand sanitizers should primarily be used only as an optional follow-up to traditional hand washing with soap and water, except in situations where soap and water are not available. In those instances, use of an alcohol gel is certainly better than nothing at all. It is also worth noting that the amount of alcohol-based hand antiseptic is important to its overall effectiveness. Failure to cover all surfaces of the hands and fingers will also greatly reduce the efficacy of alcohol-based hand antiseptics.[55]

Considerations When Selecting Skin Cleansing Products

Dermatologists and consumers are faced with a variety of choices when recommending or selecting a personal cleansing product. It has been shown that skin cleansers can impact the skin in a number of ways and produce a range of skin effects. To choose the most appropriate cleanser, consumers should consider their skin type, skin problems, and any skin allergies. First, the skin type should be determined, i.e., dry, oily, or normal for the particular body part when selecting a skin cleanser. After determining the skin type, any skin problems, such as acne, dryness, or excessive flakiness, and skin allergies should be assessed and taken into consideration.[31] Based on the individual differences in skin type and sensitivity, cleanser selection is somewhat subjective in most cases.

Here is a brief summary of the general properties of skin cleansing products that should be considered when selecting such products.

- Traditional soaps provide effective cleansing; however, they can irritate the skin due to their pH and remove the majority of the protective lipid layer from the skin. They tend to leave the skin with a tight and dry feeling.
- Facial toners are mainly used after cleansing (as a second step before applying moisturizers) to remove soap, oil, and makeup residue from the skin. As they contain astringents, they can dry the skin and leave it with a tight feeling. They are a good option for oily skin users; however, they are not recommended for consumers with dry skin.
- Exfoliating agents help physically remove dirt and cellular debris from the skin and provide a smooth skin surface. They are advantageous for aging skin, acne, and other skin conditions when high exfoliation is required. As mentioned previously, proper use is important to avoid damaging the SC barrier. Manufacturers' directions for use should be followed to avoid skin damage when using these products.

- Facial cleansing wipes are very efficient in removing makeup and debris from the skin. In addition to the chemical cleansing provided by the various ingredients deposited on the cloths, they also provide physical cleansing through the rubbing action. They may deposit emollients on the skin surface, which is beneficial for users with dry skin.
- Water temperature also impacts the interaction between the skin and cleansing products.[56] Bathing in warm rather than hot water is recommended to reduce drying and irritation.
- For consumers with dry skin and for those who do not like applying moisturizing products after taking a bath/shower, moisturizer body washes that can deposit light emollients on the skin while taking a shower are recommended.

Typical Quality Problems of Skin Cleansing Products

Although the formulation of skin cleansing products is precisely controlled at cosmetic companies, sometimes batches formulated do not meet the quality specifications set. This section reviews ●**the typical quality-related issues of skin cleansing products**, which **include sticky bath salts and bath bombs, broken bath bombs, discoloration of bar soaps, poor foaming activity of foaming formulations, separation of emulsions, cloudy solution, microbiological contamination, clumping, and rancidification,** as well as their potential causes and solutions.

Sticky Bath Salts and Bath Bombs Humidity and the hygroscopic property of the ingredients can be a major issue in the case of solid bath products, especially in the case of effervescent products. These products can take water up from the environment and start dissolving in that water, becoming sticky. In order to avoid such problems, manufacturing and storage conditions should be kept under strict control to overcome this problem.

Broken Bath Bombs Compression force influences the breaking hardness of the products. Bath bombs should be hard enough to "survive" packaging and shipping without breaking. Some of the products may contain starch as a disintegrant, which also acts as a binding agent, increasing the internal binding force between the individual particles. Salts, such as sodium chloride, can also be used as a binder. It is partially dissolved in water, then recrystallizes and forms solid bridges between the particles during drying. If low breaking hardness is an issue, either the formulation parameters or the composition should be checked. If salt is used as a binder, more liquid may be needed to provide appropriate hardness. Additionally, humidity may be an issue during storage if starch is used since it is a hygroscopic material. It acts as a disintegrant by taking up water and swelling. The swelling effect will lead to breaking of the bath bomb.

Discoloration of Bar Soaps Discoloration is a quality and aesthetic problem that may occur in part of a bar, with some parts being lighter and others being darker, or in the whole bar. Chemical changes, such as oxidation, may cause discoloration assisted with or without light. Incompatibility among the components, or other stability problems, such as pH stability, may also cause discoloration. Fragrance components are very sensitive to pH and pH changes, which may trigger oxidation and discoloration in fragranced bar soaps. Metal ions can also be the source of the problems; thus, the incorporation of chelating agents is recommended.

Poor Foaming Activity of Foaming Formulations As discussed earlier, most consumers expect cleansing formulations to generate foam as a sign of proper cleaning. However, low-foaming and non-foaming formulations are also effective cleansing alternatives to foaming cleansing aids. Due to this consumer expectation, most cleansing aids are foaming formulations. The type and amount of surfactants used have a huge influence on the foaming activity and foam quality. In addition, thickeners may also affect foam generation; therefore, their type and amount should also be taken into consideration. Defoaming agents, such as fatty acids, produced by the hydrolysis of bar soaps, as well as emollients, may also result in poor foaming characteristics.

Separation of Emulsions As discussed in Section 3 of Chapter 1, emulsions are thermodynamically unstable formulations, which tend to separate over time (it may take only a few hours to years, depending on the formulation and process parameters). Separation can be reversible and also irreversible, and the latter needs reformulation. The main mechanisms of physical emulsion instability are illustrated in Figure 3.6. Reversible changes include creaming, sedimentation, and flocculation, while irreversible changes include phase inversion, coalescence, and Ostwald ripening.[57,58]

- **Creaming and sedimentation** usually occur due to the density mismatch between the two droplets and the continuous phase. **Creaming** usually occurs in O/W emulsions. In this phenomenon, the less dense phase (i.e., oil phase) migrates to form a thin, milky layer at the top of the emulsion. **Sedimentation** usually occurs in W/O emulsions where the denser water phase migrates to form a milky layer at the bottom of the emulsion. Both creaming and sedimentation are reversible by agitation.
- During **flocculation**, the dispersed droplets aggregate; however, they do not lose their identity. Flocculation often gives a fluffy cloudy appearance, which can be reversed by agitation.
- During **phase inversion**, the emulsion inverts from one type to another, for example, a W/O emulsion inverts to an O/W. It can happen when the emulsifier becomes more soluble in the dispersed phase than in the continuous phase. This process is irreversible.
- **Coalescence** is similar to flocculation, except the droplets cluster and merge to form a larger droplet. This is an irreversible process; it leads to complete separation of the two immiscible phases.
- **Ostwald ripening** is a process in which components of the dispersed phase diffuse from smaller to larger droplets through the continuous phase. Generally, this phenomenon is found in W/O emulsions. It is an irreversible process.

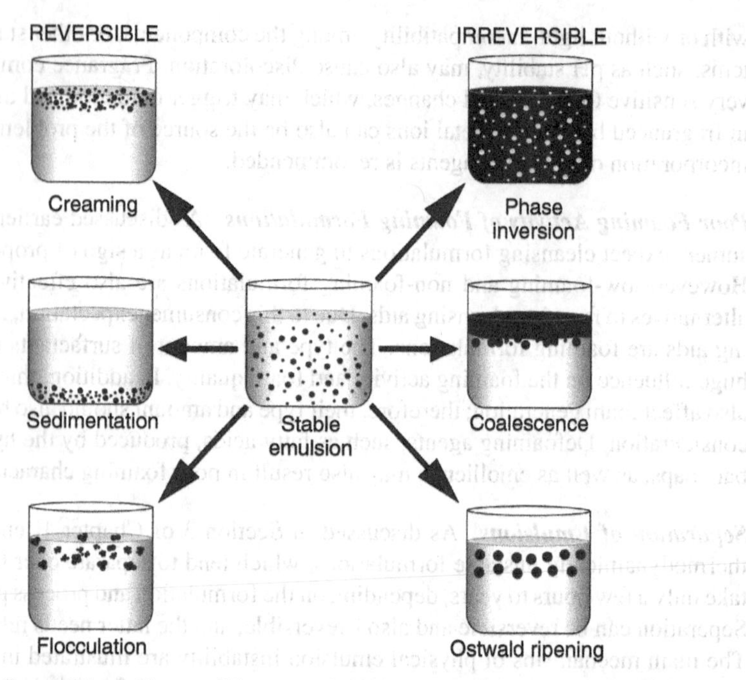

Figure 3.6 Reversible and irreversible types of physical emulsion instability. Modified with permission from Leal-Calderon, F.: Emulsified lipids: formulation and control of end-use properties. *OCL*. 2012:111–119.

DID YOU KNOW?

Although both Ostwald ripening and coalescence look similar, including the formation of larger droplets from smaller ones, these are different phenomena. In coalescence, droplets of the internal phase come into direct contact, while in Ostwald ripening, the external phase serves as a transfer medium and the droplets are enlarged through the external phase.[59]

Cloudy Solution Facial toners may contain fragrances to increase consumer satisfaction. These oily components are typically present in a solubilized state provided by surfactants; therefore, facial toners are clear solutions. If the type and/or amount of the solubilizers are not appropriate, the solution may become opaque to a certain extent due to the particle size of the oil droplets in the solution. Alternatively, the oily components may just rise to the top of the formulation and merge into a large droplet.

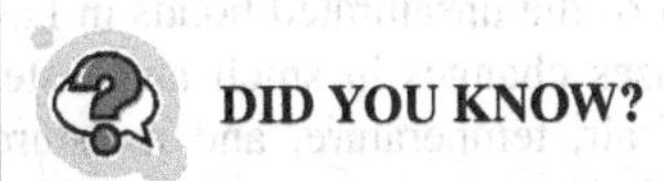

DID YOU KNOW?

The term "solubilization" can be defined as the preparation of a thermodynamically stable solution of a substance normally insoluble or very slightly soluble in a given solvent by the introduction of a surfactant. Surfactants must be used at a concentration at or above their CMC. By solubilization, oils can be incorporated into solutions (primarily water-based or hydroalcoholic), which otherwise would just float on the surface of the solution.

Microbiological Contamination The majority of cleansing aids are water-based formulations; therefore, microbiological contamination may occur in their case. Water is an optimal environment for the growth of microorganisms, and many cosmetic ingredients may actually serve as a nutrient for them. Commercial cosmetics are not expected to be completely sterile; however, they must be safe. Safety includes being free of high-virulence microbes and the total number of aerobic microbes per gram must be low. There are no widely accepted standard for the number of microbes in the US; therefore, most often, temporary guidelines are used. Microbiological contamination is not necessarily visible in cosmetic products, but can lead to severe irritation and infection if such products are used. The use of preservatives is, therefore, highly recommended to keep the number of microbes at a minimum level. Alcohol-based hand sanitizers contain alcohol, which is also a preservative; however, it may not be able to effectively prevent the contamination of the product against a wide range of bacteria and molds. Therefore, even hand sanitizers may contain preservatives.

Clumping It may occur in water-based formulations that are thickened with hydrophilic thickeners. Certain hydrophilic thickeners, for example, carbomers and gums, tend to aggregate and form clumps (i.e., powder particles stuck together having a dry core and a wet surface), which become difficult to disperse. This phenomenon can be avoided by properly dispersing and wetting out the polymer. It is important that the polymer be slowly added/sprinkled to rapidly agitated water, which may be heated beforehand if needed. Certain polymers may be prewetted with other ingredients in the formulations. For example, prewetting with glycerin may help uniformly disperse the polymer in water. Other non-solvent ingredients, lower HLB nonionic surfactants (HLB ~7–9), may also be used to wet out the polymer prior to adding it to water. This bulk dispersion of the polymer with a non-solvent delays the rate at which the polymer swells and hydrates into water, thus minimizing clumping.

Rancidification Rancidification is the chemical decomposition of fats and oils, which can lead to the formation of unpleasant odor, compromised stability, and a change in appearance of emulsions. Rancidity can be initiated by oxygen, light, and elevated temperature, which leads to oxidation of the unsaturated bonds in fats and oils. The formation of volatile chemicals causes changes in smell and taste. Hydrolytic rancidity initiated by the presence of air, temperature, and moisture leads to the hydrolysis of ester linkages in fats and oils and the formation of fatty acids. This is normally this is accomplished through enzymatic peroxidation, where enzymes found naturally in plant oils and animal fats can catalyze reactions between water and oil. The formed fatty acids can then undergo further autoxidation. Oxidation primarily occurs with unsaturated fats by a free radical-mediated process. Antioxidants can fight against these free radicals and retard the development of rancidity due to oxidation. This is why the addition of antioxidants is recommended to all formulations containing fats and oils. In addition, rancidification can be decreased by storing fats and oils in a cool, dark place with little exposure to heat, light, oxygen, or free radicals. Rancidification may also be triggered by enzymes produced by microorganisms, such as bacteria, molds, and yeast. Preservatives added to the formulation can inactivate these organisms and prevent this type of rancidification.

Evaluation of Skin Cleansing Products

This part discusses the various tests usually performed to evaluate the quality, performance (efficacy), and safety of skin cleansing products.

Quality Parameters Generally Tested

●**Parameters commonly tested to evaluate the quality of skin cleansing products include spreadability, extrudability, texture, and firmness of lotions, creams, and gels; actuation force; foaming property, foam stability, foam viscosity, foam density, and foam structure; hardness of batch bombs; disintegration time and dissolution time of bath bombs, bath salts, and bath beads; preservative efficacy; viscosity; and pH.** The range of acceptance and other limiting factors are usually determined by the individual manufacturers.

Spreadability of Lotions, Creams, and Gels Spreadability is a measure of the consistency of a lotion, cream, or gel, which refers to the ease of spreading a product on the skin. This parameter depends on various factors, such as type and amount of oils, fats, waxes, and butters; type and amount of surfactants and thickeners; as well as the water content of the formulation. A lipid-rich cream has poor spreadability with a higher viscosity, making the cream greasy, tacky, and difficult to spread. A lower viscosity cream, lotion, or gel spreads easily on the skin.

Spreadability can be determined by conventional methods, utilizing an extensometer, also known as the parallel-plate method. In this case, a defined weight of the lotion or cream is placed onto the middle of a glass plate. Another glass plate is carefully placed on top of the product, avoiding slide off of the plate. The area (diameter) is measured over which the sample spreads between the glass plates during a defined period of time (e.g., 1 min). A weight can be placed at the center of the plate to enhance spreadability.[60]

There is another method for the measurement of spreadability, which is often referred to as the penetration test. The equipment used includes a spreadability rig that measures the ease with which a product, such as a cream, can be applied in a thin, even layer. The equipment, depicted in Figure 3.7, comprises a 90° cone probe and precisely matched perspex cone-shaped product holders. The material is either deposited and allowed to set up in the lower cone holders in advance of testing or is filled with a spatula and then the surface is leveled. The sample holders can be stored in frozen, refrigerated, or ambient environments before testing of the sample. During the measurement, the rig moves down and the probe penetrates the sample at a specified rate and depth. The product is forced to flow outward at 45° between the upper cone and the product holder during the test, the ease of which indicates the degree of spreadability. When the specified penetration distance has been reached, the probe withdraws from the sample. Withdrawal of the probe from the sample provides information about adhesive characteristics that may be present.[61]

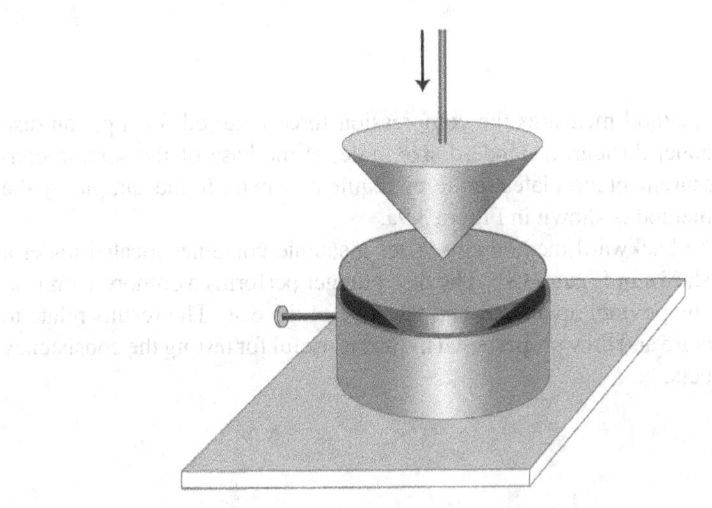

Figure 3.7 Spreadability testing of lotions, creams, and gels. Adapted from Texture Technologies Corp.

Lotion, Cream, and Gel Extrudability One desirable factor in product development is having a product that easily squeezes out of a tube and breaks off cleanly after squeezing. Extrudability test indicates the ease of extruding a cream, lotion, or gel from a tube. The equipment is shown in Figure 3.8. The extrusion rig moves forward, imitating squeezing, and the force necessary for squeezing out the product from a tube is measured. This method can only be used for products packaged into tubes.

Lotion, Cream, and Gel Texture Texture (consistency) can be measured using an extrusion rig that can move down (forward) and up (backward).

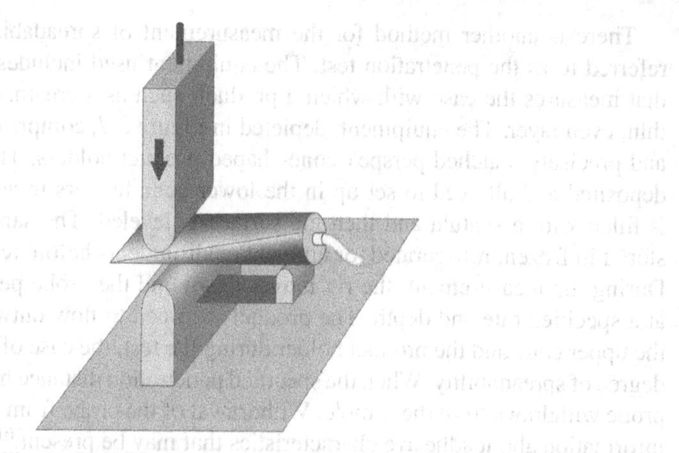

Figure 3.8 Extrudability testing of lotions, creams, and gels. Adapted from Texture Technologies Corp.

The forward method measures the compression force required for a piston disc to extrude a product through a standard size outlet at the base of the sample container. This measurement simulates the force required to extrude the sample by the consumer. The method is shown in Figure 3.9a.

The rig for the backward method comprises a sample container located under a disc plunger as shown in Figure 3.9b. The disc plunger performs a compression test, which extrudes the product up and around the edge of the disc. The results relate to the product structure and flow properties. This test is useful for testing the consistency of viscous products.

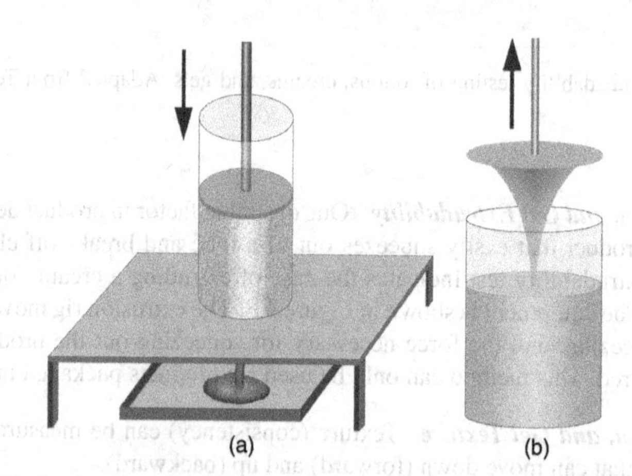

Figure 3.9 Evaluating the texture of lotions, creams, and gels. (a) Forward extrusion method and (b) Backward extrusion method. Adapted from Texture Technologies Corp.

Lotion and Cream Firmness Firmness also refers to the consistency of a product and indicates how much a product can resist an external force (e.g., application). It can be measured using a hemispherical probe. It is an imitative test simulating the ease by which a human finger will deform the sample during application of the cream. The probe shown in Figure 3.10 moves down into the sample at a specified rate to a specified depth and then it withdraws to its starting position. This test allows the consistency of creams, lotions, and gels to be assessed. The force required to penetrate the cream is a measure of the firmness of the cream.

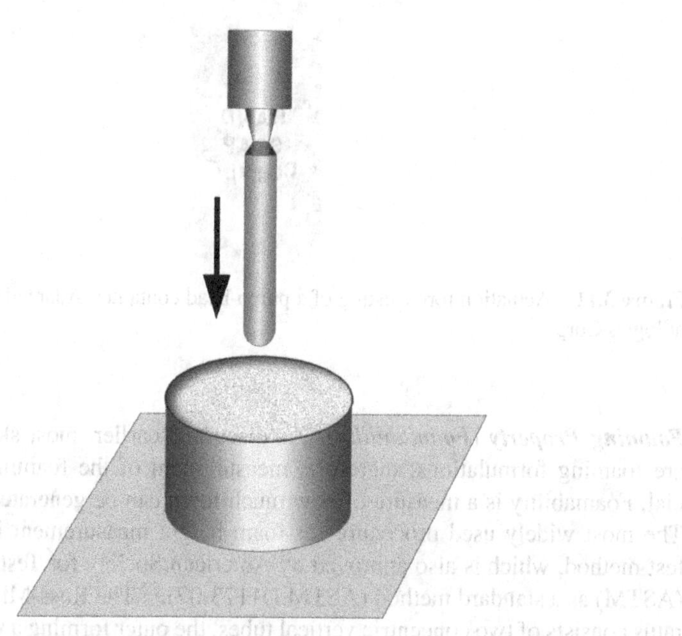

Figure 3.10 Firmness testing of lotions and creams. Adapted from Texture Technologies Corp.

Actuation Force Actuation force testing allows the measurement of the force required to release liquid cleansing products from pump-head dispensers. Manually actuated pump dispensers rely on the user to generate a pressure in order to dispense the product. The equipment used for the test is shown in Figure 3.11. It is important to test how much force is required for adults and even for children to get the product out of the container. When operating a pump-type liquid soap dispenser, a child may have difficulty applying sufficient force in the appropriate direction to operate the pump, which may cause the dispenser to move, tip, or otherwise fail to discharge the product toward the intended target. An actuation force test is an imitative test to assess such issues.

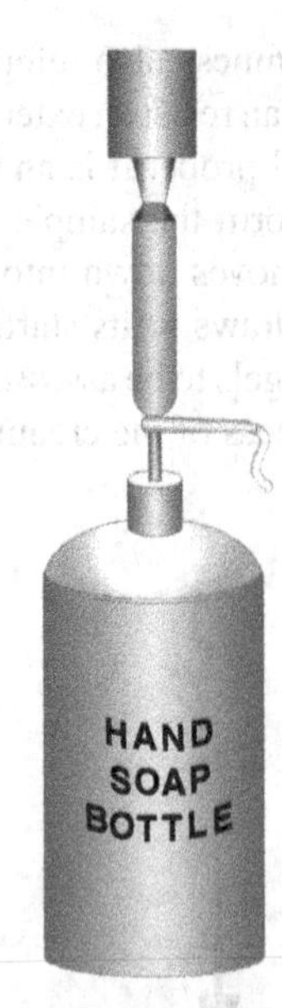

Figure 3.11 Actuation force testing of a pump-head container. Adapted from Texture Technologies Corp.

Foaming Property (Foamabililty) As discussed earlier, most skin cleansing aids are foaming formulations; therefore, measurement of the foaming activity is crucial. Foamability is a measure of how much foam can be generated from a product. The most widely used procedure for foam height measurement is the Ross-Miles test method, which is also approved by American Society for Testing and Materials (ASTM) as a standard method (ASTM D1173-07).[62] The Ross-Miles foaming apparatus consists of two concentric vertical tubes, the outer forming a warm water jacket for the inner. A volume of 200 ml of fluid is allowed to drain at a controlled rate from a glass pipette into a receiving cylinder containing 50 ml of the same fluid. The 200 ml solution generates foam when it mixes with the 50 ml solution. When all the fluid has drained out of the pipette, the foam height is measured initially and after 1 and 5 min. It also provides valuable information on foam stability.[63]

In addition to the Ross-Miles method, several other foam tests may be used to test the foaming property. A simpler method is the shake foam test. It includes placing a solution of the product into a calibrated cylinder and then inverting the cylinder a number of times or beating it with a perforated disc attached to a rod for a fixed time. The foam height generated is measured immediately after generation and after 1 min, 5 min, etc. These methods are easy ways to distinguish between solutions that do not foam from those that foam well. A better procedure consists of sparging a fixed amount of gas at a constant flow rate through a surfactant solution. The foam height in a specified container immediately after the foam has been produced gives a measure of foamability.[64]

Foam Stability Foams are thermodynamically unstable systems, and foams generated from skin cleansing products do not need to be stable for a long period of time. Consumers expect bubble bath formulations to generate and keep foam for a longer period of time (e.g., 10–15 min); however, it is not a major requirement for most other formulations.

Foam stability can be described as the ability of the foam formed to sustain itself against its breakdown or collapse. Foam stability and longevity depend on the bubble size, diffusion of the gas phase from the bubble to another or into the bulk gas phase surrounding the foam; weakening of bubble walls and bubble coalescence; and hydrodynamic drainage of liquid between bubbles, leading to rapid collapse. The presence of surfactant molecules can play an important role in strengthening the bubble walls against factors described earlier. For example, such molecules can increase the bulk viscosity of the bubbles, helping it withstand additional weight and pressure caused by the flow of fluids.[65]

A general method to determine foam stability is to determine the foam height against time and calculate the half-life of the foams. Half-life can be defined as time taken by the generated foam to reach half of its initial volume. Foam generation can be achieved in numerous ways, including the methods described earlier.

Foam Viscosity Viscosity influences the rheological behavior of skin cleansing products. Its measurement is similar to that of other cosmetics and personal care systems. The challenge is that the viscosity should be measured immediately after creating the foam since over time, it will collapse. Today, there are special viscometers and rheometers to measure the viscosity of such systems.

Foam Density Density is another factor that contributes to foam stability: the denser the foam, the more stable. This property can be determined by generating a foam, transferring the foam into a cylinder, and dropping a small item into the cylinder, for example, a rubber stopper fitting into the cylinder. The time for the item to pass between two points is measured and translated to density. New dynamic foam analyzers are also available.[66] They are able to measure the foam decay kinetics as well as traditional foam parameters, such as specific volume.

Foam Structure The volume fraction of liquid in a liquid foam is the most important parameter affecting the foam structure. There are different methods to determine the foam structure, including electrical conductivity, low-power light microscopy, electron microscopy, or video microscopy.

Hardness of Bath Bombs Hardness is an important factor of bath bombs that influences packaging and shipping. Hardness is a measure of the relative resistance of a solid product to chipping or breaking. In the test, a bath bomb is placed onto a measuring surface to which a punch comes down with a constant speed (see Figure 3.12). The force at which the bath bomb breaks down is called the breaking or crushing force. Hardness is usually inversely proportional to the disintegration/dissolution time since the harder the tablet, the longer time it takes to break down to smaller particles and dissolve in the water. On the other hand, if the hardness is too low (i.e., the bath bomb is too soft), it may not be able to withstand the handling, packaging, and shipping operations.

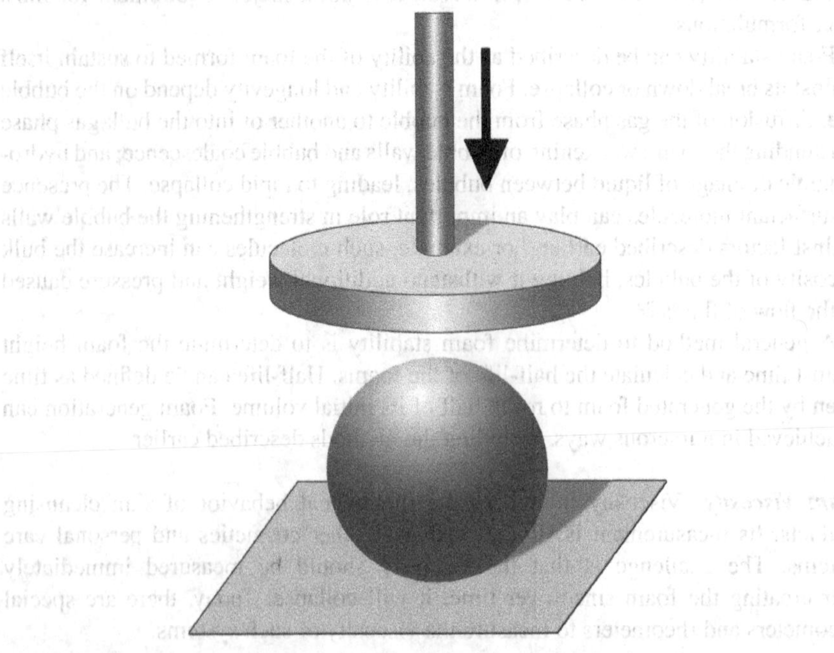

Figure 3.12 Hardness testing of a bath bomb. Adapted from Texture Technologies Corp.

Disintegration Time and Dissolution Time of Bath Bombs, Bath Salts, and Bath Beads Disintegration is the process where a compressed product, such as bath bomb, breaks down into smaller pieces in a liquid medium (i.e., hot bath water). The disintegration test is a measure of the time required under a given set of conditions for bath bombs to disintegrate. Disintegration time can be measured by placing a product sample into a sample holder, which contains water at a given temperature. The time can be measured with a stopper watch.

Dissolution is the process where a compressed and disintegrated (e.g., bath bomb) or initially uncompressed product (such as bath salt) becomes part of the dissolution medium in a molecular state (i.e., becomes invisible in the bath water). Dissolution can be simply measured by placing the sample into a reasonable amount of suitable liquid, mostly water, which is maintained at a given temperature (e.g., bath water temperature for bath bombs). The time required for dissolution can be checked with a stopper watch. As bath beads and bath salts are not compressed, they do not need to disintegrate. Additionally, they are made of water-soluble ingredients, which results in very quick dissolution in hot water. Bath bombs also consist of water-soluble materials; however, a bath bomb needs time to disintegrate and dissolve due to its compressed nature.

Preservative Efficacy As mentioned earlier, cosmetics are not expected to be completely free of microorganisms. However, they must be free of pathogenic microorganisms (i.e., those that can cause diseases) and the number of non-pathogenic microorganisms must be low. Therefore, they should be adequately preserved products containing one or more preservatives. The effectiveness of a preservative depends on all ingredients used in a product as well as the packaging. Therefore, preservative efficacy cannot be predicted based on the individual ingredients. To ensure that the preservatives used are appropriate for preserving products and preventing the growth of microorganisms, companies perform extensive microbiological testing of final formulations.

There are no widely accepted standards for the tolerable number of microorganisms in products; therefore, temporary guidelines are used. For eye-area products, counts should not be greater than **500 colony-forming units (CFU)/g**; for non-eye-area products, counts should not be greater than **1000 CFU/g.**[67] Pathogens whose incidence is of particular concern, especially in eye-area cosmetics, include *Staphylococcus aureus*, *Streptococcus pyogenes*, *Pseudomonas aeruginosa*, and *Klebsiella pneumonia*, among others.[67]

A number of methods, guidelines, and approaches have been developed over the years for preservative efficacy testing of cosmetics by regulatory agencies, standard organizations, industry and individual companies, including the PCPC, ASTM, the Association of Analytical Communities (AOAC), and various pharmacopoeias (EP/BP, JP, USP).[68] Today, the most commonly used test is the microbial challenge test. The microbial challenge testing is designed to evaluate the ability of cosmetics to prevent microbiological growth in products upon contamination with microbes that may be introduced into the products during manufacturing, storage, and/or normal consumer use.

The standard method is the United States Pharmacopeia (**USP) <51> challenge test**, which was designed for the pharmaceutical industry.[69] In the test, the cosmetic product is inoculated with five microorganisms (three bacteria: *S. aureus*, *E. coli*, and *P. aeruginosa* and two fungi: *Candida albicans* and *Aspergillus brasiliensis*), and the survival rate of the microorganisms is evaluated at specific intervals within a 28-day testing period. The effectiveness of the preservative system is determined based on the USP <51> criteria.

Many companies found the USP <51> challenge test to be a low-level challenge and prefer using other protocols and approaches. Generally, the main principles of the methods are the same: products are inoculated with microorganisms, and their survival rate is monitored over time. The tests typically pass/fail based on the preestablished criteria. The main differences are the following:

- Type of microorganism used: there are protocols that use additional microorganisms representative of species reported as manufacturing or in-use contaminants.
- Number of sampling.

- Number of inoculations: there are double challenge methods in which products are reinoculated.
- Test duration.

Although a number of cosmetic and personal care companies use the USP or PCPC tests, many companies have developed their own "in-house" methodologies. These methods are often more stringent than the standard protocols. Approaches include the following (without completeness): using more microorganisms than the recommended standard level, dilution of the samples before inoculation, longer testing periods, repeated inoculation, real-time testing, and more stringent acceptance criteria.[70]

Viscosity Viscosity (η) is a measure of the internal friction of a fluid. In other words, it is the resistance to flow. Viscosity is defined as shear stress (σ, otherwise known as shear force, i.e., the force applied per unit area, which allows the material to start flowing; shear occurs whenever the fluid is physically moved by pouring, spreading, spraying, mixing, etc.) divided by the shear rate (γ) (i.e., the velocity with which the material starts flowing upon applying a force).

$$\eta = \frac{\sigma}{\gamma}$$

Viscosity measurement is one of the most widely performed tests for all liquid and semisolid dosage forms to ensure all manufactured batches have a consistent viscosity. It is an important property since it may have a large effect on the applicability, spreadability, pumpability, and stability of products. The viscosity of a system depends on multiple factors, including the oil/wax versus water content, concentration of thickeners and emulsifiers, electrolytes, type and amount of surfactants, pH, and temperature, among others.

There are two basic types of measuring devices used for the rheological evaluation of different cosmetic systems: viscometers and rheometers. These names are not strictly defined. Viscometers, in comparison to rheometers, are usually relatively simple instruments. Most viscometers operate by a spindle rotating in one direction in the sample. Rheometers can operate by other types of movements, such as oscillation as well; therefore, they can be used to determine more complex properties.

Figure 3.13 depicts some characteristic spindles (i.e., vertical solid bodies). During the measurement, the spindle is immersed into the sample and is moved (rotated or oscillated) by the motor. The resistance to that movement is a measure of the inherent viscosity. When measuring viscosity, it is important to specify temperature since viscosity changes with temperature; for most cosmetic and personal care products, viscosity decreases as temperature is increased.

pH pH as a number refers to the degree of acidity or basicity of a solution, which depends on the hydrogen ion [H^+] concentration in a solution. The pH value is defined as the negative logarithm of the H^+ concentration in a given solution. In other words,

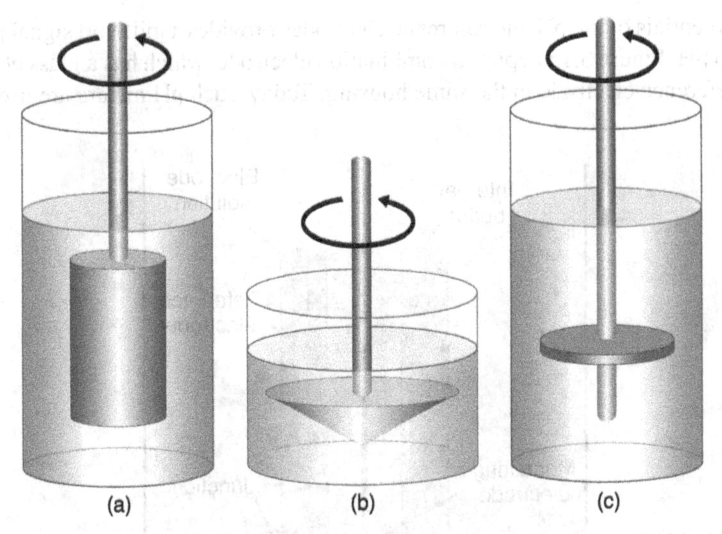

Figure 3.13 Various types of spindles used for viscosity testing. (a) cylindrical spindle, (b) conical spindle, and (c) disc spindle.

at a high concentration, e.g., 1 mol/l = 10^0, pH = 0 (acidic), while at a low concentration, e.g., 10^{-14} mol/l, pH=14 (alkaline). Different solutions are objectively compared with each other, where pH 0 is extremely acidic, pH 14 is extremely alkaline, and pH 7 is neutral (see Figure 3.14).

pH, similar to viscosity, is one of the most widely measured parameters for cosmetics and personal care products. In order to measure the pH value of a solution, a measuring electrode (pH electrode) and a reference electrode are needed, both of which are immersed in the same solution. The pH electrode uses a specially formulated pH-sensitive glass, which develops a potential (voltage) proportional to the pH of the solution. The reference electrode is designed to maintain a constant potential at any given temperature and serves to complete the pH measuring circuit within the solution. It provides a known reference potential for the pH electrode. The difference

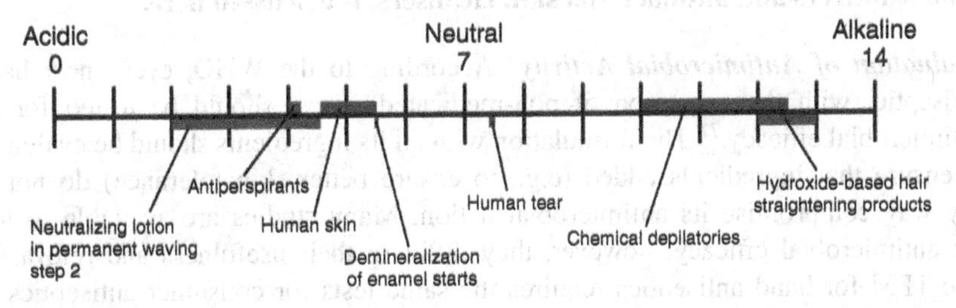

Figure 3.14 pH scale indicating the acidity/alkalinity of some common products and processes related to cosmetics and personal care products.

in the potentials of the pH and reference electrodes provides a millivolt signal proportional to pH. Figure 3.15 depicts a combination electrode, which has a glass electrode and a reference electrode in the same housing. Today, such pH meters are used.

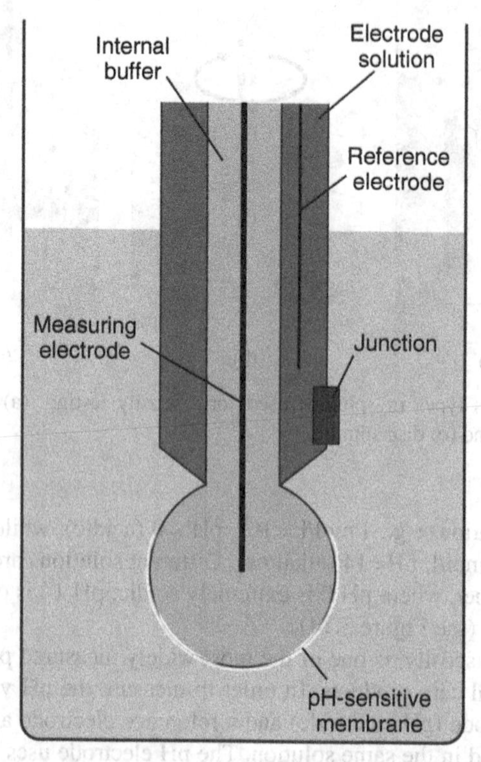

Figure 3.15 Combination electrode used in today's pH meters.

Efficacy (Performance) Parameters Generally Tested

●The most frequently tested efficacy parameter, the antimicrobial activity of hand sanitizers and antibacterial skin cleansers, is discussed here.

Evaluation of Antimicrobial Activity According to the WHO, every new hand antiseptic, with the exception of non-medicated soaps, should be tested for its antimicrobial efficacy.[71] The formulation with all its ingredients should be evaluated to ensure that ingredients added (e.g., to ensure better skin tolerance) do not in any way compromise its antimicrobial action. Many studies are available to test the antimicrobial efficacy; however, they differ in their usefulness and relevancy. The TFM for hand antiseptics requires the same tests for consumer antiseptics as healthcare personnel hand washes. The currently proposed testing consists of both *in vitro* and *in vivo* studies.

- *In vitro* studies are designed to demonstrate the product's spectrum and kinetics of antimicrobial activity, as well as the potential for the development of resistance associated with product use. These studies usually include the determination of the minimum inhibitory concentration (MIC) and time–kill studies using various bacteria and viruses. The major concern with such studies is that the efficacy is only proven in laboratory settings, and they do not reflect the human skin conditions.[72]
- *In vivo* tests and the evaluation criteria are based on the assumption that reduction in the number of bacteria translates to a reduced potential for infection and that reduction in the number of bacteria can be properly demonstrated in simulated studies. *In vivo* studies are designed to demonstrate antimicrobial efficacy in the presence of a bacterial challenge. There are different protocols and standards for different countries. In the US, hand washes and hand sanitizers are currently evaluated according to the ASTM standard E 1174.[69]

In the test, the subjects' hand and lower forearms are artificially contaminated with the selected marker bacteria. It is followed by the treatment with an antimicrobial agent and determination of the remaining viable bacteria. The before-use (baseline) data are compared to the after-use results and the reduction in the number of bacteria is determined. This contamination and hand wash procedure is repeated 10 times, and bacterial reductions are determined after the first, third, seventh, and tenth wash. This aspect of the study design is intended to mimic the repeated use of the product. After the first, third, seventh, and tenth washes, rubber gloves or polyethylene bags used for sampling are placed on the right and left hands, and the sampling solution is added to each glove, and the gloves are secured above the wrist. All surfaces of the hand are massaged for 1 min, and samples are obtained aseptically for quantitative culture. The product must achieve a specified reduction after the first and tenth washes. The current efficacy criteria are as follows: a 2 $\log_{10}$ reduction of the indicator organism on each hand within 5 min after the first use, and a 3 $\log_{10}$ reduction of the indicator organism on each hand within 5 min after the tenth use.[73,74]

Ingredients Causing Safety Concerns

❶**Ingredients causing safety concerns regarding skin cleansing products include parabens, triclosan, SLS, and microbeads.** The findings and current statements on their safety are summarized here.

Parabens Parabens, including methylparaben, ethylparaben, propylparaben, and butylparaben, among other examples, are among the most commonly used and most effective preservatives for cosmetics and personal care products. Parabens are typically combined with each other and other types of preservatives to provide a broad-spectrum protection against microorganisms. The use of mixtures of parabens allows the use of lower individual levels and higher preservative efficacy.

Concerns have recurrently arisen regarding the safe use of parabens. A summary of the main findings is provided here.

- The Cosmetic Ingredient Review (CIR) Panel initially reviewed the safety of the most commonly used parabens (methyl-, ethyl-, propyl-, and butylparaben) in 1984.[75] It concluded that they were safe for use in cosmetics at levels up to 25%. (Note: typically, these ingredients are used in lesser than 1% in cosmetic products.)
- In the early 21st century, the CIR reexamined the safety of parabens due to increasing concerns regarding their safety. A study published in 2004 linked parabens to breast cancer, based on animal experiments.[76] It discussed this activity in context of the estrogen-like properties of parabens. The study had severe limitations and was rejected by most cancer research organizations, such as the American Cancer Society (ACS)[77] and the National Cancer Institute (NCI).[78] The weak estrogen-like activity of parabens has been known (estrogens are hormones naturally found in men and women, which can be associated with the development of certain forms of breast cancer). However, studies have shown that even the most potent paraben, butylparaben, produces an estrogen-like activity 10,000–100,000 times weaker than that of naturally produced estrogens.[79] In 2006, the CIR concluded that there was no need to change its original conclusion from 1984 that parabens are safe for use in cosmetics.
- The European Commission's Scientific Committee on Consumer Safety (SCCS) also investigated the safety of parabens. In 2011, it concluded that methylparaben and ethylparaben were safe at current levels (0.4% if used alone, or 0.8% if used in combination); however, it recommended that the levels of propylparaben and butylparaben (the two most potent parabens) should be reduced so that the sum of their individual concentrations does not exceed 0.19%.[80] As for other parabens, it stated that human risk cannot be evaluated due to lack of data.[80] In Denmark, the government introduced a ban on 15 March 2011, against the use of propyl- and butylparaben in cosmetic products aimed at children under 3 years old. The Danish ban triggered a new SCCS assessment, which led to a ban on parabens in leave-on products for the nappy area in children under 6 months old.[81]
- In 2011, the PCPC requested the CIR to reexamine the review on parabens in the light of the new SCCS opinions. In 2012, the CIR carefully reviewed the SCCS opinions and reaffirmed its earlier conclusions that parabens are safe for use as cosmetic ingredients.[82]

Although no studies have confirmed the potential risk of using parabens on human health, the claims that they can cause breast cancer and endocrine disruption have been widely spread, forcing cosmetic manufacturers to remove this ingredient from their formulations and substitute it with alternative preservatives.

Triclosan Triclosan is one of the most widely used antibacterial agents in a number of personal care products, including cleansing products, deodorant, and toothpaste. One of the concerns is that it is not clearly demonstrated that triclosan is clinically beneficial for patients. Studies have found that triclosan does not have any added benefit over non-antibacterial hand soaps in reducing infectious disease symptoms or bacterial counts on the hands.[83] Difference was seen only if the hands were washed for a longer time with soap containing relatively higher concentrations of triclosan.[84] A further concern is that triclosan may contribute to the development of antimicrobial resistance and even cross-resistance in bacteria. It would lead to severe issues and additional burden as well as high cost in healthcare settings.[85] Additionally, it may also have unanticipated hormonal effects. These conclusions only apply to consumer antibacterial products and not to those used in hospitals or other clinical areas.

Due to these facts, triclosan is currently under the radar of the FDA, which is reevaluating its safety and efficacy in antibacterial cleansing products. The FDA is also collaborating with other federal agencies, such as the EPA, to study the effects of this substance on animal and environmental health.

Sodium Lauryl Sulfate SLS is an anionic surfactant that is widely used in cosmetics, and personal care products, such as skin cleaning products, toothpastes, shampoos, and shaving foams. SLS is highly effective in removing oily stains and residues and has good foaming property. There were concerns with regard to the safe use of this ingredient. It has been known that it is a highly irritative ingredient, which may cause dermatitis (i.e., inflammation in the skin) by interfering with the SC and may also irritate the eyes and mucous membranes.[86] SLS is often used as a positive control in studies evaluating the irritation potential of various chemicals.[87] Additional concerns include its heavy deposition on the skin and in the hair follicles, which may lead to damage and even hair loss; potential penetration through the skin; its comedogenic potential; carcinogenicity; and damaging potential to the immune system. The safety of SLS has been reviewed numerous times by a number of governments, including the EPA, CIR Expert Panel (in 1983 and 2005), and FDA. The major findings are summarized here:

- Studies on rat skin found heavy deposition of SLS on the skin surface and in the hair follicles. Further, it has been reported that 1–5% SLS produced significant number of comedones in rabbits. These problems along with proven irritancy should be considered in the formulation of cosmetic products.[88]
- Animal studies did not show SLS to be carcinogenic; however, it has been shown that it causes severe epidermal changes to the area of the skin to which it was applied.[89] Currently, there is no direct or circumstantial evidence that this ingredient has any carcinogenic potential.[90]
- SLS can cause skin irritation in some users, which is considered its primary side effect. This effect is dependent on the level and duration of exposure; therefore, it is important to follow the label instructions when using a cleaning product. SLS appears to be safe when used for short periods of time (e.g., cleansing the skin) followed by thorough rinsing off the skin surface. In leave-on products, concentrations should not exceed 1%.[91]

Microbeads Microbeads are tiny particles used in skin cleansers, mainly in facial cleansers as exfoliators. As discussed earlier, they can be of both natural origin and synthetic origin, i.e., plastics. In recent years, the use of plastic microbeads has increased, which lead to environmental concerns. An average consumer now has a microplastic-containing product in their home and uses it on a daily or at least weekly basis. Once used in facial washing, microbeads travel through city wastewater systems, and due to their small size, they are likely to go through the preliminary treatment screens in wastewater plants without being filtered out and finally enter the oceans.[92] These particles contribute to pollution of the ocean by potentially impacting the marine food chains.[93] These facts may question the use and need of such particles in facial cleansers in the future.

Packaging of Skin Cleansing Products

The most commonly used packaging materials for skin cleansing products include the following:

- *Plastic Bottles*: The majority of skin cleanser products are supplied in plastic bottles with a screw top, disc top, or flip-flop cap. Some facial cleansers, body washes, liquid hand soaps, and hand sanitizers are packaged into plastic jars with a special pump head. This solution provides the user with a "single dose" of the product that is generally enough for one application. This way the application may be more comfortable as well as the contamination of the products may also decrease since the bottle is not opened on a daily basis. There is a container with a hook, especially designed for shower gels; it enables the product to hang in the shower stall.
- *Foam Pump*: Liquid hand soaps and hand sanitizers can be dispensed from special dispensers, which create foam from a solution at the time of dispensing the product. There are automatic dispensers where customers do not have to touch the equipment. These products have a motion sensor that detects the user's hand and automatically dispenses a dose of foam soap.
- *Glass Bottles*: Bath oils, bath beads, and bath salts can be supplied in glass bottles. A concern with using glass packaging is that it may be dangerous if broken in the bathtub or shower and users may cut themselves.
- *Paper*: Bath bombs are usually packaged into paper boxes or supplied in paper wrap. They can be packed in a simple to a very elegant style. Facial soaps and syndets are usually also supplied in paper wrap or paper box.
- *Facial Wipe Containers*: Most cleansing wipes, including facial cloths and hand sanitizer wipes, are either individually packaged into small sachets with an alumina interior or placed into larger soft sachets with a resealable flap. This way the moisture content of the products will not decrease during continuous use, and they are easily portable.

GLOSSARY OF TERMS FOR SECTION 2

Actuation force: A measure of the force required to release liquid cleansing products from pump-head dispensers.

Antiseptic: A chemical used to prevent infections by killing or inhibiting the growth of microorganisms in or on living tissue.

Bath product: A personal care product used in the bathtub that is designed to remove dirt, perspiration, and dead cells from the body skin as well as enhance the bathing experience, soften and moisten the skin, provide a relaxing experience, and leave the skin feeling clean and fresh.

Chemical cleansing: A method for cleaning the skin by emulsifying and dissolving dirt.

CMC: Critical micelle concentration is the concentration at which micelles first form in a surfactant solution.

Disinfectant: A chemical used on inanimate surfaces or objects to destroy or irreversibly inactivate infectious microorganisms.

Extrudability: A parameter of creams, lotions, and gels packaged into tubes, which refers to the ease of extruding a cream, lotion, or gel from a tube.

Facial cleanser: A personal care product that is designed to clean the facial skin, remove dirt and makeup, provide exfoliation, and remove potentially harmful microorganisms.

Firmness: It also refers to the consistency of a product and indicates how much a product can resist an external force (e.g., application).

Foamability: A measure of how much foam can be generated from a product.

Foaming cleanser: A cleanser that produces a high amount of foam during use. It contains a significant amount of well-foaming surfactants.

Hand cleanser: A personal care product that is designed to clean the hands.

Hand sanitizer: A personal care product used to remove microorganisms from the hands with the intent of preventing infections and reducing the spread of infectious diseases. It is classified as an OTC drug–cosmetic product in the US.

Hardness: A measure of the relative resistance of a solid product to chipping or breaking.

***In vitro* test:** A test performed in a test tube, Petri dish, or elsewhere outside a living organism.

***In vivo* test:** A test performed on a living organism.

Low-foaming cleanser: A cleanser that provides a moderate amount of foam during use. It contains a lower level of well-foaming surfactants and is milder to the skin.

Non-foaming cleanser: A cleanser that does not foam during use. It contains the lowest level of well-foaming surfactants as compared to foaming and low-foaming products.

pH: A number referring to the degree of acidity or basicity of a solution.

Physical cleansing: A method for cleansing the skin by physically removing dirt via abrasion.

Rancidification: A chemical decomposition of fats and oils, which can lead to the formation of unpleasant odors, compromised stability, and changes in appearance of emulsions.

Separation of emulsions: A quality issue of emulsions, which are thermodynamically unstable systems. On the long term, emulsions tend to separate. The main reversible changes include creaming, sedimentation, and flocculation, while irreversible changes include coalescence and Ostwald ripening.

Shower product: A personal care product used in a shower stall that is designed to remove dirt, perspiration, and dead cells from the body skin as well as enhance the bathing experience, soften and moisten the skin, provide a relaxing experience, and leave the skin feeling clean and fresh.

Soap: Amphiphilic compounds made of fatty acids and alkali solutions.

Spreadability: A measure of the consistency of a lotion, cream, or gel, which refers to the ease of spreading a product on the skin.

Syndet: Amphiphilic compounds made of synthetic surfactants.

Texture: Consistency of a product.

Viscosity: A measure of the internal friction of a fluid. In other words, it is the resistance to flow.

REVIEW QUESTIONS FOR SECTION 2

Multiple Choice Questions

1. Which of the following is the unwanted effect of skin cleansers?
   a) Can solubilize lipids in the epidermis
   b) Can extract NMF
   c) Can remain in the SC after rinsing
   d) All of the above

2. Non-foaming formulations do not foam due to the following:
   a) No surfactants in the formulation + high amount of emollients
   b) Low surfactant content in the formulation + high amount of emollients
   c) Low emollient content + high amount of surfactants
   d) No emollients + higher amount of surfactants

3. What is the main drawback of nonionic surfactants?
   a) They irritate the skin
   b) They do not foam well
   c) They are unstable
   d) They smell bad

4. Facial toners are generally beneficial for the following type of skin:
   a) Oily
   b) Dry
   c) Sensitive
   d) Aging
5. Which of the following is a main issue regarding alcohol-based hand sanitizers?
   a) Intentional consumption leading to toxicity
   b) Flammability
   c) Can dry out the skin
   d) All of the above
6. According to the CDC, which of the following is the most important infection control intervention?
   a) Washing the hands with water
   b) Washing the hands with soap and water
   c) Cleaning the hands with alcohol-based hand sanitizers
   d) Cleaning the hands with non-alcohol-based hand sanitizers
7. Facial cleansing cloths remove dirt from the skin via the following mechanism:
   a) Physical cleansing
   b) Chemical cleansing
   c) Both A and B
   d) None of the above
8. Facial gels remove dirt from the skin via the following mechanism:
   a) Physical cleansing
   b) Chemical cleansing
   c) Both A and B
   d) None of the above
9. Sodium chloride can thicken anionic surfactant–based products via the following mechanism:
   a) Reducing the repulsive forces between micelles and increasing micellar size
   b) Increasing the repulsive forces between micelles and increasing micellar size
   c) Reducing the repulsive forces between micelles and decreasing micellar size
   d) Increasing the repulsive forces between micelles and decreasing micellar size

10. Antiseptics are chemicals used to ___ and are regulated by the ___.

a) Inanimate surfaces/FDA
b) Kill germs on living tissues/FDA
c) Inanimate surfaces/EPA
d) Kill germs on living tissues/EPA

11. Order the following types of surfactants based on their irritation potential. Start with the most irritative.

a) Nonionic, anionic, amphoteric
b) Anionic, nonionic, amphoteric
c) Anionic, amphoteric, nonionic
d) Amphoteric, nonionic, anionic

12. Which are the major ingredients in facial scrubs with cosmetic claims?

a) Astringents
b) Absorbents
c) Abrasives
d) Antibacterial agents

13. Which of the following refers to the ease of applying a product to the skin?

a) Spreadability
b) Extrudability
c) Texture
d) Viscosity

14. Which of the following parameters is tested usually for bath bombs?

a) Spreadability
b) Viscosity
c) Disintegration time
d) Extrudability

15. What is the main safety issue with triclosan?

a) It may contribute to the development of antimicrobial resistance
b) It is not proven that it is more effective than simple soap and water
c) It may have negative hormonal effects
d) All of the above

Fact or Fiction?

________ a) Any soap that comes in the form of a bar is harsh to the skin.
________ b) Regular soap bars have a very alkaline pH.
________ c) 100% alcohol is not as effective in killing germs as 60–90% alcohol.
________ d) Microbeads found in skin cleansers pose an environmental risk.

Matching

Match the skin cleanser ingredients in column A with their appropriate ingredient category in column B.

Column A	Column B
_____ A. BHT	1. Abrasive
_____ B. Castor oil	2. Absorbent
_____ C. Cocamide DEA	3. Amphoteric surfactant
_____ D. Coco betaine	4. Anionic surfactant
_____ E. EDTA	5. Antibacterial agent
_____ F. Kaolin	6. Antioxidant
_____ G. Polyethylene bead	7. Astringent
_____ H. Sodium lauryl sulfate	8. Chelating agent
_____ I. Triclosan	9. Emollient
_____ J. Triethanolamine	10. Hydrophilic thickener
_____ K. Water	11. Nonionic surfactant
_____ L. Witch hazel	12. pH adjustor
_____ M. Xanthan gum	13. Solvent

REFERENCES

1. Toedt, J., Koza, D., Van Cleef-Toedt, K.: *Chemical Composition of Everyday Products*, Westport: Greenwood, 2005: 4.
2. Makela, C.: *Making Natural Milk Soap*, New York: Storey Publishing, 1999: 3–4.
3. The Soap and Detergent Association: Soaps and Detergents, 1994, Accessed 12/10/2013 at http://www.cleaninginstitute.org/assets/1/AssetManager/SoapsandDetergentsBook.pdf
4. Ertel, K.: Modern skin cleansers. *Dermatol Clin.* 2000;18:561–75.
5. Partick, B., Thompson, J.: *An Uncommon History of Common Things*, Washington: National Geographic, 2011: 192.
6. Rotter, M. L.: Semmelweis' sesquicentennial: a little noted anniversary of handwashing. *Curr Opin Inf Dis.* 1998;11:457–460.
7. Semmelweis, I.; Carter, K. C. (translator and extensive foreword) (1861). *Etiology, Concept and Prophylaxis of Childbed Fever*, Wisconsin: University of Wisconsin Press, 1983;100–126.
8. Ghosh, S., Hornby, S., Grove, G., et al.: Ranking of aqueous surfactant-humectant systems based on an analysis of in vitro and in vivo skin barrier perturbation measurements. *J Cosmet Sci.* 2007;58(6):599–620.
9. Downing, D. T., Abraham, W., Wegner, B. K., et al.: Partition of sodium dodecyl sulfate into stratum corneum lipid liposomes. *Arch Dermatol Res.* 1993;285(3):151–157.
10. Gloor, M., Wasik, B., Gehring, W., et al.: Cleansing, dehydrating, barrier-damaging and irritating hyperaemising effect of four detergent brands: comparative studies using standardised washing models. *Skin Res Technol.* 2004;10(1):1–9.

11. Leveque, J. L., de Rigal, J., Saint-Leger, D., et al.: How does sodium lauryl sulfate alter the skin barrier function in man? A multiparametric approach. *Skin Pharmacol.* 1993;6(2):111-115.
12. De Jongh, C. M., Jakasa, I., Verberk, M. M., et al.: Variation in barrier impairment and inflammation of human skin as determined by sodium lauryl sulphate penetration rate. *Brit J Dermatol.* 2006;154(4):651–657.
13. Denda, M.: Epidermal proliferative response induced by sodium dodecyl sulphate varies with environmental humidity. *Brit J Dermatol.* 2001;145(2):252–257.
14. Lips, A., Ananthapadmanabhan, K. P., Vethamuthu, M.: On skin protein–surfactant interactions. Paper presented at the Society of Cosmetic Chemists Meeting, Washington, DC, May 8–9, 2003.
15. Pierard, G. E., Goffin, V., Pierard-Franchimont, C.: Corneosurfametry: a predictive assessment of the interaction of personal care cleansing products with human stratum corneum. *Dermatology*. 1994:189:152–156.
16. Dominguez, J. G., Balaguer, F., Parra, J. L., et al.: The inhibitory effect of some amphoteric surfactants on the irritation potential of alkyl sulfates. *Int J Cosmet Sci.* 1981;3(2):57-68.
17. Ananthapadmanabhan, K. P., Lips, A., Vincent, C.: pH-induced alterations in stratum corneum properties. *Int J Cosmet Sci.* 2003:25:103–112.
18. Ananthapadmanabhan, K. P., Moore, D. J., Subramanyan, K., et al.: Cleansing without compromise: the impact of cleansers on the skin barrier and the technology of mild cleansing. *Dermatol Ther*. 2004;17:16–25.
19. Imokawa, G.: Surfactant Mildness, In: Rieger, M. M., Rhein, L. D., eds. *Surfactants in Cosmetics. Surfactant Science Series*, New York: Marcel Dekker, 1997: 427–471.
20. Nyström, B.: Review impact of handwashing on mortality in intensive care: examination of the evidence. *Infect Control Hosp Epidemiol.* 1994;15(7):435–436.
21. Rotter, M., Skopec, M.: Entwicklung der Händehygiene und die Bedeutung der Erkenntnisse von Ignaz Ph. Semmelweis, In: Kampf, G., ed.: *Hände-Hygiene im Gesundheitswesen*, Berlin: Springer-Verlag KG, 2003: 1–27.
22. Larson, E. L.: Skin hygiene and infection prevention: more of the same or different approaches? *Clin Infect Dis.* 1999;29:1287–1294.
23. Zafar, A. B.: Use of 0.3% triclosan (Bacti–Stat) to eradicate an outbreak of methicillin-resistant Staphylococcus aureus in a neonatal nursery. *Am J Infect Control.* 1995;23:200–208.
24. Schneir, A. B., Clark, R. F.: Death caused by ingestion of an ethanol-based hand sanitizer. *J Emerg Med.* 2013;45(3):358-60.
25. Hand Sanitizer Intoxication, Accessed 12/12/13 at http://www.medicine.virginia.edu/clinical/departments/emergency-medicine/medtox/education/toxtalks/Apr12-HandSanitizer.pdf
26. Vonghia, L., Leggio, L., Ferrulli, A.: Acute alcohol intoxication. *Eur J Intern Med.* 2008;19(8):561–567.
27. Gormley, N. J., Bronstein, A. C., Rasimas, J. J., et al.: The rising incidence of intentional ingestion of ethanol-containing hand sanitizers. *Crit Care Med.* 2012;40(1):290–294.
28. Hoffman, L., Subramanyan, K., Johnson, A. W., et al.: Benefits of an emollient body wash for patients with chronic winter dry skin. *Dermatol Ther*. 2008;21:416-421.
29. Abbas, S., Goldberg, J. W., Massaro, M.: Personal cleanser technology and clinical performance. *Dermatol Ther*. 2004;17:35–42.

30. Woollatt, E.: *The Manufacture of Soaps, Other Detergents and Glycerin.* West Sussex: Ellis Horwood Limited, 1985.
31. Hasenoehrl, E.: Facial Cleansers and Cleansing Cloths, In: Drealos, Z. D., ed.: *Cosmetic Dermatology*, Hoboken: Wiley Blackwell, 2010:99.
32. Sharko, P. T., Murahata, R. I., Leyden, J. L., et al.: Arm wash with instrumental evaluation—a sensitive technique for differentiating the irritation potential of personal washing products. *J Dermoclin Eval Soc.* 1991;2:19–27.
33. Breen, S.: *Formulation of all Natural Body Washes*, Michigan SCC Seminar, Grand Rapids, 2013.
34. Balzer, D., Weihrauch, M.: Colloids and surfaces. *Physicochem Eng Aspects.* 1995;99:233–246.
35. Penfield, K.: A look behind the salt curve: an examination of thickening mechanisms in shampoo formulations, *AIP Conference Proceedings* 2008;1027:899.
36. Butler, H.: *Poucher's Perfumes, Cosmetics and Soaps*, New York: Springer, 2000:117–123.
37. McKetta, J. J.: *Encyclopedia of Chemical Processing and Design*, Boca Raton: CRC Press, 1982:107.
38. FDA: Aromatherapy, Last update: 3/12/2009, Accessed 11/26/2013 at http://www.fda.gov/cosmetics/productandingredientsafety/productinformation/ucm127054.htm
39. Burton, M., Cobb, E., Donachie, P., et al.: The effect of handwashing with water or soap on bacterial contamination of hands. *Int J Environ Res Public Health.* 2011;8(1):97–104.
40. CDC: Handwashing: Clean Hands Save Lives, Last update: 1/11/2013, Accessed 11/27/13 at http://www.cdc.gov/handwashing/
41. Hegde, P. P., Andrade, A. T., Bhat, K.: Microbial contamination of "In use" bar soap in dental clinics. *Indian J Dent Res.* 2006;17:70.
42. CDC: Guideline for Disinfection and Sterilization in Healthcare Facilities,2008, Last update 12/29/09, Accessed 11/29/2013 at http://www.cdc.gov/hicpac/Disinfection_Sterilization/11_0regulatory_Framework.html
43. Cosmetics and Toiletries: Comparatively Speaking: US FDA Categories of Hand Sanitizer Actives, Last update: 9/23/2009, Accessed 11/30/2013 at http://www.cosmeticsandtoiletries.com/regulatory/region/northamerica/60657047.html?page=2
44. Perencevich, E. N., Wong, M. T., Harris, A. D.: National and regional assessment of the antibacterial soap market: a step toward determining the impact of prevalent antibacterial soaps. *Am J Infect Control.* 2001;29:281–283.
45. Aiello, A. E., Larson, E. L., Levy, S. B.: Consumer antibacterial soaps: effective or just risky?, *Clin Infect Dis.* 2007;45:S137–S147.
46. Larson, E. L., Morton, H. E.: Alcohols, In: Block S. S., ed.: *Disinfection, Sterilization and Preservation*, 4th Edition, Philadelphia: Lea & Febiger, 1991: 191–203.
47. Price, P. B.: Ethyl alcohol as a germicide. *Arch Surg.* 1939;38:528–542.
48. Harrington, C., Walker, H.: The germicidal action of alcohol. *Boston Med Surg J.* 1903;148:548–552.
49. Reynolds, S. A., Levy, F., Walker, E. S.: Hand sanitizer alert. *Emerg Infect Dis.* 2006;12(3):527-529.

50. FDA: Hand Sanitizers Carry Unproven Claims to Prevent MRSA Infections, Last update: 9/3/2013, Accessed 11/29/2013 at http://www.fda.gov/ForConsumers/ConsumerUpdates/ucm251816.htm
51. Charbonneau, D. L., Ponte, J. M., Kochanowski, B. A.: A method of assessing the efficacy of hand sanitizers: use of real soil encountered in the food service industry. *J Food Protect*. 2000;63(4):495–501.
52. Kjolen, H., Anderson, B. M. Handwashing and disinfection of heavily contaminated hands – effective or ineffective. *J Hosp Infect*. 1992;21:61–71.
53. Mbithi, J. N.: Comparative in vivo efficiencies of hand-washing agents against hepatitis A virus (HM-175) and Poliovirus Type 1 (Sabin). *Appl Environ Microbiol*. 1993;59(10):3463–3469.
54. Guideline for Hand Hygiene in Health-Care Settings. Recommendations of the Healthcare Infection Control Practices Advisory Committee and the HICPAC/SHEA/APIC/IDSA Hand Hygiene Task Force 2002, Accessed 11/30/2013 at http://www.cdc.gov/mmwr/PDF/rr/rr5116.pdf
55. CDC Features: Wash Your Hands, Last update: 2/14/2013, Accessed 11/30/2013 at http://www.cdc.gov/Features/HandWashing/
56. Clarys, P., Manou, I., Barel, A. O.: Influence of temperature on irritation in the hand/forearm immersion test. *Contact Dermatitis*. 1997;36:240–243.
57. Cox, D. R.: Fundamental emulsion science, Part 1. *H PC Today*. 2012;7(3):28-31.
58. Leal-Calderon, F.: Emulsified lipids: formulation and control of end-use properties. *OCL*. 2012;19(2):111–119.
59. Weiss, J., Canceliere, C., McClements, D. J.: Mass transport phenomena in oil-in-water emulsions containing surfactant micelles: Ostwald ripening. *Langmuir*. 2000;16:6833–6838.
60. De Paula, I. C.: Development of ointment formulations prepared with *Achyrocline satureioides* spray-dried extracts. *Drug Dev Ind Pharm*. 1998.24(3):235–241.
61. Stable Micro Systems, Accessed 1/14/2014 at http://www.stablemicrosystems.com/frameset.htm?http://www.stablemicrosystems.com/TextureAnalysisAttachments.htm
62. ASTM D1173 - 07 Standard Test Method for Foaming Properties of Surface-Active Agents, Accessed 1/2/2014 at http://www.astm.org/Standards/D1173.htm
63. Rosen, M. J., Solash, J.: Factors affecting initial foam height in the Ross-Miles foam test. *J Am Oil Chem Soc*. 1969;46(8):399–402.
64. Weaire, D. L., Hutzler, S.: *The Physics of Foams*, Oxford: Oxford University Press, 2001: 51.
65. Borole, S. N., Caneba, G. T.: Foaming characteristics of vinyl acetate-acrylic acid (VA-AA) copolymer-based surfactants. *J Funct Mater Res*. 2013;1(1):6–16.
66. KRÜSS: Dynamic Foam Analyzer, Accessed 11/9/2013 at http://www.kruss.de/products/foam-analysis/dfa100/dynamic-foam-analyzer-dfa100/
67. FDA: BAM – Microbiological Methods for Cosmetics, Last update: 4/8/2013, Accessed 4/13/2014 at http://www.fda.gov/Food/FoodScienceResearch/LaboratoryMethods/ucm073598.htm
68. Yablonski, J. I., Mancuso, S. E.: Preservative efficacy testing: accelerating the process. *Cosmet Toiletries*. 2007;122(10):51–62.
69. USP Antimicrobial effectiveness testing, Section 51, 28th Edition, Rockville: United States Pharmacopeial Convention, 2005.

70. Siegert, W.: Evaluation of the microbiological safety of finished cosmetic products, *Euro-Cosmetics*. 2010;3.
71. WHO Guidelines on Hand Hygiene in Health Care: First Global Patient Safety Challenge Clean Care Is Safer Care, Accessed 12/13/2013 at http://www.ncbi.nlm.nih.gov/books/NBK144013/pdf/TOC.pdf
72. Woolwine, F. S., Gerberding, J. L.: Effect of testing method on apparent activities of antiviral disinfectants and antiseptics. *Antimicrob Agents Chemother*. 1995;39:921–923.
73. FDA Tentative final monograph for healthcare antiseptic drug products; proposed rule. *Fed Reg*. 1994:31441–31452.
74. CDC: Boyce, J. M., Pittet, D.: Guideline for Hand Hygiene in HealthCare Settings, 2002;51(RR16);1–44, Accessed 12/2/2013 at http://www.cdc.gov/mmwr/preview/mmwrhtml/rr5116a1.htm
75. Elder, R. L.: Final report on the safety assessment of methylparaben, ethylparaben, propylparaben, and butlyparaben. *J Am Coll Toxicol*. 3:149–209.
76. Darbre, P. D., Aljarrah, A., Miller, W. R., et al.: Concentrations of parabens in human breast tumours. *J Appl Toxicol*. 2004;24(1):5-13.
77. American Cancer Society: Antiperspirants and Breast Cancer Risk, Last update: 9/20/2013, Accessed 8/12/14 at http://www.cancer.org/cancer/cancercauses/othercarcinogens/athome/antiperspirants-and-breast-cancer-risk
78. Mirick, D. K., Davis, S., Thomas, D. B.: Antiperspirant use and the risk of breast cancer. *J Natl Cancer Inst*. 2002;16;94(20):1578–1580.
79. Routledge, E. J., Parker, J., Odum, J., et al.: Some alkyl hydroxy benzoate preservatives (parabens) are estrogenic. *Toxicol Appl Pharmacol*. 1998;153(1):12–19.
80. SCCS: Opinion on Parabens SCCS/1348/10, 12/14/2010, Last revision: 3/22/2011, Accessed 8/12/14 at http://ec.europa.eu/health/scientific_committees/consumer_safety/docs/sccs_o_041.pdf
81. SCCS: Clarification on Opinion SCCS/1348/10 in the Light of the Danish Clause of Safeguard Banning the Use of Parabens in Cosmetic Products Intended for Children under Three Years of Age (SCCS/1446/11), 10/10/2011, Accessed 8/12/14 at http://ec.europa.eu/health/scientific_committees/consumer_safety/docs/sccs_o_069.pdf
82. CIR Expert Panel Meeting: Parabens, 2012. Accessed 8/12/14 at http://www.cir-safety.org/sites/default/files/paraben_build.pdf
83. Allisonn E., Aiello, A. E., Larson, E. L., et al.: Consumer antibacterial soaps: effective or just risky? *Clin Infect Dis*. 2007;45(2):S137–S147.
84. Larson, E., Aiello, A., Lee, L. V., et al.: Short- and long-term effects of handwashing with antimicrobial or plain soap in the community. *J Community Health*. 2003;28:139–50.
85. Dann, A. B., Hontela, A.: Triclosan: environmental exposure, toxicity and mechanisms of action. *J Appl Toxicol*. 2011;31(4):285–311.
86. Beradesca, E., Fideli, D., Gabba, P., et al.: Ranking of surfactant skin irritancy in vivo in man using the plastic occlusion stress test (POST). *Contact Dermatitis*. 1990;23;1–5.
87. Lee, C. H., Maibach, H. I.: The sodium lauryl sulfate model: an overview. *Contact Dermatitis*. 1995;33(1):1–7.
88. Final report on the safety assessment of sodium lauryl sulfate and ammonium lauryl sulfate. *Int J Toxicol*. 1983;2:127–181.
89. De Jongh, C. M., Verberk, M. M., Withagen, C. E., et al.: Stratum corneum cytokines and skin irritation response to sodium lauryl sulfate. *Contact Dermatitis*. 2006;54(6):325–333.

90. Soap and Detergent Association: Sodium Lauryl Sulfate, Accessed 12/2/2013 at http://www.cleaningproductfacts.com/sodium-lauryl-sulfate.htm
91. CIR: SLS, Accessed 12/2/2013 at http://www.cir-safety.org/sites/default/files/imports/alerts.pdf
92. Browne, M. A., Galloway, T., Thompson, R.: Microplastic – an emerging contaminant of potential concern? *Integr Environ Assess Manage.* 2007;3:559–561.
93. Fendall, L. S., Sewell, M. A.: Contributing to marine pollution by washing your face: Microplastics in facial cleansers. *Mar Pollut Bull.* 2009;58(8):1225–1228.

SECTION 3: SKIN MOISTURIZING PRODUCTS

LEARNING OBJECTIVES

Upon completion of this section, the reader will be able to

1. define the following terms:

Aesthetics	Barrier cream	Ceramides	Emollient
Humectant	Hygroscopic	Irritant dermatitis	Moisturizer efficacy
Moisturizer product	Non-invasive method	Occlusive	Oil-in-water emulsion
Skin hydration	Skin protectant	Skin rejuvenator	Skin surface roughness
Water-in-oil emulsion	Water-in-silicone emulsion		

2. briefly discuss whether skin moisturizer products are considered cosmetics or drugs in the US;
3. explain what the term "skin protectant" refers to;
4. explain why there are specific moisturizer products for face, body, and hands;
5. list a few beneficial effects skin moisturizers can have on the human skin;
6. briefly discuss the negative effects skin moisturizers can have on the human skin;
7. list the various required cosmetic qualities and characteristics that an ideal skin moisturizer should possess;
8. list the various required technical qualities and characteristics that an ideal skin moisturizer should possess;
9. briefly discuss which factors influence the elasticity of the stratum corneum;
10. explain what the term "moisturizer" refers to;

11. differentiate between humectants and emollients;
12. briefly discuss how the following ingredients moisturize skin: humectants, emollients, and occlusives;
13. provide examples for each of the following ingredients: humectants, emollients, and occlusives;
14. briefly explain what ceramides are;
15. list the major ingredient types of skin moisturizers, and provide some examples for each type;
16. briefly discuss the main product types that skin moisturizers are usually formulated into;
17. explain what barrier creams are and what they are used for;
18. list some typical quality issues that may occur during the formulation and/or use of skin moisturizers, and explain why they may occur;
19. list the typical quality parameters that are regularly tested for skin moisturizer products, and briefly describe their method of evaluation;
20. name the efficacy parameter that is generally tested for skin moisturizers, and briefly discuss the various ways for evaluating this parameter;
21. briefly explain why aesthetic properties are especially important in the case of skin moisturizers;
22. briefly discuss how aesthetic properties of skin moisturizers products are generally tested;
23. name some ingredients that can cause safety issues in skin moisturizers;
24. list the typical containers available for skin moisturizers.

KEY CONCEPTS

1. Moisturizers are designed to improve the skin quality, maintain and/or restore the moisture content of the SC as well as keep it smooth and pliable, and aid in alleviating the symptoms of dry skin.
2. The term "moisturizer" is a generic term used to describe ingredients that add moisture to the skin. Today, four major types of moisturizers are distinguished based on their physical and chemical properties and mechanism of action, including humectants, emollients, occlusives, and skin rejuvenators.
3. It has been shown that ceramides applied externally, in the form of moisturizers, can effectively reduce dry skin symptoms.
4. Skin moisturizers are typically used after proper skin cleansing. Most of these formulations are emulsions, including both low-viscosity lotions and higher viscosity creams.
5. Barrier creams are used to avoid unpleasant reactions and prevent irritant dermatitis.

6. The typical quality-related issues of skin moisturizer products include separation of emulsions, microbiological contamination, clumping, and rancidification.
7. Parameters frequently tested by cosmetic companies to evaluate the quality of skin moisturizers include the determination of the emulsion type; spreadability, extrudability, texture, and firmness of lotions, creams, and gels; actuation force; preservative efficacy; viscosity; and pH.
8. The most commonly tested performance parameters include the moisturizing effect and aesthetic properties of the products.
9. There is a critical balance between clinical measurements and consumer perception of performance. By most accounts, consumers perform an initial assessment of a moisturizer's performance based on how the product applies to their skin, how it smells, and how it feels on the skin, in addition to perceptions later in the day of relief from dry skin symptoms.
10. Ingredients causing safety concerns regarding skin moisturizing products include fragrances, lanolin, urea, propylene glycol, and herbal extracts.

Introduction

Today, skin care products form the largest sector of all cosmetics and personal care products, out of which skin moisturizer and protective and nourishing products have the highest market share. Moisturizers are used by people all over the world, including men and women. A study reports that 75% of the young population uses moisturizers daily.[1] Over the years, with the introduction of many new raw materials plus advances in surfactant/emulsion technology, products with good functionality and aesthetic appeal have been developed.

This section reviews the various types of skin moisturizers, their ingredients, main characteristics, formulation, and testing methods. In addition, it also provides a summary on how these products may affect the skin and what consumers' general requirements are.

Types and Definition of Skin Moisturizers

❶**Moisturizers are designed to improve the skin quality, maintain and/or restore the moisture content of the SC as well as keep it smooth and pliable, and aid in alleviating the symptoms of dry skin.** Such products beautify users and help them maintain and/or restore their skin's youthful appearance without initiating any changes in its function or structure. Therefore, most moisturizer products are considered cosmetics in the US. Certain products contain active ingredients to smooth out lines and wrinkles. They are considered drugs in the US. There are additional products available and marketed as cosmeceuticals, claiming to turn back time and make users look years younger. As discussed in Chapter 1, from a regulatory standpoint, there is no meaning of the word "cosmeceutical." If a product changes the structure or function of the skin, it is considered a drug even if a company wants to sell it as a cosmetic.

Some ingredients, including mineral oil, petrolatum, and dimethicone, are listed in the OTC monograph for skin protectants. The monograph specifies the ingredients that can be used with their individual concentration. If the concentration of the ingredients falls into the specified range, the formulations are considered OTC drugs in the US.

Moisturizers for the face, body, and hands do not differ significantly in their ingredients. However, many products on the market are specifically recommended for certain body parts. The main reason for that is that the hydration state of the skin may vary among different body areas. For example, the hands of a person working in a hospital may be extremely dry due to regular hand washing and the continuous use of hand sanitizers, whereas the hydration state of his/her other body parts may be normal. In addition, the sensitivity of different body parts to certain conditions may also differ. For example, facial skin typically is more prone to acne and rosacea than other body parts, or the face and neck are more often drier than the back since these areas are more exposed to the elements, such as sun and wind. For all these reasons, it is worth selecting moisturizers according to the skin's needs.

History of Using Skin Moisturizers

Moisturizing the skin has been viewed throughout history as a compliment to beauty and hygiene. Products have been found at ancient burial sites in Egypt[2] as well as in ancient Greece and Rome.[3,4] The purpose of applying different lipids, oils, waxes and butters, and animal fats on the facial skin and the overall body was primarily to protect the skin from the weather (e.g., wind and cold) as well as to prevent dry skin, soften the skin, and slow down the formation of wrinkles. The type of moisturizer used depended on the natural ingredients available in the various geographic locations. Popular ingredients used included avocado oil, palm oil, olive oil, sesame seed oil, black seed oil, neem seed oil, rosehip oil, and animal fats. Even the Bible mentions that that lotions made of olive oil and spices were used to maintain skin smoothness and suppleness.[5] Olive oil was a popular ingredient among the ancient Greeks. The ancient Olympics were conducted by men greased in olive oil. To prepare for training, an athlete in the ancient times would cover his body with olive oil and then dust it with fine sand. This combination was believed to help regulate their body temperature and provide protection from the sun.[6] Ancient Greek and Roman women used bread and milk as antiaging ingredients. Galen, a famous physician in ancient Rome, is credited with the formulation of the forerunner of cold cream (i.e., W/O emulsion). He added as much water as could be incorporated into a mixture of beeswax and olive oil. The product provided the skin with a cooling sensation as well as moisturization.[7] For many centuries, most products were made at home, using mainly natural-sourced ingredients, such as vanilla, cloves, lemon, honey, as well as oils, fats, and waxes.

Queen Elizabeth used a cowslip cream for her skin to keep herself young and prevent wrinkling.[8] The cream became very popular afterward.

In the 19th century, pharmacies and then industrial companies manufactured and sold raw ingredients for homemade products. Later, they started to manufacture complete formulations. One of the most commonly used ingredients today, petroleum jelly was discovered in the mid-1800s in Pennsylvania by a chemist called Chesebrough. He discovered that a gooey substance was causing problems to the oil rig workers since it stuck to the drilling rigs. He also noticed that this substance had the property to heal cuts and burns. After months of testing, he successfully extracted petroleum jelly in a usable form.[9] By 1870, Chesebrough was marketing his new product made of petroleum jelly, called Vaseline®. It has been used since then for treating dry skin; healing cuts, wounds, and burns; protecting the skin; and other reasons.

The first significant advancement of simple moisturizers occurred in the 19th century when emulsifiers were developed to create stable emulsions. Pond's introduced its first skin cream in the 1840s, which had soothing and reparative benefits of hazel extract.[10] Silicone oils were started to be used in cosmetics and personal care products during the 20th century; since then, the use of these materials has been increasing.[11] Today, there are hundreds of thousands of different types of moisturizers for various skin conditions; even combination products that fulfill both cleaning and moisturizing functions are available. Many products now use special ingredients for specific reasons, such as alpha hydroxy acids (AHAs) for antiaging claims.

How Skin Moisturizers May Affect the Skin?

The main reason for using skin moisturizers is to help maintain the skin's integrity and its barrier functions. As discussed previously, the interaction of surfactants in skin cleansing products with the skin can have severe consequences. Repetitive and excessive contact with cleansers and water, or low temperature, wind, and humid air, can be very drying and irritating to the skin. On the other hand, limited contact with milder cleansers followed by the use of skin moisturizers can help maintain the skin in good condition. The beneficial effects of using moisturizers are summarized as follows.

- Moisturizer products contain ingredients that can keep the skin **hydrated**, replace the lost NMF in the corneocytes, replenish the skin with intercellular lipids, and form a protective layer over the skin. The main benefit of such formulations is that they can keep all types of skin hydrated by varying the type and amount of ingredients used. Many consumers feel that applying products to their skin, especially to the legs, back, and chest, is pointless if no symptoms indicate they have dry skin. This is not true since all skin types need moisturization. Dryness is a stage when the skin visually appears dry. Consumers should not wait with applying moisturizers until dryness becomes visible.
- Today, concerns about the potential damaging effect of the **sun** and its short-term and long-term consequences, e.g., burning, aging, and skin cancer development, are increasing. Therefore, many facial moisturizers contain

sunscreens (similar to facial foundations), which can chemically or physically filter the UV radiation reaching the skin and prevent these negative effects.

- More and more moisturizer products contain ingredients that are claimed to fight **aging**. One of the most frequently used ingredients is AHAs. They were the first very effective anti-aging ingredients introduced to the market in the late 20th century. Additional ingredients are discussed in Section 4 of this chapter.
- **Cellulite** is fat that collects in pockets under the skin surface. It forms around the hips, thighs, abdomen, and buttocks. The skin with cellulite is often described as one with "orange peel" or "cottage cheese" appearance.[12] Cellulite deposits cause the skin look dimpled.[13] Cellulite is not a serious medical condition; however, it is an aesthetically unacceptable cosmetic problem for most women.[14] It can be unsightly and may cause social and psychological problems. Today, there are specialty moisturizer formulations containing various ingredients known to be beneficial in improving cellulite.
- **Acne** often causes concerns to patients, and many consumers categorize themselves to have acne-prone skin (i.e., more susceptible to having acne than others). Many moisturizer ingredients have been categorized to be comedogenic (i.e., clog the pores in the skin) and, therefore, are not recommended by dermatologists. However, more and more ingredients thought to be comedogenic are demonstrated not to be comedogenic but beneficial for the skin. One of these ingredients is mineral oil, which has been known as one of the most irritating ingredients for years. However, animal and human studies have demonstrated that it does not to cause comedo formation (see more detail in Section 3 of Chapter 4). Another ingredient that was thought to cause acne problems is lauric acid, which can be found in high amounts in coconut oil. Recent studies indicate that this ingredient has antibacterial properties and potential to be used in antiacne products.[15]

Although moisturizers may strengthen the skin barrier function, some may also weaken it. Moisturizers are generally considered safe; however, skin reactions, such as allergic contact dermatitis, may occur.[16] Individuals with impaired barrier function are especially at risk for adverse reactions.

- Exposure of mildly irritating preparations to sensitive areas may lead to **irritation**. Facial skin has been found to be more sensitive to moisturizers than other body parts. The reason for this is a less efficient barrier, a decreased number of SC layers, and larger follicular pores in the facial skin. It has been shown that fragrances are the main irritating components in skin care products. Additional highly irritating ingredients include parabens, vitamin E, essential oils, and lanolin, among others.[17]
- AHAs are widely used ingredients in cosmetics. A concern with regard to the use of such formulations is that they **sensitize** the skin to the damaging effects of UV light. This condition is called photosensitivity. It has been shown, however, that photosensitivity is reversed within a week of terminating treatment with

products containing AHAs. It is advisable, therefore, to use daily sun protection for up to 1 week after discontinuing the use of AHA-containing products.[18] In addition, AHAs can cause mild skin irritation, redness, swelling, itching, and skin discoloration.[19]

Required Qualities and Characteristics and Consumer Needs

From a consumer perspective, a quality skin moisturizer should possess the following characteristics:

- Neutral or pleasant odor and color
- Easy to spread and a pleasant feeling during application
- Non-oily/ non-greasy after application
- Non-comedogenic
- Provides effective hydration and prevent transepidermal water loss
- Provides protection from environmental factors, for example, wind, cold temperature, UV light
- Reduces dryness, improve dull appearance
- Smoothens and softens the skin
- Well-tolerated and non-allergenic

The technical qualities of skin moisturizers can be summarized as follows:

- Long-term stability
- No microbiological contamination and growth
- Appropriate rheological properties
- Smooth texture
- Appropriate performance
- Dermatological safety

Typical Ingredients and Formulation of Skin Moisturizers

It is known that the elasticity of SC is dependent on a proper balance of lipids, NMF, and water, in conjunction with its keratin proteins. Water is a plasticizer for keratin, allowing the SC to bend and stretch, avoiding cracking and fissuring. It has been shown that corneocytes that possess the highest concentration of NMF retain more water and appear more swollen.[20] Furthermore, water increases the activity of enzymes involved in the desquamation process. The capacity of the SC to hold onto water is greatly influenced by the NMF in the corneocytes as well as the SC's lipids.

When the skin is cleaned, lipids covering the skin surface are partially or totally removed, which exposes the SC to potential damage. As discussed earlier, surfactants may penetrate the SC and remain there, causing further problems. It is, therefore, important to apply moisturizers as part of the skin care regimen. Although people may think that only dry skin needs regular moisturization, it is not true. Consumers with oily skin often tend to use products with harsher surfactants and toners to effectively remove oil from their face. In addition, they may believe that applying additional moisturizers make their skin even oilier; therefore, they would often skip the moisturization step. Over time, this regimen can lead to tightness and drying. In the light of these, it is important to keep in mind that all types of skin need moisturization to be able to maintain its barrier functions. Of course, the type of moisturizing ingredients and products should be selected based on the individual needs, similar to cleansing products.

The term "moisture" means water or other liquid. ●**The term "moisturizer" is a generic term used to describe ingredients that add moisture to the skin. Today, four major types of moisturizers are distinguished based on their physical and chemical properties and mechanism of action, including humectants, emollients, occlusives, and skin rejuvenators.** Unfortunately, these terms are often used interchangeably. However, each of these materials provides different benefits to the skin, as discussed in the following sections.

Humectants Humectants are hygroscopic ingredients that can increase the water content of the top layer of the skin by enhancing water absorption from the dermis into the epidermis (see Figure 3.16). Some also think that humectants can hydrate the SC by absorbing water from the external environment.[21] Humectants also allow the skin to feel smoother by filling the holes in the SC through swelling.[22,23]

These ingredients serve to replace the skin NMF that has been washed away or otherwise depleted. Humectants act in the same way as NMF, and indeed, some of the humectants commonly used in moisturizers are components of the skin NMF, e.g., lactic acid and urea. The key functionality of a humectant is to form hydrogen bonds with molecules of water. Examples for humectants include glycerin, AHAs (e.g., lactic acid, glycolic acid), pyrrolidone carboxylic acid, propylene glycol, urea, hyaluronic acid, and sorbitol. The most common humectant used in moisturizers is glycerin. It acts on several different parameters, which makes it the gold standard. It is an effective moisturizer, accelerates the maturation of corneocytes, reduces dryness, and enhances the cohesiveness of intercellular lipids. Overall, when combined with occlusive agents, glycerin has the ability to produce significant moisturizing effects in the skin.[24]

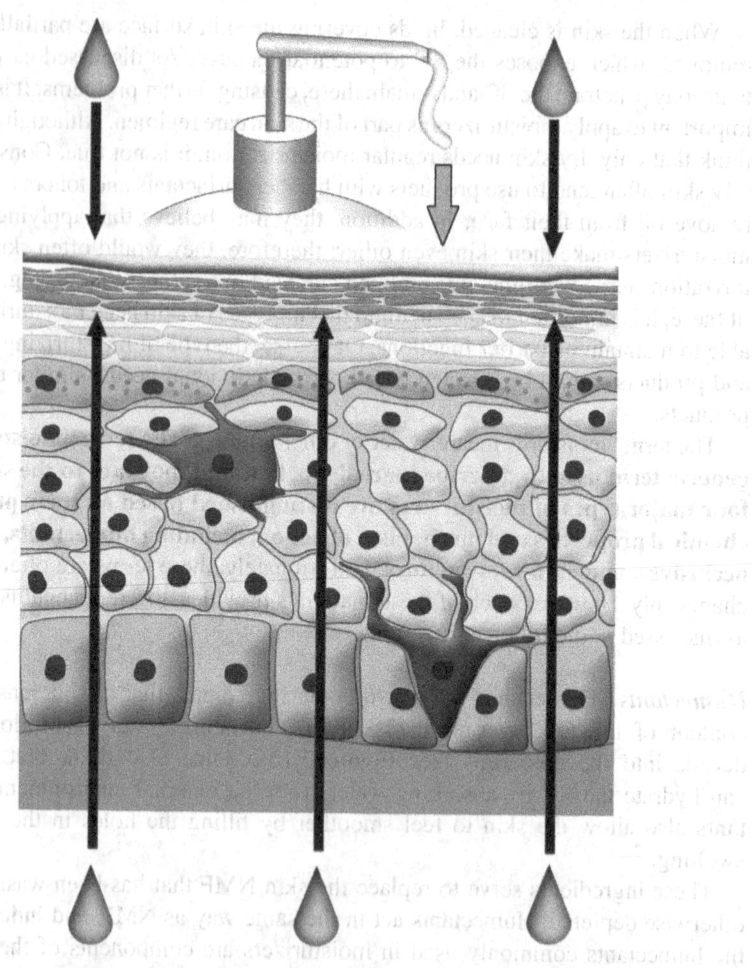

Figure 3.16 Working principle of humectants.

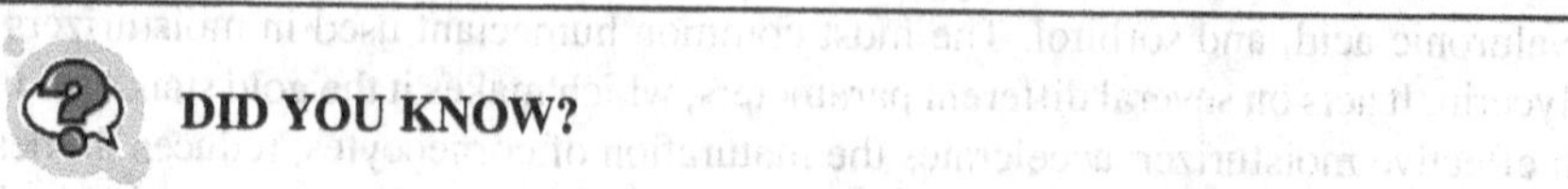

DID YOU KNOW?

Since humectants can enhance water absorption upward from the dermis, these compounds can cause excessive water loss from the dermis through evaporation into the environment if humidity is low. This may be of special concern if the skin barrier is compromised. Due to this phenomenon, moisturizer formulations often contain a mixture of humectants and occlusives that prevent water loss.[24]

Emollients Emollients are designed to plasticize, soften and smooth the skin, usually by filling in the void spaces between the corneocytes and replacing the lost lipids in the SC (see Figure 3.17). Emollients can also provide protection and lubrication on the skin surface to minimize chafing and enhance the skin's aesthetic smoothness and softness.

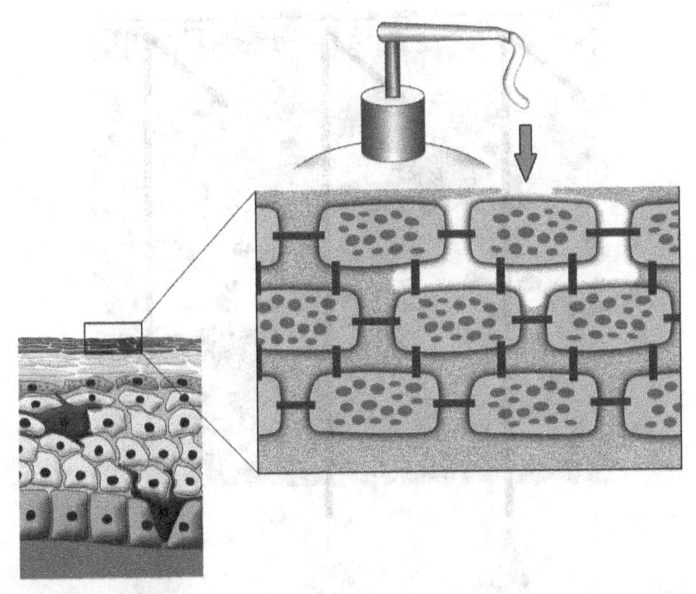

Figure 3.17 Working principle of emollients.

The most popular emollients are based on hydrocarbons, such as mineral oil and petrolatum and their derivatives; fatty acids, such as stearic acid, linoleic acid, and lauric acid and their derivatives, such as alcohols and esters; vegetable oils, such as almond oil; synthetic triglycerides; silicones; waxes, such as beeswax, carnauba wax, polyethylene wax, and cetyl alcohol; lanolin derivatives; and polymers. Essential fatty acids (i.e., C_{18} unsaturated linoleic and alpha-linoleic acids) influence the skin physiology via their effects on skin barrier functions, eicosanoid production, membrane fluidity, and cell signaling.[25]

Occlusives Occlusive agents create a hydrophobic barrier to physically block TEWL from the SC (see Figure 3.18). In moisturizer formulations, occlusives complement the water-attracting nature of humectants. As they prevent water evaporation from the skin, they can be particularly effective in the treatment of dry skin, which is already damaged. They may have additional emollient effects.

Although occlusives are not the most appealing ingredients to most consumers since they are sticky, not easy to remove, and may leave the skin with a greasy feeling, they are very effective in reducing TEWL. The most commonly used ingredient is

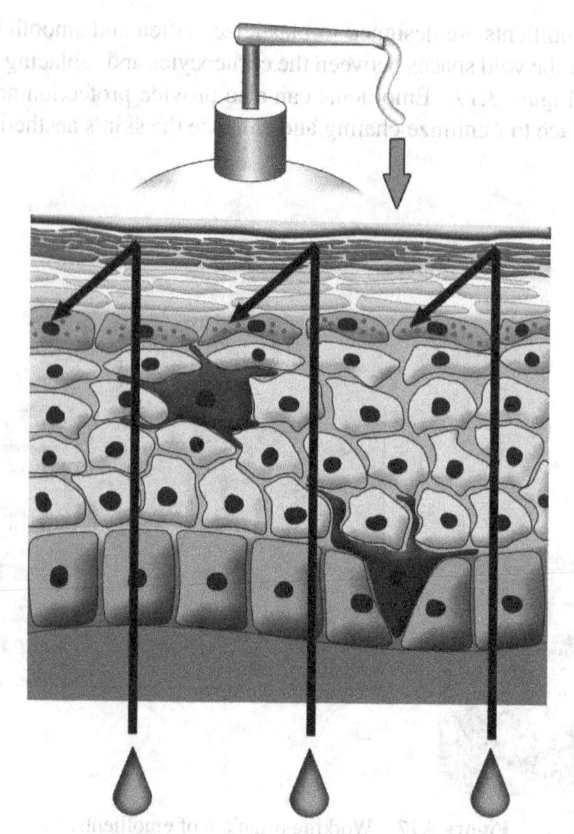

Figure 3.18 Working principle of occlusives.

petrolatum. In addition to forming an impermeable layer on the skin, it can penetrate into the skin's upper layers and initiate the production of intercellular lipids.[26] Lanolin was also very popular in the past; however, its use diminished as it is a known irritant ingredient and has an unpleasant odor. Today, there are newer occlusives that have cosmetically more appealing characteristics. These include silicone derivatives, such as dimethicone (polydimethylsiloxane). They further enhance the aesthetic quality of the formulation by imparting a "dry touch." They dry very quickly, are easy to apply and remove, do not block the pores on the skin (non-comedogenic), and are able to provide nice and shiny products. Additional examples for occlusives include vegetable oils, fatty acids, fatty alcohols, waxes, and cholesterol.

Skin Rejuvenators Skin rejuvenators, otherwise known as enhancers of the skin barrier, are claimed to restore, protect, and enhance the skin's barrier function, thereby reinforcing skin hydration. This is the newest class among moisturizer ingredients. This category includes proteins, primarily skin proteins, such as keratin, elastin, and collagen. Proteins may provide temporary relief from dry skin by filling in the irregularities in the SC. When they dry on the skin, they slightly shrink, leaving a protein film that appears to smooth the skin and stretch out some of the fine wrinkles.

Ceramides Ceramides can be naturally found in the SC. They are the most important structural elements of the intercellular lipids, which are necessary to link the protein-rich corneocytes into a waterproof barrier that is capable of protecting the underlying skin tissues and regulating body homeostasis.[27] Ceramides are orderly arranged in lamellar form to act as a membrane and fill the intercellular space in the SC. They function to help maintain the integrity of the skin barrier. There are certain diseases when lipid metabolism in the skin is abnormal, causing a deficiency of ceramide and NMF and impairment of the epidermal barrier function.[27] ●**It has been shown that ceramides applied externally, in the form of moisturizers, can effectively reduce dry skin symptoms.** These lipid molecules have been used increasingly in recent years in the treatment of dry skin and in cosmeceuticals.[28]

Additional Ingredients in Skin Moisturizers If we take a look at the products that can be found on the market, we can hardly find two similar compositions. This is not surprising since formulators can choose from numerous types of raw materials with varying characteristics. The actual composition always depends on various factors, including the application time of the products (day vs night), target age group, target skin type, product type, compatibility issues, aesthetic needs, formulation cost constraints, packaging needs, product claims, safety, whether makeup will be applied over the product, and others. Despite the large number of different formulations, the general ingredient types used are quite similar. Now we will review the major ingredient types, their characteristics and functions, and some examples for them.

- **Emulsifiers** are an essential part of the formulations since the majority of moisturizing, nourishing, and protective formulations are emulsions. Thousands of emulsifying agents are available on the world market today. Emulsifier selection is not only crucial for the stability of an emulsion but also has also a large impact on consistency and viscosity, skin feel, color, odor, and care properties of the final formulation. Generally, different types of emulsifiers are used in formulations to best tailor the formulation's properties to consumers' needs. The most commonly used types include nonionic and polymeric surfactants. Cationic emulsifiers can also be used; they have additional conditioning properties.
- **Thickening agents**, otherwise known as texturizing agents, play an important role in the skin feel of emulsions. They contribute to the stability as well as appropriate rheological property of the formulations.
  - Examples for thickeners commonly used in skin moisturizing products include hydrophilic ingredients, such as gums (e.g., xanthan gum), cellulose derivatives (e.g., hydroxyethyl cellulose), and acrylic polymers, among others, as well as liposoluble ingredients, such as waxes (e.g., cetyl alcohol); many emollients used may also have additional thickening properties. Certain nonionic emulsifiers may also have additional thickening effects.

- **Water** is a basic component of emulsions.
- **Preservatives** inhibit bacterial growth in the formulation and prevent microbiological deterioration.
  - Examples include parabens and methylisothiazolinone.
- **Antioxidants**: Fats, oils, and butters are sensitive to oxidative processes, which may trigger deterioration of these components, leading to color change, odor formation, and stability problems. Antioxidants prevent the oxidization of sensitive components and thus rancidity.
  - Examples include BHT and BHA.
- **Fragrances** improve the overall aesthetic qualities of moisturizer products, which may be important to consumers, especially for formulations applied to the face. These components may be employed to mask the natural taste of raw ingredients without adding any characteristic smell or to provide a pleasant scent.
- **Sunscreens**: Some facial formulations have additional benefits over moisturization, including sun protection. Sunscreens prevent the skin from the harmful radiation of sunlight. Due to the increasing incidence and mortality rates of skin cancer, sunscreen use as a daily protecting agent has become more important to consumers. See more detail on sunscreens in Section 5 of this chapter.
- **Coloring Agents**: Most formulations do not contain coloring agents, but if they do, they contain soft colors. The reason for this is that, on the one hand, if the formulations contained vivid and strong colors, they could leave a visible color on the skin, which is obviously undesired. On the other hand, no one would be familiar with applying strong colors (such as orange or sea blue) with the aim of moisturizing and nourishing the facial skin. However, tinted moisturizers, such as BB creams, are also available today, which are colored products (find more detail in Chapter 4).
- **Aesthetic Agents**: Pearlescent pigments, soft-feeling agents, may also be added to the products to make them unique and enhance their appearance. Formulations can also contain opacifying agents, which help cover up the formulations' color and provide a uniform appearance. The most frequently used opacifiers are titanium dioxide and glycol stearate.
- **Electrolytes**: W/Si emulsions usually contain electrolytes, such as sodium citrate, magnesium sulfate, sodium chloride, or sodium tetraborate, in approximately 1–2% concentration to improve the formulations' stability.
- **Functional Ingredients**: Many other ingredients may be incorporated into moisturizer products, which can have endless functions, including moisturizing, nourishing, protecting, improving the skin's structure, softening, vitalizing, anti-inflammatory, and others. The main types of ingredients include the following:

- **Natural Additives**: During the past several years, the demand for natural products has increased. Thus, more manufacturers have incorporated ingredients with natural origin in their formulations. Examples include fruit and vegetable extracts, vitamins, and amino acids.
- **Vitamins**: In addition to the positive nutritional effect of vitamins, there is a growing interest in their topical application. The most commonly used vitamins include E, A, and panthenol. They have an antioxidant effect, improve the skin tone, and reduce wrinkles.
- **Peptides and Proteins**: Protein derivatives such as hydrolyzed proteins, amino acids, and proteoglycans are very substantive to the skin, leaving it feeling soft and smooth. The most popular types include collagen, hyaluronic acid, milk proteins, silk proteins, and amino acids.
- **Essential Fatty Acids**: 7-Linolenic acid is an essential fatty acid, which occurs naturally in many vegetable seed oils. It improves the skin's efficiency as a barrier to TEWL. In addition, it is thought to be incorporated into the skin's structural lipids, thus increasing the suppleness and flexibility of the epidermis.
- **Hydroxy Acids**: The use of AHA and, more recently, β-hydroxy and α-keto acids in the treatment of aging skin has attracted much interest over the past decade. See more detail on the use of AHAs in Section 5 of this chapter.
- **β-Glucans**: They are polysaccharide materials derived from natural oats, wheat, and baker's yeast. They are used in medicine in the management of wound healing and aging.

Product Types As discussed earlier, skin cleansing products can contain moisturizing ingredients, which are then deposited on the skin. When using moisturizing cleansers, taking a shower may serve two functions, i.e., cleaning and moisturizing. However, it should be kept in mind that part of the moisturizers is washed away during rinsing. For this reason, it is believed that this type of moisturization is not as effective as using moisturizing products after cleaning the skin. Non-foaming skin cleansers offer a higher amount of emollients deposited on the skin surface. Moisturizing cleansing wipes also offer a pleasant way to clean and moisturize the skin at the same time.

●**Skin moisturizers are typically used after proper skin cleansing. Most of these formulations are emulsions, including both low-viscosity lotions and higher viscosity creams.** Droplet size is usually, but not necessarily, large enough to interfere with the path of light, and thus, emulsions are usually white (these are referred to as macroemulsions). Special cases exist where the particle size is so small that the liquid is clear. These are special types of emulsion, known as microemulsions and/or **nanoemulsions**.

- Most commonly, facial moisturizers are O/W and water-in-silicone (W/Si) emulsions. In O/W emulsions, water in the external phase of the emulsion helps hydrate the SC of the skin. This is desirable when one desires to incorporate water-soluble active ingredients in the vehicle. These emulsions are easy to apply and absorb quickly due to the quick evaporation of the external water phase. In addition, due to their relatively high water content, the O/W emollients exert a cooling effect as water evaporates following topical application. W/Si emulsions are desired for customers with oily and sensitive skin. Silicone is known to be noncomedogenic, nonacnegenic, and hypoallergenic. These formulations go on smoothly, dry fast, and they leave the skin with a smooth feeling.

- Products for the hand and body are more frequently water-in-oil (W/O) emulsions. In these emulsions, the oil phase forms the external phase, which is deposited on the skin surface after the water evaporates. They are very beneficial for consumers with dry skin; however, they may feel slightly greasy. Silicones may also be incorporated into the oil phase to improve its aesthetic properties.
- Recently, there has been a growing interest in alternative emulsions, including water-in-oil-in-water (W/O/W) emulsions, also referred to as multiple emulsions. Their benefit is the claimed sustained release of entrapped materials in the internal phase and the potential for using incompatible ingredients in the same formulation.

Additional product types include ointments, oils, and gels. Gels provide a rapid, cooling watery feel to the skin and are commonly used in body massage products and cellulite products.

A specialty product type is the **barrier creams**. Chronic exposure to even mild irritants, including antiseptic and cleansing products, insult the skin. **Barrier creams are used to avoid unpleasant reactions and prevent irritant dermatitis**, i.e., inflammation of the skin caused by irritants.[29] They are also often referred to as "skin protective creams." In practice, their use still remains subject to a lively debate. In general, there is a lack of evidence supporting their efficacy. Some authors even suggest that inappropriate barrier cream applications are prone to induce additional irritation rather than providing benefit.[30,31]

- Products aimed to prevent aqueous solutions are typically W/O emulsion. They are usually heavier in occlusives and emollients than regular moisturizers and serve as protective formulations. Preparations that exhibit water repellency may be based on petroleum jelly, lanolin, and silicones, such as dimethicone.
- For oil protection, non-oil-soluble film-formers such as gum acacia, tragacanth, or sodium alginate can be used. Glycerin can be utilized as a plasticizer to the gum film.

As mentioned previously, these products are considered drugs in the US since they prevent and/or treat irritation and related conditions.

Formulation of Skin Moisturizers The formulation procedure for moisturizers is emulsification in the majority of the cases. Emulsifiers are generally dispersed in the external phase of the emulsion. Water-soluble components are mixed and/or dissolved in the water phase, while oil-soluble components are mixed and/or dissolved in the oil phase. Both phases are heated, and when both are at the same temperature, the phases are mixed; typically, the internal phase is added slowly to the external phase using intensive mixing. After the mixture has cooled down and the emulsification process is complete, temperature-sensitive ingredients can be added with preservatives and fragrances.

In the case of multiple emulsions, the process is somewhat different. When preparing a W/O/W emulsion, first the W/O primary emulsion is created by combining the water phase and the oily phase that contains the lipophylic emulsifier. Next, the hydrophilic emulsifier is added to the remaining amount of water. This water phase is combined with the W/O primary emulsion at either room or higher temperature with continuous mixing to form the W/O/W multiple emulsion.

Considerations When Selecting Skin Moisturizers

Similar to skin cleansing products, a variety of skin moisturizers are available today. Skin moisturizers have different texture, speed of absorption, oil/water content, emollient properties, irritating potential based on the ingredient used, and dosage forms formulated. Therefore, they can impact the skin in a number of ways and produce a range of skin effects, similar to skin cleansers. To choose the most appropriate moisturizer, consumers should consider their skin type on the particular body part, such as dry, oily, or normal; then the skin problems and any skin allergies should also be taken into account. Based on the individual differences in the skin type and sensitivity, moisturizer selection is somewhat subjective in most cases.

Here is a brief summary of the general properties of skin moisturizing products that should be considered when selecting such products.

- To maintain the moisture content of normal skin, formulations with humectants and emollients work very well. They provide a light, non-greasy feel.
- Oily skin is often prone to breakouts and acne. Although it is oily, it still needs moisturizers, especially after using skin care products that remove oils and dry out the skin. Primarily, water-based formulations, such as lotions and gels, work the best, and they do not contain and deposit too much oil on the skin. In addition, oily skin consumers should look for non-comedogenic formulations, which are proven not to clog the pores.
- For dry skin, products with humectants only will not be enough. Dry skin users are recommended to use products that contain both humectants and occlusives, which can also be combined with emollients. They may leave the skin with a greasy feeling due to the occlusive content; however, these are the most efficient ingredients in blocking water loss through the skin.
- Fragrances may irritate and sensitize the skin; therefore, consumers with sensitive skin should avoid using products with fragrances.

- Although parabens (which are well-known and widely used preservatives) may be irritating to the skin, formulations without preservatives are more dangerous to the users. Most moisturizers are water-based products, and with all the additional ingredients, they provide a favorable environment for bacteria, yeast, and molds to growth. Users dip their fingers into the jar every day (or even twice daily if they used moisturizers in the morning and evening as well), and with all the microorganisms present on the hands, it is quite easy to contaminate products. If no preservatives are present, microorganisms may start growing in formulations within weeks or even days. In addition, parabens have been known to be one of the most effective preservatives; therefore, consumers should not worry about them.

Typical Quality Issues of Skin Moisturizer Formulations

●The typical quality-related issues of skin moisturizers products include separation of emulsions, microbiological contamination, clumping, and rancidification. Since these problems were revised in the previous section, they are not discussed here.

Evaluation of Skin Moisturizing Products

Quality Parameters Generally Tested **●Parameters frequently tested by cosmetic companies to evaluate the quality of skin moisturizers include the determination of the emulsion type; spreadability, extrudability, texture, and firmness of lotions, creams, and gels; actuation force; preservative efficacy; viscosity; and pH.** Tests that were revised in the previous section are not reviewed in detail here.

Determination of the Emulsion Type Various types of tests can be employed to determine the type of emulsion formulated. Here, a few of the most commonly used tests are discussed.

- **Conductivity Measurement**. The basic principle of this test is that water is a good conductor of electricity. In this test, an assembly is used in which a pair of electrodes connected to an electric bulb is dipped into an emulsion. If the emulsion is O/W type, the electric bulb glows, indicating that water is the outer phase. For W/O emulsions, the bulb will not glow.
- **Solubility**: By using either oil-soluble or water-soluble dyes, one of the phases of an emulsion can be stained and then it can be checked under a microscope. As an example, imagine using a poorly water-soluble red dye, such as Sudan III. It is added to the preparation and stirred until homogenous. If we can see red cells in a clear field under the microscope, it means that we have an O/W emulsion since the dye stained the oil phase, which is the internal phase in this case. On the other hand, if we see a red field with clear globules, we have a W/O emulsion.

- **Dispersion by Dilution.** Emulsions can be diluted with their external phase; however, addition of the internal phase leads to instability and separation of the phases. If the emulsion is O/W and it is diluted with water, it will remain stable and dispersed since water is the dispersion medium. However, if it is diluted with oil, the emulsion will separate.

Efficacy (Performance) Parameters Generally Tested **●The most commonly tested performance parameters include the moisturizing effect and aesthetic properties of the products.**

Moisturizing Effect As discussed earlier, the main goal of skin moisturizers is to add moisture to the skin as well as help maintain, restore, and enhance the skin's barrier function and its natural reparative processes. It is essential in the cosmetic product development process to evaluate the product's performance, i.e., whether it effectively moisturizes the skin or not. There are various ways to objectively assess the skin's barrier function of the skin based on the SC integrity. Most commonly, noninvasive methods are used as part of clinical testing. The choice of subjects, skin properties, ethnicity, age, sex, disease, and drug ingestion are important factors to be taken into consideration when performing a clinical study. Usually, inclusion and exclusion criteria are set up, that help objectively select the subjects for the study.

DID YOU KNOW?

Clinical studies usually fall into two major categories: proof-of-concept trials and full clinical trials. Proof-of-concept trials are generally small studies (10–15 subjects) aimed to determine whether the study protocol is feasible, see which measurement techniques will be used, and see if products show any effects. Full clinical studies are larger and more complex and provide opportunity for head-to-head comparisons.[32]

Measurements are typically performed by experts before applying any moisturizer products (this is referred to as the "baseline data"), and after using the product. The before and after values are compared, and the results provide a good estimate concerning the moisturizing efficacy of the product. In addition, subjects have to perform self-grading and fill in questionnaires. The functional, structural, and beauty parameters generally measured in the cosmetic industry to evaluate moisturizer products' performance include the following:

- Skin's water content (hydration) and rate of water loss (a measure of barrier integrity)

- Skin Surface Properties: roughness, friction, scaling, sebum, and elasticity
- Skin pH
- Skin color (for skin lightening, darkening, and inflammation)

Since moisturizers primarily modify the skin's water content and surface properties, the evaluation of these parameters are discussed in more detail here.

- **Skin Hydration**: Methods evaluating the SC's water content are very popular today. The two commonly used techniques include the conductance and capacitance methods.
  - **Conductance Method:** This method is based on the changes in the electrical properties of the SC. It is known that water is a conductor of electricity; therefore, the amount of current conducted in the skin is directly related to its water content. Dry SC has weak electrical conduction, whereas hydrated SC is more sensitive to the electric field, inducing an increase in the dielectric constant.
  - **Capacitance Method**: Corneometry is an example for such methods.[33] The instrument (Corneometer®) used for this test consists of a probe that is placed onto a hair-free skin surface with slight pressure. It determines the capacitance of the skin due to its behavior as a dielectric medium.[34] The measurement can detect even the slightest changes in hydration level. A disadvantage of this method is that salts or ions on the skin surface can affect the reading.
- **TEWL**: TEWL is one of the most commonly measured skin parameters and widely considered an indicator of the skin's barrier function. A damaged SC allows water to evaporate, resulting in high TEWL readings. The device measures the amount of water vapor leaving the skin. Open chamber and closed chamber devices may be used to measure TEWL.
  - **Open chamber** devices (such as Evaporimeter® and Tewameter®)[35] are open to the atmosphere. They contain two water sensors. When the probe is placed on the skin surface in a stable ambient environment, the sensors are able to determine the vapor pressure gradient above the skin surface. This gradient is translated into the amount of water leaving the SC per hour and square meter of the skin surface (g of water/h m^2). The value is an estimate of the skin's ability to retain moisture as well as an index of the extent of possible damage of the skin's water barrier function. As open chamber sensors are highly sensitive to the ambient and body-induced airflow, closed chamber systems have been developed.
  - **Closed chamber** devices (such as VapoMeter®)[36] can measure the humidity in the chamber or the diffusion gradient, similar to open chamber systems. These devices, however, are not sensitive to the angle at which the device is held and the environmental conditions.[37] The closed chamber method does not permit recording of continuous TEWL since when the air inside the chamber is saturated, skin evaporation ceases.[32]
- **Skin surface (topography)**.
  - *Roughness*: Assessing the skin's appearance is often used to determine its texture, which can be related to its hydration state. There are a number

of different ways by which the skin surface topographic features can be non-invasively analyzed for the degree of roughness. Certain instruments move over the skin surface and translate surface unevenness to an electrical signal. A widely used method is to cast a replica of the skin surface by using one of the many excellent silicone rubber impression materials, which can be analyzed later. Another method to measure the roughness (as a measure of hydration state of the skin) is the measurement of tactile acuity. The more hydrated the skin is, the softer and more supple it becomes. Consequently, it becomes more sensitive in terms of its tactile acuity, i.e., its ability to have a discriminating sense of touch. This method is often referred to as the two-point gap discrimination method. The experiment consists of the discrimination of two small metal points pressed on the subject's skin. The distance between the two points is reduced, until only one point can be discerned. It is then identified as the acuity threshold.[38]

- *Friction*: Various devices can be employed to measure skin friction, including a rotating wheel, a revolving ground glass disc, a sliding sled, and others.[39–41] In all the cases, the underlying principle is that the frictional properties of human skin *in vivo* can be assessed by determining how much force is required to drag an object across the skin surface. It may be assumed that the smoother the skin, the less force is required. In practice, however, smooth skin can actually increase friction as a result of increased contact area with the moving surface of the measuring device.[41,42] These measurements, therefore, sometimes correlate very poorly with the real smoothness of the skin.[32]
- *Desquamation*: Recently, D-squame® discs have become a very popular way to sample the skin surface and objectively determine the level of dryness and scaling. D-squame® discs are circular, clear, adhesive discs placed on the skin surface with firm pressure and then pulled away. The removed skin is observed and parameters such as the amount of skin removed, size of flakes, and coloration can be recorded. Differences between dry skin and normal moisturized skin are clearly evident upon examination of the disc, and further characterization can be carried out to differentiate levels of dryness and qualitative differences in desquamation.[43]
- *Sebum*: Excess sebum production can contribute to packing of horny cells at the follicle surface, leading to an occlusive plug or comedo. This is why its quantification and suppression are of great interest to dermatologists and cosmetic scientists. The measuring principle of this instrument (Sebumeter®) is based on the observation that a ground glass plate of certain opacity becomes translucent when its surface is covered by lipids. The translucency is proportional to the amount of lipids on the surface. The head of the device is pressed against the skin for 30 s during which the plastic tape becomes transparent due to the absorbed sebum.[44] This transparency can be translated into actual numbers, expressing the surface sebum amount of the skin.
- *Elasticity*: SC hydration state influences the skin's softness and extensibility (i.e., the extent of how quickly it regains its shape after stretching); therefore, the assessment of these mechanical properties can be related to its hydration state. There are devices available (such as the Dermal Torque Meter®[45]) that

can analyze deformation of the skin. The instrument consists of a fixed disc that is attached to the skin and a central moving (rotating) disc. A weak torque is applied to the rotating disc and the angle of rotation of the central disc is measured by a highly sensitive angle sensor.

Aesthetic Properties of the Products **There is a critical balance between clinical measurements and consumer perception of performance. By most accounts, consumers perform an initial assessment of a moisturizer's performance based on the product's aesthetics, i.e., how it applies to their skin, how it smells, how it feels on the skin, in addition to perceptions later in the day of relief from dry skin symptoms.** A formulator can formulate the best body lotion ever; however, if consumers do not like how it feels on their skin or how it smells, they will not buy it more than once. It is very important, therefore, to emphasize that, in addition to product's performance, consumers' opinion should also be taken into account.

Product aesthetics is usually tested by experts and volunteers. The description of product aesthetics may differ from person to person. Therefore, these evaluations are more subjective than those used to determine the moisturizer effectiveness. Here is a brief summary of the sensory properties typically tested for skin care products.[46,47]

- **Rub-out** refers to the easiness with which a product absorbs (spreads) upon application. This property depends on several things, including initial viscosity and viscosity under shear stress (i.e., spreading), amount of water, and type of the emulsion.
- **Appearance** refers to the physical appearance of the product, including its gloss, i.e., how shiny it is; its thickness, i.e., whether it holds its shape after dispensing it and whether it has clumps; and similar types of defects.
- **Pick-up** refers to the product's texture when it is rubbed between a finger and thumb. It is usually evaluated after removing the product from its container.
- **Greasiness** depends on the amount of emollients and occlusives used in the formulation. If their oil phase is higher, they may feel sticky, greasy, and tacky on the skin.
- **Tackiness** is determined by both the oil phase and the water phase. Certain oils may feel tacky when applied to the skin, such as lanolin or cetyl alcohol. In addition, certain hydrophilic thickeners, such as gums, or proteins may also produce a tacky feeling.
- **Break** refers to the behavior of the product on the skin and is primarily related to the components of the formulation. This property is determined when the product is being spread on the skin.
- **Slip** describes how the product glides on the skin. Products for damaged skin will have advanced slip to avoid additional damage to the skin.
- **After-feel** can be described as softness and smoothness of the skin after application. It depends on some of the earlier discussed parameters, such as greasiness, tackiness, as well as how much product is still on the skin. This property has a vital role in product selection from the consumer's viewpoint.

- **Delayed after-feel** is usually evaluated by measuring the amount of product that is still on the skin 5, 10, and 30 min after applying the product.

It is important to take into account the target group for the product when analyzing the results from of sensory evaluations. For example, tackiness may be undesired for a daily moisturizer for normal skin; however, it may be beneficial (or at least not negative) when formulated for users with dry skin.

Ingredients Causing Safety Concerns ●**Ingredients causing safety concerns regarding skin moisturizing products include fragrances, lanolin, urea, propylene glycol, and herbal extracts.** The findings and current statements on their safety are summarized here.

Moisturizers are usually free from strong irritants; however, repeated exposure of sensitive areas to mildly irritating preparations may cause various skin reactions, including dryness, redness, burning sensation, and itching. Strong irritants are typically easy to identify; however, weak irritants are less obvious to avoid. Here is a brief summary of the main ingredients that usually cause safety concerns for consumers.

- **Fragrances** are usually identified as the major sensitizers in skin care products. They may consist of more than 50 ingredients, of which many are very irritating. While these ingredients may increase consumers' willingness to buy products, their components may be mild irritants, causing concerns.
- **Lanolin** is often believed to be a main sensitizer. It is derived from the fleece of sheep and comprises hundreds of different chemicals, making it difficult to isolate the contact allergens. However, the true prevalence of positive reactions is unknown since reactions are often not reproducible and false positives and negatives commonly occur.[48] Therefore, patch testing with the patient's own product may be of value before starting to use a product containing this ingredient.
- **Urea**, an effective humectant, was found to be irritating to the skin in a few clinical studies. However, most studies evaluating urea's effect did not report severe irritation, even in higher concentrations. The CIR Expert Panel evaluated the safety of urea and concluded that urea was safe to be used in cosmetics and personal care products. The CIR Expert Panel did note that urea can increase the absorption of other ingredients through the skin and that this should be taken into account when conducting product safety assessments.[49]
- **Propylene glycol** is used in a number of products as a humectant and a solvent. Questions arose regarding the potential for adverse effects from the cosmetic use of propylene glycol in light of skin sensitization studies that reported positive results. The CIR Expert Panel evaluated propylene glycol in personal care products and concluded that propylene glycol was safe for use in cosmetic products at concentrations up to 50%.[50] It was noted that consumers with damaged skin barriers were at a higher risk of developing irritation/sensitization reactions even at low concentrations. It was also noted that in normal subjects, propylene glycol may be an irritant under occlusive patches. However, when used in cosmetics and personal care products, it is simply spread on the skin surface without applying a patch over it. Reviews from independent studies also found

that true allergic reactions to propylene glycol are uncommon and the clinical significance is probably overestimated.[51] Similar to urea, propylene glycol can also increase the cutaneous penetration of applied other ingredients, including drugs, which should be taken into account.[52]

- **Herbal extracts** may also cause sensitization and irritation. However, these reactions are rare, possibly due to the low concentration of such extracts in skin moisturizer formulations.

Packaging of Skin Moisturizers

The most commonly used packaging materials for skin moisturizers include the following:

- *Plastic Bottles*: The majority of body moisturizing products are supplied in plastic bottles. Some facial and hand moisturizers may also be packaged into plastic bottles. Many containers have a pump head that dispenses a small dose of the product. Plastic bottles may also be supplied with a flip-top, disc top, or screw-on cap.
- *Soft Tubes*: The majority of hand lotions and creams are supplied in soft tubes with a screw-on, disc top, or flip-top cap.
- *Plastic and Glass Jars*: Some body lotions and creams, and the majority of facial moisturizers, are packaged into plastic or glass jars. Glass jars are more dangerous in the sense that they can cause serious damage if broken. Manufacturers often take the target group's needs and expectations into account when designing a new package.

GLOSSARY OF TERMS FOR SECTION 3

Aesthetics: A set of properties of a product that refers to how a product looks, smells, applies to the skin, and feels on the skin.

Barrier cream: A type of moisturizing product that is used to avoid unpleasant reactions and prevent irritant dermatitis, that is, inflammation of the skin caused by irritants.

Ceramides: Structural elements of the intercellular lipids in the stratum corneum, and they help maintain the integrity of the skin barrier.

Emollient: A type of moisturizing ingredient that can soften and smoothen the skin by filling in the void spaces between the corneocytes and replacing the lost lipids in the stratum corneum.

Humectant: A type of moisturizing ingredient that can increase the water content of the top layer of the skin by enhancing water absorption from the dermis into the epidermis.

Hygroscopic: An ingredient that tends to absorb moisture from the air.

Irritant dermatitis: Skin inflammation caused by contact with various chemicals, such as surfactants, solvents, and fragrances.

Moisturizer product: A personal care product that is designed to improve the skin quality, maintain and/or restore the moisture content of the stratum corneum as well as keep it smooth and pliable, and aid in alleviating the symptoms of dry skin.

Moisturizing efficacy: The ability of a cosmetic product to hydrate the skin and increase its water content.

Non-invasive method: A test method that does not involve the introduction of instruments into the human body.

Occlusive: A type of moisturizing ingredient that can create a hydrophobic barrier to physically block transepidermal water loss from the stratum corneum.

Oil-in-water emulsion: An emulsion in which the internal phase consists of oils and oil-soluble and oil-miscible ingredients, while the external phase consists of water, emulsifier(s), and water-soluble and water-miscible ingredients.

Skin hydration: Skin's water content.

Skin protectant: An OTC drug product that temporarily protects injured or exposed skin or mucous membrane surfaces from harmful or annoying stimuli and may help provide relief to such surfaces.

Skin rejuvenator: A type of moisturizing ingredient that can restore, protect, and enhance the skin's barrier function, thereby reinforcing skin hydration.

Skin surface roughness: A quality parameter that refers to the texture of the skin surface. The state of roughness can be related to the skin's hydration state.

Water-in-oil emulsion: An emulsion in which the internal phase consists of water and water-soluble and water-miscible ingredients, while the external phase consists of oils, emulsifier(s), and oil-soluble and oil-miscible ingredients.

Water-in-silicone emulsion: An emulsion in which the internal phase consists of water and water-soluble and water-miscible ingredients, while the external phase consists of silicone oils, emulsifier(s), and silicone-soluble and silicone-miscible ingredients.

REVIEW QUESTIONS FOR SECTION 3

Multiple Choice Questions

1. Glycerin is the gold standard for ____.
   a) Humectants
   b) Emollients
   c) Occlusives
   d) Enhancer of the skin barrier

2. Which of the following is true for ceramides?
   a) They are the most important structural elements of dead cells
   b) They are the most important structural elements of the intercellular lipids
   c) They disrupt the integrity of the skin barrier
   d) They induce dry skin symptoms when applied to the skin
3. Evaluating the moisturizing effect of a skin moisturizer is a ___ test.
   a) Safety
   b) Quality
   c) Performance
   d) Stability
4. Which of the following methods can be used to evaluate the water content of the skin?
   a) Conductance method
   b) Capacitance method
   c) Both A and B
   d) None of the above
5. Which of the following is used to evaluate skin topography?
   a) Water content of the skin
   b) Water loss from the skin
   c) Rub-out of the product
   d) Roughness of the skin
6. Which of the following is used to evaluate the aesthetic properties of skin moisturizers?
   a) Appearance, friction, desquamation
   b) Appearance, pick-up, after-feel
   c) Conductance, capacitance, TEWL
   d) None of the above
7. Which of the following is used to evaluate the moisturization efficacy of skin moisturizers?
   a) Skin topography measurement
   b) Skin hydration measurement
   c) TEWL measurement
   d) All of the above

8. Which of the following is NOT true for barrier creams?
  a) They provide sun protection
  b) They are used to prevent skin irritation
  c) Products for water protection are typically W/O emulsions
  d) Products for oil protection are usually based on nonoil-soluble film-formers

9. Which of the following ingredients is known to provide a "dry touch"?
  a) Glycerin
  b) Lanolin
  c) Almond oil
  d) Dimethicone

10. What is the most common dosage form for skin moisturizers?
  a) Cream
  b) Ointment
  c) Gel
  d) Paste

Fact or Fiction?

_________ a) Lauric acid (found in coconut oil) is comedogenic.
_________ b) Alpha hydroxy acids (AHAs) make the skin sensitive to ultraviolet light.
_________ c) Only dry skin needs daily moisturization.
_________ d) Emulsions can be diluted with their internal phase.

Matching

Match the skin cleanser ingredients in column A with their appropriate ingredient category in column B.

Column A	Column B
_____ A. Collagen	1. Aesthetic agent
_____ B. Dimethicone	2. Electrolyte
_____ C. Glycol stearate	3. Emollient
_____ D. Gum acacia	4. Film-former, hydrophilic thickener
_____ E. Lanolin	5. Functional ingredient
_____ F. Methylisothiazolinone	6. Humectant
_____ G. Mineral oil	7. Occlusive
_____ H. Propylene glycol	8. Preservative
_____ I. Sodium citrate	9. Silicone
_____ J. Vitamin E	10. Skin rejuvenator

REFERENCES

1. Held, E.: So moisturizers may cause trouble! *Int J Dermatol.* 2001;40(1):12–13.
2. McDonough, E. G.: *Truth About Cosmetics*, New York: The Drug and Cosmetics Industry, 1937: 12.
3. Fernandez Rodriguez, M. C., Selles, F. E.: The splendour of the cosmetology in the ancient Greece. *Cosmetic News* 1998; XXI.
4. Gunn, F.: *The Artificial Face: A History of Cosmetics*, New York: Hippocrene Books, 1983.
5. Stepanovs, J.: *Skin Saver Remedies*. Bermagui: Harald Tietze Publishing, 1999.
6. The Olympic Games in Antiquity, Accessed 12/17/2013 at http://www.olympic.org/Documents/Reports/EN/en_report_658.pdf
7. Spelios, T.: An informal history of cosmetics II. *Drug Cosmet Ind.* 1983;133:142.
8. Encyclopaedia Britannica, *14th Ed*, 1929
9. Vaseline: Commitment to Quality: For Over 140 Years, Accessed 12/17/2013 at http://www.unileverusa.com/brands-in-action/detail/Vaseline-/298226/
10. Pond's Institute: A History of Pond's Heritage, Accessed 12/17/2013 at http://www.pondsinstitute.co.uk/history.php
11. Disapio, A., Fridd, P.: Silicones: Use of Substantive Properties on Skin and Hair, *Int J Cosmet Sci.* 1988;10:75–89.
12. Drealos, Z. D., Marenus, K. D.: Cellulite: etiology and purported treatment. *Dermatol Surg.* 1997;23:1177–1181.
13. Khan, M. H., Victor, F., Rao, B., et al.: Treatment of cellulite: Part I. Pathophysiology. *J Am Acad Dermatol.* 2010;62(3):361–370.
14. Rawlings, A. V.: Cellulite and its treatment. *Int J Cosmet Sci.* 2006;28(3):175–190.
15. Nakatsuji, T., Kao, M. C., Fang, J. Y., et al.: Antimicrobial property of lauric acid against Propionibacterium acnes: its therapeutic potential for inflammatory acne vulgaris. *J Invest Dermatol.* 2009;129(10):2480–2488.
16. Held, E., Johansen, J. D., Agner, T.: Contact allergy to cosmetics: testing with patients' own products. *Contact Dermatitis.* 1999;40(6):310–315.
17. Zirwas, M., Stechschulte, S. A.: Moisturizer allergy. diagnosis and management. *J Clin Aesthet Dermatol.* 2008;1(41):38–44.
18. Kaidbey, K., Sutherland, B., Bennett, P., et al.: Topical glycolic acid enhances photodamage by ultraviolet light. *Photodermatol Photoimmunol Photomed.* 2003;19:21–27.
19. Loden, M.: Do moisturizers work? *J Cosmet Dermatol.* 2003;2(3–4):141–149.
20. Bouwstra, J. A., de Graaff, A., Gooris, G. S., et al.: Water distribution and related morphology in human stratum corneum at different hydration levels. *J Invest Dermatol.* 2003;120:750–758.
21. Draelos, Z. D.: Active agents in common skin care products. *Plast Reconstr Surg.* 2010;125(2):719–724.
22. Robbins, C. R., Fernee, K. M.: Some observations on the swelling of human epidermal membrane. *J Soc Cosmet Chem.* 1983;37:21–34.

23. Idson, B.: Dry skin: moisturizing and emolliency. *Cosmet Toiletries*. 1992;107:69–78.
24. Nolan, K., Marmur, E.: Moisturizers: reality and the skin benefits. *Dermatol Ther*. 2012;25(3):229–33.
25. Anderson, P. C., Dinulos, J. G.: Are the new moisturizers more effective? *Curr Opin Pediatr*. 2009:21(4):486–490.
26. Grubauer, G., Feingold, K. R., Elias, P. M.: Relationship of epidermal lipogenesis to cutaneous barrier function. *J Lipid Res*. 1987;28:746–752.
27. Hon, K. L., Leung, A. K.: Use of ceramides and related products for childhood-onset eczema. *Recent Pat Inflamm Allergy Drug Discov*. 2013;7(1):12–19.
28. Hon, K. L., Pong, N. H., Wang, S. S., et al.: Acceptability and efficacy of an emollient containing ceramide-precursor lipids and moisturizing factors for atopic dermatitis in pediatric patients. *Drugs R D*. 2013;13(1):37–42.
29. Kresken, J., Klotz, A.: Occupational skin-protection products – a review. *Int Arch Occup Environ Health*. 2003;76:355–358.
30. Zhai, H., Maibach, H. I.: Barriers creams – skin protectants: can you protect skin? *J Cosmet Dermatol*. 2002;1:20–23.
31. Schliemann, S., Kleesz, P., Elsner, P.: Protective creams fail to prevent solvent-induced cumulative skin irritation - results of a randomized double-blind study. *Contact Dermatitis*. 2013;69(6):363–371.
32. Galzote, C., Suero, M., Govindarajan, R.: Noninvasive Evaluation of Skin in the Cosmetic Industry, In: Walters, K. A., Roberts, M. S., eds: *Dermatologics, Cosmeceutic and Cosmetic Development*, New York: Informa Healthcare, 2008.
33. Corneometer, Accessed 12/8/2013 at http://www.courage-khazaka.de/index.php/en/faq-en/faq-scientific-devices/61-corneometer
34. Leveque, J. L., de Rigal, J.: Impedance methods for studying skin moisturization. *J Soc Cosmet Chem*. 1983;34:419–428.
35. Grove, G. L., Grove, M. J., Zerweck, C. and Pierce, E.: Comparative metrology of the evaporimeter and the DermaLab® TEWL probe. *Skin Res Technol*. 1999;5:1–8.
36. Delfin Technologies: VapoMeter, Accessed 12/8/2013 at http://www.delfintech.com/en/vapometer/
37. Charbonnier, V., Paye, C., Mainbach, H. I.: Determination of Subclinical Changes of Barrier Function, In: Wilhelm, K. P., Zhai, H., Maibach, H. I., eds: *Dermatotoxicology*, Boca Raton: CRC Press, 2010: 563.
38. L'Oreal: Soft and Moisturizer: Water – What a History, Accessed 12/8/2013 at http://www.skin-science.com/_int/_en/topic/topic_sousrub.aspx?tc=SKIN_SCIENCE_ROOT^OBSERVING_THE_SKIN^SOFT_AND_MOISTURIZED_WATER_WHAT_A_STORY&cur=SOFT_AND_MOISTURIZED_WATER_WHAT_A_STORY
39. Comaish, S., Bottoms, E.: The skin and friction: deviation from Amonton's laws, and the effects of hydration and lubrication. *Br J Dermatol*. 1971;84:37.
40. Prall, J. K.: Instrumental evaluation of the effects of cosmetic products on skin surfaces with particular reference to smoothness. *J Soc Cosmet Chem*. 1973;24:693–707.
41. Naylor, P. F. D.: The skin surface and friction. *Br J Dermatol*. 55(67):239.
42. El-Shimi, A. F.: In vivo skin friction measurements. *J Soc Cosmet Chem*. 1977;28:37.

43. CuDerm: D-Squame®, Accessed 12/8/13 at http://www.cuderm.com/products_dSquame_rd2.php
44. Elsner, P.: Sebum, In: Berardesca, E., Elsner, P., Wilhelm, K. P., eds: *Bioengineering of the Skin: Methods and Instrumentation*, Boca Raton: CRC Press, 1995:81–85.
45. Dermal Torque Meter®, Accessed 12/8/2013 at http://cosderma.com/en/prestations/techniques/28-dermal-torque-meter
46. Aust, L. B., Oddo, L. P., Wild, J. E., et al.: The descriptive analysis of skin care products by a trained panel of judges. *J Soc Cosmet Chem*. 1987;38:443–449.
47. Barton, S.: Formulation of Skin Moisturizers, In: Leyden, J. J., Rawlings, A. V., eds: *Cosmetic Science and Technology: Skin Moisturization*, New York: Marcel Dekker, 2002:547–589.
48. De Groot AC. Sensitizing Substances, In: Lodén M, Maibach HI, eds: *Dry Skin and Moisturizers: Chemistry and Function*, Boca Raton: CRC Press, 2000: 403–411.
49. Cosmetic Ingredient Review Panel: Final report of the safety assessment of urea. *Int J Toxicol*. 2005;24(3):1–56.
50. Cosmetic Ingredient Review Panel: Final report of the safety assessment of propylene glycol and polypropylene glycols, cosmetic ingredient review, *J Am Coll Toxicol*. 1994;13(6):473–491.
51. Funk, J. O., Maibach, H. I.: Propylene glycol dermatitis: re-evaluation of an old problem. *Contact Dermatitis*. 1994;31(2):36–41.
52. Turpeinen, M.: Absorption of hydrocortisone from the skin reservoir in atopic dermatitis. *Br J Dermatol*. 1991;124(4):358–360.

SECTION 4: PRODUCTS FOR SPECIAL SKIN CONCERNS – AGING AND ACNE

LEARNING OBJECTIVES

Upon completion of this section, the reader will be able to

1. define the following terms:

Acne vulgaris	Antibiotic	Antibiotic resistance	Antioxidants
Blackhead	Botanical extract	Chronological aging	Comedo
Cyst	Extrinsic aging	Hydroxy acids	Hyperkeratinization
Intrinsic aging	Keratolytic	Liposome	Microcomedo
Nodule	Noisome	Oxidative stress	*P. acnes*
Papule	Photoaging	Photosensitivity	Pustule
Retinoids	Whitehead		

2. differentiate between intrinsic and extrinsic aging;
3. briefly discuss what changes occur in the skin during aging;
4. name some environmental factors that can accelerate the normal aging process;
5. briefly discuss whether anti-aging products are considered drugs or cosmetics in the US;
6. name some non-invasive technologies and products that can be used for the prevention and/or treatment of aging skin;
7. name some invasive technologies that can be used for the prevention and/or treatment of aging skin;
8. name some botanical extracts that can be used in the prevention and/or treatment of aging skin;
9. explain how oxidative stress contributes to aging;
10. explain how the following can prevent/treat aging skin, and name some examples for each: antioxidants, peptides and proteins, topical retinoids, and hydroxy acids;
11. list the major factors that generally influence the penetration of an active ingredient into the skin;
12. name some delivery systems that can be used for anti-aging ingredients;
13. differentiate between liposomes and niosomes;
14. briefly discuss the main formulation challenges that can occur when formulating products with common anti-aging actives;
15. briefly discuss the main safety concerns that arose in the past with regard to the use of the following ingredients: alpha hydroxy acids, topical retinoids, and peptides and proteins;
16. explain how acne vulgaris develops;
17. name the major factors contributing to the development of acne;
18. briefly discuss how the following factors contribute to the development of acne: diet, sunlight, facial cleansing, and makeup use.
19. differentiate between inflammatory and non-inflammatory acne;
20. explain the typical signs and symptoms of acne;
21. explain how noninflammatory acne can progress into inflammatory acne;
22. briefly discuss the main types of ingredients used in the topical treatment of acne, and provide some examples for each type;
23. briefly discuss the main types of ingredients used in the systemic treatment of acne, and provide some examples for each type;
24. name some alternative treatments that can be used in the treatment of acne;
25. briefly discuss the main safety concerns that arose in the past with regard to the use of the following ingredients: topical retinoids, oral retinoids, and antibiotics.

KEY CONCEPTS

1. The general aging process, which is genetically determined and occurs by the passing of time alone, is called intrinsic skin aging or the chronological aging process, whereas the skin aging process induced by environmental factors is called extrinsic skin aging.
2. Today, there is a wide range of products and procedures for the prevention and/or treatment of aging skin, ranging from non-invasive technologies to invasive methods.
3. Botanical extracts, antioxidants, proteins and peptides, retinoids, hydroxy acids, and sunscreens are claimed to prevent oxidative reactions and the formation of free radicals, resurface the epidermis, and promote the natural synthesis of collagen and elastin.
4. The majority of anti-aging formulations are emulsions since they offer cosmetically elegant formulations with the potential to include oil- and water-soluble components at the same time.
5. Concerns arose regarding the safe use of some anti-aging ingredients, including AHAs, topical retinoids, and peptides and proteins.
6. Acne vulgaris is a common dermatological disorder of the pilosebaceuos unit that has a complex pathophysiology and can be triggered by a number of factors.
7. The clinical presentation of acne can range from mild comedonal (i.e., non-inflammatory) form to severe inflammatory cystic acne.
8. Acne vulgaris is also often graded based on its severity, including the number of comedones, inflammatory lesions, total lesion count, and cysts.
9. Milder cases are best managed with OTC and prescription-only topical regimens, whereas systemic, prescription-only drugs are indicated in more severe cases.
10. The main safety issues associated with the use of acne treatments include irritation and dryness caused by many topical ingredients and teratogenicity of oral retinoids. An additional concern is the risk for antibacterial resistance when using antibiotics.

Introduction

Skin aging is a complex biological process influenced by a combination of endogenous and exogenous factors, which leads to structural and physiological alterations in the skin layers as well as changes in skin appearance, especially on the sun-exposed skin areas. Acne vulgaris is a very common skin condition, especially among teenagers and young adults. It primarily affects the face; however, in severe cases, it may also affect the chest, neck, and back. In addition to the visible symptoms it can cause, it has clear detrimental psychosocial effects on the patients' quality of life (QoL). As skin health and beauty are considered principal factors representing the overall well-being and the perception of health, skin care formulations targeting the signs of aging and acne have gained great importance over the ages.

This section reviews the skin aging process, possible preventive measures, as well as some currently used ingredients and formulation techniques. In addition, the section also provides an overview of acne, its major signs and symptoms, current treatment options, and some formulation considerations for anti-acne products.

Part 1: Anti-Aging Products

Changes in Skin Structure and Function during Aging Today, most consumers are familiar with the aging process and its visible signs. It is known that young individuals have soft, smooth, and supple skin, which becomes less elastic, tougher, uneven, and wrinkled over the years. Aging of the skin is influenced by two separate processes. ❶**The general aging process, which is genetically determined and occurs by the passing of time alone, is called intrinsic skin aging or the chronological aging process, whereas the skin aging process induced by environmental factors is called extrinsic skin aging.** Each skin aging process leads to characteristic skin aging signs as discussed here.[1]

Intrinsic aging occurs as a natural result of aging; it affects the skin on the entire body, including the photoprotected areas in a similar manner. Intrinsic aging can be considered as a slow, continuous, and irreversible tissue degradation.[2] However, people of identical chronological age may appear to have younger- or older-looking skin. The reason for this is their different genetic makeup, which plays an important role in intrinsic aging.[3] Clinically, intrinsically aged skin is smooth, thin, pale, and finely wrinkled. Histologically, the intrinsically aged skin can be characterized primarily by functional alterations, such as thinning of the dermis, degeneration of the elastin network, and loss of hydration.

- Thinning of the dermis is related to its collagen content. Collagen synthesis gradually diminishes over time, which leads to a reduction in the amount of collagen in the skin. As collagen is responsible for the strength of the skin and plays an important role in wound healing, its decreasing amount leads to reduced strength, atrophy (i.e., weakening and thinning) of the dermis, and a reduced rate of wound healing.[4]
- In addition to collagen, the synthesis of elastin also decreases, which results in a loss of skin elasticity.[5]
- The superficial muscles on the face begin to shrink, which causes lines to appear. Typical facial wrinkles form in the areas where muscles contract to make facial expressions. These include "worry lines" spanning the forehead, "laugh lines" around the eyes and mouth, and "crow's feet" radiating from the outside corners of the eyes.
- Additionally, the production of immune cells also decreases over time, which also contributes to the compromised wound healing properties.[6] The loss of strength, thickness, and healing properties causes the skin to become more vulnerable.
- There is a gradual flattening of the wavy attachment between the epidermis and dermis due to reduced production of keratinocytes. It is believed that this

flattening may contribute to the increased fragility of aging skin and may also lead to reduced nutrient transfer between the dermal and epidermal layers.[7]

- With advancing age, both the mortar and bricks of the skin are affected, which leads to a reduction in the SC lipid content, damaged barrier functions, increased epidermal water loss, and drier skin.[8,9]

Additional changes that occur during aging include changes in the skin pigmentation and size of the sebaceous glands.

- With the onset of intrinsic aging, there is a decrease in the density of melanocytes. This decrease contributes to the pale appearance of older skin, which is also related to a decrease in skin vascularity. Older patients are less able to protect themselves from the sun because of a decrease in melanin. Therefore, they are more susceptible to developing sun-induced cancers.[7]
- The number of sebaceous glands remains approximately the same throughout life, although their size tends to increase with age. As a result, the skin's pores may widen.

Extrinsic aging can be defined as the acceleration of intrinsic aging caused by exposure to various environmental factors, such as sun, wind, smoking, and air pollutants. The primary factors causing human skin aging are UV radiation from the sun as well as smoking. When aging is caused by the sun, the process is referred to as "photoaging." Photoaging is a cumulative process, which means that exposure adds up over the years. Individuals who have outdoor lifestyles, live in sunny climates, and are lightly pigmented will experience the greatest degree of photoaging.[10] It has been estimated that photoaging accounts for more than 90% of the skin's associated cosmetic problems.[11] In the vast majority of individuals, photoaging and aging due to smoking overshadow intrinsic changes, especially in the skin of the face, neck, hands, and forearms. On an individual basis, the rate and extent of extrinsic skin aging depend on the individual exposure pattern to different environmental factors and also on the individual's genetic makeup.

DID YOU KNOW?

It has been known that smoking leads to premature aging, i.e., smokers in general look older than their chronological age. It is more interesting, however, that smoking has been shown to be a great contributor to facial wrinkling than sun exposure.[12]

In contrast to intrinsic aging, the extrinsically aged skin can be characterized by more visible morphologic and physiologic changes. Clinically, extrinsically aged skin can be described as a course skin with wrinkles, hyperpigmented areas, sallowness, increased fragility, textural roughness, and dilated blood vessels that are visible under the skin (which is referred to as "telangiectasia" as a medical term). Histological features of externally aged skin include significant epidermal and dermal alterations.

- The epidermal thickness may be increased or decreased, leading to thickening or thinning, respectively. Skin type significantly influences the clinical signs of photoaging. An extreme example is fair-skinned individuals with freckles in early childhood. After chronic sun exposure, thinning of the skin and fine wrinkling develop in such persons. In contrast, in darker-skinned individuals, the skin becomes coarse and leathery over years upon sun exposure.[11] In general, the epidermis becomes thinner over the years with thicker, coarse areas. Blood vessels can be easily seen through the skin where it is thinner, and in these areas, the skin bruises and bleeds more easily than normal.
- Loss of Langerhans cells and immune cytokins can be observed in the case of photoaged skin as well, leading to compromised immune competency of the skin.
- Due to UV exposure and smoking, the skin starts to lose its normal pink glow and a sallow yellow color develops instead. In areas of the skin that are exposed to the sun, usually there is an increased number and activity of melanocytes. This is manifested by the appearance of darker spots (i.e., age spots) on the skin. Hypopigmented areas also often appear on photodamaged skin.
- Changes triggered by photodamage and smoking can also be observed in the dermis. Repeated exposure to such factors leads to the degradation of collagen and elastin and the accumulation of abnormal elastin (medically termed "elastosis"). The subcutaneous fatty layer decreases in thickness as well. All these result in decrease in the skin strength and elasticity as well as wrinkling and sagging.[13] It has been shown that sun exposure significantly speeds up the rate of degradation of collagen, leading to dramatic structural changes in the exposed areas.[14]

As you can see, age-associated skin changes and sun-exposed skin changes are similar; however, the latter are more dramatic and can potentially lead to more severe outcomes.

DID YOU KNOW?

UV light is divided into three groups based on the light's wavelength. It includes UVC, which is largely blocked by the ozone layer and has little impact on the skin. UVB penetrates only into the epidermis and is responsible for the redness, sunburn, DNA damage, hyperpigmentation, and skin cancer. UVA requires a much higher dose to cause sunburn, and for this reason, it was long considered irrelevant to skin damage. However, because it can penetrate deeper into the skin than UVB, today, UVA is thought to play a more substantial role in the photoaging. UVA is believed to be responsible for most chronic skin damage associated with photoaging; therefore, photoprotection should cover both UVA and UVB radiation.[15]

Drug or Cosmetic?

Anti-aging products can be either drugs or cosmetics. According to the FDA, "whether a wrinkle remover product should be regulated as a drug or a cosmetic depends on the claims the manufacturer makes for the product." Referring back to the definition of a drug and cosmetic in the US (Section 1 of Chapter 1), the main difference between these two categories is the intended use of the products. It is the manufacturers' responsibility to decide whether their claims and/or ingredients used in their products make the products to be considered drugs based on the regulations and act according to that.[16]

Some ingredients used in anti-aging formulations are listed in the OTC monograph as skin protectants in certain concentrations.[17] Examples were discussed in the previous sections. If they are included in anti-aging formulation in the listed concentrations, the products are considered drugs.

As discussed earlier, anti-aging products are often advertised as cosmeceuticals and claim to offer additional benefits over "simple cosmetics" (i.e., not cosmeceuticals). However, since most anti-aging products are sold as cosmetics, manufacturers cannot use the term "active ingredient" on the labels. Therefore, the main ingredients are often called "bioactive" ingredients (or functional ingredients). As most of these formulations contain one or more ingredients aimed at delaying the appearance of the first signs of aging and/or removing these signs, there is a base for claiming additional benefits. The term is widely used not only in cosmetic advertising but also in periodicals and promotional materials. However, as discussed in Section 1 of Chapter 1, it is not recognized by US law. New anti-aging ingredients are continuously developed and synthetized, leading to an increasing number of available ingredients. Their mechanism of action, however, is not precisely understood in all cases. Therefore, even if a manufacturer does not claim drug-like effects for a product and sells it as a cosmetic, it may still affect the structure and/or function of the human skin.

Typical Ingredients Used in Topical Anti-Aging Products

●Today, there is a wide range of products and procedures for the prevention and/or treatment of aging skin, ranging from non-invasive technologies to invasive methods.

- **Non-invasive** technologies include primary prevention, such as healthy lifestyle, refraining from smoking, and using sunscreens, topical cosmetics and drug products applied directly to the skin, microdermabrasion, certain laser devices, as well as systemic treatments, including hormone replacement therapy.
- **Invasive** technologies include chemical peels, topical injections of chemicals, such as botulinum toxin and various dermal fillers, dermabrasion, various laser devices, as well as corrective surgeries. This section focuses on the non-invasive treatment options, primarily on the topically applied products and their main ingredients.

A healthy skin barrier is an important protector against dehydration and penetration of various physical and chemical entities and contributes to skin regeneration, elasticity, and smoothness.[18] Therefore, its daily maintenance is an essential part of the anti-aging therapy. A further step in the anti-aging therapy is to prevent or at least slow down the degradation of skin elements, such as collagen and elastin, which naturally occurs over time. ❸**Botanical extracts, antioxidants, proteins and peptides, retinoids, hydroxy acids (HAs), and sunscreens are claimed to prevent oxidative reactions and the formation of free radicals, resurface the epidermis, and promote the natural synthesis of collagen and elastin.** Their main characteristics with commonly used examples are discussed here.

- **Botanical extracts** are very popular today, and they are frequently used as bioactive ingredients in anti-aging products. Botanical extracts are naturally sourced ingredients with various, often, multiple effects, including antioxidation (such as coffee and pomegranate),[19] photoprotection (such as black tea and olive),[20] smoothing, calming, anti-inflammatory effects (such as chamomile and various mushrooms),[21,22] skin lightening (such as blueberry and ginseng),[23] skin tightening (such as peppermint and witch hazel), and emolliating effect (such as jojoba and coconut extract). Most commonly, botanical extracts are used in combination with one another for their combined benefits.Natural products, including herbal extracts, are very popular today. However, there are some facts that should be kept in mind regarding these products. In general, herbal extracts exhibit more variation than synthetic products, regarding their ingredients. The constituents of herbs and other botanicals highly depend on multiple factors, including growing conditions, climate, weather, harvesting methods, and extraction methods. Herbal extracts are often used in very small amounts in cosmetic products, which fall far below their therapeutic ranges. Although some extracts (including green tea) have been shown to be able to influence mechanisms in the skin that are beneficial for anti-aging via antioxidant and anti-inflammatory properties, there is only a limited number of clinical studies performed with topical formulations containing such extracts. For this reason, the real usefulness of such ingredients in providing anti-aging benefits is still questionable.
- Oxidative stress is considered to be the cornerstone of the biochemical pathways, leading to both chronological aging and photoaging.[24] Human skin has an efficient antioxidant system, which is able to counteract the negative effects of oxidative stress.[25] However, in the case of chronic or severe oxidative stress, this endogenous network is not effective enough; therefore, tissue damage may occur.[26] According to the free radical theory of aging, oxidative stress increases with age and, at the same time, the endogenous antioxidant system becomes less effective.[27] Developing free radicals promote degradation of collagen, elastin, and hyaluronic acid and cause signs of aging. Therefore, supplying the skin with antioxidants is believed to be beneficial in slowing down aging. **Antioxidants** protect the cells from damage by neutralizing the free radicals. Today, antioxidants can be found in an increasing number of

cosmetic and OTC drug–cosmetic products and also in drinks and foods. Commonly used ingredients include vitamins, such as vitamin C, vitamin B_3, and vitamin E; botanical extracts, such as green tea and grape-seed extract; and other ingredients, such as resveratrol, lipoic acid, ferulic acid, and coenzyme Q10 or ubiquinone. In terms of preventing the effects of photoaging, it is not yet known which antioxidants are the most effective. It is believed that, for the best results, foods and drinks containing antioxidants should be consumed in addition to using topical products.[28] Antioxidants not only can serve as protective molecules against oxidation but also, some of them, can stimulate collagen production, preserve hyaluronic acid levels in the skin, and also exert anti-inflammatory effects.[29]

- **Peptides and proteins** are also very popular today as antiaging ingredients that are formulated into topical formulations as well as injectable products. They have multiple effects, including anti-inflammatory, wrinkle reduction by stimulating collagen production, thickening the skin and firming it; smoothing; moisturization; and skin protective effects. Examples for commonly used peptides and proteins include natural ingredients, such as collagen, elastin, hydrolyzed wheat, and soy proteins; a number of synthetic ingredients, such as matrikines, neuropeptides, including botulinum toxin; and a variety of enzymes, such as proteolytic enzymes.

DID YOU KNOW?

Hyaluronic acid is a linear polysaccharide naturally found in the dermis. It is a natural humectant that keeps the skin moisturized. Hyaluronic acid is commonly used as a dermal filler in the US to fill in facial wrinkles.

- **Retinoids** are a class of substances comprising vitamin A (retinol) and its derivatives. These ingredients are incorporated into both prescription-only medications and OTC drug–cosmetic products. Retinoic acid and its derivatives adapalene, tazarotene, and bexarotene are registered prescription drugs. Topical retinol derivatives that are used in antiaging formulations include retinyl esters, retinol, retinaldehyde, and oxiretinoids.[30]

  In general, retinoids are lipophilic molecules that can penetrate the epidermis and enter the dermis.[31] Their effects include the improvement of surface roughness, improvement of age spots, and reduction of fine lines by decreasing the amount of collagen breakdown and stimulating the production of new collagen.[32,33] Retinoids work at the molecular level through receptors in the skin cells.

  Retinoic acid was originally intended for acne treatment. Clinical studies, however, demonstrated that it is not only effective in treating acne but also prevents both skin aging and improves the already damaged skin. It has been extensively studied since then and used as an effective topical treatment for photoaging, acne, and other dermatological disorders. It is available in prescription strength only. One of its major drawbacks is that it can be irritating to the skin, and the condition developed is usually referred to as "retinoid dermatitis" or "retinoid reaction," which limits its use. The topical effects, such as the burning sensation, redness, and peeling, mainly occur in the first weeks of the retinoic acid treatment and are almost always transient and diminish with continued application.[34] However, many patients discontinue the therapy upon appearance of such symptoms. Approaches have been made to decrease the irritating potential of retinoic acid. These include the formulation of specific delivery systems as well as synthesis of new derivatives.[35] An additional side effect associated with retinoid therapy is photosensitization, which normally occurs at the beginning of the therapy. Patients on retinoid therapy are advised to avoid excessive sun exposure and take precautionary measures (such as the use of sunscreens) for sun protection. However, after a few months of therapy, the skin's response to UV radiation returns to normal.[36]

Retinaldehyde and retinol are gentler derivatives of retinoic acid, which are also scientifically proven to deliver anti-aging effects. They are available without prescription. Both retinaldehyde and retinol are precursors of retinoic acid (i.e., they are converted into retinoic acid, the biologically active form of vitamin A in the skin).[37] Retinol has been shown to be safe, effective, and better tolerated than retinoic acid in low concentrations.[38–40] Currently, retinol is the most often used retinoid, which has been shown to have positive effects not only on extrinsic aging but also on intrinsic skin aging.[41] To prevent irritation, products should be used at night in small amounts. Using moisturizers is also often recommended by dermatologists to prevent dryness. The use of sunscreens is also essential when using retinoids since these products make the skin photosensitive. Retinoids are best used at night as they are photoinactivated, meaning that direct sunlight will break them down and make them less effective.[42] When using retionoids, it has been shown that most of the improvement in aging signs occurs within the first year of treatment and improvement is maintained with continued use by up to 4 years.[43]

DID YOU KNOW?

Some AHAs are also BHAs when the molecule contains two or more carboxyl groups. In this case, the hydroxyl group is in the alpha position to one carboxyl group and, at the same time, is in the beta position to the other carboxyl group. For example, malic acid (apple acid) containing one hydroxyl and two carboxyl groups is both an AHA and a BHA.

- HAs are naturally occurring organic carboxylic acids, which have attracted much interest over the past decade in the treatment of aging. They include AHAs, such as lactic acid, glycolic acid, malic acid, tartaric acid, and citric acid; beta hydroxy acids (BHAs), such as beta-hydroxybutonic acid; polyhydroxy acids (PHAs), such as gluconolactone; bionic acids (BAs), such as lactobionic acid; and aromatic hydroxy acids (AMAs), such as salicylic acid. These organic acids are normally synthesized; but some of them can be derived from natural sources, such as sugar cane, sour milk, apples, grapes, and citrus fruits. HAs can deliver numerous skin effects, including exfoliation, moisturization, anti-inflammatory properties, and antioxidative effects. One of their major beneficial effects is the improvement of photoaged skin. Visible results include a decrease in roughness, skin discoloration and overall pigmentation, an increase in total skin thickness and in the density of collagen, and an improved quality of elastic fibers, which makes the skin look firmer.[44,45] In low concentrations (4–10%), HAs are components of nonprescription creams and lotions that are promoted as being effective for skin aging. In high concentrations (>20%), they can be used as chemical peels to treat various conditions and diseases, including calluses, acne, and photoaging.

  AHAs, such as glycolic acid and lactic acid, are widely used in skin care products to provide exfoliation (as a chemical peeling agent) and skin resurfacing. Citric acid is widely used in topical formulations as an antioxidant and pH adjustor, as well as an anti-aging ingredient.[46] Gluconolactone, a PHA, is also commonly used in cosmetic formulations since it strengthens the skin barrier function,[47] and it is a gentle moisturizing and antioxidant substance.[48] In addition, it delivers anti-aging benefits. An advantage of this ingredient is that it does not lead to an increase in sunburn after UVB irradiation, which is known to occur with glycolic acid.[49] Gluconolactone can also be formulated with oxidative drugs, such as benzoyl peroxide, to help reduce irritation potential and erythema caused by the oxidative drug.[50] BAs are hygroscopic materials, and their aqueous solution turns into a gel at room temperature since water evaporates from it. This gel can provide protective and soothing effects for inflamed skin; therefore, it can be used after cosmetic procedures that weaken the skin's barrier, such as superficial chemical peels.[51] Lactobionic acid, a BA, also provides significant benefits against sun-initiated damage of the skin, preventing wrinkle formation, skin laxity, and the formation of visible red spots.[52] Salicylic acid, an AMA, is also a widely used ingredient in OTC and prescription-only cosmetic preparations for a variety of applications. Salicylic acid is a lipophilic (i.e., fat soluble), which makes it useful in subjects with oily skin. It has keratolytic (peeling) effect and has antibacterial properties.[53]

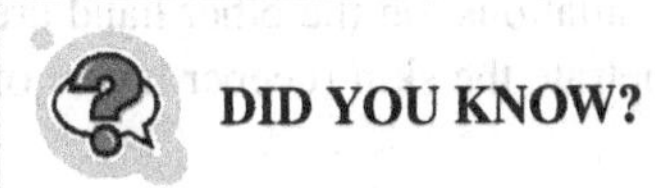

Chemical peeling is a minimally invasive procedure used for the cosmetic improvement of the skin. During this procedure, a chemical agent of a defined strength is applied to the skin, which causes a controlled destruction of the skin layers (epidermis and/or dermis) with subsequent regeneration and rejuvenation of the tissues, resulting in improvement of texture and surface abnormality.[54] Chemical peels are classified into three categories, depending on the depth of the wound created by the peel: (a) superficial peels penetrate the epidermis only, causing surface desquamation[55]; (b) medium-depth peels damage the entire epidermis and upper part of the dermis, causing coagulation of membrane proteins and destroying living cells[56]; and (c) deep peels create a wound to the level of the mid-dermis, leading to protein coagulation and complete peeling of the epidermis as well as restructure of the basal layer and restoration of the dermal structure.[56] HAs, primarily glycolic acid, lactic acid, and salicylic acid, are often used as superficial peeling agents in practice, either alone or in combination with other dermatologic devices, to provide complementary benefits and enhance outcomes of procedures, including injectable fillers, microdermabrasion, and laser therapy. Chemical peeling is usually a safe procedure when performed by qualified and experienced professionals. However, these procedures are not recommended for everyone. Patients should visit their dermatologists and find out whether this treatment is indicated for them.

- **Sunscreens** provide photoprotection by physically blocking the sunlight or absorbing and transforming it in the skin. As sunscreens are discussed in a separate section (Section 5 of this chapter), these ingredients are just mentioned here as anti-aging ingredients. Although there are several treatments available for aged skin, prevention of extrinsic aging remains the best approach to prevent aging. This includes avoiding sun exposure and using sunscreens when sun exposure cannot be avoided, avoiding smoking and pollution, preventing unnecessary stretching of the skin, eating a diet rich in vegetables and fruits, as well as consuming antioxidants and using topical antioxidants.

Formulation Considerations for Topical Anti-Aging Products

The majority of cosmetic formulations applied to the skin are not required to penetrate into the deeper skin layers as they act topically on the skin surface. Skin cleansers remove dirt and soil, moisturizers hydrate the SC and smoothens it via supplying the SC with lipids, while makeup products enhance the appearance, even out skin tone, and cover skin problems. Most anti-aging formulations, on the other hand are examples for formulations that are necessary to penetrate the skin (deeper layers of epidermis and dermis) for optimal effects.

Penetration of the actives through the skin is not an easy process as many would think. First, the actives need to be in a dissolved state in order to be absorbed into the skin (penetrate the skin). After applying the emulsion (the majority of cosmetic products) to the skin, certain ingredients, including water, start to evaporate. Evaporation changes the composition and structure of the emulsion on the skin surface. If the active is dissolved in the oil phase, water evaporation is a beneficial phenomenon since it may accelerate the delivery of the active into the skin. On the other hand, if the active is dissolved in the water phase, water evaporation can significantly affect the amount of actives remaining in the dissolved state, i.e., staying "deliverable." In such cases, water evaporation retardants have to be used to slow down the water's evaporation rate as much as possible.[57]

From the active's perspective, there are many factors that can influence its successful delivery to the skin. The SC has a lipoid characteristic due to its lipid components. The deeper layers of the epidermis and also the dermis are more aqueous compared to the SC. An ingredient has to penetrate the SC from the vehicle and be able to diffuse through the SC to the viable epidermis and upper dermis in order to be efficacious. There are a number of factors influencing the ability of a given molecule to penetrate through the SC and get down to the dermal layer. The two most important considerations are the **molecular size** and **solubility**.[57] In general, molecules with more than 500 Da do not penetrate easily through the SC. As for solubility, hydrophobic (i.e., lipophilic) components have a better chance of penetrating the lipids in the SC than water-soluble components. However, highly lipophilic components may move into the SC and stay there instead of moving to the more aqueous epidermis and dermis. Additionally, unionized molecules penetrate the skin better than an ionized molecule. As pH can significantly influence the ionization state of molecules, the formulations' pH is always critical to have optimal penetration. Beyond these prime determinants, there are additional molecular properties that can affect penetration of drugs into the skin. Such factors include binding of the active to skin components (such as lipids, proteins, and enzymes), interaction between the active and the tissue, and particle size if the active is in a suspended state in the formulation.[58]

There are a number of techniques to increase the penetration of the ingredients through the skin. Examples include formulating a supersaturated solution of the main ingredient in the vehicle using specific solvents, lowering the melting point of the main ingredient, altering the solubility of the main ingredient, using ingredient that can enhance the penetration of the main ingredient (the so-called penetration enhancers), and many others.

●**The majority of anti-aging formulations are emulsions (both lotions and creams) since they offer cosmetically elegant formulations with the potential to include oil- and water-soluble components at the same time.** When water evaporates from O/W emulsions, a thin film is left behind on the skin that contributes to the silky feeling on the skin. In the case of W/O emulsions, the film left behind is much oilier, providing a higher degree of moisturization and/or occlusive effect. However, their aesthetic properties are not as appealing to patients.

In addition to simple macroemulsions, other delivery systems have also been developed to deliver ingredients to the skin surface, including gels, foams, masks, cloths, and serums, as well as more advanced delivery systems, such as liposomes, niosomes, solid lipid nanoparticles (SLNs), and nanostructured lipid carriers (NLCs), among others. The latter systems can be also formulated as lotions, creams, and gels. The main advantage of these systems is that they can increase the stability of the sensitive, unstable ingredients, reduce the irritating potential of irritating ingredients, and also increase the aesthetics (i.e., skin feel) of formulations. A short summary of the main characteristics of some advanced delivery systems is provided here.

DID YOU KNOW?

Petrolatum is a very effective ingredient for skin healing; however, patients do not like applying petrolatum to their skin since it feels sticky and greasy and is not easy to wash off the skin. When formulated into advanced delivery systems, such as liposomes, the system could allow small amounts of petrolatum to be released onto the skin surface, avoiding the aesthetic drawbacks.[59]

- **Liposomes** are microscopic spherical vesicles that have a lipid bilayer structure resembling the natural structure of the biomembranes. In the case of unilamellar liposomes, the core (interior) is hydrophilic, while the inside of the phospholipid layers is lipophilic (see Figure 3.19a). Based on this feature, both hydrophilic and lipophilic active ingredients can be incorporated into liposomes to enhance their skin penetration and thus efficacy. In addition, the phospholipids used as

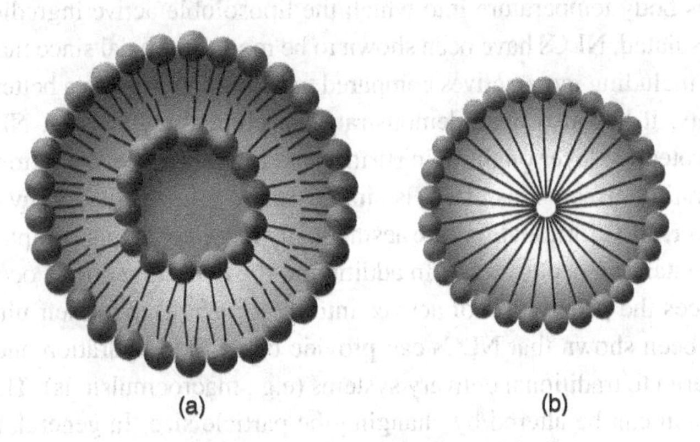

Figure 3.19 (a) Liposome and (b) micelle.

the membrane components can function as moisturizers to the skin.[59] Despite these positive features, a major problem is the limited physical stability of liposomes in emulsions. Liposomes tend to fuse with one another and with the surfactant layer of oil droplets, shortening the shelf-life of such formulations.[60] Additionally, studies show that they are rigid and cannot deeply penetrate the skin.[61]

DID YOU KNOW?

Liposomes may look similar to micelles; however, these are not similar entities (see Figure 3.19a, b). They are composed of a lipid bilayer and have one or more hydrophilic compartments (depending on the number of lipid bilayers) and lipophilic compartments. Micelles are monolayered structures, and they are surfactant aggregates. They typically have a lipophilic core and a hydrophilic surface, or a hydrophilic core and a lipophilic surface (in the case of inverted micelles).

- **Niosomes** are also spherical vesicles similar to liposomes; however, instead of phospholipids, they are made of nonionic surfactants of the alkyl and dialkyl polyglycerol ether class and cholesterol.[62] Niosomes have a structure consisting of hydrophilic, amphiphilic, and lipophilic moieties together, and therefore, they can accommodate active ingredients with a wide range of solubility (similar to liposomes).[63] Niosomes do not contain phospholipids; therefore, they do not provide the skin surface with a moisturizing effect. However, they tend to have better oxidative stability compared to liposomes.
- SLNs and NLCs are both specific delivery systems belonging to the nanomaterial scale. Both systems contain solid particles (i.e., solid lipids) at room as well as body temperature into which the liposoluble active ingredients can be encapsulated. NLCs have been shown to be more beneficial since they are capable of including more actives compared to SLNs and provide a better long-term stability. It has also been demonstrated in clinical studies that NLCs form a thin protective layer on the skin surface and lower water evaporation, thus have a hydrating effect. This effect is similar to the effect provided by occlusives; however, NLC formulations are aesthetically more appealing compared to products containing occlusives.[64] In addition to the hydrating effect, occlusion also enhances the penetration of actives into the skin.[65] Due to their ultrafine size, it has been shown that NLCs can provide better skin hydration and occlusion compared to traditional delivery systems (e.g., macroemulsions). The degree of occlusion can be altered by changing the particle size. In general, the smaller the size of the particles, the less water evaporates from the skin.[66,67]

It is well known that the vehicle plays an important role in the penetration of the actives and efficacy of formulations. In addition, using an appropriate vehicle can also decrease the irritation potential of retinoids, for example, and can increase the stability of unstable ingredients, such as vitamin C (which has oxygen sensitivity).[68] Therefore, selecting an appropriate vehicle is essential for optimal results.

Formulation Challenges for Common Anti-Aging Ingredients

When formulating anti-aging products, enhancing the penetration of the ingredients to the deeper viable layers of the skin (if needed) is just part of the challenge. Many ingredients are highly sensitive to light, oxygen, and aqueous environment, and therefore, additional precautions should be taken during product formulation. Some examples for these challenges are discussed here.

- When working with **retinoids**, safety, efficacy, and tolerability of the formulations should be taken into account at the same time. Although certain retinoids are proven to be more efficacious than other, their tolerability is worse, which leads to discontinuation of the therapy early on.[69] Therefore, selection of the active ingredient and optimization of the formulation to reduce the irritation potential, even with the inclusion of anti-inflammatory ingredients, are essential.
- **Retinoids** are unstable, especially in the presence of oxygen and light. To increase their stability, special delivery systems can be employed, as well as both formulation and packaging should be done in an environment with minimal exposure to oxygen and light.[70]
- **Vitamin C** and its derivatives are highly oxygen sensitive, leading to degradation of the active compound. Despite various stabilization strategies, such as exclusion of oxygen during formulation and packaging, using advanced delivery systems, changing pH of the formulation, and using antioxidants, ascorbate stability is still a challenge.[70] Another issue with vitamin C is its poor penetration into the skin. It is a water-soluble compound, and due to its polarity, it will not readily enter the SC. Similar to improving its stability, various approaches have been made to enhance its skin penetration, including lowering the pH and using derivatives. At pH 3.5, the ionic charge on the molecule is removed, and it is transported well across the SC.[71]
- **AHAs** have a low pH. It has been shown that the lower the pH, the greater the efficacy is and, unfortunately, also the irritation.[72] As for the formulation, this low pH may be challenging from a thickening perspective since some common thickeners, such as carbomers, are not effective at such low pH.
- Delivery of peptides and proteins can be very challenging due to their solubility, charge, and size. It has been shown that the addition of a lipophilic chain to the molecule highly increases the penetration.[73] Additionally, using advanced delivery systems, such as liposomes, may also help in penetration. Additional challenges include their limited stability in aqueous environments[74] as well as their high cost.

As seen in this short summary, there are many challenges that formulators have to overcome in order to have a stable, aesthetically appealing, safe, effective, and well-tolerated formulation.

Safety Issues Regarding the Use of Topical Non-invasive Anti-Aging Ingredients

Anti-aging ingredients, as seen earlier, can absorb and significantly alter the structure and physiological function of the skin. As most products are used for a long period of time and ingredients may have mild, moderate, or severe side effects, **concerns arose regarding the safe use of some anti-aging ingredients, including AHAs, topical retinoids, and peptides and proteins.** The major concerns are discussed here.

Alpha Hydroxy Acids (AHAs) One of the major concerns with regard to the use of AHAs is that they can increase the skin's UV sensitivity. The type and extent of additional side effects of AHA therapy depend on the type and concentration of the ingredients used, but generally, they can include burning, rash, pigmentary changes, peeling, and itching. The more serious adverse reactions appear to occur most often with products that cause the greatest degree of exfoliation, such as chemical peels. Both the FDA and the CIR investigated the safety of AHAs. The FDA concluded that sensitivity is reversible and does not last long after discontinuing the use of the AHA cream. In their study, one week after the treatments were discontinued, no significant differences in UV sensitivity among the various skin sites were found. Additional studies concluded that glycolic acid did not affect photocarcinogenesis (i.e., the development of cancer cells associated with exposure to light) in mice and that salicylic acid had a photoprotective effect (protection against the effects of light) in mice.[75] Currently, the FDA has the following labeling recommendation for topically applied cosmetics containing AHAs: "Sunburn Alert: This product contains an alpha hydroxy acid (AHA) that may increase your skin's sensitivity to the sun and particularly the possibility of sunburn. Use a sunscreen, wearprotective clothing, and limit sun exposure while using this product and for a week afterwards."[76] Its purpose is to educate consumers about the potential safety issues. The CIR Expert Panel has also evaluated the safety of AHAs. To reduce the risk of irritation, they recommended that AHAs concentration should not be higher than 10% and the pH should be at or above 3.5. In addition, the CIR Panel recommended that AHA-containing products should be formulated to avoid enhancing sun sensitivity and that consumers should be advised to use sun protection daily.

Topical Retinoids Safety concerns for the use of topical retinoids have been discussed earlier (under ingredients). The major concerns include retinoid dermatitis, which diminishes within a few weeks of use, and sun sensitivity, which also disappears after a few months of therapy. Over the past decades of research, no systemic side effects on long-term treatment with topical retinoids have been observed in young adults. This is related to the insignificant systemic absorption of the applied doses of retinoids.[77] Nevertheless, pregnant women are generally advised to cease topical retinoid treatment during pregnancy.

Peptides and Proteins Cosmetics, including anti-aging formulations, contain proteins and peptides for their beneficial effects. Certain ingredients can be found in nature, including wheat and oat proteins (in plants), as well as collagen and elastin (in animals). There were severe concerns in the past century about the use of beef proteins and hormonal extracts from bovine organs and glands in cosmetics due to their potential to cause mad cow disease (medically termed bovine spongiform encephalopathy, BSE). Mad cow disease is a fatal disorder of cattle, belonging to a group of progressive degenerative neurological diseases known as transmissible spongiform encephalopathies (TSEs). "Bovine" means that the disease affects cows, "spongiform" refers to the way the brain of a sick cow looks spongy under a microscope, and "encephalopathy" indicates that it is a disease of the brain. Currently, the most accepted theory is that the agent causing BSE is a modified form of a normal cell protein known as a prion. It is believed that BSE was spread via meat and bone meal fed to cattle. The vast majority of BSE cases (over 90%) were reported from the United Kingdom (UK), and other places included some European countries. In the US, only four cases have been reported to date.[78] There are TSEs affecting humans, and these include the classical Creutzfeldt–Jakob disease (CJD) and variant Creutzfeldt–Jakob disease (vCJD), among others. The classical CJD is an extremely rare disease, while vCJD cases were reported in a higher number in the past two decades, mainly from the UK.[79,80] It is not known whether using topical cosmetics containing BSE agents on intact, broken, or abraded skin would lead to neurodegenerative diseases in human; however, it cannot be ruled out. To prevent any infections, certain cattle-sourced materials are on the prohibited ingredient list in the US and cannot be used to manufacture cosmetics. These include ingredients sourced from specific risk organs, such as the brain, eyes, spinal cord, and others. It is important to emphasize that milk and milk products, among others, are not included in this list.[81]

Part 2: Anti-Acne Products

Introduction ●**Acne vulgaris is a common dermatological disorder of the pilosebaceous unit that has a complex pathophysiology and can be triggered by a number of factors.** It primarily affects teenagers; however, it can also affect adults. Acne is estimated to affect 40–50 million people in the US, with an 85% prevalence in those aged 12–24 years,[82] which makes it the most common skin disorder that dermatologists treat.[83,84] With regard to gender, acne is significantly higher among women than among men in all age groups.[85] Acne primarily affects the skin; however, it may also cause distress and, in some individuals, contributes to lowered self-image.

Development of Acne The pathogenesis of acne vulgaris can be attributed to multiple factors. The major factors include the following (also see Figure 3.20):

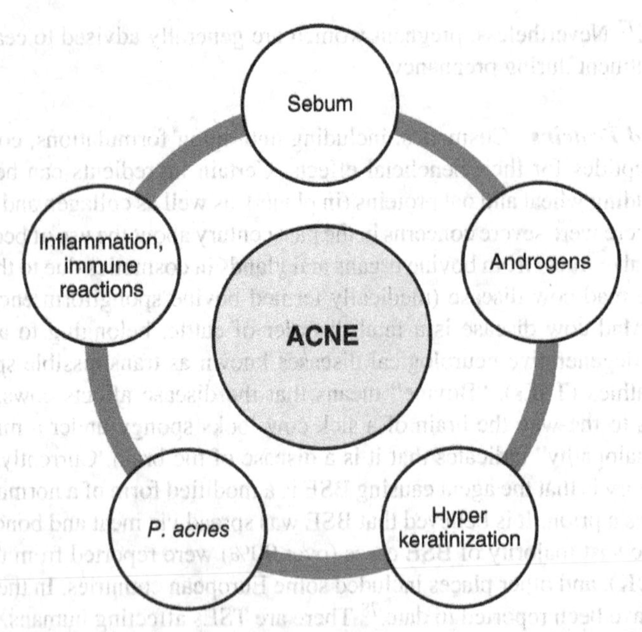

Figure 3.20 Major factors contributing to the development of acne vulgaris.

- Increased sebum production by the **pilosebaceous** unit
- Increased androgen activity
- Abnormally increased keratinocyte production and desquamation with subsequent plugging of the follicles
- Microbial colonization of the **pilosebaceous** unit by a commensal bacterial *Propionibacterium acnes* (*P. acnes*)
- Inflammation and immune reactions.[86–88]

Acne occurs when the hair follicles become plugged with oil and dead skin cells. The pathogenesis of acne is still not clearly understood, and the development process can be summarized as follows. For reasons not clearly understood, the body produces more epidermal cells (the process is referred to as "hyperkeratinization") and excess sebum. Unlike the normal case where cells that are shed within the follicle are swept out of the follicle onto the surface of the skin along with the secreted sebum, in acne, the excess amount of sebum and cells released into the follicle cannot get out to the skin surface, but they build up in the hair follicles. They form a plug, and it creates a sebum-rich, oxygen-poor environment, which is ideal for the proliferation of *P. acnes*. At the same time, the draining duct widens and the sebaceous gland grows larger and wider due to the accumulated material. These bacteria produce enzymes that hydrolyze the triglycerides in the sebum, and this leads to further hyperkeratinization and inflammation. This early stage of acne is called a **microcomedo**. Microcomedones occur under the skin and are small enough that the human eye cannot see them. However, they require special attention with regard to the development of therapeutic strategies because they represent the primary precursor lesions that will evolve into either noninflammatory or inflammatory lesions.[89]

Causes of Acne and Potential Exacerbating Factors

The exact underlying cause of acne and follicular blockage is not known. There are many factors believed to contribute to the development of acne. Here is a summary of the most important factors. Their relevance and the scientific evidence behind them are also reviewed.

- **Hormones**: Androgens are usually referred to as male hormones, which can also be found in women's blood, but in a smaller amount. Their level increases in boys and girls during puberty and causes the sebaceous glands to enlarge and produce more sebum. It is believed that androgens play a crucial role in the development of acne.[90] Hormonal changes related to pregnancy, and the use of contraceptives can also affect sebum production. Additionally, numerous other hormones, including growth hormone, and insulin-like growth factor have been identified as potential causative factors for the development of acne.[91]
- **P. acnes**: *P. acnes* is an anaerobic bacteria present in acne lesions. The significance of the involvement of *P. acnes* in acne pathogenesis is still controversial. While several studies showed that the number of *P. acnes* is higher in acne patients than in healthy individuals, other studies found no difference between the numbers of *P. acnes* in affected and non-affected follicles. Nevertheless, an abnormal colonization of *P. acnes* has been implicated in the occurrence of acne by inducing inflammation. It can promote inflammation through a variety of mechanisms. *P. acnes* stimulates inflammation by producing proinflammatory mediators that diffuse through the follicle wall.[92] Hypersensitivity to *P. acnes* may also explain why some individuals develop inflammatory acne vulgaris while others do not.[93]
- **Diet**: The influence of diet on the induction and aggravation of acne has been a matter of intense debate over the past few years. Epidemiologic, observational, and experimental evidence suggest an association between diet and acne. Major dietary factors include dairy products, mainly milk as well as foods with glycemic load, such as bread, bagels, and spaghetti.[94] This evidence, to date, does not demonstrate that diet causes acne, but can aggravate or promote it.[95–97]
- **Sunlight**: The effects of sunlight on acne and light therapy in acne treatment have also been studied; however, it is not clear whether UV radiation or visible light worsens, improves, or has no effect on acne. Some studies suggest that sunlight has a positive effect on acne,[98] while other studies did not demonstrate any

significant effects.[99] This may be a result of such studies inherently being difficult to conduct. Light therapy is increasingly used in acne treatment, and some studies have shown evidence of sunlight being beneficial to acne patients. However, the risk of development of skin cancers must be taken into account, and therefore, care should be taken in advising therapeutic sun exposure.[100] Additionally, photosensitivity is an issue with commonly used, efficacious medical treatments of acne, including tetracyclines and isotretinoin.[101]

- **Facial Hygiene and Face Cleansing**: A common perception is that poor levels of hygiene lead to the development or exacerbation of acne.[102] Due to this perception, patients tend to cleanse diligently, in addition to using topical treatments, most of which are drying and initially irritating to the skin. Aggressive cleansing, however, can lead to disruption of the skin barrier, increased TEWL, rough and irritable skin, increased bacterial colonization, increased comedonal formation, burning, and stinging. These negative effects can make patients unable to tolerate topical medications and lead to poor patient adherence.[103,104] Numerous studies have been conducted to test whether washing helps acne treatment or worsens it. Most studies had severe limitations, and therefore, the evidence is weak. In general, studies show that cleansing with facial cleansers containing beneficial ingredients for acne is helpful in its treatment.[105] The most important factors that should be taken into consideration when selecting facial washes for acne patients are the following. As discussed in Section 2 of this chapter, regular soaps can severely damage even normal skin. Therefore, they can be very damaging to acne-prone skin and should be avoided. Acne cleansers should have a pH similar to that of the skin. Facials scrubs containing abrasive particles can also traumatize the skin and can exacerbate the symptoms and, therefore, are often recommended by dermatologists not to be used for acne patients. Products containing dyes and perfumes should be avoided for acne patients since they may exacerbate acne and irritate the skin. Highly foaming liquid cleansers are beneficial for oily patients, while dry, sun-damaged, and sensitive skin requires milder cleansers that may also have added moisturizing benefits. Toners should only be used on oily areas of the skin, such as the forehead, nose, and chin. If used on other parts, they may dry out the skin due to their alcohol content and trigger oil production.[105]
- **Makeup Use**: For many years, it has been thought that facial makeup products, including facial foundations and concealers, trigger and/or worsen acne. Therefore, they were recommended to be avoided by dermatologists. However, it has been shown in new studies that acne patients can benefit from the use of color cosmetic products even using those without any active ingredients since these products can cover up blemishes, redness, and even scars and can significantly improve patients' self-image, self-esteem, and QoL.[106] Find more detail on this topic in Chapter 4.
- **Additional Factors**: In addition to the factors listed earlier, stress, genetic factors,[107] and certain medications, including corticosteroids,[108] have also been

linked to the development and/or exacerbation of acne. The role of smoking in triggering acne is still controversial.[109]

Symptoms and Types of Acne Vulgaris

●**The clinical presentation of acne can range from mild comedonal (i.e., non-inflammatory) form to severe inflammatory cystic acne.** It primarily affects the areas of skin with the densest population of sebaceous follicles; these areas include the face, the upper part of the chest, and the back.[110]

- Non-inflammatory lesions (see Figure 3.21) consist of open and closed comedones, which are not inflamed and red because the follicle walls are intact. Unlike microcomedones, these lesions are large enough to be seen by the naked eye.[111]
  - **Blackheads**, also known as open comedones, are follicles that have a wider opening filled with sebum and dead cells. They are called blackheads due to the dark appearance of the plugs in the follicles. The dark color is caused by the exposure of the top of the comedo to oxygen. Open comedones rarely develop into inflammatory lesions.
  - **Whiteheads**, also known as closed comedones, occur when the follicle's opening remains closed. They are more difficult to see; they usually have normal skin color. A closed comedo is not an inflammatory lesion; however, it is much more likely to progress into an inflammatory lesion than an open comedo.

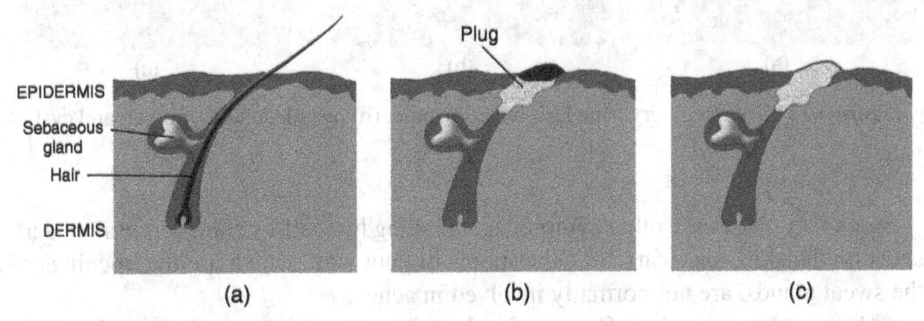

Figure 3.21 Noninflammatory acne lesions. (a) normal, intact hair follicle, (b) blackhead, and (c) whitehead.

- Inflammatory acne (see Figure 3.22) lesions are characterized by the presence of papules, pustules, nodules, and cysts. These lesions are red and inflamed due to blood penetrating the follicle after the follicle has ruptured. This classification is important since the two types are treated differently. The closed comedo becomes larger and more packed due to debris and inflammation from *P. acnes*.

P. acnes produces enzymes that make the follicle wall weaker, and eventually it ruptures. At this point, white blood cells along with red blood cells migrate to the place of rupture to fight against *P. acnes* and contain the rupture. This is when the lesion turns red. At this point, the closed comedo turns into a papule.[111]

- **Papules** are small, raised, usually red, and tender bumps under the skin. These are the primary inflammatory lesions.
- **Pustules** are red, tender bumps with white pus at their tips. While fighting the infection, white blood cells die and create pus. As the pus reaches the skin surface, a pustule is formed, also called a pimple.
- A **nodule** is a deep lesion, similar to a papule, but much deeper in the dermis. Nodules often occur when rupture of the follicle wall happens deep down at the bottom of the wall. Nodules are hard to touch, more painful than other lesion types, and deep red or purple in color. Nodules may even involve more than one follicle, forming a large pocket of infection.
- A **cyst** is a large pus-filled lesion that is the result of a severe inflammatory reaction. As it affects deeper layers of the skin, cystic acne frequently causes scarring.

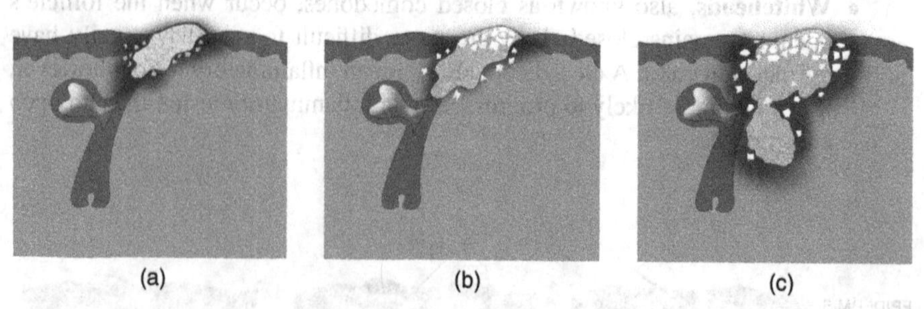

Figure 3.22 Inflammatory acne lesions. (a) papule, (b) pustule, and (c) nodule and cyst.

Acne may also cause other symptoms, including low self-esteem, depression, dark spots on the skin, and scars.[112] Other pores in your skin, which are the openings of the sweat glands, are not normally involved in acne.

❽**Acne vulgaris is also often graded based on its severity, including the number of comedones, inflammatory lesions, total lesion count, and cysts.**[113] Acne is categorized broadly into mild, moderate, and severe forms. This classification is important as it is used as a basis for selecting the treatment.[114]

- Mild acne is typically limited to the face and is characterized by the presence of noninflammatory closed and open comedones with few inflammatory lesions.
- Moderate acne is characterized by an increased number of inflammatory papules and pustules on the face and also often affects other body parts.

- Finally, acne is considered to be severe when nodules and cysts are present. In these cases, facial lesions are often accompanied by widespread disease of other body parts, including the neck, chest, and back.[115]

Treatment of Acne Vulgaris

The multifactorial etiology of acne vulgaris makes it challenging to treat. Careful assessment of the morphology and severity of acne is an important first step in management since lesion morphology largely dictates the optimal treatment approach.[116] Today, there are many effective acne treatments. However, it does not mean that every acne treatment works for everyone who has acne. Treatment should be designed to target precursor lesions (microcomedones) and active inflammatory lesions.

❶**Milder cases are best managed with OTC and prescription-only topical regimens, whereas systemic, prescription-only drugs are indicated in more severe cases.** Current treatments include topical retinoids, benzoyl peroxide, topical and systemic antibiotics, azelaic acid, systemic isotretinoin, and combined oral contraceptives for women.[117,118] Adjunctive and/or emerging approaches include chemical peels, optical treatments, as well as complementary and alternative medications.

Despite the claims, acne treatment does not work overnight. With most treatments, one may see the first signs of improvement in 4–8 weeks, and the skin may often get worse before it gets better due to the side effects of commonly used medications. The management of acne can be complex, often requiring aggressive combination therapy and a long-term therapeutic strategy.[114] Maintenance therapy is usually necessary for many acne patients, as acne lesions have been shown to return after discontinuing a successful treatment regimen.

Topical Treatment Options Topical therapies generally employ OTC actives, such as benzoyl peroxide and salicylic acid, as well as prescription-only antibiotics and retinoids. OTC active ingredients can be found in the relevant part of the OTC monograph, which lists ingredients that are generally recognized as safe and effective (GRASE) for the treatment of acne. These include the following ingredients:

- **Benzoyl Peroxide (BPO)**: It is a nonantibiotic antimicrobial agent that can kill bacteria by producing reactive oxygen species within the clogged pore. It increases cell turnover, cleans pores, desquamates the skin, and also has anti-inflammatory properties. BPO can produce rapid improvement in the inflammatory lesions.[119] It is a mainstay treatment of mild-to-moderate acne, often combined with antibiotics and/or retinoids. BPO is available as topical OTC products in a concentration of 2.5–10% in various forms, including leave-on products, such as creams, gels, and lotions, as well as rinse-off products, such as facial wash. Its main side effects include peeling, dryness, burning, and redness of the skin. Irritation typically resolves with continued use during the first month of treatment. Patients should also be warned that BPO may bleach clothing, bedding, and hair.[120]

- **Salicylic Acid (SA)**: It has been used for years for acne treatment. SA has desquamating and comedolytic properties. Studies indicate that it is less potent than topical retinoids, however, but better tolerated by patients.[119] SA is an option for patients who cannot tolerate a topical retinoid due to skin irritation.[121] It is available as OTC products in concentrations of 0.5–2% in various product forms, including leave-on products, such as lotions, creams, and foams, as well as cleansing products, such as facial wash gels, toners, and cleansing pads. Additionally, there are several OTC drug–cosmetic combination products containing SA, including foundations and concealers, available on the market. These products have an additional benefit over "simple" OTC formulations since they help cover skin imperfections. Temporary typical side effects of the topical SA treatment include skin dryness, redness, scaling, itching, and burning. They typically dissipate in a few weeks.
- **Resorcinol and sulfur** have also been used in the therapy for a long time; however, they are not considered a first-line therapy for acne. These active ingredients are generally available as creams, masks, ointments, and soap bars.[118]

DID YOU KNOW?

The term "comedolytic" means lysis (i.e., disintegration) of a comedo via lysis of plugged corneocytes, whereas keratolytic is defined as the lysis of superficial corneocytes, nails, and hair via lysis of keratin.[122]

Topical ingredients available for prescription-only drugs include the following:

- **Topical Retinoids**: They are vitamin A derivatives that normalize the abnormal desquamation pattern in sebaceous follicles, decrease the coherence of follicular keratinocytes, and prevent the formation of new microcomedones. Some retinoids also have anti-inflammatory properties.[123] Topical retinoids are effective ingredients and, therefore, are often recommended for all cases of acne, except when oral retinoids are used. Mild noninflammatory comedonal acne may be treated with retinoid monotherapy. When inflammatory lesions are present, retinoids are combined with antimicrobial therapy or BPO. As retinoids prevent the development of microcomedones, they can also be used for maintenance therapy.[88] Currently available FDA-approved topical retinoids include tretinoin, adapalane, and tazarotene; all are available for prescription-only drugs in the US.[89] These ingredients are available in cream, gel, liquid, and microsphere formulations, each at multiple concentrations. The use of retinoids is limited by their side effects, including transient skin irritation, burning sensation, redness, itching, and peeling. These negative effects can be prevented by selecting lower concentration of active ingredients or modifying the vehicle of the active (e.g., formulating a cream or foam instead of a gel). Improvement of acne symptoms usually occurs within weeks, and the maximum benefit can be typically expected after 3–4 months of treatment.[124]

- **Topical Antibiotics**: They are used for mild-to-moderate acne when inflammatory lesions are present.[125] The commonly used antibiotics for acne include clindamycin and erythromycin.[125] Antibiotics have bacteriostatic and anti-inflammatory properties.[126] They are available as gels, creams, lotions, foams, toners, and pads. An emerging issue with regard to antibiotic use is antibiotic resistance, which can decrease the antibiotic resistance over time. Therefore, antibiotic monotherapy and maintenance therapy alone are not recommended, nor the combination of topical and systemic antibiotics. Current guidelines recommend combining antibiotics with retinoids and/or BPO.[117] BPO can minimize bacterial resistance, while retinoids can provide synergistic comedolytic and anti-inflammatory properties.[127]
- Additional topical ingredients include **azelaic acid**. It is an alternative to retinoids that has comedolytic, antimicrobial, and anti-inflammatory properties.[117] It is, however, not approved by the FDA in the US.

As discussed earlier, in order to achieve optimal results, topical OTC drugs are often combined with topical antibiotics and retinoids. Benefits of such a combination therapy can be the complementary mechanism of action, reduced risk of antibiotic resistance, and improved treatment outcomes. Today, there is an increasing number of actual combination products available on the market.[128,129]

Systemic Treatment Systemic agents are typically recommended for patients with moderate-to-severe acne. These therapies are also useful in patients with larger surfaces affected, including the neck, chest, and back. In their case, topical treatment would be difficult. Systemic agents include antibiotics, hormones, and an oral retinoid, isotretinoin.

- **Oral Antibiotics**: Antibiotics have been widely used for many years in the management of moderate and severe acne.[88] Similar to topical antibiotics, systemic agents have antimicrobial and anti-inflammatory effects.[120] Most commonly, doxycycline, minocycline, tetracycline, and erythromycin are used. Typically, antibiotic selection is driven by their side effect profiles and patterns of *P. acnes* resistance. A major problem affecting antibiotic therapy of acne has been bacterial resistance, which has been increasing. As with topical antibiotics, oral antibiotics should be combined with other agents to minimize the development of bacterial resistance and improve treatment efficacy. Oral antibiotics should be used with topical retinoids or BPO.[130] For this reason, dermatologists usually recommend tapering off these medications (i.e., continuously decreasing their dose) as soon as the symptoms begin to improve or as soon as it becomes clear that the drugs do not help.[130] Antibiotics may cause side effects, such as an upset stomach, dizziness, or skin discoloration. Doxycycline can increase sun sensitivity; tetracyclines can cause teeth discoloration, while monocyline can lead to skin hyperpigmentation.

- **Hormonal Therapy**: Hormonal therapy is a useful adjunct therapy in women with moderate-to-severe acne. It is especially beneficial for those who desire oral contraception or in whom traditional therapy has failed.[131] Although hormonal therapy may help reduce and/or prevent outbreaks, it is not effective for existing lesions; therefore, it is used as an adjunct rather than a stand-alone therapy.[132] Hormonal agents commonly used for the treatment of acne are available in two major forms: combination oral contraceptives (i.e., contraceptives containing both an estrogen and a progestin), which suppress ovarian androgen production, and androgen receptor blockers, which block the effect of androgens on the sebaceous gland.[133] The available androgen receptor blockers are spironolactone, cyproterone acetate, and flutamide; however, none of these are approved by the FDA for acne treatment. In addition to being useful adjuncts, oral contraceptives can also be used as part of the maintenance therapy, usually in combination with topical retinoids.[132] Common side effects of hormonal therapy include headache, breast tenderness, nausea, and depression. The most serious potential complication is a slightly increased risk of heart disease, high blood pressure, and blood clots.[133]
- **Isotretinoin**: Isotretinoin is an oral retinoid; it is a powerful medication for the treatment of severe acne and moderate acne that does not respond to other treatments.[134] Isotretinoin targets all major components involved in the development of acne: normalizes follicular desquamation, decreases sebum secretion, inhibits the growth of *P. acnes*, and exerts anti-inflammatory effects.[135] Due to its broad effects, it is usually used as a monotherapy. The common side effects caused by isotretinoin include dryness of the skin, eyes, mouth, lips, and nose; itching; nosebleeds; muscle aches; sun sensitivity; and poor night vision. The drug may also increase the levels of triglycerides and cholesterol in the blood and may increase liver enzyme levels.[136] In addition, isotretinoin may be associated with an increased risk of depression and suicide.[137] A major concern when taking isotretinoin is the risk for malformations in the developing fetus. Women of childbearing age have to take certain precautions when taking this medication, and these are discussed under the safety concerns.

Additional Treatments Additional treatment options include the use of chemical peels, comedo removal, optical treatments, and complementary and alternative medications, among others.

- **Chemical Peels**: Both AHA-based and AMA-based peeling preparations have been used in the treatment of acne. AHAs act by desquamating the SC, which results in a smoother appearance. In addition, glycolic acid has a moderate growth inhibitory and bactericidal effect on *P. acnes*.[138] SA's properties were discussed earlier.

- **Comedo Extraction**: There are two ways to release the contents of comedones: squeezing with fingertips and using a comedo extractor. There is limited evidence published that comedo removal for acne treatment is effective. However, it can greatly improve the patient's appearance, which may positively impact compliance with the treatment program.
- **Optical therapies** that have been used to treat acne include broad-spectrum continuous-wave visible light (blue and red), intense pulsed light, pulsed dye lasers, potassium titanyl phosphate lasers, photodynamic therapy (PDT), and pulsed diode laser. Although optical therapy may improve acne initially, a standardized treatment protocol, longer-term outcomes, comparisons with conventional acne therapies, and widespread clinical experience are still lacking. Therefore, optical treatments are not included among first-line treatments.[101]
- The use of **herbal and alternative therapies** in acne is widespread. Commonly used herbal ingredients include aloe vera, fruit-derived acids, and tea tree oil. Although these appear to be well tolerated, very limited data exist regarding the safety and efficacy of such agents.[114]
- **Dietary restriction** has not been demonstrated to be of benefit in the treatment of acne.

Formulation Considerations

Similar to anti-aging formulations, anti-acne formulations also have to penetrate the skin and follicles in order to topically affect acne. All formulation considerations discussed under anti-aging products, including the solubility, molecular size, and charge of the actives, are also relevant for anti-acne products.

Formulation development is essential to the successful treatment of acne. The formulation, including the vehicle, determines how often a product should be applied; it influences efficacy and tolerability and has a huge influence on patient compliance. Selection and appropriate design of the vehicle for an anti-acne drug is the first step in formulation. When designing the vehicle, it should be kept in mind that as soon as a formulation is applied to the skin, water starts to evaporate and the formulation mixes with the lipids found on the skin surface, which slightly alters its composition. The formulation still has to be able to deliver the actives from this slightly changed composition. An effective topical formulation has to be able to provide chemical stability, deliver active ingredients at optimal concentrations for efficacy, be cosmetically acceptable, and not add side effects of its own.[139]

As discussed earlier, the currently used topical active ingredients, primarily retinoids and BPO, often cause temporary irritation, burning, redness, and itching upon application. These side effects often lead to an early discontinuation of the anti-acne therapy. Although these side effects are temporary and cease within 3–4 weeks of treatment, many patients do discontinue the therapy early on.[140] In addition to the irritation effect of the active ingredients, many components of the vehicle are also potentially irritating. Examples include surfactants, fragrances, preservatives, high levels of organic solvents, and alcohol. Today, anti-acne products are available in various forms, including solutions, gels, lotions, creams, and foams, among others. Novel drug delivery systems, including those discussed under anti-aging products, have the potential to improve the topical delivery of anti-acne ingredients while decreasing their side effects. It has been shown that liposomally encapsulated tretinoin is much better tolerated than the gel.[141] Additionally, it has been demonstrated that tazarotene foam's tolerability is better than that of the gel.[142] Additional advanced delivery systems include microspheres and microsponges that have been developed to increase the transdermal penetration and thus the efficacy of topically applied anti-acne drugs.[143]

Ingredients Causing Safety Concerns

⑩**The main safety issues associated with the use of acne treatments include irritation and dryness caused by many topical ingredients, as well as teratogenicity of oral retinoids. An additional concern is the risk for antibacterial resistance when using antibiotics.** These issues and precautions that should be taken when using such ingredients are discussed here.

Topical Ingredients BPO, SA, and topical retinoids usually cause irritation, dryness, redness, or even itching upon application. As discussed earlier, these side effects are usually transient and relieve in a few weeks upon continuous application. Irritation can be decreased by selecting lower concentration formulations or using creams or foams (as discussed previously). Additionally, aggressive skin cleansing may also irritate the skin and, therefore, exacerbate active acne lesions and limit the patient's tolerance and adherence to the anti-acne therapy.[103] Studies show that skin cleansing should be restricted to twice-daily maximum in order not to have a detrimental effect on the skin's condition.[144]

Oral Retinoids Isotretinoin is a highly effective ingredient; however, it is a known teratogenic agent (i.e., can induce malformations in an embryo or a fetus). Therefore, it cannot be safely taken by pregnant women or women who may become pregnant during the course of treatment or within several weeks of concluding treatment. Female patients must demonstrate a negative pregnancy test before and use contraception when taking this ingredient.[145] Additionally, women of reproductive age must participate in an FDA-approved monitoring program to receive a prescription for the drug.[146] Patients taking oral retinoids are recommended to use two forms of birth control 1 month before starting the treatment, during the treatment, and 1 month after the treatment. Even if a hormonal contraception is used, a second form should also be used.[147]

Antibacterial Resistance Antibiotics are widely used in the treatment of acne, both topically and systemically. These ingredients have activity against *P. acnes*, which is associated with the development of acne. Although antibiotics play an important role in acne management, resistance of *P. acnes* to antibiotics commonly used in the treatment of acne has been an increasing concern over the past decades. Antibiotic resistance typically leads to reduced clinical response to the antibiotic therapy, which occurs in the case of acne. Additionally, the pathogenicity of *P. acnes* has increased over the recent years.[148] It resulted in more severe signs and symptoms, which were more challenging to handle, leading to a prolonged use of the medications and increase in the number and severity of side effects. Additionally, and more importantly, from a public health standpoint, prolonged antibacterial therapy has resulted in the transfer of resistance to potentially pathogenic bacteria, such as certain strains of *Staphylococcus* or *Streptococcus* species.[149] These resistant organisms can potentially present clinical challenges. Current guidelines on the treatment of acne recommend strategies to limit the potential for resistance while still achieving optimal outcomes. As discussed earlier, combination of antibiotics with BPO is often used to minimize the development of resistance.[150]

GLOSSARY OF TERMS FOR SECTION 4

Acne vulgaris: A common dermatological disorder of the pilosebaceous unit that can be characterized by clogged pores.

Antibiotic resistance: The ability of bacteria or other microbes to resist the effects of an antibiotic.

Antibiotic: A drug used to fight infection caused by microorganisms.

Antioxidants: Ingredients often used in antiaging products since they can protect cells from oxidative stress by neutralizing free radicals.

Blackhead: Open comedo. It is a noninflamed acne lesion that is black due to oxidation of the sebum and bacteria in the plug.

Botanical extract: Extracts from a plant or herb, often used in antiaging products due to their antioxidant, anti-inflammatory, soothing, emolliating, skin lightening, and skin tightening functions.

Chronological aging: A synonym for intrinsic aging.

Comedo: A widened follicle filled with sebum, bacteria, and dead skin cells.

Cyst: A large pus-filled inflammatory acne lesion that is the result of a severe inflammatory reaction.

Extrinsic aging: Acceleration of intrinsic aging caused by exposure to various environmental factors, such as sun, wind, smoking, and air pollutants.

Hydroxy acids: A class of chemical ingredients that consist of a carboxylic acid substituted with a hydroxyl group. They are commonly used in anti-aging products and anti-acne products due to their exfoliating, moisturizing, anti-inflammatory, and antioxidative effects.

Hyperkeratinization: An increased production of epidermal cells.

Intrinsic aging: Aging that is genetically determined; it affects the skin on the entire body, including the photoprotected areas in a similar manner.

Keratolytic: An ingredient that helps soften and shed the stratum corneum.

Liposome: A microscopic spherical vesicle made of phospholipids that has a lipid bilayer structure resembling the natural structure of the biomembranes.

Microcomedo: The first stage of a comedo, and it is the very beginning of a pore blockage. It cannot be seen with the naked eye.

Nodule: A severely inflamed acne lesion occurring when rupture of the follicle wall happens deep down at the bottom of the wall. Nodules are hard to the touch, more painful than other lesion types, and deep red or purple in color.

Noisome: A spherical vesicle made of nonionic surfactants of the alkyl and dialkyl polyglycerol ether class and cholesterol.

Oxidative stress: It is considered to be the cornerstone of the biochemical pathways, leading to both chronological aging and photoaging.

***P. acnes*:** *Propionibacterium acnes* are anaerobic bacteria that are associated with the development of acne.

Papule: An inflamed acne lesion resembling a small, tender bump under the skin.

Photoaging: Aging caused by the sun.

Photosensitivity: Oversensitivity of the skin to sunlight.

Pustule: An inflamed acne lesion resembling a red, tender bump with white pus at its tip.

Retinoids: A class of substances comprising vitamin A (retinol) and its derivatives. They are used in both antiaging products and antiacne products.

Whitehead: Closed comedo. It is a noninflamed acne lesion that is clogged by the buildup of dead skin cells and sebum.

REVIEW QUESTIONS FOR SECTION 4

Multiple Choice Questions

1. Teratogenic ingredients can ___.
   a) Induce malformations in an embryo
   b) Induce cancer formation in human cells
   c) Induce DNA mutations in humans
   d) All of the above
2. How does benzoyl peroxide fight against acne?
   a) It kills the bacteria in the clogged pores
   b) It decreases sebum production
   c) It produces reactive oxygen species in the clogged pore
   d) All of the above

3. Which of the following is NOT a topical OTC drug for the treatment of acne?
   a) Benzoyl peroxide
   b) Salicylic acid
   c) Isotretinoin
   d) Sulfur
4. Which of the following is a type of non-inflammatory acne?
   a) Papule
   b) Nodule
   c) Cyst
   d) Closed comedo
5. Which of the following is NOT a contributory factor to acne?
   a) Increased sebum production
   b) Poor facial and body hygiene
   c) Colonization by Propionibacterium acnes
   d) Abnormal epidermal cell production and shedding
6. Why are certain cattle-sourced materials on the prohibited ingredient list in the US?
   a) Due to risks of mad cow disease
   b) Due to risks of photosensitivity
   c) To protect animal rights
   d) Due to risks of bacterial resistance
7. What is the most commonly used dosage form for anti-aging products?
   a) Ointments
   b) Solutions
   c) Emulsions
   d) Loose powders
8. Which of the following can influence the delivery of active ingredients into the skin?
   a) Molecular size of the active ingredient
   b) Solubility of the active ingredient
   c) Ionization of the active ingredient
   d) All of the above

9. Which of the following is NOT true for the aging of the human skin?

a) Intrinsic aging is normal aging accelerated by sunlight
b) Extrinsic aging is the acceleration of intrinsic aging by sunlight, wind, and smoking
c) The collagen content in the skin decreases during aging
d) The density of melanocytes in the skin decreases during aging

10. What are "bioactive ingredients"?

a) It is a synonym for active ingredients in anti-aging products
b) The main ingredients in anti-aging products that are claimed to deliver the antiaging effect
c) The official terminology required by the FDA for cosmetic ingredients in anti-aging products
d) a and c are true

Fact or Fiction?

_________ a) The more frequent the skin cleansing, the better results in acne.
_________ b) Makeup use can have positive effects on the patients' QoL in acne.
_________ c) Antiacne formulations are expected to be absorbed into the deeper layers of the skin.
_________ d) Antiaging products can be either drugs or cosmetics.
_________ e) Antiaging products labeled as "cosmeceuticals" are considered drugs in the US.

Matching

Match the ingredients in column A with their appropriate ingredient category in column B.

	Column A	Column B
_____	A. Benzoyl peroxide	1. Alpha hydroxy acid
_____	B. Blueberry extract	2. Antibiotic
_____	C. Collagen	3. Botanical extract with antioxidant benefits
_____	D. Glycolic acid	4. Natural protein found in the skin
_____	E. Isotretinoin	5. Systemic anti-acne agent
_____	F. Retinaldehyde	6. Topical anti-acne agent
_____	G. Tetracycline	7. Vitamin A derivative delivering anti-aging benefits

REFERENCES

1. Friedman, O.: Changes associated with the aging face. *Facial Plast Surg Clin North Am.* 2005;13:371–380.
2. El-Domyati, M., Attia, S., Saleh, F., et al.: Intrinsic aging vs. photoaging: a comparative histopathological, immunohistochemical, and ultrastructural study of skin. *Exp Dermatol.* 2002;11:398–405.
3. Bergfeld, W. F.: The aging skin. *Int J Fertil Womens Med.* 1997;42:57–66.
4. Uitto, J., Bernstein, E. F.: Molecular mechanisms of cutaneous aging: Connective tissue alterations in the dermis. *J Invest Dermatol Symp Proc.* 1998;3:41–44.
5. Uitto, J., Fazio, M. J., Olsen, D. R.: Molecular mechanisms of cutaneous aging: age associated connective tissue alterations in the dermis. *J Am Acad Dermatol.* 1989;21:614–622.
6. Gilchrest, B. A., Vrabel, M. A., Flynn, E., et al.: Selective cultivation of human melanocytes from newborn and adult epidermis. *J Invest Dermatol.* 1984;83(5):370–376.
7. Baumann, L.: Skin ageing and its treatment. *J Pathol.* 2007;211(2):241–251.
8. Tsutsumi, M, Denda, M.: Paradoxical effects of beta-estradiol on epidermal permeability barrier homeostasis. *Br J Dermatol.* 2007;157:776–779.
9. Elsner, P., Wilhelm, D., Maibach, H. I.: Effect of low-concentration sodium lauryl sulfate on human vulvar and forearm skin. Age-related differences. *J Reprod Med* 1991;36:77–81.
10. Fisher, G. J., Kang, S., Varani, J., et al.: Mechanisms of photoaging and chronological skin aging. *Arch Dermatol.* 2002;138(11):1462–1470.
11. Garmyn, M., Van der Oord, J.: Clinical and Histological Changes of Photoaging, In: Rigel, D. S., Weiss, R. A., Lim, H. W., et al., eds: *Dover: Photoaging*, Boca Raton: CRC Press, 2004.
12. Kennedy, C., Bastiaens, M. T., Bajdik, C. D.: Leiden: skin cancer study. Effect of smoking and sun on the aging skin. *J Invest Dermatol.* 2003;120:548–554.
13. Kang, S., Chung, J. H., Lee, J. H., et al.: Topical N-acetyl cysteine and genistein prevent ultraviolet-light induced signaling that leads to photoaging in human skin in vivo. *J Invest Dermatol.* 2003;120:835–841.
14. Griffiths, C. E., Russman, A. N., Majmudar, G., et al.: Restoration of collagen formation in photodamaged human skin by tretinoin (retinoic acid). *N Engl J Med.* 1993;329(8):530–535.
15. Gilchrest, B. A.: A review of skin aging and its medical therapy. *Br J Dermatol.* 1996;135:867–875.
16. FDA: Over-the-Counter Drug Products, Safety and Efficacy Review. Federal Register: December 31, 2003, Accessed 1/1/2014 at http://www.fda.gov/ohrms/dockets/98fr/03-32102.htm
17. CFR Title 21 Part 347.10
18. Tabata, N., O'Goshi, K., Zhen, Y. X., et al.: Biophysical assessment of persistent effects of moisturizers after their daily applications: evaluation of corneotherapy. *Dermatology.* 2000;200:308–313.

19. Viera, M., Sadegh, M. What's new in natural compounds for photoprotection. *Cosmet Dermatol.* 2008;21:279–289.
20. Cornuelle, T., Lenhart, J.: Topical Botanicals. In: *Cosmetic Formulations of Skin Care Products*, New York: Taylor and Francis, 2006:299–308.
21. Baumann, L.: Less-known botanical cosmeceuticals. *Dermatol Ther.* 2007;20:330–342.
22. Skougaard, G., Jensen, A., Sigler, M.: Effect of novel dietary supplement on skin aging in post-menopausal women. *Eur J Clin Nutr.* 2006;60:1201–1206.
23. Chanchal, D., Swarnlata, S.: Novel approaches in herbal cosmetics. *J Cosmet Dermatol.* 2008;7,89–95.
24. Wenk, J., Brenneisen, P., Meewes, C., et al.: UV-induced oxidative stress and photoaging. *Curr Probl Dermatol.* 2011;29:83–94.
25. Thiele, J. J., Schroeter, C., Hsieh, S. N., et al.: The antioxidant network of the stratum corneum. *Curr Probl Dermatol.* 2001;29:26–42.
26. Jacob, R. A., Burri, B. J.: Oxidative damage and defense. *Am J Clin Nutr.* 1996;63:985S–990S.
27. Barja, G.: Free radicals and aging. *Trends Neurosci.* 2004;27:595–600.
28. Passi, S., De Pita, O., Grandinetti, M., et al.: The combined use of oral and topical lipophilic anti-oxidants increases their levels both in sebum and stratum corneum. *Biofactors* 2003;18(1–4):289–297.
29. Nusgens, B. V., Humbert, P., Rougier, A., et al.: Topically applied vitamin C enhances the mRNA level of collagens I and III, their processing enzymes and tissue inhibitor of matrix metalloproteinase 1 in the human dermis. *J Invest Dermatol.* 2001;116(6):853–839.
30. Green, C., Orchard, G., Cerio, R.: A clinicopathological study of the effects of topical retinyl propionate cream in skin photoaging. *Clin Exp Dermatol.* 1998;23:162–167.
31. Chew, A. L., Bashir, S. J., Maibach, H. I.: Topical Retinoids, In: Elsner, P., Maibach, H. I., eds: *Cosmeceuticals: Drugs vs Cosmetics*, New York: Marcel Decker, 2000: 107–122.
32. Sorg, O., Antille, C., Kaya, G., et al.: Retinoids in cosmeceuticals. *Dermatol Ther.* 2006;19:289–296.
33. Sorg, O., Kuenzli, S., Kaya, G.: Proposed mechanisms of action for retinoids derivatives in the treatment of skin. *J Cosmet Dermatol.* 2005;4:237–244.
34. Thiboutot, D., Del Rosso, J. Q.: Epidermal barrier is acne vulgaris associated with inherent epidermal abnormalities that cause impairment of barrier functions? Do any topical acne therapies alter the structural and/or functional integrity of the epidermal barrier? *J Clin Aesthet Dermatol.* 2013;6(2):18–24.
35. MacGregor, J. L., Maibach, H. I.: Specificity of Retinoid-Induced Irritation and Its Role in Clinical Efficacy, In: Wilhelm, K. P., Zhai, H., Maibach, H. I., eds: *Dermatotoxicology*, Boca Raton: CRC Press, 2010.
36. Torras, H.: Retinoids in aging. *Clin Dermatol.* 1996;74:207–215.
37. Bailly, J., Crettaz, M., Schifflers, M. H., et al.: In vitro metabolism by human skin and fibroblasts of retinol, retinal and retinoic acid. *Exp Dermatol.* 1998;7:27–234.
38. Bellemere, G., Stamatas, G. N., Bruere, V., et al.: Anti-aging action of retinol: from molecular to clinical. *Skin Pharmacol Physiol.* 2009;22:200–209.

39. Kafi, R., Kwak, H. S., Schumacher, W. E.: Improvement of naturally aged skin with vitamin A (retinol). *Arch Dermatol.* 2007;143:606–612.
40. Rossetti, D., Kielmanowicz, M. G., Vigodman, S., et al.: A novel anti-aging mechanism for retinol: induction of dermal elastin synthesis and elastin fibre formation. *Int J Cosmet Sci.* 2011;33(1):62–69.
41. Varani, J., Warner, R. L., Gharaee-Kermani, M., et al.: Vitamin A antagonizes decreased cell growth and elevated collagen-degrading matrix metalloproteinases and stimulates collagen accumulation in naturally aged human skin. *J Invest Dermatol.* 2000;114(3):480–486.
42. Carlotti, M. E., Ugazio, E., Sapino, S., et al.: Photodegradation of retinol and anti-aging effectiveness of two commercial emulsions. *J Cosmet Sci.* 2006;57(4):261–277.
43. Bhawan, J., Olsen, E., Lufrano, L., et al.: Histologic evaluation of the long-term effects of tretinoin on photoaged skin. *J Dermatol Sci.* 1996;11:177–182.
44. Green, B.: After 30 years ... the future of hydroxyacids. *J Cosmet Dermatol.* 2005;4(1):44–45.
45. Yu, R. J., Van Scott, E. J.: α-Hydroxyacids, Polyhydroxy Acids, Aldobionic Acids and Their Topical Actions, In: Baran, R. Maibach, H. I., eds: *Textbook of Cosmetic Dermatology*, New York: Taylor & Francis, 2005, 77–93.
46. Bernstein, E. F., Underhill, C. B., Lakkakorpi, J.: Citric acid increases viable epidermal thickness and glycosaminoglycan content of sun-damaged skin. *Dermatol Surg.* 1997;23:689–694.
47. Berardesca, E., Distante, F., Vignoli, G. P.: Alpha hydroxyacids modulate stratum corneum barrier function. *Br J Dermatol.* 1997;137:934–938.
48. Edison, B. L., Green, B. A., Wildnauer, R. H.: *A polyhydroxy acid skin care regimen provides anti-aging effects comparable to an alpha-hydroxyacid regimen. Cutis.* 2004;73(2)14–17.
49. Bernstein, E. F., Brown, D. B., Schwartz, M. D.: The polyhydroxy acid gluconolactone protects against ultraviolet radiation in an in vitro model of cutaneous photoaging. *Dermatol Surg.* 2004;30:1–8.
50. Kakita, L. S., Green, B. A.: A review of the physical and chemical properties of alpha-hydroxyacids (AHAs) and polyhydroxy acids (PHAs) and their therapeutic use in pharmacologics. *J Am Acad Dermatol.* 2006;54:AB107.
51. Briden, E., Jacobsen, E., Johnson C.: *Combining superficial glycolic acid (AHA) peels with microdermabrasion to maximize treatment results and patient satisfaction. Cutis.* 2007;79(1):13–16.
52. Thibodeau, A.: Metalloproteinase inhibitors. *Cosmet Toiletries.* 2000;115:75–76.
53. Herrmann, M.: Salicylic acid: an old dog, new tricks, and staphylococcal disease. *J Clin Invest.* 2003;112(2):149–151.
54. Khunger, N.: *Step by Step Chemical Peels*, 1st Edition, New Delhi: Jaypee Medical Publishers; 2009: 280–297.
55. Fartasch, M., Teal, J., Menon, G. K.: Mode of action of glycolic acid on human stratum corneum: ultrastructural and functional evaluation of the epidermal barrier. *Arch Dermatol Res.* 1997;289(7):404–409.

56. Deprez, P.: *Textbook of Chemical Peels. Superficial, Medium, and Deep Peels in Cosmetic Practice*, London: Informa UK, 2007.
57. Abbott, S.: How the 'stuff' in formulations impacts the delivery of actives. *Cosmet Toiletries*. 2013;128(2):116–123.
58. Williams, A.: *Transdermal and Topical Drug Delivery: From Theory to Clinical Practice*, London: Pharmaceutical Press, 2003.
59. Draelos, Z. D.: Eczema Regimens, In: Drealos, Z. D., ed.: *Cosmetic Dermatology Products and Procedures*, Hoboken: Wiley-Blackwell, 2010.
60. Müller, R. H., Dingler, A.: The next generation after the liposomes: solid lipid nano particles (SLN®, LipopearlsTM) as dermal carrier in cosmetics. *Eurocosmetics*. 1998;7/8:19–26.
61. Geusens, B., Van Gele, M., Braat, S., et al.: Flexible nanosomes (SECosomes) enable efficient siRNA delivery in cultured primary skin cells and in the viable epidermis of ex vivo human skin. *Adv Funct Maters*. 2010;20:4077–4090.
62. Malhotra, M., Jain, N. K.: Niosomes as drug carriers. *Indian Drugs*. 1994;31:81–86.
63. Udupa, N.: Niosomes as Drug Carriers, In: Jain, N. K.: *Controlled and Novel Drug Delivery*, 1st Edition, New Delhi: CBS Publishers and Distributors, 2002.
64. de Vringer, T., de Ronde, H. A.: Preparation and structure of a water-in-oil cream containing lipid nanoparticles. *J Pharm Sci*. 1995;84(4):466–472.
65. Sivaramakrishnan, R.: Glucocorticoid entrapment into lipid carriers–characterization by parelectric spectroscopy and influence on dermal uptake. *J Control Release*. 2004;97(3):493–502.
66. Müller, R. H., Petersen, R. D., Hommoss, A., et al.: Nanostructured lipid carriers (NLC) in cosmetic dermal products. *Adv Drug Deliv Rev*. 2007;59:522–530.
67. Loo, C. H., Basri, M., Ismail, R., et al.: Effect of compositions in nanostructured lipid carriers (NLC) on skin hydration and occlusion. *Int J Nanomedicine*. 2013;8:13–22.
68. Raschke, T., Koop, U., Dusing, H. J., et al.: Topical activity of ascorbic acid: from in vitro optimization to in vivo efficacy. *Skin Pharmacol Physiol*. 2004;17:200–206.
69. Fluhr, J. W., Vienne, M. P., Lauze, C., et al.: Tolerance profile of retinol, retinaldehyde, and retinoic acid under maximized and long-term clinical conditions. *Dermatol*. 1999;199S:57–60.
70. Bissett, D. L.: Common cosmeceuticals. *Clin Dermatol*. 2009;27:435–445.
71. Matsuda, S., Shibayama, H., Hisama, M., et al.: Inhibitory effects of novel ascorbic derivative VCP-IS-2Na on melanogenesis. *Chem Pharm Bull*. 2008;56:292–297.
72. Slavin, J. W.: Considerations in alpha hydroxy acid peels. *Clin Plastic Surg*. 1998;25:4–52.
73. Foldvari, M., Attah-Poku, S., Hu, J., et al.: Palmitoyl derivatives of interferon alpha: potent for cutaneous delivery. *J Pharm Sci*. 1998;87:1203–1208.
74. Ruiz, M. A., Clares, B., Morales, M. E., et al.: Preparation and stability of cosmetic formulations with an anti-aging peptide. *J Cosmet Sci*. 2007;58:157–171.

75. NTP Technical Report on the Photocarcinogenesis Study of Glycolic Acid and Salicylic Acid (CAS Nos. 79-14-1 and 69-72-7) in SKH-1 Mice), Uploaded: 9/2007, Accessed 9/13/2013 at http://ntp.niehs.nih.gov/ntp/htdocs/lt_rpts/tr524.pdf
76. FDA: Cosmetics, Product and Ingredient Safety, Selected Cosmetic Ingredients: Alpha Hydroxy Acids in Cosmetics, Last update: 9/13/2013, Accessed 12/30/2013 at http://www.fda.gov/cosmetics/productandingredientsafety/selectedcosmeticingredients/ucm107940.htm
77. Krautheim, A., Gollnick, H.: Transdermal penetration of topical drugs used in the treatment of acne. *Clin Pharmacokinet*. 2003;42:1287–1304.
78. CDC: BSE (Bovine Spongiform Encephalopathy, or Mad Cow Disease), Last update: 2/21/2013, Accessed 12/30/2013 at http://www.cdc.gov/ncidod/dvrd/bse/
79. Will, R. G., Ironside, J. W., Zeidler, M., et al.: A new variant of Creutzfeldt-Jakob disease in the UK. *Lancet*. 1996;347:921–925.
80. Henry, C., Knight, R.: Clinical features of variant Creutzfeldt-Jakob disease. *Rev Med Virol*. 2002;(3):143–150.
81. FDA: Bovine Spongiform Encephalopathy (BSE) Questions and Answers, Last update: 06/18/2009, Accessed 12/30/2013 http://www.fda.gov/biologics bloodvaccines/safetyavailability/ucm111482.htm
82. White, G. M.: Recent findings in the epidemiologic evidence, classification, and subtypes of acne vulgaris. *J Am Acad Dermatol*. 1998;39:S34–S37.
83. Feldman, S. R., Fleischer, A. B.: Role of the dermatologist in the delivery of dermatologic care. *Dermatol Clin*. 2000;18:223–227.
84. Yentzer, B. A., Hick, J., Reese, E. L., et al.: Acne vulgaris in the United States: a descriptive epidemiology. *Cutis*. 2010;86:94–99.
85. Collier, C. N., Harper, J. C., Cafardi, J. A., et al.: The prevalence of acne in adults 20 years and older. *J Am Acad Dermatol*. 2008;58(1):56–59.
86. Kurokawa, I., Danby, F. W., Ju, Q., et al.: Review new developments in our understanding of acne pathogenesis and treatment. *Exp Dermatol*. 2009;18(10):821–832.
87. Thiboutot, D., Gollnick, H., Bettoli, V., et al.: New insights into the management of acne: an update from the global alliance to improve outcomes in acne group. *J Am Acad Dermatol*. 2009;60(5):S1–50.
88. Gollnick, H., Cunliffe, W., Berson, D., et al.: Management of acne: a report from a global alliance to improve outcomes in acne. *J Am Acad Dermatol*. 2003;49(1):S1–S37.
89. Thielitz, A., Gollnick, H.: Topical retinoids in acne vulgaris update on efficacy and safety. *Am J Clin Dermatol*. 2008;9(6):369–381.
90. Makrantonaki, E., Ganceviciene, R., Zouboulis, C.: An update on the role of the sebaceous gland in the pathogenesis of acne. *Dermatoendocrinology*. 2011;3(1):41–49.
91. Melnik, B. C., Schmitz, G.: Review role of insulin, insulin-like growth factor-1, hyperglycaemic food and milk consumption in the pathogenesis of acne vulgaris. *Exp Dermatol*. 2009;18(10):833–841.
92. Kim, J., Ochoa, M. T., Krutzik, S. R., et al.: Activation of toll-like receptor 2 in acne triggers inflammatory cytokine responses. *J Immunol*. 2002;169(3):1535–1541.
93. Webster, G. F.: Inflammatory acne represents hypersensitivity to Propionibacterium acnes. *Dermatology*. 1998;196(1):80–81.

94. Foster-Powell, K., Holt, S. H., Brand-Miller, J. C.: International table of glycemic index and glycemic load values. *Am J Clin Nutr*. 2002;2002(76):5–56.
95. Bowe, W. P., Joshi, S. S., Shalita, A. R.: Diet and acne. *J Am Acad Dermatol*. 2009;63(1):124–141.
96. Burris, J., Rietkerk, W., Woolf, K.: Acne: the role of medical nutrition therapy. *J Acad Nutr Diet*. 2013;113:416–430.
97. Melnik, B. C.: Diet in acne: further evidence for the role of nutrient signaling in acne pathogenesis. *Acta Derm Venereol*. 2012;92:228–231.
98. Al-Ameer, A. M., Al-Akloby, O.: Demographic features and seasonal variations in patients with acne vulgaris in Saudi Arabia: a hospital-based study. *Int J Dermat*. 2002;41:870–871.
99. Gfesser, M., Worret, W. I.: Seasonal variations in the severity of acne vulgaris. *Int J Dermat*. 1996;35:116–117.
100. Harrison, S., Hutton, L., Nowak, M.: An investigation of professional advice advocating therapeutic sun exposure. *Aust N Z J Public Health*. 2002;26:108–115.
101. Hamilton, F. L., Car, J., Lyons, C. et al.: Laser and other light therapies for the treatment of acne vulgaris: systematic review. *Br J Dermatol*. 2009;160:1273–1285.
102. Tan, J. K., Vasey, K., Fung, K. Y. Beliefs and perceptions of patients with acne, *J Am Acad Dermatol*. 2001;44:439–445.
103. Goodman, G.: Cleansing and moisturizing in acne patients. *Am J Clin Dermatol*. 2009;10(1):1–6.
104. Magin, P., Pond, D., Smith, W., et al.: A systematic review of the evidence for 'myths and misconceptions' in acne management: diet, face-washing and sunlight. *Family Practice*. 2005;22(1):62–70.
105. Choi, Y. S., Suh, H. S., Yoon, M. Y., et al.: A study of the efficacy of cleansers for acne vulgaris. *J Dermatolog Treat*. 2010;21(3):201–205.
106. Hayashi, N., Imori, M., Yanagisawa, M., et al.: Make-up improves the quality of life of acne patients without aggravating acne eruptions during treatments. *Eur J Dermatol*. 2005;15(4):284–247.
107. Bataille, V., Sneider, H., MacGregor, A. J., et al.: The influence of genetic and environmental factors in the pathogenesis of acne: a twin study of acne in women. *J Invest Dermatol*. 2002. 119:1317–1322.
108. Callender, V. D.: Acne in ethnic skin: special considerations for therapy. *Dermatol Ther*. 2004;17:184–195.
109. Schafer, T., Nienhaus, A., Vieluf, D., et al. Epidemiology of acne in the general population: the risk of smoking. *Br J Dermatol*. 2001;145:100–104.
110. Brown, S. K., Shalita, A. R.: Acne vulgaris, *Lancet*. 1998;351:1871–1876.
111. Mark Lees: *Clearing Concepts: A Guide to Acne Treatment*, Stamford: Cengage Learning, 2013: 14–18.
112. AAD: Acne: Signs and Symptoms, Accessed 01/04/2014 at http://www.aad.org/dermatology-a-to-z/diseases-and-treatments/a–d/acne/signs-symptoms
113. Lehmann HL, Robinson KA, Andrews JS, et al.: Acne therapy: a methodological review. *J Am Acad Dermatol*. 2002;47:231–240.
114. Strauss, J. S., Krowchuk, D. P., Leyden, J. J., et al.: Guidelines of care for acne vulgaris management. *J Am Acad Dermatol*. 2007;56:651–663.

115. Rinzler, C. A.: *The Encyclopedia of Cosmetic and Plastic Surgery*, New York: Infobase Publishing, 2009.
116. Dawson, A. L., Dellavalle, R. P.: Acne vulgaris, *Brit Med J* 2013;346:f2634.
117. Williams, H. C., Dellavalle, R. P., Garner, S.: Acne vulgaris. *Lancet.* 2012;379(9813):361–372.
118. GoUnick H, Cunliffe W, Berson D, et al.: Global alliance to improve outcomes in acne. Management of acne: a report from a global alliance to improve outcomes in acne. *J Am Acad Dermatol.* 2003;49(1):S1–S37.
119. Gamble, R., Dunn, J., Dawson, A., et al.: Topical antimicrobial treatment of acne vulgaris: an evidence-based review. *Am J Clin Dermatol.* 2012;13:141–152.
120. Leyden, J. J., Del Rosso, J. Q., Webster, G. F.: Clinical considerations in the treatment of acne vulgaris and other inflammatory skin disorders: a status report. *Dermatol Clin.* 2009;27:1–15.
121. Ramanathan, S., Hebert, A. A.: Management of acne vulgaris. *J Pediatr Health Care.* 2011;25:332–337.
122. Webster, G. F., Rawlings, A. V.: *Acne and Its Therapy*, Boca Raton: CRC Press, 2013.
123. James, W. D.: Clinical practice. Acne. *N Engl J Med.* 2005;352:1463–1472.
124. Haider, A., Shaw, J. C.: Treatment of acne vulgaris. *JAMA.* 2004;292:726–735.
125. Abad-Casintahan, F., Chow, S. K., Goh, C. L., et al.: Toward evidence-based practice in acne: consensus of an Asian Working Group. *J Dermatol.* 2011;38:1041–1048.
126. Tan, H. H.: Topical antibacterial treatments for acne vulgaris: comparative review and guide to selection. *Am J Clin Dermatol.* 2004;5:79–84.
127. Elston, D. M.: Topical antibiotics in dermatology: emerging patterns of resistance. *Dermatol Clin.* 2009;27:25–31.
128. Bowe WP, Shalita AR. Effective over-the-counter acne treatments. *Semin Cutan Med Surg.* 2008;27:170–176.
129. Simpson RC, Grindlay DJ, Williams HC. What's new in acne? An analysis of systematic reviews and clinically significant trials published in 2010-11. *Clin Exp Dermatol.* 2011;36:840–843.
130. Del Rosso, J. Q., Kim, G.: Optimizing use of oral antibiotics in acne vulgaris. *Dermatol Clin.* 2009;27:33–42.
131. Smith, E. V., Grindlay, D. J., Williams, H. C.: What's new in acne? An analysis of systematic reviews published in 2009–2010. *Clin Exp Dermatol.* 2011;36:119–122.
132. Ebede, T. L., Arch, E. L., Berson, D.: Hormonal treatment of acne in women. *J Clin Aesthet Dermatol.* 2009;2(12):16–22.
133. George, R., Clarke, S., Thiboutot, D.: Hormonal therapy for acne. *Semin Cutan Med Surg.* 2008;27:188–96.
134. Nast, A., Dreno, B., Bettoli, V., et al.: European evidence-based (S3) guidelines for the treatment of acne. *J Eur Acad Dermatol Venereol.* 2012;26(1):1–29.
135. Chivot, M.: Retinoid therapy for acne. A comparative review. *Am J Clin Dermatol.* 2005;6:13–19.
136. Newman, M. D., Bowe, W. P., Heughebaert, C., et al.: Therapeutic considerations for severe nodular acne. *Am J Clin Dermatol.* 2011;12:7–14.
137. Marqueling, A. L., Zane, L. T.: Depression and suicidal behavior in acne patients treated with isotretinoin: a systematic review. *Semin Cutan Med Surg.* 2007;26:210–220.

138. Takenaka, Y., Hayashi, N., Takeda, M., et al.: Glycolic acid chemical peeling improves inflammatory acne eruptions through its inhibitory and bactericidal effects on Propionibacterium acnes. *J Dermatol.* 2012;39(4):350–354.
139. Ceilley, R. I.: Advances in topical delivery systems in acne: new solutions to address concentration dependent irritation and dryness. *Skinmed.* 2011;9(1):15–21.
140. Leyden, J. J., Grossman, R., Nighland, M.: Cumulative irritation potential of topical retinoid formulations. *J Drugs Dermatol.* 2008;7(8):s14–s18.
141. Schafer-Korting, M., Korting, H. C., Ponce-Poschl, E.: Liposomal tretinoin for uncomplicated acne vulgaris. *Clin Investig.* 1994;72:1086–1091.
142. Jarratt, M., Werner, C. P., Saenz, A. B. A.: Tazarotene foam versus tazarotene gel: a randomized relative bioavailability study in acne vulgaris. *Clin Drug Investig.* 2013;33:283–289.
143. Nyirady, J., Nighland, M., Payonk, G., et al: A comparative evaluation of tretinoin gel microsphere, 0.1%, versus tretinoin cream, 0.025%, in reducing facial shine. *Cutis* 2000;66:153–156.
144. Choi, J. M., Lew, V. K., Kimball, A. B.: A single-blinded, randomized, controlled clinical trial evaluating the effect of face washing on acne vulgaris. *Pediatr Dermatol.* 2006;23(5):421–427.
145. Harper, J. C.: An update on the pathogenesis and management of acne vulgaris. *J Am Acad Dermatol.* 2004;51(1):S36–S38.
146. Dai, W. S., LaBraico, J. M., Stern, R. S.: Epidemiology of isotretinoin exposure during pregnancy. *J Am Acad Dermatol.* 1992;26:599–606.
147. FDA: REMS, The iPLEDGE Program, Last update: 4/2012, Accessed 4/3/2014 at http://www.fda.gov/downloads/Drugs/DrugSafety/PostmarketDrugSafetyInformationforPatientsandProviders/UCM234639.pdf
148. Leyden, J., Levy, S.: The development of antibiotic resistance in Propionibacterium acnes. *Cutis.* 2001;67(2):21–24.
149. Harkaway, K. S., McGinley, K. J., Foglia, A. N., et al.: Antibiotic resistance patterns in coagulase-negative staphylococci after treatment with topical erythromycin, benzoyl peroxide, and combination therapy. *Br J Dermatol.* 1992;126(6):586–590.
150. Cunliffe, W. J., Holland, K. T., Bojar, R., et al.: A randomized, double-blind comparison of a clindamycin phosphate/benzoyl peroxide gel formulation and a matching clindamycin gel with respect to microbiologic activity and clinical efficacy in the topical treatment of acne vulgaris. *Clin Ther.* 2002;24(7):1117–1133.

SECTION 5: SUN CARE PRODUCTS

LEARNING OBJECTIVES

Upon completion of this section, the reader will be able to

1. define the following terms:

Actuation force	After-sun product	Broad spectrum	Chemical sunscreen
Glide	Melting point	Nanoparticle	Pay-off
Photoaging	Photoinstability	Photostabilizer	Physical sunscreen
Primary sunscreen	Secondary sunscreen	Skin cancer	Softening point
SPF	Spray characteristics	Sunburn	Sunscreen
Tanning	UV radiation	UVA	UVAI
UVAII	UVB	Valve clogging	Vitamin D
Water-resistance	Waterproofing agent		

2. name the three main types of ultraviolet radiation, and discuss their main characteristics;
3. differentiate between UVA and UVB radiation from the perspective of tanning, sunburn, and skin cancer;
4. explain what the abbreviation "SPF" refers to;
5. list some factors that influence how much UV radiation we are exposed to;
6. explain what the term "broad-spectrum protection" refers to;
7. explain what the minimum requirements are for a sunscreen product to be labeled as "broad spectrum";
8. explain what the terms "water resistant" and "very water resistant" refer to regarding sunscreens;
9. list some beneficial effects of the UV light on the human body;
10. list some negative effects that the UV light can have on the human body;
11. differentiate between primary and secondary sunscreens;
12. differentiate between sunscreens and after-sun products from the perspective of their function and legal status in the US;
13. list various required cosmetic qualities and characteristics that an ideal sunscreen product should possess;

14. list various required technical qualities and characteristics that an ideal sunscreen product should possess;
15. differentiate between physical and chemical filters, and explain how they block the UV light;
16. name the major ingredient types that can be incorporated into sunscreens, and provide a few examples for each type;
17. name the major dosage forms in which sunscreens are generally available;
18. briefly discuss the major advantages and disadvantages of the dosage forms mentioned in the previous section;
19. name some strategies for increasing the efficacy of sunscreen products;
20. name some ingredients that can be incorporated into after-sun products as cooling and pain relieving agents;
21. list some typical quality issues that may occur during the formulation and/or use of sunscreens, and explain why they may occur;
22. list the typical quality parameters that are regularly tested for sunscreens, and briefly describe their method of evaluation;
23. list the typical performance (efficacy) parameters that are regularly tested for sunscreens, and briefly describe their method of evaluation;
24. briefly discuss the main safety issues regarding the use of organic sunscreens and nanosized physical sunscreens in sunscreen products;
25. list the typical containers available for sunscreens and after-sun products.

KEY CONCEPTS

1. The UV radiation reaching the Earth's surface is largely composed of UVA (approximately 95%) with a small UVB component (approximately 5%).
2. UVB rays mainly penetrate the epidermis.
3. UVA rays penetrate deeper into the skin, down to the dermis.
4. SPF is a measure that indicates how long it takes for UV rays to redden protected skin compared to unprotected skin.
5. The term "broad-spectrum protection" refers to protection against both UVA and UVB rays.
6. Sunscreens are considered OTC drugs in the US since they help prevent various skin conditions, including sunburn, aging, and skin cancer. After-sun products, on the other hand are considered cosmetics in the US.
7. The typical quality-related issues of sunscreens and after-sun products include valve clogging, separation of the emulsions, microbiological contamination, clumping, and rancidification.
8. Parameters commonly tested by cosmetic companies to evaluate the quality of their sunscreens include spray characteristics; aerosol can leakage; actuation force, pressure test for aerosol products; spreadability, extrudability, texture,

and firmness of lotions, creams, and gels; stick hardness, stick softening and melting point; stick pay-off and glide; preservative efficacy; viscosity; and pH.

9. The most frequently tested parameters referring to the performance (efficacy) of sunscreens include the SPF value, water resistance, and broad spectrum.
10. Safety concerns arose in the past regarding the use of sunscreens containing organic sunscreens and nanosized physical sunscreens as well as products in aerosol forms.

Introduction

Skin exposure affects the skin in many ways. In the short term, it can lead to reddening, irritation, and eventually tanning, which is the main reason for most people sunbathing. There are, however, long-term effects of UV radiation, which are irreversible and often malignant. Sun exposure is now increasingly recognized as the possible cause of premature wrinkling and various types of skin cancer. For these reasons, sun protection has become a very important issue today. UV filters, ingredients that can effectively protect the skin from UV radiation, are now incorporated not only into sunscreens but also in daily-use cosmetics, such as moisturizing creams, foundations, and lipsticks. Appropriate skin care should follow any type of sun exposure to help the skin maintain its integrity and barrier function.

This section provides a summary on the major types of UV radiation and their positive and negative effects on the human body. It reviews sunscreen ingredients, the major product types, and their characteristics. The major steps that can be taken to increase the products' efficacy are also discussed here. The section provides an overview of the sunburn (after-sun) preparations and their main ingredients. Additionally, the section reviews the major quality and performance testing methods as well as safety issues related to the use of sunscreens.

Sun Protection Basics

The Light Spectrum The sun emits a constant flow of energy in the form of electromagnetic radiation, which ranges over a wide spectrum of wavelengths. The electromagnetic spectrum is divided into categories defined by the size of the wavelength, frequency, and energy. Radio waves at the bottom of the spectrum have the lowest energy and lowest frequency; their wavelengths are long with peaks far apart (Figure 3.23). Microwave radiation has higher energy, followed by infrared waves, visible waves, UV rays, and X-rays. At the top of the spectrum, gamma rays have the highest energy and shortest wavelength with peaks closer to one another. Most electromagnetic radiation from space is unable to reach the surface of the Earth due to the Earth's atmosphere. The electromagnetic waves reaching us consist of only a portion of the UV light, visible light, a portion of the infrared rays (near-infrared light), and radio waves.[1] Consequently, the surface of our planet is not exposed to cosmic rays, gamma rays, and X-rays, each of which is potentially lethal. Of the wavelengths of radiation that reach the Earth's surface, UV has the highest energy. Therefore, it has the highest importance in relation to sun exposure.

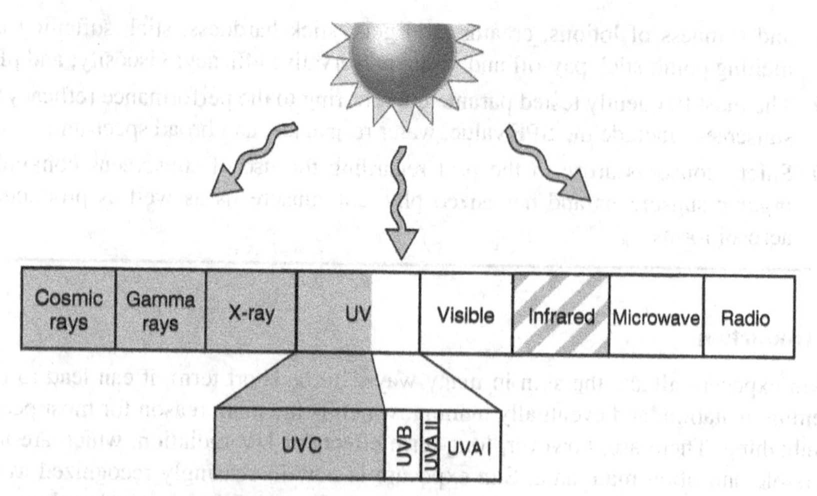

Figure 3.23 Electromagnetic spectrum. The shaded areas indicate rays that do not reach the Earth's surface. Infrared rays are partially filtered out.

UV radiation represents 5% of the total solar radiation reaching the Earth's surface.[2] As sunlight passes through the atmosphere, all UVC and approximately 90% of UVB radiation are absorbed by ozone, water vapor, oxygen, and carbon dioxide. UVA radiation is less affected by the atmosphere. Therefore, ●**the UV radiation reaching the Earth's surface is largely composed of UVA (approximately 95%) with a small UVB component (approximately 5%).**[3]

UV radiation can be divided into three categories based on wavelength. They differ in their biological activity and the extent to which they can penetrate the skin. The shorter the wavelength, the more harmful the UV radiation. However, shorter wavelength UV radiation is less able to penetrate the skin.

- **UVC** ranges from 100 to 280 nm and is blocked out by the ozone layer.
- **UVB** radiation ranges from 280 to 320 nm. ●**UVB rays mainly penetrate the** superficial skin layers, i.e., **epidermis** (Figure 3.24). UVB radiation is the major cause of sunburn, which is acute skin damage perceived as redness. Additionally, it has been identified as a leading factor in the development of skin cancer. The immediate result of UVB radiation is skin redness and thickening of the SC, which is a defense reaction of the body to reduce the UVB effect on the epidermis.[4] The redness and potential pain subside in a relatively short amount of time; however, the underlying damage accumulates over time. This leads to the formation of various types of skin cancer. Additionally, UVB contributes to photoaging and tanning and also has immunosuppressive effects. An advantageous effect of the UVB radiation is that it is responsible for the synthesis of vitamin D in the skin. The intensity of UVB varies by season, location, and time of day. The most significant amount of UVB is experienced in the US between 10 am and 4 pm during late spring and early summer.[5] However, UVB rays can

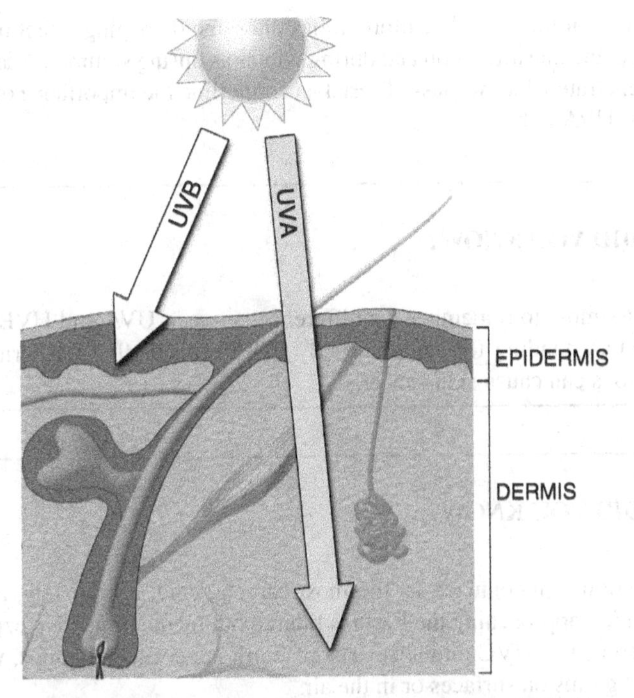

Figure 3.24 Penetration depth of ultraviolet A and B light.

burn and damage the skin all year, especially at high altitudes and on reflective surfaces such as snow or ice.[6] UVB rays do not significantly penetrate glass.

- UVA radiation ranges from 320 to 400 nm. Due to their longer wavelength, ●**UVA rays penetrate deeper into the skin, down to the dermis** (Figure 3.24). These rays can be further subdivided into UVA II (320–340 nm) and UVA I (340–400 nm). On the short term, UVA radiation leads to skin tanning (i.e., browning), which is often considered a sign of health. It is unfortunate since tanning, whether outdoor or indoor tanning, causes cumulative damage over time, leading to photoaging. UVA has long been identified as a leading factor in photoaging; however, until recently, it was not believed to cause any cancer. Studies over the past decades, however, showed that UVA damages keratinocytes in the basal cell layer where the most of the skin cancers occur. Therefore, on the long term, it can also contribute to and may even initiate the formation of skin cancer.[7–9] Additionally, UVA weakens the immune system, which also helps the development of skin cancer.[10] Photosensitivity reactions are also primarily mediated by UVA.[11] Tanning beds primarily emit UVA radiation. The lamps used in tanning salons can emit doses of UVA up to 12 times that of the sun.[12] Therefore, indoor tanning is associated with a significantly increased risk of skin cancer, and the risk is higher with use in early life (<25 years).[13] It should be kept in mind that UVA radiation retains the same energy level all day long, every day of the year. Therefore, it has the same damaging effect in the morning as in the late afternoon and during winter as during summer. Additionally, it can penetrate window glass. These facts underline the importance of protection against UVA light.

DID YOU KNOW?

To make it simple to remember the difference between UVA and UVB, think of "A" (UVA) as standing for aging and "B" (UVB) as standing for burning. Both UV radiations can cause skin cancer.

DID YOU KNOW?

UVC radiation is recognized as the most carcinogenic. Fortunately, almost all UVC radiation approaching the Earth is filtered out by the protective ozone layer. The only source of UVC on the surface of Earth is germicidal lamps, which are used to kill germs on surfaces or in the air.

Sun Protection Factor (SPF)

The term "sun protection factor" (hereinafter referred to as "SPF") was adopted by the FDA to describe the effectiveness of sunscreens. ❶**SPF is a measure that indicates how long it takes for UV rays to redden protected skin** (i.e., skin with a sunscreen) **compared to unprotected skin** (i.e., skin without a sunscreen). As the SPF value increases, sunburn protection increases. The effectiveness of a given SPF is measured in terms of redness (medically termed "erythema") that appears on the skin after sun exposure. The amount of UV energy required to produce the first visible redness on the skin is referred to as the minimal erythema dose (hereinafter referred to as "MED"). As SPF values are determined from the test that measures protection against sunburn caused by UVB radiation, SPF values only indicate a sunscreen's UVB protection.[14]

In the Federal Register, SPF is defined as the UV radiation required to produce 1 MED on protected skin after application of $2\,mg/cm^2$ of product divided by the UVR to produce 1 MED on unprotected skin.

$$\text{sun protection factor (SPF)} = \frac{\text{minimal erythema dose (MED) with sunscreen}}{\text{minimal erythema dose (MED) without sunscreen}}$$

Currently, sunscreens in the US are required to carry an SPF value, which informs users how well the product protects against UVB light. The number you see associated with SPF represents the length of time (in minutes) you can **theoretically** stay out in the sun without burning, multiplied by the corresponding number. For example, a person who would normally start to burn in 10 min could **theoretically** have 150 min of sun protection with a sunscreen that has an SPF of 15. This is only true in theory, though since SPF is not directly related to the **time** of solar exposure, but to the **amount** of solar exposure. Although the amount of solar energy is related to the exposure time, there are other factors that can impact the dose of solar energy.

DID YOU KNOW?

The FDA standard dose for SPF testing is a sunscreen application of 2 mg/cm^2. Practically, it equals approximately 30 g of sunscreen to cover an average-sized body, including arms, legs, neck, ears, and face. For the average-sized person, this equates to the amount it takes to fill a shot glass.[15]

Many other factors influence the amount of UV radiation we are exposed to, including the following:

- **Geography**: UV rays are the strongest in areas close to the equator.
- **Altitude**: Higher altitudes have greater UV exposure since there is athinner layer of atmosphere to absorb UV rays.
- **Time of the Year**: The sun's angle in relation to the Earth varies according to season. During the summer months, the sun is in a more direct angle, resulting in a greater amount of UV radiation.
- **Time of the Day**: UV is most intense at noon when the sun is at its highest point in the sky. The peak sun hours are considered from 10 am to 4 pm when the highest amount of UV exposure can be experienced during a day.
- **Weather Conditions**: Clouds can absorb UV rays; however, one can get sunburn even on a cloudy day.
- **Reflection**: Some surfaces, such as snow, sand, and water, can reflect UV radiation; therefore, UV exposure is higher under such conditions.[16]

In addition to solar intensity, there are a number of other factors that influence the amount of solar energy to which a consumer is exposed. These include the following:

- **Skin Type**: Fair-skinned consumers are likely to absorb more solar energy than dark-skinned consumers under the same conditions (for the various skin types, refer back to Section 1 of this chapter).
- The **amount of sunscreen applied** also impacts the amount of solar radiation absorbed; more sunscreen results in less solar energy absorption. The amount of 2 mg/cm^2 is used for SPF effectiveness testing[17]; however, studies show that

in reality, users usually use only 20–50% of the amount needed to achieve the labeled SFP.[18–21]

- Sunscreens wear off over time; therefore, **reapplication** frequency is also critical for optimal effectiveness. Physical activities, sweating, and running may rub off the products, while swimming may wash off the product; these conditions shorten the reapplication time.

Historically, SPF values have ranged from 2 to greater than 100. This led to the assumption that higher SPF provides significantly better sun protection (i.e., SPF 30 being twice as protective as SPF 15). This is not true, however. It has been shown in *in vitro* tests that SPF 15 sunscreens filter out 93% of UVB rays, while SPF 30 protects against 97% and SPF 50, 98%.[22–24] The higher the number, the smaller the difference in terms of UVB protection. Again, keep it in mind that the SPF value refers to the amount of solar exposure against which the sunscreen provides protection and not the length of solar exposure you can have without getting sunburn.

Broad-Spectrum Protection

●The term "broad-spectrum protection" refers to protection against both UVA and UVB rays. It has been shown in studies that using sunscreens with even a high SPF value does not provide appropriate protection against UVA radiation, which is responsible for photoaging and skin cancer formation.[25,26] Therefore, in order to prevent the long-term effects, protection is needed against both types of rays. Due to increasing concerns regarding the skin damaging effect of UVA radiation, the FDA recently issued new requirements for sunscreens. The new rules include testing and labeling requirements for protection against UVA radiation. Sunscreens that meet the FDA standard for both UVA and UVB protection can be labeled as "broad spectrum." This term has been used for a while; however, it did not have an official definition or a standard testing. Now standardized methods are available to measure UVA protection.

The new requirements also specify statements that should be used on product labels to indicate whether the product provides protection against aging and skin cancer or just sunburn. Scientific studies have demonstrated that products that are broad-spectrum with SPF 15 or higher can reduce the risk of skin cancer and early skin aging when used with other skin protection measures (e.g., sun protective clothing). According to the new rules, broad-spectrum sunscreens with an SPF 15 or higher can have the following statement on the labels: "... decreases the risk of skin cancer and early skin aging caused by the sun." However, any sunscreen that is not broad spectrum or is less than SPF 15 broad spectrum has to put the following warning statement on its label to inform consumers about the potential adverse effects of sun damage. The warning statement reads as follows: " ... Spending time in the sun increases your risk of skin cancer and early skin aging. This product has been shown only to help prevent sunburn, not skin cancer or early skin aging." Details of the UVA testing are discussed later under efficacy testing.

Water Resistance

Sweating, swimming, and other water exposure can reduce the efficacy of sunscreens by rubbing them off and washing them off from the skin. In order to clarify whether a product provides any water-resistance properties, the new sunscreen guidelines include terms that should be and can be used on the labels. Additionally, it identifies claims that cannot be made. According to the FDA monograph on sunscreens, products can be labeled as "water resistant" or "very water resistant" depending on how long they can retain the stated SPF after immersion in water. The actual length of time is indicated on the principal display panel (for more information on the principal display panel, refer back to Section 1 of Chapter 2). Testing requirements are detailed later under efficacy testing. In order to maintain the claimed sun protection, sunscreens have to be reapplied regularly, especially after swimming or heavy sweating. The labels also have to include directions to reapply the product.[27] Terms such as "sunblock," "waterproof," and "sunproof" are not allowed to be used on product labels since these may be misleading. No sunscreens today are 100% waterproof and none of them can filter out sunlight at 100%. Additionally, sunscreens cannot claim to provide sun protection for more than 2 h without reapplication or to provide protection immediately after application (e.g., instant protection) without obtaining FDA approval.

DID YOU KNOW?

There is a proposed rule under revision currently regarding the maximum allowable SPF value for sunscreens. According to the proposed regulation, manufacturers could not make a claim for an SPF above 50. The highest rating would be "50+." The reason for this is that the FDA does not have adequate data demonstrating that products with SPF values higher than 50 provide additional protection compared to products with SPF values of 50.[28]

Effects of Ultraviolet Radiation on the Human Body

UV radiation can affect human health, both positively and negatively. First, the beneficial effects are summarized.

- Its main positive effect is related to **vitamin D production** in human skin, which is catalyzed by UVB light. Vitamin D has a significant role in bone health and the prevention of osteoporosis and osteomalacia.[29] Additionally, it affects a variety of adult health problems. It lowers blood pressure in hypertensive patients[30]; the incidence and severity of cardiovascular disorders,[31] type 2 diabetes,[32] and rheumatoid arthritis[33]; and also helps prevent tooth loss.[34] Vitamin D also has a significant role in reducing the mortality from various cancers, including colon, breast, prostate, ovarian, and even – ironically – melanoma (the most severe form of skin cancer).[35–37] For most children and adults, the major source of vitamin D is sunlight,[38] which obviously creates a public health dilemma since UV radiation also causes skin cancers. Studies were conducted to determine whether the amount of vitamin D made from everyday exposure to sunlight is sufficient to acquire an optimum vitamin D blood level. The results are controversial; some studies show that even with sunscreens, a proper amount of vitamin D can be produced in the skin,[39] while other studies show that many people have insufficient levels for most of the year, especially during and after winter.[40]

- Additionally, exposure to UV radiation in the form of lasers, lamps, or a combination of these devices is helpful in certain **skin conditions** that do not respond to other methods of therapy. Such skin conditions include psoriasis (extreme skin dryness), eczema, certain fungal skin infections,[41] and acne, among others.[24] Phototherapy involves exposing a patient to a carefully monitored dose of UV radiation on a regular schedule. This type of therapy does not eliminate the negative side effects of UV exposure; however, treatment is carefully supervised by a healthcare professional to ensure that the benefits outweigh the risks.[16]
- Sun exposure has also been found helpful to reduce the severity of **depression** in patients with seasonal affective disorders.[42]

Exposure to UV radiation, whether outdoor or indoor, also carries potential risk to human health.

- Tanning is not healthy; it is a sign of **skin damage**. This damage builds up and accelerates aging and can also increase the risk for all types of skin cancer.[43,44] Both the US Department of Health and Human Services and the WHO have identified UV as a proven **human carcinogen**.[45,46] UV radiation is considered the main cause of non-melanoma skin cancers, including basal cell carcinoma and squamous cell carcinoma. Additionally, UV radiation is one of the major risks for the development of melanoma, the deadliest form of skin cancer, especially for fair-skinned people.[47,48] Skin cancer is the most common of all cancers in the US.[49] According to current statistics, more than 3.5 million non-melanoma skin cancers in more than 2 million people are diagnosed in the US annually.[49] Current estimates are that one in five Americans will develop skin cancer in their lifetime.[50,51] On the molecular level, exposure to UV radiation can result in DNA damage, leading to the formation of malignant cells as well as immunosuppression.[52]
- UV radiation is also linked with various **eye diseases**, including cataract formation and retinal degeneration.[53]
- As discussed earlier, UV radiation plays a leading role in **photoaging** and the premature appearance of lines and wrinkles as well as sun spots.
- UV radiation causes the melanocytes, located in the basal cell layer of the epidermis, to produce **melanin**, which is the pigment that makes the skin darker.

It is important to understand the relationship between skin pigmentation and photoprotection from the perspective of skin cancer. Studies show that melanin provides the skin with natural protection against solar damage by serving as a physical barrier and an absorbent.[54] However, the amount of melanin that is produced in fair-skinned people following exposure to the sun is relatively low and does not afford them adequate protection. Therefore, they must take additional precautions to prevent solar skin damage. The efficacy of melanin as a sunscreen is assumed to be about SPF 1.5–2.0, maximum SFP 4, meaning that melanin absorbs 50–75% of UVR. Dark skin, which contains more melanin than fair skin, is better protected against UV-induced damage.[55] That is why dark-skinned people often look younger than fair-skinned people of the same age. Nevertheless, even dark-skinned people should avoid excessive sun exposure. While the epidermis of African American skin allows only 7.4% of UVB and 17.5% of UVA to penetrate, 24% UVB and 55% UVA pass through Caucasian skin.[56] Claims made for the healthfulness of indoor tanning are misleading since it may make consumers believe that they can acquire effective protection when tanning. The amount of melanin produced, however, is not enough to be an effective filter against UV radiation. It should also be kept in mind that all types of UV-induced tanning, either outdoor or indoor, result in DNA damage and cellular damage and can lead to photocarcinogenesis (i.e., cancer formation).[57,58]

DID YOU KNOW?

Studies have shown that UVA oxidizes the existing melanin, causing immediate pigment darkening. UVB causes inflammation, which releases new melanin and, in about 72 h, leads to further tanning that lasts much longer than UVA-triggered tanning. All of this melanin production signals that DNA damage has already occurred: it is the body's imperfect attempt to protect the skin from further damage. Nonetheless, the damage from repeated UV exposures keeps accumulating and can ultimately lead to skin cancer.[58]

Types and Definition of Sun Care Products

Sun exposure is unavoidable in our daily lives; however, without appropriate protection, the sun can severely damage our skin on the short as well as long term. **Sunscreens are considered OTC drugs in the US since they help prevent various skin conditions, including sunburn, aging, and skin cancer. After-sun products, on the other hand are considered cosmetics in the US.**

- **Sunscreens** are designed to provide photoprotection against the harmful radiation of the sun, including both UVA and UVB radiations. The protection they provide can be through chemical and/or physical methods. Sunscreens are available in a variety of dosage forms, including creams, lotions, sticks, wipes, gels, as well as aerosols. Sunscreen products can be classified into two main categories according to their purpose:

- **Primary sunscreens** are products whose main purpose is to provide photoprotection for the skin, such as beach sunscreens and products used for outdoor activities.
- **Secondary sunscreens** are products that have a primary use other than skin protection, such as daily moisturizing creams, antiaging creams, and color cosmetics, such as facial foundations. For this category of products, sun protection is an additional benefit, but not the main purpose of their use. Nevertheless, they are also considered OTC drugs in the US since they provide photoprotection.

- **After-sun products** are designed to be used after exposure to the sun or other UV radiation. After sunbathing, even without any signs of redness, appropriate skin care is recommended. After-sun preparations help smooth and moisturize the skin as well as provide a cooling effect and relieve pain resulting from sunburn. Product forms include lotions, creams, and gels.

History of Using Sun Care Products

The use of sunscreen products dates back to the ancient times. In ancient Egypt and China, light skin tone was considered to be more beautiful than dark skin. It can be related to the fact that those working outdoor, i.e., the working class, had tanned skin due to the unavoidable sun exposure. In contrast, the higher class was able to avoid the sun's rays and maintain the fine white skin. Light skin became part of the social status and was, therefore, more desired. In the early ages, clothing was an important way of sun protection. Umbrellas were also used to provide sun protection in ancient Egypt, Mesopotamia, China, and India. Even in Victorian times, elegance and fashion dictated that skin should be white and that brown skin was indicative of hard labor and poverty. Many forms of physical protection have been used, including oils, tars and herbs, and plant extracts, which were probably used for other cosmetic reasons but also served as sunscreens.[59]

Until the 19th century, it was widely believed that sunburn was caused by heat damage. In 1801, Johann Ritter discovered UV light.[60] He postulated, correctly, that UV radiation was causing skin cells to become inflamed, and that it was this radiation, not heat, that caused sunburn. In 1820, Everard Home discovered that melanin protects the skin from sunlight and stated that darker skin is better protected than the skin of white people.[61] In the late 1920s, Karl Eilham Hausser and Wilhelm Vahle reported that sunburn is caused by specific wavelengths of the UV spectrum and realized that the skin could be protected by filtering out those wavelengths. This led to the introduction of a sunscreen in 1928 in the US, blending benzyl salicylate and benzyl cinnamate.[62] The first commercially available sunscreen product was introduced by Eugene Schueller in France in the 1930s, the founder of L'Oréal.[63] Franz Greiter in Austria created a product called Glacier Cream in the late 1930s to protect against the sun while climbing.[64] In the US, the first sunscreen product was invented in Florida in the 1940s by Benjamin Green, the future Coppertone founder. It was a red jelly-like substance, which was known as the "red vet pet," a veterinary petroleum-based product.[65] It was widely used during the Second World War; however, it had some disadvantages, such as stained fabrics, and it was sticky. Later, he developed a more consumer-friendly product. The concept of SPF was introduced in the 1960s by Franz Greiter.[66] This system has become a worldwide standard for measuring the effectiveness of sunscreens. In 1972, the FDA reclassified sunscreens from cosmetics to OTC drugs, and labeling requirements became stricter.[67] However, many consumers continued to use suntan lotions, i.e., products with low SPF, which provided only minimal protection against UV radiation. In the 1980s, consumers started to learn more about the negative effects of sunlight, and suntan lotions and similar products began to disappear. In recent decades, a number of improvements were implemented, new dosage forms were formulated, and formulations are more and more appealing, in addition to being functional. Today, many personal care products, including lipsticks, facial creams, and hand creams, contain UV filters.

Required Qualities and Characteristics and Consumer Needs

From a consumer perspective, a quality sun care product should possess the following characteristics:

- Sunscreens:
  - Provide protection against both UVB and UVA radiation
  - Allow some tanning
  - Water-resistant
- Sun damage preparations:
  - Alleviate pain
  - Sooth and moisturize the skin
  - Provide cooling effect.
- Non-odorous and non-sticky
- Non-staining for skin or clothing
- Pleasant feeling during and after application
- Quick drying time and absorption after application
- Easy spreadability
- User-friendly to encourage frequent application and provide reliable protection
- Non-toxic, non-irritant, and non-sensitizing

The technical qualities of sun care products can be summarized as follows:

- Proven efficacy and/or performance
- Long-term stability
- Appropriate texture

- Appropriate rheological properties
- Stability to heat and UV radiation (photostability)
- Dermatological safety

Sunscreens

Sunscreens represent a practical approach to photoprotection. Sunscreen ingredients are usually referred to as UV filters and are identified on product labels as active ingredients. Additional ingredients in sunscreen products are listed as inactive ingredients in the Drug Facts section on the information panel (for more information on the information panel, refer back to Section 1 of Chapter 2).

UV Filters

UV filters are classified into two groups based on their mechanism of action: physical and chemical sunscreens. A single product usually contains more than one active ingredient, both physical and chemical filters, to achieve the desired SPF value and broad-spectrum protection. It should be noted, however, that the OTC monograph includes restrictions for the combination of certain ingredients.

- **Physical sunscreens**, otherwise known as inorganic UV filters, reflect and scatter UV radiation (Figure 3.25a).[23,68] There are two approved physical filters in the US, namely titanium dioxide (TiO_2) and zinc oxide (ZnO). Both of these ingredients are white powders that are insoluble in the sunscreen product base. Therefore, they are suspended in such products. Inorganic sunscreens can only penetrate the outer layer of the skin.[69] Therefore, they have an excellent safety

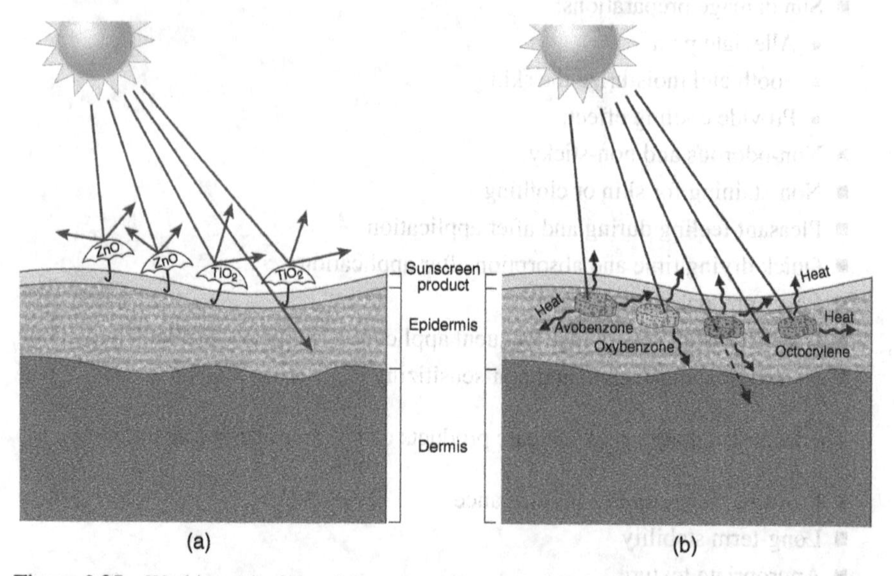

Figure 3.25 Working principle of the two main types of UV filters. (a) physical sunscreens and (b) chemical sunscreens. Modified with permission from Hagadorn J. W.

profile. Additionally, inorganic filters are photostable, independent of the sunscreen base and other ingredients. Inorganic filters provide a broad-spectrum protection since they reflect and scatter both UVA and UVB radiations (TiO_2 offers UVB and UVA II protection, while ZnO provides protection against UVB, UVA II, and UVA I radiations).[70] The main disadvantage of inorganic filters with regular particle size is that they reflect and scatter UV radiation into the visible spectrum (>400 nm), which provide a white appearance on the skin after application. This can make sunscreens cosmetically less appealing and can lead to reduced application by consumers. The reflection spectrum of TiO_2 and ZnO, and, therefore, the white appearance can be, however, modified by decreasing their particle size. The smaller the particle size, the less reflection occurs in the visible spectrum. Therefore, today, TiO_2 and ZnO are commonly used in micronized (i.e., with a particle size of 1–100 μm) and nanonized (i.e., with a particle size of 1–100 nm) forms for an aesthetically more appealing cosmetic look. Another advantage of smaller particles is that they feel lighter on the skin. Products that contain nanoparticles no longer reflect visible light and, therefore, do not appear white but transparent on the skin. However, there is an important fact that should be emphasized: particle size reduction leads to a shift in the UV radiation protection profile of inorganic sunscreens toward the UVB and UVA II range, and the UV protection becomes imbalanced.[71] Using nanosized and microsized particles in combination may improve this situation. Additionally, inorganic filters are more often used in combination with organic filters to provide broad-spectrum protection while maintaining good aesthetics.

- **Chemical sunscreens**, otherwise known as organic UV filters, are generally aromatic compounds. Their molecular structure is responsible for absorbing UV energy. Organic filters absorb UV rays, which produce excitation of the sunscreen chemical to a higher energy state. Then, they return to the ground state and convert the absorbed energy into longer, lower energy wavelengths (heat) (Figure 3.25b).[72] Currently, there are 15 organic sunscreen ingredients approved in the US. Examples for organic UVB absorbing filters include octinoxate, octisalate, and padimate O. Examples for UVA filters include avobenzone, oxybenzone, and meradimate. Organic filters are often combined with one another to achieve the desired broad-spectrum protection. According to the OTC monograph for sunscreens, a sunscreen product must have a minimum SPF of not less than the number of active sunscreen ingredients used in combination multiplied by 2.

Organic sunscreens can penetrate the skin due to their lipophilic nature, which may cause safety concerns.[73] An issue with many of the organic sunscreens is their photostability. Upon exposure to UV radiation, the structure of UV filters may be negatively affected and/or destructed. As a result, instead of returning to the ground state, they lose their absorption capacity. Therefore, most formulations contain photostabilizers (see more information under other ingredients).

As mentioned earlier, inorganic and organic UV filters are often combined to provide an optimal photoprotection. Inorganic UV filters are stable upon exposure to sunlight. However, these metal oxides may produce free radicals after UV exposure, which can degrade organic filters. Therefore, when organic and inorganic filters are combined, surface-treated inorganic filters should be used, for example, silica-coated, aluminum-coated, or silicon-coated TiO_2 or ZnO.[74]

Additional Ingredients of Sunscreens

Inactive ingredients generally found in sunscreen products depend on the dosage form. The vehicle determines which UV filters can be used based on their polarity and solubility characteristics. The most commonly used ingredient types are summarized here.

- **Waterproofing agents** are included in formulations to increase their water-resistance properties. As O/W emulsions are one of the most popular formulations, incorporation of waterproofing agents is important in their case to help the product withstand water in any form, such as sweat, sea or a swimming pool. The choice of sunscreen can also influence water resistance.
  - Examples include silicone oils, such as dimethicone 350, cyclomethicone, and dimethicone/trimethylsiloxysilicate. They are very resistant to water penetration and are easy to spread and form a continuous thin water-repellent film on the skin surface. Additionally, polymeric film-formers are gaining more interest as waterproofing agents. Alkylated polyvinylpyrrolidones (PVPs), which are closely related to the film-formers found in hair sprays, are also excellent for imparting water resistance to a formulation.
- **Photostabilizers** are ingredients that can prevent the degradation of organic UV filters, otherwise initiated by UV exposure. As discussed earlier, certain chemical filters are very sensitive to UV radiation and may suffer significant changes in their structure. Photostabilizers can help stabilize UV filters through chemical bonds, or help them dispose the UV energy more quickly, thus reducing or even eliminating the possibility of a chemical reaction.[75]
  - Examples include UV filters, such as octocrylene (for avobenzone specifically) and TiO_2, as well as other ingredients, such as polyester-8.
- **Emollients** are lipophilic ingredients that help increase water resistance since they are not washed off easily. They can also serve as solvents for lipophilic organic sunscreens. The maximum absorbance wavelength of sunscreen agents can be shifted to either longer or shorter wavelengths depending on the polarity of the sunscreen and that of the solvent. This may be either beneficial or detrimental for a formulation affecting the protection they can provide.[76]
  - Examples include mineral oil, shea butter, castor oil, cocoa butter, isopropyl myristate, isohexadecane, paraffin, and silicones.
- **Water** is an essential ingredient in O/W and W/O emulsions as well as in gels. It also serves as a solvent for water-soluble ingredients. For aerosol products, usually alcohol is used as a solvent.
- Usually, a combination of **emulsifiers** is used to provide the appropriate stability for the products.

  - Examples include glyceryl stearate, PEG-100 stearate, cetyl alcohol, polyglyceryl-3-methylglucose distearate, and cetyl dimethicone copolyol.
- **Thickeners** act as rheology modifiers in the formulations; they affect application, spreadability, and efficacy of sunscreens.
  - Examples include carbomers; cellulose derivatives, such as hydroxy propylcellulose; gums, such as xanthan gum; and other polymers, such as acrylates/C10-30 alkyl acrylate crosspolymer. Lipophilic thickeners, such as waxes, are also often used in sunscreens.
- **Film-forming ingredients** help form an even and uniform film on the skin after application and drying. This way they allow for a higher SPF factor.[77] They are sometimes referred to as "SPF boosters" for this reason.
  - Examples include hydrolyzed wheat protein/PVP crosspolymer, methylcellulose, polyester-7, and acrylates/octylacrylamide copolymer.
- **Antioxidants** The damage attributed to UVA radiation, for example, aging, is attributed to the formation of free radicals in the deeper skin layers. Although the body has natural antioxidant defenses against the reactive oxygen species, this endogenous system is quickly overwhelmed when faced with excessive oxidative stress. Antioxidants can help prevent oxidative reactions and, therefore, can be added to sunscreens, although their impact is not universally accepted.[78]
  - Examples include vitamins E and C.
- **Preservatives** are necessary when water is present in the formulations.
  - Examples include parabens, benzyl alcohol, methylchloroisothiazolinone, methylisothiazolinone, and phenoxyethanol.
- **Humectants** provide moisturization.
  - Examples include sorbitol, glycerin, and propylene glycol.
- **Chelating agents** contribute to the stability of the system by forming complexes with the metal ions that may cause premature deterioration of the products.
  - Examples include EDTA and its derivatives.
- **Propellants** are essential ingredients in aerosol sunscreens. They help expel the content of the aerosol can.
  - Examples include isobutane and dimethyl ether.
- Additional ingredients may include **pH modifiers (neutralizers)**, such as citric acid and triethanolamine; **fragrances;** and **colorants,** TiO_2 could act as an opacifier in certain cases; however, this effect is reduced with size reduction. **Natural ingredients,** such as allantoin, aloe vera, panthenol, and vitamins, can also be added to formulations.

Product Forms

Sunscreens are available in a variety of dosage forms, including O/W or W/O emulsions; anhydrous systems, such as ointments, sticks, oils, and silicone-based aerosols; wipes; and gels.

- **Emulsions** are the most popular form since they offer a variety of textures (such as sprayable lotions, thicker lotions, and creams). Additionally, another advantage of emulsions is that both oil- and water-soluble ingredients can be incorporated into them. O/W emulsions are generally more popular as the skin feel they offer is cosmetically more appealing, and are easier to manufacture at relatively low cost. Products with good spreadability are generally applied more uniformly to the skin, which results in a smooth surface with low variation in SPF and a greater application thickness.[79] W/O emulsions are greasier and stickier; however, they have been shown to have a higher efficacy. Additionally, due to the oily outer phase, they provide a higher level of water-resistance.[80] Lower viscosity emulsions may be packaged into nonaerosol spray containers. It can make application easier and more comfortable.
- **Sticks** are available as lipsticks and lip balms as well as sticks for babies. They are suitable for smaller surfaces areas only, such as the face and lips and for babies. However, applying them to a larger surface area, such as legs or back of adults, may be time consuming; therefore, sticks are usually not offered for such purposes. Due to their anhydrous nature, they provide water resistance and higher efficacy, but are often expensive to prepare, for this reason. The choice of waxes determines the melting point of the stick and its performance on the skin. Melting point can be reduced by the level and type of the emollients used, although firmness must be maintained in hot weather conditions.
- **Aerosol sprays** are one of the most popular formulations. The multiposition spray nozzles allow for quick and easy application by covering a relatively larger area on the body, so even hard-to-reach areas can be covered more easily. A common problem, however, is that they are not applied appropriately and the amount of product applied is far less than what is recommended by the FDA.[81] Aerosol spray sunscreens should be evenly spread on the skin and rubbed onto the skin after application (not just sprayed on). Concerns arose with regard to the safe use of sunscreens after severe burning cases were reported from consumers using aerosol products near a flame source. Aerosols are considered a safe dosage form; however, users should read the labels before using a product and take precautions when applying such products. Aerosol sprays are generally alcohol-based products containing alcohol-soluble organic UV filters. As physical sunscreens are not alcohol soluble, they are usually not formulated into aerosol sprays or would otherwise clog the valve. They usually contain film-forming agents to ensure the formation of an even film on the skin.
- **Ointments and oils** were quite popular years ago when many people used low-SPF products and wanted a deep-colored tan. Today, they are not as popular as other forms. Ointments and oils are greasier than W/O emulsions, which are not desired by consumers. Additionally, they are based on waxes, fats, and oils and are, therefore, more expensive than other forms.
- **Gels** provide a nice skin feel and absorb quickly; however, providing water-resistance properties for such formulations is quite difficult. Additionally, active ingredients that can be used for this dosage form are very limited. Another disadvantage may be irritation, if alcohol is used in the formulation.

- Probably, the latest sunscreen formulations are the **wipes**, which are similar to wet facial wipes. These are pre-moistened by the manufacturer. The formulators can place emollients and other moisturizers onto the cloth wipes, which may, therefore, provide additional benefits.
- **Cosmetic products** containing sunscreens (which are considered drugs for this reason), such as facial liquid foundation, face powder, BB creams, and CC creams, are also available today. These products are a convenient way for women to provide photoprotection on the facial region every day.[72]

The FDA currently considers sunscreens in the form of oils, creams, lotions, gels, butters, pastes, ointments, sticks, and sprays to be eligible for potential inclusion in the OTC sunscreen monograph; i.e., they can be marketed without individual product approvals. However, wipes, towelettes, powders, body washes, and shampoos are not eligible for the monograph. Therefore, they cannot be marketed without an approved application.[28]

Products formulated specifically for babies are worth mentioning at this point (see more information on baby care products in Section 2 of Chapter 7). Baby skin is much more sensitive than adult skin and penetration of ingredients into baby skin can be much higher and faster. Therefore, ideally, babies under 6 months should avoid sun exposure and no sunscreens should be used on their skin. The best sun protection for them is to keep them in the shade as much as possible in addition to wearing long sleeves, pants, a wide-brimmed hat, and sunglasses. For babies older than 6 months, sunscreens can be used to protect their skin from the sun. These are normally similar to adult formulations but are usually perfume-free to reduce the chance of irritation and sensitization. Additionally, irritation and sensitization potential can also be minimized using physical sunscreens, which do not penetrate the viable skin layers.[82] Products designed for babies should have a high SPF and water-resistance properties. In addition, baby skin is more susceptible to moisture loss; therefore, formulations should have additional moisturizing benefits as well to promote the protective skin barrier.[83] Sunscreen products for babies are available in a variety of dosage forms, including aerosols, creams and lotions, pump sprays, and sticks.

Optimizing Product Efficacy and Consumer Acceptance

There are several strategies commonly employed to formulate sunscreens with broad-spectrum protection, preventing both short-term and long-term effects of UV radiation. Although it seems logical to combine various UV filters in order to achieve better UV protection, active ingredients are not the only ingredients that can modify a product's properties. Inactive ingredients (i.e., cosmetic ingredients) also play a significant role in optimizing the formulation for both efficacy and consumer acceptance. Additionally, developing and marketing new cosmetic ingredients is much easier than those of UV filters, which also have to be approved by the FDA. Here is a summary of commonly used strategies:

- UVB protection is not enough. Therefore, UVB filters should be **combined** with UVA filters in order to achieve optimal protection. Combining physical and chemical UV filters allows for a better coverage of the UVB and UVA spectrums, resulting in broad-spectrum protection. Extending protection to the UVA region also helps enhance the SPF value.

- UV filters are either hydrophilic or lipophilic. When **combined**, a synergetic effect can be observed. This property can be used to obtain higher efficacy against UVB and UVA radiations.[84] Additionally, as most sunscreen formulations are emulsions, having active ingredients in both phases provides better overall efficacy, even if the product is not uniformly distributed on the skin surface.
- The **combination** of organic and inorganic filters is also advantageous. Combining nano-TiO_2 with chemical UV filters often provides better UVB protection than expected, based on the SPF of each ingredient. Studies show that as inorganic filters scatter light in the upper layers of the skin, they increase the optical path length of UV radiation and create more opportunity for absorption by chemical filters.[20]
- Applying **photostabilizers** is very important in all formulations containing chemical filters. As discussed earlier, many chemical filters, such as avobenzone, are sensitive to UV radiation. In order to prevent photodegradation, various photoprotective ingredients should be included in the formulations. Photostabilizers are able to quench the excited state of the organic filters by accepting the excited state energy, thereby returning the UV-absorbing molecule back to its ground state.[85]
- Homogenous **distribution** of the active ingredients in the product is also very important. Vehicles that dissolve and disperse the UV filters uniformly can enhance the overall UV protection by providing an even coverage on the skin, thus better protecting against sunburn.[86] As inorganic filters are insoluble particles, they will be in a suspended state in the formulation. Rheology modifiers have to be used to provide an appropriate viscosity for the formulations. Ideally, products are easy to spread; however, they do not flow off the skin surface after application but stay there and form a uniform film.

DID YOU KNOW?

Avobenzone is the only long-range chemical UVA filter (340–400 nm) widely available to sunscreen manufacturers in the US. However, the molecule is inherently unstable and, by itself, loses nearly 50% of its screening capacity after just 1 h of UV exposure.[87] Furthermore, the addition of other ingredients, even certain UVB filters, to sunscreens containing avobenzone accelerates the degradation of both compounds. This is why incorporating photostabilizers is extremely important in formulations.

Formulation The formulation of emulsions should follow the general formulation steps. Homogenous dispersion of the insoluble inorganic filters is very important for an effective and aesthetically appealing product. The formulation of gels should also follow the general steps, usually starting with hydrating the thickener(s). Sticks are formulated with the molding technique. It is discussed in more detail in Section 1 of Chapter 4. Similarly, the formulation of makeup products can be found in Chapter 4. Sunscreen wipes are clothes soaked in sunscreen lotions or solutions. Their formulation is identical to that of facial wipes.

The manufacture of aerosol formulations must be performed in a flameproof area using special equipment since they are highly flammable and/or explosive. First, the product concentrate is formulated, which is later filled into the can with the propellants using a special technique called pressure filling.[88] In this process, the product concentrate is usually filled into the can first, and then the valve is inserted and crimped into place. Finally, the liquefied propellant is forced into the can under pressure through the valve orifice after the valve is sealed. Air can be removed from the can by means of vacuum or displacement with a small amount of propellant vapor. After the filling operation is complete, the valve is tested for proper function.[89]

After-Sun Preparations

After-sun preparations, otherwise known as sunburn preparations, are used to hydrate the skin and sooth irritated, red, burnt skin. Even if no sunburn occurs, after exposure to UV radiation, the skin needs care, including soothing, smoothing, and moisturizing.

Sunburn Although tanning is believed to be healthy by many, any degree of tanning is a sign of skin damage, even without reddening. Sunburn is an acute inflammatory skin reaction for excessive exposure to UV radiation, which can come from a variety of sources, including sun, tanning beds, and phototherapy lamps. Sunburn is generally classified as a superficial or a first-degree burn. It can cause various signs and symptoms, such as tenderness, mild pain upon touch, itching, and redness, which may be followed by scaling.[90] More severe cases may proceed to edema or even blistering. Studies have shown that repeated severe burns at an early age increase the risk for melanoma and other types of skin cancers later in the life.

DID YOU KNOW?

In 1988, as few as 1% of American adults reported using indoor tanning facilities; by 2007, that number had increased to 27%.[51] At the same time, many tanning devices employed more powerful UV lamps. When combined, these results indicate a potential public health issue with tens of millions of individuals putting themselves at increased risk of developing skin cancer in the years ahead.[91]

Skin Care after Sun Exposure As a first step of treating sun-affected skin, it is usual to clean the skin with water and mild cleaners to remove any remaining product, sweat, and dirt. Thereafter, hydration therapy can take place. Vitalization and hydration of skin are very important after sun exposure to help the skin maintain or restore its healthy status. Most after-sun formulations are O/W emulsions (such as lotions, creams, and sprays) and gels containing moisturizers as well as anti-inflammatory agents and antioxidants. Gels have the advantage of an immediate cooling effect due to the high water content. The main functional ingredients in sunburn preparations are summarized here.

- **Soothing and anti-inflammatory** ingredients help alleviate pain, redness, and burning. Examples include azulene and bisabolol (from chamomile), allantoin, aloe extract, and panthenol. Witch hazel is an astringent and anti-inflammatory agent that is commonly used.
- **Cooling** agents such as menthol, eucalyptus, and alcohol provide an immediate short cold sensation. They soothe and alleviate the warm sensation and tenseness of sun-irritated skin, at least for a short period of time.
- **Antioxidants** help replenish the depleted antioxidant pool and/or boost the antioxidant defenses of the horny layer. Examples include vitamin E, vitamin A, green tea extract, and pomegranate extract, among other natural ingredients.
- **Moisturizers** help replace the water lost during sun exposure to improve elasticity and dryness by maintaining moisture levels. The use of suitable moisturizing agents also prevents, or at least postpones, the unsightly features of skin flaking and peeling. Primary ingredient types used include humectants and emollients.

The formulation should follow the general routine of the conventional formulation of emulsions and gels.

Typical Quality Problems of Sun Care Products

●**The typical quality-related issues of sunscreens and after-sun products include valve clogging, separation of the emulsions, microbiological contamination, clumping, and rancidification.** Issues discussed in Section 4 are not discussed in detail here.

Valve Clogging Valve clogging may occur in aerosol products. This phenomenon means that the product cannot be dispensed by pushing down the valve. As discussed previously, aerosol sunscreens are alcohol-based products in which the chemical filters are soluble. Valve clogging may occur due to several reasons, including inappropriate type and/or amount of the film-formers, inappropriate type and/or amount of propellant used, incompatibility between the formulation and the propellant, incompatibility between the formulation and the plastic components of the aerosol can, and inappropriate valve and actuator type.[92] This quality problem can be overcome with careful control of the composition of the formulation and container.

Evaluation of Sun Care Products

Quality Parameters Generally Tested **❽Parameters commonly tested by cosmetic companies to evaluate the quality of their sunscreens include spray characteristics; aerosol can leakage; actuation force, pressure test for aerosol products; spreadability, extrudability, texture, and firmness of lotions, creams, and gels; stick hardness, stick softening, and melting point; stick pay-off and glide; preservative efficacy; viscosity; and pH.** The evaluations that were discussed in the previous chapters are not reviewed here in detail.

Spray Characteristics The most important spray characteristics include spray rate, spray pattern, and droplet size.

- **Spray rate**, or otherwise known as the aerosol discharge rate, is usually determined by measuring the weight loss over time. This determination measures the amount of product in weight, which is dispensed during a given period of time. The aerosol can is weighed before the measurement; this value is the initial weight. Then, the product is discharged with a constant rate using a standard apparatus for a given period of time. By reweighing the container, the change in weight per time is the discharge rate, which then can be expressed as grams per second.
- **Spray pattern** refers to the area of the sunscreen spray that hits the skin (see Figure 3.26). This parameter can be determined by using an alcohol-sensitive paper (as aerosol sunscreens are alcohol-based). As the droplets are deposited on the paper, they initiate a color change (staining). The shape and diameter of the stain provide information on the spray pattern. Normally, a circle should be seen on the paper (Figure 3.26a). Others shapes (Figure 3.26b–d), such as a hollow cone or flat stream sprayed in any direction, are not acceptable.
- Sunscreen spray droplets are formed upon expansion and rapid evaporation of the propellant after expelling the product from the can. The **droplet size** depends on the pressure within the can, the type of the actuator, and the physical properties of the formulation. The particle size distribution produced during atomization relates directly to the rate of drying and the likely formation of actual droplets on the skin after application. Generally, laser diffraction technique is used to determine both the droplet size and the droplet size distribution of aerosol sprays. It is based on the principle that particles (droplets) scatter light. The intensity of scattered light can be measured as a function of the angle, which can be then used to yield a droplet size distribution.

Aerosol Can Leakage Leakage testing is a basic requirement for all aerosol containers. One of the general testing methods is the immersion method. It is performed by completely immersing the filled aerosol can in cold/hot water and examining whether air bubbles are formed or not. This testing is typically done under conditions that could cause a can to erupt, presenting safety risks to the operator or the equipment.

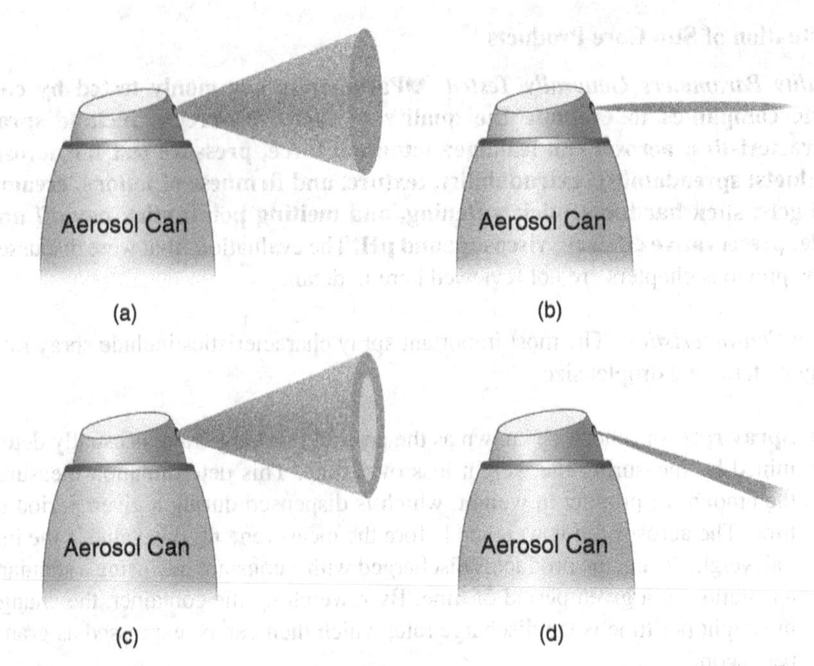

Figure 3.26 Various spray patterns of aerosol sprays. (a) normal spray pattern. (b–c–d) abnormal spray patterns.

The USP describes a method where the weight of filled aerosol cans should be measured and recorded. The method is applicable only to topical aerosols that are fitted with continuous valves. The weight should be remeasured within 3 days, and leakage can be calculated using a formula.[89]

Actuation Force for Aerosol Cans Actuation force testing allows measurement of the force required to release sunscreens from aerosol cans. The equipment used for aerosols is shown in Figure 3.27. Testing of the actuation force is important to see whether users will be able to get the product out of the container by pressing the actuator. Therefore, this test provides an imitative test and helps assess any issues related to the use of the product.

Pressure Test Pressure in the aerosol cans can be tested in several ways. An option is to subject containers to hot water. The duration of the test and the temperature should be such that the internal pressure reaches that which would be reached at 55 °C. If the contents are sensitive to heat, the temperature of the bath must be set at between 20 °C and 30 °C and one container in 2000 must be tested at the higher temperature. No leakage or permanent deformation of a container may occur during the test.[93]

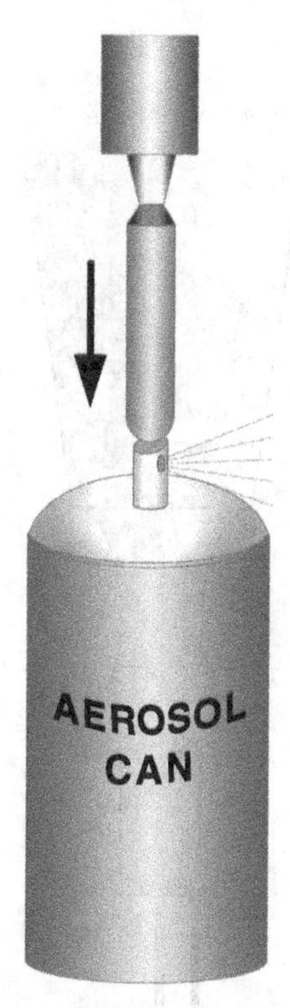

Figure 3.27 Actuation force testing of an aerosol can.

Stick Hardness Test The hardness of sunscreen sticks depends on the type and the ratio of the ingredients. Waxes provide luster, adequate firmness, and appropriate molding properties for the sticks and also significantly influence the final texture of the product. Waxes, however, differ in hardness. Measuring the hardness of a stick by varying the composition and wax content is a useful tool in quality control for assessing whether a wax of interest is suitable for producing the desired properties. The generally used test is the penetration test, adopted from the ASTM Standard Method of Test.[94] As Figure 3.28 depicts, the penetrometer contains a metal needle probe that penetrates the product. Methods used may vary; they either determine the distance of penetration when applying a specified force over a period of 5 s, or they measure the force of penetration at a defined deformation distance of 5 mm. The test can monitor the hardness over a range of temperatures that the stick may be subjected to during transportation and storage. In addition to determining the stick hardness, this test may indicate the presence of unwanted trapped air bubbles or a "grainy" texture.

Melting Point of Sticks Determination of the melting point is important as it is an indication of the limit of safe storage. This is mainly and significantly determined by the ratio and types of wax used. The melting point is the temperature at which a

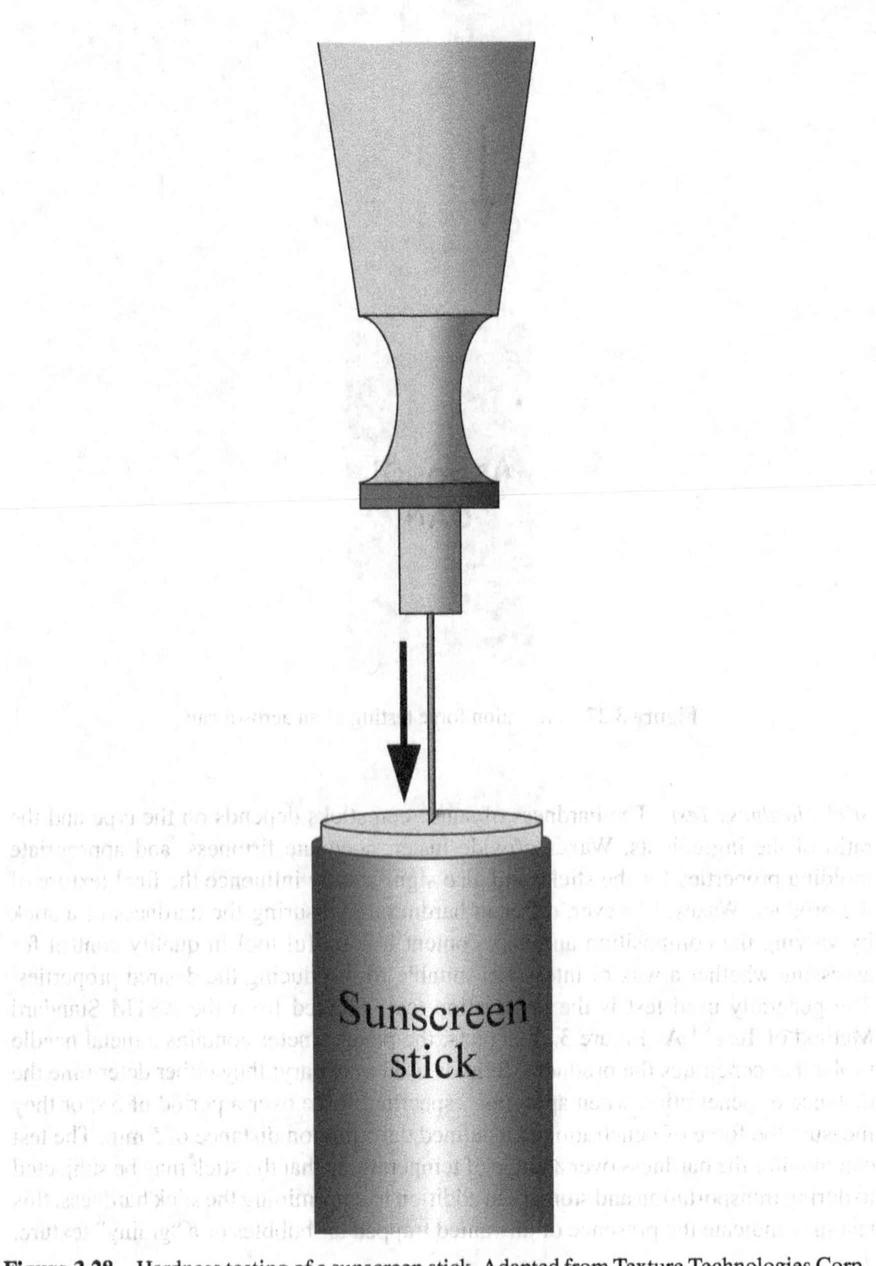

Figure 3.28 Hardness testing of a sunscreen stick. Adapted from Texture Technologies Corp.

solid phase converts to a liquid phase at 1 atm of pressure. It is usually determined by the capillary tube method in which a small sample of the substance is placed into the melting point capillary to a depth of about 0.04 in. (1 mm). The capillary is then placed into the capillary apparatus and the sample is gradually heated. The temperature at which the liquid is first seen is the lower end of the melting point range. The temperature at which the last solid disappears is the upper end of the melting point range.

Softening Point of Sticks A sunscreen stick should withstand a range of conditions to which it will be subjected in the consumers' handbag. It should be resistant to varying temperatures and be just as easy to apply in hot as in cold weather. Softening point is usually determined by the ring and ball method. The process is shown in Figure 3.29. A stick is fitted into a ring or support orifice, and the extra mass above and below the ring/orifice is removed with a sharp blade, so practically, a tablet of stick remains inside the ring/orifice. This is placed in a refrigerator for a few minutes, for example, 10 min, and then the ring is fastened onto a stand. A beaker containing 500 ml water at room temperature is placed on a hot plate having a magnetic stirrer. A steel ball is delicately placed on the stick tablet. The bar with its support is then inserted into the beaker until it submerges into it. Heating and slow stirring are then started and the temperature is monitored using a thermometer. The temperature at which the mass and steel balls are loosened and fall to the bottom of the beaker is the softening point.

Pay-off and Glide Pay-off refers to the weight of a product transferred to a surface, for example, the skin, upon application. Glide refers to the easiness of moving the stick over a surface. It can be characterized by measuring the friction as the stick is

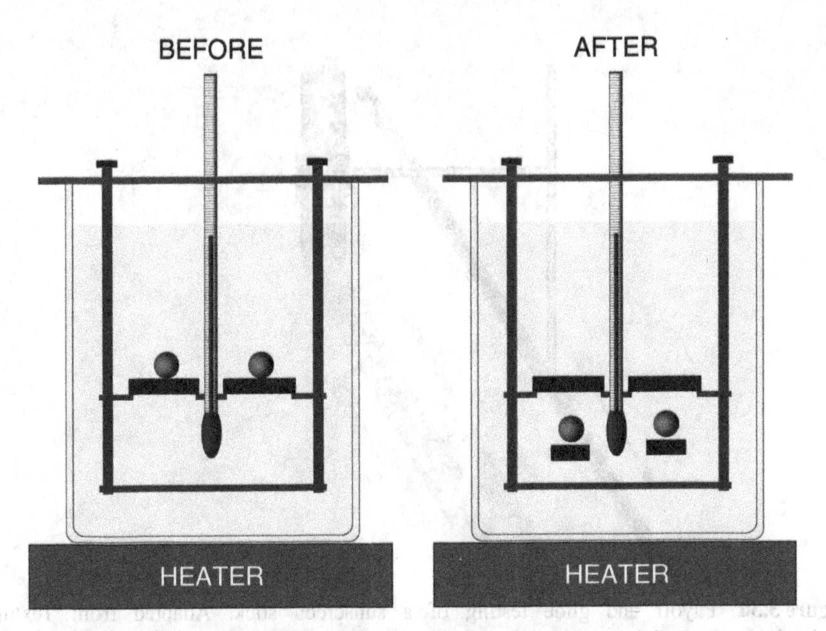

Figure 3.29 Softening point testing of sunscreen sticks.

moved on a surface.[95] Both properties depend on the wax/oil ratio of the formulation and are also in relationship with the hardness of the product. Harder and more rigid sticks have a lower level of pay-off, which is unfavorable from a consumer perspective. Inadequately, soft products (containing more oils) may, however, soften at lower temperatures and lose their shape. The equipment used for this test usually contains a piece or roll of paper and an arm for fixing the stick. The stick is moved on the surface of the paper at a constant speed with a standard pressure, mimicking the force applied by consumers (see Figure 3.30). The mass in grams transferred to the paper, i.e., pay-off, is measured afterward. Additionally, the friction coefficient is also measured during the movement, which refers to the glide.

Performance (Efficacy) Parameters Generally Tested ●**The most frequently tested parameters referring to the performance (efficacy) of sunscreens include the SPF value, water resistance, and broad spectrum.**

Evaluation of the Sun Protection Factor (SPF) Determination of the SPF value is performed *in vivo*, on human subjects, as it is required in the OTC monograph. All subjects for the SPF study are typically Fitzpatrick skin types I–III since they are the most susceptible to sunburn (for the Fitzpatrick categories and general characteristics of the major skin types, refer back to Section 1 of this chapter). First, the MED is determined for each subject for unprotected skin. The lowest dose, which produces the first definite redness, is defined as the MED for unprotected skin. Generally, the back is used for this test as usually it does not have high prior sun exposure and is not hypersensitive. To simulate the correct wavelength and strength of the UVB radiation, typically a solar simulator is used, which emits light from 290 to 400 nm, similar to

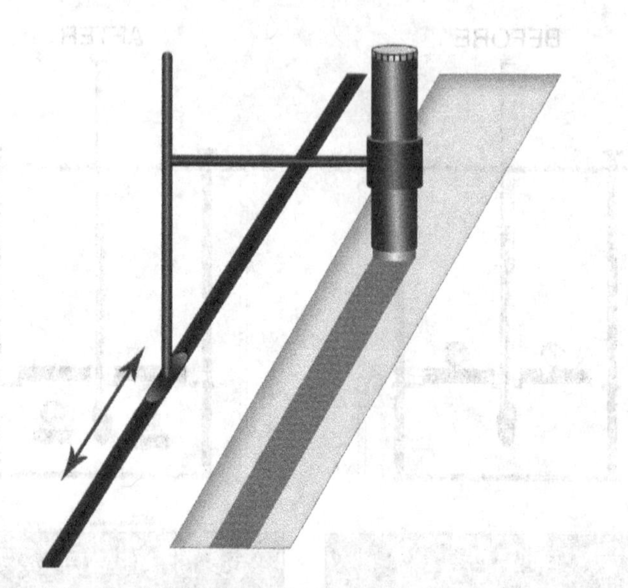

Figure 3.30 Payoff and glide testing of a sunscreen stick. Adapted from Texture Technologies Corp.

midday, midsummer sunlight. After determining the baseline MED, the sunscreen as well as a reference product are applied to the back in a dose of 2 mg/cm^2. Test sites are outlined on the back of the subjects, each of which has 4–5 subsites. These subsites receive varying dosages of UV radiation. At specific points, evaluators assess the back looking for signs of redness, evaluating it on scale. The ratio between the baseline MED value and postexposure MED value provides the basis for calculating the SPF value of the product.

DID YOU KNOW?

There were ethical issues of directly exposing the human skin directly to UV light, i.e., solar simulator. Therefore, alternative methods are being developed. One method uses a fiber-optic liquid light guide that reflects UV light from the solar simulator, down the tube, and onto the skin site.[96]

As the SPF test uses human volunteers, it is sensitive to a certain degree of biological variability among the subjects due to factors such as skin type, skin condition (e.g., sweating), and evenness of product spreading. To rule out significant variations, the OTC monograph requires a test panel with not more than 25 subjects. From this panel, at least 20 subjects must produce valid data for analysis.[97] Variability can also result from factors such as solar simulator lamp variation, environment (e.g., temperature, humidity), as well as consistency and accuracy of the MED assessment. On-site use, e.g., product application on the beach, may be also influenced by factors such as application thickness by users, reapplication rate, presence of any photosensitizing ingredients, substantivity of the product, skin temperature, and wind, among others. Therefore, SPF numbers should be viewed as a guide to protection level and not as an absolute measure.

Water-Resistance Testing of Sunscreens Water resistance refers to the property of a sunscreen to maintain its degree of protection when exposed to water, such as sweating or swimming. Intensive sweating and water immersion are common phenomena occurring when using sunscreens; therefore, it is very important that sunscreens maintain their efficacy under such conditions. Water-resistance testing is an *in vivo* test, performed similar to the SPF value determination. According to the OTC monograph of sunscreens, for a product claimed to be "water resistant," the label SPF should reflect the SPF measured after 40 min of water immersion. During the test, the subjects are immersed into water two times for 20 min separated by a 20-min rest period without toweling. The SPF is determined after the second 20-min water immersion. For a product claiming to be "very water resistant," its labeled SPF should reflect the SPF value determined after 80 min of water immersion. For the test phases of 20-min water immersion, a 20-min resting without toweling is repeated four times and the SPF is determined after the last water immersion.

Broad-Spectrum Testing The currently required test by the FDA to determine whether a sunscreen provides protection against UVA radiation, in addition to UVB protection, is an *in vitro* test. This broad-spectrum test procedure measures a product's UVA protection relative to its UVB protection. This method is referred to as the critical wavelength method, which is a pass/fail test. The critical wavelength must be greater than 370 nm as its standard for acceptable broad-spectrum protection.[98] In this test, the sunscreen product is applied to an optical grade polymethylmethacrylate (PMMA) plates suitable for UV transmittance measurements at a dose of 0.75 mg/cm^2. The UV light emitted is in the UVB–UVA range (290–400 nm range). The UV absorbance of the sunscreen measured is plotted against wavelength. The width of the curve plotted on the graph shows how broadly effective a sunscreen is across the UV spectrum. The area under the curve (AUC) is measured, and a line can be drawn at the wavelength where 90% absorption occurs (90% AUC). It is defined as the critical wavelength. Therefore, the higher the critical wavelength, the better the UVA protection provided by the sunscreen.[97,99]

Safety Concerns **Safety concerns arose in the past regarding the use of sunscreens containing organic sunscreens and nanosized physical sunscreens as well as products in aerosol form**. The main scientific findings are summarized here.

Organic Sunscreens One of the major concerns regarding the use of organic sunscreens is the possibility of skin penetration and, as a consequence, potential toxic systemic effects. Due to their lipophilic nature, certain organic sunscreens could potentially penetrate the skin.[73] Recent studies have, however, shown that, although some sunscreens may be able to absorb into the skin, including the SC and viable epidermis, their level is far below the toxic level.[100] Additionally, it has been shown that the data generated under *in vitro* conditions may overestimate *in vivo* skin concentrations of sunscreen actives.[101]

Photostability of Organic Sunscreens An additional issue with regard to many organic sunscreens is their photostability/photoinstability. As mentioned earlier, the structure of UV filters may be negatively affected and/or destructed upon UV exposure. Questions arose whether photodegradation products can cause any short-term adverse reactions, such as skin irritation and allergic reactions, or long-term adverse reactions, such as toxicity upon repeated exposure to these photodegradation products over months and years. An additional concern is an increase in UV exposure due to degradation of the sunscreen, which could lead to sunburn and chronic skin damage, such as skin cancer. However, the reported occurrence of photoirritation and photoallergic responses to sunscreens is rare compared with adverse events, such as skin irritation or sensitization, produced by cosmetics.[102] Additionally, existing *in vivo* animal studies did not identify a health hazard related to the photodegradation of UV filters.[103] Studies also showed that the long-term benefits of using sunscreens, protection against skin cancers, seem to outweigh any potential adverse consequences attributed to photodegradation. Sunscreens are thoroughly tested for efficacy as discussed earlier, both *in vivo* and *in vitro* studies. During these studies, sunscreens are exposed to UV radiation. Therefore, photoinstability and photodegradation would become obvious during these studies if it was significant. Today's sunscreens containing potentially photounstable UV filters always contain photostabilizing ingredients as well.[103]

Nanoparticles in Sunscreens TiO_2 and ZnO are frequently used in sunscreens either alone or in combination with each other as physical UV filters. Additionally, they are often combined with chemical filters since it can increase the efficacy of formulations, as discussed earlier. The main drawback of regular-sized materials is that they provide a white appearance when applied to the skin. This aesthetically negative effect, however, can be solved by decreasing the particle size under 100 nm, which is referred to as the nanorange. Nanosized TiO_2 and ZnO are transparent, which is cosmetically very appealing. However, concerns arose regarding the use of such small particles, questioning whether they can penetrate human skin and get to the viable layers. Studies have shown that nanosized TiO_2 and ZnO penetrate the upper layers of the SC only in healthy human skin and do not reach the living skin.[69,104] Additionally, topical toxicity studies demonstrated that nanosized TiO_2 and ZnO have low toxicity and are well tolerated on the skin.[105] The weight of evidence suggests that nanosized physical sunscreens currently used in sunscreens (or cosmetic products) pose no risk to human skin or human health when applied topically.[106] On the contrary, they provide a huge benefit by protecting human skin against the adverse effects of UV radiation, including skin cancer.

Aerosol Sunscreens The FDA recently requested more information on the safety and efficacy of spray sunscreens to have a better understanding whether these products are similar in efficacy and safety to other dosage forms, such as lotions and creams. As sprays may be inhaled intentionally, it is important to see whether it would pose any risk to consumers.

DID YOU KNOW?

A nanometer (nm) is 1000 times smaller than a micrometer and a 1,000,000 times smaller than a millimeter. Nanosized materials cannot be seen with the naked eye. To give a better understanding on how small nanoparticles are, here is an example: the diameter of Caucasian hair varies from 50–90 μm,[107] i.e., 50,000–90,000 nm. Just think about it; it is hard to see the width of a hair with the naked eye, without using a magnifying glass. Now you understand how tiny a nanometer is.

Packaging of Sun Care Products

The most commonly used packaging materials for sun care products include the following:

- *Plastic Bottles*: The majority of the sunscreens and after-sun preparations are supplied in plastic bottles, usually with a flip-top cap. Some low-viscosity sunscreens as well as sunbathing oils are packed into plastic bottles with a pump head.
- *Aerosol Cans*: Aerosol sunscreens are packaged into aerosol cans. These are one-compartment cans, where the formulation and the propellant are mixed together. Upon pressing the valve, some of the product is dispensed with the help of the propellant. This type of spray system produces a much smaller particle size. The multiposition spray nozzles allow for quick and easy application by covering a relatively larger area on the body, so even hard-to-reach areas can be covered much easily. As discussed earlier, however, attention needs to be paid that enough product is applied to ensure adequate protection.
- *Wipes*: Today, even sunscreen wipes are available. These are ready-to-use, prewetted formulations. These are usually packed into soft sachets with either a resealable flap or a flip-top cap. Sometimes, they are packed individually, or placed into plastic containers (similar to hand sanitizer wipes).

GLOSSARY OF TERMS FOR SECTION 5

Actuation force: The force required to release an aerosol product from an aerosol can or a liquid or a semisolid product from a pump-head container.

After-sun product: A personal care product that is designed to be used after exposure to the sun or other UV radiation. It helps smooth and moisturize the skin as well as provide a cooling effect and relieve pain resulting from sunburn.

Broad spectrum: A term used to describe sunscreens that provide protection against both UVA and UVB radiations.

Chemical sunscreen: An organic UV filter that absorbs UV radiation.

Glide: The easiness of moving the stick over a surface, for example, skin.

Melting point: The temperature at which a solid phase converts to a liquid phase at 1 atm.

Nanoparticle: A particle with a size of 1–100 nm.

Payoff: The weight of a product transferred to a surface, for example, the skin, on application.

Photoaging: Intrinsic aging accelerated by sun exposure. Primarily, UVA is responsible for photoaging.

Photoinstability: The lack of stability to UV exposure.

Photostabilizer: An ingredient that can prevent organic UV filters degrade on exposure to UV light.

Physical sunscreen: An inorganic UV filter that reflects and scatters UV radiation.

Primary sunscreen: A sunscreen product that is primarily used to provide photoprotection for the skin. An example is beach sunscreens.

Secondary sunscreen: A sunscreen product whose primary function is other than skin protection, such as skin moisturization, evening the skin tone, covering up skin problems, and prevention of aging. It has the added benefit (secondary function) of sun protection.

Skin cancer: Abnormal growth of skin cells that tend to proliferate in an uncontrolled way and, in some cases, spread. It most often develops on the skin exposed to the sun. Both UVA and UVB radiations are associated with the development of various types of skin cancer.

Softening point: The temperature at which a solid mass softens but does not melt completely.

SPF: Sun protection factor, a measure that indicates how long it takes for UV rays to redden protected skin compared to how long it takes to redden unprotected skin.

Spray characteristics: A set of parameters used to characterize sprayable products. The most commonly used parameters include spray rate, spray patterns, and droplet size.

Sunburn: Acute skin damage perceived as redness. UVB radiation is the major cause of sunburn.

Sunscreen: A personal care product that is designed to provide photoprotection against the harmful radiation of the sun, including both UVA and UVB radiations. The protection they provide can be through chemical and/or physical methods.

Tanning: Browning of the skin after exposure to the sun.

UV radiation: Ultraviolet radiation is part of the electromagnetic spectrum. Of the wavelengths of radiation that reach the Earth's surface, UV has the highest energy. Only a portion of the UV radiation reaches the surface of the Earth, including UVA and UVB. They are associated with aging, burning, tanning, and skin cancer.

UVA: Ultraviolet A radiation, ranging from 320 to 400 nm.

UVA I: A type of UVA radiation, ranging from 340 to 400 nm.

UVA II: A type of UVA radiation, ranging from 320 to 340 nm.

UVB: Ultraviolet B radiation, ranging from 280 to 320 nm.

Valve clogging: A quality problem of aerosol sunscreen products. It means that the product cannot be dispensed by pushing down the valve.

Vitamin D: A fat-soluble vitamin, which has a significant role in bone health and the prevention of osteoporosis and osteomalacia. It is produced in general by UV exposure.

Waterproofing agent: An ingredient that is added to sunscreen formulations to increase their water-resistance properties.

Water resistance: A term used to describe products that are water repellent.

REVIEW QUESTIONS FOR SECTION 5

Multiple Choice Questions

1. Which of the following is responsible for skin burning?

a) UVA radiation
b) UVB radiation
c) UVC radiation
d) All of the above

2. Which of the following is responsible for skin aging?

a) UVA radiation
b) UVB radiation
c) UVC radiation
d) All of the above

3. SPF refers to protection against ___ radiation.

a) UVA radiation
b) UVB radiation
c) UVA and UVB radiations
d) UVC radiation

4. Broad-spectrum protection refers to protection against ___ radiation.

a) UVA radiation
b) UVB radiation
c) UVA and UVB radiations
d) UVC radiation

5. Which of the following is TRUE for chemical UV filters?

a) They reflect UV light back
b) They are not soluble in either water or oils
c) They provide the skin with a white appearance, especially the larger particles
d) Some of them are photounstable and may suffer degradation upon UV exposure

6. Titanium dioxide provides a white appearance when applied to the skin. How can this effect be diminished and/or eliminated?

a) Combining it with zinc oxide
b) Combining it with avobenzone
c) Reducing its particle size
d) Dissolving it in oils

7. Which of the following techniques can be used to increase the efficacy of sunscreens?

a) Combining physical and chemical filters
b) Incorporating photostabilizers when chemical filters are used
c) Combining UVA and UVB filters
d) All of the above

8. A sunscreen with an SPF value of 10 can provide protection (theoretically) for ___.

a) 10 min
b) 10 h
c) A period 10 times longer than it takes for someone to see the first signs of redness
d) A period 10 min longer than it takes for someone to see the first signs of redness

9. Which of the following is TRUE for UV radiation?

a) UVB light penetrates the dermis
b) UVB light penetrates the epidermis only
c) UVA light penetrates the epidermis only
d) UVC light penetrates both the epidermis and the dermis

10. Which of the following is usually tested for aerosol sunscreens?

a) Spray pattern
b) Can leakage
c) Actuation force
d) All of the above

11. Which of the following parameters refers to the amount of a product transferred to a surface upon application?

a) Spreadability
b) Extrudability
c) Pay-off
d) Hardness

12. Which of the following is NOT true for the evaluation of the sun protection factor?
    a) Evaluators are looking for signs of redness
    b) It is tested *in vitro*
    c) A solar simulator is used to simulate the correct wavelength of the UVB radiation
    d) The standard sunscreen dose in the test is 2 mg/cm^2

13. What is the main purpose of application of after-sun products?
    a) To wash off sunscreens from the skin
    b) To remove sweat and dirt from the skin
    c) To hydrate the skin and relieve redness
    d) To prevent the skin from sunburn

14. Which of the following is TRUE for sunscreens?
    a) Under SPF 10, they are considered cosmetics in the US
    b) They are considered cosmetics in the US, irrelevant of their SPF
    c) They are considered OTC drugs in the US
    d) a and c are true

15. Which of the following provides a better UV protection?
    a) A sunscreen cream with an SPF value of 15
    b) A BB cream with an SPF value of 15
    c) The protection they provide is the same

Short Answers

1. The critical age under which sunscreens should not be used on babies is: ___
2. The critical wavelength used in the broad-spectrum test of sunscreens is: ___
3. Sunscreens that are broad-spectrum and have an SPF value of at least ___ can claim that they can prevent aging and skin cancer.
4. A sunscreen claimed to be "very water resistant" still provides protection after ___ minutes of water immersion.
5. A particle is usually called nanoparticle if its size is smaller than ___ nm.

Matching

Match the ingredients in column A with their appropriate ingredient category in column B.

	Column A	Column B
_____	A. Aloe extract	1. Chemical sunscreen
_____	B. Avobenzone	2. Cooling ingredient
_____	C. Citric acid	3. Emollient
_____	D. Cyclomethicone	4. Film-former
_____	E. Hydrolyzed wheat protein/PVP crosspolymer	5. pH buffer
_____	F. Isobutane	6. Photostabilizer
_____	G. Isopropyl myristate	7. Physical sunscreen
_____	H. Menthol	8. Propellant
_____	I. Polyester-8	9. Soothing and anti-inflammatory ingredient
_____	J. Titanium dioxide	10. Waterproofing agent

REFERENCES

1. NASA: Electromagnetic Spectrum, Last updated: 3/2013, Accessed 4/13/2014 at http://imagine.gsfc.nasa.gov/docs/science/know_l1/emspectrum.html
2. Brugè, F., Tiano, L., Astolfi, P., et al.: Prevention of UVA-Induced Oxidative Damage in Human Dermal Fibroblasts by New UV filters, Assessed Using a Novel *In Vitro* Experimental System. *PLoS One*. 2014;9(1).
3. WHO: Ultraviolet Radiation and Health, Accessed 4/13/2013 at http://www.who.int/uv/uv_and_health/en/
4. Gambichler, T., Künzlberger, B., Paech, V., et al.: UVA1 and UVB irradiated skin investigated by optical coherence tomography in vivo: a preliminary study. *Clin Exp Dermatol*. 2005;30(1):79–82.
5. CDC: Skin Cancer Risk, What Can I Do to Reduce My Risk?, Last update: 1/22/2014, Accessed 4/12/2014 at http://www.cdc.gov/cancer/skin/basic_info/prevention.htm
6. Moehrle, M.: Outdoor sports and skin cancer. *Clin Dermatol*. 2008;26(1):12–15.
7. Schulman, J. M., Fisher, D. E.: Indoor ultraviolet tanning and skin cancer: health risks and opportunities. *Curr Opin Oncol*. 2009;21(2):144–1449.
8. Coelho, S. G., Hearing, V. J.: UVA tanning is involved in the increased incidence of skin cancers in fair-skinned young women. *Pigment Cell Melanoma Res*. 2010;23(1):57–63.
9. de Gruijl, F. R.: Photocarcinogenesis: UVA vs UVB. *Methods Enzymol*. 2000; 319:359–366.
10. Garland, C. F., Garland, F. C., Gorham, E. C.: Epidemiologic evidence for different roles of ultraviolet A and B radiation in melanoma mortality rates. *Ann Epidemiol*. 2003;13:395–404.

11. Moyal, D., Binet, O.: Polymorphous Light Eruption (PLE): Its Reproduction and Prevention by Sunscreens, In: Lowe, N. J., Shaat, N., Pathak, M., *Sunscreens: Development and Evaluation and Regulatory Aspects*, 2nd Edition, New York: Marcel Dekker, 1997.
12. Miller, S. A., Hamilton, S. L., Wester, U. G., et al.: An analysis of UVA emissions from sunlamps and the potential importance for melanoma. *Photochem Photobiol.* 1998;68(1):63–70.
13. Wehner, M. R., Shive, M. L., Chren, M. M., et al.: Indoor tanning and non-melanoma skin cancer: systematic review and meta-analysis. *BMJ.* 2012. 345:e5909.
14. FDA: Questions and Answers: FDA Announces New Requirements for Over-the-Counter (OTC) Sunscreen Products Marketed in the U.S., Last updated: 6/23/2011, Accessed 4/15/2014 at http://www.fda.gov/drugs/resourcesforyou/consumers/buyingusingmedicinesafely/understandingover-the-counter medicines/ucm258468.htm#Q3_What_does_the_SPF
15. FDA. Sunscreen drug products for over-the-counter human use; Final Monograph. *Fed Reg.* 1999;64(98):27666–27693.
16. FDA: Radiation-Emitting Products, Last update: 9/26/2013, Accessed 4/17/2014 at http://www.fda.gov/Radiation-EmittingProducts/RadiationEmittingProductsandProcedures/Tanning/ucm116425.htm
17. CFR Title 21 Part 352.3
18. Autier, P., Boniol, M., Severi, G., et al.: Quantity of sunscreen used by European students. *Br J Dermatol.* 2001;144:288–291.
19. Diaz, A., Neale, R. E., Kimling, M. G., et al.: The Children and Sunscreen Study: a crossover trial investigating children's sunscreen application thickness and the influence of age and dispenser type. *Arch Dermatol.* 2012;48:606–612.
20. Lademann, J., Schanzer, S., Richter, H.: Sunscreen application at the beach. *J Cosmet Dermatol.* 2004;3:62–68.
21. Petersen, B., Datta, P., Philipsen, P. A., et al.: Sunscreen use and failures–on site observations on a sun-holiday. *Photochem Photobiol Sci.* 2013;12:190–196.
22. Sayre, R. M.; Agin, P. P.; Levee, G. J.; et al.: Comparison of *in vivo* and in vitro testing of sunscreening formulas. *Photochem Photobiol.* 1979;29:559–566.
23. Stechschulte, S. A., Kirsner, R. S., Federman, D. G.: Sunscreens for non-dermatologists: what you should know when counseling patients. *Postgrad Med.* 2011;123:160–167.
24. Wilson, B. D., Moon, S., Armstrong, F.: Comprehensive Review of Ultraviolet Radiation and the Current Status on Sunscreens. *J Clin Aesthet Dermatol.* 2012;5(9):18–23.
25. Moyal, D. D., Fourtanier, A. M.: Broad-spectrum sunscreens provide better protection from solar ultraviolet-simulated radiation and natural sunlight-induced immunosuppression in human beings. *J Am Acad Dermatol.* 2008;58(5):S149–S154.
26. Jean-Louis Refrégier, M.: Relationship between UVA protection and skin response to UV light: proposal for labelling UVA protection. *Int J Cosmet Sci.* 2004;26(4):197–206.
27. US FDA: Labeling and effectiveness testing; sunscreen drug products for over-the-counter human use. *Fed Regist.* 2011;76:35620–35665.

28. FDA: Administration. FDA Sheds Light on Sunscreens, Last update: 5/17/2012, Accessed 4/20/2014 at http://www.fda.gov/ForConsumers/ConsumerUpdates/ucm258416.htm
29. Holick, M. F.: Optimal vitamin D status for the prevention and treatment of osteoporosis. *Drugs Aging*. 2007;24(12):1017–1029.
30. Pfeifer, M., Begerow, B., Minne, H. W.: Vitamin D and muscle function. *Osteoporos Int*. 2002;13(3):187-194.
31. Zittermann, A., Schleithoff, S. S., Koerfer, R.: Review Putting cardiovascular disease and vitamin D insufficiency into perspective. *Br J Nutr*. 2005;94(4):483–492.
32. Lindqvist, P. G., Olsson, H., Landin-Olsson, M.: Are active sun exposure habits related to lowering risk of type 2 diabetes mellitus in women, a prospective cohort study? *Diabetes Res Clin Pract*. 2010;90(1):109–114.
33. Pelajo, C. F., Lopez-Benitez, J. M., Miller, L. C.: Review Vitamin D and autoimmune rheumatologic disorders. *Autoimmun Rev*. 2010;9(7):507–510.
34. Krall, E. A., Wehler, C., Garcia, R. I., et al.: Calcium and vitamin D supplements reduce tooth loss in the elderly. *Am J Med*. 2001;111(6):452–456.
35. Berwick, M., Armstrong, B. K., Ben-Porat, L., et al.: Sun exposure and mortality from melanoma. *J Natl Cancer Inst*. 2005;97(3):195–199.
36. Berwick, M., Kesler, D.: Ultraviolet radiation exposure, vitamin D and cancer. *Photochem Photobiol*. 2005;81:1261–1266.
37. Freedman, D. M., Dosemeci, M., McGlynn, K.: Sunlight and mortality from breast, ovarian, colon, prostate and non-melanoma skin cancer: a composite death certificate based case-control study. *Occup Environ Med*. 2002;59:257–262.
38. Glerup, H., Mikkelsen, K., Poulsen, L., et al.: Commonly recommended daily intake of vitamin D is not sufficient if sunlight exposure is limited. *J Intern Med*. 2000;247(2):260–268.
39. Lim, H. W., Gilchrest, B. A., Cooper, K. D., et al.: Sunlight, tanning booths, and vitamin D. *J Am Acad Dermatol*. 2005;52(5):868–876.
40. Godar, D. E., Pope, S. J., Grant, W. B., et al.: Solar UV doses of adult Americans and vitamin D_3 production. *Dermatoendocrinology*. 2011;3(4):243–250.
41. Boztepe, G., Sahin, S., Ayhan, M., et al.: Narrowband ultraviolet B phototherapy to clear and maintain clearance in patients with mycosis fungoides. *J Am Acad Dermatol*. 2005;53(2):242.
42. Golden, R. N., Gaynes, B. N., Ekstrom, R. D.: The efficacy of light therapy in the treatment of mood disorders: a review and meta-analysis of the evidence. *Am J Psych*. 2005;162(4):656.
43. AAD: Sunscreen FAQs, Accessed 4/13/2014 at http://www.aad.org/media-resources/stats-and-facts/prevention-and-care/sunscreens
44. FDA: Indoor Tanning: The Risks of Ultraviolet Rays, Last update: 4/23/2014, Accessed 4/24/2014 at http://www.fda.gov/forconsumers/consumerupdates/ucm186687.htm
45. US Department of Health and Human Services, Public Health Service, National Toxicology Program: Report on Carcinogens, 10th Edition, 2002, Accessed 4/20/2014 at http://www.reeglawyers.com/Portals/0/PDFs/Tenth-Report-on-Carcinogens.pdf

46. WHO: Sunbeds, Tanning and UV Exposure, Last update: 4/2010, Accessed 4/20/2014 at http://www.who.int/mediacentre/factsheets/fs287/en/
47. Wang, T., Herlyn, M.: The macrophage: a new factor in UVR-induced melanomagenesis. *J Invest Dermatol.* 2013;133(7):1711–1713.
48. Gallagher, R. P., Lee, T. K.: Adverse effects of ultraviolet radiation: a brief review. *Biophys Mol Biol.* 2006;92(1):119–131.
49. Rogers, H. W., Weinstock, M. A., Harris, A. R.: Incidence estimate of nonmelanoma skin cancer in the United States, 2006. *Arch Dermatol.* 2010;146(3):283–287.
50. Stern, R. S.: Prevalence of a history of skin cancer in 2007: results of an incidence-based model. *Arch Dermatol.* 2010;146(3):279–282.
51. Robinson, J. K.: Sun exposure, sun protection, and vitamin D. *JAMA.* 2005;294: 1541–1543.
52. Matsumura, Y., Ananthaswamy, H. N.: Toxic effects of ultraviolet radiation on the skin. *Toxicol Appl Pharmacol.* 2004;195(3):298.
53. Lucas, R. M.: An epidemiological perspective of ultraviolet exposure - public health concerns. *Eye Contact Lens.* 2011;37(4):168–175.
54. Kaidbey, K. H., Agin, P. P., Sayre, R. M., et al.: Photoprotection by melanin–a comparison of black and Caucasian skin. *J Am Acad Dermatol.* 1979;1(3):249–260.
55. Gloster, H. M., Neal, K.: Review skin cancer in skin of color. *J Am Acad Dermatol.* 2006;55(5):741–760.
56. Halder, R. M., Bridgeman-Shah, S.: Review skin cancer in African Americans. *Cancer.* 1995;75(2):667–673.
57. Brenner, M., Hearing, V. J. The protective role of melanin against UV damage in human skin *Photochem Photobiol.* 2008;84(3):539–549.
58. Miyamura, Y., Coelho, S. G., Schlenz, K., et al.: The deceptive nature of UVA-tanning versus the modest protective effects of UVB-tanning on human skin. *Pigment Cell Melanoma Res.* 2011;24(1):136–147.
59. Urbach, F.: The historical aspects of sunscreens. *J Photochem Photobiol B.* 2001;64(2–3):99–104.
60. Rik, R.: History of Human Photobiology, In: Herbert, H. W. L., Honigsmann, H., Hawk, J. L. M.: *Photodermatology*, Boca Raton: CRC Press, 2007.
61. Norlund, J. J., Ortonne, J. P.: The Normal Color of Human Skin, In: Norlund, J. J., Boissy, R. A., Hearing, V. J., King, R. A., Oetting, W. S., Ortonne, J. P., eds: *The Pigmentary System*, 2nd Edition, Lake Oswego: Blackwell Publishing, 2006.
62. Shaath, N. A.: Evolution of Modern Sunscreen Chemicals, In: Lowe, N. J., Shaath, N. A., Pathak, M. A.: *Sunscreens: Development, Evaluation, and Regulatory Aspects*, 2nd Edition, New York: Marcel Dekker, 1997.
63. L'Oreal, Accessed 4/30/2014 at http://www.lorealparisusa.com/en/brands/skin-care/sublime-sun.aspx
64. Piz Buin, Accessed 4/23/2014 at http://www.pizbuin.com/uk/our-brand/
65. Coppertone, Accessed 4/30/2014 at http://www.coppertone.com/coppertone/sunderstanding/milestones.jspa
66. Greiter, F.: Der Sonnenschutzfaktor ist nur ein Aspekt moderner Sonnenschutzmitteln, *Seife-Öle-Fette-Wachse.* 1984;110:195–197.

67. Sunscreen drug products for over-the-counter use. *Fed Regist.* 1978;43(166):38256.
68. Sayre, R. M., Killias, N., Roberts, R. L.: Physical sunscreens. *J Soc Cosmet Chem.* 1990;41:103–109.
69. Filipe, P., Silva, J. N., Silva, R., et al.: Stratum corneum is an effective barrier to TiO_2 and ZnO nanoparticle percutaneous absorption. *Skin Pharmacol Physiol.* 2009;22(5):266–275.
70. The Skin Cancer Foundation's Guide to Sunscreens, Accessed 4/24/2014 at http://www.skincancer.org/prevention/sun-protection/sunscreen/the-skin-cancer-foundations-guide-to-sunscreens
71. Smijs, T. G., Pavel, S.: Titanium dioxide and zinc oxide nanoparticles in sunscreens: focus on their safety and effectiveness. *Nanotechnol Sci App.* 2011;4:95–112.
72. Wolverton, S. E., Levy, S. B.: *Comprehensive Dermatologic Drug Therapy*, 2nd Edition, Philadelphia: Saunders, 2007.
73. Watkonson, A. C., Prediction of the percutaneous penetration of ultraviolet filters used in sunscreen formulations. *Int J Cosmeti Sci.* 1992;14:263–275.
74. Nguyen, U., Schlossman, D.: Stability study of avobenzone with inorganic sunscreens. Poster presented at a NYSSS meeting in 2001, Accessed 4/23/2014 at http://www.koboproductsinc.com/downloads/nyscc-avobenzone.pdf
75. Bonda, C., Steinber, D. C.: A new photostabilizer for full spectrum sunscreens. *Cosmet Toiletries.* 2000;115(6):37–45.
76. Agrapidis-Paloympis, L. E., Nash, R. A., Shaath, N. A.: The effect of solvents on the ultraviolet absorbance of sunscreens. *J Soc Cosmeti Chem.* 1987;38:209–221.
77. Hunter, A., Trevino, M.: Film-formers enhance water resistance and SPF in sun care products. *Cosmet Toiletries.* 2004;119(7):51–56.
78. Wang, S. Q., Osterwalder, U., Jung, K.: *Ex vivo* evaluation of radical sun protection factor in popular sunscreens with antioxidants. *J Am Acad Dermatol.* 2011;65(3):525–530.
79. Pissavini, M., Diffey, B., Marguerie, S., et al.: Predicting the efficacy of sunscreens *in vivo* veritas. *Int J Cosmet Sci.* 2012;34:44–48.
80. Latha, M. S., Martis, J., Shobha, V., et al.: Sunscreening agents. *J Clin Aesthet Dermatol.* 2013;6(1):16–26.
81. Barr, J.: Spray-on sunscreens need a good rub. *J Am Acad Dermatol.* 2005;52:180–181.
82. Cross, S. E., Innes, B., Roberts, M. S., et al.: Human skin penetration of sunscreen nanoparticles: in-vitro assessment of a novel micronized zinc oxide formulation. *Skin Pharmacol Physiol.* 2007;20(3):148–154.
83. Paller, A. S., Hawk, J. L. M., Honig, P., et al.: New insights about infant and toddler skin: implications for sun protection, *Pediatrics.* 2011;128:92.
84. L'Alloret, F., Candau, D., Seité, S., et al.: New combination of ultraviolet absorbers in an oily emollient increases sunscreen efficacy and photostability. *Dermatol Ther.* 2012;2(1):4.
85. Method of quenching electronic excitation of chromophore-containing organic molecules in photoactive compositions. US 7597825 B2
86. Schwarzenbach, R., Huber, U.: Optimization of Sunscreen Efficacy, In: Ziolkowsky, H., ed.: *Sun Protection*, Augsburg: Verlag Fur chemische Industrie, 2003.
87. Bonda, C.: The Photostability of Organic Sunscreen Actives: A Review, In: Shaath, N.: *Sunscreens: Regulations and Commercial Development*, 3rd Edition, Boca Raton: Taylor & Francis, 2005.

88. Mitsui, T.: *New Cosmetic Science*, Amsterdam: Elsevier Science, 1997.
89. USP34 <601> *Aerosols, Nasal Sprays, Metered-Dose Inhalers, and Dry Powders Inhalers*, Baltimore: United Book Press, Inc., 2010.
90. Goroll, A. H., Mulley, A. G.: *Primary Care Medicine: Office Evaluation and Management of the Adult Patient*, Philadelphia: Lippincott Williams & Wilkins, 2009.
91. Schulman, J. M., Fisher, D. E.: Indoor UV tanning and skin cancer: health risks and opportunities. *Curr Opin Oncol.* 2009;21(2):144–149.
92. Benson, A. B., Hourihan, J. C., Tripathi, U.: Aerosol hair spray composition, US5094838, 1992.
93. Title 49 Code of Federal Regulations 173.306
94. ASTM Standard D1321 10. Standard Test Method for Needle Penetration of Petroleum Waxes, DOI: 10.1520/D1321-10.
95. Payout-glide-flakeoff apparatus for characterizing deodorant and antiperspirant sticks US 8661887 B2
96. McLeod, C.: Sun protection factor. *Cosmet Toiletries*. 2013;128(9):624–628.
97. 21 CFR 352
98. Diffey, B. L.: A method for broad spectrum classification of sunscreens. *Int J Cosmet Sci.* 1994;16:47–52.
99. Diffey, B. L., Tanner, P. R., Matts, P. J., Nash, J. F.: *In vitro* assessment of the broad-spectrum ultraviolet protection of sunscreen products. *J Am Acad Dermatol.* 2000;43(6):1024–1035.
100. Hayden, C. G., Cross, S. E., Anderson, C., et al.: Sunscreen penetration of human skin and related keratinocyte toxicity after topical application. *Skin Pharmacol Physiol.* 2005;18(4):170–174.
101. Benech-Kieffer, F., Meuling, W. J. A., Leclerc, C., et al.: Percutaneous absorption of Mexoryl SX® in human volunteers: comparison with *in vitro* data. *Skin Pharmacol Appl Skin Physiol* 2003;16:343–355.
102. Gaspar, L. R., Tharmann, J., Campos, P. M., et al.: Skin phototoxicity of cosmetic formulations containing photounstable and photostable UV-filters and vitamin A palmitate. *Toxicol In Vitro*. 2013;27:418–425.
103. Nash, J. F., Tanner, P. R.: Relevance of UV filter/sunscreen product photostability to human safety. *Photodermatol Photoimmunol Photomed*. 2014;30(2–3):88–95.
104. Pinheiro, T., Allon, J., Alves, L. C., et al.: The influence of corneocytes structure on the interpretation of permeation profiles of nanoparticles across the skin. *Nuclear Instrum Methods Phys Res B* 2007;260:119–123.
105. Fabian, E., Landsiedel, R., Ma-Hock, L., et al.: Tissue distribution and toxicity of intravenously administered titanium dioxide nanoparticles in rats. *Arch Toxicol.* 2008;82(3):151–157.
106. Stern, S. T., McNeil, S. E.: Nanotechnology safety concerns revisited. *Toxicol Sci.* 2008;101(1):4–21.
107. Robbins, C. R.: *Chemical and Physical Behavior of Human Hair*, Berlin: Springer, 2012.

SECTION 6: DEODORANTS AND ANTIPERSPIRANTS

LEARNING OBJECTIVES

Upon completion of this section, the reader will be able to

1. define the following terms:

Antiperspirant	Apocrine sweat gland	Body odor	Caking
Deodorant	Eccrine sweat gland	Extrudable cream	Extrudable gel
Fabric staining	Glycine	Hyperhidrosis	Roll-on
Shrinkage	Sniff test	Sweat reduction test	Sweating

2. differentiate between a deodorant and an antiperspirant;
3. explain whether deodorants and antiperspirants are considered cosmetics or drugs in the US;
4. differentiate between eccrine and apocrine sweat glands;
5. explain how body odor develops;
6. briefly discuss the potential negative effects deodorants and antiperspirants can have on the skin;
7. list the various required cosmetic qualities and characteristics that an ideal deodorant and/or antiperspirant should possess;
8. list the various required technical qualities and characteristics that an ideal deodorant and/or antiperspirant should possess;
9. list several ways body odor can be controlled and/or reduced, and provide one example for each method;
10. explain how antiperspirants reduce the amount of sweat produced;
11. explain the function of glycine in aluminum–zirconium-based antiperspirants;
12. explain the major characteristics of the following types of deodorants and/or antiperspirants: roll-on, solid stick, extrudable gel, extrudable solid, and aerosol;
13. briefly explain the important factors that should be considered when formulating a deodorant and/or an antiperspirant product;
14. explain the major differences between deodorant and antiperspirant aerosols;
15. list some typical quality issues that may occur during the formulation and/or use of deodorants and/or antiperspirants, and explain why they may occur;

16. list the typical quality parameters that are regularly tested for deodorants and/or antiperspirants, and briefly describe their method of evaluation;
17. briefly describe the sniff test;
18. briefly explain the difference between standard effectives and extra effectiveness regarding the performance of antiperspirant products;
19. describe the method of evaluation for sweat reduction testing;
20. briefly discuss the main safety concerns that arose in the past with regard to the use of deodorants and/or antiperspirants;
21. list the typical containers available for deodorants and/or antiperspirants.

KEY CONCEPTS

1. Sweat glands found in the human skin are classified into two different types: eccrine glands and apocrine glands.
2. Sweat by itself is odorless. A characteristic odor develops by the activity of bacterial flora on the skin surface. Bacteria break down various chemicals in sweat, resulting in volatile by-products, which have an unpleasant odor.
3. Deodorants are topically applied products designed to reduce or mask unpleasant body odors by reodorization and/or antibacterial action.
4. Antiperspirants are topically applied products designed to reduce underarm wetness by limiting body transpiration.
5. There are numerous ways of masking or reducing body odor as well as reducing the production rate of perspiration.
6. Deodorants and antiperspirants are available in various dosage forms; the most common delivery systems include roll-ons, solid sticks, extrudable clear gels, extrudable soft solids, and aerosols.
7. The typical quality-related issues of deodorants and antiperspirants, which include shrinkage of extrudable gels, caking of aerosols, valve clogging, staining and fabric damage, poor pay-off, separation of emulsions, microbiological contamination, clumping, and rancidification.
8. Parameters frequently tested by cosmetic companies to evaluate the quality of their deodorant and antiperspirant products include pay-off and glide testing for sticks; drying time; spreadability, firmness, and texture of gels and creams; hardness of sticks; actuation force for aerosols; aerosol can leakage; pressure test for aerosols; spray characteristics for aerosols; preservative efficacy; pH; and viscosity.
9. The most commonly performed performance tests include *in vitro* microbiological testing, sniff test, and sweat reduction test for antiperspirants.
10. Ingredients that caused safety concerns regarding the use of deodorants and antiperspirants include aluminum salts, zirconium used to complex the aluminum salts, as well as VOCs used as propellants in aerosols.

Introduction

Deodorants and antiperspirants have been used for a long time. The first products used in the ancient ages were based primarily on fragrances that simply masked body odor. These products gradually evolved into complex chemical compounds containing aluminum and zirconium salts, which are capable of reducing the amount of sweat produced. As body odor, especially underarm odor, and excessive sweating are perceived unpleasant by most cultures worldwide, underarm care has become more important over the past decades. Today, most consumers consider deodorants and antiperspirants to be basic grooming products.[1] Therefore, today, these products make up a significant segment of personal care products.

This section reviews the anatomy and physiology of the human axilla and the mechanism of body odor formation. It also summarizes the main methods to control body odor and sweating rate and the main ingredients and product types of deodorants and antiperspirants. In addition, the section discusses the main quality issues of deodorant and antiperspirants, as well as the parameters most frequently tested to evaluate the quality and performance of such products. As the safety of these products has been in the middle of attention for years, the main safety issues are also reviewed.

Anatomy and Physiology of Human Sweat Glands

Sweating has a significant biological role for humans. It is regulated by the sympathetic nervous system and is an important body temperature regulator, especially in warm and humid weather climates, stress situations, or during heavy exercise. It also functions to remove waste and toxic by-products from the body. Sweat glands are widely distributed in the skin, and according to Gray's Anatomy, most people have several million sweat glands distributed over their bodies, providing plenty of opportunity for underarm odors to develop.[2] ❶**Sweat glands found in human skin are classified into two different types: eccrine glands and apocrine glands** (see Figure 3.31).

- **Eccrine glands** are simple, coiled tubular glands. Their secretory portion is found deep in the dermis, from which a duct leads directly onto the skin surface. These glands function continuously and are known as the "true" sweat glands since their main function is to control body temperature and electrolyte balance through the evaporation of water from sweat on the body surface.[3] Eccrine glands exist and start function from birth. These glands are found all over the body, especially on the palms, soles, axillae (underarms), and forehead.[4] They are under psychological and thermal control. Their secretion consists mainly of water with various salts, primarily sodium chloride and potassium chloride; amino acids, peptides, and proteins; and various electrolytic components, such as ammonia, calcium, uric acid, urea, copper, lactic acid, potassium, and phosphorus.[5] Warm and humid environment allows for rapid decomposition of the organic materials, which is made of primarily low-molecular-weight,

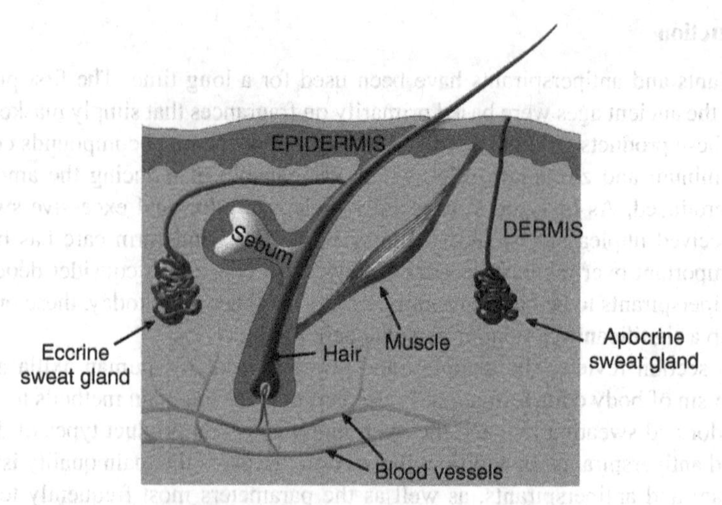

Figure 3.31 Eccrine and apocrine sweat glands.

volatile fatty acids and various steroids. These compounds produce the recognizable body odors.

- **Apocrine glands** are primarily limited to certain body parts, such as the axilla, anus, and breast. Within the axilla, apocrine glands outnumber eccrine glands by 10–1.[6] These are also found in the dermis; however, these are larger than eccrine glands and their ducts open into the hair follicle duct.[4] Apocrine glands also exist at birth; however, they become functional at puberty when the sex hormones are produced. They are usually triggered by emotions, such as excitement, anger, and fear. Apocrine glands produce an odorless viscous secretion, which acquires a distinct and unpleasant odor after bacterial decomposition.[4] This secretion primarily consists of lipids, cholesterol, proteins, sulfur-containing amino acids, volatile short-chain fatty acids, and various steroids.[7,8]

As discussed earlier, ❷**sweat by itself is odorless. A characteristic odor develops by the activity of bacterial flora on the skin surface. Bacteria break down various chemicals in sweat, resulting in volatile by-products, which have an unpleasant odor** (see Figure 3.32). The microorganisms present in the underarm area include bacteria, such as *Corynebacterium*, *Streptococcus*, *Propionibacterium* spp., and *Micrococcus*, as well as yeast, *Malassezia*. The microorganisms primarily responsible for the production of body odor include *Corynebacterium*, *Streptococcus*, and *Propionibacteria*.[9,10] The human scent is genetically controlled and systemically influenced by dietary and medicinal intake, as well as by the application of fragrance products.[11,12] Sweat collected from the skin surface contains a diverse range of metabolites, depending on the physiology status of the donor as well as the functional and developmental states of the sweat glands.

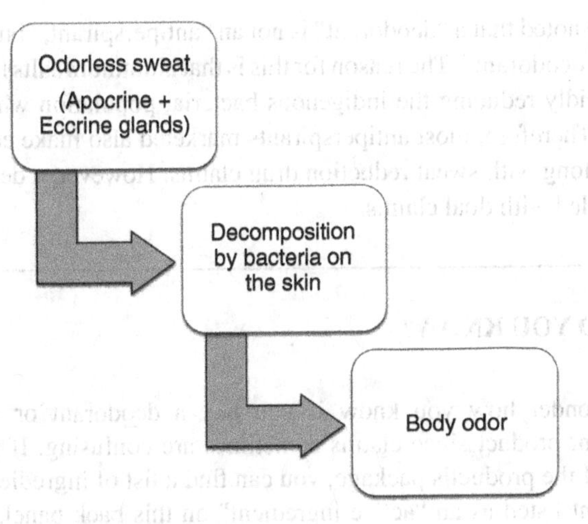

Figure 3.32 Body odor formation.

DID YOU KNOW?

The loss of excessive amounts of salt and water from the body can quickly dehydrate a person and can lead to circulatory problems, kidney failure, and heat stroke. Therefore, sweating has a cooling effect and is important to maintain the body temperature; however, it is also important that people drink fluids when exercising or when outside in high temperatures to make up for the missing water and electrolytes.[13]

Types and Definition of Products Reducing Body Odor

Most people use the terms "antiperspirant" and "deodorant" interchangeably, although they have quite distinct actions.

- **●Deodorants are topically applied products designed to reduce or mask unpleasant body odors by reodorization and/or antibacterial action.** However, they do not interfere with the delivery of sweat gland secretions. They do not have any therapeutic effect and are considered as cosmetics in the US.
- **●Antiperspirants are topically applied products designed to reduce underarm wetness by limiting body transpiration.** They affect the structure and function of the body by inhibiting perspiration that is secreted by the eccrine glands; therefore, these products are classified as OTC drugs in the US. They usually contain aluminum-based or aluminum–zirconium-based compounds that can form a temporary plug within the sweat duct and stop the flow of sweat to the skin surface.

It should be noted that a "deodorant" is not an "antiperspirant," but an "antiperspirant" can be a "deodorant." The reason for this is that aluminum salts have bactericidal properties, rapidly reducing the indigenous bacterial population when applied on a regular basis. Therefore, most antiperspirants marketed also make cosmetic deodorancy claims, along with sweat reduction drug claims. However, a deodorant product cannot be labeled with dual claims.

DID YOU KNOW?

You may wonder how you know if you buy a deodorant or an antiperspirant/deodorant product since claims sometimes are confusing. If you check the back panel of the product's package, you can find a list of ingredients. If you see any ingredient listed as an "active ingredient" on this back panel, it means you are looking at an antiperspirant/deodorant product. If you see only "ingredients" without specifying the active ingredient(s), you have a deodorant product.

DID YOU KNOW?

Some countries/regions consider antiperspirants cosmetics since they do not affect the biologic physiology of the body; therefore, in those countries/regions they are not subject to such strict standards as in the US. Examples for such countries/regions include Canada and the European Union.

History of Using Deodorants and Antiperspirants

First attempts at masking the natural body odors were made in the early ages of civilization. Ancient Egyptians used perfumed oils, such as preparations containing citrus and cinnamon.[14] As discussed in Section 5 of this chapter, Egyptians were among those cultures where people removed their unwanted hair. Hair removal also helped reduce odor since underarm hair provides a greater surface area for bacteria to grow, which eventually breaks down sweat into odorous chemicals. Alum stone (potassium alum) was first used in the ancient Roman civilization for deodorant purposes to reduce body odor.[15] Romans and Greeks used perfumed oils and aromas.[16] Other cultures were also frequent users of aromatic products, including Arabia, China, and India.[17] It was very frequent in the early centuries to trade various goods for perfumes. Alcohol-based perfumes were shipped to Europe from the Middle East in the 13th century.[14] Later, in the 16th century, Italy and France became huge centers for perfumes in Europe. Among the wealthy, the use of perfumes, spices, and aromatics became an important part of everyday life.[18]

The first modern deodorant came out in the late 19th century, and it was a cream containing zinc oxide, which had antimicrobial properties.[19] It was a waxy cream that was difficult to apply and very messy. In the early 20th century, the first modern aluminum-containing antiperspirant, EverDry®, was introduced. It was based on a very astringent aluminum chloride solution having a very acidic pH (around pH 3).[20] It led to skin irritation and damaged the fabricas well. Aluminum chlorohydrate products were introduced in the 1940s; they had an internal buffer that maintained the pH closer to the neutral pH (around pH 4). These products were less irritating with less fabric destruction than aluminum chloride.[21] Stopette®, a deodorant spray, was introduced in the 1950s. It was a liquid deodorant; but it was known as a "Spray Deodorant" due to the packaging the deodorant came in. It was a flexible plastic squeeze bottle.[22] The first roll-on deodorant was also launched in the 1950s; its design was based on the ballpoint pen. The first aerosol antiperspirants appeared on the market in the late 1950s, which soon became popular. However, in late 1970s, environmental and health concerns arose regarding to the safe use of aerosol products. Zirconium-based antiperspirants were banned in aerosol formulations,[23] and the use of chlorofluorocarbon (CFC) propellants was also banned.[24] Since then, other regulations were introduced, restricting the use of propellants in order to protect the environment and users. As a consequence, the popularity of aerosol products decreased. Today, we have a wide variety of deodorants and antiperspirants in various forms, including roll-ons, aerosols, sticks, gels, and creams. According to statistics, stick and roll-on products are the most popular product types in the US.[25]

How Deodorants and Antiperspirants May Affect the Human Skin and Body?

Sweating is a natural process to cool the body temperature. However, the odor developing via degradation of sweat by bacteria can be embarrassing, can affect self-esteem, and can have psychological consequences. Therefore, deodorants and antiperspirants are an essential part of most consumers' daily personal care routine.

- There is a condition called **hyperhidrosis** or excessive sweating. The profusion of sweat may be in the axillary sites, palms, feet, face, trunk, or a combination of any or all of these. The excessive sweat leads to unpleasant body odor that can adversely affect the person's ability to attain a normal and healthy QoL.[26] In adolescents and young adults, an incidence rate of 0.6–1% is reported for hyperhidrosis.[27] The treatment of hyperhidrosis usually starts with OTC antiperspirants. However, none of today's OTC antiperspirants are specifically designed or claimed to have a beneficial effect on excessive sweating. Additional treatment options, depending on the site of excessive sweating, include prescription antiperspirants, oral medications, topical injections, and surgery. Prescription antiperspirants contain higher doses of aluminum chloride, which, however, can cause irritation and damage clothing.[28] In 2004, the FDA approved Botox (botulinum toxin type A), a drug used to temporarily erase wrinkles for cosmetic purposes, to treat severe underarm sweating that cannot be managed by topical agents.[29] Botox is approved for the treatment of the underarms, but not for excessive sweating of other sites such as the feet and palms.

Usually, deodorants and antiperspirants have a low chance to develop side effects if used as recommended on the label. However, minor negative effects may occur. These are summarized here.

- Commonly reported negative effects include **skin irritation** and **allergies**. One of the major causes of skin irritation can be the use of products on broken skin (e.g., from shaving). This irritation can be avoided if the product is not used after such procedures. Allergies and skin sensitization are most commonly related to the fragrances present in antiperspirants.[30]
- Antiperspirants control the amount of sweat produced by eccrine glands in the underarm region. Although axillary sweating makes up less than 1% of the total body sweating rate,[31] consumers, particularly in hot countries, began to have concerns about antiperspirants that they may interfere with the body's natural cooling process, leading to **overheating**.[32] Studies show that this is not true for several reasons: the axillary region is more involved with apocrine sweating, which is triggered by emotional arousal, than eccrine sweating, which regulates the body temperature. Even when eccrine sweating happens, the sweat cannot efficiently evaporate and cool the body due to the occluded nature of the underarm region. Additionally, the surface area affected by the use of antiperspirants is relatively small.[33]
- Antiperspirants can **stain** the clothes, which is a distinct negative effect from consumers' perspective. More detail is discussed under the quality control issues of deodorants and antiperspirants.

Required Qualities and Characteristics and Consumer Needs

From a consumer perspective, a quality deodorant and/or antiperspirant should possess the following characteristics:

- Neutral or pleasant odor
- Easy to spread
- Pleasant feeling during application
- Well-tolerated and non-allergenic
- Long-term deodorization
- Quick drying properties
- Non-staining properties

The technical qualities of hair removal products can be summarized as follows:

- Fulfill long-term stability

- Appropriate texture
- Appropriate rheological properties
- No irritation potential
- Dermatological safety

Types of Ingredients, Product Forms, and Formulation of Deodorants and Antiperspirants

Before starting the discussion on the various types of ingredients found in deodorants and antiperspirants, you should be familiar with the various ways of how body odor can be controlled and/or reduced. ❺**There are numerous ways of masking or reducing body odor as well as reducing the production rate of perspiration.** Hundreds of patents, research papers, and literature articles have been published, which focus on the various and advanced solutions to this problem. Popular methods for a deodorant and/or antiperspirant action include using the following types of ingredients:

- **Odor masking** ingredients reduce the perception of odor through blending with underarm odor and masking it. Examples for such ingredients are fragrances.[34]
- **Odor neutralizing** ingredients chemically neutralize odorous compounds, yielding odorless components. Examples for ingredients acting as odor neutralizers include sodium and potassium bicarbonate and zinc carbonate.[35–37]
- **Odor quenching** ingredients bind to the odorous chemicals and form complexes with these materials. Examples for such ingredients include zinc ricinoleate as well as certain metal oxides, such as zinc oxide.[38–40] Hydroxyapatite has also been found to be effective at binding to odorous chemicals.[41]
- **Odor absorbing/adsorbing** ingredients physically neutralize odorous molecules formed in the axilla via absorption or adsorption. This results in the immobilization of those molecules, decreasing their volatility and thus decreasing the perceived odor. Attempts have been made to incorporate various resins into formulations with the aim of odor absorption.[42,43] Additionally, a range of silicones and silicates are claimed to offer odor absorption benefits.[44]
- **Esterase inhibitors** act by directly inhibiting certain enzymes of the underarm bacteria, which results in odor reduction. An example for such ingredients is zinc glycinate.[45,46] Another option to inhibit enzymes is to shift the pH optimal for the development of underarm odor (pH 6) to the acidic range. Lipophilic derivatives of citric acid are examples for such ingredients.[47]
- **Antimicrobial** ingredients are commonly used in today's deodorants to prevent underarm odor formation by inhibiting or deactivating the bacteria responsible for bad odor formation. As a result, there is no or only slight metabolism of sweat components, thus preventing/reducing the occurrence of body odor. Examples for such ingredients include ethanol; triclosan; quaternary ammonium salts; glyceryl fatty acid esters, such as diglyceryl monolaurate; and sucrose fatty acid esters, such as sucrose monostearate among others.[48,49] Essential oils, such as thyme and clove oil, can also have antimicrobial benefits,

in addition to masking bad odor. Antiperspirant active ingredients also have antimicrobial properties, which help control the bacterial growth on the skin. A popular product today is called "crystal" products. These products are made from a mineral known as potassium aluminum sulfate. Unlike aluminum salts used in antiperspirants, alum does not prohibit sweating; it only helps control the growth of bacteria that can cause underarm odor.[50]

- **Antiperspirants** reversibly block sweat gland excretion by forming a temporary, gelatinous plug in the eccrine duct that reduces, but do not stop, the flow of axillary perspiration. The process is shown in Figure 3.33.[51–54] These blockages prevent sweat from reaching the skin surface in the axilla. They can remain within the sweat duct for 7–14 days, depending on the rate of skin desquamation, user's hygiene regimen, activity type, and quality.[55] The final OTC monograph for antiperspirants lists the active ingredients, their concentrations allowed, dosage forms in which they can be formulated, as well as claims that can be made for them.

The active ingredients used in today's formulations can be broadly divided into two groups:

- **Aluminum-based** agents, such as aluminum chloride, aluminum chlorohydrate, aluminum sesquichlorohydrate, aluminum dichlorohydrate, and aluminum sulfate. Aluminum chlorohydrates can be complexed with polyethylene glycol or propylene glycol as alcohol-solubilizing adjuncts.
- **Aluminum–zirconium-based** agents, such as aluminum zirconium tri-, tetra-, penta-, and octachlorohydrate. These complexes can be buffered with glycine, an amino acid, to stabilize them and mitigate the acidic harshness, which could result when applied to underarm axilla.[56]

The final monograph also specifies whether the complexes can be formulated into aerosol and/or non-aerosol products. Antiperspirants containing zirconium

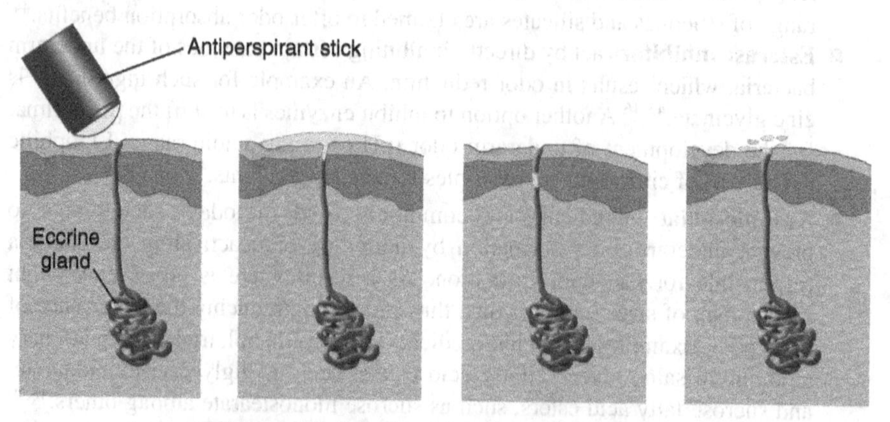

Figure 3.33 Working principle of antiperspirants.

are currently not allowed to be formulated into aerosol products due to some concerns regarding the inhalation of such complexes and potential granuloma formation.[57]

Most Common Dosage Forms for Deodorants and Antiperspirants **●Deodorants and antiperspirants are available in various dosage forms; the most common delivery systems include roll-ons, solid sticks, extrudable clear gels, extrudable soft solids, and aerosols.** Since the basic dosage forms and their main ingredient types, as well as formulation technology, are the same for deodorants and antiperspirants, these characteristics are discussed in one section.

Roll-Ons Roll-on deodorants and/or antiperspirants are very versatile and popular product types. Their popularity is due to the nonoily feel and the good spreadability of the product on the underarm skin. There are several various types of roll-ons differing in their vehicle:

- **Water-based** roll-ons are usually opaque O/W emulsions. The active ingredient is typically formulated into the external phase to provide better efficacy. These systems are usually based on nonionic surfactants, such as polyethylene glycol or propylene glycol ethoxylated alcohols, as well as emollients, humectants, hydrophilic thickeners, antioxidants, chelating agents, texturizers (such as talc or corn starch for a soft skin feel), preservatives, and fragrances.
- **Hydroalcoholic roll-ons** have a shorter drying time and offer a refreshing feeling upon application. An important factor that should be taken into consideration is that only alcohol-soluble actives can be used in this system. These products are usually thickened with a hydrophilic polymer, such as cellulose derivatives or carrageenan. Emollients and silicones can also be incorporated using proper emulsifiers.
- **Silicone-based roll-ons** can be formulated as anhydrous products where the actives are suspended in volatile silicone oils, such as cyclomethicone. To prevent sedimentation of the powder as well as thicken the silicone-based formulation, usually non-surfactant suspending agents, such as quaternium-18 hectorite, are used. An important factor to consider is the particle size of the active ingredient in order to prevent sedimentation. The smaller the particle size, the slower the settling rate and the more stable the product. Silicone-based roll-ons can also be formulated as W/Si emulsions, which are very popular due to the dry, non-tacky feel they provide.

Solid Sticks Solid sticks are the most popular product types today.

- **Deodorant sticks** are typically based on sodium stearate as the gelling agent. The deodorizing agents and fragrances are usually dissolved in the hydrophilic vehicle, which is primarily a mixture of water and propylene glycol and/or dipropylene glycol. Products with alcohol are also available; however, they are not as popular as glycol-based formulations. Nonionic surfactants can be employed to provide transparency to the formulations; examples include PPG-3 myristyl ether and isosteareth-20. Sodium stearate can also contribute to the clarity of the formulas. Additionally, solid deodorant sticks usually also contain preservatives, antioxidants, and chelating agents to improve the stability and extend the shelf-life. Additionally, neutralizing agents may be needed to adjust the product's pH, and colors can also be added to these formulations.

- **Antiperspirant sticks** are usually anhydrous suspensions containing the suspended antiperspirant actives in a silicone-based vehicle. Examples for such silicones include cyclopentasiloxane and cyclomethicone, which provide quick drying and dry skin feel without being tacky or oily. In addition, sticks usually contain various waxy and liquid emollients for a soft skin feel and glideability, such as myristyl myristate (solid) and octyldodecanol (liquid). Nonionic surfactants, such as PPG-4 butyl ether, are often employed as dispersing agents for the antiperspirant actives. Hardening agents are generally white waxy solids, such as stearyl alcohol, behenyl alcohol, and hydrogenated castor oil, which give structure to the stick, help keep ingredients intact in a formulation, and act as lubricants. Thickeners commonly used include quaternium-18 hectorite, and waxy solids can also be used to thicken formulations. These ingredients help keep the suspended particles in a suspended state and slow down settling. Talc and silica can also be used since they are effective suspending agents and can also provide lubricity for the formulation. Additional ingredients can include colorants, titanium dioxide (as an opacifying agent), fragrance, buffers (if needed), and antioxidants, such as BHT. Antiperspirant sticks are solid white formulations.

DID YOU KNOW?

Cyclic siloxanes, such as cyclopentasiloxane, are volatile and evaporate quickly. From a regulation perspective, however, they are considered non-VOCs in the US. They meet the federal and state regulatory requirements and, therefore, can be used in antiperspirants.[58]

Extrudable Clear Gels Extrudable gel sticks are transparent formulations, which are often associated by consumers with a lack of white residue on the skin. Therefore, they are also quite popular. These formulations are usually W/Si emulsions, in which the refractive indices of the water and silicone phase are identically matched. They are similar to silicone-based roll-ons with a higher viscosity. They generally contain the antiperspirants actives dissolved in water in the internal phase. The internal phase may also contain alcohol and humectants, such as propylene glycol. Commonly used silicones include cyclopentasiloxane, dimethicone, cyclomethicone, and trisiloxane. The blend of cyclomethicone and dimethicone copolyol can also help disperse and solubilize the actives.[50] For W/Si emulsions, specials silicone surfactants are required to achieve optimal stability. A commonly used surfactant in extrudable clear gels is the blend of cyclopentasiloxane and PEG/PPG-18/18 dimethicone copolymer.[59] They may also contain thickeners, electrolytes for stability, and fragrances.

Extrudable Soft Solids Extrudable soft solids are usually white anhydrous silicone suspension pastes. In these formulations, the powder active ingredient is suspended in the silicone (such as cyclopentasiloxane) and/or hydrocarbon (such as isohexadecane) vehicle. The paste is thickened to the desired viscosity with waxes, such as C18–36 acid triglyceride and tribehenin. The products may also contain additional ingredients, such as preservatives, chelating agents, antioxidants, color additives, and fragrances. These products rub in quickly, are non-tacky, leave little or no visible residue on skin, and deliver high levels of antiperspirant protection. Similar to the anhydrous, silicone-based sticks, the particle size of the powdered active ingredient should be taken into consideration. Ideally, an active with a particle size below 45 μm ensures a good skin feel.[58] Again, the smaller the particle size, the slower the sedimentation and the greater the stability of the system.

Aerosols Aerosols are popular delivery systems for consumers who prefer a hygienic and easy-to-use product form. Generally, there are some critical factors that should be taken into consideration when formulating aerosol sprays. These include the spray rate, spray shape, particle size, product concentrate and propellant product ratio, fragrance concentration, and the pressure in the aerosol can. All these factors have to be optimized in order to have a product with a dry skin feel that does not clog the valve upon application.

- **Deodorant aerosols** typically contain a solution of a deodorant ingredient, such as an antimicrobial active in an ethanol and/or a propylene glycol, which is blended with a liquefied propellant. In these systems, the deodorizer ingredient is solubilized in the vehicle. Propane, butane, and isobutane are the most commonly used propellants. They condense to form a clear, colorless, and odorless liquid. Deodorant sprays provide a dry skin feeling as they are anhydrously formulated.
- **Antiperspirant aerosols** are also anhydrous formulations, similar to deodorant aerosols. However, there is a huge difference between an antiperspirant aerosol and a deodorant aerosol: antiperspirant actives are suspended in the product concentrate and are not solubilized. The vehicle usually consists of volatile silicone oils, such as cyclomethicone, or a mixture of ester oils and silicones. Agglomeration of solid particles and settling of actives can be minimized by using suspending agents, such as amorphous silicon dioxide or clays, such as stearalkonium bentonite and hectorite. These systems generally contain the same propellants as deodorant aerosols. In addition, they can contain emollients, such as dimethicone and isopropyl palmitate; fragrances; and preservatives. As these systems are suspensions, sedimentation is a potential concern for them, despite

the application of thickeners. Therefore, most antiperspirant aerosol sprays have directions on shaking the can before use in order to homogenize the product.

DID YOU KNOW?

Clays when used as thickeners in antiperspirant aerosols form a weak gel in the presence of an oil/silicone phase. This gel can be destroyed by shaking the aerosol can before use. It will, however, reform upon standing, ensuring stability of the system. As organoclays are agglomerated, shear is needed to deagglomerate the particles. For this, usually a polar activator, such as propylene carbonate or ethanol, is used, which disperses clays and induces gelation of the oil phase.[60]

Formulation of Deodorant and Antiperspirant Products When formulating deodorant and antiperspirant products, formulators have to carefully select deodorant ingredients or active ingredients as well as vehicles, structurizing agents, and additional components in order to meet regulatory requirements and satisfy consumers. There are some general rules that should be taken into consideration when formulating antiperspirants.

- Antiperspirant actives are generally soluble in water. However, the application of a concentrated aqueous solution can leave a tacky feeling on the skin.[61] Product qualities and skin feeling can be improved by using silicone oils or ester oils.[62]
- Aluminum powders tend to leave a visible white residue on the skin or even on the clothing when formulated into anhydrous systems (such as sticks and aerosols) as powders. Emollients, such as PPG-14 butyl ether, can help minimize this problem.[51,61,63]
- Antiperspirant actives have an acidic pH value (in the range of pH 4.0–4.2) when dissolved in water. It should be taken into consideration when selecting ingredients, such as hydrophilic thickeners. As discussed previously in multiple chapters, there are some thickeners, such as carbomers, that have to be neutralized to a specific pH value in order to reach their optimum viscosity. These thickeners are not necessarily stable at such low pH and, therefore, cannot be used for such systems.[51]
- Antiperspirant aerosols contain antiperspirant actives in an undissolved suspended state. As discussed earlier, sedimentation and valve clogging can be potential problems in such systems. Sedimentation can be prevented by using proper suspending aids, which also thicken the formulation, making it more difficult for the powder to settle. In addition, using powders with a fine particle size also helps prevent this problem. Particle size is important from another perspective as well. Valve clogging can happen if particles are large enough to

be stuck in the valve upon application. Silicone oils used as a vehicle also help lubricate the valve to prevent clogging.

- Antiperspirant actives are soluble in sweat; this is how they form a viscous plug in the duct. Hydrophobic ingredients, such as emollients, can greatly influence the effectiveness of the antiperspirant actives by covering the sweat gland pores and preventing the actives from entering the eccrine duct. Therefore, the efficacy of an antiperspirant active is expected to be higher in water-based systems when compared to anhydrous formulations.[51,64]
- As discussed earlier, when formulated into water-based systems, antiperspirant actives are dissolved in water. It is expected that antiperspirant actives are much more readily available and thus effective if the water phase is the outer phase of an emulsion. Similarly, diffusion of the actives from the vehicle to the skin after application is expected to be quickest and easiest from aqueous solution. The order of effectiveness starting from the most effective type is usually as follows: aqueous solution, then low-viscosity emulsions where the water phase is the outer phase (i.e., O/W), and, finally, higher viscosity O/W emulsions.[64,65]
- In the case of extrudable creams and sticks, the suspended antiperspirant ingredient has to be solubilized in the sweat in order to act as an antiperspirant. The effectiveness highly depends on the rate of solubilization, which is influenced by the particle size of the solid and the viscosity of the formulation. The smaller the particle size, the quicker the solubilization. Similarly, the lower the viscosity, the faster the solubilization since a thick texture can act as a barrier against solubilization.[61] However, lower viscosity allows for more rapid sedimentation. Therefore, viscosity has to be optimized for stability and effectiveness.

The formulation of deodorant and antiperspirant products should follow the general formulation steps of the basic dosage forms, such as emulsions (O/W or W/Si), solutions (water-based and hydroalcoholic), gels, as well as suspensions. In the case of aerosols, usually the product concentrate is prepared as a uniform suspension first, which is then filled into the can. The propellant is usually added using the pressure filling method, similar to sunscreens.

Typical Quality Problems of Deodorants and Antiperspirants

●**The typical quality-related issues of deodorants and antiperspirants include shrinkage of extrudable gels, caking of aerosols, valve clogging, staining and fabric damage, poor pay-off, separation of emulsions, microbiological contamination, clumping, and rancidification.** The main characteristics and causes of these issues and the potential solutions are discussed here. Problems that were discussed earlier are not reviewed in detail here.

Shrinkage of Extrudable Gels Shrinkage refers to a visible decrease in size of the gel. This phenomenon can be related to the volatility of alcohol in such formulations.

The rate of water evaporation may be reduced by using alcohol evaporation; retardants may be included in the formula to reduce the amount of lost alcohol. These are generally humectants, such as glycerin or sorbitol.[66,67]

Caking of Aerosols In the case of antiperspirant aerosols, the active ingredient is suspended in the oil and silicone-based vehicle. If the viscosity of this phase is too low, it may result in settling of the active to the bottom of the can. Settling can lead to the formation of a hard-to-resuspend cake. To overcome this problem, thickeners, such as stearalkonium hectorite or bentonite, should be applied in an optimal quantity in the product.

Valve Clogging This phenomenon means that the product cannot be dispensed by pushing down the valve. Aerosol antiperspirants delivering suspended antiperspirant actives are very popular today. When formulating such systems, there are a number of factors that should be taken into consideration in order not to clog the valve. These include the basic raw materials used in the formula, the particle size of the active(s), the type of the propellant(s), and the valve itself. This problem is usually attributable to the composition of the formulation and/or type and material of the valve. If the particle size of the suspended particles in the product concentrate is too large, or fine particles form bigger aggregates, they may clog the valve. A high concentration of clay thickeners can also clog the valve. Additionally, an improper valve system can also contribute to clogging. There can be an interaction between the product and the valve, which may slowly cause clogging during continuous use.[68] Silicone oils can be also helpful in preventing clogging.

Staining and Fabric Damage When using aluminum salt–containing antiperspirants, staining of clothing, primarily in the underarm region, and fabric damage are quite frequent. The term "staining" refers to the formation of yellow-to-brown stains on lighter clothes and chalky white stains on darker clothes. The stains usually do not develop immediately after wearing clothes, but appear over an extended period of time. There are many reasons that can lead to such discoloration. One of them can be the buildup of aluminum salts from the product. If not removed properly (often pretreatment is needed before washing), it can lead to the formation of a stiff residue. Additionally, water-insoluble aluminum salts can form when dirty clothes are cleaned with alkaline detergents, which are not removable by washing. Sometimes, clothes manufacturers use an acid-sensitive dye for coloring clothes. These dyes can also lead to discolorations when exposed to the acidic antiperspirants. These changes may be reversed by soaking the affected area in a slightly alkaline solution, e.g., diluted ammonia solution.[69]

Fabric damage can also happen for a number of reasons. One of these is the acidic pH of antiperspirants, which can degrade fabric. Aluminum salts from an antiperspirant product when combined with sweat and sebum on the skin can lead to chemical damage of the fabric, mainly those made of cotton.[69] They can also lead to discoloration as well as a loss in strength of the fabric.

Poor Pay-Off When a deodorant/antiperspirant stick is drawn across the skin surface, a thin film of the stick composition is transferred to the skin surface. "Pay-off" describes the weight lost to a surface from a deodorant/antiperspirant stick upon typical application. Poor pay-off refers to the phenomenon when the amount of product transferred to the skin is not enough for an optimal antiperspirant effect. A typical cause of poor pay-off is an improper type and/or insufficiently high concentration of the hardening agents in the formulation. This problem can be overcome with proper selection of the ingredients.

Evaluation of Deodorants and Antiperspirants

Quality Parameters Generally Tested ❸**Parameters frequently tested by cosmetic companies to evaluate the quality of their deodorant and antiperspirant products include pay-off and glide testing for sticks; drying time; spreadability, firmness, and texture of gels and creams; hardness of sticks; actuation force for aerosols; aerosol can leakage; pressure test for aerosols; spray characteristics for aerosols; preservative efficacy; pH; and viscosity.** Tests that were discussed in the previous sections are not discussed here in detail.

Hardness of Sticks The hardness test for deodorant and/or antiperspirant sticks is similar to that of sunscreen sticks. However, for deodorant and antiperspirant sticks, a metal cone penetrates the product instead of a needle (see Figure 3.34).

Drying Time Solvents from liquid deodorants and antiperspirants, such as roll-ons and extrudable gels, should quickly evaporate from the skin surface after application. Slow drying time leads to a sticky feeling and may leave stains on the clothes if users do not wait until the product is completely dried on their skin. Drying time can be measured by evaluating stickiness. Stickiness is the degree to which the underarm region sticks.[70] This test is usually performed by experts.

Performance Parameters Generally Tested ❾**The most commonly performed performance tests include *in vitro* microbiological testing, sniff test, and sweat reduction test for antiperspirants.** The main characteristics of these tests are discussed here.

In vitro *Microbiological Testing* As discussed earlier, deodorants and antiperspirants can contain antimicrobial ingredients. As odor formation is related to the activity of the bacteria in the armpit, inhibition or deactivation of the bacteria can potentially reduce body odor. In formulations containing antimicrobial agents, *in vitro* antimicrobial activity is usually tested as part of the product performance testing. The microbiological test includes the cultivation of selected bacteria on agar plates. After providing proper incubation time and conditions, microbial growth is evaluated. The main disadvantage of *in vitro* tests is that they are not reliable enough to indicate the true potential of formulations for malodor control. Therefore, additional studies, such as *in vivo* sniff test, are usually also performed.

Figure 3.34 Hardness testing of a deodorant/antiperspirant stick. Adapted from Texture Technologies Corp.

Sniff Test Sensory assessment of body odor by performing a sniff test is an accepted method for evaluating the *in vivo* efficacy of deodorants. The evaluation is usually tested by experts, often referred to as sniffers.[71] The test usually starts with a conditioning phase, which lasts several days. In this phase, the subjects use a perfume-free, non-antibacterial soap to clean the underarm region and are usually recommended not to use antiperspirants and deodorants. At the end of the preconditioning phase, body odor can be easily detected on the test subjects. In the subsequent treatment phase, the subjects use the test deodorant for a few days on one armpit and the control formulation on the other. The control formulation is usually identical to the test formation with the only exception that it does not contain deodorant ingredients. The intensity of body odor is scored on a scale, typically a 0–5 scale, with 0 being no odor and 5 being very intensive odor. The test subjects perform normal activities without washing their body. The intensity is recorded a few times (e.g., every 6 h) for 1 day. The difference of the average score between the two underarms is the basis for the evaluation of performance. If there is a significant difference between the treated and the untreated side after 24 h, the efficacy is proven.[72]

Sweat Reduction Test for Antiperspirants Antiperspirants reduce the amount of sweat produced by the eccrine glands. Performance claims for such products typically refer to the actual performance, such as "maximum protection"[73] and "extra effective,"[74] or the duration of performance, such as "odor protection lasts all day"[75] or "48-h odor protection."[76] As antiperspirants are considered OTC drugs in the US, all these claims must be clinically proven as described in the relevant final monograph. The monograph differentiates between two types of effectiveness, standard and extra effectiveness, and accordingly has different tests to substantiate these claims. It is important to note that the testing procedure recommended by the FDA is published in the form of a guideline, and it is emphasized that alternative methods may also be appropriate to qualify an antiperspirant drug product as effective. Therefore, manufacturers can also use other methods. The ultimate goal is that finished products should meet the criteria established in the FDA guideline, regardless of the testing method.[77]

The guideline states that, in order for an antiperspirant to be labeled as such, it must meet a minimum of 20% sweat reduction in at least 50% of the test population for standard effectiveness or a minimum of 30% sweat reduction in at least 50% of the test population for extra effectiveness.[78]

Subjects are usually selected by a panel. In order to be eligible, test subjects must produce at least 100 mg of sweat during 20 min in a controlled environment. Additionally, the subjects should represent the greater population; therefore, the differences between the highest and lowest rates of sweating among the subjects must exceed 600 mg in 20 min per axilla.[78]

Subjects first undergo a conditioning phase, which is typically a 17-day dry-out period when no other antiperspirants or deodorants are used by the subjects, except for those issued by the testing facility. In the actual study, which is commonly referred to as the "hot room" study, perspiration is stimulated by high levels of heat (approximately 100 °F/37 °C) and humidity (30–40%). Sweat is continuously collected (every 20 min) by the gravimetric method. As its name suggests, the gravimetric method is a weight method where absorbent pads (cotton pads) are placed in the underarm and their weight is measured before and after 20 min. The antiperspirant is applied to one axilla and a placebo to the other axilla once daily. For extra effectiveness, gravimetric testing is conducted at least two times during the period of the claim. Change in perspiration is statistically calculated and expressed as a percentage change.[78]

DID YOU KNOW?

Some products claim "pH-balanced formula," which may be very reassuring for consumers. However, this claim has no scientific basis since all stable formulations are pH balanced, and the pH of antiperspirant and deodorant products is generally dictated by the pH of the active. All antiperspirant actives currently listed in the monograph have an acidic pH. Changes in the pH could negatively affect the formulation's stability.

Ingredients Causing Safety Issues Although antiperspirants are used by a large number of people every day and are deemed safe, there were a number of concerns among consumers regarding their use and long-term effect. **⓾Ingredients that caused safety concerns regarding the use of deodorants and antiperspirants include aluminum salts, zirconium used to complex the aluminum salts, as well as VOCs used as propellants in aerosols.** The main findings behind these concerns and the statement of the regulatory authorities are discussed here.

Aluminum One of the biggest concerns with regard to the use of antiperspirants was whether they can cause **breast cancer**. The breast cancer–antiperspirant concerns ("myths" as the FDA refers to them)[79] first appeared in the 1990s, and they still continue to recirculate every year. The false information suggests that antiperspirants can absorb through razor nicks from underarm shaving and deposit in the lymph nodes. As antiperspirants prevent sweating, users cannot sweat out toxins, but they remain in the body, leading to cancer.[80] Most studies performed showed no link between breast cancer and the aluminum content or antiperspirants;[81–83] however, some studies provided conflicting results.[84,85] A study looked at how much aluminum can absorb from antiperspirants applied to the underarms and found that only a tiny amount (0.012%) was absorbed.[86] The actual amount of aluminum absorbed would be much less than what would be expected to be absorbed from the foods a person eats at the same time. Additionally, it seems that breast cancer tissue does not contain more aluminum than normal breast tissue.[87] The ACS[80] and the NCI[88] state that there is no evidence linking deodorants or antiperspirants with cancer. Currently, there is no clear link between antiperspirants containing aluminum and breast cancer.

Now the FDA requires all antiperspirant products to have a warning statement that advises people with kidney disease to consult a physician before using the product.[89] The reason for this was that the kidneys play a large role in eliminating aluminum from the body, and therefore, patients with impaired kidney function may have a higher level of aluminum exposure.

Even parabens, one of the most frequently used preservatives, have been linked to breast cancer by using them in deodorants and antiperspirants. However, according to the FDA, most major brands of antiperspirants and deodorants do not even contain parabens.[90]

DID YOU KNOW?

Before undergoing a mammogram, you will be asked by your doctor not to use antiperspirant or deodorant on the day you undergo a mammogram. The reason for this is that aluminum – a metal ion – in antiperspirants can show up on the mammogram as tiny specks. These specks can look like microcalcifications, which are one of the features doctors look for as a possible sign of cancer. Not using these products helps prevent any confusion when the mammogram films are reviewed.[80]

Additional health concern arose in the 1960s regarding the use of aluminum-containing antiperspirants and **Alzheimer's disease**. This suspicion led to concerns about everyday exposure to aluminum through sources such as cooking pots, foil, beverage cans, antacids, and antiperspirants. Since then, studies have failed to replicate the results of that particular study performed in the 1960s and confirm the role of aluminum in causing Alzheimer's. There are ongoing studies investigating this topic. Currently, the research community is generally convinced that aluminum is not a key risk factor in developing Alzheimer's disease. Public health bodies sharing this conviction include the WHO, the US National Institutes of Health (NIH), the EPA, and Health Canada.[91]

Zirconium Zirconium is used as a complexing agent to form various aluminum–zirconium chlorohydrate complexes. All these complexes contain aluminum chloride and zirconium chloride.[92] As mentioned earlier, glycine can also be used to form the complex as a buffering agent. Concerns arose regarding the carcinogenic potential of aerosolized aluminum–zirconium chlorohydrate-containing products through inhalation. The FDA declared that all zirconium-based antiperspirants when formulated as aerosols are Category II (not recognized safe), which led to the removal of such products from the market.[57] Today, no zirconium-containing active ingredients can be incorporated into aerosolized antiperspirant formulations.

Propellants Aerosol products have been the target of regulatory restrictions for a long time. First, the use of CFC propellants was banned in the US in the late 1970s due to concerns regarding the ozone layer.[93] As a result, numerous manufacturers switched from CFCs to alternative propellants and mechanical pump sprays. To replace these propellants, formulators started using various hydrocarbons, which belong to the class of ingredients known as VOCs. When exposed to sunlight, VOCs are capable of reacting with oxides of nitrogen (NO_x) to form ground-level ozone, the primary component of urban "smog." Smog is known to have negative effects on health, such as causing respiratory problems in humans. Concerns about smog led to a limitation in the type of propellant that could be used for aerosols on a national as well as a state level. In 1998, the US EPA established national VOC-content limits for consumer products, including hair sprays and deodorants.[94]

The current regulation differentiates between high-volatility organic compounds (HVOCs) and medium-volatility organic compounds (MVOCs). Any organic compound exerting a vapor pressure greater than 80 mmHg when measured at 20 °C is considered an HVOC, while that exerting a vapor pressure greater than 2 mmHg but less than or equal to 80 mmHg when measured at 20 °C is considered a MVOC.[95]

States are allowed to set more stringent limits on consumer products sold within their borders, and therefore, certain states have more stringent requirements than the EPA's National Rule (with California being the most stringent). VOC-content limits for consumer products are typically set in terms of weight percent VOCs. According to the national VOC emission standards for consumer and commercial products, antiperspirant aerosols can contain up to 60% HVOCs, while deodorant aerosols can contain up to 20% HVOCs.[95] California's regulations are much stricter and state that any products manufactured for sale in California, supplied or sold in California, cannot contain more than 40% HVOCs and 10% MVOCs, while deodorants cannot contain HVOCs and can contain only up to 10% MVOCs (effective from 1/1/2001).[96]

Packaging of Deodorants and Antiperspirants

The most commonly used packaging materials for deodorants and antiperspirants include the following:

- *Stick Containers*: Antiperspirant sticks are packaged in hollow tubes with an elevator platform inside that can move up and down to dispense the product. In most packages the product is lifted by turning a screw at the bottom of the container that causes the product to travel up along a central threaded post. Plastic bottles for extrudable soft solids and gels are similar to the stick bottles, although they have an inner cap with porous openings in it. The product is extruded through these holes upon turning the screw at the bottom of the container.
- *Roll-on Containers*: Roll-on containers consist of a hollow glass or plastic body, a cap, and a roll. The roll itself is made of a hollow ball and a housing for the ball. During application, the ball freely rolls in the housing while dispensing a small amount of product, which is adhered to it.
- *Aerosol Cans*: Aerosol deodorants/antiperspirants are supplied in one-compartment cans, where the formulation and the propellant are mixed together. Aerosol containers can be fabricated from tin-coated steel, tin-free steel, or aluminum.

GLOSSARY OF TERMS FOR SECTION 6

Antiperspirant: A topically applied personal care product designed to reduce under-arm wetness by limiting body transpiration. They are considered OTC drugs in the US.

Apocrine glands: A type of sweat gland that exists at birth but starts to function only at puberty. They are limited to certain body parts, such as the axilla, anus, and breast. They are usually triggered by emotions. Their secretion primarily consists of lipids, cholesterol, proteins, sulfur-containing amino acids, volatile short-chain fatty acids, and various steroids.

Body odor: A characteristic odor that develops after odorless sweat produced by the sweat glands is decomposed by bacteria on the skin surface.

Caking: A quality problem of aerosol antiperspirants, which are oil- and silicone-based suspensions. Caking refers to the formation of a hard-to-resuspend aggregate of insoluble particles at the bottom of the container.

Deodorant: A topically applied personal care product designed to reduce or mask unpleasant body odor by reodorization and/or antibacterial action.

Eccrine gland: A type of sweat gland that exists and starts to function from birth. These glands are found all over the body; their main function is to control body temperature. Their secretion consists mainly of water with various salts.

Extrudable cream: A deodorant and/or an antiperspirant in the form of a thick cream that is applied from a container through a cap with porous openings.

Extrudable gel: A deodorant and/or an antiperspirant in the form of a gel that is applied from a container through a cap with porous openings.

Fabric staining: The development and appearance of yellow-to-brown stains on lighter clothes and chalky white stains on darker clothes as a long-term consequence of using antiperspirants.

Glycine: An amino acid that can be used for aluminum–zirconium-based antiperspirants as a stabilizer, and it can also mitigate the acidic harshness of the products.

Hyperhidrosis: A medical condition that can be characterized by excessive sweating even at room temperature or lower or at rest.

Roll-on: A deodorant and/or an antiperspirant that is applied by means of a rotating ball in the neck of the container.

Shrinkage: A visible decrease in the size of a system, for example, gel.

Sniff test: An *in vivo* test used to evaluate the performance (efficacy) of deodorant products.

Sweat reduction test: An *in vivo* test used to evaluate the performance (efficacy) of antiperspirants.

Sweating: The process of producing and releasing a salty liquid from the sweat glands.

REVIEW QUESTIONS FOR SECTION 6

Multiple Choice Questions

1. Which of the following ingredients is used to complex aluminum–zirconium-based antiperspirant actives to stabilize them and reduce their harshness?
   a) Glycerin
   b) Glycine
   c) Glycol
   d) Glycerol

2. Which of the following is used to evaluate the performance of antiperspirants?
   a) Sniff test
   b) Viscosity test
   c) Extrudability
   d) Sweat reduction test

3. Which of the following is used to evaluate the performance of deodorants?
   a) Sniff test
   b) Viscosity test
   c) Extrudability
   d) Sweat reduction test
4. Which of the following has a quicker effect and higher antiperspirant efficacy?
   a) W/Si emulsion–based extrudable antiperspirant gel
   b) Water-based antiperspirant roll-on
   c) Anhydrous antiperspirant stick
   d) Powder active ingredient by itself
5. What is the pH of antiperspirant actives in aqueous solution?
   a) Approximately 2
   b) Approximately 4
   c) Neutral
   d) 8–9
6. In which dosage form is zirconium not recognized safe by the FDA?
   a) Gel
   b) Cream
   c) Aerosol
   d) Stick
7. What is the reason for staining of clothes when using antiperspirants?
   a) The formation of colored sweat
   b) Buildup of insoluble aluminum salts
   c) Waxes used as thickeners stain the clothes
   d) All of the above
8. Which of the following is TRUE for deodorant aerosol sprays?
   a) The deodorizer ingredient is solubilized in the vehicle
   b) The deodorizer ingredient is suspended in oils and/or silicones
   c) Caking can occur on standing
   d) They decrease the amount of sweat produced
9. Which of the following is usually NOT anhydrous?
   a) Antiperspirant sticks
   b) Extrudable clear gel antiperspirants
   c) Extrudable cream antiperspirants
   d) Aerosol antiperspirants

10. Antiperspirants block ___ sweat glands.

a) Apocrine
b) Eccrine
c) Both apocrine and eccrine

Fact or Fiction?

_______ a) Deodorants only cover up body odor and do not change the amount of sweat produced.
_______ b) Sweat becomes odorous when it is decomposed by bacteria on the skin.
_______ c) Antiperspirants can lead to overheat since they block perspiration.
_______ d) Antiperspirants cause breast cancer.

Matching

Match the ingredients in column A with their appropriate ingredient category in column B.

Column A	Column B
_____ A. Aluminum–zirconium tetrachlorohydrate	1. Antimicrobial agent
_____ B. Corn starch	2. Antiperspirant active
_____ C. Cyclomethicone	3. Hardening agent in sticks
_____ D. Ethanol	4. Humectant
_____ E. Hydrogenated castor oil	5. Nonionic surfactant
_____ F. Isosteareth-20	6. Odor-quenching deodorant ingredient
_____ G. Propylene glycol	7. Opacifying agent
_____ H. Quaternium-18 hectorite	8. Silicone oil
_____ I. Titanium dioxide	9. Suspending agent
_____ J. Zinc ricinoleate	10. Texturizing agent

REFERENCES

1. Market Research: Deodorants and Antiperspirants, US, Mintel International Group Ltd. Accessed 3/14/2014 at http://www.marketresearch.com/Mintel-International-Group-Ltd-v614/Deodorants-Antiperspirants-7086246/
2. Standring, S.: *Gray's Anatomy: The Anatomical Basis of Clinical Practice*, 39th Edition, London: Churchill Livingstone, 2004.

3. Venkataramm M.: *ACSI Textbook on Cutaneus and Aesthetic Surgery*. Clayton: JP Medical Ltd, 2012.
4. Eroschenko, V. P., di Fiore, M. S. H.: *DiFiore's Atlas of Histology with Functional Correlations*, Philadelphia: Lippincott Williams & Wilkins, 2012.
5. Udby, L., Cowland, J. B., Johnsen, A. H., et al. An ELISA for SGP28/CRISP-3, a cysteine-rich secretory protein in human neutrophils, plasma and exocrine secretions. *J Immunol Methods*. 2002;263:43–55.
6. Quinton, P. N., Elder, H., Jenkinson, D. M., Bovell, D. L.: Structure and Function of Human Sweat Glands, In: Laden, K., ed.: *Antiperspirant and Deodorants*, 2nd Edition, New York: Marcel Dekker, 1999;19.
7. Labows, J. N., Preti, G., Hoelzle, E., et al. Steroid analysis of human apocrine secretion. *Steroids*. 1979;34:249–258.
8. Jacoby, R. B., Brahams, J. C., Ansari, S. A., et al.: Detection and quantification of apocrine secreted odor-binding protein on intact human axillary skin. *Int J Cosmet Sci*. 2004;26:37–46.
9. Kanlayavattanakul, M., Lourith, N.: Body malodours and their topical treatment agents. *Int J Cosmet Sci*. 2011;33(4):298–311.
10. Taylor, D., Daulby, A., Grimshaw, S., et al.: Characterization of the microflora of the human axilla. *Int J Cosmet Sci*. 2003;25(3):137–145.
11. Bhutta, M. F.: Sex and the nose: human pheromonal responses. *J R Soc Med*. 2007;100:268–274.
12. Preti, G., Willse, A., Labows, J. N., et al. On the definition and measurement of human scent: comments on Curran et al. *J Chem Ecol*. 2006;32:1613–1616.
13. Rados, C.: FDA antiperspirant awareness: it's mostly no sweat. *FDA Consum*. 2005;39(4):18–24.
14. Partick, B., Thompson, J.: *An Uncommon History of Common Things*, Washington: National Geographic, 2011;193.
15. Plinius. *Naturkunde*. 35,52
16. Rimmel, E. *The Book of Perfumes*, London: Chapman & Hall, 1865.
17. Crawford, T. H., Nagarajan, T. S. *J Soc Cosmet Chem*. 1954;5:202.
18. Barraclough, G. *The Times Atlas of World History*, London: Times Books, 1978.
19. MUM: Accessed 3/13/14 at http://www.mum-deo.com/about-mum/history/
20. IFSCC *Monograph No 6. Antiperspirants and Deodorants, Principles of Underarm Technology*, Weymouth: Micelle Press, 1998.
21. Cuzner, B., Klepak, P.: Antiperspirants and Deodorants, In: Butler, H.: *Poucher's Perfumes Cosmetics and Soaps*, 9th Edition, London: Chapman & Hall, 1993:3–26.
22. *Life*, 1949;25.
23. 42 FR 41374, 1977.
24. EPA: Protecting the Stratospheric Ozone Layer Last Update: 3/6/2012, Accessed 3/10/2014 at http://www.epa.gov/air/caa/peg/stratozone.html
25. Statista 2014: U.S. Households: Which Forms of Deodorant and Antiperspirant Do You Use Most Often in 2013, Accessed 3/12/14 at http://www.statista.com/statistics/275489/us-households-form-of-deodorants-and-anti-perspirants-used/

26. Ruchinskas, R.: Hyperhidrosis and anxiety: chicken or egg? *Dermatology*, 2007;214(3):195–196.
27. Adar, R., Kurchin, A., Zweig, A., et al.: Palmar hyperhidrosis and its surgical treatment: a report of 100 cases. *Ann Surg*. 1977;186(1):34–41.
28. Rados, C.: Antiperspirant Awareness: It's Mostly No Sweat. *FDA Consumer Magazine*. 2005. Accessed on 2/3/14 at www.medicinenet.com/script/main/art.asp?articlekey=53207
29. Hornberger, J., Grimes, K.: Recognition, diagnosis and treatment of primary focal hyperhidrosis. *J Am Acad Dermatol*. 2004;51:274–286.
30. Johansen, J. D., Rastogi, S. C., Bruze, M., et al.: Deodorants: a clinical provocation study in fragrance-sensitive individuals. *Contact Dermatitis*. 1998;39(4):161–165.
31. Sato, K., Kane, N., Soos, G., et al.: The eccrine sweat gland: basic science and disorders of eccrine sweating. *Derm Found*. 1995;29:1–11.
32. Can too much antiperspirant cause me to overheat? *Popular Science* 2013;282(1):74.
33. Burry, J. S., Evans, R. L., Rawlings, A. V., et al.: Effect of antiperspirant on whole body sweat rate and thermoregulation. *Int J Cosm Sci*. 2003;25:189–192.
34. Geria, N.: Fragrancing antiperspirants and deodorants. *Cosmet Toiletries*. 1990; 105:41–45.
35. Lamp, J. H.: Sodium bicarbonate: an excellent deodorant. *J Invest Dermatol*. 1946; 7:131–133.
36. Berschied, J. R.: Antiperspirant-deodorant cosmetic stick products containing active agent particles in organic matrix, which matched densities for homogeneous products. Patent WO 9413256.
37. Winston, A. E. Microporous alkali metal carbonate powder—comprises particles of average particle size of 0.1 to 50 microns, surface area of 5 to 20 sq.m/f, average pore size of 10 to 500 nm and total pore volume of 0.1 to 2 cc/g and is useful as lightweight deodorant ingredient. Patent WO 9424996.
38. Zekorn, R., *Deowirkstoff auf Basis Zinkricinoleat. Parf Kosmet*. 1996;77:682–684.
39. Zekorn, R.: Zinc ricinoleate. *Cosmet Toiletries*. 1997;112:37–40.
40. Kanda, F., Yagi, E., Fukuda, M., et al.: Quenching short chain fatty acids responsible for human body odors. *Cosmet Toiletries*. 1993;108:67–72.
41. Yagi, E., Fukuda, M., Ota, T., et al.: Deodorants containing hydroxylapatite and synthetic resins, Patent JP 62135411, 1987.
42. Thurmon, F. M., Ottenstein B.: Studies on the chemistry of human perspiration with especial reference to its lactic acid content. *J Invest Dermatol*. 1952;18:333–339.
43. Ikai, K.: Deodorizing experiments with ion-exchange resins. *J Invest Dermatol*. 1954;23:411–422.
44. Nakahara, Y., Shimai, Y., Yoshikawa, T.: Body deodorant powder containing spherical, porous, hollow microparticles. Patent JP 62212315, 1987.
45. Charig, A., Froebe, C., Simone, A., et al.: Inhibitor of odor producing axillary bacterial exoenzymes. *J Soc Cosmet Chem*. 1991;42:133–145.
46. Use of monoalkyl or dialkyl ester(s) of aliphatic, satd. dicarboxylic acids. DE 4343265 A1
47. Process of suppressing odors employing deodorants containing esters of citric acid US 4010253 A

48. Cox, A. R.: Efficacy of the antimicrobial agent triclosan in topical deodorant products. *J Soc Cosmet Chem*. 1987;38:223–231.
49. Dillenburg, H., Jakobsonm G., Klein, W., et al.: Cosmetic deodorant preparations containing di- or triglycerin esters. Patent No. EP 666732 Al/Bl.
50. Abrutyn, E. S.: Antiperspirants and Deodorants, In: Drealos, Z. D.: *Cosmetic Dermatology*, Hoboken: Wiley Blackwell, 2010.
51. Abrutyn, E. S.: Antiperspirants. *Cosmet Toiletries*. 2011;126(11):780–786.
52. Swaile, D. F.: Variation in efficacy of commercial antiperspirant products based on differences in active composition and manufacturing method. In: 61st Annual Meeting of the American Academy of Dermatology, San Francisco, 2003:P337.
53. Reller, H. H., Luedders, W. L.: Mechanism of action of metal salt anti-perspirants. *Adv Mod Toxicol*. 1977;4:18–54.
54. Quatrale, R. P., Coble, R. P., Stoner, K. L., et al.: The mechanism of antiperspirant action of aluminum salts. III. Histological observations of human eccrine sweat glands inhibited by aluminum zirconium chlorohydrate glycine complex. *J Soc Cosm Chem*. 1981;32:195–222.
55. Wild, J. E., Lanzalaco, A. C., Swaile, D. F.: Antiperspirants, In: Drealos, Z. D., Thaman, L. A.: Cosmetic Science and Technology, Vol 30. Cosmetic Formulation of Skin Care Products, New York: Taylor & Francis, 2006.
56. Abrutyn, E.: Antiperspirant and Deodorants, In: Reiger, M. M.: *Harry's Cosmetology*, 8th Edition, New York: Chemical Publishing Company Inc., 2000.
57. FDA: Title 21, Chapter 1, Subchapter D, Part 310, Subchapter G, Part 700: Aerosol drug and cosmetic products containing zirconium, final rule, *Fed Reg*, 1977;42(158):41374–41376.
58. Abrutyn, E. S.: Antiperspirant sticks, soft solids and gels, *Cosmet Toiletries*. 2009;124(5):24–35.
59. A Quick Guide to Dow Corning® brand Silicone Emulsifiers, Accessed 3/10/2014 at http://www.dowcorning.com/content/publishedlit/27-1369.pdf?wt.svl=BPC_pFace_FaB2
60. Schreiber, J.: Antiperspirants, In: Barel, A. O., Paye, M., Maibach, H. I.: *Handbook of Cosmetic Science and Technology*, 3rd Edition, New York: Informa Healthcare, 2009.
61. Abrutyn, E. S., Bahr, B. C.: Formulation enhancements for underarm applications. *Cosmet Toiletries*. 1993;108:51–54.
62. Alexander, P.: Monograph antiperspirants and deodorants. *SOFW*. 1994;120:117–121.
63. ICI Speciality Chemicals: A new emollient for antiperspirant sticks. *HAPPI*. 1989;50–51.
64. Osborne, G. E., Lausier, J. M.: Statistical evaluation of vehicle effect on antiperspirant activity with a limited number of subjects. *J Soc Cosmet Chem*. 1982;33:179–191.
65. Majors, P. A., Wild, J. E.: The evaluation of antiperspirant efficacy – influence of certain variables. *J Soc Cosmet Chem*. 1974;25:139–152.
66. Juneja, P. S.: Deodorant gel sticks containing 1-hydroxy pyridinethione active EP 0662817 A1
67. Geria, N.: Formulation of stick antiperspirants and deodorants. *Cosmet Toiletries*. 1984;99:55–66.
68. Chadwick, J.: Aerosol anatomy, aerosol product litigation – Part 2, *Spray*, 2007;11:20–22.
69. Andrasik, M. J.: Deodorants and Antiperspirants, *Int Fabr Instit Bullet, 626*, Accessed 3/12/2014 at http://70.88.161.72/ifi/BULLETIN/TOI/Toi626.pdf

70. Meilgaard, M. C., Carr, B. T., Civille, G. V.: *Sensory Evaluation Techniques*, 4th Edition, Boca Raton: CRC Press, 2006.

71. Maxeiner, B., Ennen, J., Rützel-Grünberg, S., et al.: Design and application of a screening and training protocol for odour testers in the field of personal care products. *Int J Cosmet Sci.* 2009;31(3):193–199.

72. Smits, J., Senti, B., Herbst, N.: usNeoTM - a Naturally Effective Deodorant and Antibacterial Active. *Cosmet Sci Technol.* 2012;1–8.

73. Rexona Men Maximum Protection, Accessed 3/8/2014 at http://www.unilever.com/brands-in-action/detail/Rexona/292102/

74. Lady Speed Stick Clinical Proof, Accessed 3/8/2014 at http://dailymed.nlm.nih.gov/dailymed/lookup.cfm?setid=6137b038-dcfd-4ef6-9a59-45e19dc3807a

75. Old Spice Fresh Invisible Solid, Accessed 3/8/2014 at http://www.oldspice.com/en-US/products/product/20/old-spice-high-endurance-antiperspirantdeodorant-invisible-solid-fresh/

76. Secret Invisible Protection Outlast, Accessed 3/8/1014 at http://www.secret.com/Outlast-InvisibleSolid.aspx

77. CFR Title 21 Part 350.60

78. Guidelines for Effectiveness Testing of OTC Antiperspirant Drug Products, Accessed 3/10/2014 at http://www.fda.gov/downloads/AboutFDA/CentersOffices/CDER/ucm106437.pdf

79. Rados C.: Antiperspirant Awareness: It's Mostly No Sweat. *FDA Consumer Magazine.* 2005, Accessed 2/3/2014 at www.medicinenet.com/script/main/art.asp?articlekey=53207

80. American Cancer Society: Antiperspirants and Breast Cancer Risk, Last update: 9/20/2013, Accessed 3/10/2014 at http://www.cancer.org/cancer/cancercauses/othercarcinogens/athome/antiperspirants-and-breast-cancer-risk

81. Mirick, D. K., Davis, S., Thomas, D. B.: Antiperspirant use and the risk of breast cancer. *J Natl Cancer Inst.* 2002;94(20):1578–1580.

82. Gikas, P. D., Mansfield, L., Mokbel, K.: Do underarm cosmetics cause breast cancer? *Int J Fertil Womens Med.* 2004;49(5):212–214.

83. Namer, M., Luporsi, E., Gligorov, J., et al.: The use of deodorants/antiperspirants does not constitute a risk factor for breast cancer. *Bull Cancer.* 2008;95(9):871–880.

84. Darbre, P. D.: Aluminium, antiperspirants and breast cancer. *J Inorg Biochem.* 2005;99(9):1912–1919.

85. McGrath, K. G.: An earlier age of breast cancer diagnosis related to more frequent use of antiperspirants/deodorants and underarm shaving. *Eur J Cancer Prev.* 2003;12(6):479–485.

86. Flarend, R., Bin, T., Elmore, D., et al.: A preliminary study of the dermal absorption of aluminium from antiperspirants using aluminium. *Food Chem Toxicol.* 2001;39(2):163–168.

87. Rodrigues-Peres, R. M., Cadore, S., Febraio, S., et al.: Aluminum concentrations in central and peripheral areas of malignant breast lesions do not differ from those in normal breast tissues. *BMC Cancer.* 2013;13:104.

88. Fact Sheet: Antiperspirants/Deodorants and Breast Cancer, Last update: 1/4/2008, Accessed 2/27/2014 at http://www.cancer.gov/cancertopics/factsheet/Risk/AP-Deo

89. CFR Title 21 Part 350.50(c) Labeling of antiperspirant drug products.

90. FDA: Parabens, Last update: 3/20/2014, Accessed 2/27/2014 at http://www.fda.gov/cosmetics/productandingredientsafety/selectedcosmeticingredients/ucm128042.html
91. Alzheimer News 2/02/2005, Accessed 3/11/2014 at http://www.alz.org/news_and_events_alzheimer_news_02-02-2005.asp
92. D. F. Williams, W. H. Schmitt: *Chemistry and Technology of the Cosmetics and Toiletries Industry*, Springer, 1996:321.
93. 42 FR 24,536
94. EPA: Cleaning Up Commonly Found Air Pollutants, Accessed 2/12/2014 at http://www.epa.gov/airquality/peg_caa/cleanup.html
95. 40 CFR Part 59—National Volatile Organic Compound Emission Standards for Consumer and Commercial Products (Current as of 12/12/2013)
96. California Air Resourced Board: Final Regulation Order, Regulation for Reducing Volatile Organic Compound Emissions from Antiperspirants and Deodorants, Subchapter 8.5. Consumer Products, Article 1. Antiperspirants and Deodorants

4

COLOR COSMETICS

INTRODUCTION

Today, color cosmetics are part of our regular culture and fashion; they have a major role in satisfying personal desires for self-improvement, self-adornment and grooming for one's own sense of well-being, and for the general attention or attraction of others. However, this was not always the case. Color cosmetics were created a few thousand years ago. The purposes of their application in the early ages were beautification, maintenance of a young appearance, protection of the body from the environment, preparation for religious rituals, and even medical treatment. Although many of the ingredients used were highly poisonous, makeup products were very popular in Egypt, the Middle East, and Asia. However, when cosmetics started to spread in Europe, there was resistance in some cultures. Color cosmetics were believed as extravagant, abhorrent, and unnecessary by many cultures. There were countries where the use of cosmetics was not allowed and only women without makeup were considered beautiful. With the development of cinema, photography, and television, the popularity of makeup started to rise again in all classes. Today, the cosmetic industry is a multibillion-dollar business worldwide. Formulators always find new ways to keep and increase the interest of women and men to buy these products. Latest developments in color cosmetics include products that resist abrasive removal (known as transfer-resistant products), long-lasting products that will stay on the application surfaces for an extended period of time without reapplying them, and multifunctional products that combine a few products into one to satisfy consumers' increasingly growing desires.

Introduction to Cosmetic Formulation and Technology, First Edition. Gabriella Baki and Kenneth S. Alexander.
© 2015 John Wiley & Sons, Inc. Published 2015 by John Wiley & Sons, Inc.

The four sections of this chapter focus on the various types of color cosmetics, including products applied to the human lips, eyelids and eyelashes, facial skin, and nails. After reading these sections, you will understand the use and function of these products as well as how they are formulated and tested.

SECTION 1: LIP MAKEUP PRODUCTS

LEARNING OBJECTIVES

Upon completion of this section, the reader will be able to

1. Define the following terms:

Aeration	Bleeding	Break strength	Cracking
Cratering	Deformation	Extrusion	Flaming
Laddering	Lip balm	Lip gloss	Lip liner
Lip plumping lipstick	Lipstick	Mold	Molding
Sweating	Seams	Streaking	Rigidity
Vermilion zone			

2. describe the distinct anatomical regions of the human lips;
3. list some of the causes for the dryness of human lips;
4. describe the changes during aging of human lips;
5. differentiate between lipstick, lip gloss, lip liner, and lip balm;
6. explain why the majority of lip balms are considered OTC drug–cosmetic products in the US;
7. list various required cosmetic qualities and characteristics that an ideal lip makeup product should possess;
8. list various required technical qualities and characteristics that an ideal lip makeup product should possess;
9. list the typical ingredient types of lip makeup products, and provide some examples for each type;
10. list the major lipstick types, and describe their characteristics;
11. describe the method of "molding" used to prepare lipsticks and name its four basic steps;
12. explain why premilling of the applied pigments is important before mixing them into the lipstick base;
13. describe how special effect pigments should be handled during molding;
14. differentiate between molding and flaming;

15. describe the simple mixing process, and explain which types of products can be formulated with this technique;
16. describe the extrusion process, and explain which types of products can be formulated with this technique;
17. list some of the typical quality issues that may occur during the formulation of lip makeup products, and explain why they may occur;
18. list the typical quality parameters that are tested for lip makeup products, and briefly describe their method of evaluation.
19. name the efficacy parameter that is generally tested for lip makeup products claiming moisturization, and describe the method of evaluation.
20. briefly discuss the potential safety issues regarding lead contamination of lipsticks;
21. list typical containers available for lip makeup products.

KEY CONCEPTS

1. Human lips have a complex anatomy consisting of mucosa and skin.
2. The following products belong to the category of cosmetics in the US: lipsticks, lip glosses, and lip liners. They are applied to enhance or color the lips and are purely for increasing beauty and attractiveness, which meet the definition of a cosmetic product.
3. Although lip balms are also applied to the lips, similar to the types of lip care products previously, the majority of lip balms, also called lip protectants, are considered OTC drug–cosmetic products in the US.
4. The most frequently applied technique to produce lipsticks and lip balms is the process known as molding. The usual steps of this technique include (i) pigment premilling, (ii) melting and mixing, (iii) molding and packaging, and (iv) flaming.
5. Additional techniques include molding without flaming, simple mixing without heating, as well as extrusion.
6. The typical quality-related issues of lip care products include aeration, laddering, chipping, deformation, cratering, streaking, sweating, mushy failure, seams, bleeding, and poor pay-off.
7. Parameters commonly tested to evaluate the quality of lip care products include break strength, hardness, melting point, softening point, color matching, color brightness and streakiness, pay-off and glide, rigidity, viscosity, and pH.
8. The most frequently tested efficacy parameter is the moisturizing effect of lip balms.
9. The main ingredient that caused safety concerns regarding lipsticks is lead.

Introduction

Lip makeup products are primarily used by women to enhance their attractiveness and femininity; however, there are products available for men as well. Products for men usually include uncolored sticks offering hydration and sun protection.

This section reviews the main anatomical and physiological characteristics of human lips. In addition, it provides an overview of the various cosmetic products and OTC drug–cosmetic products applied to the lips, including their typical ingredients, formulation technology, testing methods, and packaging materials. It is also discussed how these products may affect the lips and what the consumers' general requirements are.

Anatomy and Physiology of Human Lips

❶ **Human lips have a complex anatomy consisting of mucosa and skin.** The outer surface of the lips is covered by epidermis and hair, sweat glands, and sebaceous glands.[1] The skin of the face meets the mucous membrane of the mouth in a transition zone called the vermilion border. This transition zone is characterized by lightly keratinized epithelium; it does not contain hair or sweat glands. The inner surface of the lips is continuous with the oral mucosa, which is covered with non-keratinized epithelium and contains numerous tiny salivary glands.[2] In the red area of the lip (i.e., vermilion border), cornification and pigmentation of the epithelium diminish. The red color of the border area is thought to be the result of the decreased density keratin along with the translucency of the tissue, which allows the observer to see easily the small capillary vessels.[3] This is the zone where makeup products are applied to. Figure 4.1 depicts the different parts of the lips, and Figure 4.2 shows the basic histological characteristics of the human lips.

Human lips have a specific shape due to the muscles and soft connective tissue of which they are made. The muscle, the so-called orbicularis oris muscle, makes a hooked curve toward the exterior at the edge of the vermilion area, which gives the lips their shape. The physical appearance of the Cupid's bow is also caused by the configuration of the underlying muscle.[4]

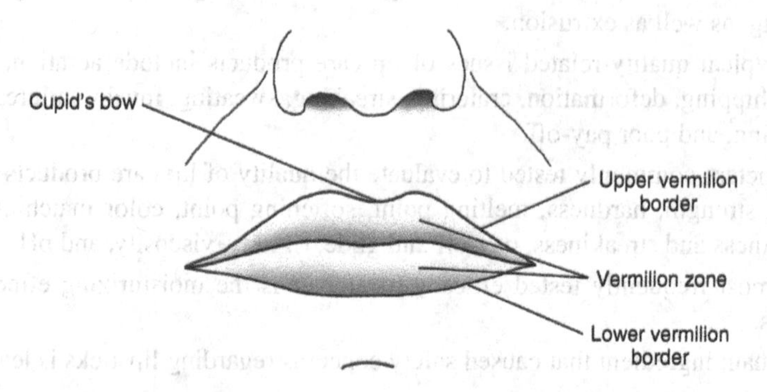

Figure 4.1 Structure of the human lip.

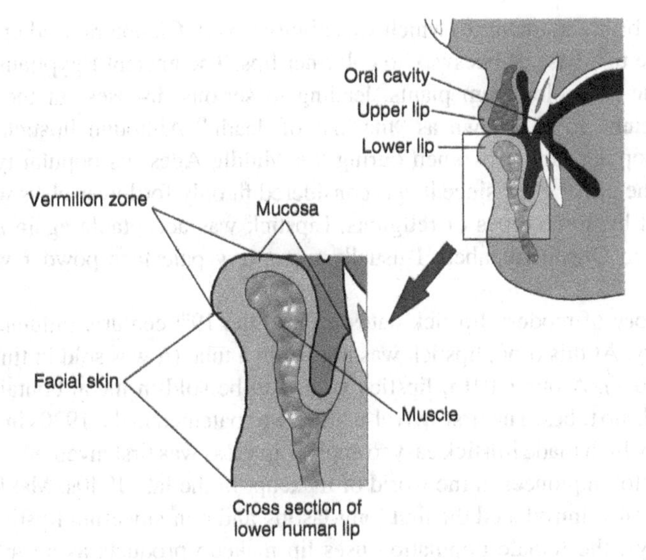

Figure 4.2 Cross section of the lips.

The production of natural emollients is very low in the lips. In addition, it also appears to contain less water than other areas of the face and lose water faster than the cheeks.[5] As a result, human lips can become dry and chapped easily.

An interesting characteristic of the vermilion border is that it has very rapid cell turnover. This is why the lips recover very quickly if you hurt or burn them. Another specific property of the lips is that they have a lot of sensory receptors and have great tactile sensitivity. According to the literature, the sensitivity of the lips is similar to that of the tongue and the fingertips.[6] In addition to being sensitive to touch, lips are very sensitive to chemical, physical, and microbial damages as well. The fact that lips are not rich in keratinized cells and melanin significantly contributes to their sensitivity. Their prolonged exposure to sunlight may lead to visible damage, and they can be more easily infected with yeast and bacteria than other parts of the skin.

The quality and structure of our lips change over time. As we grow older, thinning skin, loss of muscle support, changes in bone structure, bone resorption, and reductions in soft tissue volume lead to a number of changes in the perioral region (i.e., the area around the lips).[7] The lips become flat and thin and less shapely. In addition, fine lines and deeper wrinkles develop on the upper and lower lips that may cause lip makeup products to flow into these lines from the red area of the lips; this phenomenon is usually referred to as "bleeding." The tactile sensitivity of the lips decreases with age as well.

History of Using Lip Makeup Products

Most of the history of lipstick dates back to the prehistoric times. Women in ancient Mesopotamia as far back as 3000 BC tinted their lips with crushed gemstones and inorganic ingredients such as red clay, iron oxide (rust), or even with henna, seaweed, iodine, and bromine mannite (which was highly toxic). Cleopatra used crushed ants and carmine in a base of beeswax to color her lips. The ancient Egyptians squeezed out a purple-red color from plants, leading to serious diseases. In the course of time, this came to be known as "the kiss of death." Although lipstick remained generally popular among women during the Middle Ages, its popularity began to decline in the upper class since it was considered fit only for lower-class women and was banned by some types of religions. Lipstick was acceptable again in the 16th century when Queen Elizabeth I usually wore very pale face powder with bright red lips.

The history of modern lipstick dates back to the 19th century, influenced by the film industry. At this time, lipstick was not yet in a tube (it was sold in tinted papers or paper tubes). Around 1915, lipstick started to be sold in metal containers, with various push-up tubes. The first swivel-up tube was patented in the 1920s in Nashville, Tennessee, which made lipstick easy to apply. Lip gloss was first invented in the 1930s by Max Factor, a pioneer in the world of makeup. In the late 1940s, Ms. Bishop, an organic chemist, introduced the first long-lasting and non-smearing lipstick.

Nowadays, the female population uses lip makeup products as an integral part of their daily life worldwide. We have a great and continuously increasing variety of shades, shapes, and flavors as well as advanced technologies and multifunctional products.

Types and Definition of Lip Makeup Products

There are several different types of products available on the market that can be applied to the lips for various reasons. Let us review the most common types of lip products and their characteristics. ❶ **The following products belong to the category of cosmetics in the US: lipsticks, lip glosses, and lip liners. They are applied to enhance or color the lips and are purely for increasing beauty and attractiveness, which meet the definition of a cosmetic product.**

- **Lipsticks** are designed to enhance the appearance of the lips by imparting color and gloss. They consist of waxes, butters, fats, oils, and hydrocarbons, which are usually referred to as the base, as well as pigments for the color. Additionally, lipsticks can contain flavors and fragrances as well as specific ingredients for ultraviolet (UV) protection and a plumping effect.
- **Lip glosses** are designed to give the lips a glossy luster and, sometimes, subtle color. They usually have a lower viscosity than traditional lipsticks and are more transparent. Lip glosses consist of a higher ratio of oils and lower ratio of waxes; therefore, they apply smoothly, have a greater shine, but do not wear as long as a lipstick does. They can be clear or translucent or exhibit various shades of opacity with various finishes, such as metallic, glittered, glassy, and frosted.

- **Lip liners** are designed to redefine the outline of the lips. They consist of blends of waxes, butters, fats, and oils, similar to lipsticks; however, the finished formulation is harder and the level of pigments is slightly lower. Lip liners are most often slim pencils, or fluids encased in special "pens," to which is attached a fine brush through which the product is dispensed.

There is an additional product type available on the market, it is called a lip balm. ❻ **Although lip balms are also applied to the lips, similar to the types of lip cosmetics previously, the majority of lip balms, also called lip protectants, are considered OTC drug–cosmetic products in the US.** The reason for this is that they temporarily prevent dryness of the lips and help relieve chapping. This action, i.e., prevention, meets the definition of drugs. The OTC monograph for skin protectant drugs specifies the active ingredients and their concentration that qualify as OTC drugs.[8] If a product contains at least one of the following (without completeness), it is considered an OTC drug in the US:

- Allantoin in 0.5–2%
- Cocoa butter in 50–100%
- Dimethicone in 1–30%
- Glycerin in 20–45%
- White petrolatum in 30–100%.

FYI

For more information, visit the Code of Federal Regulations (CFR) Title 21, Part 347 about "Skin protectant drug products for over-the-counter human use."

The same monograph also contains approved claims, such as "helps prevent" (or optionally "temporarily protects" or "helps relieve") "chapped or cracked lips." Optionally, the following statement can also be added: "helps prevent and protect from the drying effects of wind and cold weather."[8]

Lip balms are made of waxes, oils, fats, butters, and hydrocarbons as well as additional ingredients to prevent chapping and drying of the lips. These products are usually not or only slightly colored.

It should be noted that today many lip makeup products are multifunctional, which means that they fulfill several functions at the same time. One of the most frequent extra functions of a lip product, in addition to increasing the attractiveness or providing protection against dryness, is its protective effect against sunlight. It should be noted that a lip product containing a sunscreen (UV filter) is also considered a drug in the US under the current sunscreen regulations (see more details in Section 5 of Chapter 3).

How Lip Makeup Products May Affect the Human Lips?

Lip makeup products are primarily used due to their coloring. In addition, they can have further advantages and some disadvantages as well.

- Certain lipsticks, mainly long-wearing lipsticks, tend to **dry** the lips. It is attributed to the silicones found among the ingredients. These ingredients seal the color to the lips and make the lipstick "kiss-resistant." Lip plumping lipsticks containing menthol and camphor may also cause dryness. Other factors that might contribute to the development of dry lips include nasal obstruction, promoting mouth breathing, and dentures that might not fit properly (commonly seen in the elderly). Dry lips may also be seen in children who are thumb suckers. As discussed, dryness can be prevented and/or treated with the use of lip balms.

DID YOU KNOW?

Continuous wetting of the lips with the saliva does not help chapped lips and provides only a temporary relief. After wetting the lips, water evaporates from the skin's surface, which has a cooling effect. In addition, the evaporating water drags additional water from the outer layers of the skin. As a result, wetting the lips frequently with the saliva has actually a drying effect. Dryness can be prevented and treated with the application of moisturizing ingredients.

- Recent research has shown that **lip cancer** occurs in men more frequently than in women.[9,10] This difference is attributed to the widespread use of lipsticks and lip balms with UV filters by women. UV filters block the sun's rays and prevent penetration of them into the skin. Unfortunately, many people remain unaware how important consistent lip protection is. It is also important when using these photoprotective products to reapply them over time in order to provide and maintain effective protection. Lip balms can be applied from a container similar to that of lipsticks or by fingers from metal or plastic pots and tins.
- Lip balms containing sunscreens may also be used to prevent the reoccurrence of **herpes simplex**, since the virus is photoreactivated.[11] In the case of viral infections, e.g., herpes, lip balms and lipsticks should not be shared as they can transmit the virus as long as the blister fluid remains moist.

Required Qualities and Characteristics and Consumer Needs

From a consumer perspective, a quality lip cosmetic and OTC drug–cosmetic product should possess the following characteristics:

- Attractive shades
- Homogeneous color when applied

- Good coverage
- Long-lasting effect
- Pleasant taste and smell
- Easy to apply
- UV protection (common requirement)
- No staining or bleeding into the fine lines surrounding the lips
- Lip glosses: provide a wet and shiny look
- Lip liners: have high pigment content to accent the lines of the lips, firm enough not to run into the lines around the lips
- Lip balms: moisturize the lips and prevent chapping.

The technical qualities of lip makeup products can be summarized as follows:

- Long-term stability
- Dermatological safety
- Creamy gliding without a greasy sensation
- Firm adherence to lips without being brittle and tacky
- High retention of color intensity without any change in shade
- Appropriate rigidity and hardness
- Resistance to temperature from 39 °F to 104 °F (4–40 °C) without hardening or running.

Typical Ingredients of Lip Makeup Products

Lip makeup products, including cosmetics and OTC drug–cosmetic products, can contain a dozen of different ingredients. Despite the potentially high number of ingredients and a vast array of colors, ingredients can be categorized into several basic groups. This section reviews these typical ingredient types as well as their main characteristics and functions.

- **Waxes** function as structuring agents, providing lipsticks with rigidity and solidity. In addition, they stabilize the sticks and allow them to be molded into shape. By combining waxes with different properties, such as high shine, flexibility, and brittleness, optimal cosmetic performance can be achieved. The melting points of waxes vary widely depending on their unique composition and chain lengths. They are usually mixed with oils to achieve the desired softness; waxes alone would be too rigid.
    - Examples for waxes used in lip makeup products include beeswax, candelilla wax, carnauba wax, paraffin wax, ozokerite wax, microcrystalline wax, polyethylene, and lanolin alcohol.
- **Oils, fats, and butters** provide a slippery and soft texture to the formulations. They also have a moisturizing effect and act as emollients, i.e., prevent drying

and chapping of lips. Oils are also used to disperse pigments and pearls. A fact that should be taken into consideration when adding oils to a formulation is that a high concentration (typically above 50%) may result in a greasy and sticky feeling. Butters and fatty acid esters can improve the adhesiveness for the formulations to the lips. Low molecular weight silicones may also be used, and since they are volatile, they can provide excellent carrier properties for transfer-resistant lipsticks.

- Examples for oils, fats, and butter used in lip makeup products include plant oils, such as castor oil, grape-seed oil, almond oil, meadowfoam oil, olive oil, coconut oil, palm oil, and triglyceride; butters, such as avocado butter, shea butter, and cocoa butter; fatty acid esters, such as isopropyl myristate, isopropyl palmitate, isostearyl isostearate, and butyl stearate; and hydrocarbons, such as polyisobutene, mineral oil, petrolatum, isododecane, and isoeicosane. Additionally, silicones, such as dimethicone and cyclomethicone, can also be used.

- **Color additives** are the most important components from a commercial and appearance point of view since colors set lipsticks apart in most consumers' minds. The application of colorants is strictly controlled by the FDA, and color additives are specifically approved for the mouth area. Typically, the following types of color additives are used in colored lip care formulations: organic colors, inorganic pigments, lakes, and specific or the so-called effect pigments. As discussed in Chapter 1, dyes are readily soluble, whereas pigments are insoluble in the medium/vehicle in which they are used. Effect pigments may include particles that provide pearlescent effect, sparkle, luster, glitter, matte, metallic, and other types of effects to lipstick formulations. Effect pigments used in lip makeup products are generally available in various colors and particle sizes. By varying the particle size of these ingredients, different effects can be achieved. Smaller particles (in the range of 20 μm) create silky and satin effects and opacify the mass. Larger-sized particles create (in the range of 120 μm) high luster effects, sparkling, or glittering, combined with high brilliance and transparency. Special pigments usually do not add too much color to formulations; therefore, they are usually combined with regular color additives. However, in low-color products, such as lip glosses or lip balms, they can be used alone.
  - Examples for inorganic pigments used in lip makeup products include iron oxides (red, yellow, brown, and black), titanium dioxide, and zinc oxide; organic colors include reds (such as Red 6, 7, and 21), yellows (such as Yellow 6), oranges (such as Orange 5 in a concentration $\leq$5%), and lakes such as Red 7 Lake and Yellow 5 Lake. The most frequently used special effect pigments include micas coated with iron oxides and titanium dioxide (which can provide a sparkly or glittery effect, a soft luster, or a pearlescent effect depending on the coating), and bismuth oxychloride (which provides a pearlescent effect).

DID YOU KNOW?

Although pigments are not soluble in the medium to which they are added, e.g., a lipstick base made of waxes and oils, they still can be dispersed in these liquids or vehicles. A dispersion is a mixture of a substance (e.g., a red pigment) **separated** from another substance (e.g., lipstick base). They are separated because they do not dissolve in each other. To us, a melted, colored lipstick mix may seem as a continuous solution; however, under the microscope, you would see two phases **dispersed** in each other. It means that dispersions have components that are not soluble in each other but still form a stable and homogenous mixture.

- **Antioxidants** are added to prevent rancidity and the oxidation of sensitive ingredients.
  - Examples for antioxidants used in lip makeup products include vitamin E, butylated hydroxyanisole (BHA), and butylated hydroxytoluene (BHT).
- **Preservatives** provide protection against microbiological contamination.
  - Examples include parabens and phenoxyethanol.
- **Fragrances** can be used in lip makeup products to mask fatty or wax odor. They should be neither irritant nor toxic by ingestion, and their taste must be agreeable as well. In addition, they should be stable at higher temperatures since the formulation of lip makeup products might take place at a higher temperature.
- **Flavoring agents** can also be added to formulations to provide a pleasant taste. Lip makeup products may get in contact with the taste buds as opposed to the majority of cosmetic products, which are not ingested. Therefore, product taste should also be taken into consideration. Any FDA-approved food flavoring agent can be used for flavoring. Taste can also be adjusted by using sweeteners, such as sodium saccharin. Lip plumping lipsticks containing cinnamon or menthol are also popular today. These ingredients add a specific flavor to the formulations.
- **Texturizing ingredients**, such as talc, silica, and mica, may also be used to improve the texture, application, and stability of products. Additional examples include titanium dioxide, which provides a soft focus effect (spherical ingredients are able to scatter light in different directions, making the skin appear soft and blurred, which is referred to as a soft focus effect), and bismuth oxychloride, which gives a satin, shimmering effect to the products.
- Certain products, mainly long-lasting formulas, contain **fixatives** that prevent colors from bleeding on the lips; they help seal the lipsticks on the lips.
  - Primarily, silicone resins are used as fixatives in lipsticks.
- **Active ingredients** can also be used in lip makeup products.

- Examples include the lip protectant ingredients and UV filters, both of which are listed in the relevant OTC monographs.

Typical Lipstick Types

Lipstick manufacturers created numerous words for describing their products; some of them are used interchangeably, while others refer to more specific categories. Here is a short summary of the basic types currently available on the market.

- **Matte** lipsticks are rich in pigments and waxes but lighter in emollients. Since they contain less oil, they deflect light and are less shiny. Mattes can often be longer wearing than shinier types of lipstick.
- **Glossy** or **glaze** lipsticks are heavy in oil and often contain flavors and scents as well. They usually add shine and volume to the lips. Generally, glossy lipsticks come in lighter shades. The coverage they provide lasts longer than that of lip for glosses, however, not as long as that of with crème lipsticks. Their main disadvantage is that they wear off quickly.
- **Crème** lipsticks as well as **satin** and **sheer** lipsticks usually fall between mattes and glosses. They contain a high concentration of emollients; therefore, they create a shiny, glossy finish. Crème lipsticks typically contain pigments with a smaller particle size to achieve a silky effect. They also tend to wear off quickly and, therefore, must be reapplied frequently.
- **Shimmer** lipsticks, also known as **frosted** lipsticks, contain light-reflecting particles, such as coated micas with a larger particle size, to add luster to the color. They are more often used for special occasions and come in lighter colors.
- **Long-wearing** lipsticks are usually two-part systems consisting of a colored base formula and a colorless cover. The base contains pigments; therefore, it looks like a regular lipstick. It may also contain silicones and hydrocarbons, which seal in the color. Once the base dries, consumers can apply the colorless cover formulation for the shine. While this may sound like an appealing option to keep a lipstick on all day long, the formula can be very drying.
- **Lip plumping** lipsticks are designed to make lips appear fuller by slightly irritating the delicate skin on the lips to make them swell. Ingredients used to provide this effect include ginger, cinnamon, cayenne, camphor, and menthol.

Formulation of Lip Makeup Products

There are various techniques available for the formulation of lip makeup products. Most lipsticks are formulated using the molding technique. It involves four steps and is usually a lengthy process. Lip balms are formulated using a similar technique; however, not all steps of the classic molding technique are applied in their case. Lip glosses do not necessarily need heating, since the ingredients used are liquid and semisolid at room temperature. Lip pencils can be produced via molding and extrusion. These techniques are discussed in the following sections.

Molding ● **The most frequently applied technique to produce lipsticks and lip balms is the process known as molding. The usual steps of this technique include (i) pigment premilling, (ii) melting and mixing, (iii) molding and packaging, and (iv) flaming.**

1. **Pigment premilling** is a step where agglomerates in the powder color are broken up. In order to provide a homogenous, smooth, and even color to lipsticks, pigments should be finely ground. This step is generally performed by preparing a "pigment grind" using an oil, such as castor oil, with a triple roll mill, bead mill, or similar conventional mill, or pestle and mortar for the laboratory-scale preparation. Today, pre-micronized pigments are also available, which do not require grinding as such; however, some degree of shear may be required to break down any agglomerated particles. Ready-to-use liquid pigment dispersions are also widely used. These can be added directly to the oils and waxes without pre-processing.

   An important rule to be noted is that special effect pigments, such as pearlescent pigments, should never be ground. These particles are prepared by coating the cores with one or more layers of metal oxides. By grinding them, the coat may be damaged or removed from the core, which could result in a loss of luster and change in color. They should be added when the lipstick mass has been fully mixed, and only light stirring should be used.
2. **Melting and Mixing** Since waxes are solid at room temperature, they cannot be mixed with the other components and poured into the lipstick molds. In order to make these steps possible, waxes should be heated and melted. They can be usually mixed with the oils and melted with them. To facilitate mixing, various types of mixers/dispersers can be used. When the lipstick base is melted, the pigment dispersion can be added to the melted phase and mixed until a homogeneous product is achieved. The mixture is slightly cooled and then specific pigments, preservatives, fragrances, and other ingredients can be added and mixed until homogenous.
3. **Molding** is the actual step where the melted lipstick mix is poured into plastic or metal molds. The mix should be poured while it is still hot and liquid; however, it is beneficial to let the mixture cool slightly (without starting to solidify, the mixture's temperature should be about 10 °C above its melting point). The reason for this is that as the mixture cools, its viscosity increases, which helps avoid settling of suspended pigments on the tip of the lipstick and thus assures uniform pigment distribution. In addition, this allows air entrapped in the mix to surface before molding. The slightly cooled mixture should be poured in an uninterrupted stream into a mold. If filling is interrupted, a layered stick may form, which will break apart upon use. The mix is typically poured in excess since its base (which is the top during molding) shrinks as it cools. Then the mold is placed on a cooling table or into a refrigerator until it hardens. Any excess material should be sliced from the base of the stick. When the sticks have hardened, they can be removed from the mold by slight pressure and can

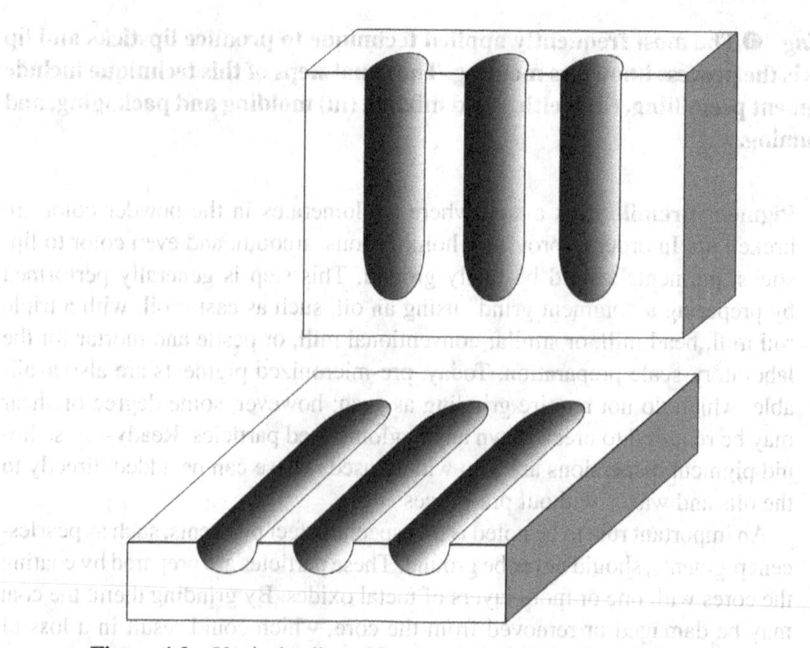

Figure 4.3 Vertical split mold commonly used for lipstick formulation.

be put into the lipstick holder. If needed, the tip of the lipstick can be cut off to provide easier application.

Most commonly, plastic or metal vertical split molds are used; for an example, see Figure 4.3. Their inner surface must be smooth to produce non-sticking sticks. If they are made of metal, they can be previously heated and used while hot. It may also be helpful to grease the mold with oil before molding.

4. **Flaming** is the last step of lipstick formulation where sticks are passed through flame, e.g., using a gas burner, to produce a glossy finish to lipsticks (see Figure 4.4). The flames are adjusted to a level hot enough to just melt the surface of the stick. Lipsticks are typically held and twisted in the flame for up to a second and then removed to avoid melting and losing shape. Finally, the caps are put on the lipstick holder and are labeled.

An important point, which should be considered during formulation, is the **minimization of air incorporation** into the mass. Air in the bulk is difficult to remove and will lead to unsightly pinholes in the stick during molding, resulting in slow production and increased rejection rate. It can be minimalized by avoiding overmixing of the mix. In addition, relatively slow but still uninterrupted filling can also help eliminate this unwanted phenomenon.

Most modern lipsticks are produced on automatic lipstick machines. One of the most common automatic machines consists of a table of metal molds, which rotates through various heating and cooling chambers. The warm molds are first lubricated

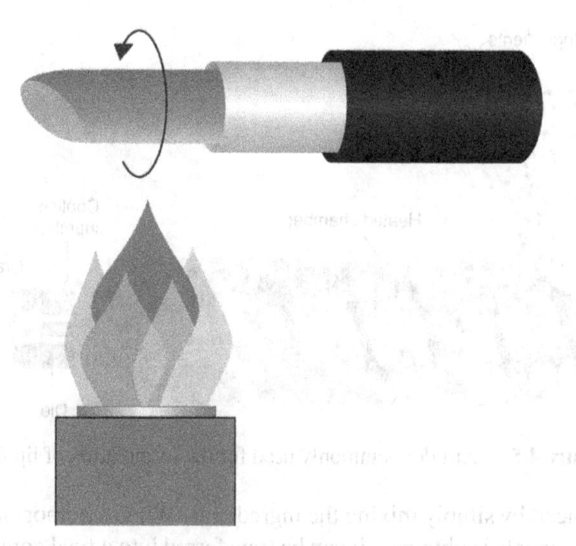

Figure 4.4 Flaming of a lipstick.

by spraying in a small amount of oil; they then are passed under the filling head where the molten lipstick is injected into them. As they move around the table, they are cooled by refrigerated compartments and, finally, ejected out of the molds by a small jet of compressed air straight into the cases. The full holders are then passed through a tunnel while spinning in front of jets of hot air, which flames the surface to give the required glossy finish.[12]

Variations for the Molding Technique ❺ **Additional techniques include molding without flaming, simple mixing without heating, as well as extrusion.**

- **Lip balms** are typically formulated using a base similar to that of lipsticks (i.e., mixture of waxes and oils). However, instead of pigments, they contain additional moisturizers, sunscreens, or other active ingredients. Lip balms in a stick form are usually formulated with the molding technique. However, the melted and homogenous mixture can be directly poured into the final lip balm containers, which may be plastic tubes, pots, or metal tins. Lip balms are typically not subject to flaming. Sometimes, the tip of the lip balms is flamed to remove any holes that may be present after molding and provide a shiny surface to the stick's tip.
- As discussed previously, **lip glosses** generally contain a smaller amount of waxes and larger amount of oils; therefore, their viscosity is significantly lower than that of sticks. If they contain waxes, melting is still necessary to mix the ingredients. Therefore, they will also be formulated via molding. As the lip gloss mixture cools down, it will not completely solidify (what would be expected for a lipstick); however, its viscosity will significantly increase. Lip glosses that do not contain waxes or other ingredients that should be melted can

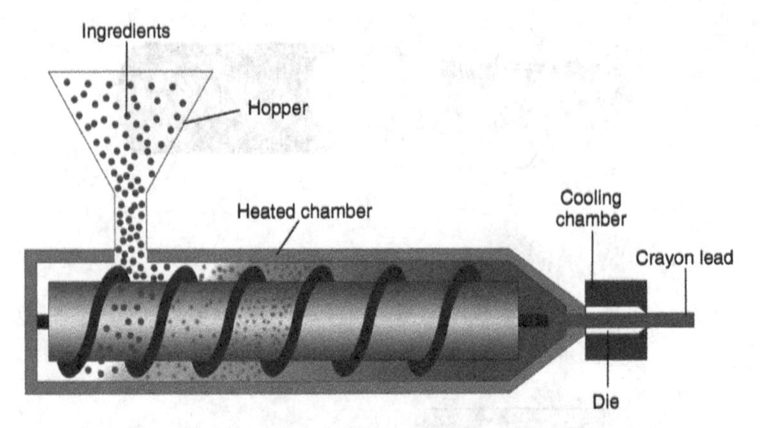

Figure 4.5 Extruder commonly used for the formulation of lip liners.

be produced by simply mixing the ingredients. When a homogenous mixture of the components is obtained, it can be transferred into a final container, typically into a vial using a pipette or syringe. Although melting is not necessary, gentle warming may be needed for easier packaging.

- **Lip liners** in the form of pencils can also be made via molding. In this case, the melted mix is injected directly into the hollow barrels using a syringe. The tip of the syringe can be stuck into one end of the pencil and the melted and still hot mix can be slowly injected into the barrel without interruption. The pencil is filled completely when the filling starts dripping at the other end of the pencil. After injecting, the pencil should be held horizontally to avoid dripping. The product can be placed into a refrigerator until the filling solidifies. Before use, the pencil should be sharpened.
- **Extrusion** Lip liners and some lip glosses are available in pencil form. These products can be formulated by a special technique called extrusion. Figure 4.5 depicts the main parts of an extruder. During this process, the lipstick mix is loaded into a hopper and is then forced through a narrow opening, called the die, which is usually round shaped. First, the crayon (i.e., the long, thin extruded, waxy material) is generally soft and flexible and must undergo a tempering phase for a couple of days before it reaches its optimum texture. The sticks are then placed in wooden slats and shaped, pointed, painted, and sharpened. Due to the heavier pigment/wax concentration of these products, the bulk is thicker and usually requires kneading to ensure proper mixing. As a general practice, a roller mill is used to mill the whole mixture. There is a rotating screw within the equipment that also helps mix the ingredients and formulate a smooth, clump-free product.

Typical Quality Problems of Lip Makeup Products

❻ The typical quality-related issues of lip care products include aeration, laddering, chipping, deformation, cratering, streaking, sweating, mushy failure, seams,

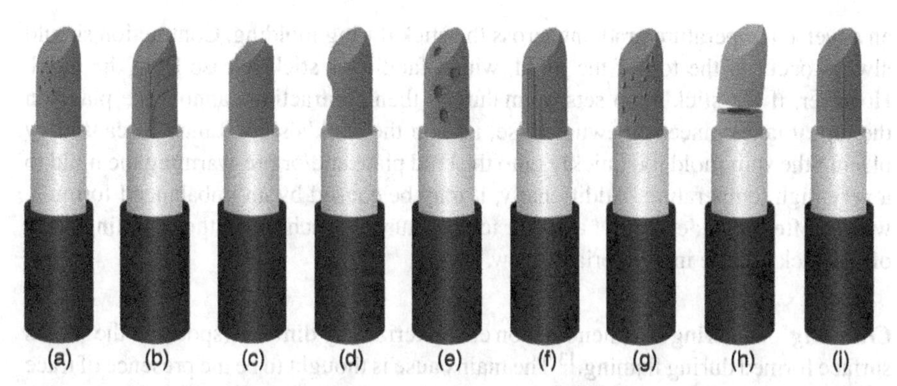

Figure 4.6 Common quality problems of lipsticks: (a) aeration, (b) laddering, (c) chipping or cracking, (d) deformation, (e) cratering, (f) streaking, (g) sweating, (h) mushy failure, and (i) seams. Modified with permission from Reference 13 of Section 1.

bleeding (Figure 4.6 depicts these problems), **and poor pay-off.** Their potential causes as well as solutions are also discussed here. As the problem of pay-off was discussed previously under, it is not reviewed here again.

Aeration Aeration, also known as pinholing, refers to the formation of tiny holes on the surface of the stick.[13] It may occur if the melted ingredients are mixed together quickly using a mixer or dispenser and are immediately molded. During mixing, air is also mixed into the melted mix, which should leave the mass prior to molding. However, if there is no time for the air to leave the melted mix; it will appear in the form of small holes within the stick. In addition, if too much oil is used to grease the mold, it may form a thin layer on the surface of the stick. When the stick is subjected to flaming, the oil on its surface immediately burns, leaving a tiny hole on the stick's surface.

Laddering Laddering of lipsticks refers to a multilayered, i.e., ladder-like appearance. This may occur due to any of the following: the mold being kept at a very low temperature, the lipstick mix is not hot enough or the filling rate is too slow. In such cases, the first layer of the mass just fills the mold's cavity and sets; the next layer comes over the top of the initial layer and sets.[13] This process is repeated until a series of layers are formed, resulting in a ladder-like appearance. This quality problem is most noticeable in pearl formulations.

Chipping or Cracking Chipping or cracking may occur if the lipstick mix is too brittle.[13] They may be caused by an imbalanced wax/oil ratio and/or a faulty cooling technique. In such cases, modification of the composition and/or cooling technique is necessary.

Deformation Lipstick deformation is a phenomenon where the shape of the lipstick looks abnormal.[13] It is more noticeable in softer formulas. Deformation is caused by an adverse temperature gradient across the stick during molding. Contraction should always occur at the top of the mold, which facilitates stick release from the mold. However, if the stick's top sets prematurely, then contraction cannot take place on the top; it takes place somewhere else, i.e., on the stick's side. It may be caused by placing the split mold too quickly onto the cold plate and/or pre-warming the mold to a very high temperature. Additionally, it may be caused by an imbalanced formula, which softens and deforms at a higher temperature. In such cases, the softening point of the stick will be inappropriately low.

Cratering Cratering is a phenomenon characterized by dimples (spots) on the stick's surface formed during flaming.[13] The main cause is thought to be the presence of trace amounts of oils either in the formula or in the formulation process. Allowing the stick to set for a few days before flaming can improve the condition.

Streaking Streaking refers to the appearance of discontinuities and streaks, i.e., a thin line or band of different color substances on the stick's surface.[13] There are a few hypotheses on the cause of the problem. These include milling of surface-treated titanium dioxide, which damages the surface and creates uneven wetting. It may lead to floating of the particles on the waxy mix's surface, which results in separation of the suspended colored particles. Another theory assumes that the particle size and wettability are the major reasons for this phenomenon.[14] An additional theory is related to the surface tension of the lipstick base. This would require reformulation of the base, which could solve the problem but might potentially generate new ones as well.[15]

Sweating Sweating of lipsticks refers to the appearance of oil droplets on the stick's surface. This may be caused by an incompatibility between individual ingredients in the formulation and imbalanced composition, such as high oil content or inferior oil-blending capacity of the wax composition. This problem is directly related to the composition and not to the manufacture and handling of the lipstick mix; therefore, the solution is to reformulate the product.[13]

Mushy Failure Mushy failure refers to a phenomenon where the central core of the stick lacks structure and breaks. This problem is usually related to the speed and/or temperature of molding.[13] If pouring of the mix is too fast and/or interrupted for several times, or happens below the optimal temperature, the lipstick mix flows down along the wall of the molding cavity and hardens at the top of the cavity, enclosing air into the middle of the stick. To avoid this problem, the mix should be molded in an uninterrupted stream with a constant speed at the proper temperature.

Seams Seams, also known as vertical marks, may form on the lipstick, mainly when split molds are used. If attachment of the two parts is not precise, a thin line may be left that can leave a mark on the sticks. To avoid such a quality problem, proper attachment should always be checked before molding.

Bleeding The term "bleeding" may refer to two things. First, it may refer to the separation of colored liquids from the waxy base, which leads to an extremely uneven color distribution. This may be a solubility issue or another type of incompatibility issue between the ingredients. Secondly and more frequently, it refers to the phenomenon when the lipstick flows into the fine lines around the lips. Usually, an unbalanced formulation is the cause of this problem, and it needs fine-tuning of the ingredients' ratio.

Evaluation of Lip Makeup Products

Quality Parameters Generally Tested ❼ **Parameters commonly tested to evaluate the quality of lip care products include break strength, hardness, melting point, softening point, color matching, color brightness and streakiness, pay-off and glide, rigidity, viscosity, and pH.** The range of acceptance and other limiting factors are usually determined by the individual manufacturers. Tests that were discussed in the previous sections are not reviewed in detail here.

Break Strength An important property of a lipstick is that it must not bend, crumble, crack, or break during application. The breaking action is induced by a hemispherical-edged blade (see Figure 4.7), which simulates the bending action

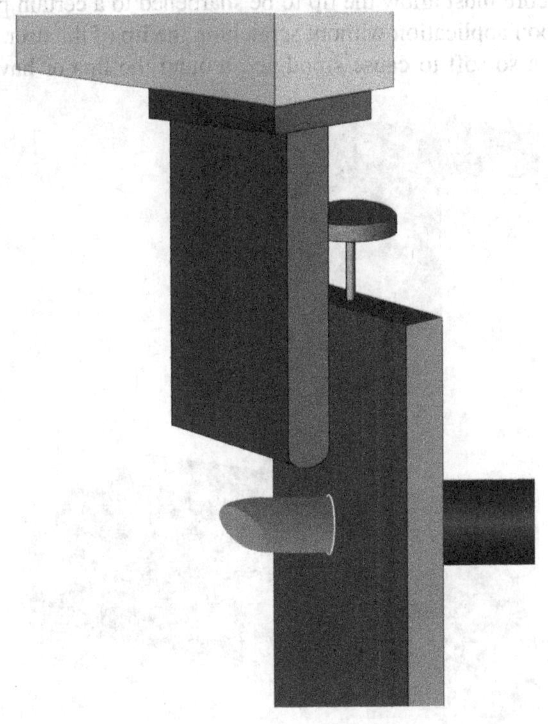

Figure 4.7 Break strength testing of a lipstick. Adapted from Texture Technologies Corp.

caused during application. The lipstick is usually held horizontally or at an angle of 45° in a socket. The weight is gradually increased by a specific value at a specific interval, and the weight at which it breaks is considered the breaking point.

Color Matching Color matching is the process when dispersed pigments and the finished product are compared to a previously approved standard color. It is essential to produce the same shade of a lipstick every time a company manufactures it. The test can be done visually or using spectrophotometers that provide quantitative information on the shade. If done visually, the testing environment and lighting should be standardized since many artificial light sources might distort the color seen.

Color Brightness and Streakiness Color brightness and streakiness of lip makeup products can be evaluated visually, and using a spectrophotometer or colorimeter. Spectrophotometers and colorimeters provide quantitative information on the color; therefore, they can identify the match with the previous batches. Color on an application surface can be evaluated by image analysis. With this technique, it is possible to study the skin areas of various sizes, take measurements without any contact with the skin, and analyze the image pixel by pixel, providing quantitative information according to localization.[16]

Rigidity of Lip Liner Pencils and Lip Gloss Pencils The rigidity of a lip liner or a lip gloss pencil core must allow the tip to be sharpened to a certain point to provide a defined line upon application without scratching the lip of the user. An ideal product should not be so soft to cause smudging around the lips or have the tendency

Figure 4.8 Rigidity testing of a lip liner pencil. Adapted from Texture Technologies Corp.

to become crumbly or brittle after prolonged or low-temperature storage. The test equipment depicted in Figure 4.8 has an arm for holding the sharpened pencil at 45°. It measures the force (load) needed to break the tip of the pencil. This technique enables to quantitatively measure the hardness/rigidity of the pencil's tip.

Efficacy Parameter Generally Tested ❽ **The most frequently tested efficacy parameter is the moisturizing effect of lip makeup products.**

Moisturizing Effect Lip balms and other lip makeup products claiming to moisturize the lips should truly provide moisturization on application. The moisturizing effectiveness of lip makeup products is usually evaluated similar to that of skin moisturizing products. Methods include noninvasive techniques, such as conductance and capacitance. Hydration state is generally measured at the beginning of the experiment to obtain a baseline value, and then it is continuously measured during the study. At the end of the study, the baseline values are compared to the in-study values and are analyzed statistically.[17]

Ingredient Causing Safety Concerns ❾ **The main ingredient that caused safety concerns regarding lipsticks is lead.** The findings and current statements on its safety are summarized here.

Lead Content Lead can be found in color additives used for lip makeup products, especially in red colors. Lead contamination of lipsticks is considered a safety issue since lipsticks may be ingested during wear and lead absorbed into the blood is highly poisonous. It can build up in the body over time and can cause learning, language, and behavioral problems. Lead contamination of lipsticks may sound scary, indeed. However, certain things should be taken into consideration. Lead has to be absorbed in order to build up and have side effects. In reality, a large portion of lead that is swallowed does not absorb into the blood, only passes through the body. The FDA limits lead in color additives to maximum specified levels, typically no more than 20 parts per million (ppm) for color additives approved for use in cosmetics. In addition, batch certification also includes testing each batch for lead.[18]

The first concerns arose with regard to the safety of lipsticks back in the 1990s. At that time, reports of analytical results from a commercial testing laboratory suggested that traces of lead in lipstick might be of concern. In 2007, additional reports were sent to the FDA regarding the lead contamination of lipsticks. FDA scientists developed and validated a highly sensitive method for the analysis of total lead content in lipsticks. They concluded that the lead levels found (0.09–3.06 ppm) were within the range that would be expected from lipsticks formulated with permitted color additives and other ingredients that had been prepared under GMP conditions.[19] The FDA did a follow-up evaluation in 2010, when it examined a wide variety of shades, prices, and manufacturers. In 2010, 400 lipsticks available on the US market were tested for total lead content. The FDA selected lipsticks based on the parent company's market share; in addition, it included some lipsticks from niche markets. The FDA found that the average lead level (1.11 ppm) was very close to the initial results.[20]

Lipstick, a product intended for topical use with limited absorption, is ingested only in very small quantities. The FDA states that lead levels found in lipstick should not be considered a safety concern.[20]

Packaging of Lip Makeup Products

The most commonly used packaging materials for lip makeup products include the following:

- *Lipstick Cases*: Classical packaging for lipsticks and lip balms includes the cases, also called tubes. The tubes have a swiveling, twisting, or pushing base that lifts the lipstick up from the hollow cylinder. Lipsticks are placed in the cases after molding and cooling, while lip balms are directly molded into the tubes.
- *Lip Gloss Tubes*: Lip glosses are usually dispensed from clear glass or plastic tubes to which is fitted a cap that incorporates a wand with an applicator attached to the bottom. Products with a lower viscosity, e.g., lip glosses, can be packaged into soft tubes that are fitted with a roller ball in the housing. Medium-viscosity formulas that are still not sticks, e.g., softer lip balms at room temperature, can be dispensed in soft tubes that are fitted with a slant-tip applicator.
- *Jars and Pots*: Alternatively, lip glosses and lip balms can be packed into small jars, pots, or thin metal tins, which are then applied with fingers.
- *Pencil Cases*: Lip liners and some lip glosses are packaged into wooden or plastic pencil cases. The plastic cases are similar to the lipstick cases; they can twist and push the lead up during application.

GLOSSARY OF TERMS FOR SECTION 1

Aeration: A quality problem characterized by the formation of tiny holes on the surface of lipsticks.

Bleeding: A quality problem leading to separation of the colored liquids from the waxy base in lipsticks.

Break strength: A measure of a solid product's ability to withstand an increasing tension without breaking.

Cracking: A quality problem resulting in chipped lipsticks.

Cratering: A quality problem characterized by the formation of dimples on the lipsticks' surface during flaming.

Deformation: A quality problem resulting in lipsticks of abnormal shape.

Extrusion: A process of forcing a mixture through a dye of the desired cross section. It can be used to formulate lip liners.

Flaming: The process of passing lipsticks through a flame to produce a glossy finish to their surface.

Laddering: A quality problem of lipsticks resulting in the formation of a multilayered, ladder-like appearance.

Lip balm: A personal care product that can temporarily prevent dryness of the lips and help relieve chapping. Lip balms can also be colored.

Lip gloss: A color cosmetic designed to give lips a glossy luster and, sometimes, a subtle color.

Lip liner: A color cosmetic designed to redefine the outline of the lips.

Lip plumping lipstick: A type of lipstick designed to make the lips appear fuller by slightly irritating the delicate skin on the lips to make them swell.

Lipstick: A color cosmetic designed to enhance the appearance of the lips by imparting color and gloss.

Mold: A hollow form used to give shape to the molten lipstick mix when it cools and hardens.

Molding: A formulation technique typically used for lipsticks and lip liners.

Mushy failure: A quality problem characterized by the formation of a hole in the center of lipsticks leading to improper hardness.

Rigidity: The measure of relative stiffness of a lip liner that allows it to resist bending, stretching, twisting, or other deformation under a load.

Seams: Vertical marks on lipsticks.

Streaking: A quality problem characterized by the formation of discontinuities and streaks on the lipsticks' surface.

Sweating: A quality problem characterized by the appearance of oil droplets on the lipsticks' surface.

Vermilion zone: A transition zone between the skin of the face and the mucous membrane of the mouth, which is the application surface for lip care products.

REVIEW QUESTIONS FOR SECTION 1

Multiple Choice Questions

1. Which of the following determine the hardness of sticks?
   a) Colors
   b) Sunscreens
   c) Fixatives
   d) Waxes

2. What is the aim of premilling during the molding process?
   a) To break up the agglomerates that may be present in the powder color additive
   b) To grind pigments, including special effect pigments
   c) To mill the waxes
   d) To mill and mix pigments with the waxes

3. Which of the following provide slip and softness to sticks?
   a) Waxes
   b) Texturizing agents
   c) Oils and fats
   d) All of the above

4. Which of the following is usually evaluated as part of the quality control tests for lipsticks?
   a) Flaming
   b) Break strength
   c) Premilling
   d) Actuation force

5. Which of the following lists the steps for the molding process in proper order?
   a) Premilling, flaming, melting and mixing, molding
   b) Flaming, molding, melting and mixing, premilling
   c) Premilling, melting and mixing, molding, flaming
   d) Molding, premilling, flaming, melting and mixing

Matching

Match the typical quality issues of lip care products listed in column A with their appropriate description in column B.

	Column A	Column B
_____	A. Bleeding	1. A lipstick flows into the fine lines around the lips.
_____	B. Cratering	2. Dimples appear on the stick's surface.
_____	C. Deformation	3. Molded products have a hole in the center and break.
_____	D. Laddering	4. Molded products have a ladder-like appearance.
_____	E. Mushy failure	5. Molded products have an abnormal shape.
_____	F. Pinholing	6. Oil droplets appear on the stick's surface.
_____	G. Seams	7. Streaks appear on the stick's surface.
_____	H. Streaking	8. Tiny holes appear on the stick's surface.
_____	I. Sweating	9. Vertical marks appear on the stick after molding.

Fill in the Blank

Fill in the blanks using the words from the box. There are extra words provided that do not have to be used.

Beeswax	Titanium dioxide	Lipsticks	Flaming
Lead	Coconut oil	Molding	Antimicrobial
Softening point	Melting point	Lip balms	Arsenic

1. __________ is often used in lip care products to provide the necessary rigidity.
2. Preservatives are incorporated into lip care products for their __________ effect.
3. UV protection can be provided with the application of __________.
4. __________ is a procedure that provides a shiny surface for lipsticks.
5. Lip care products consisting of waxes, oils, and other components applied to the lips with the purpose of protecting them from drying are referred to as __________.
6. A metal ion causing safety concern in lipstick is __________.
7. The temperature at which a solid phase (e.g., a stick) converts to a liquid phase under 1 atm pressure is referred to as the __________.

Matching

Match the ingredients in column A with their appropriate ingredient category in column B.

	Column A	Column B
_____	A. Bismuth oxychloride	1. Antioxidant
_____	B. Carnauba wax	2. Color additive
_____	C. Castor oil	3. Dispersing agent
_____	D. Menthol	4. Hardening agent
_____	E. Phenoxyethanol	5. Lip plumping ingredient
_____	F. Red iron oxide	6. Pearlescent pigment
_____	F. Silica	7. Physical UV filter
_____	G. Sodium saccharin	8. Preservative
_____	H. Vitamin E	9. Sweetener
_____	I. Zinc oxide	10. Texturizing agent

REFERENCES

1. Rohrich, R. J., Pessa, J. E.: The anatomy and clinical implications of perioral submuscular fat. *Plast Reconstr Surg*. 2009;124(1):266–271.
2. Fritsch, H., Kühnel, W.: *Color Atlas of Human Anatomy*: in 3 volumes. Internal Organs. Vol. 2, New York: Thieme, 2008: 144.
3. Zugerman, C.: The lips: anatomy and differential diagnosis. *Cutis*. 1986;38(2):116–120.
4. Mulliken, J. M., Pensler, J. M., Kozakewic, H. P. W.: The anatomy of vermilion bow in normal and cleft lip. *Plast Reconstr Surg*. 1993;92:395–404.
5. Kobayashi, H., Tagami, H.: Functional properties of the surface of the vermilion border of the lips are distinct from those of the facial skin. *Br J Dermatol*. 2005;150:563–567.
6. Stevens, J. C., Choo, K. K.: Spatial acuity of the body surface over the life span. *Somatosens Mot Res*. 1996;13:153–166.
7. Lévêque, J. L., Goubanova, E.: Influence of age on the lips and perioral skin. *Dermatology*. 2004;208:307–314.
8. CFR Title 21 Part 347.10
9. Czerninski, R., Zini, A., Sgan-Cohen, H. D.: Lip cancer: incidence, trends, histology and survival: 1970–2006. *Br J Dermatol*. 2010;162(5):1103–1109.
10. Pogoda, J. M., Preston-Martin, S.: Solar radiation, lip protection, and lip cancer risk in Los Angeles County women (California, US). *Cancer Causes Control*. 1996;7:458–463.
11. Whitley, R. J.: Herpes simplex virus infections. In: Goldman, L., Schafer, A., eds: *Cecil Medicine*, 24th Edition, Philadelphia: Saunders Elsevier, 2011.
12. Mitsui, T.: *New Cosmetic Science*, Elsevier, 1997.
13. Dweck, A. C., Burnham, C. A. M.: Moulding techniques in lipstick manufacture: a comparative evaluation. *Int J Cosm Sci*. 1980;2:143–173.
14. Dweck, A. C., Burnham, C. A. M.: Lipstick moulding techniques—comparison and statistical analysis. *Cosmet Toiletries*. 1981;96:61–72.
15. Rhein, L. D., Schlossman, M., O'Lenick, A., et al.: *Surfactants in Personal Care Products and Decorative Cosmetics*, 3rd Edition, Boca Raton: CRC Press, 2010.
16. Korichi, R., Provost, R., Heusèle, C., et al.: Quantitative assessment of properties of make-up products by video imaging: application to lipsticks. *Skin Res Technol*. 2000;6:222–229.
17. Batisse, D., Giron, F., Léveque, J. L.: Capacitance imaging of the skin surface, *Skin Res Technol*. 2006;12:99–104.
18. FDA: Lipstick and Lead: Questions and Answers, Last update: 06/05/2013, Accessed 7/2/2013 at http://www.fda.gov/cosmetics/productandingredientsafety/productinformation/ucm137224.htm#ref1
19. Hepp, N. M., Mindak, W. R., Cheng, J.: Determination of total lead in lipstick: Development and validation of a microwave-assisted digestion, inductively coupled plasma–mass spectrometric method, *J Cosmet Sci*. 2009;60:405–414.
20. Hepp, N. M.: Determination of total lead in 400 lipsticks on the U.S. market using a validated microwave-assisted digestion, inductively coupled plasma-mass spectrometric method. *J Cosmet Sci*. 2012;63(3):159–176.

SECTION 2: EYE MAKEUP PRODUCTS

LEARNING OBJECTIVES

Upon completion of this section, the reader will be able to

1. Define the following terms:

Binder	Breaking	Bulk density	Cake mascara
Cake strength	Color extension	Draize eye irritation test	Eye makeup remover
Eyeshadow	Eyebrow liner	Eyelash perming	Eyelash tinting
Eyeliner	Filler	Isotonic	Mascara
Smudging	Tapped density	Tonicity	Transfer resistant
Waterproof	Water-resistant		

2. briefly discuss the function of the human eyelids and eyelashes;
3. differentiate between a waterproof and a water-resistant mascara;
4. differentiate between a mascara, an eyeliner, and an eyeshadow;
5. list various required cosmetic qualities and characteristics that an ideal eye makeup product should possess;
6. list various required technical qualities and characteristics that an ideal eye makeup product should possess;
7. list various required qualities and characteristic that an ideal eye makeup remover should possess;
8. briefly explain why cake mascara tends to be water sensitive and smudge;
9. list typical ingredient types of liquid mascara and provide some examples for each type;
10. differentiate between a liquid eyeliner and eyeliner pencil;
11. differentiate between loose and pressed powder eyeshadows;
12. list the basic ingredient types found in powder eyeshadows;
13. briefly describe the formulation method used for pressed eyeshadows;
14. list the basic types of eye makeup removers;
15. name the basic ingredient types found in eye makeup removers;
16. briefly discuss the factors that should be taken into consideration during the formulation of eye makeup removers;
17. list some typical quality issues that may occur during the formulation of eye makeup products, and explain why they may occur;
18. list the typical quality parameters that are tested for eye makeup products, and briefly describe their method of evaluation;
19. briefly discuss the potential safety issues regarding the use of eye makeup products;

20. describe the safety test currently used to test eye makeup products;
21. list some typical containers available for eye makeup products.

KEY CONCEPTS

1. The main application surfaces for eye makeup products include the outer layer of the eyelids, eyelashes, and the base of the eyelashes.
2. There are several different types of products available on the market, including mascara, eyeliner, eyebrow liner, eyeshadow, and eye makeup remover, which can be applied to the eyelids, eyelashes, and eyebrows. All these products belong to the category of cosmetics in the US.
3. Eye makeup removers require specific precautions since the eyes and eye contours are particularly sensitive to irritations.
4. The typical quality-related issues of eye makeup products include breaking, sticking, smudging, drying up, poor pay-off, separation of emulsions, microbiological contamination, clumping, and rancidification.
5. Parameters commonly tested to evaluate the quality of eye makeup products include cake strength for pressed cakes; flow properties of powder mixtures; bulk and tapped densities of powder mixtures; compressibility of powder mixtures; glazing and pay-off; water resistance, transfer resistance, color uniformity, and dispersion of pigments; viscosity; pH; rigidity for eyeliner and eyebrow liners; color matching; preservative efficacy; as well as spreadability, texture, and firmness of creams and gels.
6. Eye cosmetics have always been considered more dangerous to the human body since they are in direct contact with the eye and the cornea.

Introduction

Today, cosmetic products for the eyes are an essential feature of women's facial makeup. These formulations are applied to the delicate skin around the eyes as well as the eyebrows and eyelashes.

This section reviews the main anatomical and physiological characteristics of the human eyelashes and eyelids. In addition, it discusses the various eye makeup products, as well as their ingredients, formulation technology, testing methods, and packaging materials. Although the purpose for the application of eye makeup removers is not to beautify users, they are also applied to the thin-skinned eye area. Therefore, they are also included in this section. Potential safety issues relating to eye makeup products as well as consumers' general requirements for eye makeup products are also discussed in this section.

Anatomy and Physiology of Human Eyelids and Eyelashes

● **The main application surfaces for eye makeup products include the outer layer of the eyelids, the eyelashes, and the base of the eyelashes.** The eyelids' function is to protect the eyeball from local injury. Additionally, they aid in regulating the light reaching the eye, in tear film maintenance, in tear flow, and in distributing the tear film over the eye's surface (called the cornea) during blinking. Regular blinking (20–30 blinks per minute) protects the eyes from drying out by evenly distributing the tear. Mechanical irritants (such as grains of sand) evoke the blink reflex.[1] Humans have upper and lower eyelids, which can be considered analogous structures; the only difference is in their muscle arrangement.[2] Figure 4.9 depicts the anatomy of human eyes. The skin of the eyelids is the thinnest of the human body (<1 mm). It is basically identical to the skin of the rest of the face. The eyelid's outer layer is in contact with the environment, while its inner layer is in direct contact with the eyeball's surface. The eyelids' outer layer consists of the eyelid skin, sweat glands, ciliary glands (i.e., modified sweat glands), sebaceous glands, muscles, and nerve endings.[3]

Beyond mere aesthetics, eyelashes help protect our vision by defending the eyes against debris and signaling the eyelids to close when something gets to their close proximity. The eyelashes are terminal hairs; the hair follicles where they grow from are located in the eyelids. Similar to other terminal hairs, eyelashes are coarser, longer, and more pigmented than vellus hairs (i.e., the fine hairs that cover most of the body of children and adults; see more detail on the various types of hair in Chapter 5). Like all hair on the human body, the eyelashes are made up of keratinized dead cells bound together as well as melanin, minerals, lipids, and a small amount of water. There are some specific properties of the eyelashes and hair follicles for eyelashes, such as they do not usually lose pigmentation and turn gray with age. Additionally, they lack arrector pili muscles, which make other types of hair stand up.[4] They are not influenced by androgens, which make some scalp hairs fall out. The eyelashes belonging to the upper lid and lower lid differ from each other. The upper lid has twice the number of follicles, resulting in 100–150 lashes in the upper lid, and there are about half as many in the lower eyelid.[5]

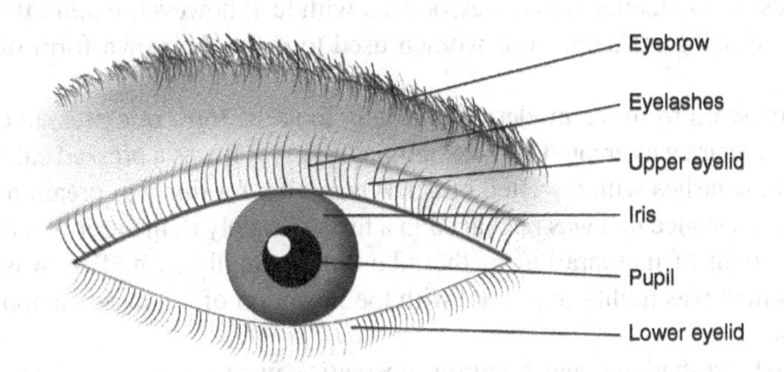

Figure 4.9 Structure of the human eye.

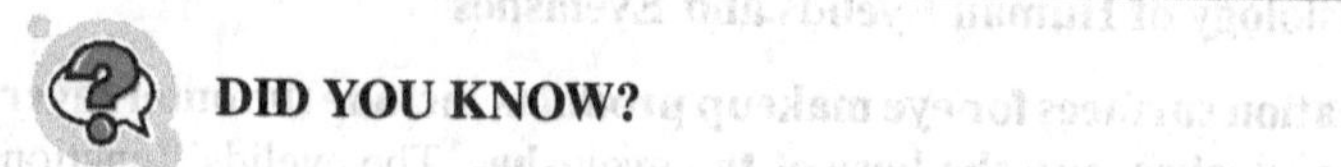

DID YOU KNOW?

The normal growth cycle for eyelashes is typically 5–6 months, in contrast to the scalp follicles (6–7 years). This is one of the reasons that scalp hair is so much longer. Additionally, eyelashes grow much slower, 0.12 mm/day on average,[6] which is about half the growth rate of scalp hair.

History of Using Eye Makeup Products

In the history of makeup use, highlighting the eyes has been controversial. It may be due to the fact that the eyes have been considered mysterious, expressive, and evocative of lust and sin in different cultures and times. In modern times, eye makeup was one of the last face cosmetic products to become popularized and mass marketed.

In ancient Egypt, eye makeup was a very important cosmetic implement. They used mainly black and green paints to darken their lashes, eyebrows, and eyelids, which were probably made of galena, lead, copper, burnt almond, ochre, crushed antimony, ash, malachite, and charcoal or soot. Apart from enhancing the appearance, kohl was also thought to be beneficial for the prevention of eye infections; it was used for religious purposes, as protection from the glare of the sun.[7] In ancient Greece, respectable women were forbidden from using cosmetics, and eye makeup was considered particularly abhorrent. The Romans were more liberal with the use of cosmetics than the Greeks; they allowed the use of eye makeup. The Iraqis believed that eyeshadow would protect them from "the evil eye." Kohl has been used extensively in India to darken the upper and the lower eyelid and even used as an eyeshadow since the Bronze Age. It was used to ward off the evil eye in infants and adults and also to protect the eyes from the harsh sun. Apart from India, kohl has also been used in places such as North Africa and Morocco.

As with lip cosmetics, eye makeup was not popular and widely accepted in Europe during the Middle Ages. Even in Elizabethan England, eye makeup was not fashionable. The fashion of eye makeup in Italy and France was significantly more elaborate. The real evolution in the history of makeup began during the early 20th century. By then, the use of cosmetics was not associated with lust; however, mainly the wealthy could afford using makeup. Poor women used to make their own form of makeup products.

Early mascara from the modern era usually took the form of a pressed cake. The first cake mascara was created by Maybelline in 1917.[8] It was a pressed cake and was applied to the lashes with a wetted brush. A variation for this was cream mascara, a lotion-like substance that was packaged in a tube. To apply it, the user would squeeze a small amount of mascara out of the tube onto a small brush. This was a messy process, which was highly improved with the invention of the mascara applicator in the 1960s.

The first eyeshadows and eyebrow cosmetics were made in the early 1930s; however, they became popular only in the 1960s. By the 1970s, waterproof mascara was marketed and became more popular than the regular ones. During the past 40 years, improvement in the variety and performance of colors as well as polymers has resulted in more durable and more appealing eye cosmetics.

Types and Definition of Eye Makeup Products

❷ **There are several different types of products available on the market, including mascara, eyeliner, eyebrow liner, eyeshadow, and eye makeup remover, which can be applied to the eyelids, eyelashes, and eyebrows. All these products belong to the category of cosmetics in the US.** They are applied to define, enhance, and emphasize the eyes or clean the sensitive eye area, which meet the definition of a cosmetic product.

- **Mascara** is designed to produce an intense look and make the eyelashes thicker, longer, and darker. It brings out the contrast between the iris (i.e., colored part of human eyes) and the sclera (i.e., the white of human eyes) and highlights and dramatizes the eyes. Mascara typically consists of a blend of waxes, pigments, texturizers, emulsifiers, and solvents. Today, it is primarily used in liquid form; however, the old-fashioned cake form is also available on the market.
- **Eyeliners** are designed to help draw a precise line at the base of the eyelashes to contour the eyes. They give the illusion of a larger or smaller eye, bringing out the contrast between the iris and the white of the eye. Eyeliners are typically made as emulsions consisting of water, texturizers, pigments, emollients, and emulsifiers. Often, polymers are also incorporated to increase the viscosity and add texture to the formula as well as increase the adhesiveness of the product to the eyelids. Eyeliners can be found in pencil, liquid, gel/crème, and cake forms.
- **Eyebrow liners** are designed to accentuate the natural line and natural hair, create fuller brows, or cover areas that have no hair. With regard to their composition and formulation, they are identical to eyeliners and lip liners; however, they are usually harder than lip liners or eyeliners.
- **Eyeshadows** are designed to add depth and dimension to the eyes and thus draw attention to the eye appearance or eye color. They are applied to the eyelids. Their predominant form is powders, both pressed and loose powders; however, they are also available in other forms such as gel, crème, and stick.
- **Eye makeup removers** are designed to remove any eye makeup from the eyelashes and eyelids. The most popular forms are creams, lotions, and remover pads. An important requirement for them is to be compatible with the delicate skin of the eyelids and eye area.

How Eye Makeup Products May Affect the Eye Area?

Women apply various types of eye makeup products every day to emphasize their eyes and draw attention to them. However, these products can cause mild to severe irritation, inflammation, and other types of unwanted reactions in consumers with sensitive skin or if applied improperly. Here is a summary of the major problems that might happen when using and removing various eye cosmetics.

- Eyelash and eyebrow **tinting** are popular treatments in all parts of the world in order to enhance the color of lashes and brows. It should be known, however, that no color additives are approved by the FDA for permanent tinting of the eyelashes and eyebrows. Dyes used to perform such treatments have been known to cause severe eye injuries, including swelling of the eyelids, inflammation of the eyelids, eye infections, and even blindness.[9] Users often use their hair colorants on their lashes and eyebrow to match the color of their eyebrow and eyelashes to their hair. However, it should be kept in mind that the FDA does not allow the use of hair dyes on body parts other than scalp hair.
- Liquid mascara can easily become **contaminated** with bacteria, yeast, and molds. Dermatologists recommend discarding mascara after 3 months of use.[10] The FDA states that users should never use saliva to moisten the dried up mascara since bacteria from the mouth may grow in the mascara and cause infection. Similarly, it warns users not to share their mascara since bacteria from another user's skin might be hazardous to them.[11]
- Eyelash **extensions** are also popular to increase the length and volume of natural lenses. Artificial lashes are typically held in place by methacrylate-based adhesives, which are part of the kit consumers can get in the stores. It should be known, however, that these glues can result in irritation, redness of the eyes, and allergic reactions. Similarly, the solvent used to remove artificial eyelashes can also cause allergic reaction in some individuals.[12] Therefore, eyelash extensions, if necessary, should be applied with caution by professionals.
- People with **contact lenses** should be careful when applying eye makeup products since contact lenses can be soiled or damaged as a result of contamination from various eye cosmetics.[13] The improper use of cosmetics can even cause damages to the eyes. The Association of Contact Lens Manufacturers and the American Optometric Association have recommendations on the safe use of eye cosmetics for people with contact lenses. They recommend that contact lenses are inserted before applying eye makeup, because it will help see what the user is putting on to the eyelashes or eyelids and will also reduce the risk of soiling the lenses. For similar reasons, they recommend that lenses be removed before removing eye makeup. They suggest staying away from eye makeup, cleansers, and other types of products, such as hand creams that can leave a greasy film on the lenses. None of these groups suggests using lash-extending mascara since the fibers may flake off and get into the eyes, causing irritation. They recommend that users avoid waterproof mascara, which is difficult to remove and may stain contact lenses.[14,15]

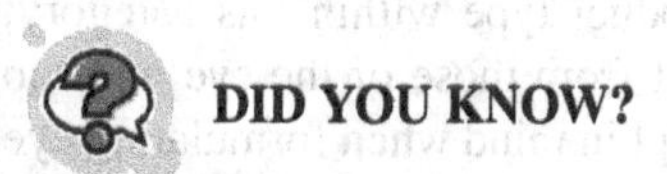

DID YOU KNOW?

Kohl is a color additive that provides a dark black color. It has been used since ancient times in various parts of the world to enhance the appearance of the eyes; however, its use is not approved for cosmetics in the US. Kohl consists of salts of heavy metals, such as antimony and lead, which are not safe ingredients. You might find eye cosmetics labeled with the word "kohl" in the US; however, it indicates only the shade and does not mean that the product actually contains this ingredient. The FDA warns people from using kohl since its use has been linked to lead poisoning in children.[16–18]

Required Qualities and Characteristics and Consumer Needs

From a consumer perspective, a quality eye makeup product should possess the following characteristics:

- Attractive shades
- Non-allergic and non-irritant
- Homogeneous color when applied
- Good coverage
- Long-lasting effect
- Easy to apply and remove without hurting the delicate skin round the eye
- Firm adherence to eyelids and eyelashes without being brittle and tacky
- Water resistance
- Quick drying after application
- Mascara: no clumping and flaking
- Eyeshadow sticks and eyeliners: good gliding, not too soft to run into the small lines around the eyes.

The technical qualities of eye makeup products can be summarized as follows:

- Fulfill long-term stability
- Dermatological safety
- Appropriate rheological properties
- High retention of color intensity without any change in shade
- Pressed powders: good pay-off
- Pressed/bulk powders: free-flowing nature to enable easy filling of the pans or godets and prevent adherence to the punches.

Eye makeup removers represent a different product type within this category; therefore, their required characteristics are different from those of the eye makeup formulations. The following properties should be kept in mind when formulating eye makeup removers:

- Good tolerance, no irritation
- pH equal to that of human tear
- Isotonicity to avoid irritation
- Sufficient cleansing power to remove eye makeup without insisting too much
- Pleasant or neutral odor
- Non-greasy
- Non-tacky
- Dermatological safety.

Typical Ingredients and Formulation of Eye Makeup Products

Mascara Colored pigments mixed with oils and waxes have been used since ancient times to enhance the appearance of the eyelashes. Today, the color, thickness, and length of the eyelashes are enhanced by using suspensions of permitted pigments in a film-forming medium to which various ingredients, e.g., lengthening ingredients, can be added.

Today, two main types of mascara are marketed: cake mascara and the more popular liquid mascara. Mascaras can also be classified as water-resistant and waterproof types. Water-resistant mascara resists smearing and smudging but is not fully waterproof, i.e., it does not entirely prevent the penetration of water. Waterproof, however, means that the formula will not smudge or smear when subjected to water (such as sweating or swimming) or tears in tests.

Cake Mascaras Cake mascara was the first type of eye makeup product to appear on the market in the 1920s and is still available today. It has a waxy texture and is applied to the eyelashes by wetting an applicator brush, rubbing it onto the cake to pick up the product, and then using the brush to transfer the product to the lashes. Most formulations of this type tend to have little water sensitivity and will smudge if the wearer cries or rubs the eyes. This is because cake mascara is based on a soap/wax/pigment blend, which is emulsified when a wet brush is applied to its surface. The proportion of these materials can be varied to control the required degree of hardness and water repellency. Waxes increase the glossy appearance of the cake, whereas soaps give it a dull appearance. An ideal formula contains a low soap and higher wax and other fatty material concentration to provide water resistance to the mascara film.

Basic ingredients for cake mascara include the following: soaps, such as glyceryl monostearate and triethanolamine stearate; emollients, such as isopropyl myristate and lanolin; waxes, such as carnauba wax and beeswax; pigments; and antioxidants.

Cakes are commonly prepared by incorporating pigments into the hot wax and soap blend and heating and mixing it thoroughly. The cold mass is transferred to a three-roll mill and ground to form a smooth homogenous mass. This mass is either compressed into metal or plastic pans or remelted and poured into warm molds.

Liquid Mascaras Water-resistant liquid mascara (or otherwise known as cream mascara) is typically formulated in the form of O/W emulsions delivering waxes, polymers, pigments, and other ingredients. It is an advanced form of the cake mascara, which already incorporates water. The amount of water is determined by the formulator and not the consumer.

Waterproof liquid mascaras are typically anhydrous formulations made in the form of via a dispersing waxes in non-aqueous solvents. These solvents provide a waterproofing effect and contribute to the quick drying properties of the formulation at the same time. They provide a long-wearing film on the lashes, which is resistant to water, smudging, and smearing. The major disadvantage of this type of mascara is that the waterproof property makes them difficult to remove from the lashes and necessitates the use of special eye makeup remover.

Liquid mascara can contain a dozen or more ingredients; however, these ingredients can be categorized into several basic groups. A summary of these typical ingredient types is provided here, including their main characteristics, functions, and some examples. The ingredient types for water-resistant and waterproof mascaras are similar; the main difference is that waterproof mascaras are based on non-aqueous solvents and anhydrous raw materials.

- **Solvents** act as a vehicle and help deliver other ingredients.
  - In water-resistant mascara, water is used as the primary solvent. Additional solvents include glycerin, propylene glycol, and sorbitol.
  - Waterproofing solvents quickly evaporate right after application and make the mascara dry quicker. Examples include hydrocarbons, such as isododecane, isoeicosane, and polyisobutene; silicones, such as cyclomethicone; and paraffin distillates, such as C8-9 isoparaffin.
- **Structurants** provide an optimal, creamy texture for the formulations that glide onto the lashes. They are also known as consistency ingredients.
  - Examples for structurizing ingredients include primarily waxes, such as beeswax, candelilla wax, carnauba wax, ozokerite wax, and cetyl alcohol.
- **Thickeners and stabilizers** increase the viscosity and provide a good texture to the formulations.
  - Examples include waxes; clays, such as bentonite clay; cellulose derivatives, such as hydroxyethyl cellulose; gums, such as gum arabic and xanthan gum; acrylates copolymers; and certain emulsifiers that have thickening properties, such as stearic acid. Additionally, talc, kaolin, silica, and starch may also be used as texturizing agents.
- **Emulsifiers** are essential ingredients in formulations. They stabilize the two immiscible parts of the formulation and prevent separation.
  - Examples include mainly nonionics, such as steareth-2, glyceryl stearate, and isoceteth 20.

- **Color additives** are very important components in every mascara formulation since consumers buy mascara to color their lashes. As mentioned in Chapter 1, not all color additives are approved for the eye area. According to the CFR, the "area of the eye" includes "the area enclosed within the circumference of the supra-orbital ridge and the infra-orbital ridge, including the eyebrow, the skin below the eyebrow, the eyelids and the eyelashes, and conjunctival sac of the eye, the eyeball, and the soft areolar tissue that lies within the perimeter of the infra-orbital ridge."[19] The most frequently used colors include black, brown, and blue.
- **Film-formers** act as adhesion promoters and enhance film formation on the lashes.
  - Examples for film-formers used in mascara include cellulosic polymers, such as hydroxyethyl cellulose; gums (see examples under thickeners); acrylates copolymers; and other types of polymers, such as polyvinylpyrrolidine (PVP, also known as polyvidone, povidone), PVP/VA (vinyl alcohol) copolymer, PVA (polyvinyl alcohol), and carboxy methyl chitosan.
- **Preservatives** provide protection against microbial contamination, especially in water-based formulations.
  - Examples include parabens, potassium sorbate, and phenoxyethanol.
- **Antioxidants** prevent rancidification of waxes and oils.
  - Examples include vitamin E, BHT, and BHA.
- **Chelating agents** contribute to product stability by catching metal ions.
  - Examples include ethylenediaminetetraacetic acid (EDTA) and its derivatives, such as disodium EDTA and tetrasodium EDTA.
- **Emollients** are typically used in combination with waxes to provide the required consistency for the formulations.
  - Examples for emollients used in mascara include jojoba oil, palm oil, castor oil, provitamin B5, and panthenol.
- **Lash-elongating synthetics** provide the illusion of false lashes. They can build on the lashes and extend beyond the natural end of the lashes.
  - Examples include rayon, silk, and nylon fibers.
- **Additional ingredients** can include hollow particles to create thicker lashes or pearlescent pigments for extra effects.

Formulation of Mascaras Formulating water-based mascaras is a general O/W emulsification process. The water-soluble thickeners are hydrated in water and neutralized if necessary. After the hydration process is complete, all water-soluble components are added to the water phase and the phase is heated. Waxes, oils, and other emollients are mixed and heated to melt the ingredients. When both phases are at the same temperature, they can be blended in smaller portions with continuous mixing. Color additives are either powders that have to be dispersed in oils, or premixed dispersions. Color additive dispersions are usually added to the emulsion in the next step. After the mix starts to cool, preservatives and other temperature-sensitive components can be added and properly mixed.

Trends in Eyelash Enhancement Longer, fuller eyelashes are frequently considered a desirable physical attribute by women. Therefore, there are a lot of products available today on the market, in addition to mascara, to enhance the appearance of the eyelashes.

- **False lashes** intensify the length of the lashes. There are products available as entire strips or individual groups to be adhered to the eyelids with a non-permanent adhesive. Mascaras with false lash ingredients, such as nylon or rayon fibers, are also marketed.
- A frequently applied technique is called **lash tinting**. This process is similar to semi-permanent hair coloring (see Chapter 5). The lash tint is mixed with hydrogen peroxide, and the mixture is then applied to the eyelashes. During the development time, some of the dye penetrates into the lashes. The effects of tinting generally last for up to 6 weeks. An important fact that should be kept in mind is that hairdressing products should not be used for tinting the eyelashes since hair dyes contain strong chemicals, which are not suitable for use in the eye area.
- **Eyelash perming** is another example of today's trends. Perming provides a long-lasting effect similar to scalp hair perming (see more details on this topic in Chapter 5). First, eyelashes are rolled around a thin cotton tube. Then they are coated with a perming lotion, which is a high-pH lotion that penetrates into the lashes and breaks disulfide bonds holding the keratin protein strands inside the lashes together. After about 10–15 min, a neutralizing product is applied to the lashes to reduce the high pH (neutralize) and reform bonds in their new position after the cotton tube is removed.

Eyeliners Eyeliners are typically applied after the eyeshadow in order to give the eyes a more striking appearance. They are generally available in pencil, liquid, gel/cream, and cake forms.

Liquid Eyeliners Liquid liners can create the most perfectly defined eye, if applied correctly, and provide longer wear than other types. They contain colorants dispersed in solvents and can be applied with a fine pen-like applicator. In the package, a wick is soaked with a liner formulation containing ultrafine pigments. This liquid is fed by capillary action to an applicator nib.

As for their dosage form, liquid eyeliners are deeply pigmented emulsions in an aqueous (water-resistant) or non-aqueous (waterproof) base, which also contains film-forming agents similar to those used in mascaras. The emulsion must be viscous enough to avoid running and should dry fast to a smooth, strong film. Therefore, polymers as thickeners are typically incorporated. They have the additional benefit of increasing the adhesiveness of the formula to the eyelids. The emulsion can be patterned after the emulsions used in mascaras.

Eyeliners contain similar types of raw materials as mascaras; however, the eyeliners are less viscous. Typical ingredients for liquid eyeliners include solvents, such as water; glycerin; hydrocarbons such as isododecane or polyisobutene; thickening agents, such as xanthan gum; surfactants/emulsifiers, such as polysorbates, polyethylene glycol, lecithin, or glyceryl monostearate; structuring agents, such as waxes; pigments; film-formers, e.g., PVP, acrylic derivatives, PVA, or PVP/PVA copolymer; and preservatives. Since these products are similar to water-resistant liquid mascaras, their formulation is also very similar. For a brief description, refer to liquid mascara.

Pencil Eyeliners (Soft Crayon Pencils) Pencil eyeliners are supplied in either wooden cases or mechanical (plastic) cases. The latter ones are usually softer and deliver the mass with minimal pressure. The leads are blends of waxes, hardened fats, oils, pigments, and pearls evenly dispersed throughout (similar to lip liners; refer to Section 1 of this chapter). Similar to lip liners, the softness of the lead can be modified by the amount and types of the ingredients used; for example, higher levels of hard, high-melting point waxes result in harder leads. The leads are typically formulated by extrusion or molding as lip pencils. An important difference that should be kept in mind is the softness of the products. Eye liners are typically softer since the skin around the eyes is more sensitive than that of the lips.

Eyebrow Liners Eyebrow liners are prepared in a range of colors, from black through brownish-black and brown to blue. The wooden-cased type or pencil type is typically fairly hard, while the plastic-cased type is softer. The composition of eyebrow pencils is similar to that of lip pencils, with a higher proportion of waxes to increase the hardness. The skin under the eyebrow is not considered a highly sensitive area; therefore, these pencils can be harder. For a detailed list of ingredients, see the previous section of this chapter about lip care products. The formulation is also identical to that of the lip liners and eyeliners (extrusion or molding).

Eyeshadows Eyeshadows are available as pressed and loose powders, creams, sticks, and pencils.

Powder Eyeshadows Loose and pressed powder eyeshadows are the most popular forms of this product category. They are typically applied to the upper eyelid by lightly stroking a soft sponge-tipped applicator or a fine brush across the skin. Pressed powders represent the "on-the-go" form of loose powders. Although they are similar with regard to their ingredients, there is a distinct difference between the two product types: pressed powders contain binders to hold the powder particles together in a pressed form. Other ingredients are quite similar; therefore, they are discussed in another section.

Both pressed and loose powders are primarily composed of powder ingredients, such as fillers, pigments, and pearls.

- **Fillers** provide a base (bulk) for the pigments. They contribute to the slip and consistency of powders. As they are typically white powders, they also help dilute the colors, making them less intensive.
  - Examples for fillers used in eyeshadows include talc, magnesium stearate, starch, bismuth oxychloride, and micas.
- **Absorbents** are typically dense powders that increase the overall density of the eyeshadow powders, making them easier to compress. In addition, they provide a matte finish surface effect to the eyeshadows as well as to the applied surface. Absorbents can be used to absorb liquids, such as fragrance prior to mixing them into the eyeshadow powder. Due to this property, they can absorb sweat and oil on the face and make the skin velvety.
  - Examples include kaolin, starch, and calcium carbonate (chalk).
- **Binders** help the cake stick together in the godet, add some water repellency to the formulas, and provide adherence to the skin. In addition, liquid binders are often used as pigment dispersing agents and emollients.
  - Binders used in eyeshadows are usually divided into the following two categories: solid and liquid. Examples for solid binders include starches. Examples for liquid binders include primarily oils such as mineral oil, isopropyl myristate, and silicone oils.
- **Colorants** As with other types of eye makeup products, the number of approved color additives for this application surface is limited. Color additives used for the eye area include mainly inorganic colorants (e.g., iron oxides and ultramarines) since most organic colors are prohibited by FDA. Special effect pigments, such as pearlescent pigments, can also be used—with restrictions—in the eye area. These make products unique and provide dramatic effects.
- **Preservatives** are added to powder eyeshadow formulations to keep microorganisms away from the products.

Formulation of Powder Eyeshadows The initial steps in the manufacture of loose and pressed eyeshadows are the same. However, the latter requires the addition of binder and compression. The basic steps involved in the formulation are the following: color extension, base powder preparation, blending, milling, filling, and sieving; for a compressed powder, compression is the last step. The actual steps depend on the ingredients; repetition of milling and blending may be necessary in order to obtain complete homogeneity.

Loose powders are identical to other bulk powders, such as talcum powders. A general consideration in the case of loose powders is the particle size and particle size distribution. In order to avoid separation over time, the particle size of the various components should be very similar. It can be achieved by grinding the ingredients prior to and during mixing them with each other. Grinding also breaks up the lumps that may be present in the powders and thus fastens the mixing process. A key stage in the processing of pigmented powder products is to achieve a homogeneous dispersion of the pigments in the mixture of other, typically white or off-white, ingredients (which are often referred to as the "base"). Homogeneity can be attained by adequate grinding and blending of the pigments with one of the base ingredients, typically talc. This process is referred to as "extension." After the pigments are properly extended, they are mixed with the rest of the white base. Fragrances, if part of the formula, are added after a homogenous mix is formed. They are typically sprayed onto the powder with continuous mixing. In the case of loose powders, they are subject to filling.

In the case of pressed powders, liquid binders are generally added after a homogenous dispersion is formed. They may be added with the fragrances as well. Mixing and grinding will be continued until the mixture is completely developed. If pearls are used, they are added and gently mixed in the last step. The powder is then ready for compression, which usually takes place in a powder press machine. The type, e.g., pneumatic, hydraulic, or ram type, and working principle, for example, whether the punch is coming down onto the powder or it remains fixed and the metal pans are pushed from beneath, may be different. Pressed eyeshadows are typically compressed directly into metal pans (called godets). However, other solutions, such as pressing them into suitably designed cavities and removing them afterwards, are also possible.

Cream and Gel Eyeshadows Eyeshadows in cream form are mainly anhydrous emulsions based on oils thickened with either clay gelling agents or waxes. Anhydrous eyeshadows are often called "cream-to-powder" eyeshadows since they glide onto the eyelids as a cream and instantly transform into a supersoft powder. Cream eyeshadows usually have high viscosity; therefore, the pigments and pearls do not sink or float. However, they can be still easily applied due to their rheology. The formulation of such products includes heating and mixing of ingredients, homogenously distributing the pigments and pearls in the hot mixture, cooling, and filling it into an appropriate container.

Cream eyeshadows can also be water-based emulsions containing oils, emollients, and thickeners to provide appropriate rheological behavior for products as well as pigments, pearls, and preservatives. Their manufacturing basically includes an emulsification process with heating if necessary.

Eyeshadows in gel form can be water-free and water-based formulations, containing appropriate solvents (either water or quickly evaporating hydrocarbons or silicone oils), thickeners, emollients, emulsifiers, preservatives, and pigments.

Eyeshadow Sticks Eyeshadow sticks are typically based on waxes, oils, and texturizing ingredients into which the colors are dispersed. They have a creamy texture and glide easily onto the eyelids. The main types of ingredients and the formulation process are similar to those of lipsticks (for more information, refer to Section 1 of this chapter). However, eyeshadow sticks are generally softer than lipsticks.

Eye Makeup Removers ● **Eye makeup removers require specific precautions since the eyes and eye contours are particularly sensitive to irritations.** Today, three main types of eye makeup removers are available on the market: cleansing milks based on O/W emulsions, cleansing waters in the form of aqueous solutions with ultramild surfactants, and cleansing pads and wipes impregnated with a makeup remover solution.

Ingredients for such products include the following:

- Solvents, such as water (in water-based formulations)
- Emollients, such as mineral oil, isodecyl oleate, and cyclohexasiloxane
- Extra mild surfactants, such as poloxamer, sulfosuccinate, and PEG-40 stearate
- Humectants, such as glycerin
- Thickeners, such as carbomers
- pH buffers, such as disodium phosphate or triethanolamine
- Preservatives, such as the parabens, DMDM hydantoin, and iodopropynyl butylcarbamate
- Chelating agents, such as calcium disodium EDTA and citric acid
- Color additives; fragrances; proteins; and natural extracts, such as aloe extract and cucumber extract.

Eye makeup removers are made according to the solution formulation or emulsification process. All hydrophilic ingredients are dispersed/dissolved in water under slow stirring; some surfactants may need to be dispersed in hot water (40 °C). Preservatives and surfactants are also added to the water phase. The oil phase is added in the last step when thorough agitation and mixing is necessary to complete the emulsification process. If necessary, pH is adjusted.

Special Considerations Considering the sensitivity and special requirements of the eye area, there are some factors that should be taken into consideration when formulating eye makeup remover products. Ideally, the pH of eye makeup removers should be equivalent to that of human tears, which is slightly alkaline, pH = 7.4 (the usual range is from 7.3 to 7.7).[1–22] Therefore, eye makeup removers should contain appropriate ingredients to establish the same pH.

Additionally, products should contain tonicity agents, if needed, to establish isotonicity with tears (i.e., the saline concentration of the product should be equal to that of the tears) to avoid discomfort and irritation. The osmotic pressure in tears is equivalent to 0.9% sodium chloride solution.[23] It should be noted that the eyes can usually tolerate solutions equivalent to 0.5–1.8%. However, the optimal value should be kept in mind when formulating such products.

Typical Quality Problems of Eye Makeup Products

❶ **The typical quality-related issues of eye makeup products include breaking, sticking, smudging, drying up, poor pay-off, separation of emulsions, microbiological contamination, clumping, and rancidification.** Their characteristics, potential causes, as well as solutions are discussed here. Quality issues that were discussed in previous sections are not reviewed here.

Breaking Breaking is a typical quality problem of pressed powders, such as pressed eyeshadows. The main cause of breaking without dropping the cake is typically air remaining in the powder mixture. If air is not removed completely during pressing, it will expand as soon as the pressing tool is released, causing the cake to break apart or break into several layers. In such cases, the pressing machine has to be adjusted to let the air escape before the compression process is complete. Another source of this problem might be the inappropriate breaking force, meaning that the cake is not firm and hard enough to withstand normal processing. If a very low force is used to press the powder into a compact form, it may result in breaking.

Sticking Sticking of the powder blend to the pressing tools may result from inadequate lubrication. If the mixture does not contain any lubricating ingredients, or does not contain them in an appropriate concentration, it may stick to the pressing tools' surface. This phenomenon can be easily avoided by properly designing the formulation. However, it should be noted that it is a relatively infrequent problem since pressed eyeshadow formulations typically contain ingredients responsible for providing slip and lubrication to the powder. Slip is one of the major requirements of eyeshadows, and the ingredients used for slip during application will help avoid sticking to the pressing tools at the same time.

Smudging Smudging of an applied eye makeup product is a typical drawback. Water-resistant products, as discussed previously, cannot completely prevent the penetration of water. However, they should not smudge after application. Smudging may be caused by high oil or glycerin concentrations. These may be overcome to some extent by basing the products with volatile silicones rather than traditional oils; therefore, once applied, the silicones evaporate, leaving a powdery product behind on the eyelids, which have a lesser tendency to smudge. On the other hand, certain types of silicones never completely dry out; thus, they may also increase the risk of smudging. Waxes may be substituted with conventional film-formers if the composition is appropriate for their incorporation. If the glycerin concentration is the cause of the problem, its quantity should be decreased.

Drying up Drying up of makeup formulations means that a hard cake forms at the bottom of the container. Quick drying of eye makeup products on the eyelashes and eyelids is a very important customer expectation; however, this property may also be one of the disadvantages if drying occurs in the container. Dried products will not be able to perform as expected, since mascara will form clumps on the lashes. If the container is left open, the solvents might easily evaporate and the product dries up. It may eventually happen during application, which is one of the reasons for the limited period of application (e.g., several months in case of mascaras).

Evaluation of Eye Makeup Products

Quality Parameters Generally Tested ● **Parameters commonly tested to evaluate the quality of eye makeup products include cake strength for pressed cakes;**

flow properties of powder mixtures; bulk and tapped densities of powder mixtures; compressibility of powder mixtures; glazing and pay-off; water resistance, transfer resistance, color uniformity, and dispersion of pigments; viscosity; pH; rigidity for eyeliner and eyebrow liners; color matching; preservative efficacy; as well as spreadability, texture, and firmness of creams and gels. Manufacturers usually set up their own range of acceptance for each characteristic. Tests that were discussed in previous sections are not discussed here.

Cake Strength Eyeshadow cakes must be compacted hard enough to have the necessary stability that prevents the cake from crumbling, flaking, and dusting during both shipment and use. This property is judged by measuring the hardness or "cake strength." There are two types of tests used to investigate this property.

- In the so-called **drop test**, the filled godet is dropped onto a standard surface (usually a wooden floor or rubber matte) from a set height to investigate its ability to withstand shock. The number of drops before breaking is counted and a decision, whether the product passes the test or not, is based on the standard acceptance range. This test simulates the accidental drop of the product by the consumer.
- Another way to measure cake strength is demonstrated in Figure 4.10. The test is called **penetration test,** and it is similar to that used for lipsticks. A needle probe

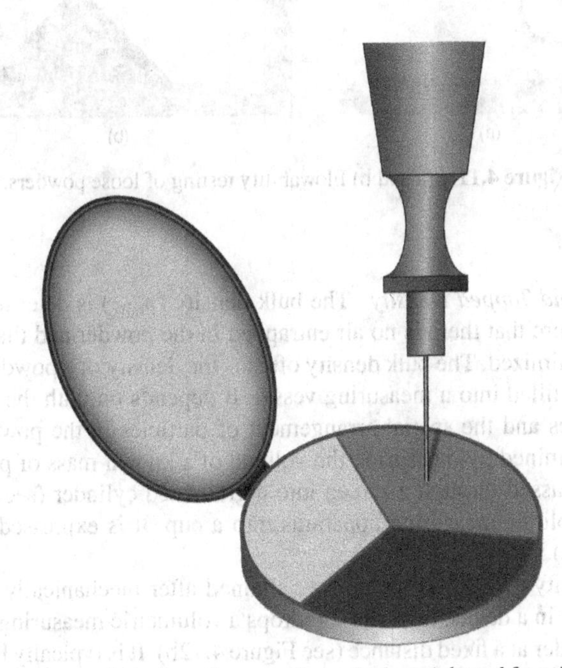

Figure 4.10 Cake strength testing of a pressed eye shadow. Adapted from Texture Technologies Corp.

penetrates into the pressed cake with a set force and the distance of penetration is registered, or the probe penetrates from a defined distance and the force of penetration is determined. This test can even indicate the presence of air pockets beneath the sample surface, which is undesirable.

Flow Properties of Powders During the filling process (either into the final container in the case of loose powders or into the feeder of the cake press machine in the case of compressed powders), the final powder formulation is volumetrically dispensed. Considering this fact, it is obvious that control of the flow and density of the product is vital to achieve high product quality. Flow properties are measured by the flow time of a given volume of powder (e.g., 100 mL) and its angle of repose (Figure 4.11): the smaller the angle, the better the flow; the larger the angle, the greater the stickiness of the powder and an increased resistance to flow.

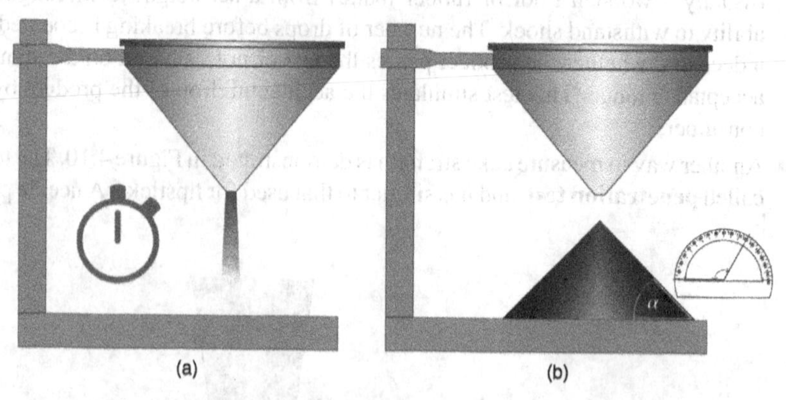

Figure 4.11 (a and b) Flowability testing of loose powders.

Bulk Density and Tapped Density The bulk density (ρ_{bulk}) is determined for loose powders to ensure that there is no air entrapped in the powder and that incorrect fill weights are minimized. The bulk density often is the density of a powder "as poured" or as passively filled into a measuring vessel. It depends on both the density of the powder particles and the spatial arrangement of particles in the powder bed. Bulk density is determined by measuring the volume of a known mass of powder sample that has been passed through a screen into a graduated cylinder (see Figure 4.12a) or through a volume-measuring apparatus into a cup. It is expressed in grams per milliliter (g/mL).[24]

Tapped density (ρ_{tapped}) is the density attained after mechanically "tapping" the powder, usually in a device that lifts and drops a volumetric measuring cylinder containing the powder at a fixed distance (see Figure 4.12b). It is typically higher than the bulk density. Tapped and bulk densities provide information on the compressibility of a powder mixture (see in the following section).

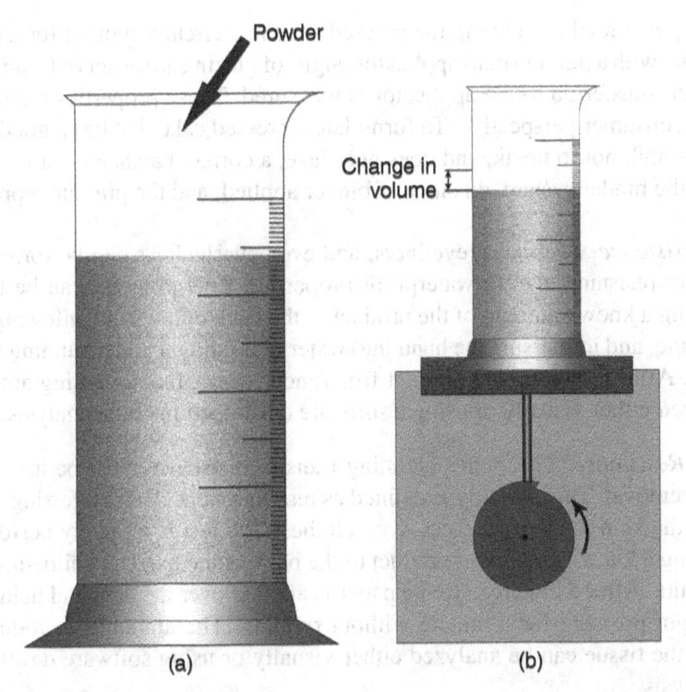

Figure 4.12 (a and b) Measuring bulk and tapped density of loose powders.

Compressibility of Powder Mixtures Interparticular interactions might significantly influence powder flow; a comparison of bulk and tapped densities can give a measure of the relative importance of these interactions in a given powder. Such a comparison can be used to predict the ability of a powder to flow. In free-flowing powders, such interactions are generally less significant, and the bulk and tapped densities will be closer in value. For poor-flowing materials, there are frequently greater interparticle interactions, and a greater difference between the bulk and tapped densities will be observed. The differences are typically reflected in the compressibility index and Hausner ratio. Smaller values refer to better flow. Higher numbers refer to a higher level of interaction between the particles. The equations used to calculate these factors are as follows:[24]

$$\text{Compressibility index} = 100 \times \left(\frac{\rho_{\text{tapped}} - \rho_{\text{bulk}}}{\rho_{\text{tapped}}} \right)$$

$$\text{Hausner ratio} = \frac{\rho_{\text{tapped}}}{\rho_{\text{bulk}}}$$

Glazing and Pay-off—Pressure Testing Pay-off refers to the weight of the material transferred to the eyelids when applying an eye shadow. Glazing refers to the appearance of dark, oily surface hardening within the pressed cake. These measurements are typically performed by rubbing the pressed cake in a circular motion for a predetermined time with a puff or small applicator. Signs of glazing are observed, and the mass of product transferred to the applicator is measured. These properties are important from the consumer perspective. To formulate a pressed cake that has a good pay-off, is hard enough not to break, and does not glaze, a correct balance must be achieved between the binder system, quantity of binder applied, and the pressure applied.

Water Resistance Mascaras, eyeliners, and even eyeshadows can be formulated to have water-resistant or even waterproof properties. This property can be tested by transferring a known amount of the product to the back of the hand, allowing it to set for a minute, and immersing the hand into water or holding it under running water for a minute. After removing the product from/under water, the remaining amount can be analyzed either visually or using a software developed for such analysis.

Transfer Resistance Cosmetics claiming transfer resistance must be able to resist abrasive removal. This property is defined as resistance against transferring the product from the skin to other surfaces, e.g., clothes. The test is typically performed by transferring a known amount of product to the back of the hand and allowing it to set for a minute. After a minute, a tissue paper is applied over the area and held in place under slight pressure for a minute without rubbing. The amount of product transferred to the tissue can be analyzed either visually or using software developed for such analysis.

Dispersion of Pigments and Color Uniformity Homogenous dispersion of the pigments is critical for high-quality formulations. Pigments are often used in a high amount for powder eyeshadows; therefore, any undispersed pigment would normally show up as streaks when applied to the skin. Color uniformity and streakiness can be evaluated visually or using a spectrophotometer or colorimeter or image analysis (similar to lip cosmetics).

Safety Concerns ● **Eye cosmetics have always been considered more dangerous to the human body since they are in direct contact with the eye and the cornea.** Microbiological contamination can occur in a number of cosmetic products; therefore, all cosmetics should be adequately preserved to prevent microbiological insults. This is especially true for eye cosmetics where the use of a contaminated product could lead to severe eye inflammations, infections, and serious complications. In the early 1930s, an untested eyelash dye was introduced into the market. It caused severe allergic reactions, even loss of vision.[25] This was one of many cases in the US that led to the FDA overseeing cosmetics. In particular, it led to a positive list of colorants that could be used for eye-area cosmetics.[26]

Eye irritation testing, therefore, is of crucial importance in the case of eye cosmetics to make sure they will not cause any allergy or other irritation to consumers. The Draize eye irritancy test, which is still used today, was developed by an American scientist in 1944 to assess eye irritation caused by various chemicals.[27] In this test, a substance is placed in one eye, with the other eye serving as the control. The eyes are evaluated after 1 h and then at 24-h intervals for up to 14 days. Sometimes, the eyes are continuously evaluated for up to 3 weeks or more. The level of irritation to the eyes is scored numerically by observation of the three major tissues of the eye (cornea, conjunctiva, and iris). This method has been criticized several times as being unreliable, subjective, and cruel to test animals. Two non-animal alternatives, i.e., the bovine corneal opacity and permeability test method and isolated chicken eye test method, have been created to allow for partial replacement of animal tests. At this time, however, none of these tests has been accepted by the scientific community as total replacement for the Draize test.[28] They can be used in the process of hazard identification, and they allow for the elimination of severe eye irritants, but fail to detect mild irritants. Therefore, live animals must still be used to confirm the negative results. Several tests, including human reconstructed tissue models, are currently under evaluation and validation.

Packaging of Eye Makeup Products

The most commonly used packaging materials for eye makeup products include the following:

Liquid Mascara Containers: Mascara is typically marketed in unique containers consisting of a slim bottle and an applicator attached to the cap. In the past, the applicator was made of nylon; however, today, a new category of mascara applicators, known as molded applicators, has been introduced into the market. The advantage of such applicators is that the mascara load on the applicator is smaller. Due to the consistent gaps between the bristles, these applicators enable the bristles to penetrate deeper into the lashes to increase mascara transfer and to achieve more efficient and regular separation of the lashes. The applicator is inserted through a small aperture (called the wiper), which is located in the neck of the container to remove the excess amount of product. The applicator and wiper must work together so that the excess mascara is removed from the applicator. However, if too much is removed, it will result in poor application. The variety of applicator colors, textures, and shapes found today is countless.

Pencil Cases: Eyeliners, eyebrows liners, and some eyeshadows are available as pencils, which can be sharpened to allow a line to be drawn. They also can be found in plastic (mechanical) pencils that can twist and push the lead up during application (as in the case of lip liners). Several types of eyebrow pencil caps have a small brush included, which provides a natural finish. There is a special pencil type: one of its ends is a pencil with which a fine line can be drawn, while the other end is a pointed rubber end, which makes smudging easy.

Liquid Eyeliner Containers: Liquid eyeliners are marketed in slim bottles supplied with a thin pointed brush. Felt-tip eyeliners are also available, which are similar to felt-tip pens but have a more precise tip.

Godets: Compressed powders and cake mascaras are usually packed into metal or plastic godets (the final package housing). They are marketed with or without an applicator included in the package. Many packaging companies also provide a mirror.

Plastic Containers with Applicator: Cream and loose powder eyeshadows are often packaged into plastic bottles that may even include an applicator. Tubes with applicators attached to the cap of the tube are also commonly used (these are very similar to lip gloss tubes).

Glass Jar: Cream and gel eyeliners are usually supplied in tiny glass jars and are applied with a brush.

GLOSSARY OF TERMS FOR SECTION 2

Binder: An ingredient added to pressed powders to help the particles stick together.

Breaking: A quality problem of pressed powders leading to crumbling and the formation of smaller pieces without dropping the cake.

Bulk density: The density of a powder "as poured" or as passively filled into a measuring vessel.

Cake mascara: Mascara in a solid form that is applied with a wet brush.

Cake strength: A measure of the resistance of a pressed product (cake) to breaking.

Color extension: The process of mixing pigments with an ingredient of the white base, typically talc, to prepare a homogenous mixture.

Draize eye irritation test: An animal test typically performed to evaluate the safety of eye makeup products.

Eye makeup remover: A personal care product designed to remove any eye makeup from the eyelashes and eyelids.

Eyeshadow: A color cosmetic designed to add depth and dimension to the eyes, thus drawing attention to the eyes or eye color.

Eyebrow liner: A color cosmetic designed to accentuate the natural line and natural hair, create fuller brows, or cover areas that have no hair.

Eyelash perming: The process of curling the eyelashes using cotton tubes and a perming lotion.

Eyelash tinting: The process of coloring the eyelashes with a dye; the color usually lasts for a few weeks.

Eyeliner: A color cosmetic designed to help draw a precise line at the base of the eyelashes to contour the eyes.

Filler: An ingredient used in loose and pressed powders to provide a bulk for the pigments.

Isotonic: A solution with the same osmotic pressure as another solution. It is often used with reference to a physiological fluid.

Mascara: A color cosmetic designed to produce an intense look and make the eyelashes thicker, longer, and darker.

Smudging: A term used to describe eye makeup products that smear and crease on wear, run into the fine lines around the eyes, and/or flake around the lashes on rubbing.

Tapped density: The density attained after mechanically tapping the powder.

Tonicity: The measure of the osmotic pressure between two solutions.

Transfer-resistant: Able to resist abrasive removal.

Waterproof: Impervious to water, a product that does smudge or smear when subjected to water.

Water-resistant: Able to resist but cannot entirely prevent the penetration of water.

REVIEW QUESTIONS FOR SECTION 2

Multiple Choice Questions

1. Which of the following is considered a drug in the US?
- a) Mascara
- b) Eye makeup remover
- c) Eye shadow
- d) None of the above

2. Which of the following is TRUE for kohl?
- a) It is an approved color additive for cosmetics in the US
- b) The term "kohl" can be used on labels to indicate the color only
- c) The term "kohl" can be used on labels to feature the main ingredient
- d) It has a deep blue color

3. What does the term "water-resistant" refer to?
- a) A product that is fully waterproof
- b) A product that does not smudge or smear when subjected to water
- c) A product that resists smudging and smearing to a certain extent, but is not fully waterproof
- d) A product that smudges and smears as soon as it gets into contact with water

4. What is the function of binders in pressed eyeshadows?
- a) To absorb sweat and oil on the face
- b) To provide a base for the pigments
- c) To help the cake stick together
- d) All of the above

5. What does the term "extending the powders" refer to regarding eyeshadows?
- a) To gradually mix the pigments with a white base powder, for example, talc
- b) To mix powders with oils to form a smooth dispersion
- c) To sieve the powder
- d) To compress the powder into the desired form

6. Which of the following factors is important to consider for eye makeup removers?
   a) pH
   b) Tonicity
   c) Mildness
   d) All of the above

7. The penetration test is usually used to evaluate the quality of ____.
   a) Eye makeup remover wipes
   b) Liquid mascaras
   c) Liquid eyeliners
   d) Pressed eyeshadows

8. What is the pH of the human tear?
   a) 2.5
   b) 4
   c) 7.4
   d) 9

9. What is the Draize eye irritation test?
   a) A test performed on human volunteers to test the safety of eye makeup products
   b) A test performed on animals to test the safety of eye makeup products
   c) A test performed on human volunteers to test the quality of eye makeup products
   d) A test performed on animals to test the quality of eye makeup products

10. Which ingredients are approved in the US to permanently tint the eyelashes?
    a) All ingredients that are approved for temporary tinting
    b) Black iron oxide only
    c) No ingredients are approved
    d) Hair colorants

Fact or Fiction?

_____ a) Since the pH of the human tear is acidic, the pH of eye makeup removers should be alkaline.
_____ b) Mascara is used to tint the eyelashes temporarily.
_____ c) Eyeliner pencils are usually softer than lip liner pencils.
_____ d) Women with contact lenses should not use any eye makeup products.

Matching

Match the ingredients in column A with their appropriate ingredient category in column B.

	Column A	Column B
_____	A. Isododecane	1. Absorbent
_____	B. Isopropyl myristate	2. Filler for loose eyeshadow
_____	C. Kaolin	3. Film-former
_____	D. Mica	4. Hydrophilic thickener
_____	E. Ozokerite wax	5. Lash-elongating synthetic
_____	F. PVP/VA	6. Liquid binder
_____	G. Rayon	7. Nonionic emulsifier
_____	H. Steareth-2	8. Solvent
_____	I. Water	9. Structurant
_____	J. Xanthan gum	10. Waterproofing solvent

REFERENCES

1. Evinger, C.: *Eyelid Anatomy and the Pathophysiology of Blinking, Encyclopedia of the Eye*, Oxford: Academic Press; 2010: 128–133.
2. Matsuo, T., Takeda, Y., Ohtsuka, A..: Stereoscopic three-dimensional images of an anatomical dissection of the eyeball and orbit for educational purposes. *Acta Med Okayama*. 2013;67(2):87–91.
3. Reid, R. R., Said, H. K., Yu, M., et al.: Revisiting upper eyelid anatomy: introduction of the septal extension. *Plast Reconstr Surg*. 2006;117(1):65–72.
4. Krstic, R. V.: *Human Microscopic Anatomy: An Atlas for Students of Medicine and Biology*, New York: Springer, 1997: 538.
5. Elder, M. J.: Anatomy and physiology of eyelash follicles and its relevance to lash ablation procedures. *J Plast Reconstr Surg*. 1997;13:21–25.
6. Thibaut, S., De Becker, E., Caisey, L., et al.: Human eyelash characterization. *Br J Dermatol*. 2010;162(2):304–310.
7. Kleiner, F. S.: *Gardner's Art through the Ages: A Global History*, Boston: Cengage Learning, 2010:56.
8. Maybelline New York: Our History, Accessed 8/19/2013 at Maybelline website http://www.maybelline.co.nz/ABOUT_US/Our_History.aspx
9. Teixeira, M., de Wachter, L., Ronsyn, E., et al.: Contact allergy to para-phenylenediamine in a permanent eyelash dye. *Contact Dermatitis*. 2006;55(2):92–94.
10. Draelos, Z. D.: Cosmetics, Accessed 9/28/2013 at Medscape website http://emedicine.medscape.com/article/1066966-overview#a30

11. FDA: Eye Cosmetic Safety, Last update: 7/10/2013, Accessed 9/28/2013 at FDA website http://www.fda.gov/Cosmetics/ProductandIngredientSafety/ProductInformation/ucm137241.htm
12. O'Donoghue, M. N. Eye cosmetics. *Dermatol Clin.* 2000;18:633–639.
13. Tlachac, C. A.: Cosmetics and contact lenses. *Optom Clin.* 1994;4:35–45.
14. The Association of Contact Lens Manufacturers, Contact Lens Guides, Lens and Makeup Care, Accessed 9/14/2013 at ACLM website http://www.aclm.org.uk/index.php?url=03_Contact_Lens_Guides/06_Makeup_Care/default.php
15. American Optometric Association: Contact Lenses and Cosmetics, Accessed 9/28/2013 at AOA website http://www.aoa.org/patients-and-public/caring-for-your-vision/contact-lenses/contact-lenses-and-cosmetics
16. Alkhawajah, A.M.: Alkohl use in Saudi Arabia: Extent of use and possible lead toxicity. *Trop Geogr Med.* 1992;44(4):373–377.
17. Nir, A., Tamir, A., Nelnik, N., et al.: Is eye cosmetic a source of lead poisoning? *Isr J Med Sci.* 1992;28(7):417–421.
18. Rahbar, M. H., White, F., Agboatwalla, M., et al.: Factors associated with elevated blood lead concentrations in children in Karachi, Pakistan. *Bull World Health Organ.* 2002;80(10):769–775.
19. CFR Title 21 Part 70.3(s)
20. Fischer, F. H., Wiederholt, M.: Human precorneal tear film pH measured by microelectrodes. *Graef Arch Clin Exp.* 1982;218(3):168–170.
21. Abelson, M. B., Udell, I. J., Weston, J. H.: Normal human tear pH by direct measurement. *Arch Ophthalmol.* 1981;99(2):301.
22. Andrés, S., García, M. L., Espina, M., et al.: Tear pH, air pollution, and contact lenses. *Am J Optom Physiol Opt.* 1988;65(8):627–631.
23. Garg, A.: Tear Film Physiology, In: Agarwal, S., Agarwal, A., Apple, D. J., eds: *Textbook of Ophthalmology*, New Delhi: Jaypee Brothers Publishers, 2002:43–44.
24. US Pharmacopoeia (USP 34) and National Formulary (NF 29), <616> Bulk density and tapped density of powders, Rockville The US Pharmacopeial Convention, Vol. 1, 2010: 241–243.
25. McCally, A. W., Farmer, A. G., Loomis, E. C.: Corneal ulceration following use of lash lure. *JAMA.* 1933;101:1560–1561.
26. CFR Title 21 Part 700, Subchapter G.
27. Draize, J. H., Woodard, G., O'Calvary, H.: Methods for the study *of* irritation and toxicity of substances applied topically to the skin and mucous membranes. *J Pharmacol Exp Ther.* 1944;82:377.
28. AAVS: Testing Alternatives, Accessed 8/31/2013 at American Anti-Vivisection Society (AAVS) website http://www.aavs.org/site/c.bkLTKfOSLhK6E/b.6457299/k.82D0/Types_of_Animal_Testing.htm#.UiIF2j-2bko

SECTION 3: FACIAL MAKEUP PRODUCTS

LEARNING OBJECTIVES

Upon completion of this section, the reader will be able to

1. Define the following terms:

BB cream	Blush	Compact powder	Concealer
Cream-to-powder foundation	Foundation	Mineral powder	Non-transfer foundation
Oil-control product	Oil-free product	Primer	Two-way foundation

2. differentiate between a foundation and a concealer;
3. differentiate between a compact and a mineral facial powder;
4. explain whether facial makeup products are considered cosmetics or OTC drug–cosmetic products in the US;
5. list various required cosmetic qualities and characteristics that an ideal facial makeup product should possess;
6. list various required technical qualities and characteristics that an ideal facial makeup product should possess;
7. list the main ingredient types typically found in powder facial makeup products, and provide some examples for each type;
8. explain the importance of using binders in facial powder formulations;
9. name the four major types of liquid and semisolid facial makeup products, and briefly discuss their main characteristics;
10. explain what the term "oil-free" refers to;
11. differentiate between oil-free and oil-control products;
12. list the main ingredient types typically found in liquid and semisolid facial makeup products, and provide some examples for each type;
13. describe the potential ways how color additives can be added to the white base of liquid and semisolid foundations;
14. differentiate between a stick foundation, a two-way foundation, and a cream-to-powder foundation;
15. list some typical quality issues that may occur during the formulation and/or use of facial makeup products and explain why they may occur;
16. list the typical quality parameters that are tested for face makeup products and briefly describe their method of evaluation;

17. briefly discuss the potential safety issues and the FDA's statement on talc used to make facial makeup products;
18. list typical containers available for facial makeup products.

KEY CONCEPTS

1. Facial makeup products are used to cover up small imperfections on the facial skin and make it more refined, finer textured, and look still natural.
2. There are several different types of facial makeup products available on the market, including foundations, concealers, and blushes. These products can be considered either cosmetics or OTC drug–cosmetic products in the US, depending on their ingredients and claims.
3. Facial powder makeup products, including facial foundation, concealer, and blush, are available as loose powders and compact powders.
4. Liquid and semisolid makeup formulations are primarily emulsions in the form of lotions, creams, and mousses.
5. Additional product types available on the market today include sticks, two-way foundations, cream-to-powder and cream-in-powder foundations, transparent facial powders, and primers.
6. The typical quality-related issues of facial makeup products include breaking, sticking, drying up, and poor pay-off for compact powders; separation for emulsions; clumping, microbiological contamination, rancidification, and quality issues specific for sticks (including sweating, streaking, laddering, deformation, mushy failure, cratering, seams, cracking or chipping, aeration, and pinholing).
7. Parameters commonly tested to evaluate the quality of facial makeup products include spreadability, extrudability, texture, and firmness of lotions, creams, and gels; break strength, hardness, and melting and softening points of sticks; cake strength of pressed powders; bulk and tapped densities of loose powders; compressibility; flow properties of powders; glazing and pay-off of pressed powders; transfer resistance; color matching; dispersion of pigments and color uniformity; preservative efficacy; viscosity; and pH.
8. The main ingredient causing safety concerns regarding facial makeup products is talc.

Introduction

❶ **Facial makeup products are used to cover up small imperfections on the facial skin and make it more refined, finer textured, and look natural.** To achieve this purpose, they unify the color of the skin; improve a dull and tired complexion; give a matte finish; and mask the dark spots, small wrinkles, dark rings under the eyes, and pores of the skin surface.

This section reviews the various facial cosmetic products available today, including facial foundation, concealer, and blush, as well as their main characteristics, typical ingredients, formulation technology, testing methods, and packaging materials. It also provides a brief overview on how these products may affect the skin and what consumers' general requirements are for such products.

Types and Definition of Facial Makeup Products

❶ There are several different types of facial makeup products available on the market, including foundations, concealers, and blushes. These products can be considered either cosmetics or OTC drug–cosmetic products in the US, depending on their ingredients and claims. The main facial makeup products include the following:

- **Facial foundations** are designed to create a uniform color, provide a basic coverage to the skin, as well as blend uneven facial color for all skin tones. They are typically applied to the total facial area. To satisfy various consumer needs, foundations can be found in a large number of types and forms. Foundations are available as liquids, which represent the most popular form of facial makeup products, as well as compact products, sticks, creams, mousse, and loose powders.
- **Concealers**, also known as color correctives, are very similar to foundations with respect to their ingredients. However, they are designed to hide minor skin problems, such as small visible blemishes, pimples, black marks, or dark circles under the eye. Concealers are much thicker and heavier in pigments and hence are more noticeable than foundations. They are primarily applied to areas of concern on the face and not the whole facial skin. They are available in various forms, including powders, creams, sticks, and liquids.
- **Blush,** also known as rouge, is designed to add color to the cheeks. However, to many customers, blush denotes a powdered product, which is more popular, while rouge denotes a creamy product. They are usually applied over a foundation, to emphasize and highlight the cheekbones and provide structure to the face. Powder blush is formulated similar to facial powder, except more vivid pigments are used in blush. Cream rouges are similar to anhydrous foundations, which contain light esters, waxes, mineral oil, titanium dioxide, and pigments.

As mentioned previously, today, many makeup products are formulated as multifunctional products, including facial makeup products. Examples for today's breakthrough formulations are BB creams and CC creams that fulfill several functions at the same time. These products are typically tinted moisturizers, i.e., mixtures of moisturizers and foundations. In addition, many of them contain sunscreens and anti-aging ingredients.

History of Using Facial Makeup Products

The history of facial makeup preparations dates back to the prehistoric times. Women in ancient Mesopotamia, as far back as 3000 BC, used cosmetics consisting of a mixture of mineral pigments based on "talak" (derived from the word "talc"). In general, it was desirable for women to have a white skin without facial problems, e.g., blemishes, since these were indications of a privileged life of leisure. Throughout history, white skin tone was often considered a sign of a life of leisure, while tan skin indicated a life of outdoor labor. People used various chemicals, many of them being highly toxic and even lethal, to cover up their skin and hide their imperfections. Priests in ancient Egypt used plaster (white powder) to cover and whiten their faces. Additional ingredients used in ancient Egypt included chalk, tin oxide, pearls, and lead carbonate mixed with fats and waxes. Similarly, a white matte facial complexion symbolized purity in ancient Greece and Rome as well. It was typically obtained through ingredients similar to those used in Egypt, such as plaster, chalk, kaolin, and lead carbonate.

During the European Middle Ages, pale skin was a sign of health. In the 6th century, women sought drastic measures to achieve that look by bleeding themselves. During the Italian Renaissance, lead paint was used to lighten the face, which was very damaging to the wearers' skin. Cosmetics were seen as a health threat in Elizabethan England, although women wore egg whites and lead over their faces for a glazed look. Lead was mixed with vinegar to form a paste called ceruse. The white lead often made the hair fall out, and its extensive use is linked to the presence of high foreheads in this era.[1] During the French Restoration in the 18th century, red rouge and lipstick were used to give the impression of a healthy, fun-loving spirit. During the Regency era (1810–1820s), the most important item was rouge, which was used by almost everyone. In order to maintain a pale complexion, which was again indicative of a life of leisure, women wore bonnets, carried parasols, and covered all visible parts of their bodies with whiteners and blemish removers. Unfortunately, more than a few of these remedies were lethal. Victorians abhorred makeup and associated its use with prostitutes and actresses.

The real evolution actually began during the 1910s. In 1914, Max Factor introduced its pancake makeup. Pancake makeup, a mixture of stearate, lanolin, and dry powders, was not easy to apply. Pressed powder blush followed soon after. Although women were more inclined to wear cosmetics, makeup was still not a part of everyday life. In the 1950s, a real boom occurred in the number of products on the market. Compact makeup was made available in a cream-based form; foundation became a fluid cream. Today's trend seems to have reverted to the more natural look with a blend of styles from the past.

How Facial Makeup Products May Affect the Skin?

Today, facial cosmetics are part of most women's daily routine and are applied every day to provide an even color to the face and cover up skin problems. When applied properly, facial cosmetics can significantly improve women's self-confidence and quality of life. This is especially true for situations when these products are used to cover otherwise visible skin problems, such as acne, rosacea, scars, or even pigmentation problems.

- **Acne** is a common dermatological condition, affecting mainly the face, neck, back, and chest. As discussed previously, in acne, skin pores become clogged with overproduced cells and sebum in which bacteria begin to grow, causing various symptoms.[2,3] Acne can significantly affect patients' quality of life if left untreated. Many dermatologists consider makeup as an exacerbating factor for acne. In spite of this, many women being treated for acne continue to use makeup. Recent studies have shown that decorative cosmetics have a favorable effect on the quality of life and interpersonal relationships of female acne patients when skin care and makeup products are used as recommended by their dermatologist.[4–6]
- **Rosacea**, another common disease, mainly affects the face, including the cheeks, nose, chin, and forehead. It is characterized with redness, or, in more severe cases, bumps and pimples.[7] It has also been shown that decorative cosmetics can be used to conceal the effects and symptoms of these diseases and improve the patients' quality of life. Therefore, such products can be essential parts of the treatment.[4] For rosacea, green makeup and green-tinted foundations are recommended since green and red are complementary colors.

FYI

The National Rosacea Society has detailed recommendations on the use of makeup in rosacea. For more information, visit its website.

- The term "acne cosmetica" was developed in the early 1970s to describe the association between cosmetic use and acne breakouts.[8] Many ingredients were and still are considered to be "acnegenic" (i.e., able to trigger acne) or "comedogenic" (i.e., leads to clogged pores), including surfactants, such as sodium lauryl sulfate; emollients, such as isopropyl myristate and octyl palmitate; and occlusive ingredients, such as mineral oil and lanoline.[9] It should be noted that many of these ingredients are found in almost all types of facial products, since they enhance the products' aesthetics and increase their performance. Due to the general belief that these ingredients trigger or worsen acne, patients with acne-prone skin tend to look for gentle products on the market and often end up buying products labeled as "non-comedogenic," "oil-free," and "hypoallergenic." As discussed in Chapter 1, the term "hypoallergenic" has no accepted definition, and it is the same in the case of the other two terms. Recent studies in humans have demonstrated that some cosmetics may cause acne in human; however, not as many as were thought in the past. Studies also concluded that the concept "acne cosmetica" is outdated.[9] It has also been shown that cosmetics containing ingredients considered comedogenic are not necessarily comedogenic.[10] Additionally, it has been proven that ingredients thought to be potentially comedogenic, including petrolatum and mineral oil, are not comedogenic in humans.[11,12]

- Multifunctional makeup products, such as **BB creams** and **CC creams,** are one of the most successful developments today, combining cosmetic and skin care benefits to cover blemishes on the face. BB originally stood for "blemish balm;" however, today, companies create their own meaning for this abbreviation, including "beauty balm" and "beauty benefit crème."[13] CC stands for "color correction" or "color control." These products combine facial foundations, serum, moisturizing and nourishing ingredients, sunscreens, and, in many cases, anti-aging ingredients in a single formulation.[14]

Required Qualities and Characteristics and Consumer Needs

From a consumer perspective, a quality facial makeup product should possess the following characteristics:

- Pleasant and easy application
- Soft and comfortable
- Guaranteed evenness and concealment of flaws
- Good coverage
- Hides wrinkles and pores
- Consistent color and matt and smooth finish
- Non-settling, non-tacky, non-greasy
- Improves appearance, but not artificially
- Long-lasting effect
- Non-transfer properties
- Provides a final homogenous and unifying film
- Moderately quick drying to allow an even application
- Pleasant, neutral odor.

The technical qualities of facial makeup products can be summarized as follows:

- Long-term stability
- Good adhesive property
- Dermatological safety
- Resistance to heat and humidity
- Appropriate rheological properties
- Appropriate texture
- Good performance (good pay-off for pressed powders).

Typical Ingredients and Formulation of Facial Makeup Products

Facial makeup products can be classified in several ways, based on their color, e.g., brown tones versus pink tones; the application surface, such as the whole facial area versus spots on the face and cheekbones; as well as dosage forms, such as sticks, lotions, creams, and powders. From a formulation standpoint, classification based on dosage form is the most reasonable, since the ingredients and formulation technology are similar for the same dosage forms. Therefore, facial makeup products are discussed in this context in the following section.

Powder Makeup Products for the Face

❸ **Facial powder makeup products, including facial foundation, concealer, and blush, are available as loose powders and compact powders.** These products are identical to powder eye shadows in terms of the ingredient types and formulation technology. Liquid foundations have largely replaced facial powders. However, for those who wish sheer coverage with excellent oil control, facial powders perform excellently. They may also be used over a liquid foundation to provide a professional finish.

DID YOU KNOW?

The commonly used term "mineral powder" refers to facial foundations in loose powder form.

As with eyeshadows, pressed (compact) facial powder (otherwise known as pancake foundation) represents the "on-the-go" form of loose powders. Pressed powders typically contain binders to hold the particles together.

Ingredients The basic components of powder facial makeup products are similar; therefore, all of them are discussed in a single section. The main difference is in the applied percentage of each component and the shade of the pigments obtained within each product.

- **Fillers** provide a base (bulk) for the pigments. They contribute to the slip and consistency of powders. Since they are typically white powders, they also help dilute the colors, making them less intensive. Fillers, especially talc, are often used as extenders for colors. In addition, they may also increase the adhesiveness of facial powders to the skin.
  - Examples for fillers used in facial powders include talc, magnesium stearate, starch (rice starch, corn starch, and, today, modified starches), boron nitrides, bismuth oxychloride, and micas.

DID YOU KNOW?

Starch is often used in cosmetic powders since it provides a peach-like bloom and a smooth surface on the skin. However, its major disadvantages are that it tends to cake when moistened, and when wet, it provides an optimal environment for bacterial growth.

- **Absorbents** are typically dense powders that increase the overall density of facial powders. They also improve the aesthetics of the formulation and can absorb excess oil from the skin. Due to this property, they are often used in oil-control formulations. Additionally, they impart a certain degree of skin adhesion to the finished product.
  - Examples for absorbents used in facial powders include kaolin, starch, and calcium carbonate (chalk).
- **Binders** help the cake stick together, provide some water repellency to the formulas, and provide adherence to the skin. They are used in pressed powders.
  - Binders used in facial powders are usually divided into the following two categories: solid and liquid. Examples for solid binders include zinc and magnesium stearate and starches. Examples for liquid binders include primarily oils such as mineral oil, isopropyl myristate, and silicone oils.
- **Colorants** Mainly inorganic pigments are used, including iron oxides (yellow, red, and black), ultramarine (blue), and chrome hydrate and chrome oxide (green). These inorganic pigments are dull in hue. In addition to inorganic pigments, colored and uncolored pearls can also be added to provide a lustrous effect to the skin. For improved pigment dispersion and formula stability, treated or coated pigments can be used. For example, particles can be coated with silicone to provide slip and allow for easier dispersion. Coating is beneficial since they eliminate the need to mill pigments, which is commonly needed for uncoated pigments to break up the agglomerates. Titanium dioxide and zinc oxide are also often used in facial powders, but not for their color. These are white powders and are utilized in such formulations as covering agents. In addition, they may also be employed as sunscreen agents to provide protection against the damaging effect of UVA and UVB rays. In the case of blush, coverage is not a priority; therefore, it typically does not contain titanium dioxide or zinc oxide.
- **Preservatives** Due to the types of ingredients used in powder systems, microbiological contamination will generally not occur in them. However, certain ingredients, such as starches, unfortunately, are ideal media for microbiological growth. Thus, preservatives are often used in powders as well.
  - Most often, phenoxyethanol is used in facial powders as a preservative.

- **Antioxidants** The use of antioxidants may also be required to protect some of the ingredients from degradation and, consequently, rancidity.

  Low levels of BHA, BHT, or vitamin E can be used when necessary.

- **Additional Ingredients** Fragrances can also be incorporated in facial powders. Absorbents can be used to absorb these ingredients before mixing them with the other ingredients in the formulations. Polymers, such as nylon, polyethylene, polypropylene, silica beads, silicone powders, Teflon, microcrystalline cellulose, and acrylates copolymers, can be utilized in facial powders as texture enhancers. They can help other ingredients mix more uniformly, provide additional slip, and improve transfer resistance and oil resistance as well.

Formulation of Facial Powders The manufacturing process of facial powders is identical to that of powder eyeshadows. The basic steps involved are the following: color extension, base powder preparation, blending, milling, and filling, and if it is a compressed powder, the last step is compression. Again, two important considerations are the particle size and the air remaining in the powder mixture.

Liquid and Semisolid Makeup Products for the Face ● **Liquid and semisolid makeup formulations are primarily emulsions in the form of lotions, creams, and mousses.** They can be oil-based, water-based, oil-free, or water-free formulations. The main characteristics of these formulations are discussed here.

- **Oil-based** formulations are W/O emulsions containing suspended pigments. They also contain emollients, occlusives, water, and emulsifiers and may include additional components, such as vitamins, sunscreens, and moisturizing agents. These foundations are easy to apply since the external phase is the oil phase, and due to this, the pigment can be spread over the face for up to 5 min prior to setting. Water evaporation from the foundation just after application leaves the pigments in the oil on the face. This creates a moist skin feel, which is especially desirable in patients with a dry complexion. Oil-based foundations do not shift color as they mix with sebum since the color is fully developed in the oily phase of the formulation. Their main disadvantage results from the type of the emulsion. They may feel greasy and heavy if not used by consumers with dry skin.
- **Water-based** products are typically O/W emulsions containing a small amount of oil in which pigments are suspended with a relatively large quantity of water (usually 50–60%). These foundations are appropriate for minimally dry to normal skin. Since pigments are already developed in the oil, this foundation type is not subject to color drift either. The application time is shorter than that of oil-based foundations due to the lower oil content.
- **Oil-free** formulations typically do not contain animal, vegetable, or mineral oils, as their name implies. Instead of oils, they are generally based on silicones, such as dimethicone or cyclomethicone. Thus, these products are water-in-silicone (W/Si) emulsions. Consumers with oily skin or with skin

problems, such as acne, often look for oil-free formulations, not to trigger further oiliness and clogging of skin pores. However, the definition of "oil-free" is more complicated than it first seems. Oil-free formulations typically do not contain any ingredients whose name contains the actual word "oil," such as mineral oil. However, there are plenty of ingredients, including emollients and occlusives, that do not have the literal word "oil" in their name, such as isopropyl myristate or lanolin. Sometimes, they are formulated into oil-free products. Due to their oily nature, however, they should not be included in oil-free products. Since there is no rule for the use of this term, the skin care industry is not always strict in avoiding these ingredients in oil-free formulations.

Silicone-based foundations are usually designed for individuals with oily complexion, as they leave the skin with a dry feeling. Silicone is non-comedogenic (i.e., will not trigger the formation of clogged pores in the skin),[15] non-acnegenic (i.e., will not trigger acne development), and hypoallergenic, accounting for a tremendous popularity of this type of facial foundation formulation. W/Si foundations go on smoothly and dry fast; therefore, they must be blended quickly for even coverage. Pigments are typically dispersed in the silicone phase; therefore, standard, oil-dispersible pigments may not be suitable for this emulsion type. It is recommended to use silicone-treated pigments. This type of emulsions usually contains an electrolyte, e.g., sodium citrate, magnesium sulfate, sodium chloride, or sodium tetraborate, in approximately 1–2% concentration. It has been shown that these salts tend to further reduce the interfacial tension that helps stabilize the emulsion.[16]

DID YOU KNOW?

The terms "oil-free" and "oil-control" are frequently used claims for liquid foundations. They may sound alike; however, they do not refer to the same properties. Oil-control formulas are designed to reduce shininess of oily face; they contain oil-absorbing ingredients, such as talc, kaolin, and starch. However, they may contain silicones and oils as well. In contrast, oil-free foundations are based on silicones and do not contain any oil. Therefore, an oil-control formulation is not the same as an oil-free formulation.

- **Water-free** (i.e., anhydrous) products offer waterproof properties. These consist of oils, such as vegetable oil; hydrocarbons, such as mineral oil; synthetic esters; silicones, such as dimethicone; as well as waxes into which higher concentrations of pigments can be mixed. They are generally more opaque compared to the other types. Water-free foundations and concealers are well suited for use in patients with facial scarring who desire camouflaging. The main disadvantage of this type of foundation is that it cannot be easily removed; consumers need special makeup remover.

Ingredients The basic ingredient types for liquid and semisolid formulations are the same. The main difference between these product types is in the applied amount of ingredients. Liquid and semisolid foundation, blush, and concealer all consist of a powder dispersion and an emulsion base containing the rest of the ingredients. A summary of the main components is provided here.

The powder dispersion typically consists of pigments and extenders for the pigments.

- **Fillers** act as matting and texturizing agents, and they are used to extend and fully develop colors. They can act as absorbents (for sebum and sweat) so as to make the skin velvety and fix the color of the skin and also contribute to the spreadability of the product.
  - Examples include talc, magnesium stearate, starches, mica, and bismuth oxychloride. Sometimes, they are superseded by different varieties of silica, polymers (such as nylon), and Teflon.
- **Pigments** The same types of pigments are used as in facial powders, including iron oxides, ultramarine, chrome hydrate, and chrome oxide, as well as titanium dioxide and zinc dioxide. Blush formulas more likely contain pink-purple or brown-orange shades.

The emulsion base contains the rest of the ingredients, including the following:

- **Emollients** serve as the oil phase in emulsions. They provide a slippery and soft texture and also have a moisturizing effect. Waxes function as structuring and thickening agents as well as emulsifiers in some formulations. Over the years, silicones have been introduced, and they overtook the role of many conventional oils in facial foundations. As volatile silicones evaporate, the tinted film concentrates on the skin. Adhering while drying on the skin surface, the tinted film withstands friction and does not stain clothes. These are the so-called non-transfer foundations.
  - Examples for emollients used in liquid and semisolid facial makeup products include waxes, such as beeswax; vegetable oils; hydrocarbons, such as mineral oil and isoeicosane; synthetic esters of fatty alcohols and fatty acids, such as isopropyl palmitate and glyceryl stearate; and silicones, such as cyclomethicone, cyclopentasiloxane, and dimethicone.
- **Water** is a basic component of water-based emulsions, serving as a vehicle.
- **Emulsifiers** stabilize the two immiscible parts of the formulation to prevent separation. Emulsifiers with low HLB value can also act as pigment-wetting agents.
  - Examples for emulsifiers used in foundations and blushes include anionic emulsifiers, such as stearic acid, and nonionic emulsifiers, such as propylene glycol stearate, sorbitan sesquioleate, sorbitan laurate, and polysorbate 20, among others.

- **Thickening agents** increase the viscosity of the formulations. Pigments have to be uniformly suspended in the emulsion and should not sediment over time, and thickeners contribute to this.
  - Examples include hectorite, cellulose derivatives, gums, and acrylate copolymers. Oil-phase thickeners are also used, especially for W/O emulsions. These include long-chain wax esters.
- **Preservatives** Water-based emulsions have a high susceptibility for microbiological contamination; thus, the application of preservatives is essential.
- **Antioxidants** prevent rancidity of the oily components.
- **Chelating agents** bind to metal ions and block the pro-oxidant action of metal ions, e.g., iron oxides in suspension. In addition to the traditional chelating agent, EDTA and its derivatives, polyphosphonic acids have also shown the ability to improve dispersion and aid in pigment wetting.
- **Additional ingredients** can also be incorporated into foundations and blushes for additional benefits, which can include antiaging ingredients, such as antioxidants, botanical extracts, and proteins; sunscreens; fragrance; and humectants. Since humectants are hygroscopic ingredients, they can control the rate of water evaporation from the skin's surface, letting the foundation shade onto the skin smoothly. Additionally, film-formers may be necessary to provide adequate water resistance and long-lasting wear for formulations. Polyacrylate derivatives are often used due to their advantageous properties, e.g., nontacky and wear resistant.

Formulation of Liquid and Semisolid Facial Makeup Products Since liquid and semisolid facial makeup products are emulsions (O/W, W/O, and W/Si emulsions), their manufacturing is an emulsification process. Pre-milled pigments are typically dispersed with an oil and are then added to the oil phase. Even with maximum process accuracy, shade correction is frequently needed.

Coloration of the emulsion base may be handled in different ways, such as using direct pigments, pigment blend, or monochromatic color solutions.[17] Each method has its advantages and disadvantages.

- In the **direct pigment** method, pigments are weighed directly into one of the phases and dispersed using a mill, such as a colloid mill. The emulsion is then formed in the usual manner. The major problem is that there are too many color adjustments needed and accurate color matching is very difficult.
- When using **pigment blends**, pigments are premixed and pulverized with the extender(s) and the color of the blend is matched to a standard. The pigment blend is then dispersed in one of the phases of the emulsion, and the emulsion is formed as usual. The finished shade is color-matched at the blending stage. This reduces the number of color corrections needed. However, it takes extra time to prepare making the dispersions.
- **Monochromatic color solutions** are color concentrates of individual pigments. These solutions are made in a ready-to-use, finished form. It is easy to color

match using monochromatic solutions; however, large storage space is needed, and the possibility for contamination is increased.

Other Facial Makeup Products ❸ **Additional product types available on the market today include sticks, two-way foundations, cream-to-powder and cream-in-powder foundations, transparent facial powders, and primers.** These product types are discussed in this section.

- **Stick** foundations, concealers, and blushes are molded foundations, similar to lip liners, but are typically larger. They are anhydrous containing a higher proportion of low-viscosity oils and solid esters and waxes. Sticks are formulated by the molding process, which was discussed in detail under lip care products (see Section 1 of this chapter). However, there is a significant difference between the formulations of lip balms and foundation sticks since facial makeup formulations usually contain extenders instead of oils to provide homogenous pigment dispersion.
- **Two-way foundations** are a type of compact powder that can be applied to the skin by the use of either a wet or a dry sponge. The overall function is to provide a natural-looking smooth finish. In many ways, they combine the properties of a cream foundation with that of a face powder. Using it dry gives a make-up similar to a facial powder, whereas using it wet gives a more even coverage. This dual use requires the majority of ingredients to be hydrophobic; so, the formulation will not clump and cake when wet. Mostly treated pigments, mainly those treated with silicones, are used.
- **Cream-to-powder** foundations are supplied in a compact form. They are typically applied with a sponge. As their name implies, they transform from a velvety cream to a microfine powder when they come in contact with the skin.
- A new type of product on the market is "cream-in-powder" (or powder-to-cream) formulation, which is basically a powder that is turned into a cream during application. It is a loose powder that contains a high proportion of liquid phase. These products are based on the combination of a treated pigment, a cosmetic powder, and a liquid phase. Thickeners also play an important role in their formulation.[18]
- **Transparent** facial powders (also known as finishing powders) are also popular today. They provide a sheer coverage and provide oil-blotting properties of a previously applied foundation. They typically contain less pigment as color, and coverage are not a priority. They can also be used for touch-up during the day.
- Makeup **primers** are either colorless or slightly tinted semisolid formulations. They are used to smoothen the skin surface. They also prevent the facial makeup from feathering and creasing. They are typically silicone-based formulations that evaporate very quickly after application and leave a dry finish on the skin. Green primers are typically used to cancel redness, while purple tones are for yellow and sallow tones. Peach and salmon colors cancel the blue color under the eye.

Typical Quality Problems of Facial Makeup Products

The typical quality-related issues of facial makeup products include breaking, sticking, drying up, poor pay-off for compact powders, separation for emulsions, clumping, microbiological contamination, rancidification, as well as the quality issues specific for sticks (including sweating, streaking, laddering, deformation, mushy failure, cratering, seams, cracking or chipping, aeration, and pinholing). Since these quality issues were discussed in the previous sections, they are not reviewed here in detail.

Evaluation of Facial Makeup Products

Quality Parameters Generally Tested **Parameters commonly tested to evaluate the quality of facial makeup products include spreadability, extrudability, texture, and firmness of lotions, creams, and gels; break strength, hardness, and melting and softening points of sticks; cake strength of pressed powders; bulk and tapped densities of loose powders; compressibility and flow properties of powders; glazing and pay-off of pressed powders; transfer resistance; color matching; dispersion of pigments and color uniformity; preservative efficacy; viscosity; and pH. The** range of acceptance and other limiting factors are usually determined by individual manufacturers. Since these tests were discussed in previous sections, they are not reviewed here in detail.

DID YOU KNOW?

The performance of colored facial makeup products should always be assessed when applied to the skin. The inner forearm is often used by formulators for this test. The reason for this is that the color of a thin film of pigment may be different from the effect given by the product when viewed in bulk. The thin film color effect is known as the **undertone,** and the bulk effect as the **mass** tone. The color of a product on the skin depends on the opacity of the white and colored pigments used, their particle size, degree of dispersion, thickness of the applied film, moisture content, oiliness of the application surface, and the skin color.

Ingredients Causing Safety Concerns **The main ingredient causing safety concerns regarding facial makeup products is talc.** The findings and current statements on their safety are summarized here.

Talc Based on the scientific literature, going back to the 1960s, it was believed that talc can cause cancer. Talc is a naturally occurring mineral; chemically, it is a hydrous magnesium silicate. There are many grades of talc, each of which is categorized according to its level of purity. At the top of this purity scale is cosmetic-grade talc. Only talc that meets very high levels of quality and purity is permitted for use in cosmetics. It should be noted, however, that it is not the same as industrial talc, which may frequently contain impurities. A major impurity that caused most safety concerns is asbestos. It is also a naturally occurring silicate mineral, but with a different crystal structure. Unlike talc, however, asbestos is a known carcinogen. For this reason, the FDA considers it unacceptable for cosmetic talc to be contaminated with asbestos.[19]

The use and safety of talc as a cosmetic ingredient have been evaluated numerous times in the past. A few highlights summarized are as follows:

- In 1994, the FDA and the International Society of Regulatory Toxicology and Pharmacology (ISRTP) held an open workshop to review all the available data on talc safety. They concluded that no health hazards had been demonstrated in connection with the normal use of cosmetic talc.[20,21]
- In 2004, the National Toxicology Program (NTP) in the US considered listing talc in its report on carcinogens after a review in 2000 found considerable confusion over the mineral nature and consequences of exposure to talc. In October 2005, however, the NTP stated that the existing scientific data were insufficient to identify talc as a cancer-causing agent and withdrew it from their review process.[22]
- Because safety questions about the possible presence of asbestos in talc are raised periodically, the FDA conducted an exploratory survey in 2009–2010 of the cosmetic-grade raw material talc, as well as some of the cosmetic products containing talc, marketed at that time. They analyzed 4 cosmetic-grade talc and 34 cosmetic products containing talc. The survey found no asbestos fibers or structures in any of the samples of cosmetic-grade raw material talc or cosmetic products containing talc.[21]
- The CIR Expert Panel released its report on the safety assessment of talc used in cosmetics in April 2013. It concluded that it is safe in the current practices of use and concentration; talc is reported to be used at up to 100% in cosmetics.[23]

To date, it can be stated that talc is safe to be used in cosmetics.

Packaging of Facial Makeup Products

The most commonly used packaging materials for facial makeup products include the following:

Bottles and Soft Tubes: The majority of liquid foundations are available in glass or plastic bottles or in soft tubes. Bottles often have a pump head to provide convenient dispensing. In addition, it reduces the chance of microbial contamination. Liquid foundations are usually applied with the fingers, a brush, or a sponge. Some cosmetic companies prefer tubes with an applicator for concealers. These are identical to lip gloss containers.

Plastic Powder Containers: Loose face powder (mineral powder) is typically available in transparent plastic containers. It is usually applied directly from the container with a puff or a large brush. To prevent leakage, a sifter is placed on the jar, under the cap. It will also help dispense the powder. There is a state-of-the-art dosing system, in which the brush is attached to the plastic bottle. It makes the application of the product easier and more comfortable.

Glass and Plastic Jars: Cream, mousse, and cream-to-powder foundations and some cream blushes are usually marketed in glass or plastic jars.

Stick Cases: Cream blush and concealers can be found in the form of sticks packaged into stick cases (similar to lipsticks).

Godet: Compact facial powders, two-way foundations, and compressed blushes are usually available in a godet identical to containers of cake mascara and compact eyeshadow. It usually comes with a small brush or a sponge, and in the majority of the products, a mirror is also part of the package.

Pencil: A small group of concealers are found in the form of soft leaded crayons or pencils. These are identical to eyeshadow pencils.

GLOSSARY OF TERMS FOR SECTION 3

BB cream: A color cosmetic designed to serve multiple functions, including facial moisturization, foundation, sunscreen, and even antiaging.

Blush: A color cosmetic designed to add color to the cheeks.

Compact powder: Pressed powder.

Concealer: A color cosmetic designed to hide minor skin problems, such as small visible blemishes, pimples, black marks, or dark circles under the eye.

Cream-to-powder foundation: A color cosmetic formulated as a cream that transforms into a microfine powder on application.

Foundation: A color cosmetic designed to create a uniform color, provide a basic coverage to the skin, as well as blend uneven facial color for all skin tones.

Mineral powder: A term often used for loose facial powders.

Non-transfer foundation: A foundation that can resist mechanical abrasion.

Oil-control product: A product designed to reduce shininess of oily face; they contain oil-absorbing ingredients, such as talc, kaolin, and starch.

Oil-free product: A term often used on product labels to indicate the absence of oily components. There is no standard definition for this term. Sometimes, it refers to the absence of mineral oil; in other cases, it refers to the absence of ingredients that have the word "oil" in their name.

Primer: A colorless or slightly tinted semisolid cosmetic product that is used before the foundation to smoothen the skin surface and provide a base for the foundation.

Two-way foundation: A color cosmetic formulated as a compact powder that can be applied using either a wet or a dry sponge.

REVIEW QUESTIONS FOR SECTION 3

Multiple Choice Questions

1. What is the function of absorbents in facial makeup products?
   a) To contribute to slip of the product
   b) To provide a shiny finish
   c) To take up oil and sweat on the face
   d) All of the above
2. Fillers are used in liquid foundations for the following reason:
   a) To help the cake stick together
   b) To provide a bulk for the pigments
   c) To absorb moisture from the air
   d) To moisturize the skin
3. Talc caused concerns with regard to the safe use of facial makeup products. Which of the following ingredients was the major cause of these concerns?
   a) Asbestos
   b) Silicones
   c) Hydrous magnesium silicate
   d) Formaldehyde
4. Green pigments are used in concealers due to the following reason:
   a) To cancel the blue color under the eye
   b) To cancel the red color on the face
   c) To emphasize the color of the eye
   d) To provide oil-blotting properties
5. Which of the following is a distinct advantage of facial powders over liquid or semisolid foundations?
   a) Powders can absorb oil from the face, but semisolids cannot
   b) Powders can supply the face with emollients, but semisolids cannot
   c) Powders have a much longer shelf-life
   d) Powders do not cause acne
6. What is the purpose of the application of concealers?
   a) To create a uniform coverage to the facial skin
   b) To add color to the cheeks
   c) To blend uneven facial color
   d) To cover up minor skin problems

7. Usually, which of the following is included in a BB cream?
   a) Sunscreen
   b) Moisturizing ingredients
   c) Facial foundation
   d) All of the above

8. Which ingredients can provide a foundation with oil-control properties?
   a) Absorbents
   b) Abrasives
   c) Emollients
   d) Preservatives

Fact or Fiction?

______ a) Oil-free formulations do not contain any ingredients that have an oily texture.
______ b) Products not labeled "non-comedogenic" are all comedogenic.
______ c) Cosmetic-grade talc is the not the same as industrial-grade talc.
______ d) Primers prevent facial makeup from creasing.

Matching

Match the ingredients in column A with their appropriate ingredient category in column B.

	Column A	Column B
______	A. Cyclopentasiloxane	1. Absorbent
______	B. EDTA	2. Chelating agent
______	C. Hectorite	3. Color additive
______	D. Isoeicosane	4. Emulsifier
______	E. Kaolin	5. Filler
______	F. Phenoxyethanol	6. Hydrocarbon emollient
______	G. Polysorbate 20	7. Preservative
______	H. Talc	8. Silicone emollient
______	I. Ultramarine	9. Solvent
______	J. Water	10. Thickener

REFERENCES

1. Toedt, J., Koza, D., Van Cleef-Toedt, K.: *Chemical Composition of Everyday Products*, Westport: Greenwood Publishing Group, 2005:33.

2. Zouboulis, C. C.: Propionibacterium acnes and sebaceous lipogenesis: a love-hate relationship? *J Invest Dermatol.* 2009;129(9):2093–2096.
3. Makrantonaki, E., Ganceviciene, R., Zouboulis, C. C.: An update on the role of the sebaceous gland in the pathogenesis of acne. *Dermatoendocrinology.* 2011;3(1):1–49.
4. Boehncke, W. H., Ochsendorf, F., Paeslack, I., et al.: Decorative cosmetics improve the quality of life in patients with disfiguring skin diseases. *Eur J Dermatol.* 2002;12(6):577–580.
5. Hayashi, N., Imori, M., Yanagisawa, M., et al.: Make-up improves the quality of life of acne patients without aggravating acne eruptions during treatments. *Eur J Dermatol.* 2005;15(4):284–287.
6. Matsuoka, Y., Yoneda, K., Sadahira, C., et al.: Effects of skin care and makeup under instructions from dermatologists on the quality of life of female patients with acne vulgaris. *J Dermatol.* 2006;33:745–752.
7. National Rosacea Society: *All About Rosacea*, Accessed 9/28/2013 at NRS website http://www.rosacea.org/patients/allaboutrosacea.php
8. Kligman, A. M.: Acne cosmetica. *Arch Dermatol.* 1972;106(6):843–850.
9. Draelos, Z. D., DiNardo, J. C.: A re-evaluation of the comedogenicity concept, *J Am Acad Dermatol.* 2006;54:507–512.
10. Singh, S., Mann, B. K., Tiwary, N. K.: Acne cosmetica revisited: a case-control study shows a dose-dependent inverse association between overall cosmetic use and post-adolescent acne, *Dermatology.* 2013;226:337–341.
11. Kligman AM: Petrolatum is not comedogenic in rabbits or humans: a critical reappraisal of the rabbit ear assay and the concept of 'acne cosmetica'. *J Soc Cosmet Chem.* 1996;47:41–48.
12. DiNardo, J. C.: Is mineral oil comedogenic? *J Cosmet Dermatol.* 2005;4:2–3.
13. Matthews, I.: Multifunctional BB Creams Versus Tailored Approaches in Beauty, Posted 11/6/12, Accessed 9/29/13 at Cosmetics and Toiletries website: http://www.cosmeticsandtoiletries.com/formulating/category/skincare/Multifunctional-BB-Creams-Versus-Tailored-Approaches-in-Beauty-177499591.html?page=1
14. Rigano, L.: BB creams. *Cosmet Toiletries.* 2013;128(2):88–91.
15. Fulton, J. E., Comedogenicity and irritancy of commonly used ingredients in skin care products. *J Soc Cosmet Chem.* 1989;40:321–333.
16. Dahms, G. H., Zombeck, A. New formulation possibilities offered by silicone copolyols. *Cosmet Toiletries.* 1995;110(3):91.
17. Dweck, A. C., Foundations—a guide to formulation and manufacture. *Cosmet Toiletries.* 1986;4:41–44.
18. Desmarthon, E.: Cream in powder form: a new concept in makeup. *Cosmet Toiletries.* 2007;122(12):71–74.
19. FDA: Selected Cosmetic Ingredients, Talc in Cosmetics, Last update: 9/3/2013, Accessed 9/8/2013 at FDA website: http://www.fda.gov/Cosmetics/ProductandIngredientSafety/SelectedCosmeticIngredients/ucm293184
20. CFR Title 21 Part 347
21. Carr, C. J.: Talc: consumer uses and health perspectives. *Regul Toxicol Pharm.* 1995;21(2):211–215.

22. National Toxicology Program (NTP): Report on Carcinogens. Talc (Cosmetic & Occupational Exposure), Last update: 4/25/2007, Accessed 9/8/2013 at NTP website: http://ntp.niehs.nih.gov/index.cfm?objectid=03CA6E02-FBD5-5C52-9699F9DD00863ED7
23. CIR: Safety Assessment of Talc as Used in Cosmetics, Accessed 9/8/2013 at CIR website: http://www.cir-safety.org/supplementaldoc/safety-assessment-talc-used-cosmetics

SECTION 4: NAIL CARE PRODUCTS

LEARNING OBJECTIVES

Upon completion of this section, the reader will be able to

1. Define the following terms

Abrasion resistance	Artificial nail	Base coat	Brushability
Bubbling	Chipping	Cracking	Cuticle
Cuticle remover	Cuticle softener	Formaldehyde	Nail hardener
Nail moisturizer	Nail plate	Nail polish	Nail polish remover
Resin	Thixotropy	Top coat	

2. describe the distinct anatomical regions of the human nail unit;
3. explain what determines the hardness and flexibility of human nails;
4. differentiate between functional and decorative nail care products;
5. differentiate between a nail hardener and a nail polish;
6. explain how nail cosmetics can be used in the management of brittle nails;
7. explain how the use of nail care cosmetics can lead to inflammations and infections;
8. briefly discuss how cuticle removers work and why the cuticle should not be overmanipulated;
9. discuss whether nail care products are considered cosmetics in the US;
10. list various required cosmetic qualities and characteristics that an ideal nail care product should possess;
11. list various required technical qualities and characteristics that an ideal nail care product should possess;
12. name ingredients that are typically used in nail hardeners;
13. explain what the purpose of the application of nail moisturizers is;
14. differentiate between a top coat, a base coat, and a regular nail polish;
15. list the main types of ingredients typically used in nail polish, and provide some examples for each type;

16. explain the function of resins and thixotropic agents in nail polish;
17. briefly discuss the differences between traditional nail polish and shellac nail polish;
18. describe how nail polish is formulated;
19. differentiate between preformed nails and sculptured nails;
20. list various ingredients generally found in nail polish removers;
21. explain how nail polish removers remove nail polish;
22. list some of the typical quality issues that may occur during the formulation and/or use of nail care products, and explain why they may occur;
23. list the quality parameters that are typically tested for nail care products, and briefly describe their method of evaluation;
24. briefly discuss the potential safety issues related to the use of nail care products;
25. list typical containers available for nail care products.

KEY CONCEPTS

1. The human nail unit is composed of the nail plate, nail matrix, nail folds, nail bed, cuticle, and the hyponychium.
2. Most nail care products belong to the category of cosmetics in the US.
3. Functional nail care products aid in maintaining nail hardness. They can be used by both genders since these products do not necessarily leave a shiny film on the nail plate. This category also includes chemical cuticle removers.
4. Decorative care products include colored formulations that enhance the appearance of human nails. This category includes various types of nail polish and artificial nails.
5. Nail polish removers remove the lacquer by redissolving the resins.
6. The typical quality-related issues of nail care products include bubbling, cracking, chipping, and thickening.
7. Parameters commonly tested to evaluate the quality of nail care products include abrasion resistance, gloss, film flexibility and hardness, drying time, adhesion test, brushability, color, dispersion of pigments, and viscosity.
8. The main ingredients causing safety concerns regarding nail care products include toluene, phthalates, and methylene glycol.

Introduction

Nail cosmetics represent a significant segment of color cosmetics. They are more popular among females; however, the market for male manicures is also rapidly growing.

Nail care is an increasingly important segment of normal body care since aesthetically pleasing healthy nails reflect one's general health status, and, to some extent, the condition of nails may also reflect a person's social status.

This section reviews the main anatomical and physiological characteristics of human nails. Additionally, it provides an overview of the various products applied to the nails, their usual ingredients, formulation technology, testing methods, and packaging materials. The section also reviews how these products may affect the nails and what consumers' general requirements are.

Anatomy and Physiology of Human Nails

The function of human nails is to protect the fingers, aid in picking up small objects, and assist in delicate manual activities. Toenails are equally important for protection. ❶ **The human nail unit is composed of the nail plate, nail matrix, nail folds, nail bed, cuticle, and the hyponychium.**[1] Figures 4.13a and b depict the basic parts of the human nail.

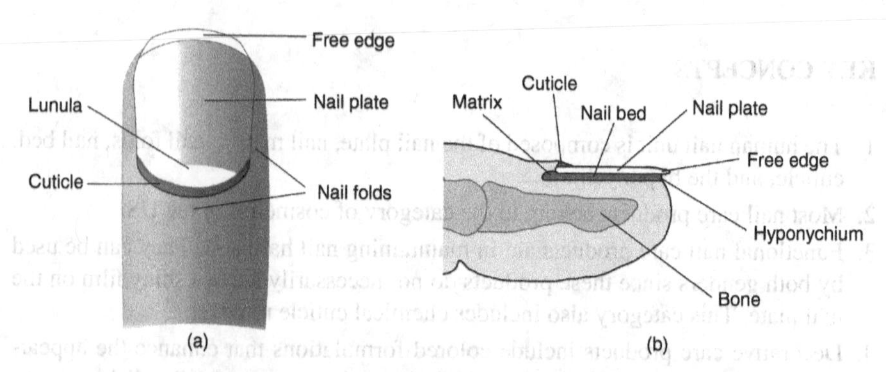

Figure 4.13 (a) Structure of the nail unit and (b) cross section of the nail.

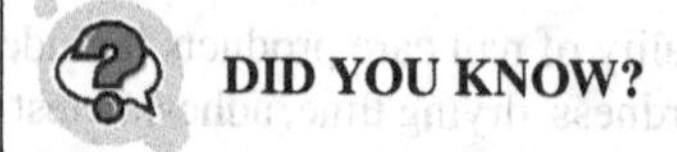

DID YOU KNOW?

The term "matrix" refers to a material that is the origin of something enclosed within it.[2] Nail matrix is a thickened epithelium at the base of the fingernail or toenail from which a new nail plate develops.

Nail plate is the external visible part of the nail. **Nail matrix** refers to the root of the nail. This is the living part of human nails. Its cells are in the area of lunula and under the skinfold at the base of the nail base. Matrix cells continuously divide; therefore, nails continuously grow. Nail matrix cells are formed at birth and cannot regenerate following injury. For this reason, any injury or trauma to the nail matrix can result in a permanently deformed nail that cannot repair and will not grow normally.[3] **Lunula** is the pale structure at the base of the nail. It represents the visible front part of the nail matrix from which the nail grows. **Nail folds** surround and protect the nail unit by sealing out environmental irritants and microorganisms through tight attachment to the nail plate.

Nail bed is a thin soft tissue that underlies the nail plate. It contains living epidermis and is pink because of the visible network of capillaries in the dermis. The actual nail plate (as can be seen at the edge of the nail that protrudes beyond the edge of the finger) is white. **Cuticle** is the skinfold at the base of the nail. It forms a watertight seal between the non-living nail and the skin of the fingertip. Damage to the cuticle results in water, chemicals, molds, and bacteria reaching the nail matrix cells, which may lead to severe infection. For this reason, dermatologists recommend that the cuticle should not be dislodged, pushed back, trimmed, or manipulated in any way.[4–6] The **hyponychium** is the cutaneous margin underlying the free edge of the nail plate. Its function is similar to that of the cuticle and acts as an adherent seal to protect the nail unit. The hyponychium should not be overmanipulated during nail grooming to avoid separation of the nail plate from the nail bed.

Human nails are composed mainly of keratin, the same protein that makes up the hair and skin surface; however, this is a special type of keratin. Nails are made up of "hard keratin," which contains much more sulfur than the normal skin keratin.[7]

DID YOU KNOW?

A common belief is that the strength of the nails is related to their calcium content. However, it is not true[8] since their hardness, and therefore strength, depends on the number of keratin fibers and disulfide bonds gluing the cysteine components of keratin together.

In addition to keratin, nails also contain water, lipids, and small amounts of elements, such as calcium, iron, aluminum, copper, zinc, and others. The normal water content of human nails is in the range of 10–15%, with an average of 12%.[9] It has been found that the nail plate becomes soft and tends to be double layered when its water content exceeds 20% and tends to split when the water content is below 10%.[10] Lipids are present only in a small amount (about 5%), which is significantly lower than that of the skin.[11] They contribute to the flexibility of nails. Due to the low concentration of lipids, the nail plate is about 1000-fold more permeable to water than skin.[12] As a result, it can be rapidly hydrated and dehydrated. If its water content decreases, it becomes brittle.

The average fingernail growth is 0.1 mm/day or 3 mm/month. Fingernails grow out completely in 6 months. Toenails grow at one-third to half of the rate of fingernails and take 12–18 months to grow out completely.[13] Nail growth is greatly influenced by the weather: they grow slower at night and during winter.

History of Using Nail Care Products

The use of nail cosmetics is well rooted in history. There is no exact record of the invention of fingernail polish; however, it is believed that it originated in China around 3000 BC. The first nail polish was made of egg white, beeswax, vegetable dyes, gelatin, and, sometimes gum arabic. At around 600 BC, gold and silver were preferred by royalty; later, red and black become their favorite colors.[14] Ancient Egyptians and Indians used henna and other natural plant stains to color their nails. In Egypt, bright red was allowed for the highest classes, while lower ranking women could wear only pale shades. During the Ming Dynasty of China (14th to 17th century), noblewomen wore very long artificial nails as a status symbol, indicating that, unlike commoners, they did not have to do manual labor. Purity and hygiene were important during the Victorian era; clean nails were a must. The most popular type of manicure was simple, merely buffering the nails and tinting them with red oil.

In the 19th and early 20th centuries, "nail polish" was a colored oil or powder. It was used to rub and buff the nail, literally polishing and coloring the nail simultaneously.[15] As for who invented nail polish in the form that we know today, records show that it was Michelle Menard, a French makeup artist in the 1920s. Menard worked for a company established by Charles Revson and his brother. Menard took her inspiration from the enamel used to paint cars, which was very similar to the substance that she came up with for painting nails. She managed to create a modern lacquer made of the same nitrocellulose dissolved in solvent that was used on cars (except not with the same strength).[16] The Charles Revson Company became Revlon (they added "L" in the middle of the name for the other co-founder, the chemist). The first Revlon nail polish went on sale in 1932. It was Hollywood that made the new nail polish big. It helped that it was relatively cheap. Painting nails and buying matching lipstick gave a bit of Hollywood glamor. The formula for nail polish remained similar to that invented by Michelle Menard. It has been slightly amended to make it longer lasting and quicker drying; however, essentially, it is the same.

The use of acryl as a material for nail extension started accidentally. Fred Slack, a dentist, was its inventor in the 1950s in Philadelphia. He damaged his fingernail while working. He used materials that he had in his clinic, a piece of aluminum foil and dental acrylic. He was surprised at the result: the nail looked natural and the pain disappeared. He experimented with different acrylic monomers and polymers and soon patented the invention. The product he invented helped strengthen nails as well as extend them. It proved to be very popular and still is. The Slack family went on to develop the first non-yellowing, cross-linked formulations used in all traditional acrylic systems today. They introduced the first fiberglass-reinforced polymers, light-activated liquid and powder acrylic, UV-cured gels, as well as air- and heat-activated polish sealants. Today, Fred Slack, Director of R&D, and his son Rick Slack, President, lead NSI (Nail Systems International) to continually meet the demands for innovative products distributed all over the globe.[17]

Today, nail polish products are available in any hue you can think of. In addition to painting their nails, many people are also taking it a step further by practicing nail art, or the painting of intricate designs onto their nails using different colors of nail polish. Some even add 3D elements to their nails such as tiny beads, glitters, and other embellishments.

Types and Definition of Nail Care Products

The classification of nail care products includes functional nail products and decorative nail products. Functional products promote the healthy growth of human nails, help their normal maintenance and removal of cuticle. This category involves nail hardeners and cuticle removers. Decorative nail cosmetics are applied to color both natural and artificial nails. This class of products includes various types of nail polish and artificial nails. Nail polish removers are not considered decorative cosmetics; however, since they are also used on the nail plates as nail polishes, they are discussed in this section.

● **Most nail care products belong to the category of cosmetics in the US.** Although their application may prevent nail plate dryness and brittleness, they are still "just" cosmetics, if they do not contain active ingredients. Products used to treat infections, such as fungal nail infections, are, however, considered drugs in the US.

- **Nail hardeners** are designed to develop a protective layer on the nail plate; they increase the hardness and strength of the nails. They typically contain a chemical cross-linking agent. Otherwise, their composition is similar to that of the ordinary clear nail polish. Since the main aim of their application is not coloring the nail, they are more of functional nail care products. Certain types might have a shiny finish; these also provide decorative purposes.
- **Nail moisturizers** are designed to increase the hardness of nails by supplying them with moisturizing ingredients. They are typically lotions or creams directly applied to the nail plates.
- **Nail polish** is designed to color and enhance the aesthetics of nails, giving them a more attractive aspect. Nail polish basically consists of pigments suspended in a volatile non-aqueous solvent to which film-formers are added. Addition of pearlescent and metallic ingredients as well as shimmers can provide special effects.
- **Cuticle removers** are designed to chemically remove the cuticle. Chemical cuticle removers are formulated as liquid or cream and contain alkali ingredients to destroy the cuticle keratin.
- **Artificial nails** are nails created using products designed to enhance nail length. Products in this category include preformed plastics, formed acrylics, and a combination of both. Preformed plastic nail tips are stuck to the surface of the nails with the application of glue. Formed acrylic gels are mixtures of acrylic monomers and polymers, which harden on the nails' surface.
- **Nail polish removers** are liquids; generally organic solvents designed to strip nail polish from the nail plate without removing layers of nail plate. Additional

components, such as oils and fragrances, can also be added to in the formulations. They are usually applied with cotton pads; however, recently, a new sponge design is also available on the market.

How Nail Care Products May Affect the Human Nails?

Nail cosmetics can be utilized in the management of mild problems and diseases causing cosmetic issues.

- As an example, nail care products can be used for the treatment of **brittle,** soft, and/or splitting **nails** using nail cosmetics. The term "brittle nails" usually refers to a cosmetic issue; however, it might become severe, affecting the nail function (which is more of a therapeutic issue).[18] Nail brittleness is characterized by weak inelastic nails that split, flake, and crumble. Environmental and occupational factors that lead to progressive dehydration of the nail plate play an important role in the development of nail brittleness.[19] Excessive exposure to detergents (such as dishwashing liquid) and overfrequent use of nail polish removers can damage the keratin and decrease the water content of the nails making them more brittle. Even simple washing of the hands, if carried out too frequently, can lead to brittleness. Nail polish and moisturizers might help maintain nail hydration by sealing in moisture that would otherwise evaporate.[20] Nail polish, nail hardener, gels, shellacs, and elongators can serve to physically thicken and strengthen and protect soft, weak, or otherwise fragile nails by providing an external, durable shell.[21] Nail hardeners can help brittle nails; however, their regular use may paradoxically cause brittle nails over time.
- The beneficial effect of using nail lacquers is controversial among scientists. Some believe that nail polish is protective from environmental irritants and seals the moisture in the nail plate by preventing rapid evaporation. There are some concerns, however, that brittleness and nail dryness may be enhanced by overusing nail polish removers.[22] According to the recommendations, manicures should be used only once a week or even less frequently.

Unfortunately, when nail care products are used improperly, they can lead to discoloration, inflammation, infections, and nail diseases. Here are some examples for unwanted issues related to the use of nail cosmetics.

- Cuticle serves a protective function. Its chemical removal or physical **overmanipulation** is generally not recommended by dermatologists, since these procedures can lead to damage, inflammation, and infection. In addition, cuticle removers can cause nail plate softening and are a common cause of irritant contact dermatitis, especially when left on for prolonged period of time.[23]
- **Filing** the nail plates' surface is usually performed before applying artificial nail extensions to the nail plate to facilitate adherence of the product. The nail plate is approximately 100 cell layers thick. If filing must be done, only 5% of the nail plate thickness, or approximately 5 cell layers, should be removed, which

is just enough to remove the shine and increase the attachment of the nail care product.[23]

- Plastic tips and other artificial nail extensions can reinforce and cover brittle nails; however, they are not as flexible as natural nail plates and are more likely to become **oncholytic** (i.e., a disease when nail plate separates form the nail bed) following a mechanical trauma, especially with long nails.[5,24] Oncholytic nails are also more susceptible to secondary infections.
- **Yellow staining** of the nails often occurs following continued wear of nail polish, especially if it contains Red 6 and 7 or Yellow 5 Lakes.[25] Nail staining typically fades about 2 weeks after the polish has been removed. However, if the nail polish is left on for a prolonged period of time, the colorant can deeply penetrate the nail plate and continuously leach out. In such cases, nail staining may only be resolved after the discolored nail has grown out.

Required Qualities and Characteristics and Consumer Needs

From a consumer perspective, a quality nail care product should possess the following characteristics:

- Even color and shine
- Pleasant, easy application, and easy spreadability
- Non-streaking
- No staining when removed
- Will not fade and become dull
- Easy removal with a nail lacquer remover
- Long-lasting effect (at least 4–5 days)
- Resistance to chipping, cracking, and peeling
- Moderately quick drying to allow an even application
- Good coverage
- Good adhesiveness.

The technical qualities of nail care products can be summarized as follows:

- Fulfill long-term stability
- Color uniformity
- High-shine finish film
- Sufficient thixotropic characteristics
- Sufficiently hard and flexible film
- Good drying
- Light stability
- Dermatological safety.

Functional Nail Care Products

● Functional nail care products aid in maintaining nail hardness. They can be used by both genders since these products do not necessarily leave a shiny film on the nail plate. This category also includes chemical cuticle removers.

Nail Hardeners Nail hardeners are used to harden brittle nails, as their name implies. Nail hardeners are essentially a modification of nail enamels with different solvent and resin concentrations. They are applied to the clean nail plate and function as a base coat. Many modern nail hardeners contain a chemical cross-linking agent, such as formalin (which is mistakenly equated with "formaldehyde" under current labeling rules) and dimethyl urea (DMU). These ingredients react with the keratin in the nail and create more cross-links, hardening the nails. Although more cross-links mean a harder nail, too many links can cause the nails to become dry, brittle, and yellow.[26] These products should only be applied to the free edge of the nails while the skin is shielded.[27]

DID YOU KNOW?

Formaldehyde is a highly reactive gas and is, therefore, not used in cosmetics in such form. When formaldehyde is added to a cosmetic, it is almost certainly formalin (i.e., a 37% solution of formaldehyde in water). Formalin comes from the reaction of formaldehyde and water and contains primarily methylene glycol and water, as well as a small amount of methanol and methylene glycols and only a tiny amount of free formaldehyde (approximately 0.02–0.1%). Since both formaldehyde and methylene glycol are present in the aqueous solution, they are usually referred to as formaldehyde equivalents.[28]

Originally, nail hardeners were formulated as 10% or greater solutions of formaldehyde. However, the FDA recalled these products due to reports on adverse reactions.[29] Some of them still contain free formaldehyde since its use in concentrations of 1–2% is still permitted. In recent years, it has been mainly replaced by other substances, such as polyesters, polyamides, and acrylate polymers. Substances likely to be present in nail hardeners include toluene, nitrocellulose, acrylate polymers, resins, acetates, as well as ingredients to further strengthen nails, such as proteins, biotin, nylon, glycerin, propylene glycol, and salts of various metals. DMU is reported to cause fewer side effects as formaldehyde in a concentration of 2%.

Products that do not contain cross-linking agents are usually referred to as "non-hardening" nail polish or base coat. Due to the presence of resins and polymers, they can also increase nail hardness. However, overuse of these products does not lead to brittleness or yellowing.

Nail Moisturizers The purpose of using nail moisturizers is to add moisture to the nail plate. They can also help patients with brittle nails. Nail moisturizers usually come in cream or lotion form. They contain various moisturizing ingredients, including occlusives, emollients, and humectants, and proteins may also be added. In addition, ingredients that increase the nails' water-binding capacity may also be added, and examples for such ingredients include urea and lactic acid.

There is no scientific evidence that topical gelatin, vitamin B12, calcium, and botanical extracts are effective in treating nail dehydration. Biotin has been shown to be beneficial for patients with brittle nails, when taken orally.[30] However, there is no data on whether topically applied biotin would have the same effect. It has also been shown that silicone taken orally may be beneficial in treating brittle nails.[31] Nail moisturizers should be applied under cotton gloves or other protective and occlusive aids, typically during nights.

Chemical Cuticle Removers Chemical cuticle removers dissolve the excess cuticular tissue on the nail plate by attacking the disulfide bonds of cysteine in keratin. They are usually water-based formulations, which are applied with a brush or the tube itself and left on the nail plate for 5–10 min followed by removal. After the cuticle is softened, it is usually pushed back from the nail using specially designed fine instruments and cuticle retractors, such as an orange stick or rubber-ended stick. The fibrous cuticular ridge should not be removed; however, sometimes, it is slightly damaged. Incorrect use may cause mechanical damage to the nail structure, called paronychia, and produce permanent deformation.

Cuticle removers are available as liquids, gels, or creams. Typical ingredients include cuticle-dissolving agents, such as sodium hydroxide and potassium hydroxide (in a concentration of 2–5%). These products typically have a highly alkaline pH and can be irritating to the skin. Milder preparations, which are less effective at the same time, contain trisodium phosphate or tetrasodium pyrophosphate. Additional ingredients of these formulations include water, emollients, humectants, thickeners, and preservatives.

DID YOU KNOW?

Cuticle softeners are a lighter subtype of cuticle removers, which only wear down the cuticle for easy mechanical removal by subsequent cutting or trimming. Cuticle softeners contain quaternary ammonium compounds in a 3–5% concentration, and they are sometimes combined with urea.

Decorative Nail Care Products ❶ **Decorative care products include colored formulations that enhance the appearance of human nails. This category includes various types of nail polish and artificial nails.** This section reviews the characteristics, ingredients, and formulation methods of these products.

Nail Polish Nail polish forms the largest group of manicure preparations. It is used for coloring, both for beautification and camouflaging surface irregularities or discolorations and strengthening weak brittle nails. There are a few synonym names for this term, including nail varnish, nail enamel, nail paint, and nail lacquer.

There are several subsets of nail polish, which are different in their color, basic function, and concentration of ingredients. Let us review these types before discussing the major ingredients of nail polish.

- **Base coats** are clear solutions, usually applied to the clean nail plate before the nail polish. These preparations usually enhance the normal, healthy growth of the nails, and they contain nutrients and moisturizers. They also prevent the nail polish from staining the nails, and they create a smooth nail surface to which the nail lacquer has better adhesion. They usually contain less primary film-formers and more secondary film-formers and are less viscous than colored nail polish.
- **Top coats** are also clear solutions applied over nail polish. They provide a protective layer and prevent the applied polish from fading and chipping as well as enhance gloss and reduce drying time. They contain increased amount of primary film-formers, more plasticizers, and less secondary film-formers. They are also less viscous than color nail polish for the same reason as base coats.
- **Nail polish** is a colored suspension in volatile solvents containing viscosity-modifying agents.

Ingredients Nail polish can contain a dozen or more different ingredients. Despite the potentially high number of ingredients and a vast array of colors, ingredients can be categorized into several basic groups. The raw materials applied in base coats, top coats, and nail polish are the same; generally, the percentage of the ingredients is different.

- **Resins**, otherwise known as film-formers, hold the ingredients of the lacquer together while forming a strong film on the nails. Chemically, resins are polymers that are solid or gummy in their pure state. They improve adhesion of the product to the nail. They also give the polish its characteristic glossy appearance. Two types of resins—hard, glossy resins (otherwise known as primary film-formers) and soft, pliable resins (otherwise known as secondary film-formers)—are used in various concentrations depending on the desired effect of the nail polish.
    - **Hard, glossy resins** provide the lacquered nail with a hard, brittle film. Examples include nitrocellulose, vinyl polymers, methacrylate and acrylate polymers or copolymers, acrylate esters, acrylamide, as well as cellulose derivatives, such as cellulose acetate proprionate. Nitrocellulose is the most commonly used primary film-forming agent in nail polish. The film is somewhat oxygen permeable, allowing gas exchange between the

atmosphere and the nail plate; this gas exchange is important for ensuring nail plate health.[32]

- **Softer, more pliable resins** enhance adhesion, gloss, flexibility, and resistance performance. For a long time, the most popular resin was tosylamide/formaldehyde resin (toluenesulfonamide/formaldehyde resin, TSFR); however, it might potentially cause allergic contact dermatitis. Today, mainly, other classes of ingredients are used instead of TSF resin. Examples include toulenesulfonamide/epoxy resin, polyester resins, acrylate and metacrylate copolymers, and polyvinyl butyral. Base coats incorporate a higher proportion of pliable resins.

A proper ratio of the two types of film-formers should be provided in formulations to obtain films that have appropriate drying time and are flexible. Excessive amount of secondary film-formers can lead to films that dry slowly and are too soft, whereas insufficient amount can lead to films that dry quickly and become very hard and brittle.

- **Solvents** Solvents act as carriers of lacquer. They dissolve resins, suspend pigments, and evaporate, leaving a smooth film. They also prevent the product from drying out in the bottle. They help regulate viscosity, application, flow, leveling, drying time, hardness, gloss, and stability. Typically, a blend of solvents is used to achieve an optimal level of the above-mentioned properties. The boiling point of the solvents influences the viscosity of the lacquer solution.
  - Examples for solvents used in nail polish formulations include alkyl esters, such as ethyl acetate and *n*-butyl acetates; glycol ethers, such as propylene glycol monomethyl ether; alcohols, such as isopropyl alcohol; and alkanes, such as hexane and heptane. Formerly, toluene was a commonly used solvent; however, the industry trend is to refrain from using it in response to expressed health concerns (i.e., exposure for both nail technicians and consumers).
- **Plasticizers** improve resin flexibility and chip resistance.
  - Frequently used plasticizers include camphor, castor oil, glyceryl tribenzoate, glycerol, triphenyl phosphate, trimethyl pentanyl diisobutyrate, acetyl tributyl citrate, ethyl tosylamide, sucrose benzoate, ethyl toluene sulfonamide, and polymer plasticizers called NEPLAST (a polyether-urethane). Dibutyl phthalate was a popular plasticizer in the past; however, it has been banned (by California and the EU); therefore, it is no longer used widely.
- **Color Additives** Today, we have an endless variety of shades. Of the many colorants available, the choice is limited to those that have good permanence and are insoluble in solvents in order to avoid staining and discoloration of the nails, as well as to avoid any chemical reaction to the lacquer. Most often, lakes are used for nail polish. A shimmer effect is typically created by incorporating powdered aluminum, mica flakes, and bismuth oxychloride.
- **Thixotropic agents**, otherwise known as thickeners or suspending agents, provide flow control and keep the color additives dispersed. They increase the nail polish viscosity at rest, which, however, becomes fluid as soon as a mechanical

constraint is exerted, such as shaking the bottle or brushing. Uncolored formulations, such as base coats and top coats, do not require thickeners since they do not contain suspended particles.

- Most commonly, clay derivatives, such as stearalkonium bentonite or stearalkonium hectorite, are used. Silica can also be used as a thickener.

Thixotropy refers to the non-Newtonian flow nature of materials, which is characterized by a reversible gel-to-sol formation with no change in volume or temperature. The term thixotropy means "to change by touch." Systems having thixotropic properties usually exhibit a reversible time-dependent decrease in apparent viscosity. Figure 4.14 shows this special rheological behavior. Therefore, when such a material is subjected to a particular shear rate, the viscosity will decrease over time. Once the shear stress has been removed, the structure reforms. The process is not immediate; viscosity increases over time as the molecules return to the original state under the influence of Brownian motion. The presence of the hysteresis loop indicates that a breakdown in structure has occurred, and the area within the loop may be used as an index of the degree of breakdown.

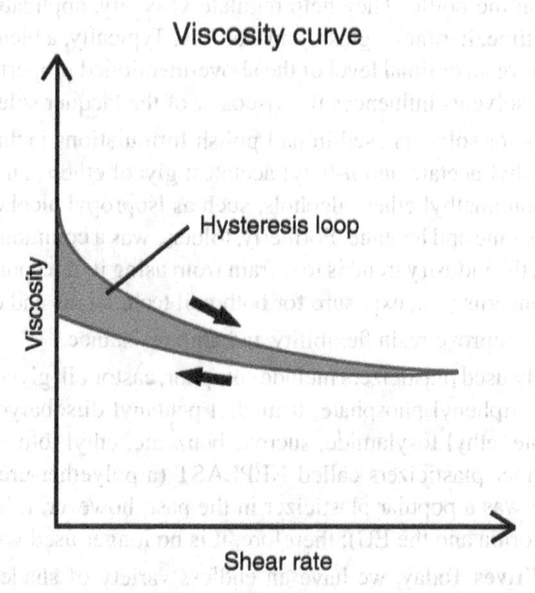

Figure 4.14 A viscosity curve characteristic of thixotropic materials.

- **Color stabilizers/UV absorbents** prevent color shifting of the nail polish on exposure to UV light. Some top coats also contain UV filters to prevent fading over time.
  - Examples for color stabilizers include benzophenone-1 and etocrylene.

DID YOU KNOW?

Some nail polish formulations and top coats contain UV filters, otherwise known as sunscreens. Although these are the same ingredients as those used in sunscreen lotions, the aim of their application in nail care products is not to prevent cancer but simply to provide color protection.

- **Nail treatment ingredients** can include compounds that strengthen the nails and enhance nail health. Examples include vitamins, minerals, vegetable oils, herbal extracts, and fibers, such as silk. There is a new substance called poly-ureaurethane, which was approved by the FDA in 2012 for managing the signs and symptoms of nail dystrophy, i.e., nail splitting and nail fragility.[33] In addition, antifungal agents can also be incorporated into nail polish formulations to serve as a treatment for fungal infection of the nails. Typically, tolnaftate is used as an OTC active ingredient in a 1% concentration.

Additional Types of Nail Polish In addition to organic solvent-based nail polish, other product types are also available, including the following:

- **Water-based nail polish** is now available on the market. Since the evaporation rate of water and, therefore, the drying time of products are much slower than those of organic solvents, they are unlikely to replace solvent-based products in the foreseeable future. Their advantages include that they are cheap, non-flammable, and odorless. In this case, incorporation of preservatives is also recommended to prevent microbial growth over time.
- Newly advertised **magnetic nail polish** contains iron powder. The products are supplied with a tiny magnet built into the cap of the nail polish container. It should be held over the nail before solvents evaporate, which causes the iron powder in the formulation to gravitate toward the magnet and form a desired pattern. It should be kept in mind that nail polish containing magnetic materials must be removed prior to magnetic resonance imaging (MRI).

DID YOU KNOW?

When you go to the dentist office to have your cavities filled, you will experience that they use UV light during filling. The reason for this is the filling material used hardens with UV light. Shellac nail polish is very similar to this polymer; it also hardens with UV light.

- Photocured, otherwise known as UV-cured nail polish or **shellac nail polish**, is a relatively new, innovative formulation. These products contain the same pigments as those in the conventional nail polish formulations. However, instead of a solvent/resin base, they consist of methacrylate or acrylate oligomers and monomers. They also contain a photoinitiator, which causes polymerization on exposure to UV light, leaving a polymer/pigment coat. Photocured products are designed to provide a thin coat, similar to conventional products; however, in this case, the film developed is much harder, shinier, longer lasting, and more resistant to chipping. UV curing takes about 10 min, there is no need for additional drying, and consumers usually do not experience smudging and/or chipping when using these products.

Formulation of Nail Polish The formulation of various types of nail polish involves intensive mixing of highly flammable and volatile materials, and therefore, it involves a high risk. Both the buildings and the manufacturing equipment must be flame proof and carefully monitored for harmful vapors.

Colored nail polish formulations are formulated as suspensions. The process generally consists of the following steps: formulation of the pigment blend, formulation of the nail polish base, and coloring of the lacquer base. As with all color cosmetic formulations, pigment preparation is the most important step. This is particularly true for nail polish; as the finer the pigment is ground, the higher the gloss and the more stable the product. Pigments for nail polishes are usually prepared in "chip form." They are mixed with nitrocellulose and plasticizer using a high-shear mixer. The resultant mixture is then passed through a mill, such as a triple roll mill, dried, and "chipped" (i.e., split up into solid fragments). In small-scale size, the pigments are prewetted by making a 1:1 or 1:2 premix in the carrier (oil or solvents) under slow stirring conditions. The polish base is made by blending all components under intensive stirring. In the last step, the color mixture is dispersed in the nail polish base using controlled speed stirrers. Other solvents and additional ingredients are added when a uniform color has been achieved with continuous mixing. Following this, the thixotropic agent is added and the viscosity is properly adjusted. As noted previously, base coats and top coats are formulated without suspension bases and pigments.

Artificial Nails Artificial nails became very popular during the last century. They are longer lasting than natural nails and highly resistant to cracking and chipping. They can be utilized for patients with brittle nails or nail discoloration to hide these imperfections. In the past, artificial nails were almost exclusively applied by professionally trained manicurists in nail salons. Today, artificial nail kits designed for home use are now available in many shops. This section reviews the frequently used artificial nail types, their ingredients, and how they are applied to the nail plates.

Nail Tips Nail tips are popular since they are easy to use. They are available in a variety of designs, shapes, and colors; typically, they are made of nylon or plastic. Nail tips are either preglued or have to be glued by the user before applying them. The glue is an adhesive, typically methacrylate-based or cyanoacrylate-based glue.

While applying, the glued plastic tip is placed on the nail plate for 5–10 s to ensure proper adhesion. The tips can then be reshaped and colored, if needed.

Sculptured Nails An increasingly popular method of obtaining stronger, longer, and more attractive nails is nail sculpturing. The nails made with this technique are also known as acrylic nails and porcelain nails. The word "sculpturing" refers to the process of building custom-made nails on a template attached to the natural nail plate. These nails fit perfectly, and, if applied well, they are almost impossible to distinguish from natural nails. The basic steps include cleansing and filing, putting on the sculpting form, priming, mixing, sculpting, and finishing process. A short description of the basic steps is provided, which is as follows:

- First, a solvent, such as isopropyl alcohol, is applied to the nail plate to remove any nail polish residue and oils. Then, the nails' surface is abraded with a pumice stone or file to further clean the nail plate and create the optimal surface for adhesion of the sculpted nail. Some dermatologists suggest the application of antifungal and antibacterial products to prevent bacterial and fungal infections. Care must be taken when applying these solvents because they may dry out the skin surrounding the cuticle.
- In the next step, a flexible template (e.g., made of Teflon) is placed under the natural nail plate, which provides additional surface beyond the natural nail plate.
- Primers may also be applied to the nail plate before sculpting. Primers are substances that improve adhesion. Most types act as a double-sided tape; one side sticks to the nail enhancement, while the other side holds tightly to the nail plate.
- Sculpturing is performed by applying layers of acrylic polymers onto the natural nails' surface and template using a brush. This is done by dipping the brush into a liquid monomer and drawn through a polymer powder, and a small bead forms at the end of the brush. After sculpting is done, the hardened polymers can be shaped into the desired length and width. The template is removed from the fingers.
  - The majority of monomer liquids is composed of ethyl methacrylate and other methacrylate monomers to provide cross-linking; inhibitors, such as hydroquinone, to slow down polymerization (i.e., hardening of the polymers on the nail plate, since if hardening is too fast, there may be not enough time to level the polymer layers on the nails); UV stabilizers; catalysts; flexibilizing agents; and other additives.
  - The powder polymer is made of methyl and/or ethyl methacrylate polymer coated with benzoyl peroxide, which acts as a polymerization initiator. It may also contain color additives.
- The final nail sculpture is sanded to a high shine. Nail polish, jewels, decals, and decorative metal strips may be added, depending on the customer's request.

These nails harden at room temperature due to a chemical reaction between the liquid and the powder polymer. Since these artificial nails grow with the natural nail, a gap will develop between the cuticle and the acrylic nail, which has to be filled in regularly. Removal of these products is not easy and takes a long time (even up to an hour). Typically, fingers are placed into a small bowl and immersed into solvents, such as acetone. The removal process can greatly be accelerated by filing the nails before soaking them into solvents, which removes the bulk of the artificial nail.

DID YOU KNOW?

A common belief is that artificial nails should be regularly removed to allow the nails to breathe. However, these nails are not designed to be taken off frequently. Even when done carefully, complete removal can damage and dry the nail plate.

Variations for the Sculpturing Technique Instead of using a flexible template, plastic tips can also be used for sculpturing nails. Such tips are similar to those adhered to the entire nail surface (discussed previously); however, the difference is that, in this case, the tips are glued only to the tip of the nails. The nail plate is filled with acrylic polymers (liquid and powder). The application of tips highly reduces the sculpturing time.

There are some artificial nails that cure under UV light; they are a variation of sculptured nails. They are also called UV gels. The sculpturing process is similar to that of the two-part liquid and powder systems; however, unlike liquid and powder systems, UV gels will not cure at room temperature. After application, the nails are placed under a UV lamp for a few minutes. Another difference is that these are one-phase systems, typically including polymerization photoinitiators, urethane methacrylate oligomers, cross-linking monomers, and catalysts.

Nail Polish Removers As discussed previously, when regular liquid nail polish dries on the nail plate, resins are deposited onto the nail plate as solvents evaporate. Therefore, when solvents—similar to those in nail polish—are added to the dry nail polish film, resins redissolve in the solvents. ● **Nail polish removers remove the lacquer by redissolving the resins.** These liquid formulations can be wiped off from the nails.

Ingredients Mainly organic solvents are used in nail polish removers; examples include acetone, ethyl acetate, butyl acetate, methyl ethyl ketone, toluene, and ethanol. California banned the use of ethyl acetate and methyl ethyl ketone and most other acetone alternatives for safety reasons.[34] Other states and countries are also considering the same actions.

A significant disadvantage of all types of nail polish removers is that they dehydrate the skin, can cause irritation to the eyes, and make nails dry and brittle. They also have a distinct chemical smell and are highly flammable. In certain cases, water may also be used in a smaller concentration (up to 10%). However, in this case, the application time (removal time) is expected to be longer. These products are less damaging to the skin barrier.

In addition to solvents, nail polish removers typically contain emollients to counter the dehydration and brittleness effects of solvents and condition the skin. In some cases, vitamins may be also incorporated into the formulations. Additional components may be fragrances to counteract the distinct smell of the organic solvents, color additives to provide a slight color to the preparation, and preservatives.

Types of Nail Polish Removers Mainly, three different dosage forms are available on the market: including solutions, cleaning pads, and sponges.

- Solutions are applied with a cotton ball or tissue and wiped over the nail to strip away the fingernail polish on it.
- Cleaning pads are pads prewetted with a nail polish remover solution. They are applied the same way as cotton balls.
- A special type is the sponge, which is also impregnated with a nail polish remover solution. It consists of a piece of foam sponge soaked with the solution in a container with a hole at its center. The fingers are dipped into the sponge through the hole to remove the polish. In this case, the cleaning time may take a little longer than in the other types of formulations.

The formulation of these products is a simple solution preparation.

Typical Quality Problems of Nail Care Products

● **The typical quality-related issues of nail care products include bubbling, cracking, chipping, and thickening.** Their potential causes as well as solutions are discussed here.

Bubbling Bubbling is a visible process on the surface of the newly painted nails. It usually happens during application. It is considered to be caused by air that may come from "overshaked" bottles. Many cosmetic technicians say that the bottles should never be shaken, only rolled between the palms. It may also occur if nail polish layers applied to the nail plate are not completely dried before the application of the next layer. Oil residue on the nail plate can cause bubbling since the nail polish will not adhere at points that are oily. An additional cause may be the thickening of the product. When applied, the microstructure collapses and the solvents cannot escape fast enough during drying, and they remain captured in the final film. Bubbles can lead to chipping when dry.

Cracking Cracking means that the applied nail polish film is uneven and there are some cracking on its surface. It may happen if the flexibility of the film is inappropriate, resulting from an abnormally low plasticizer concentration. As discussed previously, plasticizers make the film formed on the nails more flexible, decreasing its rigidity. If their concentration is too low, a rigid film develops, which cannot adjust itself to the nails' natural curvature.

DID YOU KNOW?

There is a new nail polish trend called "cracked" or "crackle" nail polish. It is a type of nail polish that shatters as it dries to provide an artistic chipped effect. The process of making cracked nails involves two steps: using a regular nail polish as a base color and applying the crackle nail polish on top of that. Unlike regular nail polish that dries into a solid, uniform coat, crackle nail polish formula forces the layer to separate into randomly placed cracks. The effect is caused by the addition of ethanol to the formulation that causes very quick drying after application.

Chipping Chipping is characterized by small, broken, and/or missing nail polish film pieces from the nail plates. It usually happens after a few days of wearing, mainly at the tip of the nail plates. The source of the problem is the inappropriate flexibility of the nail polish film. It might also may happen if too many layers or too thick layers are applied that make the final film rigid. Plasticizers in appropriate concentration can prevent this phenomenon up to a certain extent.

Thickening Thickening means that the viscosity of a product is increased compared to the starting viscosity. A correct consistency should allow for the application of at least one coat of enamel before redipping the brush. The efficacy of solvents (diluents) usually decreases over time, which may cause the thickening of the product. Another cause is that, during continuous use, part of the volatile solvents may evaporate, causing thickening.

Evaluation of Nail Care Products

Quality Parameters Generally Tested ❼ **Parameters commonly tested to evaluate the quality of nail care products include abrasion resistance, gloss, film flexibility and hardness, drying time, adhesion test, brushability, color, dispersion of pigments, and viscosity.** The range of acceptance and other limiting factors are usually determined by the individual manufacturers. Tests that were discussed in the previous sections are reviewed in detail here.

Abrasion (Wear) Resistance Nail polish films come into direct contact with many objects during wear. The abrasion resistance of the film (i.e., its ability to resist mechanical wear) is an important property from a consumer perspective, since it determines the longevity of films on the nails. Abrasion resistance can be tested on a group of test subjects having their nails painted. Any signs of abrasion can be

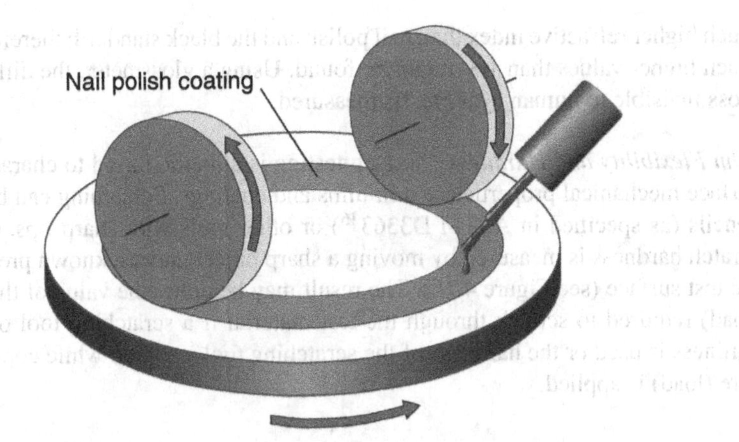

Figure 4.15 Abrasion resistance testing of nail polish films.

tested visually or using appropriate software. However, this method is not too reliable as abrasion depends on the conditions the nail polish film is exposed to, including mechanical factors and chemical agents. It may give a general idea of the wear resistance; however, without standardization (which is quite difficult), it cannot be used for quantitative analysis.

There are some other methods to test the abrasion resistance of films. One of the most frequently used equipment, called the "abraser," consists of two abrasive wheels that produce characteristic rub-wear action. Figure 4.15 shows the basic working principle of the equipment. The nail polish is applied to the steel panel. The wheels rotate while sliding on the sample film, forming a characteristic pattern on the film. There are various techniques to interpret the results, including a decrease in film thickness after a certain number of cycles (which is usually referred to as depth of wear), weight loss after a certain number of cycles (which is usually referred to as the wear index), or the number of abrasion cycles required to wear through a coating of known thickness.

Gloss Gloss is an optical property of a surface, e.g., nail polish film, characterized by its ability to reflect light. It is as important as the color itself when considering the psychological impact of products on consumers. Gloss is measured by shining a known amount of light at a surface and quantifying the reflectance (i.e., amount of light reflected). The equipment is called glossmeter, which directs light at a specific angle to the test surface and simultaneously measures the amount of reflection. The intensity is dependent on the material and angle of illumination. Many industries have adopted the 20/60/85° geometries (the numbers refer to the angle of the incident light) as specified in ISO2813/ASTM D523.[35] The measurement results of a glossmeter are compared to the amount of reflected light from a black, highly polished glass standard with a defined refractive index. This standard's reflectance is considered 100 gloss units (GUs). This scale is suitable for non-metallic films (e.g., plastics and paints) since they generally fall within this range. However, it is important to keep in mind that this number is not a value in percentage. Certain materials, such as mirrors, have much higher refractive index than nail polish and the black standard; therefore, values much higher values than 100 might be found. Using a glossmeter, the differences in gloss invisible to human eyes can be measured.

Film Flexibility and Hardness Scratch testing is a method used to characterize the surface mechanical properties of thin films and coatings. Scratching can be done by pencils (as specified in ASTM D3363[36]) or other tools with sharp tips. Generally, scratch hardness is measured by moving a sharp object under a known pressure over the test surface (see Figure 4.16). The result may be either the value of the pressure (load) required to scratch through the test material if a scratching tool of constant hardness is used or the hardness of the scratching tool is varied while constant pressure (load) is applied.

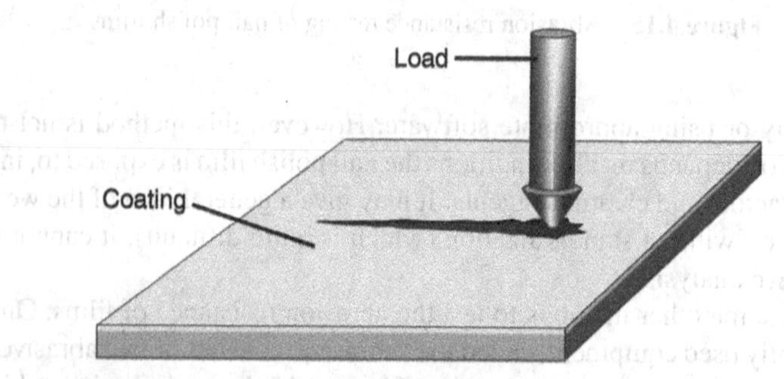

Figure 4.16 Flexibility and hardness testing of nail polish films.

Drying Time Application, performance properties, and drying time of nail polish greatly depend on the volatility of the solvents. If drying is too fast, there may be not enough time to evenly spread and level the film on the nails. However, if drying is too slow, the film does not set on the nails and can be transferred to the surfaces the customer touches. Drying time can be measured on a clean glass/metal surface by applying a thin film coating and measuring the time necessary for the complete drying of the film.

Adhesion Test For nail polish coatings to perform satisfactorily, they must adhere to the nails. A variety of recognized methods can be used to determine how well a nail polish coating is bonded to the nail plate. The commonly used measuring techniques use a tape, a knife, or a pull-off adhesion tester.

- In the tape test, a pressure-sensitive tape is applied diagonally across a coated surface, which has a specific cut pattern on its surface (see Figure 4.17). After applying the tape, it is pulled back slowly. The amount of coating removed from the surface refers to the adhesives of the nail polish to the nail or primer (base coat). This method is specified in ASTM D3359.[37]

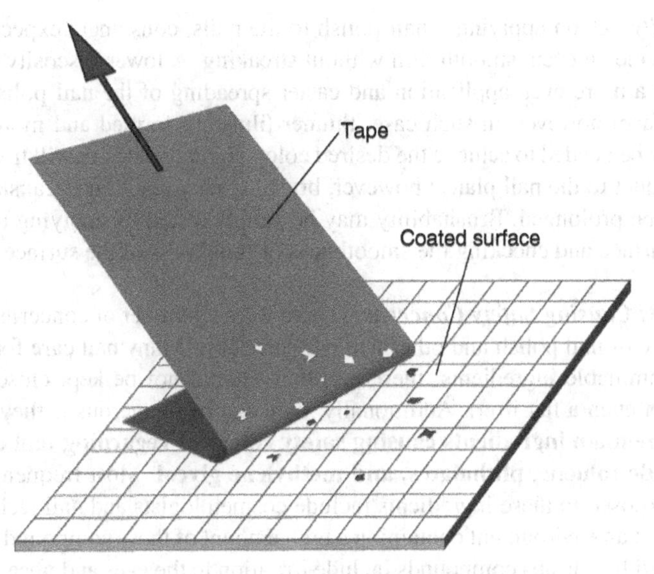

Figure 4.17 Tape testing of nail polish films.

- The knife test is a simple test in which a knife is used to remove the film from a surface. A standard method for the application and performance of this test is available in ASTM D6677.[38]
- The pull-off test is a more quantitative test where a loading fixture, commonly called a dolly, is affixed by an adhesive to a coating (see Figure 4.18). Loads are increasingly applied to the surface until the dolly is pulled off. The force required to pull the dolly off, or the force the dolly withstands, yields the tensile strength.[39]

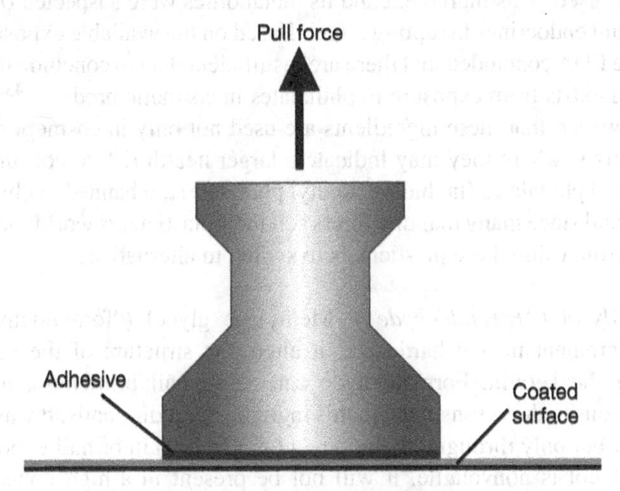

Figure 4.18 Pull-off testing of nail polish films.

Brushability Upon applying a nail polish to the nails, consumers expect the product to provide an even, smooth film without streaking. A lower-viscosity nail polish allows for a more even application and easier spreading of the nail polish over the nails' surface; however, in such case, thinner films are formed and more than two layers may be needed to achieve the desired color. Higher viscosity will transfer more of the product to the nail plates; however, brushing may result in streaks and drying time may be prolonged. Brushability may be simply tested by applying the product to a test surface and checking the smoothness or roughness of the surface.

Ingredients Causing Safety Concerns There were a number of concerns regarding the safe use of nail polish and other nail care products. Many nail care formulations contain flammable ingredients; therefore, they should not be kept close to a heat source (not even a flat iron). Additionally, they can be dangerous if they get in the eyes. ❽ **The main ingredients causing safety concerns regarding nail care products include toluene, phthalates, and methylene glycol.** Most frequently, people who are exposed to these ingredients include cosmetologists and nail technicians as they work in an environment containing a high amount of these compounds.[40] Symptoms caused by various compounds include irritation to the eyes and nose, headache, dizziness, and drowsiness, etc.

Toluene Toluene is used as a solvent in nail polish, nail hardener, and even nail polish removers. This solvent was reviewed by the CIR twice and was found to be safe for cosmetic use (in lesser than 50%) by the FDA and CIR.[41,42] However, consumer perceptions necessitate the replacement of this solvent and the use of more environmentally friendly alternatives.

Phthalates Phthalates are used in nail care products as plasticizers to make the nail polish films more flexible and resistant against chipping. Most frequently, dibutyl phthalate was used. This ingredient and its metabolites were suspected of producing teratogenic and endocrine-disrupting effect. Based on the available exposure and toxicity data, the FDA concluded that there are insufficient data to conclude that a human health hazard exists from exposure to phthalates in cosmetic products.[43,44] It should be noted, however, that these ingredients are used not only in cosmetics but also in plastics and toys, where they may indicate a larger health risk to consumers. Some frequently used phthalates (including dibutyl phthalate) are banned in children's toys in the EU[45], and since many manufacturers sell their plasticizers worldwide, the trend is to refrain from using these plasticizers to switch to alternatives.

Methylene Glycol ("Formaldehyde") Methylene glycol ("formaldehyde") is an important ingredient in nail hardeners; it alters the structure of the nail plate by cross-linking the keratin. Formaldehyde can cause nail brittleness, irritation, or allergic reactions to those sensitized to this ingredient. Additionally, it was associated with cancers, but only through inhalation and not from skin or nail exposure. Since methylene glycol is nonvolatile, it will not be present in a high concentration in the air.

The CIR Expert Panel concluded in 2012 that formaldehyde and methylene glycol are safe for use in cosmetics when formulated to ensure use at the minimal effective concentration; however, in no case should formalin concentration exceed 0.2% by weight. This would be 0.074% by weight calculated as formaldehyde or 0.118% by weight calculated as methylene glycol. In addition, the CIR also concluded that formaldehyde and methylene glycol are safe in nail care products in the current practices of use and concentration in nail hardening products.[28]

Another misunderstanding regarding formaldehyde is related to a resin called TSFR. Since it has "formaldehyde" in its name, it is assumed that it contains formaldehyde. The resin is obtained by reacting tosylamide and formaldehyde; however, the end product, the resin, essentially does not contain any formaldehyde. Allergies are still reported, which are more likely to be in connection with side products of the synthesis.[46] However, the industrial trend is to replace this resin with alternatives.

FYI

The FDA provided a detailed discussion on the safety of nail care formulations and, in particular, that of resins. For more detail, visit the FDA website (Home, Cosmetics, Product and Ingredient Safety, Product Information and "Nail Care Products").

Packaging of Nail Care Products

The most commonly used packaging materials for nail care products include the following:

Glass Bottles: Almost all types of nail polish and nail hardeners are supplied in glass bottles with varying shape. Liquid cuticle removers can also be packaged into glass bottles. The applicator brush is typically attached to the cap of the bottle. Some nail polish removers are packaged into simple glass bottles.

Plastic Bottles: The majority of nail polish removers are marketed in simple plastic bottles.

Soft Tubes: Cuticle softener, cuticle remover, and nail moisturizer products are generally available in soft tubes fitted with a nozzle applicator.

Plastic Jar: As mentioned previously, there is a special type of nail polish removers available as a piece of sponge soaked with a nail polish remover solution. These products are supplied in jars. There is a hole in the middle of the sponge, and the consumers dip their finger into the sponge through this hole. Nail polish remover pads are also prewetted with a nail polish remover solution and are generally supplied in plastic jars. Certain manufacturers package these pads individually, which are convenient for traveling, and place them in paper boxes. Nail moisturizers may also be packaged into plastic jars.

Nail Tips: Nail tips are available in "ready-to-use" form, usually supplied in a kit with all the necessary tools and ingredients.

GLOSSARY OF TERMS FOR SECTION 4

Abrasion resistance: A measure of the resistance of nail polish films to wearing off.

Artificial nails: Nails created using products designed to enhance nail length.

Base coat: A clear nail polish usually applied to the clean nail plate before the nail polish. It prevents the nail polish from staining the nails and creates a smooth nail surface to which the nail polish will have better adhesion.

Brushability: The easiness of applying nail polish to the nails with a brush.

Bubbling: A quality problem leading to the formation of bubbles on newly painted nails.

Chipping: A quality problem leading to the loss of nail polish film pieces from the nails.

Cracking: A quality problem resulting in some cracking in nail polish films.

Cuticle remover: A personal care product designed to chemically remove cuticle.

Cuticle softener: A personal care product designed to soften the cuticle for easy mechanical removal by subsequent cutting or trimming.

Cuticle: The skinfold at the base of the nail.

Formaldehyde: A highly reactive gas. Its aqueous solution is often used in nail hardeners due to its cross-linking properties.

Nail hardener: A cosmetic product designed to develop a protective layer on the nail plate; they increase the hardness and strength of nails.

Nail moisturizer: A personal care product designed to increase the hardness of the nails by supplying them with moisturizing ingredients.

Nail plate: The external visible part of the nail. It is the application surface for nail polish.

Nail polish remover: A product designed to strip nail polish from the nail plate without removing layers of nail plate.

Nail polish: A color cosmetic designed to color and enhance the aesthetics of the nails, giving them a more attractive aspect.

Resin: A film-forming ingredient used in nail polish formulations to hold the ingredients together and form a strong film on the nails.

Thixotropy: A type of non-Newtonian rheological behavior characterized by a reversible gel-to-sol formation with no change in volume or temperature.

Top coat: A clear nail polish applied over nail polish. It provides a protective layer and prevents the applied polish from fading and chipping as well as enhances gloss and reduces drying time.

REVIEW QUESTIONS FOR SECTION 4

Multiple Choice Questions

1. Which of the following is NOT true for human nails?
   a) The nail matrix is pink due to the capillaries underneath it
   b) The nail matrix cannot regenerate after injury
   c) The nail plate is made of keratin, water, lipids, and minerals
   d) The nail folds surround and protect the nail unit
2. Which of the following can lead to brittle nails?
   a) Overuse of nail hardeners that do not contain cross-linking agents
   b) Overuse of nail polish removers
   c) Overdose of biotin
   d) All of the above
3. What is the function of plasticizers in nail polish formulations?
   a) To increase the viscosity of the formulations
   b) To keep resins soft and flexible
   c) To dissolve resins
   d) To improve the adhesion of the products to the nails
4. Which of the following is used to determine the adhesion of nail polish films to the nail plates' surface?
   a) Abrader
   b) Pull-off test
   c) Scratch test
   d) Glossmeter
5. How do nail polish removers remove nail polish?
   a) By redissolving the resins
   b) By reemulsifying the resins
   c) By solubilizing the resins
   d) By physically removing the film

6. How do UV gels harden on exposure to UV light?
   a) An acid–base reaction starts, which results in hardened nails
   b) An oxidation–reduction reaction starts, which results in hardened nails
   c) A polymerization reaction starts, which results in hardened nails
   d) UV gels harden due to the heat from the UV lamp

7. Hard resins provide the nail polish film with ____.
   a) Flexibility
   b) Hardness
   c) Softness
   d) Color

8. Which of the following can lead to cracking of nail polish films?
   a) Low plasticizer concentration
   b) Low resin concentration
   c) Low color additive concentration
   d) Low thickener concentration

9. An abraser evaluates the ___ of nail polish films.
   a) Adhesion
   b) Gloss
   c) Brushability
   d) Wear resistance

10. Which of the following is the application surface for nail polish?
    a) Cuticle
    b) Nail matrix
    c) Nail plate
    d) Nail bed

Fact or Fiction?

______ a) Keratin is responsible for the strength of human nails.
______ b) Magnetic nail polish has to be removed before an MRI.
______ c) UV filters in nail polish provide nail cancer prevention.
______ d) Formaldehyde is a gas.

Matching

Match the ingredients in column A with their appropriate ingredient category in column B.

	Column A		Column B
_____	A. Benzoyl peroxide	1.	Active ingredient for fungal infection
_____	B. Camphor	2.	Cross-linking agent
_____	C. Ethyl acetate	3.	Cuticle-dissolving agent
_____	D. Formaldehyde	4.	Opacifier
_____	E. Nitrocellulose	5.	Plasticizer
_____	F. Polyester resin	6.	Polymerization initiator
_____	G. Sodium hydroxide	7.	Primary film-former
_____	H. Stearalkonium bentonite	8.	Secondary film-former
_____	I. Titanium dioxide	9.	Solvent
_____	J. Tolnaftate	10.	Thickener

REFERENCES

1. Tosti, A., Piraccini, B. M.: Biology of Nails and Nail Disorders, In: Wolff, K., Goldsmith, L.A., Katz, S. I., et al., eds: *Fitzpatrick's Dermatology In General Medicine*, New York: McGraw-Hill Companies Inc., 2007:778–794.
2. Barnhart, C. A.: *The Facts On File Student's Dictionary of American English*, New York: Facts On File, Inc., 2008:392.
3. Bharathi, R. R., Bajantri, B.: Nail bed injuries and deformities of nail. *Indian J Plast Surg*. 2011;44(2):197–202.
4. Sterry, W., Paus, R., Burgdorf, W. H. C.: *Dermatology*. Stuttgart: Thieme, 2005:32.
5. Lawry, M., Rich, P.: The nail apparatus: a guide for basic and clinical science. *Curr Probl Dermatol*. 1999;11:202–204.
6. American Academy of Dermatology: Tips for Healthy Nails, Accessed 9/14/2013 at http://www.aad.org/dermatology-a-to-z/health-and-beauty/nail-care/tips-for-healthy-nails
7. Seshadri, D., De, D.: Nails in nutritional deficiencies. *Ind J Dermatol Venerol Leprol*. 2012;78:237–241.
8. Ohgitani, S., Fujita, T., Nishio, H., et al.: Nail calcium and magnesium content in relation to age and bone mineral density. *J Bone Miner Metab*. 2005;23:318–322.
9. Barba, C., Méndez, S., Martí, M., et al.: Water content of hair and nails. *Thermochim Acta*. 2009;494(1–2):136–140.
10. Sugawara, T., Kawai, M., Suzuki, T.: The relationship between moisture content of human fingernails and the mechanical properties of the fingernail. *J Soc Cosmet Chem Jpn*. 1999;33:283–289.

11. Cashman, M. W., Sloan, S. B.: Nutrition and nail disease. *Clin Dermatol.* 2010; 28:420–425.
12. Paus, R., Peker, S.: Biology of Hair and Nails, In: Bolonia, J., Jorizzo, J., Rapini, R., eds: *Dermatology*, London: Mosby, 2003;1:1007–1032.
13. Yaemsiri, S., Hou, N., Slining, M. M., et al.: Growth rate of human fingernails and toenails in healthy American young adults. *J Eur Acad Dermatol Venereol.* 2010;24(4):420–423.
14. Patrick, B. K., Thompson, J. M.: *An Uncommon History of Common Things*, Washington: National Geographic Books, 2009:203–204.
15. Gorton, A.: *History of Nail Care*, Nails, Torrance: Bobit Business Media, 1993.
16. Discovery Fit and Health: Nail Polish, Accessed at 9/14/2013 at http://health.howstuffworks.com/skin-care/nail-care/tips/nail-polish.htm
17. NSI: Fred Slack, Accessed 9/14/2013 at http://www.nsinails.com/aboutus/staffDetail/name_stripped/fred-slack.html
18. Daniel, C. R., Elewski, B.: Simple brittle fingernails. *J Cosmet Dermatol.* 2001; 14:53–54.
19. Iorizzo, M., Pazzaglia, M., Piraccini, B. M., et al.: Brittle nails, *J Cosmet Dermatol.* 2004;3:138–144.
20. Rich P.: Nail Cosmetics: The Benefits and Pitfalls, In: Scher, R. K., Daniel, C. R., Tosti, A., et al., eds: *Nails: Diagnosis, Therapy, and Surgery*, China: Elsevier Saunders, 3rd Edition, 2005:221–227.
21. Rich, P.: Nail cosmetics. *Dermatol Clin.* 2006;24:393–399.
22. Stern, D. K., Diamantis, S., Smith, E., et al.: Water content and other aspects of brittle versus normal fingernails. *J Am Acad Dermatol.* 2007;57(1):31–36.
23. Rich, P., Kwak, H.: Nail Physiology and Grooming, In: Draelos, Z. D., ed.: *Cosmetic Dermatology Products and Procedures*, 1st Edition, Hoboken: Wiley-Blackwell, 2010:197–205.
24. Rich, P.: Nail Cosmetics. In: Rich, P., Scher, R., eds: *An Atlas of Diseases of the Nail.* New York: Parthenon Publishing, 2003;97–100.
25. Baran, R.: Nail cosmetics: allergies and irritations. *Am J Clin Dermatol.* 2002; 3(8):547–555.
26. Bryson, P. H., Sirdesai, S. J.: Colored Nail Cosmetics and Hardeners, In: Draelos, Z. D., ed.: *Cosmetic Dermatology Products and Procedures*, 1st Edition, Hoboken: Wiley-Blackwell, 2010:206–214.
27. Baran, R., Schoon, D.: Nail fragility syndrome and its treatment. *J Cosmet Dermatol.* 2004;3:131–137.
28. CIR Expert Panel Meeting Minutes, 2012, Accessed 9/25/2013 at http://www.cir-safety.org/sites/default/files/formy_build.pdf
29. Helsing, P., Austad, J., Talberg, H. J.: Onycholysis induced by nail hardener. *Contact Dermatitis.* 2007;57:280–281.
30. van de Kerkhof, P. C., Pasch, M. C., Scher, R. K., et al.: Brittle nail syndrome: a pathogenesis-based approach with a proposed grading system. *J Am Acad Dermatol.* 2005;53(4):644–651.
31. Scheinfeld, N., Dahdah, M. J., Scher, R.: Vitamins and minerals: their role in nail health and disease. *J Drugs Dermatol.* 2007;6(8):782–787.
32. Medscape: Draelos, Z. D.: Nail Cosmetics, Last update: 9/19/2012, Accessed 9/15/2013 at http://emedicine.medscape.com/article/1067468-overview#aw2aab6b3

33. Nasir, A., Goldstein, B., van Cleeff, M., et al.: Clinical evaluation of safety and efficacy of a new topical treatment for onychomycosis. *J Drugs Dermatol.* 2011;10(10):1186–1191.
34. San Francisco Department of the Environment Regulations, Accessed 9/21/2013 at http://www.sfenvironment.org/sites/default/files/fliers/files/sfe_th_hnsrp_reg_12-01.pdf
35. ASTM D523-08: Standard Test Method for Specular Gloss, DOI: 10.1520/D0523-08
36. ASTM D3363: Standard Test Method for Film Hardness by Pencil Test, DOI: 10.1520/D3363-05R11E02
37. ASTM D3359: Standard Test Methods for Measuring Adhesion by Tape Test, DOI: 10.1520/D3359-09E02
38. ASTM D6677: Standard Test Method for Evaluating Adhesion by Knife, DOI: 10.1520/D6677-07R12
39. ASTM D4541: Standard Test Method for Pull-Off Strength of Coatings Using Portable Adhesion Testers, DOI: 10.1520/D4541-09E01
40. Tsigonia, A., Lagoudi, A., Chandrinou, S., et al.: Indoor air in beauty salons and occupational health exposure of cosmetologists to chemical substances. *Int J Environ Res Public Health.* 2010;7:314–324.
41. Alexopoulos, E. C., Chatzis, C., Linos, A.: An analysis of factors that influence personal exposure to toluene and xylene in residents of Athens, Greece. *BMC Public Health.* 2006;6:50.
42. McNary, J. E., Jackson, E. M.: Inhalation exposure to formaldehyde and toluene in the same occupational and consumer setting. *Inhalat Toxicol.* 2007;19:573–576.
43. Hubinger, J. C., Havery, D. C.: Analysis of consumer cosmetic products for phthalate esters. *J Cosmet Sci.* 2006;57(2):127–137.
44. Koo, H. J., Lee, B. M.: Estimated exposure to phthalates in cosmetics and risk assessment. *J Toxicol Environ Health A.* 2004;67(23–24):1901–1914.
45. Directive 2005/84/EC of the European Parliament and of the Council of 14 December 2005 amending for the 22nd time Council Directive 76/769/EEC on the approximation of the laws, regulations and administrative provisions of the Member States relating to restrictions on the marketing and use of certain dangerous substances and preparations (phthalates in toys and childcare articles). Europa: Phthalate-containing soft PVC toys and childcare articles, Accessed 9/28/2013 at http://europa.eu/legislation_summaries/internal_market/single_market_for_goods/technical_harmonisation/l32033_en.htm
46. Hausen, B. M., Milbrodt, M., Koenig, W. A.: The allergens of nail polish. (I). Allergenic constituents of common nail polish and toluenesulfonamide-formaldehyde resin (TSFR) *Contact Dermatitis.* 1995;33(3):157–164.

5

HAIR CARE PRODUCTS

INTRODUCTION

The hair care market is one of the largest personal care markets all over the world. Washing the hair and scalp has become a near-universal practice. In the past, the main aim of using hair care products was to clean the hair by removing soilage and dirt. Today, hair care products are desired to provide additional benefits, such as beautifying the hair, making it easy to handle, or repairing damages. In addition to basic hair care formulations, there are special products for the treatment of certain hair problems, such as dandruff. A huge section of the hair care market is focused on styling products that include fixatives, coloring aids, as well as products that change the natural physical and/or chemical nature of the hair fiber, known as straightening and waving products. Hair is an extremely important factor in personal appearance; therefore, appropriate care must be taken. Despite its great strength, hair is prone to damage by weathering, chemical attack, heat, and abrasion; thus, appropriate products should be used to achieve the best results.

Hair care, color, and style play an important role in people's physical appearance and self-perception. Hair defines an individual's gender, age, sexual attitude, and social status. There are no significant differences in the number of hair follicles between men and women or among different races.

Introduction to Cosmetic Formulation and Technology, First Edition. Gabriella Baki and Kenneth S. Alexander.
© 2015 John Wiley & Sons, Inc. Published 2015 by John Wiley & Sons, Inc.

This chapter provides a basic understanding of the structure of human hair and its main functions as well as various products that are applied to the hair. These include products used to remove dirt and hair styling aids from the hair, improve hair quality, and/or temporarily or permanently change its shape or color. It also reviews the main characteristics, ingredients, formulation technology, testing methods, and packaging materials for hair cleansing products, hair conditioners, hair styling products, and hair coloring products.

SECTION 1: HAIR ANATOMY AND PHYSIOLOGY

LEARNING OBJECTIVES

Upon completion of this section, the reader will be able to

1. define the following terms:

Anagen phase	Catagen phase	Cortex	Cuticle
Eumelanin	Hair bulb	Hair follicle	Hair matrix
Hair papilla	Hair root	Hair shaft	Hair weathering
Lanugo	Medulla	Pheomelanin	Porous hair
Strong bonds	Telogen phase	Terminal hair	Vellus hair
Weak bonds			

2. differentiate between the hair root and the hair shaft;
3. name the part of the hair root where new hair cells form;
4. explain how a hair follicle receives oxygen and nutrients;
5. explain to what the term "pilosebaceous unit" refers;
6. name the three main layers of a hair shaft, and explain their function;
7. name the three phases of the hair growth cycle;
8. briefly discuss what is considered normal hair loss;
9. name some factors that can influence hair growth;
10. name the main constituents of hair fibers;
11. differentiate between strong and weak bonds in the hair fibers;
12. explain why strong bonds in the hair fibers are called strong;
13. briefly discuss the hair's surface charge and its importance;
14. explain what the strength of hair fibers depend on;
15. briefly discuss how water changes the hair fibers' structure;
16. name some of the factors that may contribute to hair weathering;

17. classify the hair based on the following factors: thickness, color, condition, shape and size, curliness, and greasiness;
18. differentiate between eumelanin and pheomelanin.

KEY CONCEPTS

1. Hair is a flexible thin keratin thread with great strength and elasticity. It is present on almost all surfaces of the human skin.
2. The cross-section of a hair shaft has three major components, from the outside to the inside: the cuticle, cortex, and medulla.
3. Hair growth is a unique and complex process that involves continuous cycles of growth and regeneration (anagen phase), transition (catagen phase), and resting (telogen phase).
4. Shedding is part of the normal process of the replacement of old hair with new.
5. Hair fibers are primarily composed of various types of keratins. The keratin fibers consist of long molecular chains intertwined and firmly attached through various bonds.
6. Bonds found in hair fibers are generally classified as strong bonds, including disulfide bonds, and weak bonds, including van der Waals forces, salt bonds, and hydrogen bonds.
7. Hair usually carries a negative charge.
8. The physical properties of human hair are related to its shape and internal constituents.
9. Weathering of hair is the progressive hair damage primarily affecting the end of the hair fibers. It results from external factors.
10. Hair can be classified in a number of different ways, based on its color, thickness, shape, texture, length, and curliness.

Introduction

In humans, hair has an aesthetic function influencing our appearance. For centuries, decoration and styling of the scalp hair have been means of social communication as well as display of social identity and status. Today, it also has social, psychological, and sexual significance. Any change in the pattern of the hair, such as hair loss, hair overgrowth, or color change, may negatively affect the self-esteem of individuals and has emotional consequences.[1] In addition, hair serves as an aid in camouflage and protection from the sun and provides sensory, tactile information about the environment.

Structure and Function of Human Hair

● **Hair is a flexible thin keratin thread with great strength and elasticity. It is present on almost all surfaces of the human skin,** except for the palms, soles, vermilion zone of the lips, and certain genital parts. Each hair consists of a **root** embedded in the dermis and a **hair shaft** protruding above the surface of the skin. The hair root is surrounded by a tube-like sheath made of epithelial cells that form a downward extension of the epidermis into the dermis. This is called the **hair follicle.**[2] The base of the root and hair follicle is slightly larger than the rest of the root; this onion-shaped structure is called the **hair bulb**. The hair bulb receives a cluster of blood vessels from the dermis which pushes into the bulb to form the **hair papilla** (otherwise known as the dermal papilla). Oxygen and nutrients via the blood vessels supply the actively growing cells in the hair follicle around the hair bulb known as the **hair matrix**. These cells are the only source of new hair. The structure of human hair is shown in Figure 5.1.

The hair shaft itself arises from the bulb region of the root and is produced by rapidly multiplying matrix keratinocytes.[3] The growth of hair is similar to that of skin cells; as cells divide and grow, they push older cells upward away from the blood supply, resulting in gradual cell death and keratinization. About midway to the surface, all cells forming the hair root die and complete the process of keratinization. The shaft is, therefore, made up entirely of dead cells composed mainly of keratin. These cells remain attached to each other by an intercellular cement-like substance.

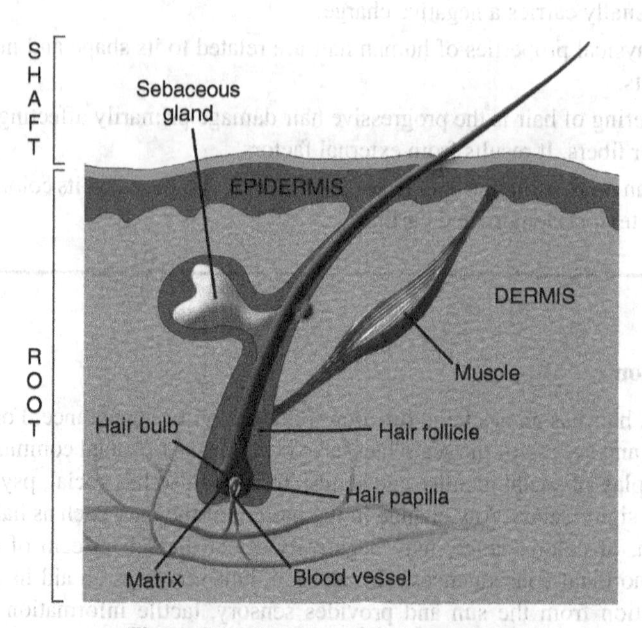

Figure 5.1 Structure of the human hair.

DID YOU KNOW?

In contrast to the common belief, shaving or cutting does not affect hair growth rate since the portion being cut or shaved off is already made up of dead cells.

As mentioned previously, hair follicles are part of the **pilosebaceous unit**, which also includes one or more sebaceous glands and a small muscle. Sebaceous glands, which produce sebum that coats the hair and skin, open into the hair follicle. An arrector pili muscle is attached to the hair follicle; when this muscle contracts, it causes the hair to stand up.

● **The cross-section of a hair shaft has three major components, from the outside to the inside: the cuticle, cortex, and medulla** (shown in Figure 5.2).

The outermost structure is the **cuticle**. It is composed of multiple layers of keratinized, flattened cells (scales), which overlap in a roof-tile formation with an intercellular cement to bind them together. It is translucent, allowing light to penetrate the cortex pigments.[1] It is composed primarily of keratin. The cuticle protects the underlying cortex, acts as a barrier, regulates the water content of the hair fiber, and is responsible for the luster and the texture of hair. The cuticle is the target for hair conditioning products, since it is responsible for many of the observable characteristics of hair, including its texture.[4] Shiny and soft hair reflects a healthy cuticle, while dry, brittle hair is the result of damaged cuticle cells.

The **cortex** is the major component of the hair shaft. It lies below the cuticle and contributes to the mechanical properties of the hair fiber, including strength, elasticity, and curliness. The cortex consists of elongated cortical cells rich in keratin filaments as well as an amorphous matrix of sulfur proteins. In addition, variable accumulations of air spaces are present in the cells and intercellular spaces.[5] It is the

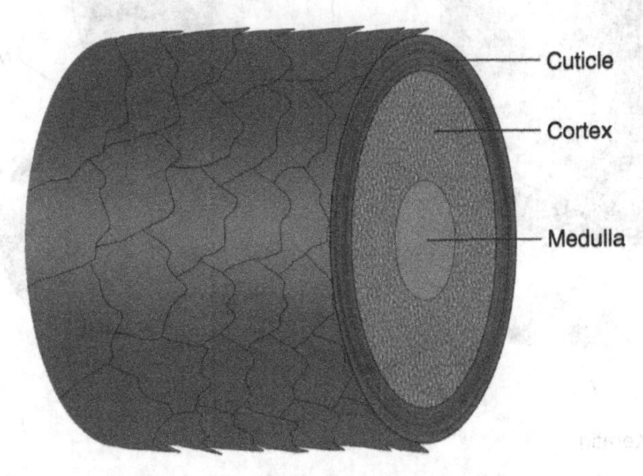

Figure 5.2 Cross section of the human hair.

presence of melanin in the cortex that gives hair color; otherwise, the fiber would not be pigmented. During aging, the amount of pigment produced is less and the number of air spaces increases. These are responsible for graying of the hair. In addition, the changes involved in oxidative hair coloring, permanent waving and straightening, and thermal styling all take place in the cortex. The cortex, similar to the cuticle, has a great cosmetic importance, as its optical properties strongly affect the color and shine of the hair fiber.[6]

DID YOU KNOW?

Cortical cells found in the hair cortex have a very unique microstructure (see Figure 5.3), which provides the hair with its great strength and elasticity. Each cortical cell is formed from bundles of keratin macrofibrils. Each macrofibril is, in fact, a collection of smaller elements called the microfibrils. Mircofibrils themselves are made of 11 protofibrils tightly assembled in a cable-like formation. Protofibrils are still not the basic elements of hair cortical cells since each protofibril is made up of 4 keratin chains twisted together in a cable-like format.[7,8]

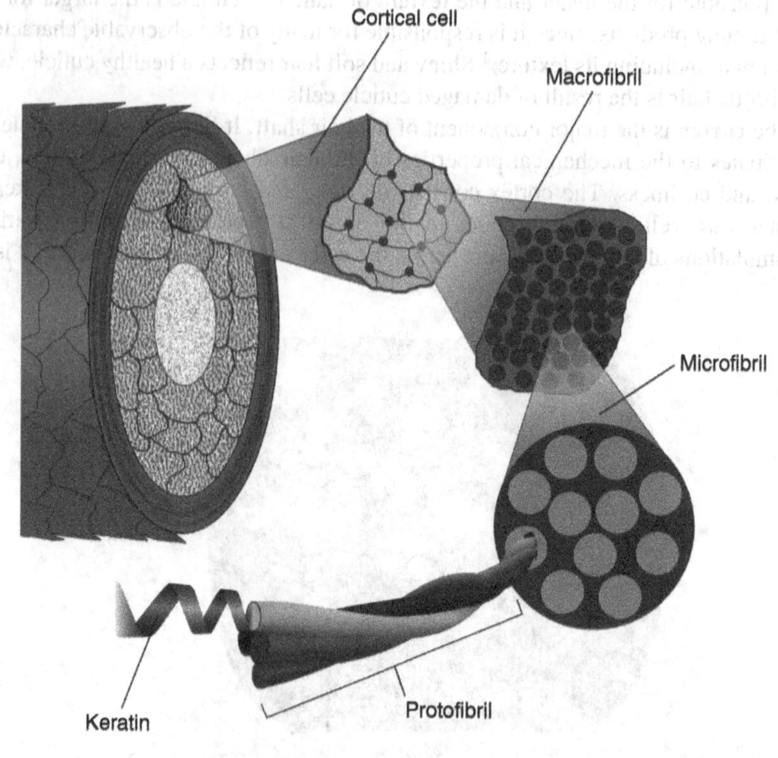

Figure 5.3 Microstructure of the human hair cortex.

The **medulla** (often referred to as the "core") is composed of flattened, cornified cells. It appears as continuous, discontinuous, or absent structure under microscopic examination of human hair fibers. It is viewed as a framework of keratin supporting thin shells of amorphous material bonding air spaces of variable size.

Hair Growth Cycle and Hair Loss

❸ **Hair growth is a unique and complex process that involves continuous cycles of growth and regeneration (anagen phase), transition (catagen phase), and resting (telogen phase)** (depicted in Figure 5.4). The cyclic activity continues throughout life; but the phases of the cycle change with age.[9]

- During the **anagen** phase, new hair is produced in the lower part of the hair follicle. Normally, most of the scalp hair (approximately 85–90%) are in their anagen phase at any time, while the remaining 10% is in the telogen or catagen phase.[10] On the scalp, the anagen phase can last 2–6 years[11]; however, in some cases, it may be longer (even 8 years).[12] The longer the anagen phase, the longer the hair is able to grow. The difference in individual's hair length can be related to the varying length of the anagen phase. Scalp hair grows at a normal rate

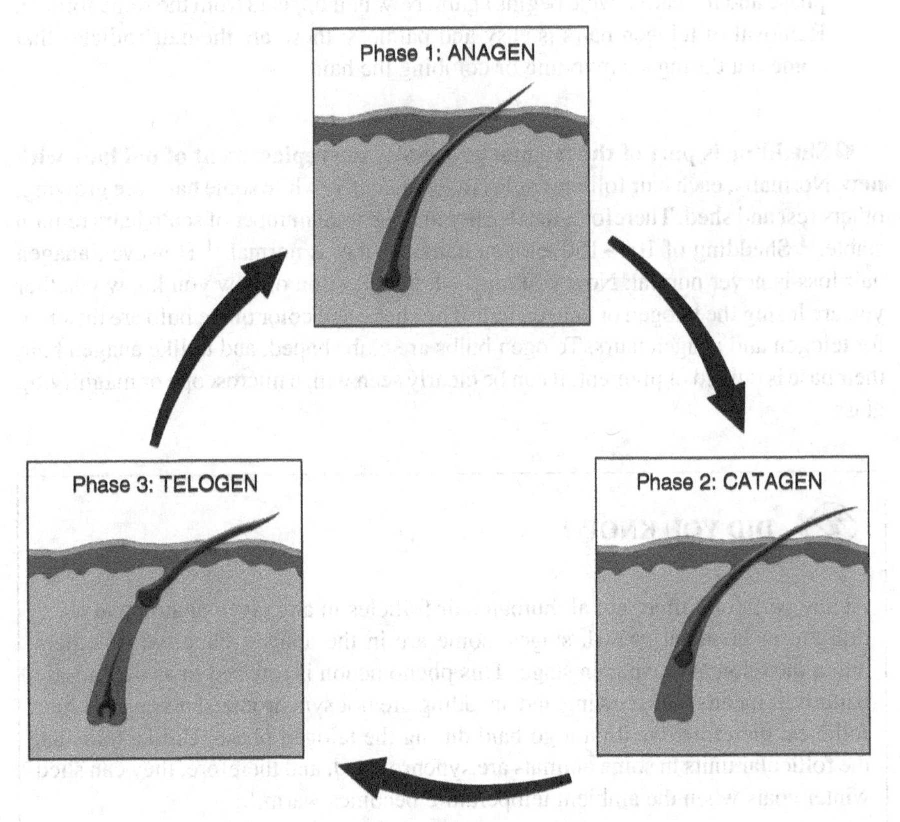

Figure 5.4 Hair growth cycle.

of about 1 mm every 3 days.[2] The hair on the arms, legs, eyelashes, and eyebrows have a much shorter anagen phase and a slower growth rate, explaining why it is much shorter than the scalp hair (the general length is in the range of 1–3 cm).

- The **catagen** phase is a brief transition phase between the growth and the resting phases, which marks the end of the growth phase. On the scalp, the catagen phase usually lasts between 2 and 3 weeks.[3] During this phase, cell division stops, the follicle tube shrinks and detaches from the dermal papilla, and the base of the follicle moves upward toward the surface of the skin. Melanin production stops in this phase, leading to a non-pigmented lower end in the hair (which is under the scalp until it falls off).
- The **telogen** phase is the final phase and lasts until the fully grown hair is shed. Although the telogen phase is called the resting phase, many activities occur during this phase, which allows the hair shaft to be shed and stimulates the conditions essential for regrowth.[3] The hair either shed during the telogen phase or remains in place until the next anagen phase, when the new hair growing in pushes it out. On the scalp, the telogen phase usually lasts for approximately 2–3 months.[11] As soon as the telogen phase ends, the hair returns to the first phase and the entire cycle begins again. New hair appears from the same follicle. Removal of telogen hairs is easy and painless; these are the hair follicles that come out during shampooing or combing the hair.

Shedding is part of the normal process of the replacement of old hair with new. Normally, each hair follicle cycles independently; while some hairs are growing, others rest and shed. Therefore, the density and the total number of scalp hairs remain stable.[13] Shedding of 100–150 telogen hairs per day is normal.[11] However, anagen hair loss is never normal. Now you may ask the question of how you know whether you are losing the telogen or anagen hair. The shape and color of the bulb are different for telogen and anagen hairs. Telogen bulbs are club shaped, and unlike anagen hair, their base is devoid of pigment. It can be clearly seen with a microscope or magnifying glass.

DID YOU KNOW?

At any particular time, not all human hair follicles in any given anatomical location are in identical growth stages; some are in the anagen stage, while others are in the telogen or catagen stage. This phenomenon is referred to as the mosaic pattern. It means that growing and shedding are not synchronized among the hair follicles; therefore, we do not go bald during the telogen phase. Unlike humans, the follicular units in some animals are synchronized, and therefore, they can shed winter coats when the ambient temperature becomes warm.[1,14]

Among the factors influencing hair growth mentioned are systemic factors, such as hormones, including androgens, estrogens, and thyroid hormones; growth factors and cytokines; as well as external factors linked to the environment, such as toxins, and deficiencies of nutrients, vitamins, and energy.[1]

Chemical Composition of Human Hair

❺ **Hair fibers are primarily composed of various types of keratins (proteins). The keratin fibers consist of long molecular chains intertwined and firmly attached through various bonds.** Additional ingredients include water, lipids, melanin, as well as trace amounts of elements, such as aluminum, chromium, calcium, copper, iron, manganese, magnesium, and zinc. Deep within the hair structure, cross-bonds and linkages create a network of strength that reinforces the hair fibers and allows them to remain responsible to various styling techniques.

❻ **Bonds found in hair fibers are generally classified as strong bonds, including disulfide bonds, and weak bonds, including van der Waals forces, salt bonds, and hydrogen bonds** (see Figure 5.5).[15,16]

- **Strong bonds**
  - *Disulfide Bonds*: Hair keratin is made up of amino acids, where cysteine is one of the most important among them. Cysteine molecules with their sulfur

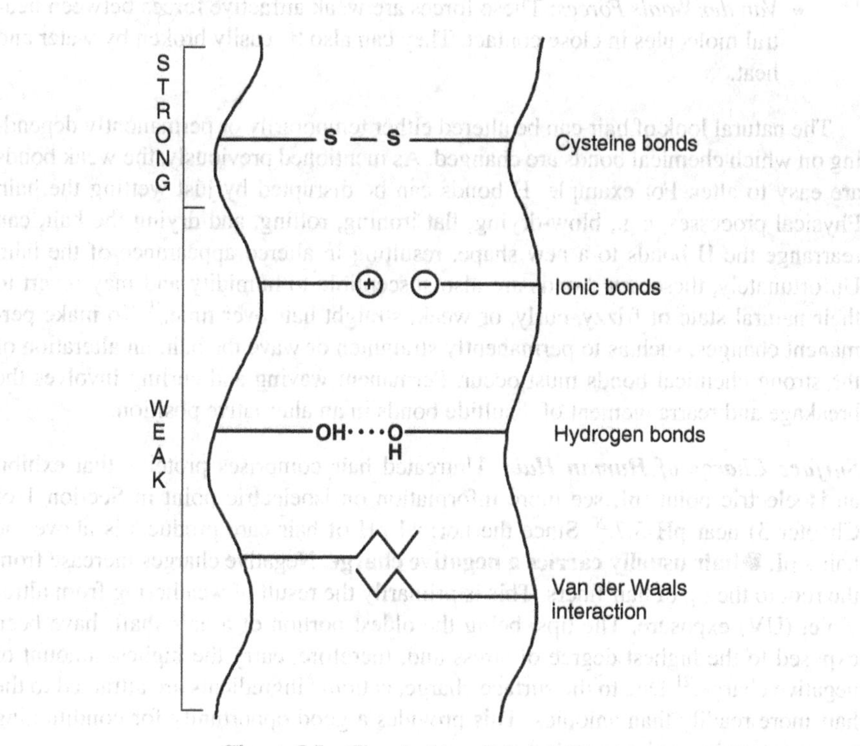

Figure 5.5 Chemical bonds in hair fibers.

atoms are able to form a very strong bond known as the disulfide bond.[17] Hair fibers contain a large number of cysteine bonds, which results in a significant effect on the physical properties of hair fibers. They contribute much to the shape, stability, and texture of the hair. When two cysteine molecules are bound together through a disulfide bond, a molecule called cystine is formed. Disulfide bonds are called strong bonds, since they cannot be broken up by heat or water, only chemically. These bonds remain intact when the hair is wet, allowing the hair to resume its original shape.

- **Weak bonds**
  - *Hydrogen Bonds (H-bonds)*: H-bonds are relatively weak and can be easily broken by water and heat. They are primarily responsible for changing the hair's overall shape. Although individual H-bonds are weak, they are present in the highest number of all types of bonds; therefore, they significantly contribute to the strength of hair fibers.
  - *Salt Bonds*: Salt bonds are formed between the positive end of one amino acid chain and the negative end of the adjacent amino acid chain.[18] They are also known as ionic bonds or coulomb interactions. These bonds are sensitive to pH; therefore, they are easily broken by strong alkaline or acidic solutions. Although these are weak bonds, together they account for a significant part of the strength of hair fibers.
  - *Van der Waals Forces*: These forces are weak attractive forces between neutral molecules in close contact. They can also be easily broken by water and heat.

The natural look of hair can be altered either temporarily or permanently depending on which chemical bonds are changed. As mentioned previously, the weak bonds are easy to alter. For example, H-bonds can be disrupted by just wetting the hair. Physical processes, e.g., blow-drying, flat ironing, rolling, and drying the hair, can rearrange the H-bonds to a new shape, resulting in altered appearance of the hair. Unfortunately, these new bonds are also susceptible to humidity and may revert to their natural state of frizzy, curly, or weak, straight hair over time.[19] To make permanent changes, such as to permanently straighten or wave the hair, an alteration of the strong chemical bonds must occur. Permanent waving and curling involves the breakage and rearrangement of disulfide bonds in an alternative position.

Surface Charge of Human Hair Untreated hair comprises proteins that exhibit an isoelectric point (pI, see more information on isoelectric point in Section 1 of Chapter 3) near pH 3.7.[20] Since the normal pH of hair care products is above the hair's pI, ● **hair usually carries a negative charge.** Negative charges increase from the root to the tip of hair fibers. This is primarily the result of weathering from ultraviolet (UV) exposure. The tips, being the oldest portion of a hair shaft, have been exposed to the highest degree of stress and, therefore, carry the highest amount of negative charge.[21] Due to the surface charge, cationic ingredients are attracted to the hair more readily than anionics. This provides a good opportunity for conditioning treatments using cationic molecules.

Physical Properties of Human Hair

❽ **The physical properties of human hair are related to its shape and internal constituents** (including the cuticle and cortex). Changes in the geometric shape and damage to the cuticle and cortex can significantly affect the hair's elasticity, strength, and other properties.

- Hair fibers are very **durable and strong** primarily due to keratin located in the cortex. Healthy hair fibers have a tensile strength similar to that of a copper wire of the same diameter. However, to resist externally applied forces, a healthy cuticle is also necessary. Damage to the cuticle significantly weakens the overall strength of the hair and may result in splitting and breakage.
- **Elasticity** is another important property of hair fibers. This property lets hair spring back to its original form after physical stress, such as grooming, without damage. Due to this property, hair can withstand forces that could change its shape, volume, or length. When healthy hair is wetted and stretched, it can increase in length by up to 30% and still return to its original length when it is dried. However, more than this extent of stretching can lead to irreversible changes, such as permanent lengthening and even breaking. The elastic properties of both wet and dry hair are related to the diameter of the hair shaft. The thicker the hair, the more it will tend to resist stretching.
- The **water content** of the hair varies according to the relative humidity of the surrounding atmosphere. For a healthy appearance, hair fibers ideally need to retain approximately 17% humidity; however, water retention up to 35% is possible.[18,22] When hair is wet, the cortex swells and the edges of the cuticle scales tend to lift. The hair surface temporarily loses its smoothness, and there is more friction when wet hair is rubbed. This can lead to matting and tangles developing during overvigorous shampooing or combing when wet.
- Hair has a high **electrical resistivity** and a fairly **low dielectric constant**. In practice, it means that it is easy to generate electrostatic charges by brushing and combing hair. It is especially noticeable in hot, dry weather. Charged hair is referred to as "flyaway" hair, which stands out from the head.

Hair Weathering

❾ **Weathering of hair is the progressive hair damage primarily affecting the end of the hair fibers. It results from external factors,** including environmental factors, such as UV radiation, cosmetic treatments, combing and brushing, which cause friction; blow-drying or straightening with flat irons using excessive heat; permanent straightening and waving using chemicals; as well as coloring and bleaching using harsh chemicals. The results of these insults are mechanical abrasion of the cuticle; damage of the cortex, altering the tensile strength of the fibers; and depletion of amino acids from keratin.[23] The visible signs of weathering include damaged cuticle, split ends, and fracture along the fibers. Generally, all hair exhibits some degree of weathering; however, longer hair, subjected to repeated insults, inevitably shows more severe changes of weathering.

Hair Types

⓾ **Hair can be classified in a number of ways, based on its color, thickness, shape, texture, length, and curliness.** The main classifications are discussed in the following sections.

Classification Based on Thickness Hair thickness refers to the diameter of individual hair shafts. Generally, hair can be classified as coarse, medium, or fine. A person may have a mixture of these types with one being the most prevalent.[24]

- **Coarse** hair is the thickest. It is often more resistant to chemical processes, such as hair perming, than medium or fine hair and usually requires more time for completing such procedures.
- **Medium** hair is usually considered the standard to which other types are compared.
- **Fine** hair is the thinnest and, therefore, is more fragile and more susceptible to damage.

Classification Based on Color Natural color is the result of melanin present in the cortex. There are two types of melanin: **eumelanin**, which provides dark brown and black color to hair, and **pheomelanin**, which provides red to blond tones. Natural hair color ranges from black, brown, and blond to red with subtle hues in each category. The diversity of hair pigmentation results mostly from the quantity and ratio of eumelanin and pheomelanin.[25]

Classification Based on Condition The hair's condition refers to the smoothness and softness of the hair by touch, which is in direct relationship with the cuticle's intactness. This is commonly called porosity, which reflects the hair fibers' ability to absorb moisture.

- **Healthy** hair with a compact cuticle layer is naturally resistant to moisture.
- If the cuticle is **damaged,** hair has a course texture and loses its shine and smoothness. In such cases, the cuticle is raised, meaning that the layers are not tightly packed. This type of hair is referred to as porous or overporous hair if damaged severely. Porous hair can absorb liquids very easily and quickly and tends to develop split ends. The damaged cuticle is fragile, and the damage worsens over time, leading to gradually weakening hair.

The hair fibers' condition should be considered before undergoing chemical processes, such as hair coloring since the needs of damaged and healthy hair are different. Damage is usually more pronounced at the tips of the hair fibers as it represents age progression along the fiber.

Classification Based on Shape, Size, and Color Based on the structure and shape of hair fibers, gross size, and time of appearance on the body, three major hair follicle types can be differentiated, including lanugo hair, vellus hair, and terminal hair.

- From about the fifth month of being a fetus, the body is covered with a very fine soft hair called **lanugo**. It is mostly shed at birth and is replaced by vellus hair.
- **Vellus** hairs are short, soft, colorless hair fibers; they are often referred to as "down" or "fuzzy" hair.[3] In adults, vellus hair is usually found on surfaces normally considered hairless, including the forehead, eyelids, and bald scalp, as well as many other hairy body parts, except for the palms and soles. Vellus hairs help in efficient evaporation and perspiration.
- **Terminal hairs** are pigmented, have a larger cross-sectional diameter, and are longer as compared to other hair fibers. Terminal hairs are located on the scalp, face, chest (mainly for men), arms and legs, underarm, and pubic area and may also appear on the back. The size and shape of terminal hair varies with body location and function. For example, terminal hair on the scalp plays an important role in UV protection and insulation against cold, while the eyelashes protect the eyes from dust.

Some hair follicles have the ability to move from one category to another at various times.[1] For example, hair follicles on an adult scalp that normally produce terminal hairs can undergo a gradual conversion and start producing vellus hair. This process results in baldness. Similarly, hair follicles on the male beard, which until puberty produce only vellus hairs, are capable of producing terminal hair thereafter.

Classification Based on Curliness Waviness (or curliness) refers to the shape of the hair shaft. There are a variety of categories describing the curliness of hair, including straight, wavy, curly, and kinky, among others. In the past, the curly nature of hair was thought to be related to the cross-section of the hair follicle. This theory claimed that hair with a round cross-section is straight, while hair with an oval cross-section is wavy. Recent studies, however, suggest that curliness depends on the shape of the hair follicle, meaning that a straight hair follicle with even an elliptical cross-section could give a straight hair.[26] In addition, new research shows that the curliness also correlates with the distribution of cells in the cortex, independent of the ethnic origin.[27]

The conventional classification of curliness distinguishes three types of hair: African, Caucasian, and Asian. Usually, Asian hair is the thickest, followed by African hair, and then Caucasian hair. An additional feature is the cross-sectional shape of the hair fibers: Caucasian and Asian hairs typically has a similar round shape, with the Asian hair being more cylindrical, while the African hair has a highly elliptical shape. These three categories, however, cannot account for the complexity of human biological diversity. Today, we can find hair types that would not fit into these three basic categories. Therefore, there are newer classifications developed that categorize hair into more than three groups, regardless of its ethnical origin (similar to Fitzpatrick's skin classification system).[28]

Classification Based on Greasiness Based on greasiness, two major hair types are usually differentiated: dry and greasy hair.[29]

- **Dry hair** does not contain enough moisture. It is usually a result of damaged (weathered) cuticle and cortex. Porous cortex cannot retain water; therefore, hair fibers have dull, unhealthy appearance. These signs are more visible in longer hair as compared to short hair, since long hair grows for a longer time and weathers more. It is more common in women, as they are more likely to have longer hair. Dryness is usually aggravated by excessive shampooing and chemical treatments, such as permanent waving or bleaching. Extremely dry hair needs special care and mild cleansing agents not to worsen the damage.
- **Greasy hair** is usually caused by overactive sebaceous glands, which produce more sebum than normally. Since sebum production is highly influenced by hormones, many consumers experience greasy hair when changes take place in their hormone levels, for example, during puberty. In such cases, the hair may become greasier and need to be washed more often. Greasy hair needs special care because if not cleaned adequately, it can lead to various scalp conditions, such as dandruff.

GLOSSARY OF TERMS FOR SECTION 1

Anagen phase: The active growth phase of hair cycle during which new cells form.

Catagen phase: The transition phase of the hair growth cycle which marks the end of the anagen phase.

Cortex: The inner layer of the hair shaft located below the cuticle.

Cuticle: The outermost structure of the hair shaft consisting of multiple layers of overlapping scales.

Eumelanin: A type of melanin found in the hair, which provides dark brown and black color to the hair.

Hair bulb: The base of the hair root, having a characteristic onion shape.

Hair follicle: A tube-like sheath that forms a downward extension of the epidermis into the dermis and surrounds the hair root.

Hair matrix: Actively growing cells in the hair follicle around the hair bulb.

Hair papilla: A cluster of blood vessels at the base of the hair root, which supplies the hair bulb with oxygen and nutrients.

Hair root: Part of the hair under the skin.

Hair shaft: Part of the hair protruding above the skin's surface.

Hair weathering: Progressive hair damage primarily affecting the end of the hair fibers. It results from external factors, including environmental factors and chemical treatments.

Lanugo: A very fine, soft hair that covers the body and limbs of a human fetus or newborn. It is mostly shed at birth.

Medulla: The innermost layer of the hair shaft located below the cortex; sometimes, it is missing.

Pheomelanin: A type of melanin found in the hair that provides red to blond tones.

Porous hair: Damaged hair that has a coarse texture and loses its shine and smoothness. It can absorb liquids very easily and quickly and tends to develop split ends.

Strong bonds: A type of bond found in the hair that can only be broken through a chemical reaction. An example of strong bonds is disulfide bonds.

Telogen phase: The final, resting phase of the hair growth cycle.

Terminal hair: Pigmented hair with a larger cross section than vellus hair. Terminal hairs are located on the scalp, face, chest (mainly for men), arms and legs, underarm, and pubic area and may also appear on the back.

Vellus hair: Short, soft, colorless hair fiber usually found on surfaces normally considered hairless, such as the forehead, eyelids, and bald scalp.

Weak bonds: A type of bond found in the hair that can be modified even by wetting the hair. Weak bonds include hydrogen bonds, salt bonds, and van der Waals forces.

REVIEW QUESTIONS FOR SECTION 1

Multiple Choice Questions

1. Which of the following is TRUE for shedding of human hair?
a) Hair loss up to 100 anagen hairs per day is considered normal
b) When hair sheds, only the hair shaft sheds, but not the root
c) Hair loss up to 100 telogen hairs per day is considered normal
d) Hair shedding is identical to skin shedding, and the outer layer sheds when the new layers are pushed upward

2. How do you know if a hair fiber you lost is in its anagen or telogen phase?
a) Color, shape
b) Shape, diameter
c) Color, diameter
d) Color, shine

3. Which of the following bonds can be modified by wetting the hair?
a) Melanin bonds
b) Hydrogen bonds
c) Disulfide bonds
d) Peptide bonds

4. Hair curliness depends on the ___.
a) Color of the hair
b) Cross-section of the hair
c) Type of the hair, i.e., terminal or vellus
d) Shape of the hair follicle

5. Which of the following is TRUE for porous hair?
a) It is healthy
b) It is resistant to moisture
c) It can absorb water more quickly than non-porous hair
d) Has a compact cuticle

6. In which layer of the skin is the root of a hair fiber located?
a) Epidermis
b) Dermis
c) Hypodermis
d) Stratum corneum

7. Which of the following is the application surface of hair conditioners?
a) Cuticle
b) Cortex
c) Medulla
d) All of the above

8. Hair's texture and luster are determined by the healthiness of the ___.
a) Cuticle
b) Cortex
c) Medulla
d) All of the above

9. What is the hair's overall charge typically?
a) Positive
b) Neutral
c) Negative

10. Hair weathering is ___
a) Mild hair damage
b) Progressive hair damage
c) A synonym for being exposed to the sun and wind
d) Temporary change in the medulla

Fact or Fiction?

______ a) Cutting the hair makes it grow faster.
______ b) Thick hair is more resistant to stretching than thin hair.
______ c) Wet hair is more vulnerable than dry hair.
______ d) Negative molecules are more attracted to the hair than positive molecules.

REFERENCES

1. Blume-Peytavi, U., Whiting, D. A., Trüeb, R. M.: Hair Growth and Disorders, New York: Springer, 2008.
2. Wingerd, B.: The Human Body: Concepts of Anatomy and Physiology, Philadelphia: Lippincott Williams and Wilkins, 2013.
3. Preedy, V. R.: Handbook of Hair in Health and Disease, Wageningen: Wageningen Academic Publishers, 2012.
4. Bhushan, B, Wei, H., Haddad, P: Friction and wear studies of human hair and skin. *Wear.* 2005;259:1012–1021.
5. William, K. J.: Krause's Essential Human Histology for Medical Students, Sydney: Universal Publishers, 2005.
6. Kharin, A., Varghese, B., Verhagen, R., et al.: Optical properties of the medulla and the cortex of human scalp hair. *J Biomed Opt.* 2009;14:1–7.
7. Agache, P., Agache, P. G., Humbert, P.: Measuring the Skin, New York: Springer, 2004: 244.
8. Voet, D., Voet, J. G.: Biochemistry, Hoboken: John Wiley and Sons, 2011.
9. Kligman, A. M.: The human hair cycle. *J Invest Dermatol.* 1959;33:307–316.
10. Wortsman, X., Wortsman, J., Matsuoka, L., et al.: Sonography in pathologies of scalp and hair. *Br J Radiol.* 2012;85(1013):47–655.
11. Herskovitz, I., Tosti, A.: Female pattern hair loss. *Int J Endocrinol Metab.* 11(4):e9860.
12. Paus, R., Cotsarelis, G.: The biology of hair follicles. *N Engl J Med.* 1999;341:491–497.
13. Harrison, S., Bergfeld, W.: Diffuse hair loss: its triggers and management. *Cleve Clin J Med.* 2009;76(6):361–367.
14. Randall, V. A., Thornton, M. J., Messenger, A. G., et al.: Hormones and hair growth: variations in androgen receptor content of dermal papilla cells cultured from human and red deer (Cervus elaphus) hair follicles. *J Invest Dermatol.* 1993;101:114S–120S.
15. Robbins, C. R.: Chemical and Physical Behavior of Human Hair, 4th Edition, New York: Springer, 2002: 33–134.
16. Wella, A. G., Schwan-Jonczyk, A.: Haar Umformung, 1st Edition, Darmstadt: Procter & Gamble Service, 2004: 10–11.
17. Bhushan, B. Introduction: Human Hair, Skin, and Hair Care Products, In: Bhushan, B., ed.: Biophysics of Human Hair: Biological and Medical Physics, Biomedical Engineering, Berlin/Heidelberg: Springer-Verlag, 2010:1–19.
18. Wolfram, L. J.: Human hair: a unique physicochemical composite. *J Am Acad Dermatol.* 2003;48:S106–S114.

19. Davis-Sivasothy, A.: The Science of Black Hair: A Comprehensive Guide to Textured Hair, Stafford: SAJA Publishing Company, 2011.
20. Wilkerson, V. A.: The chemistry of human epidermis: II. The isoelectric points of the stratum corneum, hair and nails as determined by electrophoresis. *J Biol Chem.* 1935; 112:329–335.
21. Lai, K. Y.: Liquid Detergents, Boca Raton: CRC Press, 2012: 418.
22. Mercelot, V.: Application of a tensile-strength test method to the evaluation of hydrating hair products. *Int J Cosmetic Sci.* 1998;20:241–249.
23. Trüeb, R. M.: Female Alopecia: Guide to Successful Management, New York: Springer, 2013;170–171.
24. Frangie, C. M., Botero, A. R., Hennessey, C., Lees, M., Sanford, B., Shipman, F., Wurdinger, V., eds: *Milady Standard Cosmetology*, New York: Cengage Learning, 2012:237.
25. Ito, S., Wakamatsu, K.: Human hair melanins: what we have learned and have not learned from mouse coat color pigmentation. *Pigment Cell Mel Res.* 2011;24:63–74.
26. Thibaut, S., Gaillard, O., Bouhanna, P., et al.: Human hair shape is programmed from the bulb. *Brit J Dermatol.* 2005;152(4):632–638.
27. Thibaut, S., Barbarat, P., Leroy, F., et al.: Human hair keratin network and curvature. *Int J Dermatol.* 2007;46(1):7–10.
28. de la Mettrie, R., Saint-Léger, D., Loussouarn, G.: Shape variability and classification of human hair: a worldwide approach. *Hum Biol.* 2006:79.
29. P&G: Dry or Greasy Hair, Accessed 1/22/2014 at http://pgbeautyscience.com/dry-or-greasy-hair.php

SECTION 2: HAIR CLEANSING AND CONDITIONING PRODUCTS

LEARNING OBJECTIVES

Upon completion of this section, the reader will be able to

1. define the following terms:

Antidandruff shampoo	Combability	Dandruff	Eye irritation
Film-forming ingredient	Hair conditioner	Hydrolyzed protein	Leave-in conditioner
Malassezia	Quats	Rinse conditioner	Seborrheic dermatitis
Shampoo	Split end		

2. differentiate between shampoos and hair conditioners;
3. explain whether shampoos are regulated as drugs or cosmetics in the US;
4. briefly discuss how shampoos can negatively affect the scalp and hair;
5. list various required cosmetic qualities and characteristics that an ideal shampoo should possess;
6. list various required technical qualities and characteristics that an ideal shampoo and hair conditioner should possess;
7. explain how shampoos clean the hair;
8. list the major types of ingredients found in shampoos, and provide some examples for each type;
9. differentiate between water-based and dry shampoos;
10. explain the main difference between the various types of shampoos available today;
11. list the main types of shampoos available today;
12. differentiate between dandruff and seborrheic dermatitis;
13. list some typical signs and symptoms of dandruff;
14. explain why flaking is visible when someone has dandruff;
15. name some active ingredients that can be used in antidandruff shampoos for mild dandruff;
16. explain how hair conditioners work;
17. list some commonly used conditioning ingredients and explain their working principle;
18. list the main types of hair conditioners available today;
19. list some typical quality issues that may occur during the formulation and/or use of shampoos and hair conditioners, and explain why they may occur;
20. list the typical quality parameters that are regularly tested for shampoos and conditioners, and briefly describe their method of evaluation;
21. explain how hair conditioners' performance is usually tested;
22. name the efficacy parameter i.e. generally tested for antidandruff shampoos, and describe the method of evaluation;
23. list typical containers available for shampoos and hair conditioners.

KEY CONCEPTS

1. As hair is one of the most important factors influencing our appearance, shampoos and hair conditioners are widely used personal care products among both men and women.
2. The majority of hair cleansing and conditioning products provide cosmetic benefits to consumers; therefore, they are considered cosmetics in the US. However, products used to prevent and/or treat specific problems such as dandruff are OTC drugs with cosmetic benefits.

3. Shampoos are surfactant-based preparations. Therefore, their cleaning principle is emulsification.
4. Dandruff is one of the most common skin diseases of the scalp, which presents as dry, scaly patches on the scalp.
5. Conditioners are applied to the hair after shampooing and are designed to smooth the hair, improve gloss and luster, as well as recondition chemically damaged hair, mechanically damaged hair, and weathered hair.
6. The typical quality-related issues of shampoos and hair conditioners include separation of emulsions, microbiological contamination, clumping, rancidification, and poor foaming activity of shampoos.
7. Parameters commonly tested to evaluate the quality of shampoos and hair conditioners include spreadability, extrudability, texture, and firmness of lotions, creams, and gels; actuation force; foaming property; foam stability; foam viscosity; foam density; foam structure; preservative efficacy; viscosity; and pH.
8. The most frequently tested efficacy parameters include combability and antimicrobial activity of antidandruff shampoos.
9. Adverse reactions to shampoos and hair conditioners are rare. However, eye irritation (burning sensation) still frequently occurs primarily caused by the sulfates (such as sodium lauryl sulfate and sodium laureth sulfate) used as primary surfactants.

Introduction

➊ **As hair is one of the most important factors influencing our appearance, shampoos and hair conditioners are widely used personal care products among both men and women.** Although shampooing has been the most common form of cosmetic hair treatment, primarily aimed at cleansing the hair and scalp, the present-day consumer expects more options.[1] Environmental factors, such as UV radiation, wind, and humidity; cosmetic manipulation, such as coloring, perming, and bleaching; and even basic grooming can cause hair to lose its strength, elasticity, and shine. Current shampoos and hair conditioners are adapted to the variations associated with hair quality, hair care habit, and specific problems related to the superficial condition of the scalp.[2]

This section reviews the various types of shampoos and hair conditioners, their common ingredients, formulation technology, testing methods, and packaging materials. It also provides an overview of how these products may affect the hair and scalp and what consumers' general requirements are.

Types and Definition of Hair Cleansing and Conditioning Products

➋ **The majority of hair cleansing and conditioning products provide cosmetic benefits to consumers; therefore, they are considered cosmetics in the US. However, products used to prevent and/or treat specific problems such as dandruff are OTC drugs with cosmetic benefits.**

- **Shampoos** are designed to remove all kinds of soilage, including sebum, sweat, environmental dirt, and hair conditioners, as well as to beautify the hair and make it easy to handle.
- **Antidandruff products** are essentially shampoos that contain valuable active ingredients for the prevention and treatment of dandruff. Dandruff is a common condition of the scalp characterized by flaking, itching, and redness on the scalp.
- **Hair conditioners** are designed to repair chemical and environmental damage, replace natural lipids removed by shampooing, and facilitate managing and styling hair. They are usually applied after cleaning the hair in the form of rinsing or leave-in preparations. Certain manufacturers also produce two-in-one shampoo and conditioner formulations.

History of Using Shampoos and Hair Conditioning Products

Taking care of one's hair has always been important for women and men since ancient times. Excavations from Egyptian tombs also have revealed combs, brushes, mirrors, and razors made of tempered copper and bronze. Products for cleaning the hair date as far back as soap. Early Egyptians washed their hair with a mixture of citrus juice and a small amount of soap to help remove oil from the hair.[3] As for conditioning, they used castor oil and other oils. By the Middle Ages, more refined hair products had been developed by combining soap with soda, from which soap itself is made.[3] By the late 18th century, British salons were offering customers a hair-washing massage called a "shampoo." The word "shampoo" and the massage originated from India, where it was introduced as a head massage, usually consisting of alkali, natural oils, and fragrances.[4]

By the late 1900s, it was common to have shampoos of soaps mixed with aromatic herbs. The original soap, however, left a dull film on the hair, which made it uncomfortable and irritating. With the invention of "soapless" detergents in the early 20th century, cleaning the hair without leaving a residue on it became possible. The first shampoo made of detergents was introduced after the First World War, which became very popular soon. The first hair conditioners were developed in the early 1930s, when self-emulsifying waxes became available. These waxes were combined with protein hydrolyzates and silicones to give the hair an improved feel and texture. Early sources of protein were gelatin, milk, and egg protein.[5]

Today, there is a wide variety of shampoos and hair conditioners available with specific products developed for dry and damaged hair, oily hair, normal hair, and colored hair. Many products also offer additional styling claims, such as volumizing, straightening, or curl enhancing.

How Hair Cleaning and Conditioning Products May Affect the Hair and Scalp?

Hair cleansing products are one of the most widely and frequently used personal care products today. When selected properly for the individual's needs and used appropriately, they help maintain the hair's healthy appearance. The advantages of using such products are summarized here:

- The main benefit of using shampoos is the **removal of dirt** from the hair. Similarly to dirt on the skin, dirt on the hair consists of sweat, sebum and its breakdown products, dead skin cells, residues of cosmetics and personal care products, dust, and other environmental impurities carried in the air. Most of these compounds are not soluble in water; therefore, washing the hair with simple water would not be sufficient to remove dirt. Shampoos contain surfactants (similar to skin cleansers), which are able to remove oily particles from the hair.
- **Dandruff** is more of a cosmetic problem than a disease, which, if left untreated, can have significant emotional distress. Shampoos containing antidandruff active ingredients can be effectively used for the prevention and treatment of dandruff.
- Chemical treatments, such as hair perming, can cause severe damage to the hair. Damaged hair looks dry without shine and luster, and it is characterized by split ends and also fractures along the hair fiber. Hair conditioning products deposit various ingredients on the hair that **help** temporarily **repair damages** and increase the hair's gloss and manageability.

Adverse reactions to shampoos and hair conditioners are generally rare; however, some negative effects can still occur in some cases. The most common negative effects of these products are summarized here:

- Generally, shampoos are not a common cause of **skin irritation** as they are in contact with the skin for a brief time before being rinsed off.[6] Ingredients that may be allergens in shampoos include fragrances, triclosan, propylene glycol, benzophenones, parabens, and other preservatives.[7,8]
- The same is true for hair conditioners. Although hair conditioners can be left on the hair (such as leave-in conditioners), they are usually applied to the ends of hair fibers and not the scalp; therefore, adverse reactions are rare.
- In the case of shampoos, **eye irritation**, however, is a general problem. It is usually caused by the primary surfactants used in shampoos (i.e., the surfactants responsible for cleansing), such as sodium lauryl sulfate. In order to reduce the irritation, potential shampoos contain a variety of ingredients, such as amphoteric surfactants, silicone derivatives, protein derivatives.
- In addition to eye irritation, anionic surfactants can **damage the stratum corneum** (SC). As discussed under skin cleansing products, harsh surfactants can remove barrier lipids and the water-soluble NMF from the SC, leading to dryness and changed enzyme activity in the SC.[9,10] Collectively, these changes in SC composition impact the overall barrier quality, impair desquamation, and

promote flaking. A common approach to reduce these negative effects is to use milder cleansing systems by combining anionic and amphoteric surfactants.[11]

- Moreover, as harsh surfactants are excellent components for removing sebum and dirt from the hair, their excessive use can lead to significant changes in the **appearance of the hair**. They can leave the hair dull, susceptible to static electricity, and difficult to comb. Excessive sebum removal is advantageous for oily hair, but would make dry hair even worse. This is one of the reasons why selecting the appropriate type of shampoo is important in maintaining hair with healthy appearance.
- As discussed under skin care products, the use of traditional **bar soaps** diminished in the past decades due to the appearance of more sophisticated and gentle products. However, they may still be used, mainly by the older generation, for cleansing the body and hair. Similar to their effects on the skin, they are harsh to the hair. Bar soaps leave behind a soap scum when mixed with hard water, which is difficult to rinse off from the hair and scalp. If it remains on the hair, it may be one of the aggravating factors for seborrheic dermatitis.[6]

Required Qualities and Characteristics and Consumer Needs

From a consumer perspective, a quality hair cleansing and/or conditioning product should possess the following characteristics:

- Gentle to the hair and scalp, do not dry or damage it
- Natural or pleasant odor and color
- Long-lasting effect
- Easy to spread on the hair
- Easy to rinse off from the hair
- Enhances gloss and luster of the hair and make it easy to comb
- Well-tolerated and non-allergenic
- Shampoos: remove sebum and other dirt from the hair and scalp, good foaming property, nonirritating to the eyes, act as a vehicle for the deposition of beneficial materials onto the hair and scalp.

The technical qualities of hair cleansing and/or conditioning products can be summarized as follows:

- Antidandruff shampoos: efficacious
- Shampoos: appropriate foaming activity
- Appropriate rheological properties
- Appropriate pH
- Long-term stability
- Dermatological safety

Hair Cleansing Products

Cleansing Basics Cleansing principles discussed under skin cleansing products (in Section 1 of Chapter 3) also apply to shampoos. From the aspect of their chemical nature, today, ● **shampoos are surfactant-based preparations. Therefore, their cleaning principle is emulsification.** Surfactants surround and trap tiny droplets of fat, which in this form can be rinsed off from the hair and scalp. Insoluble particulate soil can be removed by electrostatic repulsion between the soil and the hair fiber assisted by repulsion between the surfactant molecules adsorbed onto the hair fiber and those dissolved in the soil.[12]

Typical Ingredients of Shampoos Today, shampoos are available as liquids, gels, emulsions (lotions and creams), and powders. Most commonly, shampoos are colloidal dispersions of various surfactants in water. For the most part, this basic type will be covered in this section. Basic components of a classical shampoo are cleansing agents, thickeners, and water. Usually, various additives are also incorporated in the formulations in order to help the cleansing process, enhance the aesthetic properties, increase foaming, and make the hair shine.

There are countless formulations available on the market, and we hardly can find two of them whose composition is the same. Manufacturers can choose from a great variety of ingredients; however, the basic ingredient types used are the same. This section reviews these ingredient types, including their basic characteristics and functions.

- **Surfactants** aid in cleaning and foaming by reducing the surface tension between two phases. In addition, surfactants can also act as foam boosters and foam stabilizers. Typically, several surfactants are combined to achieve the desired result. For example, shampoos for oily hair contain surfactants with strong sebum removal qualities, unlike those for colored hair that are gentler to the hair. Different surfactants, however, have different characteristics and effects on the hair and scalp. Anionics can make the hair extremely clean, but will leave it with a rough, harsh feeling; while nonionics can increase luster and shine, but they do not foam as well as anionics. Selecting the appropriate type and amount of surfactants is, therefore, critical.
  - *Anionic* surfactants have good cleansing properties; therefore, they are commonly found in most shampoos. Examples of anionic surfactants include lauryl sulfates, such as sodium lauryl sulfate; laureth sulfates, such as sodium laureth sulfate; sarcosines, such as sodium lauroyl sarcosinate (they are excellent conditioners but do not efficiently remove sebum); and sulfosuccinates, such as sodium dioctyl sulfosuccinate.
  - *Cationics* are not as popular as anionics since they do not foam well and do not remove grease as efficiently as anionics. In addition, they are normally incompatible with anionics, which is another drawback. They are mainly used in formulations where minimal cleansing is required; however, softness and manageability should be increased (such as everyday shampoos for colored hair).[13]

- *Amphoterics* are compatible with all classes of surfactants. These detergents are nonirritating to the eyes, foam moderately well, and increase manageability of hair. They are often used in combination with anionics. Examples include betaines, such as cocamidopropyl betaine, and alkylamino acids.
- *Nonionics* are popular surfactants, and they are very mild. Therefore, they are often used in combination with ionic surfactants as co-surfactants, rheology modifiers, and solubilizers for insoluble components such as fragrance oils.[14] Examples include poloxamers; amine oxides, such as cocamidopropylamine oxide; and polyglucosides, such as lauryl glucoside.

DID YOU KNOW?

Soaps are usually not used in shampoos due to their negative properties. A major drawback of soaps is that they have a high pH, which may damage the hair and the skin as well. Additionally, a normal soap forms calcium salts that adhere to the hair when used with tap water. These salts make the hair look and feel dull and brittle and difficult to comb.

DID YOU KNOW?

Similar to skin cleansers, most consumers look for well-lathering formulations as they associate this property with the cleansing power of a shampoo. This is not the case, however; low-foaming formulations can be just as effective in removing sebum from hair as highly foaming formulations. Another common belief is that the thicker the shampoo, the richer it is in active ingredients and conditioners. Most ingredients added to a shampoo will not significantly affect its viscosity; the ingredients added for thickening, however, do not have conditioning effect. They simply provide better shampooing experience to the users.

- **Thickeners** provide the necessary rheological properties for the systems. A shampoo with a viscosity similar to that of water would not be favorable as it would run off the hands and not stay on the scalp, but run into the eyes instead. Thickeners increase viscosity and influence the product feel. As discussed under skin cleansing products, solutions of anionic surfactants can be thickened with sodium chloride (see Section 2 of Chapter 3 for more detail).
  - Examples of thickeners include sodium chloride, gums, celluloses, and other polymers, such as polyvinyl alcohol and acrylates copolymer.
- **Water** is the vehicle for shampoos.

- **Preservatives** prevent the growth of microorganisms in the formulations.
  - Examples include parabens; urea derivatives; isothiazolones, such as methylchloroisothiazolinone; as well as benzalkonium chloride, a cationic surfactant.
- **Opacifiers and pearlescent agents** play an aesthetic role by providing a unique pearly, shimmering effect or a creamy appearance for the formulations.
  - Examples for such ingredients include polyglycol esters, latex opacifiers, and pearlescent color additives.
- **Conditioners** make the hair soft, shiny, and easier to manage. Although the main purpose of using shampoos is to clean the hair, overcleaned hair looks dull and has less shine. Conditioners can be incorporated into shampoos; products containing such ingredients are usually referred to as two-in-one shampoo and conditioner formulations. Conditioners are particularly important in dry hair shampoos and shampoos for colored and bleached hair as these hair types are dry by themselves, which is further aggravated by using shampoos.
  - Examples for commonly used ingredients include quats (a type of cationic surfactant); humectants, such as glycerin; proteins; silicones, such as dimethicone among others (see more detail on conditioners under hair conditioner products).
- **pH buffers** adjust the pH of products. Surfactants usually provide an alkaline pH to formulations; it can lead to swelling of the cuticle, which makes it more vulnerable. This is especially a concern in dry and chemically treated hair. By shifting the pH closer to the neutral range provides less damage to the hair.
  - Examples include citric acid and glycolic acid.
- **Chelating agents**, also known as sequestering agents, contribute to the stability of the product by binding to metal ions. Metal ions, such as magnesium and calcium ions, present in tap water and can form insoluble soaps with shampoos, which, if deposited on the hair, make it dull and less manageable.
  - Examples include EDTA and its derivatives.
- **Additional ingredients** include compounds that provide a unique feel or appearance for the products but do not influence their functional (i.e., cleaning) property. Such ingredients include color ingredients; perfumes; botanical extracts, such as tea tree oil; and vitamins, such as vitamin B5 (panthenol).
- **Active ingredients** can also be incorporated into shampoos, which make them to be considered drugs. Most frequently, ingredients that are capable of preventing and/or treating dandruff are added to shampoos. These ingredients are discussed later under antidandruff shampoos.

Types of Shampoos The variety of shampoos available today in the market is endless. There are, however, some basic types that are worth discussing. When buying shampoos, we can usually see the type of hair for which a certain shampoo is recommended, including "normal hair," "oily hair," "dry hair," "colored hair," and "damaged hair." The basic ingredients are similar in these formulations; however, the concentration of the ingredients used in different types of shampoos can significantly vary. This is why users should pay attention to selecting the best type for their individual needs.

- **Normal hair shampoos** are designed to clean the hair of persons with moderate sebum production and who do not have chemically processed hair. These shampoos offer good cleansing by using sodium or ammonium lauryl sulfate with minimal conditioning.[6]
- **Oily hair shampoos** are designed to remove excess sebum from the hair and scalp. This can be accomplished by using strong surfactants, such as lauryl sulfates, with no or minimal conditioners. These shampoos are harsher to the hair than normal hair shampoos due to the stronger surfactants. Conditioning is not needed for oily hair due to the excess sebum production.
- **Dry hair shampoos** provide gentle cleansing by incorporating gentle surfactants, such as sulfosuccinates, and good conditioning. Dry scalp and hair needs special attention due to the presence of less sebum. Some companies recommend the same products for dry and damaged hair as their main characteristics are similar. These products are often referred to as two-in-one shampoos and conditioners. Dry hair shampoos provide a thin coat over the hair fibers and thus reduce the static electricity and increase manageability of fine hair.
- **Everyday shampoos** are formulated as gentle formulations that can be used every day without drying the hair or depositing too much oil on it. Normally, it is not necessary to shampoo every day unless sebum production is high.
- **Deep cleansing shampoos** are designed to thoroughly clean the hair. These products are generally used to remove retained hair styling products, such as hair gels, hair sprays, and mousse. These shampoos contain stronger surfactants, such as sodium or ammonium lauryl sulfate, similar to oily hair shampoos, to efficiently remove dirt. These shampoos are typically used once weekly to keep the hair free of hair styling product buildup.[6]
- **Baby shampoos** are usually milder, based on amphoteric surfactants, such as betaines. They offer nonirritating properties and minimal sebum production.
- **Gray Hair Shampoos**: One new significant consumer segment is for gray hair and consists of products containing blue dyes to make the gray hair color brighter and less yellowish. Overdosing or very frequent use may cause a bluish appearance of the hair.[15]
- **Hair dyeing shampoos** are special formulations that are designed to be used after permanent hair dyeing. These shampoos contain cationic surfactants and have an acidic pH, which neutralizes any residual alkalinity from the chemicals used for hair dyeing. By shifting the pH to the normal pH, swelling of the hair cuticle decreases. The cuticle should be tightly adherent to the cortex for optimum hair functioning and appearance.[16]
- **Medicated shampoos** are designed to deliver extra benefits to the hair and scalp in addition to cleansing and conditioning. Most medicated shampoos

contain active ingredients to relieve itching and scaling. These products are considered OTC drugs in the US. They are discussed in more detail later on.

- **Dry shampoos** were the earliest types of hair cleaning products. They have primarily historical importance; today, their use has diminished. The word "dry" in their name refers to their dosage form (i.e., powder or powder-based aerosol) and not the type of hair they should be used on. Dry shampoos contain powders with good oil-absorbing capacity, such as starch, silica, magnesium stearate, kaolin, and talc. These are cleansing formulations that work without soap and water. Dry shampoos are mainly used as touch-ups if customers do not have to time to wash their hair.

Antidandruff Shampoos Dandruff and seborrheic dermatitis are common diseases of the scalp, which are considered the same basic condition,[17] but differing in the severity of symptoms, such as flaking and inflammation. Dandruff is a milder variant of seborrheic dermatitis. ❶ **Dandruff is one of the most common skin diseases of the scalp,[18] which presents as dry, scaly patches.** It is not contagious.

Dandruff results from an increased rate of cell turnover, which may be even double the normal rate.[19] Rather than taking 28–30 days to migrate from the basal cell layer to the SC with dandruff, skin cells reach the horny layer in about 2 weeks. In dandruff, dead skin cells are ordered in an abnormal manner, which build up on the scalp and then flake off. As long as the rate of this turnover is reasonable and normal, shedding of dead cells is invisible. If, however, the rate of turnover increases, there is an increased number of cells produced; this also leads to increased rate of exfoliation. When cells are produced more rapidly, they do not have enough time to appropriately mature and die[20]; they stick to each other and form larger aggregates, which become visible to the naked eye as they are shed from the skin (see the factors contributing to the visible aggregates summarized in Figure 5.6).[21]

Dandruff can appear in normal (neither dry nor oily) hair. Sometimes, if the scalp is particularly dry, and the dry skin peels, this can resemble dandruff. Nevertheless, dandruff is more common in oily hair.[22] Most common symptoms include gray or yellow oily-looking flakes on the scalp and shoulders, along with itching, redness, and scaling. The exact causes of dandruff are not clearly known. Potential factors include hormones, seasonal effects (more frequent during winter), emotional stress, sebum, increased alkalinity of the skin, occlusion of the scalp,[23] and a yeast normally found on the human scalp and skin, called *Malassezia* (formerly known as *Pityrosporum*).[24] The scalp is a unique environment combining a high level of sebum production with a physical covering of hair. Under the appropriate conditions, the *Malassezia* yeasts can colonize the scalp and produce by-products that enhance lipid secretion.[25] These circumstances lead to inflammation and hyperproliferation.

Although dandruff is not a serious disease, more of a cosmetic problem, it should be taken care of because it has uncomfortable symptoms, can be unsightly and embarrassing, and thus can significantly affect a person's self-esteem and confidence.[26] The symptoms of mild dandruff can usually be controlled by shampooing with antidandruff shampoos and thoroughly cleaning the scalp. For more severe dandruff, patients

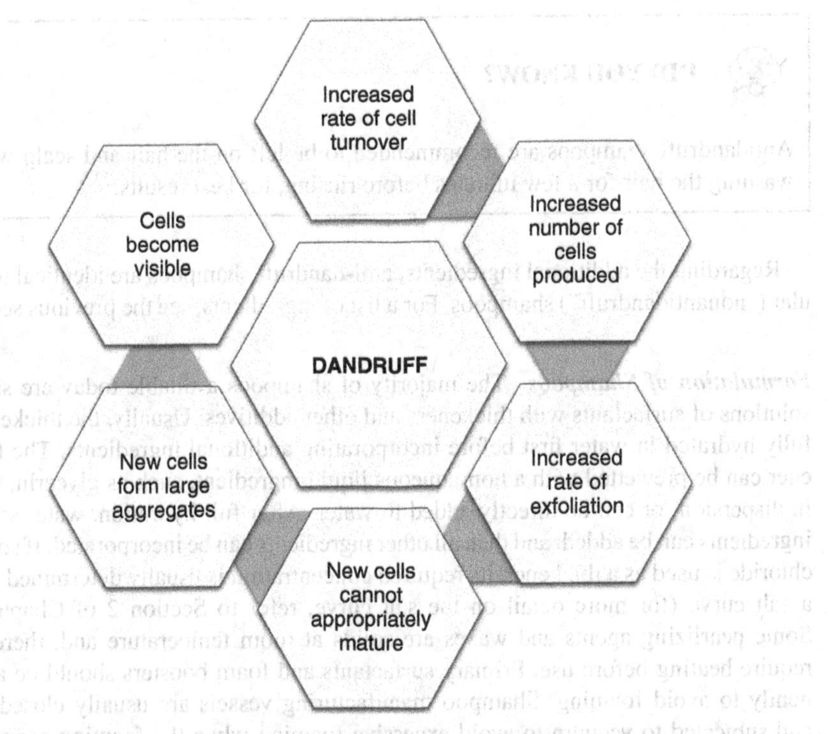

Figure 5.6 Major factors contributing to the development of dandruff.

should visit a dermatologist. Depending on the patient's age (more common in young adolescents), cell turnover tends to return to an abnormal rate once treatment is discontinued, and dandruff will occur again.

Antidandruff shampoos are basic shampoos with active ingredients; therefore, they deliver both cosmetic and drug benefits. They clean the hair and leave it in an aesthetically appealing condition. In addition, they reduce scaling, decrease the rate of cell turnover, and also have an antimicrobial effect. Commonly used monographed active ingredients include zinc pyrithione, ketoconazole, coal tar, salicylic acid, selenium sulfide, and sulfur.

- Zinc pyrithione slows down cell turnover and is an effective antifungal ingredient.
- Sulfur and salicylic acid have a keratolytic effect, which means that they dissolve the keratin of dead cells and thus prevent the formation of visible flakes. In addition, sulfur has antimicrobial properties.
- Tar slows down the rate of epidermal turnover and also has antiseptic activity.
- Ketoconazole is an antifungal ingredient, which controls flaking and itching.

DID YOU KNOW?

Antidandruff shampoos are recommended to be left on the hair and scalp when washing the hair for a few minutes before rinsing, for best results.[27]

Regarding the additional ingredients, anti-dandruff shampoos are identical to regular ("nonantidandruff") shampoos. For a list of ingredients, see the previous section.

Formulation of Shampoos The majority of shampoos available today are simple solutions of surfactants with thickeners and other additives. Usually, the thickener is fully hydrated in water first before incorporating additional ingredients. The thickener can be prewetted with a non-aqueous liquid ingredient, such as glycerin, to aid in dispersion, or can be directly added to water. After full hydration, water-soluble ingredients can be added, and then all other ingredients can be incorporated. If sodium chloride is used as a thickener, its required concentration is usually determined using a salt curve (for more detail on the salt curve, refer to Section 2 of Chapter 3). Some pearlizing agents and waxes are solids at room temperature and, therefore, require heating before use. Primary surfactants and foam boosters should be added gently to avoid foaming. Shampoo manufacturing vessels are usually closed type and subjected to vacuum to avoid excessive foaming when the foaming agents are added.

The formulation of dry shampoos is a powder blending process, which usually starts with grinding the ingredients to provide a uniform particle size. After careful blending, the mixtures may be subject to sieving to better homogenize them and again provide a uniform particle size.

Hair Conditioners

● **Conditioners are applied to the hair after shampooing and are designed to smooth the hair, improve gloss and luster, as well as recondition chemically damaged hair** (by permanent waving, hair bleaching, or hot blow-drying), **mechanically damaged hair** (by excessive brushing), **and weathered hair** (by sunlight, salty seawater, chlorinated water, or swimming pools). Conditioners act by reducing static electricity generated after combing dry hair, improving manageability by filling in the gaps around and between the cuticle scales, increasing hair shine by coating hair shafts with a thin layer,[16] decreasing split ends, and improving hair flexibility.[28]

Natural sebum is the ideal conditioner. Excessive removal of sebum leads to a harsh and dull appearance of the hair, and it necessitates the use of synthetic sebum-like products.

DID YOU KNOW?

Hair becomes negatively charged after brushing. These negative charges repel each other and make the hair look frizzy, which is often referred to as "flyaway" hair. Conditioners deposit positive charges on the hair, reducing its negative charges and minimizing frizzy hair.

DID YOU KNOW?

Split ends occur when the cuticle is damaged and removed from the underlying cortex and medulla. Keratin found in the cortex is then directly exposed to chemical, physical, and environmental damage, which cannot withstand most of these damaging factors without the cuticle on top.[29]

DID YOU KNOW?

Conditioners provide the hair with only temporary improvement. Products applied to the surface of hair fibers, are washed off with their beneficial conditioning effect when washing the hair.

Ingredients Conditioners are available as liquids, creams, or gels. The main ingredients in hair conditioners are the conditioning ingredients. There are various types of conditioning agents available, including lipids, silicones, quats, protein derivatives, silicones, and glycols, among others. Of these types, the following conditioning agents are the most widely used:[30]

- **Quaternary conditioners** are cationic detergents, which were discussed under shampoos. Due to their positive charge, cationic compounds are attracted to the negatively charged hair fiber and can remain on their surface even after rinsing.[31] They neutralize the negative charges and make hair less susceptible to static electricity and shinier. Cationic ingredients are beneficial for permanently colored or waved hair where the cuticle is damaged. The more the cuticle is damaged, the more negative electric charges its surface carries and the stronger bond it forms with the conditioner. They are usually applied after shampooing and rinsed before drying the hair. Examples for widely used cationic conditioners include stearalkonium chloride, cetrimonium chloride, quaterniums, and polyquaterniums (such as polyquaternium-10).

- **Film-forming conditioners** coat hair fibers with a thin polymer layer. In addition, they fill in defects in the cuticle to create a smooth surface.[32] The most common film-forming agent used in such conditioners is polyvinylpyrrolidone (PVP). The thin film creates a smoother surface, which translates to shinier hair, and also reduces static electricity and improves hair manageability. Film-formers are ideal conditioners for curly and kinky hair, and they can even make hair straightening easier for such hair types if applied in a higher amount. However, they can make fine straight hair limp and difficult to style. Film-forming conditioners are usually applied to towel-dried hair and are left on the hair.
- **Protein-containing conditioners** contain a small amount of proteins that can penetrate the holes in the hair shaft and increase its fracture strength. The proteins, derived from animal tissues, silk, and plants, are hydrolyzed (i.e., broken down) to smaller fragments in order to be able to penetrate the hair shaft.[33] The ability of these ingredients to strengthen hair shafts depends on the contact time. The longer they are left on the hair, the deeper proteins can penetrate the hair. Therefore, proteins can be used as rinse-off conditioners for minimal penetration and leave-in conditioners for deeper penetration.
- **Silicones** form a thin film on the hair without creating the appearance of greasy and limp hair. They are very popular conditioning ingredients. Some silicones are water-resistant and, therefore, can remain on the hair shaft even after washing the hair.[34–36] Examples for silicones include cyclomethicone, dimethicone, and amodimethicone.

Additional ingredients in hair conditioners include water, thickeners to provide viscosity, botanical extracts, vitamins, preservatives, color additives, and fragrances.

Types of Conditioners

Based on their application, there are several types of hair conditioners available today, including instant products, hair rinses, deep conditioners, and leave-in products.[37]

- **Instant conditioners** are usually formulated as lotions and are used on wet hair after shampooing. They are left on the hair for a few minutes and then rinsed off. These conditioners usually contain quats as the main ingredients. Instant conditioners improve wet combing and are primarily recommended for consumers who shampoo frequently and/or have minimally damaged hair.
- **Hair rinses** are also applied to towel-dried hair and rinsed after a few minutes. The main ingredients in these products are quats, such as stearalkonium chloride. Hair rinses are usually formulated as liquids and are generally intended

for fine oily hair, which needs less conditioning. Their main function is to aid in hair detangling.

- **Deep conditioners**, also known as hair masks, are usually recommended for chemically damaged hair and dry hair. They are applied to wet hair and are left on the hair for 20–30 min before rinsing. They are generally available as creams or oils containing quats and hydrolyzed proteins.[38]
- **Leave-in products** are typically applied to towel-dried hair, and as their name implies, they are designed to remain on the hair. The most popular leave-in conditioners are oily products based on petrolatum; mineral oil and silicones are designed for thick, curly, or kinky hair. They can moisturize the hair while aiding in hair styling. A popular category of leave-in conditioners are the blow-drying lotions, which are designed to coat the hair shaft and protect the hair protein from heat damage during the drying process.[39]

DID YOU KNOW?

There are hair conditioners containing sunscreens available to provide photoprotection.[40] Unlike the skin where skin cancer can develop as a result of UV radiation, the hair cannot undergo carcinogenesis. Hair photoprotection aims to preserve the cosmetic value of hair, including its color, shine, and healthiness.[41]

Similar to the skin that needs moisturization regardless of its type, most hair types also need regular conditioning as well. Fine hair is generally more manageable than curly hair; however, it is also more vulnerable to chemical and physical damage. There are specific products designed for straight hair, wavy hair, and curly or kinky hair, as well as oily hair and extremely dry hair, as all these types possess different characteristics. Therefore, hair type should be taken into consideration when selecting conditioners.

Formulation of Hair Conditioners Conditioners are available in various dosage forms, including mists, oils, lotions, creams, and gels. The formulation of these products can be attributed to the general dosage forms; their formulation technology was discussed in the previous chapters.

Typical Quality Problems of Shampoos and Hair Conditioners

❶ **The typical quality-related issues of shampoos and hair conditioners include separation of emulsions, microbiological contamination, clumping, rancidification, and poor foaming activity of shampoos.** Since these problems were discussed in the previous sections, they are not reviewed in detail here.

Evaluation of Shampoos and Hair Conditioners

Quality Parameters Generally Tested ● **Parameters commonly tested to evaluate the quality of shampoos and hair conditioners include spreadability, extrudability, texture, and firmness of lotions, creams, and gels; actuation force; foaming property; foam stability; foam viscosity; foam density; foam structure; preservative efficacy; viscosity; and pH.** The range of acceptance and other limiting factors are usually determined by the individual manufacturers. Since these tests were discussed in the previous sections, they are not reviewed here again.

Efficacy (Performance) Parameters Generally Tested ● **The most frequently tested efficacy parameters include combability and antimicrobial activity of antidandruff shampoos.** These evaluations are discussed here.

Conditioning Effect (Combability) As discussed previously, hair conditioners improve hair manageability via various mechanisms. There is no standard testing protocol for evaluating the conditioning effect of hair conditioners; however, there are some approaches developed by the industry. The most widely used approach involves instrumental combing experiments. This methodology involves the use of a mechanical testing device that measures the friction generated between the comb and hair during grooming. Therefore, lubrication is evaluated through a decrease in grooming forces for product-treated hair relative to unconditioned hair (control). In the test (depicted in Figure 5.7), a hair tress is suspended from a load cell i.e. connected to the crosshead of a tensile testing device. The hair tress is held vertically, close to the comb. When instructed, the arm of the equipment brings the comb down through the hair, recording the necessary force as it moves. At the end of a stroke, the comb is mechanically disengaged and moves back to its starting position to begin the next combing cycle.[42] The testing device measures the decrease in movement of the comb down through the hair after conditioning. This test can be set up in various ways; it can be used on both wet and dry hair samples.

Efficacy of Antidandruff Shampoos As antidandruff products are considered OTC drug-cosmetic products in the US, they must have a proven efficacy. The possible types of tests that can be used include the following: (i) *in vitro* tests carried out in a laboratory by cultivation of yeast in the presence of the active agent(s) and (ii) *in vivo* test, in which the product is tested on humans. The subjects can be divided into groups (one group using the antidandruff shampoo and another using a placebo; i.e., a shampoo identical to the antidandruff shampoo with the only difference being that it does not contain active ingredient), and the results of the two groups are compared. A better approach is when the subjects are their own controls, meaning that they first use a non-medicated shampoo and then an antidandruff shampoo. This way the same person can compare the two formulas on themselves. A variation of this approach is when subjects use the placebo on one half of their scalp and the medicated shampoo on the other half (known as the "half head" technique).[43,44] It

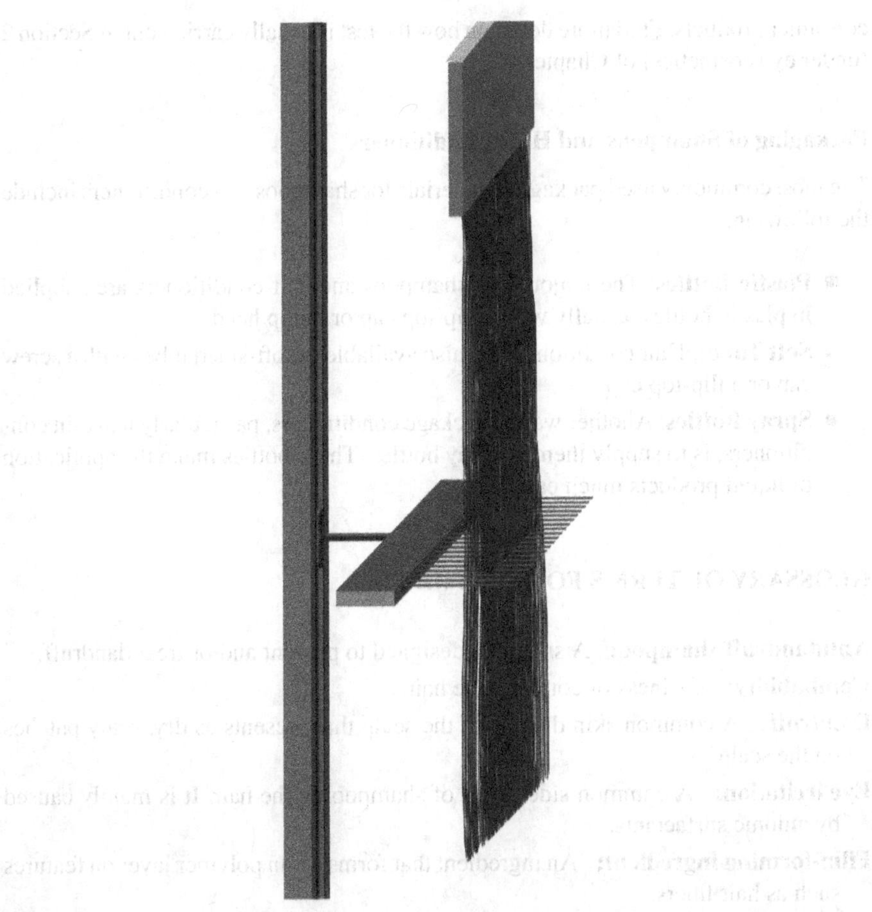

Figure 5.7 Combability testing of hair conditioners. Adapted from Texture Technologies Corp.

makes comparison easier and more realistic. At the end of the study, the efficacy is compared to the control products and the baseline results, and a statistical analysis is carried out to determine whether the difference is significant or not. The following parameters are usually checked and associated with the efficacy of the formulations: subjective parameters (itching and burning) and objective parameters (weight of flakes, size, number, and area of scalp scales determined by image analysis).[45]

Safety Parameter As discussed previously, ❶ **adverse reactions to shampoos and hair conditioners are rare. However, eye irritation** (burning sensation) **still frequently occurs, primarily caused by the sulfates** (such as sodium lauryl sulfate and sodium laureth sulfate) **used as primary surfactants.** The assessment of ocular irritation potential is, therefore, an important part of safety testing for cosmetic and consumer products. Find more detail on how the test is usually carried out in Section 2 (under eye cosmetics) of Chapter 4.

Packaging of Shampoos and Hair Conditioners

The most commonly used packaging materials for shampoos and conditioners include the following:

- **Plastic Bottles**: The majority of shampoos and hair conditioners are supplied in plastic bottles, usually with a flip-top cap or pump head.
- **Soft Tubes**: Hair conditioners are also available in soft-sided tubes with a screw cap or a flip-top cap.
- **Spray Bottles**: Another way to package conditioners, particularly leave-in conditioners, is to supply them in spray bottles. These bottles make the application of liquid products much easier.

GLOSSARY OF TERMS FOR SECTION 2

Antidandruff shampoo: A shampoo designed to prevent and/or treat dandruff.

Combability: Easiness of combing the hair.

Dandruff: A common skin disease of the scalp that presents as dry, scaly patches on the scalp.

Eye irritation: A common side effect of shampooing the hair. It is mainly caused by anionic surfactants.

Film-forming ingredient: An ingredient that forms a thin polymer layer on features such as hair fibers.

Hair conditioner: A personal care product designed to repair chemical and environmental damage, replace natural lipids removed by shampooing, and facilitate managing and styling of hair.

Hydrolyzed protein: A protein that has been broken down to smaller segments.

Leave-in conditioner: A type of hair conditioner i.e. designed to remain on the hair after application.

***Malassezia*:** Nonpathogenic yeast normally found on the skin. It is associated with the development of dandruff.

Quat: Cationic surfactant, quaternary ammonium compound.

Rinse conditioner: A type of hair conditioner designed to be rinsed shortly after application.

Seborrheic dermatitis: A more severe variant of dandruff.

Shampoo: A personal care product designed to remove all kinds of soilage, including sebum, sweat, environmental dirt, and hair conditioners, as well as to beautify the hair and make it easy to handle.

Split end: A hair tip i.e. damaged and splits into strands.

REVIEW QUESTIONS FOR SECTION 2

Multiple Choice Questions

1. Which of the following is the causing factor of dandruff?
   a) Yeast
   b) Bacteria
   c) Virus
   d) Prion
2. What happens when a person has dandruff?
   a) Cell production rate decreases
   b) Cell production rate increases
   c) Bacterial by-products trigger inflammation
   d) Less sebum is produced than the normally
3. Which of the following are used commonly in dandruff treatment?
   a) Zinc pyrithione, acetylsalicylic acid, tar
   b) Tar, sulfur, acetylsalicylic acid
   c) Zinc pyrithione, tar, sulfur
   d) Salicylic acid, silica, ketoconazole
4. Which of the following is TRUE for hair conditioners?
   a) Quats have a negative charge that strongly bonds to the negative hair fibers
   b) Film-formers coat the hair fiber and fill in the holes
   c) Protein-based conditioners contain large-sized proteins because the larger the molecules, the better is their penetration to the hair
   d) Silicones have a positive charge that strongly bonds to the negative hair fibers
5. Which of the following generally contains the highest amount of surfactants?
   a) Everyday shampoo
   b) Baby shampoo
   c) Dry shampoo
   d) Dry cleansing shampoo

6. How do shampoos clean the hair?
   a) Via emulsification
   b) Via abrasion
   c) Via dissolution
   d) All of the above

7. Which of the following is the main cause of eye irritation when using shampoos?
   a) Cationic surfactants
   b) Anionic surfactants
   c) Silicone conditioners
   d) Preservatives

8. Sodium chloride can only thicken shampoos that contain ___.
   a) Preservatives
   b) Anionic surfactants
   c) Nonionic surfactants
   d) Cationic surfactants

9. Which of the following is TRUE for instant hair conditioners?
   a) They are designed to be left on the hair for days
   b) They are used on dry hair
   c) They are used on wet hair
   d) They protect the hair from heat damage

10. Which of the following is NOT true for combability testing?
    a) Living animals are used for the test
    b) Artificial or human hair tresses are used for the test
    c) The test evaluates the change in grooming force
    d) Both wet and dry hair can be used for the test

Fact or Fiction?

_____ a) Conditioners provide a temporary aid to the hair.
_____ b) Shampoos that do not foam well do not clean well.
_____ c) Dandruff is a noncontagious infection.
_____ d) Dry shampoos work without soap and water.

Matching

Match the ingredients in column A with their appropriate ingredient category in column B.

	Column A	Column B
_____	A. Citric acid	1. Amphoteric surfactant
_____	B. Cocamidopropyl betaine	2. Anionic surfactant
_____	C. Dimethicone	3. Antidandruff active ingredient
_____	D. EDTA	4. Cationic hair conditioner
_____	E. Methylchloroisothiazolinone	5. Chelating agent
_____	F. Polyquaternium-10	6. Film-forming hair conditioner
_____	G. PVP	7. pH buffer
_____	H. Sodium chloride	8. Preservative
_____	I. Sodium laureth sulfate	9. Silicone hair conditioner
_____	J. Zinc pyrithione	10. Thickener for shampoo

REFERENCES

1. Trüeb, R. M.: Dermocosmetic aspects of hair and scalp. *J Investig Dermatol Sympos Proc.* 2005;10:289–292.
2. Bouillon, C.: Shampoos. *Clin Dermatol.* 1996;14:113–121.
3. Partick, B., Thompson, J.: An Uncommon History of Common Things, Washington: National Geographic, 2011:192.
4. Smith, V.: Clean: A History of Personal Hygiene and Purity, Oxford University Press, 2007.
5. Draelos, Z. D.: Skin and Hair Cleansers, Last update: 7/13/2013, Accessed 1/19/2014 at http://emedicine.medscape.com/article/1067572-overview#showall
6. Draelos, Z. D.: Essentials of hair care often neglected: hair cleansing. *Int J Trichol.* 2010;2(1):24–29.
7. Zirwas, M., Moennich, J.: Shampoos. *Dermatitis.* 2009;20(2):106–110.
8. Katugampola, R. P., Statham, B. N.: A review of allergens found in current hair-care products. *Contact Dermatitis.* 2005;53(4):234–235.
9. Schepky, A. G., Holtzmann, U., Siegner, R.: Influence of cleansing on stratum corneum tryptic enzyme in human skin. *Int J Cosmet Sci.* 2004;26:245–253.
10. Ananthapadmanabhan, K. P., Moore, D. J., Subramanyan, K., et al.: Cleansing without compromise: the impact of cleansers on the skin barrier and the technology of mild cleansing. *Dermatol Ther.* 2004;17(1):16–25.
11. Lips, A., Ananthapadmanabhan, K. P., Vetamuthu, M., et al.: Role of Surfactant Micelle Charge in Protein Denaturation and Surfactant Induced Skin Irritation, In: Rhein, L. D., Schlossman, M., O'Lenick, A., eds: Surfactants in Personal Care Products and Decorative Cosmetics, Boca Raton: CRC Press, 2006: 177–187.

12. Mottram, F. J., Lees, C. E.: Hair Shampoos, In: Butler, H., ed.: Poucher's Perfumes, Cosmetics and Soaps, Boston: Kluwer Academic Publishers, 2000.
13. Bouillon, C.: Shampoos and hair conditioners. *Clin Dermatol.* 1988;6:83–92.
14. Powers, D. H.: Shampoos, In: Balsam, M. S., Gershon, S. D., Reiger, M. M., et al., eds: Cosmetics Science and Technology, 2nd Edition, Hoboken: Wiley-Interscience, 1972:73–116.
15. Krause, T., Rust, R., Kenneally, D. C.: Hair Styling – Technology and Formulations, In: Drealos, Z. D., ed.: Cosmetic Dermatology, Hoboken: Wiley Blackwell, 2010.
16. Zviak, C., Bouillon, C.: Hair Treatment and Hair Care Products, In: Zviak, C., ed.: The Science of Haircare. New York: Marcel Dekker, 1986: 134–137.
17. Faergemann, J.: Seborrheic Dermatitis (Dandruff), In: Elsner, P., Maibach, H., eds: Cosmeceuticals: Drugs vs Cosmetics, New York: Marcel Dekker, 2000: 197–202.
18. Gupta, A. K., Bluhm, R.: Seborrheic dermatitis. *J Eur Acad Dermatol Venereol.* 2004;18(1):13–26.
19. Pray, W. S.: Nonprescription Product Therapeutics, 2nd Edition, Philadelphia: Lippincott Williams & Wilkins, 2006.
20. Tang, E.: Management of an embarrassing problem: dandruff. *Hong Kong Pharm J.* 2002;11:106.
21. Trüeb, R. M.: Female Alopecia: Guide to Successful Management, New York: Springer, 2013.
22. Harding, C. R., Moore, A. E., Rogers J. S.: Dandruff: a condition characterized by decreased levels of intercellular lipids in scalp stratum corneum and impaired barrier function. *Arch Dermatol Res.* 2002;294:221.
23. Gupta, A. K., Nicol K. A.: Sebborheic dermatitis of the scalp: etiology and treatment. *J Drugs Dermatol.* 2004;3:155.
24. Hay, R. J.: Malassezia, dandruff and seborrhoeic dermatitis: an overview. *Br J Dermatol.* 2011;165(2):2–8.
25. Piérard-Franchimont, C., Hermanns, J. F.: From axioms to new insights into dandruff. *Dermatology.* 2000;200:93.
26. Rippon, J. W.: Medical Mycology. The Pathogenic Fungi and Pathogenic Actinomycetes, Philadelphia: WB Saunders Co, 1984:154–168.
27. Johnson, B. A., Nunley, J. R.: Treatment of seborrhoeic dermatitis. *Am Fam Physician.* 2000;61:2703.
28. Klein, K.: Formulating hair conditioners: hope and hype. *Cosmet Toiletries.* 2003;118:28–31.
29. Swift, J.: Mechanism of split-end formation in human head hair. *J Soc Cosmet Chem.* 1997;48:123–126.
30. Braida, D., Dubief, C., Lang, G.: Ceramide. A new approach to hair protection and conditioning. *Cosmet Toiletries.* 1994;109:49–57.
31. Idson, B., Lee, W.: Update on hair conditioner ingredients. *Cosmet Toiletries.* 1983;98:41–46.
32. Finkelstein, P.: Hair conditioners. *Cutis* 1970;6:543–544.
33. Fox, C.: An introduction to the formulation of shampoos. *Cosmet Toiletries.* 1988;103: 25–58.
34. Berthiaume, M., Merrifield, J., Riccio, D.: Effects of silicone pretreatment on oxidative hair damage. *J Soc Cosmet Chem.* 1995;46:231–245.

35. Reeth, I., Caprasse, V., Postiaux, S., et al.: Hair shine: correlation of instrumental and visual methods for measuring the effects of silicones. *IFSCC*. 2001:21–26.
36. Starch, M.: Screening silicones for hair luster. *Cosmet Toiletries*. 1999;114:56–60.
37. Hordinsky, M., Avancici Caromori, A. P., Donovan, J. C.: Hair Physiology and Grooming, In: Drealos, Z. D., ed.: Cosmetic Dermatology and Cosmetic Procedures, Hoboken: Wiley and Blackwell, 2000.
38. Bouillon, C.: Shampoos and hair conditioners. *Clin Dermatol*. 1988;6:83–92.
39. Drealos, Z.: Hair Care, New York: Taylor and Francis, 2005.
40. Gao, T., Bedell, A.: Ultraviolet damage on natural gray hair and its photoprotection. *J Cosmet Sci*. 2001;52:103–118.
41. Ruetsch, S., Kamath, Y., Weigmann, H.: Photodegradation of human hair: an SEM study. *J Cosmet Sci*. 2000;51:103–125.
42. Texture Analyzer, Accessed 2/10/2014 at http://www.stablemicrosystems.com/frameset.htm?http://www.stablemicrosystems.com/CosmeticsAndSkincareTestingAttachments.htm
43. Wessman, L.: The antidandruff efficacy of a shampoo containing piroctone olamine and salicylic acid in comparison to that of a zinc pyrithione shampoo. *Int J Cosmet Sci*. 2000;22(4):285.
44. Futterer, E.: Evaluation of efficacy of antidandruff agents. *J Cosmet Chem*. 1981;32:327–338.
45. Clinical Study to Evaluate the Anti-Dandruff Efficacy of Shampoos, Accessed 1/26/2014 at http://www.fda.gov/ohrms/dockets/dailys/04/oct04/101904/04n-0050-rpt0001-E-28-proDerm-Institute-2000-vol8.pdf

SECTION 3: HAIR STYLING PRODUCTS, HAIR STRAIGHTENING PRODUCTS, AND HAIR WAVING PRODUCTS

LEARNING OBJECTIVES

Upon completion of this section, the reader will be able to

1. define the following terms:

Brazilian blowout	Curl retention	Emollient-based hair fixative	
Fixing lotion	Flaking	Hair fixative	Hair lanthionization
Hair straightening	Hair waving	Heat protection spray	Lye
Perming	Perming lotion	Polymer-based hair fixative	
Pressure filling	Protein denaturation	Stiffness	Tackiness
Valve clogging	VOCs		

2. differentiate between hair styling products and hair straightening/waving products;
3. briefly discuss whether the following are considered cosmetics or drugs in the US: hair styling products, hair straightening products, and hair waving products;
4. briefly discuss how hair styling products and hair straightening/waving products can negatively affect the hair and scalp, respectively;
5. list various required cosmetic qualities and characteristics that an ideal hair styling aid, as well as hair straightening and waving product, should possess;
6. list various required technical qualities and characteristics that an ideal hair styling aid, as well as hair straightening and waving product, should possess;
7. differentiate between polymer-based and emollient-based hair styling aids;
8. list some of the major types of ingredients found in both types of hair styling aids, and provide some examples for each type;
9. list a few examples for both polymer-based and emollient-based products;
10. differentiate between temporary and permanent hair waving/straightening;
11. briefly discuss the possible methods used for temporary hair waving and hair straightening, respectively;
12. briefly explain the process of permanent hair waving;
13. list some of the major types of ingredients found in hair waving products and provide examples for each type;
14. briefly discuss the possible methods used for permanent hair straightening;
15. list some of the typical quality issues that may occur during the formulation and/or use of hair styling aids and hair straightening and hair waving products, and explain why they may occur;
16. list the typical quality parameters that are regularly tested for hair styling aids and hair straightening and hair waving products, and briefly describe their method of evaluation;
17. name the performance parameters generally tested for hair styling products, and describe the method of evaluation;
18. name the performance parameters generally tested for hair waving products, and describe the method of evaluation;
19. explain in which types of products the following ingredients are used and why they caused safety concerns: propellants, thioglycolate, hydroxides, and formaldehyde;
20. list typical containers available for hair styling aids and hair straightening and hair waving products.

KEY CONCEPTS

1. Hair styling aids, also known as hair fixatives, are designed to reshape the hair to a desired arrangement, improve its volume, increase its shine, soften it, or stiffen it – depending on the products used – and maintain the style for a shorter or longer period of time.
2. Hair waving and straightening products help alter the natural configuration of hair fibers and modify the hair's natural straightness/curliness.
3. Polymer-based hair styling aids help build interactive forces among the hair fibers, which prevent smooth sliding of fibers over each other.
4. Emollient-based hair styling products are either anhydrous formulations or emulsions. They are often used for thicker, curly, or kinky hair i.e. hard to handle and can be very dry.
5. Hair can be temporarily curled or straightened using nonelectric and electric tools, which restructure and reshape the appearance of the hair by affecting the weak bonds, resulting in a change in the curl pattern of the hair.
6. In order to make permanent changes in the hair fibers, the strong disulfide bonds have to be denatured and reset into a new shape. Here, denaturation is a reversible process controlled by various chemical agents.
7. The typical quality-related issues of hair styling, hair straightening, and hair waving products include valve clogging, poor foaming of hair mousse, unstable foam, separation of emulsions, microbiological contamination, clumping, and rancidification.
8. Parameters commonly tested by cosmetic companies to evaluate the quality of their hair styling products include spray characteristics; aerosol can leakage; actuation force; pressure test for aerosol products; foamability, foam stability, foam viscosity, foam density, and foam structure; spreadability, extrudability, texture, and firmness of lotions, creams, and gels; preservative efficacy; viscosity; and pH.
9. Commonly tested performance parameters for hair styling products include stiffness of hair fibers, fixative tackiness, dynamic hair spray analysis, curl retention, humidity resistance, and flaking. As for hair waving products, the most common tests include deficiency in tightness (DIT) value, the curl length, and the 20% index.
10. Ingredients of hair styling, hair straightening, and hair waving products that caused major safety issues in the past include propellants, thioglycolates, hydroxides, and formaldehyde.

Introduction

Hair can be straight, wavy or curly, blonde, black, brown, red, gray, and white, and its natural variations are important to our identity. Hairstyle can be used to express personality and make a social statement. In order to achieve the desired appearance, various types of styling products and/or tools can be used. Manipulation of the normal structure of the hair shaft is universal, and it is dictated by multiple factors, including culture, religion, fashion, and even celebrities. Today, sales of hair care and styling products add up to a multibillion-dollar industry.

In this section, we will review the various types of products and tools that can be used to style the hair by altering its natural arrangement, straightness/curliness, and volume. We will review the main ingredients of these products, their formulation technology and testing, as well as consumers' needs. In addition, the main effects of such products on the hair and scalp will also be discussed.

Types and Definition of Hair Styling Products, Hair Straightening Products, and Hair Waving Products

● **Hair styling aids, also known as hair fixatives, are designed to reshape the hair to a desired arrangement, improve its volume, increase its shine, soften it, or stiffen it – depending on the products used – and maintain the style for a shorter or longer period of time.** Hair fixatives do not affect the strong bonds in hair fibers, and they reform the weak bonds only; thus, their effect is temporary (see more detail on various bonds in hair fibers in Section 1 of this chapter). Styling products can be applied to either wet or dry hair after cleaning and/or conditioning it. All hair styling aids are considered cosmetics in the US as they simply help in reshaping the hair and make consumers more beautiful.

Due to the many different hair structures, needs, and fashion trends, today, a wide variety of hair styling aids are available. In general, hair styling aids can be divided into two groups based on the major ingredients used.

- **Polymer-based products** help improve the hair volume and height by increasing the hair strand stiffness and hair fiber interactions. They coat hair fibers with a thin film layer and form links between the fibers. Hair spray, styling gel, mousse, styling gum, and liquid setting products work based on this principle.
- **Emollient-based products** can help smooth out frizz and increase hair shine by aligning fibers and reducing friction and can also have a conditioning effect on hair fibers. They work on the principle of depositing waxes and emollients on hair fibers. Hair cream, wax, pomade, brilliantine, and oil spray work based on this principle.

● **Hair waving and straightening products help alter the natural configuration of hair fibers and modify the hair's natural straightness/curliness.** Depending on the technique and ingredients used, they can affect both the weak and strong bonds, having a temporary or permanent effect. All types of hair styling aids are considered cosmetics in the US.

Hair waving products and techniques are used to make straight hair curly or wavy. They typically involve the use of curling rods for temporary waving and products applied in two separate steps for permanent waving.**Hair straightening products** and techniques are used to make curly and wavy hair straight. They typically involve the use of flat irons for a temporary effect and various products, usually in two separate steps, for a permanent effect.

DID YOU KNOW?

Permanent styling procedures involve denaturation of the structural disulfide bonds and reforming them into a new shape. Although you may argue that these products can change the appearance of the human body, i.e., the hair, these products are considered cosmetics in the US. The main reason for this is that hair shafts consist of dead cells; therefore, causing any changes in these dead cells will not result in any physiological changes.

History of Using Hair Styling Products

Hair styling began early in the human history. Hair usually does not grow naturally in a wide variety of hairstyles desired by men and women; therefore, manipulation of the hair through combing, brushing, setting and using hair styling aids and techniques is necessary to arrange the hair into the desired shape and format.

Mummy analyses found that the ancient Egyptians styled their hair using a fat-based gel.[1] Ancient Egyptians were also known to have short hair or even shave their hair for comfort and wear wigs for ceremonial occasions. These wigs were made of human hair, wool, palm fiber, and flax and were styled with beeswax.[2] The use of thermal hair appliances also dates back to the Egyptian civilization. They used to wrap their hair with wooden rollers, followed by baking in the sun, which created a temporary wave in their hair. In ancient Greece, men favored curls all over their head and often used an early form of curling iron to achieve a uniform look. Women used to pull their hair back into a chignon style, and it was usually grown long. In ancient Rome, hair styles became an expression of a person's identity as much as it is today. For women, their hair indicated how attractive and wealthy they were. Artificial curls did not appear to play an important role in the Middle Ages when hair was worn long and natural. However, during the Renaissance and especially during the Baroque period, curls were back in fashion. During this time, and even until the 19th century, hair was mainly curled using hot irons.

A new type of curved curling tong was invented in the late 19th century by Marcel Grateau, which was named "Marcel wave." He heated the tong and then manipulated it to thermally and temporarily curl the hair. It was a great hit at that time and ingenious at the same time, as the practice of thermally curling hair still exists today.[3] Nessler, a London hairdresser, produced permanent waves by heating the hair after it had been wound on a mandrel and moistened with an aqueous solution of alkali. It is generally accepted that modern permanent waving began with Nessler's method.[4,5] Nessler's heating device had individual heaters for each curl, which were heavy and uncomfortable, not to mention that the processing time lasted for hours.[6]

Additionally, these early perms were harsh and very damaging to the hair and left the hair in tight frizzy curls that were difficult to manage. The idea of the cold permanent wave, i.e., the method for waving hair without externally applied heat, was introduced in the 1930s.[7] The cold waving solution contained ammonium thioglycolate and free ammonia at a controlled pH. The use of such solutions, with slight variations, is still popular today.

The first hair straightener was invented intentionally by Garrett Morgan in the early 20th century. While preparing a chemical solution that could be used to polish sewing machine needles so their friction would not scorch and damage the fabrics as they sewed, he accidently discovered that this solution he made straightened a pony-fur cloth. He tried the solution on a dog and then on himself, and it worked. In 1913, he founded his company and sold his hair straightening cream.[8] The first permanent hair straighteners were developed around 1940 and consisted of sodium hydroxide or potassium hydroxide mixed into potato starch. These were very irritating to the scalp and hair. Later, petrolatum and fatty alcohols were added to decrease the irritation. The first no-base relaxers were available in the 1960s. Today, a great variety of hair styling products and techniques are available on the market for various hair types.

How Hair Styling Products and Procedures May Affect the Human Hair and Scalp?

Hair has an important role in influencing our appearance; it can reflect culture, socioeconomic status, political views, marital status, and even emotions. Hair styling products, procedures, and various tools are commonly and widely used by both genders and all hair types. When used properly, hair styling aids, various devices, and chemical processes can help improve the hair's appearance, make the hair more manageable and cosmetically elegant, and improve the customer's self-esteem. The advantages of using hair styling products are summarized as follows:

Styling aids not only assist in increasing volume and shine but are also useful in improving the appearance of hair dermatologic conditions that result in **thinning hair, hair damage, or hair loss.** People can damage and lose hair for various reasons, including illness, certain medications, and genetics; even harsh chemical treatments can also lead to severe damage or loss. In addition, thinning of hair can severely affect people's self-confidence and impact their quality of life and, in many cases, lead to significant psychological and emotional distress.[9] Hair styling aids, such as hair spray and gels, can be used to lift thinning hair and create the illusion of fullness.[10] Oil and silicone sprays can help restore shine and improve the appearance of very dry or even splitting hair.

The ethnic hair care market is a blooming area in the cosmetic industry. **Ethnic hair** structure is unique and warrants special attention and care. The diameter of tight, curly, (or kinky) ethnic hair is much different from those of other hair types (see more detail on the differences in Section 1 of this chapter). Generally, African hair is coarser, very dry, frizzy and fragile, and more difficult to handle than Caucasian or Asian hair. Hair styling aids and various tools and techniques can be efficiently used to make the hair more manageable and smoother.

DID YOU KNOW?

"Ethnic hair care" traditionally refers to products designed for consumers of African American origin with tight, curly, or kinky hair.

Hair styling aids are, in general, safe for the skin and hair and can have many benefits. Consumers, however, often use more styling products than what they actually need, or they do not use the proper type of styling aid. Temporary and permanent hair waving and straightening are commonly used hair styling techniques, which are considered safe only if used carefully, selected according to the consumers' hair type, and following the manufacturer's directions. Even the simplest tools, such as a curling iron, can lead to significant and permanent hair damage if used improperly.[11] Both overuse and improper use of hair styling products and/or hair styling procedures and tools can have a negative effect on the hair's appearance. Here is a summary of the major negative effects of such products and tools on the hair and scalp, as well as the main factors that should be considered before altering the hair's natural appearance.

- Overuse of hair styling products based on film-forming polymers can create a weighted down effect, reduction in luster, or even appearance of white **flakes**. Flakes are tiny polymer particles peeling off from the hair fibers, which are similar to dandruff in appearance.
- Overuse of emollient-based hair styling products can create a **greasy, lump** appearance, the opposite of what users are trying to achieve.
- Regular and proper removal of hair styling aids is essential as these products **accumulate** on the hair fibers and provide a surface for additional dust and dirt. All these can lead to a dull, greasy appearance.
- **Irritation** and **allergic reactions** to hair styling products are rare, similar to shampoos and conditioners. The main ingredients used in styling aids are generally not recognized as irritants. However, fragrances and certain preservatives added to these formulations may cause irritation reactions in sensitive skin patients.

Heat can be efficiently used to temporarily reshape the hair into a desired form; however, very high temperature may lead to significant, irreversible **damage** of hair fibers. It has been observed that blow-drying and other heat treatments, such as curling irons or flat irons, can lead to cracks and bubble formation on the cuticle due to rapid dehydration of the hair fiber. These changes can permanently damage the hair, leading to visible changes, for example, loss in shine, weathering, and split ends. Recent improvements in temperature control promote thermal stability and coatings, including ceramic and titanium, and provide durability and reduced friction. Reduced friction is critical to maintain a smooth cuticle surface and reduce breakage during thermal processing.[12] Additionally, heat protection products are also available on the market. These are typically applied before the heat treatment so they can coat the hair shafts and protect them from severe damage.

Permanent straightening and waving are chemical processes that involve breaking of the strong bonds in the hair and reforming them into a new shape. For waving or straightening the hair, a number of key factors should be considered. These include the hair's initial state, curliness/straightness, coarseness, porosity, and previous chemical treatments (e.g., coloring, bleaching, waving, and straightening). These factors aid in selecting the appropriate type of waving/straightening ingredient and its concentration. Instructions found on the package of waving/straightening products should be carefully read and followed in order to help maintain the integrity of the hair and prevent unnecessary damage. Overprocessing, improper removal of the reducing agents or hydroxide relaxing agents or oxidizing agents, and improper neutralization are the problems that are commonly associated with the misuse of permanent waving and straightening products. Improper steps in these potentially damaging chemical procedures can lead to significant **hair damage**, including loss of elasticity [13], loss of hydrophobicity; increased porosity; split ends; hair breakage; rough, dry, dull, and brittle hair; and even loss of the cuticle.[14–16]

In addition to hair damage, **skin irritation** can also commonly occur during permanent relaxing and waving of the hair. Many products used have a highly alkaline pH (around pH 12), which can lead to severe skin damage.

Required Qualities and Characteristics and Consumer Needs

From a consumer perspective, a quality hair styling product should possess the following characteristics:

- Easy to apply
- Easy to remove by shampooing
- Polymer-based products
  - Non-flaky
  - Non-tacky even at high humidity
  - Rapid drying time
  - Strong hold
- Emollient-based products
  - Non-greasy, does not make hair look oily and lump
  - Aids in hair manageability
  - Improves shine
- Good smell

From a consumer perspective, a quality hair waving and hair straightening product should possess the following characteristics:

- Easy to apply
- Provides a long-lasting effect
- Does not damage the hair or scalp

The technical qualities of hair styling, hair straightening, and waving products can be summarized as follows:

- Compatible with the other ingredients in the formulation
- Appropriate rheological properties
- Appropriate texture
- Appropriate pH
- Long-term stability
- Dermatological safety

Hair Styling Formulations

Hair styling products are not meant to be permanent, and the majority of these products are designed for easy removal by shampooing. They are usually finishing products and are applied in the very last step of styling on dry hair.

Polymer-Based Hair Styling Formulations ❶ **Polymer-based hair styling aids help build interactive forces among the hair fibers, which prevent smooth sliding of fibers over each other.** In order to create these forces, the formulations provide a type of "roughness" on the hair surface. Thus, they can help hold the style longer and support it. Although there are a wide variety of polymer-based products available, their basic types of ingredients are the same. The currently available products vary in viscosity, polymer type, and its concentration. Therefore, the main ingredients are reviewed together.

- **Polymers** are the main ingredients in all polymer-based formulations. They are responsible for holding the hair in the styled configuration and stiffening hair.
  - The first polymer used was PVP, which is compatible with water and alcohol and has good film-forming properties. The main disadvantage of PVP is that it is sensitive to humidity, becomes sticky, and loses holding power when exposed to water.[17] New polymers include PVP mixed with vinyl acetate (VA), which makes it more resistant to water. It is a valuable characteristic from styling perspective as it offers a stronger holding power; however, it can make hair more difficult to clean. Additional examples for polymers used today include other copolymers (i.e., a polymer made by reacting two different monomers), such as vinylpyrrolidone/dimethylaminoethylmethacrylate (VP/DMAEMA) copolymer, octylacrylamide/acrylates/butylaminoethyl methacrylate copolymer, and vinylmethylether and maleic acid hemiesters (PVP/MA).[18,19]
  - Mousses also contain polymers that act as both film-formers and conditioners. These ingredients are usually known as polyquaterniums, which are polymers with cationic cross-links. Due to their positive charge, they have greater substantivity to the negatively charged hair fibers.[20,21]
- **Thickeners** provide viscosity control to the formulations; however, they are usually poor film-formers. They are most commonly hydrophilic ingredients

applied primarily in gels. The concentration of these gelling agents is usually kept low in order to prevent them from coating the hair fibers and obstructing the adherence of the film-forming polymers. Additionally, thickeners used in such products are usually hygroscopic, which if used in high concentration can make the formulation absorb moisture from the air and cause softening of the polymer film and may also cause stickiness of hair fibers.[22]

- Examples include polyacrylates, such as carbomer; cellulose derivatives, such as hydroxyethylcellulose; and polysaccharides.

- **Solvents** dissolve the polymers and act as vehicles for the formulations.
  - Examples of solvents generally used in polymer-based formulations include water and alcohol. Traditionally, aerosol products are based primarily on alcohol, while non-aerosol products contain water as well. Water was first added to lower the production cost; however, its use became more important when regulations on propellants were introduced. First, the increased level of water caused significant problems in product performance, including slower drying time and tackiness upon application. However, today's polymer systems are developed to work well with lower alcohol and higher water contents.[23]
- **Propellants** are essential ingredients in aerosol formulations, such as aerosol hair sprays and aerosol mousses. They help expel the content of the aerosol can.
  - Examples include dimethyl ether, propane/butane, and fluorinated hydrocarbons.
- **Plasticizers** make polymer films more flexible.
  - Examples include dimethicone, castor oil, and mineral oil.
- **Sunscreens** are often employed in hair styling products. They provide protection against UV radiation, help maintain the attractive appearance of the hair (as UV radiation can lead to weathering in the long term), and can also stabilize the viscosity of gels.
  - Examples include benzophenone-4 and octyl methoxycinnamate.
- **Chelating agents** provide protection against metal ions, improve the stability of gels, and increase the efficiency of preservative systems.
  - The most commonly used chelating agents are EDTA and its derivatives.
- **Preservatives** provide protection against microbial contamination in water-based systems.
  - Examples include the parabens and phenoxyethanol.
- **Additional ingredients** can include humectants that contribute to the wet look provided by gels; conditioners, such as proteins and quaterniums; fragrances; vitamins; and herbal extracts.

Product Types The major types of polymer-based products include hair sprays, setting lotions, mousses, hair styling fiber gum, and hair gels. Their main characteristics are reviewed here.

- **Hair sprays** are available in two major forms: aerosol products and non-aerosol products. Both types are usually applied to dry hair as a final step of styling the hair. Aerosol hair sprays are one of the most widely used hair styling products by both consumers and hair stylists. Aerosol hair sprays are released from the container with the help of a propellant, which results in very fine particles (such as a mist). Due to the small particle size of the droplets, they dry quickly. Non-aerosol hair sprays are dispersed from a plastic container supplied with a pump head. The distribution of the hair spray is provided mechanically by the pump head, which forms bigger droplets and, therefore, can wet the hair more rapidly. Hair sprays, especially those formulated to provide strong hold, tend to form flakes on the hair.
- **Setting lotions**, also known as wave sets, are applied from a small bottle with a nozzle to towel-dried hair rolled onto curling rollers. Hair is usually dried when rolled up. Setting lotions coat the hair fibers and set them in a curled shape. After removing the curling rollers, the hair is combed gently into the desired style. Setting lotions do not work by sticking the hair fibers together, but by coating each hair fiber, creating greater interfiber friction, and reducing moisture uptake, thus conferring greater control to the hair.
- **Hair mousses** are available in aerosol and non-aerosol forms; however, the aerosol types are far more popular. Both types are typically applied to towel-dried hair followed by drying and styling. They can also be applied to dry hair to create a wet, spiky look. Products applied to wet hair should provide easy combability, and therefore, mousses usually contain conditioner polymers as well. Mousses are designed to add volume, improve texture, and help control the hair mass. They can even be colored to provide natural or unnatural highlights temporarily.
- **Hair styling fiber gums**, known as putties, are relatively new products designed to add increased hold to the hair. They are usually gum-like opaque emulsions. Fiber gums are more popular among younger customers with shorter hair. These products are bendable and moldable styling products that can be used to create various hairstyles, including spikes or a smoother look. Hair fiber gums are usually scooped from a jar and applied with the fingers. Fiber gums can be applied to both towel-dried and dry hair, depending on the look the user wants to create.
- **Hair gels** are also very popular hair styling products. They can offer both a smooth hold and an extreme look with spikes, depending on whether the product is applied to wet or dry hair and whether the hair is blow-dried or air-dried. Gels are usually applied from a tube or jar; however, spray gels are also available. Sprayable products usually have a lower viscosity, and they are easy to evenly apply to the hair and provide a weaker hold. Tube and jar gels have a higher viscosity, provide a stronger hold, and, therefore, allow for creating "gravity-defying" hairstyles. These products are also often referred to as "sculpturing" gels as they allow for creating spikes that naturally would not be able

to hold its shape. Generally, the amount of film-forming polymer usually determines the hold level. Removal of a gel requires shampooing – in some cases, a second lathering is recommended. Dry, brittle gels with high hold levels may tend to generate flakes when touched, similar to hair sprays.

Formulation of Polymer-Based Hair Styling Products The formulation of low-viscosity products is relatively simple. Generally, ingredients are blended with agitation and mixed until homogeneous. During the formulation of a product containing thickening ingredients, dispersion of the thickener is a critical step. Thickeners are hygroscopic and tend to form clumps if added too fast to the dispersing medium or if no or slow agitation is used. As discussed in the previous chapters, certain thickeners need pH adjustment (neutralization) to reach their optimal viscosity. In such cases, the neutralizer should be added after proper hydration of the thickener in order to avoid clumping.

The formulation of aerosol formulations must be performed in a flameproof area using special equipment. Formulation of the product concentrate is a solution preparation process. The concentrate contains polymers, neutralizers, plasticizers, fragrance, and other ingredients in alcohol or a mixture of water and alcohol as the vehicle. If the formulations contain water, it is usually added in the last step. Filling of the product concentrate and propellants in the aerosol cans is usually done via pressure filling (for more detail, refer to aerosol sunscreens, Section 5 of Chapter 3).

Emollient-Based Hair Styling Formulations ❶ **Emollient-based hair styling products are either anhydrous formulations or emulsions. They are often used for thicker, curly, or kinky hair, which is hard to handle and can be very dry.** Their main types of ingredients are summarized as follows:

- The **anhydrous base** provides a waxy, oily mass for the formulations. Ingredients of the anhydrous base also provide a hair conditioning effect, help reduce friction, and prevent damage.
  - Examples for such ingredients include beeswax, petrolatum, lanolin, vegetable oils, mineral oils, and silicones.[24]
- **Solvents** are used in emulsion-based products to provide a vehicle for the formulations. Usually, water is used as a solvent.
- **Surfactants** help create emulsions as well as aid in removal of the product.
- **Polymers** can also be added to these products to increase the hold level of the products.
- **Thickeners** can be used to change the product's viscosity and texture.
- **Additional ingredients** may include sunscreens, vitamins, fragrances, antioxidants, and preservatives.

Product Types The major types of emollient-based products include hair pomades, hair brilliantines, sprayable oils and silicones, and hair creams. Their main characteristics are reviewed here.

- **Hair pomades**, also known as cream brilliantines, are anhydrous products based on petrolatum and various waxes. These formulations are generally very thick and are designed to straighten, condition, and moisturize the hair as well as add shine to hair while reducing frizziness. Pomades can be applied to both towel-dried and dry hair. As the products are quite thick, they are usually warmed up in the palms before being applied to the hair.
- **Hair brilliantines** are liquids based on oils and silicones. These products are designed to allow for easy styling and add shine to the hair without providing a greasy appearance.[25]
- **Sprayable Oils and Silicones**: Curly, kinky African-American hair needs hair sprays that contain a higher amount of oils and silicones to condition the hair and decrease the combing friction. These products are also known as oil sheen sprays. These are usually aerosolized oils that may also contain additional conditioning agents, such as proteins. These products are usually lighter on the hair; therefore, they can be applied on a daily basis.
- **Hair creams** are emulsion products providing modest hold and high gloss. They may also be O/W or W/O emulsions, which break down easily on application.

Formulation of Emollient-Based Hair Styling Products The formulation of pomades is relatively simple. Liquid components are mixed with waxes and are melted to obtain a uniform mixture. All other ingredients are added to this mixture, and the resulting mixture is then is poured into appropriate containers. Hair brilliantines are generally mixtures of various liquid conditioning agents. Their formulation process consists of blending the various components until a uniform mixture is formed and filling them into containers. Hair creams are generally formulated as emulsions.

Hair Styling Procedures

Hair styling is an important part of our life. After washing and/or conditioning the hair, it is usually dried using a blow-drier or other tools. Today, various non-electric and electric devices are available to make temporary changes in the appearance of the hair, i.e., to make a wavy hair straight or a straight hair wavy for a couple of days. For longer lasting effects, a combination of chemical treatments and nonelectric or electric devices can be used. These may save time for consumers as they do not have to restyle their hair every time they wash it; however, these procedures can be more damaging to the hair.

Temporary Hair Curling and Straightening Tools ❺ **Hair can be temporarily curled or straightened using non-electric and electric tools, which restructure and reshape the appearance of the hair by affecting the weak bonds, resulting in a change in the curl pattern of the hair.** Since these processes affect only the weak bonds, they are easily modified by humidity (such as perspiration and moisture from the environment) and completely removed by washing the hair.[25]

- **Non-electric tools** include curling rollers, which are available in various sizes and types; brush rollers can be used without clips, while smooth plastic rods are usually used with clips to hold the hair. In addition, simple curling or straightening brushes can also be used – typically while blow-drying – to alter the natural shape of the hair.
- **Electric styling tools** include individually heated plastic-coated curling rods called electric curlers, curling irons, and flat irons, most of which can be thermostatically controlled today. The higher the temperature, the more long lasting the changes; however, more hair damage may occur. When using a curling iron, the hair is either clipped to the rod and wound around it, or it is held and wound on the rod by hand. For the flat iron, tresses of the hair are held between the heated metal plates of the tool while the tool is moved down to the tips of the hair fibers. These techniques remain popular as they are inexpensive and provide an easy way to change the natural curliness/straightness of the hair and improve its manageability even at home.

DID YOU KNOW?

Protein denaturation is a process in which proteins are irreversibly altered by an external stimulus. In the case of hair, denaturation can result in decreased fiber integrity. The heat sensitivity to denaturation of human skin and hair is very different. Human skin proteins start degrading at a temperature of 40 °C, which may be reversible if the temperature is lowered. Starting at 45 °C, proteins are permanently denatured in the skin, reflected by degradation of the local tissue.[26] Interestingly, human hair shows the first signs of thermal denaturation around 150—220 °C, depending on the water content of the hair. Generally, dry hair can withstand a much higher temperature as compared to wet hair; water decreases the threshold of hair for thermal injury.[27]

The popularity of straightening and curling irons has created a large market for heat protection products for the hair, such as **heat protection sprays**. These sprays are designed to protect the hair from heat damage, provide protection against friction, as well as enhance the hair's flexibility to reform and hold shape. They are typically sprayed onto the hair after blow-drying and immediately before applying a flat iron or curling iron. Usual ingredients of heat protection products include silicones, which have good thermal stability and can effectively form a film on the fibers that reduces friction and conditions.[28] Other examples include cationic conditioning polymers, such as quaternium-70 and polyquaternium-11; film-forming polymers; and proteins, such as hydrolyzed wheat protein.[29]

DID YOU KNOW?

Today, heat protection sprays are available in two main forms: water-based and hydroalcoholic products. A recent study has shown that hydroalcoholic products provide better heat protection, resulting in less hair damage.[30]

Permanent Hair Waving and Straightening Procedures **In order to make permanent changes in the hair fibers, the strong disulfide bonds have to be denatured and reset into a new shape. Here, denaturation is a reversible process controlled by various chemical agents.** Permanent styling methods are actually a combination of physical and chemical processes, since while the hair is being chemically treated, it is physically shaped into the desired format.[13,31] It should be noted, however, that the newly developing hair will not be affected by any of these alterations and will grow in its natural shape. Therefore, hair perming and straightening procedures need to be repeated regularly in order to maintain the altered appearance.

Perming Permanent waving, also known as perming, is a chemical process combined with mechanical manipulation of the hair used to make straight hair curly. Permanent waving is usually a two-step process (depicted in Figure 5.8), including the following steps:[32]

- First, the hair is washed; then it is sectioned into smaller areas, and the end of the hair in each area is wrapped into a thin sheet of tissue paper to avoid irregular wrapping around the rod and frizziness. Finally, the hair tresses are wound on rods that have holes inside to allow the perming lotion to contact

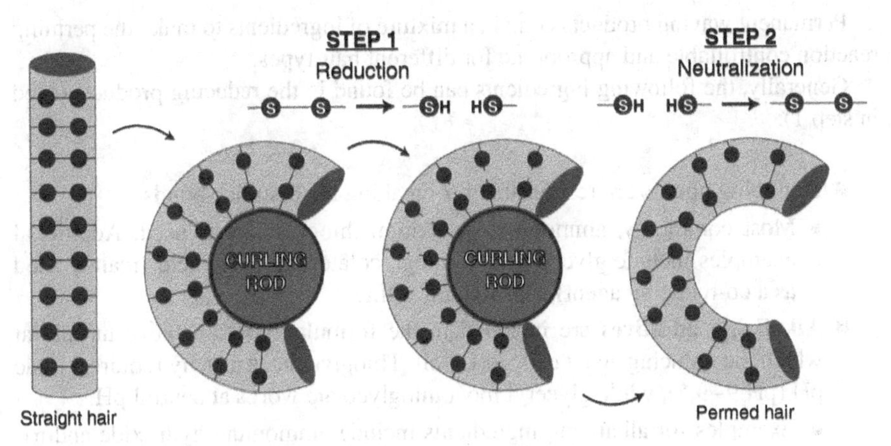

Figure 5.8 Permanent waving process.

all the surfaces of the hair shaft. The size of the rod determines the size of the curl produced: the smaller the rod, the tighter the curl. Hair should be tightly wound on the rods; however, too much tension can result in damaged, brittle hair. The perming lotion is applied to the hair after wounding, and it is allowed to remain on the hair for about 10–30 min (depending on the hair's condition). During the process, the hair is usually covered with a plastic cap to keep it in a warm environment and accelerate the disulfide bond breaking process. To make spreading and penetration of the perming lotion easier, perming lotions are usually liquids. In this step, the hair fibers are softened by breaking about 20–40% of the disulfide bonds in the hair fibers.

- Next, a fixing lotion, otherwise known as neutralizing lotion, containing an oxidizing agent is applied to the hair fibers, which are still wound on the curling rods. This lotion stops the cleavage of disulfide bonds and oxidizes the cysteine residues to cystine in a new configuration within about 5–10 min. Often, one portion of the neutralizing lotion is applied to the hair when it is still on the rods, and the other portion is applied after the rods are removed. The neutralizer is removed from the hair by thorough rinsing with water after unwinding. Usually, only about 85% of the cleaved disulfide bonds are reformed during the neutralization step.[33] Most companies recommend avoiding shampooing the hair for 1–2 days after perming to ensure long-lasting curls.

DID YOU KNOW?

When permanent waving or straightening is performed in a hair salon, an irritating rotten egg smell can be observed. This is actually the smell of sulfur, which escapes from the hair when the disulfide bonds break.

Permanent waving products contain a mixture of ingredients to make the perming reaction controllable and appropriate for different hair types.

Generally, the following ingredients can be found in the reducing products (used in step 1):

- **Reducing agents** are responsible for breaking the disulfide bonds.
  - Most commonly, ammonium or sodium thioglycolate is used. Additional examples include glycerol monothioglycolate, thiolactic acid (mainly used as a co-reducing agent), and sodium sulfite.
- **Alkalizing additives** are included in the formulations to achieve the pH at which the reducing agent can work well. Thioglycolates usually require a basic pH (pH 9–9.5), while glyceryl monothioglycolate works at neutral pH.
  - Examples for alkalizing ingredients include ammonium hydroxide and triethanolamine.

- **Buffers** help adjust the pH of the product, which could decrease when the product comes into contact with the hair.
  - An example is ammonium carbonate.
- **Chelating agents** remove metals and prevent product instability.
  - Examples include EDTA and its derivatives.
- **Wetting agents**, i.e., surfactants, help wetting the hydrophobic hair surface and, therefore, improve spreading of the perming lotion.
  - Examples include sodium lauryl sulfate, sodium laureth sulfate, and sulfosuccinate.
- **Antioxidants** prevent oxidative reactions and, therefore, increase product stability.
  - An example is vitamin E.
- **Conditioners** can be included in the formulations to aid in hair manageability and protect the hair during perming.
  - Examples include humectants, proteins, and cationic surfactants.
- **Solvents** are used as vehicle for the ingredients in the formulation. Most often, water is used.
- **Penetration enhancers** increase the penetration of the reducing actives into the hair and, therefore, speed up the perming process.
  - Examples include propylene glycol and urea.
- **Thickeners** are used to provide viscosity control to the formulations and prevent dripping of the product from the hair during application.
  - Examples include cellulose derivatives and acrylates polymers.
- **Additional ingredients** may include opacifiers, solubilizing agents, and fragrances.

The neutralizing products used in step 2 of the perming process usually contain the following ingredients:

- **Neutralizing ingredients** neutralize the alkaline pH and reform broken disulfide bonds.
  - Most commonly, hydrogen peroxide is used at concentrations of roughly 0.5–3% at a pH of 2–4.5. Additionally, potassium or sodium bromate solution can be used.
- **Buffers** are used to set the pH to an acidic level.
  - Examples include citric acid, acetic acid, and lactic acid.
- **Stabilizers** prevent hydrogen peroxide from breakdown before it gets to the hair.
  - Examples include ammonium phosphate and phosphoric acid.
- **Additional ingredients** include solvents, thickeners, conditioners, wetting agents, chelating agents, fragrances, and opacifiers with the function described under perming products.

DID YOU KNOW?

Curliness of the hair after perming depends on the rod used and not on the strength of the perming lotion. The extent of hair damage (seen as brittle hair that has lost its shine), however, depends on the strength of the perming lotion. A stronger lotion may penetrate deeper into the hair fibers, especially at the ends, making the hair dull. Therefore, the hair length and perming lotion strength should always be taken into consideration; longer hair may need a weaker perming lotion than short hair.[26]

Various Perming Product Types Major differences between various perming products are the type and amount of reducing agent used, pH of the product, and the product form. The majority of commercial waving and neutralizing lotions are water-based solutions; however, gels, creams, or aerosols are also available.[34] Depending on the type of the reducing agent, perming products may have different pH values. Thioglycolate-based perming lotions have an alkaline pH (pH 9–9.5) as thioglycolates are only effective at an alkaline pH. These products, however, can damage the hair cuticle and irritate the scalp. To reduce their irritation potential, buffers can be included in the formulations to adjust the pH to 7–8. These are considered milder perms. Perming lotions with an acidic or neutral pH (pH 6.5–7) are also available. They usually contain glycerol monothioglycolate as the reducing agent. They are less irritating and less effective as well. The so-called thermal waves are also available on the market. These products produce heat during the perming procedure due to a chemical reaction between hydrogen peroxide and thioglycolic acid. These products were introduced to provide a pleasant feel on the scalp.

DID YOU KNOW?

Both the skin and the hair contain keratin, and you may wonder how it is possible that perming products are able to break the disulfide bonds in the hair but leave the skin relatively intact. The reason for this is that the type of keratin found in the skin and in the hair is not the same. The soft keratin found in the skin has less cysteine than the hard keratin of the hair. The skin, therefore, has fewer disulfide bridges and is less susceptible to the attack by thioglycolates.[35]

Formulation Consideration for Permanent Hair Waving Products Thioglycolate forms a red complex with iron; therefore, metal contaminants should be avoided during the formulation of products containing thioglycolate. Equipment is better lined with glass or non-reactive plastic, such as Teflon, to avoid chemical reactions. The neutralizing ingredient, hydrogen peroxide, is also highly reactive; therefore, the formulation of products containing this ingredient also requires rigorous control as well.

Permanent Hair Straightening Permanent hair straightening, also known as relaxing or lanthionization, is a common practice among individuals with highly curly (kinky) hair to make the hair more manageable.[36] There are a few product types that can be used to straighten the hair, including hydroxide-, thioglycolate-, and formaldehyde-based products.[37]

- **Hydroxide-based straighteners**, also known as lye relaxers (the term "lye" refers to sodium hydroxide, which is the most commonly used hydroxide), contain various metal hydroxides, such as sodium, lithium, potassium, or guanidine hydroxide. These products usually have a highly alkaline pH (pH 12–13)[38] and are, therefore, very irritating both to the hair and to the scalp. Lye-based relaxers are available as "base" and "no-base" products. The "base" is usually an occlusive ingredient, most commonly petrolatum or a mixture of such ingredients, which is applied to the scalp and hairline before the chemical treatment to protect the scalp. Products requiring basing may contain sodium hydroxide in a concentration of 3.5%. "No-base" hair relaxers contain less sodium hydroxide (usually 1.5–2.5%), and they require base application to the hairline only (and not to the scalp), which can save a significant amount of time.[39] "No-lye" relaxers are also available, which contain alternative hydroxides, such as guanidine hydroxide. These are less irritating, do not require any base application to either the scalp or the hairline, and are, therefore, more popular.[40] Hydroxide-based relaxers are generally used as creams, which makes the application and spreading easier on the hair.

DID YOU KNOW?

No-lye hair relaxers are usually available in kits, which include all ingredients needed for hair relaxation. As the main ingredients, the kits usually contain calcium hydroxide in a cream form and guanidine carbonate as a liquid. These two are mixed right before application to form calcium carbonate and guanidine hydroxide, the "active" ingredient. The application of such products is easy and convenient since no basing is required.

After relaxing the hair with hydroxide-based products, the hair is rinsed thoroughly with water, and then a neutralizer cream or shampoo is applied to it. Neutralizer products, primarily containing hydrogen peroxide, bromates, and perborates, reset the pH to a neutral level, minimize hair shaft swelling, and complete the relaxing treatment.[41]

The hair relaxer treatment has a defatting action on the hair; therefore, various post-conditioning treatments are available to lubricate, moisturize, and add luster to the hair. Ingredients commonly used include various conditioning agents such as polyquaternium-6, behentrimonium chloride, and hexadimethrine chloride, as well as ceramides and panthenol.[42]

- **Reducing agents**, similar to hair perming products, are also available.[43] Thioglycolate-based straighteners work based on the same principle as thioglycolate-based hair perming products (for more detail, see Section 2 of this chapter); however, some differences exist. In the case of hair straightening, the hair is not washed before the treatment. The reason for this is that straightening is more damaging to the hair than waving, and shampooing would remove the protective sebum from the scalp. Additionally, the hair is not wound on rods, but it is combed straight while the reducing agent is in contact with the hair. To make spreading and application easier, reducing agents are usually formulated as thick creams. Sulfite-based products are also available. Their pH is around 7; therefore, these products are milder to the skin and hair. Their application procedure is similar to that of thioglycolate straighteners, except that no oxidant is used in step 2. Instead of oxidants, a neutralizing solution containing sodium carbonate or bicarbonate is usually used, whose effectiveness is weaker.
- In addition to hydroxides and thioglycolates, there are other ingredients used as permanent hair straighteners. A group of hair straightening agents, known as "Brazilian Keratin Treatment" or "Brazilian Blowout," contains **formaldehyde** as the main straightening ingredient. The procedure consists of the application of a straightening solution containing keratin followed by the application of high-temperature flat iron (around 150–220 °C). After flat ironing, the hair is rinsed and conditioners are applied to the hair.[44] There were a number of reports about severe adverse effects from both professionals and clients when using such formulations. The use of these formulations has been declared unsafe by the CIR Expert Panel in 2012. Both Oregon's Occupational Safety and Health Division (OSHA) (hazard alert) and FDA recommend that customers limit the exposure to such formulations.[45]

FORMULATION OF PERMANENT HAIR STRAIGHTENING PRODUCTS Similar to the formulation of hair waving products containing thioglycolates, the same precautions should be kept in mind, as discussed earlier. Relaxing products are generally formulated as creams, mainly O/W emulsions. Humectants are often incorporated to prevent excessive water evaporation during the dissolution process of lye, which is an exothermic reaction. Addition of the active phase to the emulsion is a critical step due to the highly reactive nature of alkalis, especially in the case of sodium hydroxide. Otherwise, the formulation process of these products follows the general steps of emulsification.

Typical Quality Problems of Hair Styling Products

● **The typical quality-related issues of hair styling, hair straightening, and hair waving products include valve clogging, poor foaming of hair mousse, unstable foam, separation of emulsions, microbiological contamination, clumping, and rancidification.** Issues discussed in the previous sections are not discussed here in detail.

Valve Clogging Aerosol hair sprays delivering high hold and/or fast drying time are very popular today. While hair sprays having high holding power can be achieved by increasing the level of polymers, such increase in the polymer level typically results in imparting an undesirable sticky/tacky feel to the hair and possible clogging of the valve in the aerosol can. In the case of valve clogging, the product cannot be dispensed by pushing down the valve. It may occur for various reasons, including the type and amount of polymer used, type and amount of propellant used, compatibility between the formulation and the propellant, the degree to which the polymer is neutralized, spray rate, and appropriate aerosol valve and its actuator.[46] This problem can be overcome with careful control of the rheology and composition of the formulation.

Poor Foaming of Hair Mousses Consumers expect hair mousses to create a dense foam that can stand on the palms and can be easily spread on the hair without collapsing. If the amount and/or types of the applied surfactants, i.e., foam-generating ingredients, are not optimal, the product may have a poor foaming capability. It should be kept in mind that highly purified surfactants are not able to create a viscous surface film required for foam stability. Therefore, selecting a proper impurity can help in obtaining the desired foam. Such impurities can be parts of commercial surfactants or added by manufacturers as foam boosters.[47] In the case of foams, propellants also influence foamability. Aerosol foams are created *in situ* when propellants are expelled out of the can and expand. If the amount, type, or dispersion of propellants is not carefully planned and applied, it may also lead to inappropriate foaming properties. Thickeners can also affect foam generation; therefore, their type, amount, and effect should also be taken into consideration.

Unstable Foam Hair mousse should be stable foams that will hold their internal structure after creation. Unstable foams refer to foams whose structure breaks down faster than usual, and they become flowing "liquid-like" foams after creating them. In addition to influencing foam generation, emulsifiers, propellants, and thickeners may also play a significant role in foam stabilization.

Evaluation of Hair Styling Products

Quality Parameters Generally Tested ● **Parameters commonly tested by cosmetic companies to evaluate the quality of their hair styling products include spray characteristics; aerosol can leakage; actuation force; pressure test for aerosol products; foamability, foam stability, foam viscosity, foam density, and**

foam structure; spreadability, extrudability, texture, and firmness of lotions, creams, and gels; preservative efficacy; viscosity; and pH. Since testing of all these parameters was discussed in the previous chapters, it is not reviewed here in detail.

Performance Parameters of Hair Styling Products **● Commonly tested performance parameters for hair styling products include stiffness of hair fibers, fixative tackiness, dynamic hair spray analysis, curl retention, humidity resistance, and flaking. As for hair waving products, the most common tests include DIT value, the curl length, and the 20% index.** The major characteristics of these tests are reviewed in the following sections.

Stiffness of Hair Fibers Hair fixatives can significantly change the stiffness/flexibility of hair fibers, which can be measured with a tensile meter. Stiffness is usually determined by the bending test (most commonly the three-point bending test). In this test, a hair tress is dipped into a hair spray solution or a diluted gel and is allowed to dry in a flat position. After drying, the bending stiffness is measured by fixing the hair tress horizontally on two arms and moving the central arm of the equipment vertically (applying the bending force to the hair tress) (see Figure 5.9).

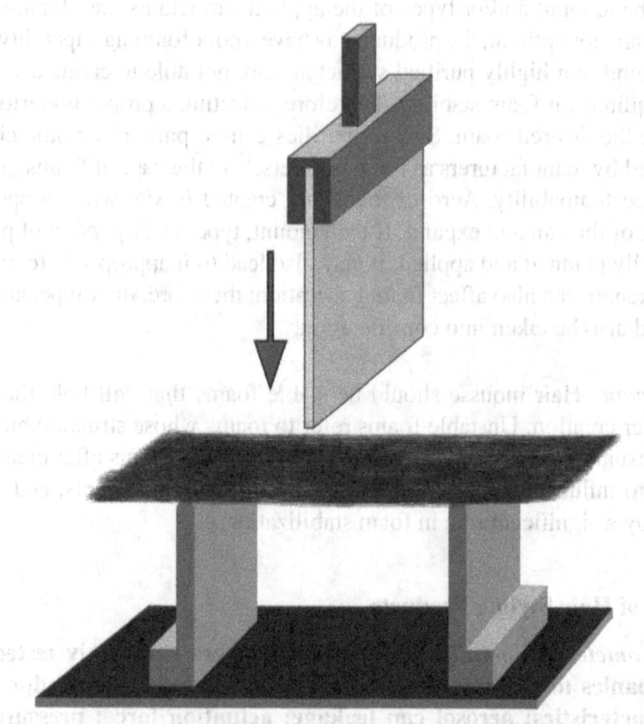

Figure 5.9 Bending test. Adapted from Texture Technologies Corp.

In the test, the force required to bend or break the polymer film is recorded, which refers to the bending stiffness force. Usually, stiffness is proportional to the viscosity of polymer solutions and the total amount of polymer deposited on the hair.[48]

Fixative Tackiness Non-aerosol polymer-based hair styling products, particularly those with a higher polymer concentration, often form actual droplets on the hair fibers, which stay tacky for a longer period of time and need longer drying time. Aerosol dispensers form fine particles, which tend to dry more quickly and produce a less tacky feel after application, making the hair spray easier to use. A slight amount of tackiness is usually expected by the users as a sign of proper working of the products. Too much tackiness will, however, cause the hair to stick to the brush/hands to a point of making it difficult to style. In the tackiness test, hair tresses are treated with a hair spray or other hair fixative and are evaluated by panelists. The time needed to indicate the absence of tack on hair is usually measured, which also refers to the drying time of the polymer film.[49]

Dynamic Hair Spray Analysis A new approach to measure multiple parameters of hair tresses is the dynamic hair spray analysis. This technique generally employs a commercial instrument, the texture analyzer. In the test, the hair tresses are shaped into an omega loop (i.e., formation resembling the Greek omega letter, Ω), while the two sides of the hair tresses are glued onto the surface (see Figure 5.10). A probe

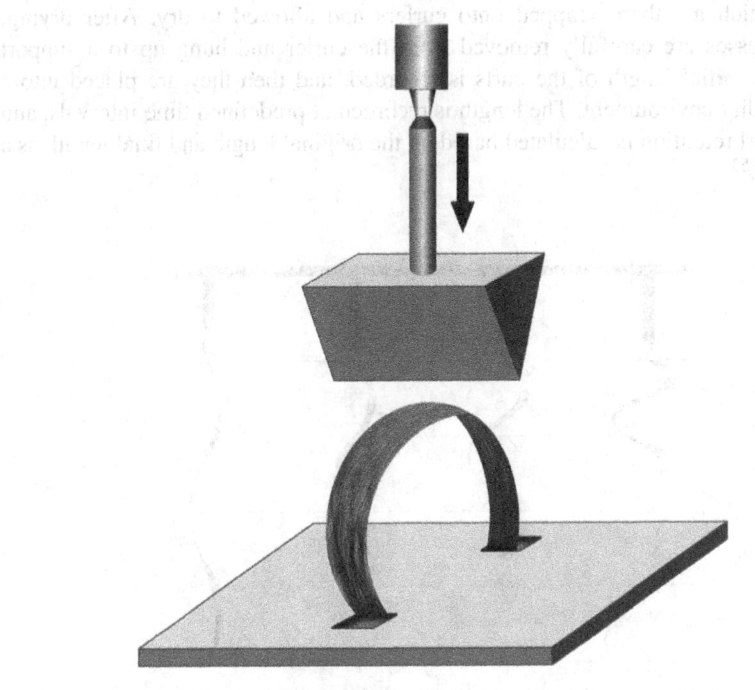

Figure 5.10 Omega loop test.

moves down to touch the hair tress and deforms it to a predefined extent (i.e., 25%) and moves back; this movement is repeated a predetermined number of times. Initially, the untreated hair is evaluated, which is then sprayed with the test product and the hair tress is deformed again. The equipment measures stiffness and increase of stiffness after applying the hair spray, as well as tackiness and drying time. Various temperature and humidity values can also be employed in the measurement to see how these would modify the performance of the formulation.[48,50,51]

Curl Retention for Long-Lasting Styling Hold Polymer-based hair fixatives may claim "long-term curl retention," which has to be substantiated. The test generally used is called the curl snap test. In this test, the hair fixative is applied to the hair tresses, which are then wrapped onto curlers and allowed to dry (see Figure 5.11). After drying, the hair tresses are carefully removed from the curler and hung up to a support board. The initial length of the curls is recorded and then a wide-toothed comb is passed through the tresses. The length (lowermost portion) of the tresses is recorded, which can be used to calculate the curl retention of the hair fixative. Curl retention is calculated as follows[52]:

$$\%\text{Curl retention} = \frac{\text{length of uncurled hair tress} - \text{length of the hair at the time of reading}}{\text{length of uncurled hair tress} - \text{length of the curl at the start of the experiment}} \times 100$$

Humidity Resistance of Treated Hair Tresses A number of hair fixative formulations claim high holding power even in humid environment. The test usually used is similar to the curl retention test. First, a hair fixative is applied to the hair tresses, which are then wrapped onto curlers and allowed to dry. After drying, the hair tresses are carefully removed from the curler and hung up to a support board. The initial length of the curls is recorded, and then they are placed into a high-humidity environment. The length is recorded at predefined time intervals, and then the curl retention is calculated based on the original length and final length as a percentage.[53]

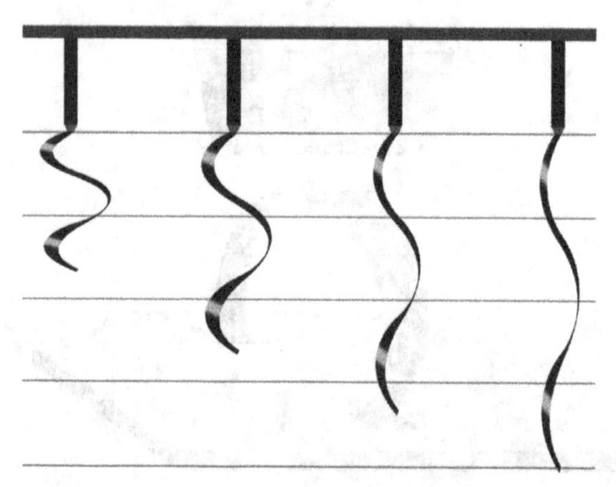

Figure 5.11 Curl retention test. Adapted from Texture Technologies Corp.

Flaking As discussed previously, overuse of polymer-based hair fixatives can lead to flaking. To test flaking, usually panelists are involved. The fixative is applied to the hair tresses and allowed to dry. The hair tresses are hung up onto a supporting board and a small-toothed comb is then passed through the hair tresses. The hair, comb, and area under the support stand are inspected and graded by panelists to evaluate flaking.[54]

Performance Parameters of Permanent Waving Products In order to make sure that permanent waving products work properly without causing significant damage, their performance must be tested. The most common industry evaluation techniques are the DIT value, the curl length, and the 20% index.[55]

The DIT value is a measure of the tightness of the curl. Generally, the higher the DIT value, the better the performance of the perming solution and the tighter the curl. It is usually calculated using the following formula: [56]

$$\text{DIT} = \frac{\text{diameter of the curl (mm)} - \text{diameter of the rod (mm)}}{\text{diameter of the rod (mm)}} \times 100$$

Curl length is a measurement of the spring of the curl. Optimally, curls should be springy after being removed from the curling rod. The curl length is usually evaluated by fixing a fresh curl and observing its spring. The degree of laxity is proportional to the degree of hair shaft damage. Thus, longer curls are indicative of greater hair shaft damage.[56]

The 20% index is used to evaluate the strength of the permanently waved hair. This parameter is determined by stretching the freshly permed hair with a uniformly increasing load. The 20% index refers to the ratio of the load required to stretch a standardized wet hair tress to 20% of its original length after and before perming. The higher the index, the lower the reduction in hair strength after perming. Hair with a low 20% index will tend to break easily upon applying minimal trauma, such as combing.[56]

Ingredients Causing Safety Concerns ⓾ **Ingredients of hair styling, hair straightening, and hair waving products that caused major safety issues in the past include propellants, thioglycolates, hydroxides, and formaldehyde.** These safety issues are reviewed in this section.

Propellants As discussed earlier, aerosols have been the target of regulatory restrictions for a long time. Aerosol hair fixatives can contain volatile organic compounds (VOCs); however, their amount is limited by the US Environmental Protection Agency (EPA) on a federal level as well as by the states. Currently, under the federal rule, hair sprays may contain no more than 80% VOCs on a weight-percent basis. The corresponding California limit is 55%. This limit is usually referred to as 80/55. Currently, the rules state that hair mousse may contain no more than 16% VOCs (6% in California), hair shine (55% in California), and hair styling gels (6% in California) (since no federal limit is set up for these two styling products).[57] It should be mentioned that VOC limits are mass-based, which means all ingredients considered VOCs are treated equally, regardless of their reactivity in the atmosphere. In a 55% VOC hair spray, 55% w/w of the formula can be made of VOCs. The individual ingredients are irrelevant as long as the mass requirement is not more than 55%.[58] There are a small number of ingredients that do not contribute to ozone formation and are, therefore, classified as "VOC exempt." An example for such propellants used in hair sprays is hydrofluorocarbon (HFC) 152A.[59]

Inhalation toxicity is an important consideration for aerosol hair sprays. Aerosol hair spray creates a fine mist of droplets that may be inhaled, especially over long-term use. The safety of an inhaled aerosol (liquid droplets in the case of hair sprays) depends on several key factors, including the chemical and physical properties of the aerosol, its concentration in inhaled air, the duration of inhalation exposure, size of the inhaled droplets, and the deposition and clearance of the aerosol in the respiratory tract.[60] There is a broad scientific consensus that the probability of penetration of droplets with a diameter of >10 μm into the lower regions of the respiratory tract is essentially zero; only the smaller droplets are considered to be the so-called respirable.[61,62] Hair sprays (both aerosol and pump sprays) have a droplet size larger than 10 μm. Therefore, even if inhaled, they will not get into the lungs; they may only be deposited in the upper respiratory tract.[63,64] Hair spray manufacturers carefully test the droplet size and size distribution of their products in order to ensure that the products are safe for at-home and in-salon use.

Thioglycolates Thioglycolates are capable of breaking the disulfide bonds in hair fibers, and therefore, these are the main ingredients in thioglycolate-based permanent hair waving and straightening products. These products are considered safe; however, various adverse reactions, such as irritant contact dermatitis, have been reported.[65,66] The CIR Expert Panel evaluated the effects of thioglycolates on the hair and skin. They determined that these chemicals are safe for use in straighteners, permanent waves, tonics, dressings and other hair grooming aids, wave sets, other non-coloring hair products, and hair dyes and colors, at concentrations up to 15.4% (as thioglycolic acid).[67] In order to reduce irritation, a layer of petrolatum can be applied to the margins of the scalp, which provides protection to the skin areas not covered by hair.[68] For sensitive patients, petrolatum can also be applied to the scalp as well, which, however, may affect the effectiveness of the perming treatment in hair parts close to the scalp.

Although glyceryl monothioglycolate works at a neutral pH as discussed earlier, which is thought to be less irritating, it is an ingredient commonly involved in the development of allergic contact dermatitis.[69,70] It is an inflammatory skin reaction occurring after direct contact with this ingredient. The reaction can occur immediately after permanent waving or later due to an allergen in the hair of those who had their hair permed using this ingredient. The hair may continue to be allergenic even after multiple times of shampooing the hair.[71]

Hydroxides Hydroxide-based hair relaxers are commonly used in salons and at home as well. These products have a highly alkaline pH, which causes swelling of the hair fibers, and they can lead to severe hair damage, such as cuticle abrasion, decreased hydrophobicity, increased porosity, loss in tensile strength, as well as breakage of hair shafts.[71] Additional chemical processing of the hair, such as coloring or bleaching, should be avoided for a while to prevent further damage of the hair shafts. In addition, hydroxide-based hair relaxers are well-known skin irritants, which can cause redness, itching, and chemical burning as well. In order to decrease skin irritation, petrolatum can be applied to the base of the scalp and hairline.

Formaldehyde As discussed previously, Brazilian hair straightening products use formaldehyde as the main straightening ingredient. As discussed under nail care products (see Section 4 of Chapter 4), formaldehyde is a gas at standard temperature and pressure. Therefore, it is normally supplied as an aqueous solution, known as formalin. In solution, there is much more methylene glycol present as compared to formaldehyde and water. In recent years, there were a number of warnings and bans introduced in the US (and other parts of the world), regarding the safety of products containing high levels of formaldehyde. Formaldehyde is considered a probable human carcinogen by the EPA.[72] Formaldehyde can lead to cause a number of health concerns via inhalation, including nose and eye irritation, allergic reactions, and neurologic effects.[73] The first safety issues surfaced in 2010 when workers at several beauty salons in Portland, Oregon, complained of illness, including headaches, nosebleeds, and breathing problems, after using the product Brazilian Blowout on clients.[74] Oregon's OSHA tested samples of Brazilian Blowout and found levels of formaldehyde ranging from 6.3 to 11.8%, including those labeled "formaldehyde free."[75] The Oregon OSHA immediately released a "Hazard Alert," warning salon owners, workers, and clients about the laboratory findings and advising them to either discontinue the use of the products or install air monitoring devices to protect their workers.[76] Similarly, the federal OSHA and agencies in other states have released hazard warnings, alerting hair salons and clients concerning hair smoothing products that could release formaldehyde into the air.[77] Additionally, FDA issued a warning letter in 2011 to the importer and distributor of Brazilian Blowout product, identifying the product as adulterated and misbranded because it contains methylene glycol, which can release formaldehyde during the normal conditions of use and because the label makes misleading statements ("Formaldehyde Free" or "No Formaldehyde").[78] In 2012, the CIR Panel reassessed formaldehyde/methylene glycol and concluded that formaldehyde and methylene glycol are unsafe in the current practices of use and concentration in hair straightening products.[79]

DID YOU KNOW?

In the US, formaldehyde/methylene glycol is recommended to be used in cosmetics only with the function of preserving with a maximum limit of 0.2% and as a nail hardener at a maximum limit of 5%.[80]

FYI

The FDA has a detailed discussion on the safety of Brazilian hair straighteners. For details, visit the FDA web site (Home, Cosmetics, Product and Ingredient Safety, Product Information and "*FDA, OSHA Act on Brazilian Blowout*").

DID YOU KNOW?

Workplace safety, in general, is regulated by OSHA. Salons, such as hair and nail salons, are also generally subject to state and local authorities, which may specify safety practices. FDA does not have authority over the operation of salons or the practice of cosmetology.[81]

Packaging of Hair Styling Products The most commonly used packaging materials for hair styling products include the following:

- **Plastic Bottles**: Many hair fixatives are supplied in plastic bottles, usually with a flip-top cap or a pump head.
- **Plastic and Metal Jars**: Hair styling formulations, such as gels, creams, and waxes, may also be supplied in plastic or metal jars.
- **Soft Tubes**: Gels and creams can also be filled into soft-sided tubes with a screw cap or a flip-top cap.
- **Aerosol Cans**: Aerosol hair sprays and hair mousse are supplied in one-compartment cans, where the formulation and the propellant are mixed together. Aerosol containers can be made of tin-coated steel, tin-free steel, or aluminum. The material of the can is an important factor to be taken into consideration as the product with a certain pH, or certain solvents, may have an interaction with the can. Valve type is also an important factor to be considered; valve selection should address sprayability, clogging, and evacuation.

GLOSSARY OF TERMS FOR SECTION 3

Brazilian Blowout: A personal care product used to permanently straighten the hair. It contains formaldehyde as the main straightening ingredient, which caused severe safety concerns.

Curl retention: The ability of polymer-based hair fixatives to hold curls in hair.

Emollient-based hair fixative: A type of hair fixative designed to smooth out frizz and increase hair shine by aligning the fibers and reducing friction and have a conditioning effect on the hair fibers. Products in this category include hair pomades, hair brilliantines, sprayable oils and silicones, and hair creams.

Fixing lotion: A product containing a neutralizing ingredient, most commonly hydrogen peroxide, used in step 2 of permanent hair curling to stop cleavage of disulfide bonds and neutralize the alkaline pH.

Flaking: Tiny film particles falling off the hair after using polymer-based hair fixatives.

Hair fixative: A personal care product designed to reshape the hair to a desired arrangement, improve its volume, increase its shine, soften it, or stiffen it – depending on the products used – and maintain the style for a shorter or longer period of time.

Hair lanthionization: Permanent hair straightening.

Hair straightening: The process of temporarily or permanently straightening the hair.

Hair waving: The process of temporarily or permanently curling the hair.

Heat protection spray: A personal care product designed to protect the hair from heat damage, provide protection against friction, as well as enhance the hair's flexibility to reform and hold shape.

Lye: Sodium hydroxide.

Perming lotion: A product containing a reducing agent, such as ammonium thioglycolate, in an alkaline environment used in step 1 of permanent hair curling to break the disulfide bonds.

Perming: Permanent hair curling.

Polymer-based hair fixative: A type of hair fixative designed to improve hair volume and height by increasing hair strand stiffness and hair fiber interactions. Products in this category include hair sprays, setting lotions, mousses, hair styling fiber gum, and hair gels.

Pressure filling: A technique used to fill a product concentrate and propellant(s) into an aerosol can.

Protein denaturation: A process in which the proteins are irreversibly altered by an external stimulus and lose their quaternary, tertiary, and secondary structures.

Stiffness: Rigidity of hair fibers after applying polymer-based hair fixatives.

Tackiness: Stickiness.

Valve clogging: A quality problem of hair sprays most commonly occurring due to the film-formers used.

VOCs: Volatile organic compounds are organic chemical compounds whose composition makes it possible for them to evaporate under normal indoor atmospheric conditions of temperature and pressure. They contribute to the formation of urban smog, and therefore, their use is limited in aerosols. They are regulated by the EPA.

REVIEW QUESTIONS FOR SECTION 3

Multiple Choice Questions

1. Which of the following changes the hair's structure permanently?
   a) Curling rollers
   b) Flat iron
   c) Perming lotion
   d) None of the above
2. In step 1 of the hair perming process ___ is used to ___.
   a) Thioglycolate lotion/break the S–S bonds
   b) Thioglycolate lotion/reform broken S–S bonds
   c) Hydrogen peroxide/break the S–S bonds
   d) Hydrogen peroxide/reform broken S–S bonds
3. In step 2 of the hair perming process ___ is used to ___.
   a) Thioglycolate lotion/break the S–S bonds
   b) Thioglycolate lotion/reform broken S–S bonds
   c) Hydrogen peroxide/break the S–S bonds
   d) Hydrogen peroxide/reform broken S–S bonds
4. Which of the following is NOT an example for a polymer-based hair fixative?
   a) Mousse
   b) Hair fiber gum
   c) Hair spray
   d) Hair brilliantine
5. Which of the following regulates the amount of VOCs that can be used in aerosol products?
   a) FDA
   b) CIR
   c) EPA
   d) VOC

6. Which of the following is the main ingredient in the product called "Brazilian Blowout"?
   a) Formaldehyde
   b) Sodium hydroxide
   c) Thioglycolic acid
   d) Petrolatum

7. Which of the following parameters is typically evaluated as a performance parameter for polymer-based hair fixatives?
   a) Curl retention
   b) Flaking
   c) Tackiness
   d) All of the above

8. What is the pH of the most effective hair perming products?
   a) Alkaline (approximately 9)
   b) Acidic (approximately 3)
   c) Neutral

9. Which of the following is NOT true for emollient-based hair styling products?
   a) They are typically used to make the hair more manageable
   b) They are typically used to hold the hair in a desired shape
   c) Hair pomade is an example for such products
   d) They provide a temporary effect

10. UV filters in hair styling products provide protection against ___.
   a) Skin cancer
   b) Hair cancer
   c) UV damage and weathering
   d) Tanning

Fact or Fiction?

______ a) Wet hair can stand a much higher temperature than dry hair.
______ b) Skin irritation caused by hydroxide-based relaxers can be prevented/decreased by using petrolatum prior to the relaxing treatment.
______ c) Perming products affect only the hydrogen bonds in hair fibers.
______ d) Emollient-based hair styling aids can lead to flaking.

Matching

Match the ingredients in column A with their appropriate ingredient category in column B.

	Column A	Column B
_____	A. Alcohol	1. Film-forming polymer
_____	B. Ammonium carbonate	2. Heat-protecting cationic surfactant
_____	C. Carbomer	3. Lye relaxer
_____	D. Castor oil	4. Penetration enhancer
_____	E. Glycerol monothioglycolate	5. pH buffer
_____	F. Hydrofluorocarbon 152A	6. Plasticizer
_____	G. Polyquaternium-11	7. Propellant
_____	H. Sodium hydroxide	8. Reducing agent
_____	I. Urea	9. Solvent
_____	J. VP/DMAEMA	10. Thickener

REFERENCES

1. McCreesh, N. C., Gize, A. P., David, A. R.: Ancient Egyptian hair gel: new insight into ancient Egyptian mummification procedures through chemical analysis. *J Archaeol Sci.* 2011;38(12):3432–3434.
2. Lucas, A., Harris, J. R.: Ancient Egyptian Materials and Industries, Courier Dover Publications, 1999: 332.
3. Marsh, M.: Compacts and Cosmetics: Beauty from Victorian Times to the Present Day, Havertown: Casemate Publishers, 2009.
4. Nessler, C.: Br. Pat. 2931; 1910.
5. Wall, F. E.: The Principles and Practice of Beauty Culture, 2nd Edition, New York: Keystone Publications, 1946: 26.
6. Lee, A. E., Bozza, J. B., Huff, S., et al.: Permanent waves: an overview. *Cosmet Toiletries.* 1988;103:37–56.
7. Frangie, C. M., Botero, A. R., Hennessey, C., Lees, M., Sanford, B., Shipman, F., Wurdinger, V., eds: *Milady's Standard Cosmetology*, New York: Cengage Learning, 2008.
8. House-Soremekun, B.: Confronting the Odds: African American Entrepreneurship in Cleveland, Ohio, Kent: Kent State University Press, 2002.
9. Williamson, D., Gonzalez, M., Finlay, A. Y.: The effect of hair loss on quality of life. *J Eur Acad Dermatol Venereol.* 2001;15(2):137–139.
10. Rushton, D. H., Kingsley, P., Berry, N. L., et al.: Treating reduced hair volume in women. *Cosmet Toiletries.* 1993;108:59–62.
11. Gray J.: Human Hair Diversity, Oxford: Blackwell Science, 2000.
12. McMullen, R., Jachowicz, J.: Thermal degradation of hair. I. Effect of curling irons. *J Cosmet Sci.* 1998;49:223–44.
13. Feughelman, M.: A note on the permanent setting of human hair. *J Soc Cosmet Chem.* 1990;41:209–212.

14. Wilkinson, J. D., Shaw, S.: Adverse Reactions to Hair Products, In: Boullion, C., Wilkinson, J., eds: The Science of Hair Care, Boca Raton: CRC Press 2005:521.
15. Bergfeld, W. F.: Side Effects of Hair Products on the Scalp and Hair, In: Orfanos, C. E., Montagna, W., Stüttgen, G., eds: Hair Research, Berlin: Springer Verlag, 1981:507–512.
16. Ishihara, M.: Some Skin Problems Due to Hair Preparations, In: Orfanos, C. E., Montagna, W., Stüttgen, G., eds: Hair Research, Berlin: Springer Verlag, 1981:536–542.
17. Lochhead, R.: The history of polymers in hair care (1940–present), *Cosmet Toiletries.* 1988;103.
18. Zviak, C.: The Science of Hair Care, New York: Marcel Dekker, 1986, 153–65.
19. Stutsman, M. J.: Analysis of Hair Fixatives, In: *Newburger's Manual of Cosmetic Analysis*, Newburger, S. H.; Senzel, A. J. eds., 2nd Edition, Washington, DC: Association of Official Analytical Chemists, 1977:72.
20. Johnson, S. C.: Acetylene-derived polymers, *Cosmet Toiletries.* 1984;99.
21. Martiny, S.: Acetylenic polymers for hair styling products, *Intl J Cos Sci.* 2002;24:125–134.
22. Laba, D.: Rheological Properties of Cosmetics and Toiletries, Cosmetic Science and Technology Series, Volume 13, New York: Marcel Dekker, 1993.
23. Pfaffernoschke, M.: Formulating low VOC aerosol hairsprays. *Int J Aerosol Spray Pack Technol.* 2005.
24. Goode, S. T.: Hair Pomades, *Cosmet Toiletries.* 1979;94:71–74.
25. Wells, F. Wells, F., Lubowe, II.: Hair grooming aids: Part II, *Cutis.* 1978;22:270–301.
26. Robbins, C. R.: Chemical and Physical Behaviour of Human Hair, 4th Edition, Berlin: Springer Verlag, 2002.
27. Jeschke, M. G., Kamolz, L. P., Shahrokhi, S.: Burn Care and Treatment, New York: Springer, 2013.
28. Robbins, C. R.: Load elongation of single hair fiber coils. *J Soc Cosmet Chem.* 1983;34:227–239.
29. Dussaud A., Fieschi-Corso, L.: Influence of functionalized silicones on hair fiber-fiber interactions and on the relationship with the macroscopic behavior of hair assembly, *J Cosmet Sci.* 2009;60:261-271.
30. McMullen, R., Jachowicz, J.: Thermal degradation of hair. II. Effect of selected polymers and surfactants, *J Cosmet Sci.* 1998;49:245–256.
31. Christian, P., Winsey, N., Whatmough, M., Cornwell, P. A.: The effects of water on heat-styling damage. *J Cosmet Sci.* 2011;62:15–27.
32. Heilingotter, R.: Permanent Waving of Hair, In: de Navarre, M. G., ed.: The Chemistry and Manufacture of Cosmetics, Wheaton: Allured Publishing, 1988, 1167–1227.
33. Wickett, R. R.: Permanent waving and straightening of hair. *Cutis.* 1987;39:496–497.
34. Inoue, T., Ito, M., Kizawa, K.: Labile proteins accumulated in damaged hair upon permanent waving and bleaching treatments. *J Cosmet Sci.* 2002;53:337–344.
35. Wickett, R., Savaides, A.: Permanent Waving of Hair, In: Knowlton, J., Pearce, S., ed.: Handbook of Cosmetic Science and Technology, 1st Edition, Oxford: Elsevier Science, 1993, 511–534.
36. Robbins, C. R., Fernee, K. M.: Some observations on the swelling of human epidermal membrane. *J Soc Cosmet Chem.* 1983;34:21–34.
37. Bernard, B.: Hair shape of curly hair. *J Am Acad Dermatol.* 2003;48:S120–S126.

38. Cannell, D. W.: Permanent waving and hair straightening. *Clin Dermatol.* 1988;6:71–82.
39. Obukowho, P., Birman, M.: Hair curl relaxers: a discussion of their function, chemistry, and manufacture. *Cosmet Toiletries.* 1995;110:65–69.
40. Khalil, E. N.: Cosmetic and hair treatments for the black consumer. *Cosmet Toiletries.* 1986;101:51-58.
41. Syed, A. N., Ayoub, H., Kuhajda, A.: Recent advances in treating excessively curly hair. *Cosmet Toiletries.* 1998;113:47–56.
42. Bernard, B. A., Franbourg, A., Francois, A. M., et al.: Ceramide binding to African-American hair fibre correlates with resistance to hair breaking. *Int J Cosmet Sci.* 2002;24:1–12.
43. Wong, M., Wis-Surel, G., Epps, J.: Mechanism of hair straightening. *J Soc Cosmet Chem.* 1994;45:347–352.
44. Westman, M.: Demi-perm hair dyes – important and misunderstood, *Cosmetiscope.* 2008;14(5):1–7.
45. FDA: Hair Smoothing Products That Could Release Formaldehyde, Accessed 2/10/2014 at http://www.fda.gov/Cosmetics/ProductandIngredientSafety/ProductInformation/ucm228898.htm
46. Benson, A. B., Hourihan, J. C., Tripathi, U.: Aerosol hair spray composition, US5094838, 1992.
47. Prud'homme, R. K.: Foams: Theory, Measurements, Applications, Boca Raton: CRC Press, 1995: 386.
48. Jachowicz, J., Yao, K.: Dynamic hairspray analysis I. Instrumentation and preliminary results. *J Soc Cosmetic Chem.* 1996;47:73–84.
49. Lubrizol: Tack Test Method, Last update: 5/26/2002, Accessed 1/29/2014 at http://www.lubrizol.com/Personal-Care/Documents/Test-Procedures/Fixate%E2%84%A2-Polymers-Test-Procedures/TP-004_Tack.pdf
50. Jachowicz, J., Yao, K.: Dynamic hairspray analysis. II. Effect of polymer, hair type, and solvent composition. *J Cosmet Sci.* 2001;52:281–295.
51. Jachowicz, J.: Dynamic hairspray analysis. III. Theoretical considerations. *J Cosmet Sci.* 2002;53:249–261.
52. Lubrizol: Curl Snap Test, Last update: 5/26/2002, Accessed 1/29/2014 at http://www.lubrizol.com/Personal-Care/Documents/Test-Procedures/Fixate%E2%84%A2-Polymers-Test-Procedures/TP-003_Curl_Snap.pdf
53. Lubrizol: Humidity Resistance of Treated Hair Tresses, Last update: 5/26/2002, Accessed 1/29/2014 at http://www.lubrizol.com/Personal-Care/Documents/Test-Procedures/Fixate%E2%84%A2-Polymers-Test-Procedures/TP-005_Humidity_Resistance.pdf
54. Lubrizol: Flaking Test Method, Last update: 5/26/2002, Accessed 1/29/2014 at http://www.lubrizol.com/Personal-Care/Documents/Test-Procedures/Fixate%E2%84%A2-Polymers-Test-Procedures/TP-001_Flaking.pdf
55. Heilingotter, R.: Permanent Waving of Hair, In: de Navarre, M. G., ed.: The Chemistry and Manufacture of Cosmetics. Wheaton: Allured Publishing, 1988: 1167–227.
56. Draelos, Z.: Hair Care, New York: Taylor and Francis, 2005: 142.
57. IISA: Summary of State and Federal VOC Limitations for Institutional and Consumer Products, 2013, Accessed 2/13/2014 at http://www.issa.com//data/File/regulatory/VOC%20Limits%20Summary%2010-25-13.pdf

58. Rigoletto, R., Mahadeshwar, A., Foltis, L., et al.: Advances in Hair Styling. *Cosmet Toiletries.* 2012;127(5):372–382.
59. EPA: Summary of Substitute Aerosol Propellants Listed in SNAP Notice 25, Accessed 2/13/2014 at http://www.epa.gov/ozone/snap/aerosol/Notice25SubstituteAerosols.pdf
60. Jensen, P. A., O'Brien, D.: Industrial Hygiene, In: Willeke, K., Baron, P. A., eds: Aerosol Measurement: Principals, Techniques, and Applications, Hoboken: John Wiley & Sons, 1993: 537–559.
61. World Health Organization (WHO). Hazard Prevention and Control in the Work Environment: Airborne Dust. Geneva, Switzerland, 1999. Report No. WHO/SDE/OEH/99.14. pp. 1-246.
62. Oberdorster, E. G., Oberdorster, J.: Nanotoxicology: an emerging discipline evolving from studies of ultrafine particles. *Environ Health Perspect.* 2005;113(7):823–839.
63. Bower, D. Unpublished Information on Hair Spray Particle Sizes Provided at the September 9, 1999 CIR Expert Panel Meeting.
64. Johnson, M. A.: The influence of particle size. *Spray Technol Market.* 2004;24–27.
65. Ishihara, M.: Some Skin Problems Due to Hair Preparations, In: Orfanos, C. E., Montagna, W., Stüttgen, G., eds: Hair Research, Berlin: Springer Verlag, 1981: 536–542.
66. Orfanos, C. E., Sterry, W., Leventer, T.: Hair and Hair Cosmetic Treatments, In: Orfanos, C. E., ed.: Hair and Hair Diseases, Stuttgart: Fischer Verlag, 1979: 853–885.
67. Cosmetics and Toiletries: CIR Assesses Aminomethyl Propanol and More, Last update: 6/27/2007, Accessed 2/14/2014 at http://www.cosmeticsandtoiletries.com/regulatory/region/northamerica/8202152.html
68. Lee, A. E., Bozza, J. B., Huff, S., et al.: Permanent waves: an overview. *Cosmet Toiletries.* 1988;103:37–56.
69. Morrison, L. H., Storrs, F. J.: Persistence of an allergen in hair after glyceryl monothioglycolate containing permanent wave solutions. *J Am Acad Dermatol.* 1988;19:52–59.
70. Storrs, F. J.: Permanent wave contact dermatitis: contact allergy to glyceryl monothioglycolate. *J Am Acad Dermatol.* 1984;11:74–85.
71. Shansky, A.: The osmotic behavior of hair during the permanent waving process as explained by swelling measurements. *J Soc Cosmet Chem.* 1963;14: 427.
72. EPA: Formaldehyde, Accessed 2/14/2014 at http://www.epa.gov/ttnatw01/hlthef/formalde .html
73. ATSDR. ToxFAQs™ for Formaldehyde, CAS# 50-00-0, September 2008. Atlanta, GA: Agency for Toxic Substances & Disease Registry, U.S. Centers for Disease Control and Prevention, Last update: 3/3/2011, Accessed 2/14/2014 at http://tinyurl.com/4yembxu
74. Muldoon, K.: Hair Product Tests Find High Formaldehyde, *The Oregonian*, 2010.
75. Oregon OSHA and CROET, Oregon Health & Sciences University, Accessed 2/8/2014 at http://www.ohsu.edu/xd/research/centers-institutes/croet/emerging-issues-and-alerts.cfm
76. See Oregon OSHA, Department of Consumer and Business Services: Hazard Alert, Hair Smoothing Products and Formaldehyde, September 2010, Accessed 2/15/2014 at http://www.orosha.org/pdf/hazards/2993-26.pdf
77. U.S. Occupational Health and Safety Administration: Hazard Alert: Hair Smoothing Products that Could Release Formaldehyde, Accessed 2/10/2014 at https://www.osha.gov/SLTC/formaldehyde/hazard_alert.html
78. FDA: Brazilian Blowout 8/22/11, Last update: 9/6/2011, Accessed 2/12/2014 at http://www.fda.gov/ICECI/EnforcementActions/WarningLetters/2011/ucm270809.htm

79. CIR: Formaldehyde and Methylene Glycol. CIR Expert Panel Meeting Minutes, 2012, Accessed 9/25/2013 at CIR website: http://www.cir-safety.org/sites/default/files/formy_build.pdf
80. FDA: Guide to Inspections of Cosmetic Products Manufacturers, Last update: 3/13/09, Accessed 3/11/2014 at http://www.fda.gov/ICECI/Inspections/InspectionGuides/ucm074952.htm
81. FDA: OSHA Act on Brazilian Blowout, Last update: 03/08/2014, accessed 4/11/2014 at http://www.fda.gov/Cosmetics/ProductsIngredients/Products/ucm228898.htm

SECTION 4: HAIR COLORING PRODUCTS

LEARNING OBJECTIVES

Upon completion of this section, the reader will be able to

1. define the following terms:

Booster	Coal-tar hair dye	Coupler	Demi-permanent hair dye
Developer	Dye intermediate	Hair bleach	Henna
Non-oxidative dye	Oxidative dye	Permanent hair dye	Progressive hair dye
Semi-permanent hair dye	Skin patch testing	Temporary hair dye	Touch-up

2. list several reasons for coloring the hair;
3. differentiate between hair dyes and hair bleaches;
4. differentiate between temporary and permanent hair dyes;
5. explain how hair coloring products can positively contribute to people's lives;
6. list some of the negative effects hair coloring and bleaching products can have on the scalp and hair;
7. list various required cosmetic qualities and characteristics that an ideal hair coloring product should possess;
8. list various required technical qualities and characteristics that an ideal hair coloring product should possess;
9. briefly discuss whether hair coloring products are considered cosmetics or drugs in the US;
10. explain what types of color additives can be used in hair coloring products in the US;

11. briefly discuss when and why warning statements are required to appear on the label of hair coloring products in the US;
12. differentiate between oxidizing and non-oxidizing hair coloring products;
13. explain and/or illustrate how temporary and semi-permanent hair dyes color the hair;
14. explain how progressive hair dyes color the hair;
15. explain and/or illustrate how oxidizing products color the hair;
16. differentiate between demi-permanent and permanent hair dyes;
17. list the main ingredient types found in permanent hair dyes;
18. explain why regular touch-up is necessary for permanent hair dyes;
19. explain how hair bleaches remove hair color;
20. list some of the typical quality issues that may occur during the formulation and/or use of hair coloring products, and explain why they may occur;
21. list the typical quality parameters that are regularly tested for hair coloring products, and briefly describe their method of evaluation;
22. name the performance parameters generally tested for hair coloring products, and describe the method of evaluation;
23. briefly discuss the main safety concerns that arose in the past with regard to the use of oxidative hair dyes;
24. list the typical containers available for hair coloring products.

KEY CONCEPTS

1. Hair colorants are widely used by both genders to alter their natural hair color by removing some of the existing color and/or adding a new color.
2. In the US, both approved and not approved color additives can be used in hair coloring products.
3. Hair coloring products can be categorized based on the presence or absence of the chemical reaction (known as oxidation) involved in hair coloring process. Non-oxidative products include temporary dyes and semi-permanent dyes, while demi-permanent dyes, permanent dyes, and hair bleaches fall into the category of oxidative products. An additional product type is known as progressive hair dyes, which are permanent but not oxidative.
4. Hair bleaching is a chemical process that involves the removal of the natural hair pigment or artificial color from the hair.
5. The typical quality-related issues of hair coloring products include valve clogging, poor foaming of hair mousse, unstable foam, separation of emulsions, microbiological contamination, clumping, and rancidification.
6. Parameters commonly tested to evaluate the quality of hair coloring products include spray characteristics; aerosol can leakage; pressure test; spreadability,

extrudability, texture, and firmness of lotions, creams, and gels; actuation force; color; preservative efficacy; viscosity; and pH.

7. The most frequently tested parameters related to the performance of hair coloring products are shine and color intensity.
8. Oxidative hair dyes have repeatedly come to the attention of dermatologists due to their allergic potential and potential carcinogenic characteristics.

Introduction

Colored hair has become a common statement of individuality, youth, and fashion. Today, according to some estimates, more than 60% of women in the US color their hair, as do a growing number of men.[1] Additionally, an estimated 50% of American women above the age of 25 color their hair; this market is expected to continue to grow.[2] ❶ **Hair colorants are widely used by both genders to alter their natural hair color by removing some of the existing color and/or adding a new color.** While most consumers have their individual reasons, the main motivator for coloring the hair is to cover up gray hair. The hair color can be changed temporarily (for 1 time up to 24 times of shampooing) or permanently.[3] The hair coloring procedure can be done at home with kits sold commercially or in salons by professionals.

This section provides an overview of the various techniques and products available to alter the natural hair color. It reviews the main product types, their ingredients and formulation technique, as well as their required qualities and consumer needs. The safe use of hair coloring products has always been a hot topic; therefore, the main issues and the scientific evidence behind them are also discussed here.

Types and Definition of Hair Coloring Products

Today's hair color products can remove (lift) natural hair color, add (deposit) a new color to the natural color, or accomplish both processes at the same time. The size of the coloring molecule, swelling of the hair at the time of application, and alkalinity of the dye product determine whether the dye penetrates the cortex and colors the hair for a long period of time or precipitates on the cuticle and provides a temporary effect.[4] Hair coloring products can be classified in several ways. Based on the action of hair coloring products, we can distinguish between hair dyes and hair bleaches.

- **Hair dyes** add color to the hair, which can be lighter or darker depending on the type of hair coloring product used.
- **Hair bleaches** only lighten the hair without adding a new color to it.

  Based on the permanency of the new color, the following hair coloring product types are available on the market today: temporary, semi-permanent, demi-permanent, and permanent dyes, as well as hair bleaches.

- **Temporary** hair dyes adhere to the outside of hair fibers by weak chemical bonds and, therefore, are washed out by the first shampooing.
- **Semi-permanent** dyes adhere to the outside of hair fibers and partially penetrate the cuticle layers, making the hair dye longer lasting.
- **Demi-permanent** hair dyes penetrate the cuticle and cortex. They contain an oxidizing agent, have a more alkaline pH and are, therefore, significantly longer lasting than the previous types. However, they are still not completely permanent and cannot lighten the original hair color.
- **Permanent** hair dyes also penetrate both the cuticle and cortex. They contain a higher amount of oxidizing agents and have a highly alkaline pH (compared to demi-permanent dyes), which make them permanent. They can be used to change the original hair color into two shades: either lightening it or darkening it. Due to their composition, they are the most damaging chemicals among the hair coloring products. Permanent hair coloring can also be achieved without oxidizing agents. These products employ metal ions and are called progressive hair dyes.
- **Hair bleaches** remove the hair color through a chemical reaction. They can be used alone or in combination with permanent hair colors.

All hair coloring products are considered cosmetics in the US, as they are used to enhance the appearance of the hair, which is a cosmetic benefit.

DID YOU KNOW?

The reason for going gray is not the production of a new gray melanin pigment. However, its underlying mechanism is still not totally understood. Genetic factors seem to play an important role in graying. It is also believed that the death of some melanocytes within a hair follicle triggers a chain reaction, resulting in the death of the rest of the melanocytes in the same follicle in a relatively short period of time.[5,6]

History of Using Hair Coloring Products

Fashion in hair color and preferences for certain hues have changed throughout history. The use of various substances to alter the hair color dates back thousands of years. Although there is a wide variety of natural hair color shades, and tones, humans have attempted to enhance or change their natural hair color even in the ancient ages. The earliest dyes were made of plant products, including fruits, flowers, and vegetables, such as chamomile, indigo, logwood, henna, and walnut hull extract.[7]

People in ancient Mesopotamia and Persia dyed their hair, which was usually long. Ancient Egyptians favored henna, which gave reddish hues to the hair, or wore wigs to alter their hair color.[7] Later on, in ancient Greece, women dyed their hair using a liquid made from potash water and yellow flowers or dusted it with yellow pollen, flour, and even gold dust to achieve blond shades.[8] The early Roman preferred dark hair. They usually started to color their hair as they went gray. Other early societies favored unusual colors, such as red, green, orange, or blue, which was common among the Saxons. From the 8th to the 13th century, blond hair was quite popular among women. Some tried to lighten their hair with mixtures of flowers, eggs, saffron, and other ingredients. A coppery red color was popular in the 16th century in Italy, which was achieved by applying caustic soda to the hair and exposing it to the sunlight.[7] In the same century in England, Queen Elizabeth popularized natural red color. Women used saffron and sulfur powder to duplicate this color, which caused numerous side effects, including headaches and nosebleeds.[7]

In the 17th century, people used lead combs, which gradually darkened their hair to black. At this time, they did not realize how toxic lead was. In the late 19th century, a London chemist E. H. Thiellay and Parisian hairdresser L. Hugot discovered that hydrogen peroxide removes hair pigment from the hair, leaving it yellow. It provided a better way to lighten the hair than alkaline solutions used before. Their discovery became widely recognized and still remains the basis of bleaching preparations today.[9]

The whole hair coloring industry began in France when Hoffman in the mid-19th century noticed that *para*-phylenediamine (PPD) produced a brown-black coloration when oxidized. This led to the birth of the synthetic hair dye industry in which PPD still dominates today. The first hair coloring product was marketed in the early 20th century by Eugène Schueller, a French chemist and the founder of L'Oréal.[10] After the 1920s, oxidative hair dyes were greatly improved and the use of hair coloring products greatly increased. Clairol, one of the leaders in hair coloring in the US, launched its first salon colors in the 1930s.[11] Originally, hair coloring was performed at professional salons; however, in the mid-20th century, hair dyeing products became available for home use as well. Today, consumers can choose from a variety of products that produce temporary or permanent results. Before 1960s, only about 7% of women dyed their hair, which is approximately 75% today.[7]

How Hair Coloring Products May Affect the Scalp and Hair?

Hair frames the face, and therefore, it is one of the most important features in terms of first impressions. Following a person's smile, eyes, and skin, the hair is often the next feature people notice on first encounters. Additionally, it is among the top features used when describing others. Attractiveness is strongly connected to confidence and positive self-esteem. Styled, well-kept hair gives the external appearance of being well managed, which in turn can contribute to feeling the same way internally.[12] Hair coloring products can significantly contribute to a healthy and well-managed appearance. This is what makes hair dyeing products one of the fastest growing segments of the entire hair care market.

As mentioned earlier, the key motivator for coloring the hair is graying. Graying can have a significant effect on both men and women, including emotional and psychological stress as well as lowered self-esteem. Hair coloring products can cover up the gray hair shafts and can have a **life-transforming** potential. Statistics show that people feel more attractive and younger after having their hair colored as well as more confident in their private life and work environment.[13]

Additionally, people are rarely satisfied with their natural hair color. A wide variety of hair coloring products offers an option to everyone to alter their hair color, which can also contribute to self-esteem and confidence.

Hair coloring products, however, can have potential negative effects on the hair and scalp even under recommended conditions of use. These should be considered before undergoing a hair coloring treatment.

- Oxidative hair dyes, including demi-permanent, permanent, and bleaching products, contain hydrogen peroxide and have an alkaline pH, which can have a significant effect on the hair shaft's **structure** and its **physical state**. The higher the pH, the more damaging the procedure. Generally, oxidative hair dyeing procedures can result in damaged cuticle, porous hair, decreased tensile strength, and increased hair breakage. All these can lead to undesirable sensorial attributes, such as poor shine, poor feel, coarse hair, which also lacks luster.[14] Additionally, hair that has been permanently colored or bleached is more sensitive to physical and environmental damage.[15] It should be emphasized that these changes are more prominent with frequent use and inappropriate application technique.
- Hair dyes can cause **allergic reactions**, mainly at the site of application. Allergic reactions to hair dyes are well known; however, the number of adverse reactions is estimated to be less than 0.5% of the general population,[16] which is still relatively rare, taking into account the number of people using such products. A key hair dye ingredient (dye intermediate), also known as a skin sensitizer, is *para*-phenylenediamine (PPD). PPD is part of the standard patch tests. Main symptoms of allergy to this ingredient include scalp redness and itching. Progressive and temporary hair dyes present minimal risk for allergic reactions, as they do not contain PPD. Permanent hair dyes contain the highest amount of this ingredient and, therefore, pose a potential risk for clients as well as hairdressers.[17] For permanent and even for semi-permanent hair colorants, consumers are advised to conduct a skin sensitivity test with the product to be used 48 h before hair coloring, by following the manufacturer's recommendations. In case of any allergic reaction, the hair dyeing product should be removed immediately, and users should contact their dermatologist. In order to prevent such reactions, hairdressers are usually advised to wear gloves during the hair dyeing process.
- Hair bleaching has also been reported to cause **skin irritation**, including scalp burns and allergic dermatitis.[18–20]

Required Qualities and Characteristics and Consumer Needs

From a consumer perspective, a quality hair coloring product should possess the following characteristics:

- Gentle to the hair and scalp, does not dry or damage it

- Good coverage for gray hairs
- Does not color skin
- Permanent dyes: long-lasting coloring effect
- Easy to spread on the hair
- Easy to rinse off from the hair
- Well-tolerated and non-allergenic.

The technical qualities of hair coloring products can be summarized as follows:

- Strong coloring power
- Appropriate rheological properties
- Appropriate pH
- Long-term stability
- Dermatological safety.

Current US Regulation on Hair Dyes

As discussed in Chapter 1, the regulation of color additives is quite complex in the US. Under the FD&C Act, all cosmetic color additives must be approved by the FDA, and the act prohibits any cosmetic product to be marketed in the US if it contains a color additive that has not been previously approved by the FDA. There is, however, an exception to this rule, which is for coal-tar hair dyes.

In addition to the approval, the FDA can subject a color additive to certification depending on the source of the color ingredient. The aim of this certification is to ensure that each batch of color additives manufactured conforms to the approved specifications in order to protect the public's health. Most approved color additives, including both certified and non-certified colors, have restrictions as to what they can be used for. For example, henna cannot be used for the eye area or the body; it is allowed to color only the hair on the scalp.

The name "coal-tar" comes from the original source of synthetic color additives. When the first synthetic organic dyes were discovered, they were named "coal-tar colors" as they were first produced from by-products of coal processing. Today, certifiable color additives are still called coal-tar dyes; however, they are synthesized mainly from raw materials obtained from petroleum and not from coal.

In the US, both approved and not approved color additives can be used in hair coloring products.

- **Approved** color additives, including both certifiable and non-certifiable color ingredients, can be used to color the hair as long as they are not specifically restricted from use. The restrictions can be found on the FDA's web site.
- **Not approved** synthetic organics, often called coal-tar hair dyes, can also be used to color the hair. In such cases, however, a special warning has to appear on the label as required by the FDA. The following warning statement should

> appear on the label: "Caution – This product contains ingredients, which may cause skin irritation on certain individuals and a preliminary test according to accompanying directions should be first made. This product must not be used for dyeing the eyelashes or eyebrows; to do so may cause blindness."[21] In addition, the label should contain adequate directions for conducting the aforementioned preliminary test.

It should be noted that this exception is only for coal-tar hair dyes; thus, hair dyes from other sources do not fall under this rule[22], and then they need to be approved by the FDA before using as hair dyes. If the label of a coal-tar color-containing hair dye product does not bear the above quoted caution statement and the directions on the preliminary test, it may be subject to regulatory action if it is determined to be harmful under customary conditions of use.[23]

Types, Typical Ingredients, and Formulation of Hair Coloring Products

Today, a wide range of products for changing the color of hair is available to consumers. ● **Hair coloring products can be categorized based on the presence or absence of the chemical reaction (known as oxidation) involved in the hair coloring process. Non-oxidative products include temporary dyes and semi-permanent dyes, while demi-permanent dyes, permanent dyes, and hair bleaches fall into the category of oxidative products. An additional product type is known as progressive hair dyes, which are permanent but not oxidative.**

Non-Oxidizing Products The products that belong to this category do not contain oxidizing agents, as their name implies. As a result, non-oxidizing dyes are not able to produce lighter shades than the originally presenting shade and cannot significantly darken the originally presenting color.

Temporary Dyes Temporary dyes or color rinses usually contain molecules that are too large to penetrate the hair cortex and also have low affinity to hair, meaning that the binding forces between the hair cuticle and the dye molecules are low.[24] As a result, temporary dyes provide a weak coating on the hair cuticle (see Figure 5.12) and are easily washed out after the first shampooing. Temporary coloring agents include azo compounds, triphenylmethane-based dyes, indoamines, and indophenols.[25] It should be kept in mind, however, that people who had permanent straightening, permanent waving, or even previous permanent coloring may have damaged cuticle, making the hair less resistant to the dyes. In such cases, the dyes can enter the cortex. Under these conditions, it may take more than one shampooing to remove the color. Temporary hair coloring products are often used to add a slight tone, brighten the hair, refresh the already colored hair, or try out a hair color before permanently dyeing the hair.

Typical product forms include liquids, shampoos, hair mousses, gels, and hair sprays.

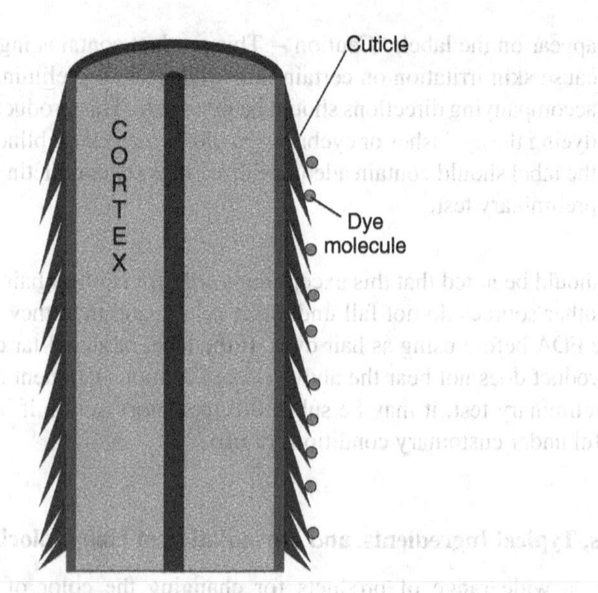

Figure 5.12 Coloring mechanism of temporary hair dyes.

- Color rinses are usually liquid products that are applied after shampooing, with the excess dyestuff being removed by rinsing. It is quite popular in the older generation to highlight the existing color or add color to the hair. It helps customers to obtain a glowing platinum color.
- Color-enhancing shampoos combine the action of a shampoo with that of a color rinse. They usually contain certified colors, which help produce highlights and slightly impart color tones.
- Hair color sprays used for parties also contain temporary dyes, which makes them easy to remove. These are usually aerosol products, which are applied to dry hair.
- Mousses are also available for parties and in natural colors as well.
- Gel products are similar to hair styling gels, except for their color. In mousses and gels, in addition to the dyes, film-forming ingredients are used. Therefore, these products serve as both styling and coloring products. Products with glitters are also popular, which give the hair a shimmery effect.[26]

Temporary hair dyes are ideal for patients with less than 15% gray hair.[27] These formulas are popular since they are easy to use and carry little risk of contact dermatitis.[28] However, these dyes can readily stain the scalp and skin.

Semi-Permanent Dyes Semi-permanent hair dyes usually employ dyes that are small enough to penetrate the hair cuticle to some degree (see Figure 5.13), in addition to staining it from the outside, and, hence, remain on the healthy hair through 6–8 shampooing.[29] Semi-permanent coloring agents usually include

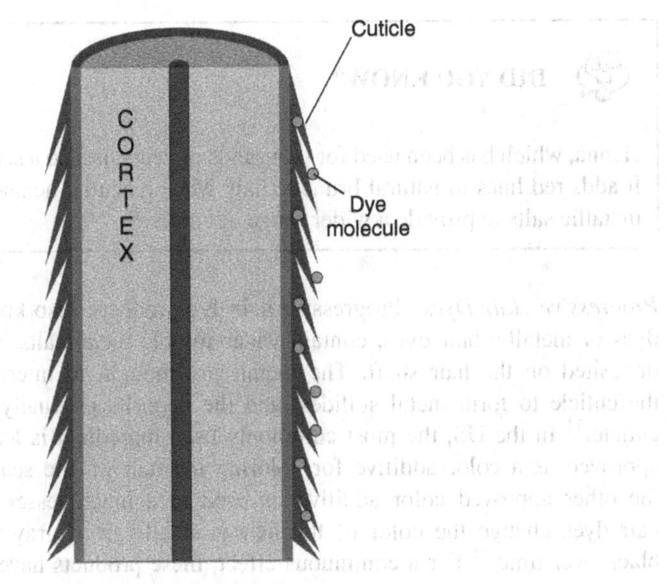

Figure 5.13 Coloring mechanism of semi-permanent hair dyes.

nitrophenylenediamines, nitroaminophenols, and azo dyes.[25] Similar to temporary dyes, their effect may be longer, if applied to damaged, porous hair.

They are usually used on natural, unbleached hair to cover gray, add highlights, and cover up unwanted tones.[30] The pH of semi-permanent dyes is slightly alkaline (7.0–9.0), making the cortex to swell and rise. It allows a certain degree of penetration of the dye into the deeper layers of the cuticle. Washing the hair opens the cuticle, allowing the color to escape over time because of the solubility of the dyes in water.

The formulation of a typical semi-permanent hair dye includes a dye, an alkalizing agent, a solvent, a surfactant, a thickener, a fragrance, and water.[29] Usually, 10–12 dyes are mixed to obtain the desired shade.[31] Typically, semi-permanent hair colorings are available as lotions, shampoos, gels, creams, and mousses.

- Shampoos are usually applied to hair and left on it for 20–30 min. Thickness of such products is important as low-viscosity products would run off the scalp.
- Mousses incorporate the dye in an aerosol form. It is also applied to wet hair and left on the hair for 20–30 min before rinsing.

Semi-permanent dyes are ideal for patients with less than 30% gray hair who want to restore their natural color.[32] Semi-permanent dyes have the potential to cause allergic contact dermatitis, and according to FDA regulations, statutory warnings and instructions to perform a skin patch test before use should be present on the packaging.[28]

DID YOU KNOW?

Henna, which has been used for thousands of years, is also a semi-permanent dye. It adds red hues to natural brunette hair. Most recently, henna is combined with metallic salts to provide a wider range of colors.[28]

Progressive Hair Dyes Progressive hair dye products, also known as gradual hair dyes or metallic hair dyes, contain water-soluble metal salts, which are gradually deposited on the hair shaft. The metals are thought to interact with cysteine in the cuticle to form metal sulfides, and the deposits gradually accumulate on the cuticle.[33] In the US, the most commonly used ingredient is lead acetate, which is approved as a color additive for coloring the hair on the scalp. Bismuth citrate, the other approved color additive, is used to a much lesser extent. Progressive hair dyes change the color of the hair gradually from gray to yellow-brown to black over time.[34] For a continuous effect, these products have to be continuously applied. Progressive hair dyes are inexpensive and can be applied at home, which contributes to their popularity. However, they are not popular as oxidative products. Sometimes, trace metals left on the hair can interact with bleaching or permanent waving products. Therefore, it is recommended to allow the hair grow out before undergoing other coloring or waving procedures.[35] These products are more popular among men, who usually do not undergo permanent waving and have their hair cut more frequently; therefore, the potential damaging effect in their case is relatively lower.

Oxidizing Products Oxidizing hair dyes are two-component systems: one component contains colorless dye intermediates (such as PPD; *p*-toluenediamine, PTD; and *p*-aminophenols) and couplers (such as resorcinol and *m*-aminophenols) in a highly alkaline formulation. The other component contains hydrogen peroxide (the oxidizing agent, otherwise known as the developer or activator). These products are mixed right before application, which generates a chemical reaction. The alkaline agent swells the hair cuticle and thus helps the penetration of the relatively small dye intermediates into the cortex. In addition, it also destabilizes hydrogen peroxide to liberate oxygen. The oxygen released destroys the hair's natural melanin (i.e., lightens hair) and also oxidizes the dye intermediates and allows them to react with the couplers within the hair shaft to form a colored molecule.[36] The final color molecule is too large to be removed by shampooing, which makes the color resistant to shampooing.[37,38] This process is depicted in Figure 5.14. It is important to emphasize that couplers do not produce hair color but alter the color of the oxidized dye intermediates. The color depends on the type and amount of dye intermediates and couplers used.

In the category of oxidizing hair coloring products, we can distinguish between two main groups: demi-permanent dyes and permanent dyes. The primary

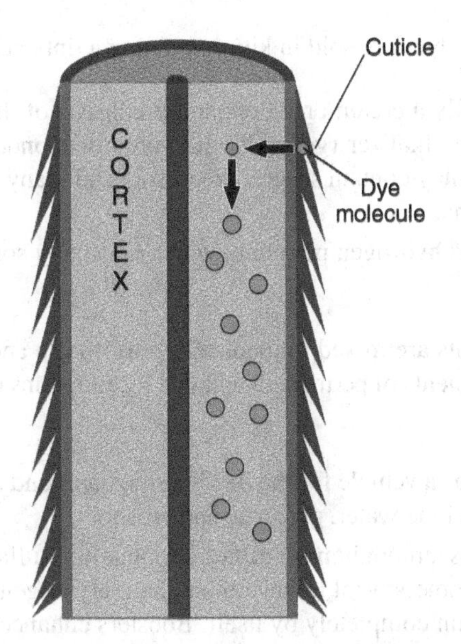

Figure 5.14 Coloring mechanism of oxidizing hair dyes.

distinctions between these two are the type and level of alkalizing agent and the concentration of the peroxide. These differences result in different color, coverage, lasting power, and lightening performance.

Demi-Permanent Hair Dyes Demi-permanent products are a newer category of hair coloring products. They typically employ 2% hydrogen peroxide and lower levels of alkalizers (usually monoethanolamine).[39] Their effect is longer lasting than that of semi-permanent colors; however, it is still not permanent due to the use of non-ammoniacal alkaline agent. They are gradually washed out, usually during 20–24 shampooing.[39] Demi-permanent colors can be used to enhance and brighten the natural hair color; however, due to the lower amount of hydrogen peroxide, they do not lighten the hair. For the same reason, they are milder to the hair than permanent hair colors and have a less disturbing smell during application. Most commonly, they are used to add red highlights to brown hair. Demi-permanent dyes can cover up gray hairs up to 50%. Demi-permanent hair dyeing products are available as gels, creams, or liquids.

Permanent Hair Dyes Permanent hair coloring agents are the most popular today. Their popularity is due to the wide variety of shades available as well as their ability to both lighten and darken the hair. Permanent hair dyes can be used to entirely cover gray hair and produce a completely new color.

Permanent colorants use up to 6% hydrogen peroxide and contain ammonia as an alkalizer to bring the pH of the final product to 9.0–10.5.[39] This allows complete penetration across the cortex.

All permanent hair dyes are sold in kits containing a tint and a developer.

- The **tint** is usually a cream or lotion, and it consists of the dye intermediates, couplers, and an alkalizer (generally ammonia or monoethanolamine). More complex colors may contain several precursors and many couplers and involve multiple reactions.
- The **developer** is hydrogen peroxide, which is usually sold as a liquid, gel, or lotion.

The two components are mixed immediately prior to use and then applied to the hair. Additional ingredients of permanent hair dye formulations include the following ingredients:

- **Solvents**, serve as a vehicle for the dye intermediates and couplers. Commonly used solvents include water, glycerin, and ethanol.
- **Boosters**, such as ammonium persulfate or potassium sulfate, can be used when dark-haired customers want to have blond hair. Hydrogen peroxide is not able to remove melanin completely by itself. Boosters enhance its effectiveness.
- **Conditioners**, such as quaternary compounds, proteins, and emollients.
- **Surfactants**, in which helps wetting the hair during the coloring process, removing the dye formulation from the hair after application, as well as help stabilize the formulations. Examples include anionic, amphoteric, and nonionic surfactants.
- **Buffers** stabilize the formulation and help neutralize the polymer, if needed.
- **Thickeners** build viscosity into the formulations and keep the formulation at the site of application.
- **Antioxidants, preservatives,** and **chelating agents** to contribute to the stability of the formulations.

Redyeing (touch-up) is usually necessary every 4–6 weeks as new hair appears at the scalp (often called the "root," which, as discussed earlier, is not the real root, as the hair roots are located deep down within the scalp). Permanent hair dyes are formulated to give hair long-lasting and intensive colors; however, several factors, including improper application of the dye; mechanical factors, such as shampooing and permanent chemical procedures damaging the cuticle; and environmental factors, such as UV light and water exposure, can accelerate color changes and fading.

DID YOU KNOW?

Red shades tend to fade fastest as they have a relatively small molecule size, which can diffuse from the hair and wash away more quickly than other shades.[9]

DID YOU KNOW?

The main difference between permanent hair dyes and hair bleaches is that permanent hair dyes act as both bleaching agents (lighten the hair) and coloring agents, while hair bleaches do not add a new color. However, permanent hair dyes cannot dramatically lighten the hair color. For example, a brown hair cannot be dyed to a golden blond color using permanent dyes only. In such cases, first a bleaching process is needed and when the color is removed, a permanent dye can be applied to color the hair.

Hair Bleaches ❶ **Hair lightening**, often referred to as **"bleaching," is a chemical process that involves the removal of the natural hair pigment or artificial color from the hair.** Hair bleaches react with the hair melanin in the cortex, removing the color in an irreversible chemical reaction. The bleach oxidizes the melanin molecule, and although the melanin is still present, the oxidized molecule is colorless. The hydrogen peroxide breaks chemical bonds in the hair, which releases sulfur that accounts for the characteristic odor of the hair coloring process. As melanin is located in the cortex, the cuticle has to be opened for optimal penetration of the bleaching agent. Therefore, bleaches are also alkaline solutions containing hydrogen peroxide, often with added boosters to accelerate the bleaching efficacy. Examples for boosters include ammonium persulfate or potassium sulfate. Since hydrogen peroxide is not stable at an alkaline pH, it is usually supplied as a diluted (6–12%)[40] aqueous solution or cream, and it is combined with an alkaline ammonia solution or cream immediately prior to use. The concentration of hydrogen peroxide is usually characterized in terms of volume. The higher the volume used, the more dramatic the achievable hair lightening.[41] Ammonia serves to speed up the oxidation reaction. The reaction usually occurs more rapidly at the scalp due to the presence of body heat. Therefore, the bleaching product is usually applied to the hair tips first and then to the part close to the scalp.[41] After leaving on the hair for a desired amount of time, the product is removed with an acidic shampoo to reset the pH to the normal level and minimize hair damage. The resulting color is often flat and difficult to control, and toners (dilute solutions of dyes) are used to make the color more aesthetically acceptable.[33] The toner can be selected from either the permanent or semi-permanent family of dyes.

Hair bleaching is commonly used by itself or to remove pigment before color is deposited. Similar to permanent hair dyes, regular reapplication is necessary to prevent visible regrowth of the naturally darker hair.

Formulation of Hair Coloring Products As discussed previously, hair dyes are available in a wide variety of product forms, including hair sprays, shampoos, lotions, creams, and gels. The formulation technique for these product types can be attributed to the basic dosage forms, such as aerosols, gels, creams, and lotions, which were discussed the in previous chapters.

Due to the reactivity of hydrogen peroxide, special precautions should be taken when formulating products containing this ingredient.

Typical Quality Problems of Hair Coloring Products

● **The typical quality-related issues of hair coloring products include valve clogging, poor foaming of hair mousse, unstable foam, separation of emulsions, microbiological contamination, clumping, and rancidification.** As these quality issues were discussed in the previous chapters, they are not reviewed here in detail.

Evaluation of Hair Coloring Products

Quality Parameters Generally Tested ● **Parameters commonly tested to evaluate the quality of hair coloring products include spray characteristics; aerosol can leakage; pressure test; spreadability, extrudability, texture, and firmness of lotions, creams, and gels; actuation force; color; preservative efficacy; viscosity; and pH.** The range of acceptance and other limiting factors are usually determined by the individual manufacturers. As these tests were discussed in the previous sections, they are not reviewed here in detail.

Performance Parameters Generally Tested ● **The most frequently tested parameters related to the performance of hair coloring products are shine and color intensity.** These tests are discussed in the following sections.

Shine and Color Intensity Hair shine and color intensity are important features of hair appearance and key consumer objectives in the hair care market. Measurement of these characteristics is quite challenging since their perception depends on various factors, including the light, hair color, surface smoothness, hair morphology, hair mass density on the scalp, and hair fiber alignment.[42] Recently, a fast polarimetric video camera called SAMBA® (by Bossa Nova) with a high polarization contrast was introduced, which can effectively separate specular and diffuse light. It has been shown that the instrumental data are in good correlation with the subjective assessments of panelist as well as consumer's subjective evaluation.[43] The experiments involve capturing two images of the hair. The first is taken with parallel polarizers and allow all reflected light to be captured. The second uses perpendicular polarizers, which eliminate the specular reflection, and capture the diffuse reflection only. Therefore, by subtraction, it is possible to quantify the specularly reflected light. These two parameters can then be used in conjunction with different shine equations from the technical literature.[44] In addition, atomic force microscopy can also be used to study hair luster and other hair surface properties, such as cuticle height, cuticle angle, and surface roughness.[45,46]

Safety Concerns ● **Oxidative hair dyes have repeatedly come to the attention of dermatologists due to their allergic potential and potential carcinogenic**

characteristics. As discussed previously, permanent hair dyes contain the highest amount of "para" dye components (such as phenol, diamines, and aminophenols). They may also contain components that can interact with "para" dyes. **Allergic contact dermatitis** and also facial edema can develop in sensitive individuals after applying permanent hair dyes.[47] The allergic potential of PPD and chemically similar compounds remains a significant cause of occupational allergy, particularly in cosmeticians and hairdressers.[48] The prevalence of contact allergy to PPD appears to have decreased in recent years in the US, despite the increasing use of hair dyes in the industrialized world. This trend has been attributed to the successful introduction of risk management measures; i.e., precautionary labeling; improved compliance with occupational safety measures, such as wearing protective gloves and minimizing the contact of the hair dye with the scalp during application; and the increased risk awareness among consumers and hairdressers.[49] The CIR Panel continuously reviews the safety of various dye intermediates and couplers and publishes its conclusion in the International Journal of Toxicology.

An additional safety concern with regard to the use of oxidative hair dyes was whether they cause cancer. Oxidative hair dye ingredients belong to the large chemical family of arylamines, which includes known human carcinogens, such as 2-naphthylamine. The carcinogenic activity of some arylamines was recognized in the late 19th century, showing an increased incidence of bladder cancer in occupationally exposed workers in the dye industry.[50] Although some arylamines are known to be carcinogenic in humans and other mammals, many substances of this large chemical class do not have carcinogenic activity. However, given the potential class hazard of arylamines, the **carcinogenic potential** of hair dye ingredients has been of major concern and has been investigated by the dye and hair dye industries, the US National Toxicology Program (NTP), independent investigators, and the International Agency for the Research on Cancer (IARC) of the World Health Organization (WHO).[51,52] Numerous studies have been conducted on the safety of hair dyes; the vast majority of these studies concluded no association of hair dye use and an increased cancer risk. In 2008, leading cancer experts of the IARC reviewed all relevant studies and scientific papers published to date and concluded that there is no evidence that personal hair dye use is associated with any increased cancer risk.[53] In 2011, the CIR Panel came to the same conclusion, stating that the available epidemiology studies are insufficient to conclude there is a causal relationship between the hair dye use and cancer and other endpoints, based on the lack of strength of the associations and inconsistency in the findings.[53]

Packaging of Hair Coloring Products

The most commonly used packaging materials for hair coloring products include the following:

- **Aerosol Can**: Temporary hair coloring sprays and mousse can be supplied in aerosol cans, similar to non-coloring hair sprays and mousses.

- **Plastic Bottles**: Shampoos, gels, and lotions are usually supplied in plastic containers, similar to non-coloring products.
- **Hair Coloring Kits**: Most oxidative hair dyes come in kits in which the two components are supplied separately in metal or plastic tubes or bottles. In addition to the actual coloring products, a set of gloves and plastic bottles with a special nozzle head are also supplied to make application easier. Certain kits may also include a small brush to help touch up the "roots."

GLOSSARY OF TERMS FOR SECTION 4

Booster: An ingredient used in permanent hair dyes and hair bleaches to enhance the effectiveness of hydrogen peroxide in removing the hair color.

Coal-tar hair dye: Synthetic organic hair dye named "coal-tar" due to historical reasons. In the past, coal-tar dyes were derived from coal; however, today, these dyes are obtained from petrolatum.

Coupler: An ingredient used in oxidative hair dyes that binds colorless dye intermediates together inside the hair fibers. This reaction allows the dye to become larger in volume and cannot easily be washed out of the hair.

Demi-permanent hair dye: An oxidative hair coloring product that penetrates the cuticle and cortex; therefore, it provides a long-lasting coloring effect (up to 20–24 shampooing).

Developer: An ingredient, usually hydrogen peroxide, used in permanent hair dyes to develop the color from the colorless dye intermediates inside the hair fibers.

Dye intermediate: Small colorless molecules used in oxidative hair dyes that, once entered into the hair cortex, react with hydrogen peroxide and form large color molecules that cannot be removed by shampooing.

Hair bleach: A product designed to remove the hair color through a chemical reaction.

Henna: An approved non-certifiable color additive for cosmetics in the US, which is only approved to color the hair (and not the skin, eyelashes, or eyebrows).

Non-oxidative hair dye: A hair coloring product that does not contain oxidizing agents.

Oxidative hair dye: A hair coloring product that works based on an oxidative reaction.

Permanent hair dye: An oxidative hair coloring product that penetrates the cuticle and cortex and provides a permanent hair coloring effect.

Progressive hair dye: A non-oxidative hair coloring product that gradually colors the hair by depositing metal ions on it.

Semi-permanent hair dye: A non-oxidative hair coloring product that adheres to the outside of hair fibers and partially penetrates the cuticle layers; therefore, it colors the hair for a few days/weeks (usually 6-8 shampooing).

Skin patch testing: A safety test required to be performed by the consumers before using hair coloring products that contain not approved synthetic organic dyes. In such cases, the product labels give detailed instructions on how to perform the test.

Temporary hair dye: A non-oxidative hair coloring product that adheres only to the outside of hair fibers and, therefore, provides a short coloring effect (one shampooing).

Touch-up: A quick restoration of natural hair color or gray hair to the desired color. Touch-up is usually needed close to the scalp.

REVIEW QUESTIONS FOR SECTION 4

Multiple Choice Questions

1. Which of the following is NOT true for progressive hair dyes?
   a) They have a temporary effect
   b) They gradually color the hair if used continuously
   c) They contain metal salts
   d) They are more popular among men

2. What is the pH of permanent oxidative hair dyes?
   a) 2–4
   b) 5–6
   c) 7–8
   d) 9–10

3. Temporary hair dyes cannot penetrate into the hair cortex due to their ___.
   a) Size
   b) Smell
   c) Viscosity
   d) Density

4. Which of the following is the main factor influencing the tightness/openness of cuticle layers?
   a) Temperature
   b) Viscosity
   c) pH
   d) Pressure

5. Permanent hair dyes work based on a chemical reaction called ___.
   a) Reduction
   b) Oxidation
   c) Hydrolysis
   d) Isomerization

6. Which of the following washes out during 6–8 shampooing?
   a) Temporary hair dyes
   b) Hair bleaches
   c) Permanent hair dyes
   d) Semi-permanent hair dyes

7. Which of the following is a necessary component of permanent hair dyeing products?
   a) Couplers
   b) Colorless dye intermediates
   c) Alkalizer
   d) All of the above

8. A warning statement has to be put on the label of hair coloring products if any ___ hair dyes are used.
   a) Approved
   b) Not approved
   c) Certified
   d) All of the above

Fact or Fiction?

_____ a) Henna is only approved to color the hair (and not the skin).
_____ b) Bleaching temporarily lightens the hair color.
_____ c) Red colors fade fastest.
_____ d) Graying is caused by the production of a gray pigment.

Matching

Match the ingredients in column A with their appropriate ingredient category in column B.

	Column A	Column B
_____	A. Ammonia	1. Alkalizer in demi-permanent dyes
_____	B. Henna	2. Alkalizer in permanent hair dyes
_____	C. Hydrogen peroxide	3. Booster
_____	D. Lead acetate	4. Developer in oxidative hair dyes
_____	E. Monoethanolamine	5. Dye intermediate
_____	F. Potassium sulfate	6. Progressive hair dye
_____	G. PPD	7. Semi-permanent hair dye

REFERENCES

1. Happi Magazine: A Colorful History, Last update: 1/18/2013, Accessed 2/15/2014 at http://www.happi.com/issues/2013-01/view_features/a-colorful-history/
2. Lin, F., Murphy, K.: Translating Trends for Hair Care Product Development, Last update February 11, 2013, Accessed 2/12/2014 at http://www.gcimagazine.com/marketstrends/segments/hair/Translating-Trends-for-Hair-Care-Product-Development-190739661.html
3. Corbett, J.: Hair coloring processes. *Cosmet Toiletries.* 1991;106:53–67.
4. Wojnarowska, F. T.: Hair Cosmetics, In: Rook, A., Dawber, R., eds: Diseases of the Hair and Scalp, 2nd Edition, Oxford: Blackwell Science; 1991:465–479.
5. Cesarini, J. P.: Hair Melanin and Hair Colour, In: Orfanos, C. E., Happle, R., eds: Hair and Hair Diseases, Berlin: Springer-Verlag, 1990: 166–197.
6. Nishimura, E. K., Granter, S. R., Fisher, D. E.: Mechanisms of hair graying: incomplete melanocyte stem cell maintenance in the niche. *Science* 2005;307:720–724.
7. Sherrow, V.: For Appearance's Sake: The Historical Encyclopedia of Good Looks, Beauty and Grooming, Westport: Greenwood Publishing, 2001.
8. Partick, B., Thompson, J.: *An Uncommon History of Common Things*, Washington: National Geographic, 2011: 204.
9. Bouillon, C., Wilkinson, J.: The Science of Hair Care, 2nd Edition, Boca Raton: CRC Press, 2005: 252–253.
10. L'Oreal: The First Steps, Constructing a Model, Accessed 2/15/2014 at http://www.lorealusa.com/group/history/1909-1956.aspx
11. Clairol Professional: About Us, Accessed 2/15/2014 at http://www.clairolpro.com/Details/ContentArticlePage.aspx?ArticleID=4
12. Vivian Diller: The Challenges of Aging in Today's Culture, Last update: 2/18/2012, Accessed 2/18/2014 at http://www.psychologytoday.com/blog/face-it/201202/the-psychology-behind-good-hair-day
13. Gray, J. The World of Hair Colour, London: Thomson, 2005.
14. Jachowicz, J.: Hair damage and attempts to its repair. *J Soc Cosmet Chem.* 1987;38: 236–286.
15. Takada, K., Nakamura, A., Matsua, N., et al.: Influence of oxidative and/or reductive treatment on human hair (I): analysis of hair damage after oxidative and/or reductive treatment. *J Oleo Sci.* 2003;52:541–548.
16. Schnuch, A.: Data Presented During a Recent BfR Symposium on the Safety of Hair Dyes, 2009, Accessed 2/15/2014 at http://www.bfr.bund.de/cd/31861
17. Goldberg, B. J., Herman, F. F., Hirata, I.: Systemic anaphylaxis due to an oxidation product of p-phenylenediamine in a hair dye. *Ann Allergy.* 1987;58:205–208.
18. Forster, K., Lingitz, R., Prattes, G., et al.: Hair bleaching and skin burning. *Ann Burns Fire Disasters.* 2012;25(4):200–202.
19. Lee J. M. S., Jeong, C. M., Kim, W. J., et al.: Significant damage of the skin and hair following hair bleaching. *J Dermatol.* 2010;37(10):882–887.
20. Jensen, C. D., Sosted, H.: Chemical burns to the scalp from hair bleach and dye. *Acta Dermato Cenerologica.* 2006;86:461–462.
21. FD&C Act Section 601(a)
22. CFR Title 21 Part 70.3(u)

23. FDA: Hair Dyes: Fact Sheet, Last update: 03/19/2014, Accessed 3/20/2014 at http://www.fda.gov/Cosmetics/ProductsIngredients/Products/ucm143066.htm
24. Corbett, J. F.: Hair coloring, *Clin Dermatol.* 1988;6:93–101.
25. Salvador, A., Chisvert, A.: Analysis of Cosmetic Products, Amsterdam: Elsevier, 2007.
26. Corbett, J. F.: Hair Dyes, In: Venkataraman, K., ed.: The Chemistry of Synthetic Dyes, Vol. 5, New York: Academic Press, 1971:475–534.
27. Drealos, Z.: Hair Care, New York: Taylor and Francis, 2005.
28. Draelos, Z. K.: Hair cosmetics. *Dermatol Clin.* 1991;9:19–27.
29. Corbett, J. F.: Hair coloring processes. *Cosmet Toiletries.* 1991;106:53.
30. Spoor, H. J.: Semi-permanent hair color. *Cutis.* 1976;18:506–508.
31. Robbin, C. R.: Chemical and Physical Behavior of Human Hair, 2nd Edition, New York: Springer-Verlag, 1988: 185–188.
32. Zviak, C.: Hair Coloring, Nonoxidation Coloring, In: Zviak, C., ed.: The Science of Hair Care, New York: Marcel Dekker, 1986: 235–261.
33. Brown, K. C.: Hair Colouring, In: Johnson, D. H., ed.: Hair and Hair Care, 1st Edition, New York: Marcel Dekker, 1977: 191–215.
34. Pohl, S.: The chemistry of hair dyes. *Cosmet Toiletries.* 1988;103:57–66.
35. O'Donoghue, M. N. Hair cosmetics. *Dermatol Clin.* 1987;5:619–625.
36. Corbett, J. F.: Hair Colorants: Chemistry and Toxicology , Cosmetic Science Monographs, Weymouth: Micelle Press, 1998.
37. Westman, M.: Demi-perm hair dyes – important and misunderstood. *Cosmetiscope.* 2008;14(5):1–7.
38. Tucker, H. H.: Formulation of oxidation hair dyes. *Am J Perfum Cosmet.* 1968;83:69.
39. Neuser, F., Schlatter, H.: Hair Dyes, In: Drealos, Z. D., ed.: Cosmetic Dermatology, Hoboken: Wiley Blackwell, 2010: 227–235.
40. Bolduc, C., Shapiro, J.: Hair care products. Waving, straightening, conditioning, and coloring. *Clin Dermatol.* 2001;19:431–436.
41. Wolverton, S. E.: Comprehensive Dermatologic Hair Therapy, Amsterdam: Elsevier Health Sciences, 2012
42. Gao, T., Pereiara, A., Zhu, S.: Study of hair shine and hair surface smoothness. *J Cosm Sci.* 2009;60:187–197.
43. Lim, J. M., Chang, M. Y., Park, M. E., et al.: A study correlating between instrumental and consumers' subjective luster values in oriental hair tresses. *J Cosm Sci.* 2006;57:475–485.
44. McMullen, R., Jachowicz, J.: Optical properties of hair - detailed examination of specular reflection patterns in various hair types. *J Cosm Sci.* 2004;55(1):29–47.
45. Swift, J. A., Smith, J. R.: Atomic force microscopy of human hair. *Scanning.* 2000;2(2):310–318.
46. Ruetsch, S. B., Kamath, Y. K., Kintrup, L., et al.: Effects of conditioners on surface hardness of hair fibers: an investigation using atomic force microscopy. *J Cosm Sci.* 2003;54:579–588.
47. Thyssen, J. P., White, J. M.: European Society of Contact Dermatitis. Epidemiological data on consumer allergy to p-phenylenediamine. *Contact Dermatitis* 2008;59(6):327–343.
48. Schnuch, A., Geier, J., Uter, P. J.: National rates and regional differences in sensitization to allergens of the standard series. *Contact Dermatitis.* 1997;37:200–209.

49. DeGroot, A. C.: Fatal attractiveness: the shady side of cosmetics. *Clin Dermatol.* 1998;16:167–179.
50. Rhen, L.: Blasengeschwülste bei Anilinarbeitern. *Archiv für klinische Chirurgie,* 1895;50:588–600.
51. International Agency for the Research of Cancer, World Health Organization. Occupational exposure of hairdressers and barbers and personal use of hair colorants. Some hair dyes, cosmetic colorants, industrial dyestuffs and aromatic amines. IARC Monographs on the Evaluation of Carcinogenic Risk to Humans. Volume 57, Geneva, Switzerland: World Health Organisation.
52. Baan, R., Straif, K., Grosse, Y., et al.: Carcinogenicity of some aromatic amines, organic dyes and related exposures. *Lancet Oncol.* 9,2008;322–323.
53. Hair Dye Epidemiology, Accessed 2/18/2014 at http://www.cir-safety.org/sites/default/files/hairdyeepidemiology-2011.pdf

6

ORAL AND DENTAL CARE PRODUCTS

INTRODUCTION

Keeping the teeth and oral cavity clean and fresh has been part of the human daily routine for a long time. In the past, dental care products were primarily cleansing aids, which were used to remove debris and polish the teeth. Today, the majority of dental care products offer additional benefits in the prevention of common dental problems such as caries or gingivitis.

Dental care products represent a smaller section of the cosmetic industry when compared to skin and hair cosmetics. However, their role is essential in our everyday life. There is a wide range of products, and their availability is excellent worldwide. It is important and worthwhile to emphasize that, despite the wide range of products and significant investments in advertising made by manufacturers, the per- capita use of oral care products is still lower than recommended by health professionals.

This chapter reviews the main anatomical and physiological characteristics of the human teeth and gums. It also discusses the various types of cosmetics and OTC drug–cosmetic products that are used in the oral cavity, including their ingredients, formulation technology, testing methods, and packaging materials. Additionally, the chapter provides an overview on how these products may affect the health of the human teeth and oral cavity and what the consumers' general requirements are.

Introduction to Cosmetic Formulation and Technology, First Edition. Gabriella Baki and Kenneth S. Alexander.
© 2015 John Wiley & Sons, Inc. Published 2015 by John Wiley & Sons, Inc.

LEARNING OBJECTIVES

Upon completion of this section, the reader will be able to

1. define the following terms:

ADA	ADA seal of acceptance	Bingham plastic	Cariogenic
Cementum	Demineralization	Dental caries	Dental floss
Dentin	Enamel	Fluorosis	Gingivitis
Gum disease	Halitosis	Mouthwash	Plaque
Pseudoplastic	Remineralization	Tartar	Tooth sensitivity
Toothpaste	Yield stress		

2. describe the distinct anatomical regions of the human teeth;
3. differentiate between enamel, dentin, and cementum;
4. explain to what the terms "plaque" and "tartar" refer;
5. briefly summarize what tooth decay is and which factors play an important role in its development;
6. briefly summarize the development and different stages of gum disease;
7. briefly summarize what may cause dental stains and how they can be treated;
8. list the possible causes of bad breath;
9. differentiate between toothpaste and mouthwash;
10. briefly discuss whether toothpaste and mouthwash are considered drugs or cosmetics in the US;
11. explain how dental fluorosis can develop and how it can be prevented;
12. list various required cosmetic qualities and characteristics that an ideal oral and dental care product should possess;
13. list various required technical qualities and characteristics that an ideal oral and dental care product should possess;
14. list the major ingredient types found in toothpaste and provide some examples for each type;
15. briefly discuss how the particle size of abrasives may influence the cleaning power of toothpastes;
16. explain why anionic surfactants are primarily used in toothpaste formulations;
17. explain what the function of thickeners is in toothpastes;
18. briefly discuss why there are preservatives in toothpastes that already contain antimicrobial ingredients;
19. list various active ingredient types that may be incorporated into toothpastes and name some examples for each type;
20. briefly discuss how fluoride can prevent caries;

21. explain and provide an illustrative viscosity curve that shows how the viscosity of toothpaste and toothpaste gel changes under increasing shear stress (i.e., squeezing the tube);
22. describe the general formulation method of toothpastes;
23. list the various ingredient types that are used in mouthwash formulations and provide some examples for each type;
24. list the various types of tooth whitening products currently available in the US;
25. summarize the major concerns with regard to tooth whitening products;
26. differentiate between dental floss and dental tape;
27. list some of the typical quality issues that may occur during the formulation and/or use of dental and oral care products, and explain why they may occur;
28. list the typical quality parameters that are regularly tested for oral and dental care products, and briefly describe their method of evaluation;
29. briefly discuss the potential safety issues with regard to the use of hydrogen peroxide in toothpaste and mouthwash;
30. describe how the antimicrobial effect of oral and dental care products is evaluated;
31. list the typical containers available for oral and dental care products.

KEY CONCEPTS

1. The human oral cavity includes the lips, buccal mucosa, salivary glands, gums, teeth, palate, tongue, and floor of the mouth. It is the most complex and most accessible microbial ecosystem of the human body.
2. There are many oral and dental diseases that may result from poor dental hygiene. The most common problems include plaque formation, dental caries, tartar formation, gum disease, tooth sensitivity, dental stain, and bad breath, which could be prevented with regular use of dental and oral care products.
3. Toothpaste and mouthwash may be classified as either cosmetics or OTC drug–cosmetic products, depending on the claims made as well as the concentration of active ingredients present.
4. Toothpaste is a mild cosmetic detergent for cleaning the teeth. Initially intended to freshen the breath and remove deposits from teeth, the evolution of toothpaste has also made it a vehicle for the protection of the teeth from cavities, calculus formation, and gum diseases.
5. Toothpaste is a non-Newtonian fluid that has a specific rheological behavior and is called a Bingham plastic.
6. Mouthwash is a clear, most of the times, colored solution that is aimed to refresh the breath by swishing the product around the mouth, followed by spitting it

out. Similar to toothpaste, it may also have additional benefits, such as prevention against tooth decay, gingivitis, plaque formation, or tartar formation, or a combination of these.

7. The typical quality-related issues of oral and dental care products include hardening, inappropriate viscosity, grittiness, cloudy solution, microbiological contamination, and clumping.
8. Parameters commonly tested to evaluate the quality of oral and dental care products include spreadability and extrudability of toothpastes and toothpaste gels, foaming property, flavor, preservative efficacy, viscosity, and color and pH.
9. The most frequently tested efficacy parameters include antimicrobial activity, whitening activity, and hypersensitivity reduction.
10. Ingredients causing safety concerns with regard to oral and dental care products include fluoride and hydrogen peroxide.

ANATOMY AND PHYSIOLOGY OF THE HUMAN ORAL CAVITY

❶ **The human oral cavity includes the lips, buccal mucosa, salivary glands, gums, teeth, palate** (which forms the roof of the mouth), **tongue, and floor of the mouth.**[1,2] **It is the most complex and most accessible microbial ecosystem of the human body.** Its parts all provide different surfaces for microbial colonization. The constant production of saliva and the presence of sugars and amino acids from foods provide nutrients for microbial growth. Generally, in a single subject, it is usual to find between 20 and 50 species of bacteria at healthy oral sites. At the diseased sites, there is a tendency for higher numbers of different species to be present, perhaps 200 or more.[3]

Humans have 32 permanent teeth that start to appear at about the age of 6 and slowly replace the initial deciduous ("milk") set of teeth. The teeth are arranged in two rows, anchored in the upper and lower jaws. They play an important role in biting, tearing, and grinding of solid foods as well as in speaking. In addition, they play an important role in an individual's self-esteem and quality of life.

As for its anatomical structure, a human tooth can be considered to consist of three major parts: the **crown,** which is visible and extends above the gum (gingiva); the **root,** which sits in the upper or lower jawbone and remains invisible until recession of the gum exposes some of the root surface; and the **neck,** which is the boundary between the root and the crown.[2] Figure 6.1 depicts the structure of the human teeth.

Gingiva is the mucosa that covers the upper and lower jaws and surrounds the neck of the teeth. The **dental cavity** (pulp cavity) contains the pulp, a soft connective tissue containing nerves and blood vessels. The pulp cavity receives blood vessels and nerves through a narrow canal, the root canal located at the base of the tooth. Dentin-forming cells are also found here, on the boundary of the dental cavity and dentin.

Each tooth consists of three hard substances that resemble a bone: enamel, dentin, and cement.

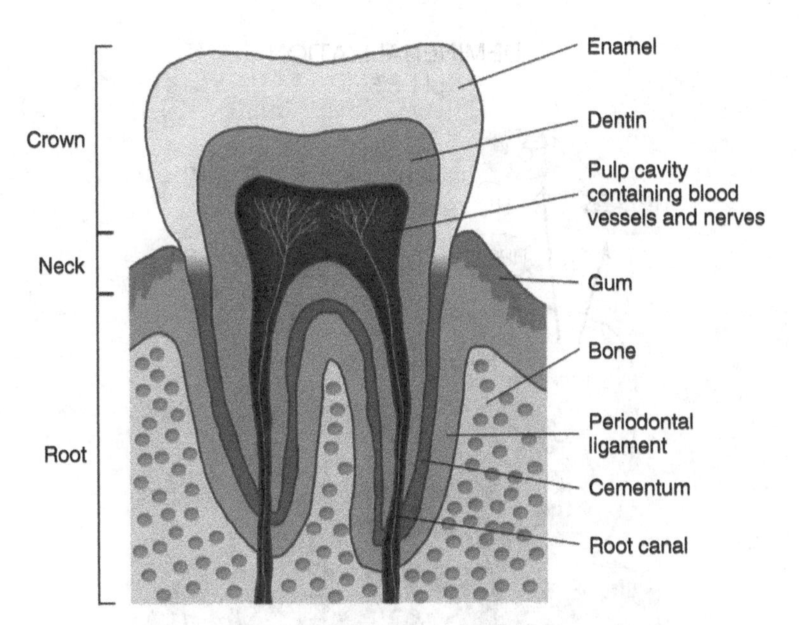

Figure 6.1 Structure of the human teeth.

- **Enamel** is the hardest substance in the human body; it is the white outer part of the tooth. Its function is to prevent the tooth from wearing away under the pressure of chewing. The calcium in our enamel, which is present in the form of hydroxyapatite, starts dissolving below pH 5.5. The pH of fruit juices and many carbonated drinks is known to be below this pH. Extensive consumption of such drinks may change the pH in the oral cavity and facilitate the chemical dissolution (i.e., demineralization) of dental calcium and phosphate. Saliva can buffer the pH up to a certain extent; however, repeated exposure to such acidic drinks can greatly increase demineralization. Under normal conditions, closer to pH 7 (neutral), such as periods between meals, the enamel tends to reacquire mineral ions (remineralize) from the saliva and minor carious lesions may be repaired. The processes of demineralization and remineralization are shown in Figures 6.2 and 6.3, respectively. These processes can have a crucial impact on the hardness and strength of the tooth enamel.
- **Dentin** forms the largest portion of the tooth; it is a yellowish bone-like layer surrounding the dental cavity. At the crown, the dentin is covered by the enamel. Dentin is less dense than enamel; thus, it is more porous, softer, more sensitive, and more susceptible to decay and wear if exposed.
- At the root, dentin is covered by the **cementum**, which is a special connective tissue. It binds the roots of the teeth firmly to the gums and jawbone. It is anchored in the jaw by the periodontal ligament.[2]

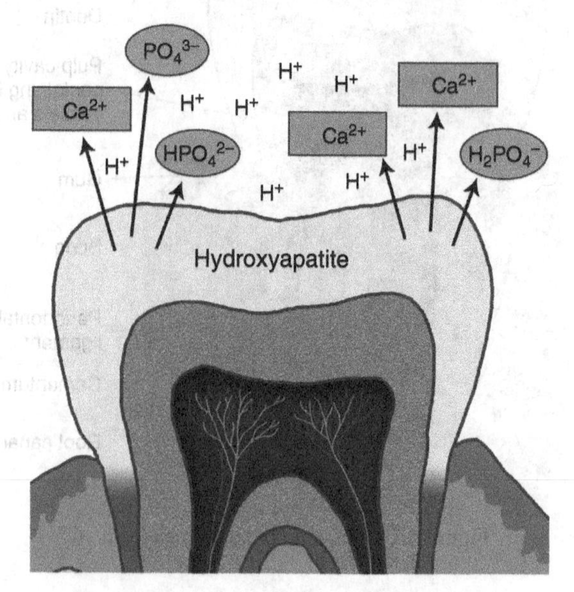

Figure 6.2 The process of demineralization.

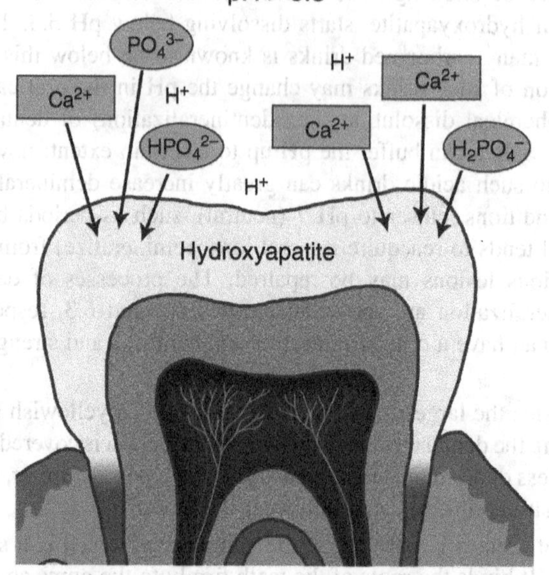

Figure 6.3 The process of remineralization.

BRIEF REVIEW OF THE MOST COMMON ORAL AND DENTAL CARE PROBLEMS

There are many oral and dental diseases that may result from poor dental hygiene. The most common problems include plaque formation, dental caries, tartar formation, gum disease, tooth sensitivity, dental stain, and bad breath, which could be prevented with regular use of dental and oral care products.

- **Plaque** is a sticky, colorless film of bacteria and sugars that constantly forms on our teeth. It is the main cause of cavities and gum disease and can harden into tartar if not removed daily.[4]
- **Dental caries**, also known as tooth decay and cavities, is a universal problem affecting all ages and all geographic locations around the world. According to the World Health Organization (WHO) report,[5] dental caries remains the most prevalent chronic disease both in children and in adults in most industrialized countries, including the US. It can be defined as a disease resulting in the breakdown and destruction of the enamel due to demineralization (i.e., loss of calcium and phosphate from the enamel).[6] It appears as tiny brown holes on the surface of the teeth. When eating sugars, the cariogenic bacteria (i.e., caries-causing), particularly *Streptococcus mutans*, in plaque produce acids that attack enamel.[7] The firm adherence of the bacteria to the teeth keeps these acids in contact with

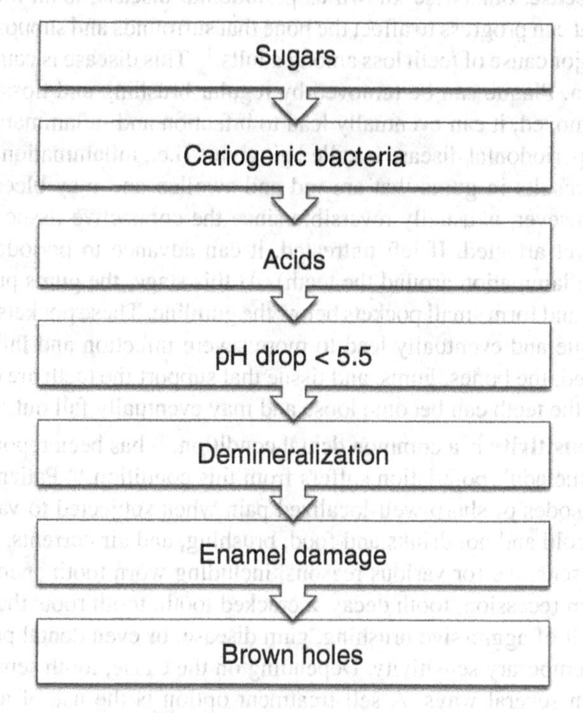

Figure 6.4 Development of tooth decay.

the teeth. Over time, the acids lead to breakdown of the enamel, and this is when caries can form (see Figure 6.4).

It is widely accepted that regular use of fluoride, i.e. in toothpaste and drinking water, is extremely effective in preventing dental caries. The US Center for Disease Control (CDC) issued a statement that water fluoridation is one of the 10 most important public health measures of the 20th century.[8] Because water fluoridation is not available in many countries, toothpaste is globally considered to be one of the most important sources of fluoride.[9] Today, the majority of toothpastes contain fluoride in an amount that is high enough to prevent the formation of cavities. Its incidence significantly decreased with the widespread availability of fluoride.[10,11] There are, however, recent studies that show an alarming increase in caries, especially in young children.[12–14] It is very important, therefore, to emphasize the importance of practicing adequate oral hygiene with the use of fluoride toothpaste.

- **Tartar**, otherwise known as dental calculus, is a plaque that is hardened (calcified) on the teeth. It can also form beneath the gums and can irritate the gum tissues. It gives plaque more surface area to grow and can be a causative factor in other dental diseases. In the long term, it causes a yellowish-brownish discoloration to the teeth. As tartar is hardened on the teeth, it cannot be removed by simple brushing; it has to be mechanically removed by dental professionals.[15]
- **Gum disease**, otherwise known as periodontal disease, is an infection of the gums that can progress to affect the bone that surrounds and supports the teeth. It is the major cause of teeth loss among adults.[16] This disease is caused by plaque formation. Plaque can be removed by regular brushing and flossing; however, if not removed, it can eventually lead to infection and inflammation. The early stage of periodontal disease is called gingivitis (i.e., inflammation of the gums). It often results in gums that are red and swollen and may bleed easily. This stage, however, is usually reversible since the connective tissue and the bone are not yet affected. If left untreated, it can advance to periodontitis (which means inflammation around the teeth). At this stage, the gums pull away from the teeth and form small pockets below the gumline. These pockets can trap food and plaque and eventually lead to more severe infection and inflammation. If not treated, the bones, gums, and tissue that support the teeth are destroyed. As a result, the teeth can become loose and may eventually fall out.[17]
- **Tooth sensitivity** is a common dental condition; it has been reported that up to 57% of the adult population suffers from this condition.[18] Patients experience brief episodes of sharp well-localized pain when subjected to various stimuli, such as cold and hot drinks and food, brushing, and air currents. The teeth can become sensitive for various reasons, including worn tooth enamel, worn fillings, gum recession, tooth decay, a cracked tooth, tooth roots that are exposed as a result of aggressive brushing, gum disease, or even dental procedures can lead to temporary sensitivity. Depending on the cause, tooth sensitivity can be treated in several ways. A self-treatment option is the use of anti-sensitivity toothpaste.

- **Dental stains**, or tooth discoloration, are spots or small areas on the teeth contrasting with the rest of the teeth color. They may be caused by multiple local and systemic conditions, which are typically classified as being either intrinsic or extrinsic.[19] Intrinsic discoloration means that the enamel and/or dentin in the tooth darken, and therefore, the teeth get a yellow tint. Discoloration may be caused by dental materials, caries, trauma, infections, medications (such as tetracyclines),[20] and even excessive fluoride intake while the teeth are still developing. Extrinsic stains are caused by certain molecules and metal ions found in our diet.[21] Such ingredients include coffee, tea, wine, and tobacco. If tooth discoloration is not treated, it can affect the appearance of a person's smile causing temporary, as well as permanent, social and psychological problems.
- **Bad breath**, otherwise known as oral malodor or halitosis, is a condition i.e. commonly experienced in the general population. It has been estimated that up to 50% of the adult population suffers with this problem at least occasionally.[22] Bad breath may be an important social problem that often is a cause for embarrassment. While malodor can be the result of a number of various factors, including certain foods, tobacco, alcohol, certain medications, gum disease, digestive, or metabolic disorders,[23,24], it has been shown that, for 9 out of 10 people, the cause of their oral malodor originates in the mouth.[25,26] The bacteria in the oral cavity produce a range of malodors. These bacteria are influenced by eating, drinking, oral hygiene, and sleep. Saliva production in the mouth facilitates the clearing away of bacteria. During sleep, when there is a reduced flow of saliva in the mouth, bacteria build up, leading to oral malodor first thing in the morning.[27]

HISTORY OF USING ORAL AND DENTAL CARE PRODUCTS

Dental and oral cleaning has its roots back to 5000 BC, when the Egyptians started using a paste to clean their teeth. This was even before toothbrushes were invented. Toothbrushing tools date back to 3500–3000 BC; the first toothbrush was made of twig by the Babylonians and the Egyptians. Early Chinese, Greek, and Roman writings also describe numerous mixtures of both pastes and powders. The purpose of their application was to clean the teeth and gums, whiten the teeth, and refresh the breath. Early toothpaste ingredients included powdered pumice stone and vinegar[28] as well as powdered fruit, burnt shells, talc, ginseng, herbal mints, salts, honey, ground shells, and dried flowers. The Romans used urine; its ammonia (an ingredient in some of the modern toothpastes) cleaned the teeth. The Romans and Greeks also tried other abrasive ingredients, such as crushed boned and shell. The Romans also used urine as a mouthwash. In the Middle Ages, fine sand and pumice were the primary ingredients in teeth-cleaning formulas. The Europeans went to barber-surgeons to have their teeth filed and painted with nitric acid, which destructed the enamel. It resulted in beautiful white teeth, which had a short life though. Later on, toothpastes using brick and china had similar effects.[28]

The development of modern toothpastes started in the 1800s. British chemists added sodium bicarbonate (for whitening) and strontium (for strengthening) to the toothpastes. In the early 1800s, Italian dentists noticed that, although their patients had stained teeth, they were cavity-free. It was the result of an interaction between the enamel and a high level of fluorides in local soil and water. By the 1840s, both in Italy and in France, dentists suggested that people regularly use lozenges made of fluoride.[29] In the 1940s, fluoride was added to drinking water, and in 1955, Procter & Gamble was the first to put it in toothpastes.[30]

In the early ages, toothpaste was actually a powder. From the 1850s, toothpaste was manufactured in the form that we use today. In 1873, Colgate started the mass production of toothpaste in jars. In 1892, Dr. Sheffield first manufactured toothpaste into a collapsible tube and called the product Dr. Sheffield's Crème Dentifrice.[31] In 1896, Colgate also started the mass production of toothpaste in collapsible tubes.[32] Specific active ingredients started to be incorporated into the formulations in the second half of the 20th century. Since then, there was a significant development in active ingredients, polymers, and even packaging materials.

The first dental floss was introduced in 1815 by a New Orleans dentist. It was a piece of silk thread.[33] Mass production of silk dental floss was started in 1882; nylon floss (which is used today) in the 1940s.

TYPES AND DEFINITION OF ORAL AND DENTAL CARE PRODUCTS

❶ **Toothpaste and mouthwash may be classified as either cosmetics or OTC drug–cosmetic products, depending on the claims made as well as the concentration of active ingredients present.** The primary function of these products is to clean and refresh the teeth and the oral cavity, which are considered pure cosmetic benefits. Tooth whitening is also considered a cosmetic claim; therefore, "simple" whitening products are considered cosmetics in the US. Mouthwashes, exclusively for bad breath, temporarily control (mask) bad breath and refresh the mouth. Since they do not kill bacteria that cause bad breath, or do not inactivate chemically odor-causing compounds, they are considered cosmetics in the US. Dental floss is considered a cosmetic product in the US since it is used to clean the teeth and interdental areas. Drug claims may include reduction, prevention, and/or treatment of oral and dental care diseases. When products have a combination of cosmetic and drug claims, for example, whitening and prevention against gingivitis, they are considered OTC drug–cosmetic products in the US.

Let us review the definition of the basic product types, irrespective of their legal status.

- **Toothpaste,** also known as oral dentifrice, is a paste or gel designed to help remove plaque and stains from the teeth and keep the breath fresh. It improves the mechanical brushing and cleaning power of a toothbrush. It typically contains abrasive ingredients, coloring, flavoring, sweetener, as well as other ingredients that make the toothpaste a smooth paste, foam, and stay moist.
- **Mouthwash,** also known as oral or mouth rinse, is a liquid i.e. designed to refresh the breath and enhance oral hygiene. It is intended for use after brushing to provide a more effective cleaning effect.

- **Tooth whitening products**, also known as bleaching products, are utilized to get teeth brighter in dental office settings, salons, shopping centers, and at home. There are a number of product types available on the market, including toothpastes, gels, mouthwashes, trays with a solution, and strips. These products contain various ingredients to remove discoloration of the teeth.
- **Other Products:** There are other products on the market for cleaning the teeth, including tooth powder and dental floss. Tooth powder is an alternative to toothpaste. It is a mildly abrasive powder i.e. used with a toothbrush. Its effect is similar to that of toothpaste; however, the dosage form and mouthfeel are different. Dental floss is a thin filament used to remove debris and bacterial film caught between the teeth and between the teeth and the gums. They are designed to clean the areas in the oral cavity that are difficult to reach with a brush.

HOW ORAL AND DENTAL CARE PRODUCTS MAY AFFECT THE TEETH AND THE ORAL CAVITY?

Dental care products play an important role in maintaining healthy teeth and preventing diseases; however, if swallowed, used improperly, or overused, they may cause diseases.

- **Taste** has a great impact on toothpaste and mouthwash consumption and patients' tendency of buying a product more than once. Sugar can serve as a masking agent and flavoring agent for different types of formulations. However, it has been known for a long time that sugar consumption increases the incidence of dental caries. Therefore, companies manufacturing toothpaste do not use sugars for providing the formulation with a sweet taste. Instead of sugars, they use sugar derivatives, which have a very intensive sweet taste without contributing to the development and progression of dental caries. Ingredients that have been used for a long time include sugar alcohols, such as xylitol, erythritol, and sorbitol, as well as other, non-sugar-like ingredients, including aspartame, acesulfame, and saccharin.[34]
- One of the major concerns related to the use of fluoride toothpaste in children is the development of **fluorosis**. Dental fluorosis is a change in the appearance of the enamel. These changes can vary from barely noticeable white spots in mild forms to staining and pitting in the more severe forms. Dental fluorosis can occur only when younger children consume too much fluoride, from any source, over long periods when permanent teeth are developing under the gums. Only children aged 8 years and younger can develop fluorosis because this is when permanent teeth are developing under the gums.[35] Fluorosis does not develop after the teeth have erupted into the mouth. Most cases of fluorosis result from taking fluoride supplements or swallowing fluoride toothpaste when the water they drink is already fluoridated. There are anti-cavity toothpastes on the market for children 2 years of age and older. However, until about age 6, children have poor control of their swallowing reflex and frequently swallow most of the toothpaste placed on their brush. Parents should provide kids

with only a small (pea-sized) amount of toothpaste and make sure that children do not swallow it. Use of flavored toothpaste is not recommended since it may encourage swallowing. It should be emphasized that fluorosis is not a disease; it does not affect the function of the teeth. The spots and stains left on the teeth are permanent and may also darken over time; therefore, fluorosis is considered a cosmetic condition. Concerns about appearance can be addressed by whitening to remove the surface stains and/or other procedures to cover the discoloration.[36]

- **Teeth whitening** products are very popular today and widely used at home without any supervision from dental professionals. They may be helpful in removing discoloration; however, they may also cause side effects. Bleaching materials can affect the filling materials and may also result in color mismatch of the teeth with the existing fillings or crowns. In addition, they can lead to damage, including sensitivity, enamel loss, translucency, erosion, and gum damage.[37] In addition, bleaching discolored teeth in which the color change is the only visible indication of an underlying abnormality may change the tooth color, but will not remove any underlying abnormality. This masking effect can result in tooth loss or other complications, depending on the underlying condition since the true cause of discoloration may go untreated.

Many dental and oral care products are freely accessible in the US. Therefore, when consumers have dental issues, they will typically buy something and try to treat themselves instead of visiting a dental professional. It should be kept in mind that, while these products may provide temporary help, they may also hide the underlying problems. If a consumer has to apply dental products to maintain acceptable breath, for example, it is advisable to visit a dental office to see what the source of the dental issue may be.

REQUIRED QUALITIES AND CHARACTERISTICS AND CONSUMER NEEDS

From a consumer perspective, a quality oral and dental care product should possess the following characteristics:

- Pleasant taste and smell
- Provides a fresh and clean sensation after every use
- Removes debris, plaque, and stains
- Foams well
- Pleasant mouthfeel: non-sticky, and good texture
- Attractive appearance: with appropriate gloss, bubble-free, and homogenous color.

The technical qualities of oral and dental care products can be summarized as follows:

- Long-term stability
- Appropriate texture
- Appropriate rheological properties (the toothpaste can stand up on the toothbrush after it is extruded from the tube)
- Safety.

TOOTHPASTE

● Toothpaste is a mild cosmetic detergent for cleaning the teeth. Initially intended to freshen the breath and remove deposits from teeth, the evolution of toothpaste has also made it a vehicle for the protection of teeth from cavities, calculus formation, and gum diseases.

DID YOU KNOW?

The seal of acceptance provided by the American Dental Association (ADA) is an important symbol of a dental product's safety and effectiveness. It is evidence that the product has been objectively evaluated in extensive laboratory and clinical studies for its safety and effectiveness.[38]

Ingredients

There are countless formulations on the market, and we hardly can find two of them whose composition is the same. Manufacturers can choose from a great variety of ingredients; however, the basic ingredient types used are the same. This section reviews these ingredient types and their basic characteristics and functions.

Inactive Ingredients

- **Abrasives** are typically insoluble inorganic ingredients that clean and polish the teeth to remove debris and residual surface stains. When selecting abrasives, there are a number of factors that should be taken into consideration, including their hardness, toughness, chemical inertness, and particle size and shape. Abrasives should not damage the gums, enamel, and dentin. The hardness of the enamel is 5 on the Mohs scale, while that of the dentin is 3–4 (on this scale, talc is rated as 1 and diamond as 10).[39] Ideally, an abrasive's hardness should be below that of the dentin and enamel. Toughness refers to the ability of an abrasive to withstand forces during brushing. Some ingredients are too hard to be used as abrasives; however, if they break down to smaller pieces under the force during brushing, they may also be used. Abrasives should be chemically inert and compatible with additional ingredients. The degree of abrasivity and cleaning properties are greatly influenced by the particle size and shape. Larger particles feel gritty in the mouth and may damage the gums and enamel. A particle's shape should also be taken into consideration. Ideally, abrasives should be spherical with no sharp edges. The ability of an abrasive to remove stained particles will generally increase with its concentration in the toothpaste, with larger particle size and a higher level of hardness. The abrasivity of a given formulation, however, also varies according to the external forces, including the toothbrush, vigor with which it is used, and the characteristics of the individual dentition.

- Examples for abrasives used in toothpaste include hydrated silica; calcium phosphates, e.g. dicalcium phosphate dihydrate, and anhydrous dicalcium phosphate; calcium carbonate; disodium pyrophosphate; hydrated alumina (however, its use significantly decreased in the past 20 years); and sodium bicarbonate (baking soda).

DID YOU KNOW?

Calcium-based abrasives should be avoided in toothpastes containing sodium fluoride since the formation of an insoluble calcium salt decreases the active fluoride concentration. Calcium-containing abrasives are significantly more compatible with sodium monofluorophosphate than with sodium fluoride.[40]

- **Surfactants** aid in cleaning and foaming. Typically, anionic surfactants are used in toothpaste formulations since they have a powerful cleaning efficacy and are generally compatible with the other ingredients.
  - Examples for commonly used surfactants include sodium lauryl sulfate (the most widely used), sodium dodecylbenzene sulfonate, sodium lauroyl sarcosinate, sodium laureth phosphate, magnesium lauryl sulfate, sodium lauryl sulfoacetate, dioctyl sulfosuccinate, and monoglycerides. Originally, soap was used; however, it is strongly alkaline and, therefore, incompatible with some other components, and it has an unpleasant odor and a bitter taste.
- **Thickeners,** also known as binders, are hydrophilic colloids that disperse and swell in the water phase of the toothpaste. They are used to maintain the integral stability of the paste, prevent sedimentation of the abrasives, and provide toothpaste formulations with a specific rheological property. Due to the presence of thickeners, the toothpaste can stand on the brush and does not run into the bristles.
  - Examples include celluloses, such as sodium carboxymethyl cellulose, hydroxyethyl cellulose; alginates, such as sodium alginate; gums, such as xanthan gum; carrageenan (which is the generic name of thickening

agents derived from seaweed); carbomer; polyacrylates, clays (both natural processed bentonite and synthetic clays).

- **pH regulators** are typically used to neutralize thickeners.
  - An example is sodium hydroxide.
- **Humectants** prevent the paste from drying out (i.e., they retain moisture), locking of the cap to the nozzle, and hardening to an unacceptable level. At the same time, they provide shine and some plasticity to the paste.
  - Examples for humectants used in toothpaste include glycerin, sorbitol, and propylene glycol.
- **Water** is the vehicle of all toothpaste formulations. It is used to disperse the thickeners, dissolve the soluble components, mix the miscible liquids, and act as a carrier for all other ingredients.
- **Preservatives** prevent microbiological growth in these water-based formulations. Some preserving effects can be obtained by the flavoring oils; however, it is advisable to use preservatives.
  - Examples include parabens, sodium benzoate, and phenoxyethanol.
- **Sweetening agents** are important for product acceptance; the final product should be neither too sweet nor too salty or bitter. It is worth noting that sweeteners should always be considered in partnership with flavors because of their combined impact. Sweeteners are usually applied in a very small concentration due to their highly sweet taste. They can be utilized to mask the taste of the raw ingredients and not necessarily to provide only a sweet taste.
  - Examples include sodium saccharin (the most widely used), acesulfame, aspartame, and sorbitol.

DID YOU KNOW?

Although sweeteners may be hundreds to thousands of times sweeter than table sugar, they do not contribute to the development of cavities. The reason for this is that, unlike sugar, sweeteners are not fermented by the microflora of the dental plaque.[41]

- **Flavoring agents** are extremely complex, one of the most expensive parts and probably the most crucial part of a toothpaste because of consumer preferences. Conventionally, mint flavors tend to predominate in adult toothpastes, while bubble gum and fruit aromas in children products. Flavor is typically a blend of many suitable ingredients, including peppermint and spearmint oil, thymol, menthol, clove oil, cinnamon oil, eucalyptol oil, wintergreen oil, bubble gum as well as fruit aromas, such as mango, strawberry, apple, and banana flavor. Many others may also be used to achieve the desired flavor.

- **Coloring agents:** Color also has a huge impact on the consumers and their purchase intent. Their use is regulated by the FDA. A small amount of color may be added to the paste as a whole to provide a pleasant appearance, or it can be added to only a small part of the formulation to make colored striped products.
  - Examples include inorganic pigments, such as titanium dioxide and mica, as well as organic pigments, such as Blue 1 and Red 33.
- **Natural Ingredients:** Today, an increasing number of people are attracted to products that do not contain synthetic ingredients, such as sodium lauryl sulfate, fluorides, dyes, preservatives, or sweeteners. A number of natural products are available on the market, which include ingredients of natural source. These may be ingredients from the earth in purified form, such as hydrated silica; may be derived from herbs, such as echinacea, myrrh, calendula, chamomile, rosemary, witch hazel as well as essential oils; they may be of homeopathic origin or can be by-products of bees, such as propolis. Natural ingredients also have a lot of beneficial effects, including antibacterial, antiseptic, and antiinflammatory effects, cleansing, whitening, and masking of bad breath.
- **Anticaries components** are used to prevent cavities; they strengthen the tooth enamel and remineralize tooth decay. Two main types of anticavity actives are available on the market: fluoride and nonfluoride components, from which the fluoride-containing products' market share is about 95%. Fluoride components include sodium monofluorophosphate, sodium fluoride, and stannous fluoride. The mechanism of action of fluorides includes initiating and speeding up remineralization and slowing down demineralization.[42–44] Saliva in itself is able to supply the enamel with lost minerals, e.g. calcium and phosphate ions, when the pH is above 5.5 (to a certain extent). However, when fluoride is present in the saliva, this process is enhanced. In the presence of fluoride (from water or toothpaste), fluorohydroxyapatite forms in addition to hydroxyapatite, and this creates supersaturated solution outside the enamel. Therefore, calcium and phosphate ions are more efficiently recovered if fluoride is present in the saliva and bacterial biofilm.[42] This process is illustrated in Figure 6.5. In addition, fluorohydroxyapatite is less soluble than hydroxyapatite and is, therefore, more resistant to subsequent demineralization when acid challenged. As mentioned earlier, it is very important to keep the fluoride ions in a stable state. If insoluble salts are formed, the anticaries activity is lost.

Active Ingredients Active ingredients are used for additional benefits in toothpaste formulations. They prevent (or at least slow down) the development and progression of common dental problems.

In addition to these ingredients, there are non-fluoride caries ingredients available on the market. These include agents containing calcium, phosphorous (such as

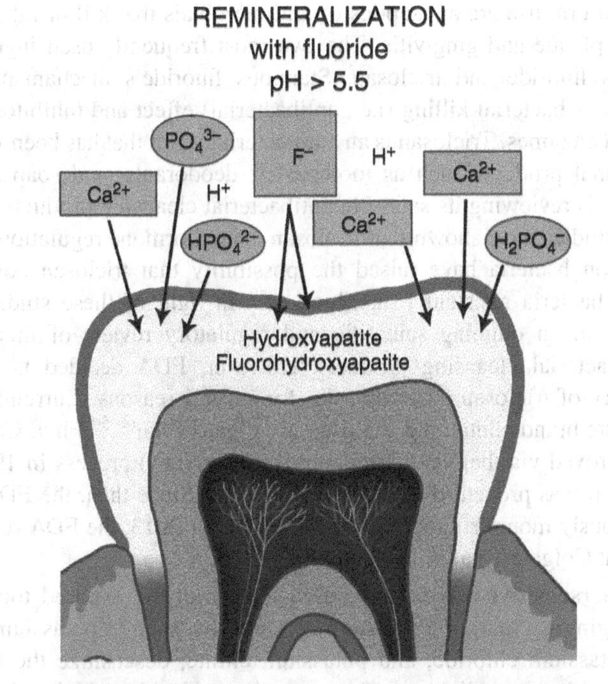

Figure 6.5 The process of remineralization with fluoride.

phosphates, trimetaphosphates, pyrophosphates, and glycerophosphates), as well as metals (such as zinc, tin, aluminum, iron, manganese, and molybdenum).

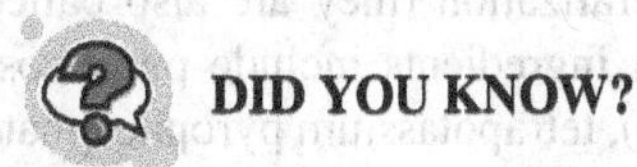

DID YOU KNOW?

Although there are some anti-cavity toothpastes that do not contain fluoride components, it should be kept in mind that, currently, all ADA-accepted toothpastes contain fluoride.[45]

The FDA allows toothpaste containing 850–1150 ppm theoretical total fluoride for children 2 years of age and older and adults, and those containing 1500 ppm total fluoride for children 6 years of age and older and adults.[46] Some prescription-strength fluoride toothpastes, which must be prescribed by a dentist, may contain even 5000 ppm of fluoride.

- **Antiplaque and antigingivitis components** are regulated according to the Advance Notice for Proposed Rulemaking (ANPR) by the FDA. It means that they do not have a final monograph yet (see more information on various types of monographs in Section 2 of Chapter 1). Most antiplaque and antigingivitis agents in oral use are antiseptics or antimicrobials that kill or inhibit bacteria causing plaque and gingivitis. The two most frequently used ingredients are stannous fluoride and triclosan. Stannous fluoride's mechanism of action involves its bacterial killing (i.e., antibacterial) effect and inhibitory effect on bacterial enzymes. Triclosan is an antibacterial agent that has been widely used in personal products such as toothpastes, deodorants, and soaps. Currently, the FDA is reviewing its safety in antibacterial cleansing products since some animal studies have shown that triclosan alters hormone regulation.[47,48] Other studies on bacteria have raised the possibility that triclosan contributes to making bacteria resistant to antibiotics.[49] In light of these studies, FDA is engaged in an ongoing scientific and regulatory review of this ingredient in antibacterial cleansing products. However, FDA decided not to review the safety of triclosan in toothpastes for several reasons. Currently, the only toothpaste brand containing triclosan is Colgate Total®.[50] First, Colgate Total was approved via the New Drug Application (NDA) process in 1997. In that process, it was proven to be safe and effective. Since then, the FDA has been continuously monitoring its safety. In December 2013, the FDA reaffirmed its view that Colgate Total is safe and effective.[51]

- **Antihypersensitivity agents** are used to protect the exposed tooth surfaces and the gingival margin. Potassium components, such as potassium nitrate (in 5%), potassium chloride, and potassium citrate, desensitize the tooth nerve endings. Other ingredients, such as strontium chloride (which is incompatible with fluoride though), stannous fluoride (in 0.454%), and calcium sodium phosphosilicate,[52] reduce the permeability of dentin and prevent the nerve endings to be stimulated.
- **Anticalculus/tartar control ingredients** prevent calculus formation on the surface of the teeth through controlling mineralization (they are also called crystal growth inhibitors). Examples for such ingredients include pyrophosphates, e.g., tetrasodium pyrophosphate (TSPP), tetrapotassium pyrophosphate (TKPP), disodium dihydrogen pyrophosphate; phosphonates, e.g., sodium hexametaphosphate (SHMP); zinc salts, e.g., zinc citrate, zinc chloride, and zinc lactate; as well as a copolymer of methyl vinyl ether and maleic anhydride (PVM/MA).
- **Whitening Ingredients:** Over the past two decades, tooth whitening or bleaching has become one of the most popular esthetic dental treatments. Whitening can be achieved in two ways: by using non-bleaching ingredients and bleaching ingredients.
  - **Non-bleaching** whitening products contain agents that work by physical or chemical action to help remove surface stains only. Any toothpaste containing abrasives provide whitening action as they remove stains from the teeth. For more efficient results, chemical action may also be utilized where positively charged ingredients (e.g., SHMP) bind to the negatively charged stain molecules.

- **Bleaching** agents actually change the natural tooth color. They contain peroxides, such as hydrogen peroxide and carbamide peroxide, that help remove deep (intrinsic) and surface (extrinsic) stains.

■ **Antimalodor Ingredients:** As mentioned previously, regular brushing of the teeth and tongue, flossing, and using a mouthwash greatly improve bad breath. When these basic hygienic aids are not enough, specific formulations can be used. Ingredients used include flavors to freshen breath by masking the bad odor. They have only a temporary action since as saliva relatively quickly washes away the flavoring ingredients. Antibacterial agents are also used to kill the bacteria causing bad breath. They offer a longer-term benefit, and, opposed to flavors, they can actually treat the source of the problem. Such antibacterial ingredients include triclosan, zinc, and stannous fluoride.

Rheology of Toothpaste Formulations

❺ **Toothpaste is a non-Newtonian fluid that has a special rheological behavior, and is called a "Bingham plastic."** As you may have noticed, you have to press the toothpaste tube to get the toothpaste out of the tube, and if you just simply turn it upside down, it will not flow. Bingham plastics behave as a solid (which does not flow) at rest. However, when a minimum force is applied on them, they start moving as a viscous fluid. The minimum force needed to flow is referred to as the "yield stress." As soon as the force (i.e., squeezing the tube) is stopped, they turn back into a solid form. This property makes it possible that the toothpaste does not flow out from an open container if no pressure is used and can "stand up" on the brush without running into the bristles of the brush or rolling off the sides. A rheogram characteristic for Bingham plastics is shown in Figure 6.6. It can be seen that the curve does not pass through the origin but intersects with the shear stress axis. This implies that a plastic material does not flow until such a value of shear stress has been exceeded. These formulations usually have a high viscosity with a relatively high yield point, although the necessary pressure to "extrude" the paste is not too high; even a child should be able to squeeze the paste out. The rheology of the paste greatly influences the dispersibility of the paste in the mouth, the generation of foam, and, above all, the release of the flavor components. However, this property may cause difficulties from a manufacturing point of view since as emptying the containers or starting to pump the "liquid" (which is really a solid) is quite difficult.[53]

It should be kept in mind that this phenomenon is different from thixotropy that we have seen in nail polish formulations (refer back to Section 4 of Chapter 4). In the case of thixotropy, the system is able to flow at any time; however, on higher shear stress, the viscosity decreases as the structure is gradually broken down, and it tends to increase after the shear stress stops and the structure reforms. However, a Bingham plastic acts as a solid material, which does not flow at low shear stress. It starts moving as the yield value is reached, which means that the structure has been destroyed. The yield value is like a borderline between the solid and the liquid state.

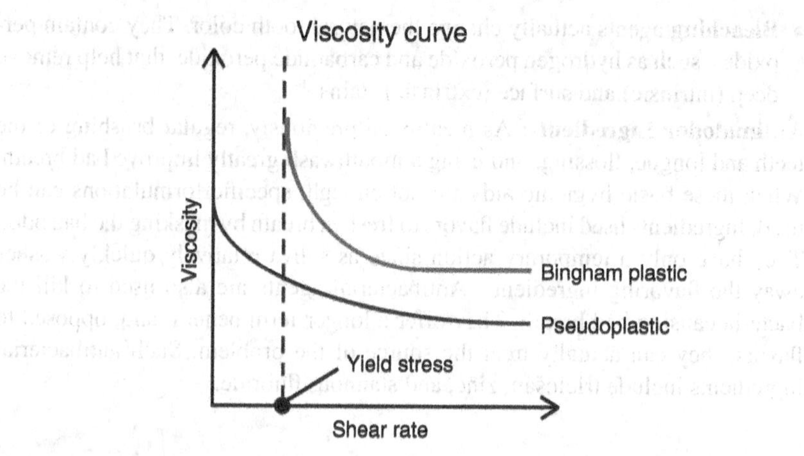

Figure 6.6 Viscosity curve characteristic of Bingham plastic materials and pseudoplastic materials.

There are gel toothpaste formulations available on the market that start flowing at any time. These are **pseudoplastic** (also known as shear thinning) materials, similar to most cosmetics and personal care products (see Figure 6.6). A shear thinning material stays viscous until you apply a force (such as squeezing the bottle) that makes it less viscous and flowable. As soon as the shear force is removed, the viscosity reverts to its original high value. The reason for this shear thinning behavior is that an increase in shear rate deforms and/or rearranges the particles, resulting in lower flow resistance and, consequently, lower viscosity. They recover their viscosity virtually and instantaneously when shear is eliminated; therefore, this behavior is called a time-independent behavior. The shear rate dependency of the viscosity can differ substantially between products as well as for a given liquid, depending on the temperature and concentration. In such cases, there is no yield stress value that should be reached and exceeded to make the material flow.

Formulation of Toothpaste

Toothpastes can be either white (or colored opaque) pastes or clear (typically colored) gels. In generally, pastes contain a higher concentration of abrasives (40–50%) and a lower concentration of humectants, whereas gels contain a higher concentration of humectants and a smaller amount of abrasives (10–15%). Gels can be translucent or transparent, depending on how close the refractive indices of the abrasive and humectant are to each other and on other factors.[54]

The processing methods vary depending on the product type and ingredients used. A method could be outlined as follows[55]: the gelling agent is fully hydrated in water. Heating may be necessary depending on the type of thickener used. The thickener can be prewetted with a nonaqueous liquid ingredient such as glycerin or a flavoring oil to aid dispersion, or can be directly added to water. It should be kept in mind that certain thickeners need neutralization in order to reach their optimal viscosity. Water-soluble ingredients, such as sweeteners and actives (except for the surfactants), are dissolved in part of the water and are then mixed with the liquid phase. All additional liquid compounds (except for the flavors) are added to the thickened liquid phase after hydration is completed. Abrasives, which are generally non-water-soluble powder ingredients, may also be prewetted with humectants, e.g., glycerin or sorbitol, and are then mixed with the thickened liquid phase until a uniform paste is formed. It should be taken into account that some ingredients, e.g., hydrated silica, have a low density and are very difficult to incorporate and wet out. At this stage, air incorporation should be avoided, which may lead to the formation of air bubbles in the paste. Surfactants and flavors are usually added in the last step under slow-speed agitation to minimize foaming.

Another method can be used when all powder components, including the abrasives, gelling agents, actives, sweeteners, and colors, are mixed to obtain a uniform mixture. In the following step, this powder mix is blended with all liquids in the toothpaste, including water and humectant. Following complete hydration of the gelling agent, a homogenous paste will be obtained.[56]

It is also feasible for certain formulas to first dissolve the surfactant in water. Prewetting the surfactant with the humectant may be a good option to help the dissolution process. Sodium lauryl sulfate needs heat in order to dissolve, and if it is added first, formulators do not have to be concerned about the heat sensitivity of other ingredients. Abrasive powders may be added next in vacuum to remove any air present in the powder, avoiding bubble formation. Thickeners can be added in a final step, preferably after preblending them with the flavor that will shorten their hydration time.[57]

DID YOU KNOW?

Sodium lauryl sulfate, one of the most frequently used surfactants, can be highly irritative to the eyes, oral cavity, and lungs. Therefore, extreme care should be provided when this ingredient is added to the toothpaste formulation.

A primary concern when formulating a dentifrice is the need to prevent inactivation of the therapeutic agents by other ingredients. Other concerns involve the degradation or inactivation of ingredients by water. Certain ingredients may be hydrolyzed by water into individual ingredients that will not perform as expected. Aeration is an additional concern in toothpaste manufacture. All powders contain a certain amount of air, and the surfactant can exacerbate the problem. Mixing is usually carried out in vacuum to overcome this; however, deaeration of the system takes a long time. If air is not removed, a mousse-like consistency will be produced, which is not preferred.

MOUTHWASH

Mouthwash is a clear, most of the times, colored solution that is aimed to refresh the breath by swishing the product around the mouth, followed by spitting it out. Similar to toothpaste, it may also have additional benefits, such as prevention against tooth decay, gingivitis, plaque formation, or tartar formation, or a combination of these. It is usually a hydroalcoholic solution in which flavors, essential oils, and other agents are combined to provide long-term breath deodorization. Palatability can be improved by including a polyhydric alcohol, such as glycerin or sorbitol. Anionic and nonionic surfactants can also be used to help solubilize the flavors and help remove debris and bacteria from the mouth.

Ingredients

Mouthwash is the simplest type of formulation, and it is a simple solution. The ingredients may vary depending on the particular formulation and claims; however, the basic ingredients are the same.

Inactive Ingredients

- **Solvents** function as a vehicle for the other ingredients. Generally, two main solvents are used in mouthwash formulations, water and alcohol (i.e., ethanol). Alcohol has an antibacterial activity, acts as an astringent, and contributes to the fresh feeling provided by the product. It may also help stabilize the product by solubilizing the flavoring oils.
- **Humectants** increase the viscosity of the product and result in a good mouthfeel. They also inhibit "crystallization" around the closure. Without these ingredients, the products would have a harsh chemical-like taste/feel. They may also contribute to the sweetness.
  - Examples for frequently used humectants include glycerin and sorbitol.

DID YOU KNOW?

Have you experienced that a solid aggregate forms around the neck of the mouthwash bottle while using it for weeks or months? It is due to the evaporation of water and alcohol from the formulation and all the remaining ingredients, including sugar crystals.

- **Surfactants** are used in mouthwash formulations to solubilize the flavoring oils and stabilize the products. They can provide foaming action on use, with this action they contribute to the cleansing effect of the product.

  - Examples include nonionic surfactants, such as poloxamers (polyoxyethylenated polyoxypropylene nonionic block polymers) and polysorbates, such as polysorbate 80, as well as anionic components, such as sodium methyl cocoyl taurate and sodium lauryl sulfate.
- **Astringents** can serve as temporary deodorizers that mask bad breath.
  - Examples include zinc chloride, ethanol, and witch hazel.
- **Preservatives:** Although alcohol has an antibacterial activity, its antibacterial profile may not cover all microorganisms that frequently contaminate such formulations. Mouthwashes are basically water/humectants systems, which is a perfect environment for microbial growth. Therefore, most formulations contain additional preservatives as well.
  - Examples include ethanol; benzoates, such as benzoic acid and sodium benzoate; and parabens.
- **Flavoring Agents:** As with toothpastes, taste has a key importance from a consumer perspective. It also contributes to the refreshing effect of the product and provides a pleasant note over the breath aroma. Some flavors may also have antibacterial action.
  - Typically, flavor blends are used for flavoring. Examples for ingredients include mint, menthol, peppermint oil, eucalyptol, methyl salicylate, thymol, and bubble gum. Many flavors listed under toothpaste formulations could also be taken as examples for mouthwash formulations.
- **Sweeteners** are also usually added to adjust the taste of the formulations. See examples under toothpaste formulations.
- **Colorants** are also an important part of mouthwashes. Only a tiny amount of water-soluble dyes are added. Colors may vary from blue to green to purple or can also be yellow and red.
- Similar to toothpastes, **natural ingredients** can also be added for extra benefits or for substituting synthetic ingredients.

Active Ingredients For therapeutic claims, various active ingredients can be incorporated into mouthwashes. The major types are the same as those discussed under toothpastes, including antibacterial, anticavity, antihypersensitivity, antiplaque, antitartar, and whitening ingredients. The only condition is that they must be water-soluble.

Examples for the main functional ingredients are summarized as follows:

- **Anticaries Agents:** sodium fluoride, stannous fluoride, and sodium monofluorophosphate
- **Antihypersensitivity Agents:** potassium nitrate and strontium chloride
- **Antiplaque/Antigingivitis Agents:** Chlorhexidine (which is a very effective agent to reduce plaque formation and gingivitis; its disadvantages include staining of the teeth, taste modification, and increased calculus formulation); essential oils; cetylpyridinium chloride (its disadvantage is that anionic surfactants

inactivate it, so it is usually not recommended to use after brushing); and triclosan

- **Antitartar Agents:** Pyrophosphates, PVM/MA
- **Whitening Agents:** Non-bleaching agents, such as SHMP, and bleaching agents, such as peroxides.

Formulation of Mouthwash

As mouthwash is a simple flavored, colored solution, which may contain essential oils in a solubilized form, formulation is a simple solution preparation process. Ingredients are dissolved in either water or alcohol, based on their solubility, and then the water and alcohol phases are mixed together.

OTHER PRODUCTS

In addition to toothpaste and mouthwash, there are other products with cosmetic benefits available on the market. In this section, we will review the basic ingredients and main characteristics of whitening aids, tooth powders, and dental flosses.

Tooth Whitening Aids

The use of hydrogen peroxide for tooth whitening can be traced back to more than a century.[58] Tooth bleaching is an oxidative process that alters the light absorbing or light reflecting nature of the tooth structure, increasing its perception of whiteness. The generally accepted mechanism involved in tooth bleaching is similar to that in textile and paper bleaching: free radicals produced by hydrogen peroxide interact with pigment molecules to produce a whitening effect.[59] Enamel dehydration during bleaching may also result in a temporary whitening effect as enamel dehydration alone is capable of producing a significant, visible tooth shade reduction.[60]

Over the past decades, the tooth whitening market has evolved into the following four categories in the US:

- Professionally applied products (in the dental office)
- Dentist-prescribed/dispensed products (used at home by patients)
- Consumer-purchased/OTC products (applied by patients)
- Other non-dental options, such as treatment offered in salons, spas, and even cruise ships.

Additionally, dentist-dispensed bleaching materials are sometimes used at home after dental office bleaching to maintain or improve the whitening results. Currently, all surface tooth bleaching products remain unclassified by the FDA; all of them are regulated as cosmetics. This includes all peroxide-based products used in the in-office, dentist-dispensed products for at-home use, patient-purchased products, as well as products used in non-dental settings.

Whitening products are available in the US in various dosage forms, such as gels, solutions, chewing gums, toothpaste, and strips.

- Gels and solutions are typically applied using a custom bleaching tray. It is inserted into the mouth after brushing the teeth for a shorter (half an hour) or a longer (even 8–10 h) period of time.
- Strips are usually made of polyethylene and polypropylene polymers coated with an adhesive gel on one side. As the strip is placed on the teeth, the gelling agents, such as carbomer, polyvinylpyrrolidine (PVP), and acrylate polymers, start swelling, providing adhesiveness and appropriate coverage on the tooth surface. The strips are placed on the teeth for about half an hour.

Most OTC bleaching products in the US are hydrogen peroxide based, although some contain carbamide peroxide. Hydrogen peroxide is a highly reactive chemical; it is a strong oxidant and can form free radicals. It is available on the market as an aqueous solution containing up to 33–37% pure hydrogen peroxide and other additives to prevent product decomposition. Carbamide peroxide decomposes to release hydrogen peroxide in an aqueous medium: 10% carbamide peroxide yields approximately 3.5% hydrogen peroxide.[61] In-office bleaching materials contain high hydrogen peroxide concentrations (typically 25–40%), while the hydrogen peroxide content in at-home bleaching products usually ranges from 3% to 10%; however, there have been home-use products containing up to 15% hydrogen peroxide.

DID YOU KNOW?

In the EU, oral hygiene products may be sold freely to consumers only if they contain no more than 0.1% hydrogen peroxide, whereas in the US, whitening products sold to consumers may contain much higher concentrations. In the EU, products containing hydrogen peroxide in a concentration higher than 0.1% can only be sold to dentists. The highest hydrogen peroxide concentration permitted in the EU is 6%, while in the US products may contain up to 40% hydrogen peroxide US.[62]

Although published studies tend to suggest that most in-office, dentist-prescribed, and at-home bleaching techniques are relatively safe and effective procedures, there are some concerns on the long-term safety of unsupervised bleaching procedures. Adverse reactions on the tooth's hard tissue, soft tissue, mucosal tissues of the mouth, and restorative materials are still reported.[63,64] Most frequent side effects of teeth whitening include transient mild-to-moderate tooth sensitivity, gingival irritation, and release of some components of fillings.[64] Tooth sensitivity and gum irritation are mild to moderate and transient in most of the cases. Gels of high hydrogen peroxide concentrations, such as those for in-office bleaching, should be used with appropriate gingival protection to avoid severe damage.

DID YOU KNOW?

Skin contact with moderate concentrations of hydrogen peroxide will cause whitening of the skin and stinging sensations. The whitening is due to the formation of gas bubbles in the epidermal layer of the skin. Stinging, in most cases, subsides quickly after thorough washing, and the skin gradually returns to normal without any damage.

Tooth bleaching in non-dental settings is of particular concern.[65] Bleaching offered in a mall kiosk, salons, spas, and cruise ships may present the image of a dental practice and professional supervision without providing the benefits of care from fully trained and licensed oral health-care providers. It is advisable to consult with a dental professional before starting any whitening procedure, and it is best to have tooth bleaching performed under professional supervision.

Tooth Powders

Tooth powder was the original commercial product sold for cleaning the teeth. It is still available today and can be used as an alternative for toothpaste. Tooth powder is a physical mix of dental abrasive, flavor, and foaming aid, which is used with a toothbrush. Although efficacious and perhaps cost-effective, powders are certainly not easy to use since they are difficult to apply to the brush, and they have several drawbacks both from a consumer perspective and in terms of the incorporation/dosage of ingredients such as fluoride. Many customers say that a tooth powder does not provide a pleasant brushing experience and the foaming activity is not appropriate either. As a result of all these factors, tooth powder application has diminished compared to toothpastes.

The method of manufacture is a simple bulk powder formulation mixing process. If a coloring material is used, it is usually added as a solution.

Dental Floss

Dental floss is a thin filament used to clean the areas in the oral cavity, which are more difficult to reach with a toothbrush (such as the surface between the teeth as well as the teeth and the gums). Dentists recommend flossing as a critical part of any daily oral care routine. Dental floss is generally recommended to use before brushing as the fluoride (from this way the toothpaste) has a better chance of reaching between the teeth.

Dental floss was once made from silk fibers twisted to form a long strand. Today, floss is usually made from waxed/non-waxed nylon, Teflon filaments, or plastic (e.g., polyethylene) monofilaments. It may be treated with flavoring agents, such as mint, to make flossing more pleasant. Waxes help introduce the floss between the teeth. Examples include beeswax, carnauba wax, and jojoba wax. Typical flavors include peppermint and spearmint. Even active ingredients may be deposited on the strings, including antitartar ingredients and fluoride. In addition, natural ingredients, such as acacia, propolis, and myrrh, may also be deposited on the floss.

Dental tape is similar to dental floss. It is also a strip of waxed/non-waxed nylon or polyethylene fiber. The only difference is in the shape of the filaments. Floss has a round cross section, and the tape has a very thin, almost rectangular cross section that helps clean wider spaces between the teeth.

The manufacture of dental floss and dental tape requires special equipment and a unique technique. First, the filaments are formed, stretched, and then put on large bobbins. Correct stretching is a vital part of the process since it adds strength to the floss as well as greatly reduces fraying or breaking. As dental floss is used in the mouth, it has to be hygienic. First, it is washed to remove any dirt. Coating takes place after the filaments are washed. First, a layer of wax is applied to the filaments, and then peppermint powder and other desired ingredients are sprayed onto the surface of the filaments. Then the filaments are put back onto the large bobbins. The floss is wrapped around tiny spools. The inner plastic rolls are placed inside the plastic containers, and then floss bobbins are inserted into the container.

TYPICAL QUALITY PROBLEMS OF ORAL AND DENTAL CARE PRODUCTS

❼ **The typical quality-related issues of oral and dental care products include hardening, inappropriate viscosity, grittiness, cloudy solution, microbiological contamination, and clumping.** The potential causes as well as solutions of these issues are discussed here. The quality issues that were discussed previously are not reviewed here.

Hardening

Hardening means that extruding toothpaste from the tube is more difficult than usual, and it needs an abnormally high pressure, and the consistency of the paste is not adequate. The cause of this problem is usually that the concentration of the humectant incorporated is not high enough. It cannot prevent water evaporation, which leads to its drying up during everyday use. In such cases, the formulation should be reconsidered and redesigned.

Inappropriate Viscosity

Viscosity is an important factor for appropriate use of toothpastes. If thickeners are not applied in a necessarily high concentration, the product cannot "hold" its shape on the toothbrush and may run into the bristles of the brush. On the other side, if the toothpaste is too thick, it cannot be easily extruded from the tube and will not spread easily in the mouth. To solve such problems, the type and amount of ingredients affecting the viscosity, including thickeners and abrasives, should be reconsidered.

Grittiness

If the hardness, particle size, and shape of abrasives are not appropriate, they may feel gritty in the mouth and may even damage the gums and mucosa. This problem can be addressed by monitoring the abrasivity of ingredients used and using abrasives with a recommended hardness and particle size.

Cloudy Solution

Mouthwash may contain essential oils as antiseptics and flavoring ingredients. These oily components are typically in a solubilized state provided by surfactants; therefore, the mouthwash can still be clear. If the type and/or amount of the solubilizers is not appropriate, the solution may be opaque to a certain extent.

EVALUATION OF DENTAL AND ORAL CARE PRODUCTS

Quality Parameters Generally Tested

❽ **Parameters commonly tested to evaluate the quality of oral and dental care products include spreadability and extrudability of toothpastes and toothpaste gels, foaming property, flavor, preservative efficacy, viscosity, color, and pH**. The range of acceptance and other limiting factors are usually determined by the individual manufacturers. The tests that were reviewed in the previous sections are not discussed here.

Flavor Flavor is one of the most important factors having a great impact on costumers' general impression as well as the choice of the product. The taste of a toothpaste, tooth powder, or mouthwash is usually tested organoleptically by a group of people, who fill in a form after brushing their teeth with the test toothpaste.

Viscosity Viscosity is an important property since it may have a large effect on the applicability of the product. It may be influenced by several factors, including the binder concentration, solid/liquid ratio, and humectant/water ratio. Evaluation of the yield stress is a critically important part of quality testing of toothpaste. Testing is usually performed using rheometers; for more detail, refer back to Section 2 of Chapter 3.

Efficacy (Performance) Parameters Generally Tested

❾ **The most commonly tested efficacy parameters include antimicrobial activity, whitening activity, and hypersensitivity reduction.** These tests are discussed here.

Claims made for the therapeutic effect or performance of toothpaste and mouthwash formulations can include whitening, caries prevention, prevention of tartar formation, killing bacteria causing bad breath, desensitizing, freshening, and many others. Depending on the mechanism of action of the ingredients used, various types of tests can be performed.

- **Antimicrobial activity** can be tested *in vitro* and *in vivo*. *In vitro* test is typically performed using oral pathogens, such as *Streptococcus mutans, Peptostreptococcus micros, Lactobacillus casei, Streptococcus salivarius*, and *Streptococcus oralis*. A generally used antimicrobial assay is the agar well diffusion method. It includes the preparation of the culture media, agar in this case. The media is then inoculated with the sample microorganism of interest. Wells are made on the agar gel's surface, and the properly diluted toothpaste or mouthwash sample is filled into the holes. The plates are incubated at a higher temperature (e.g., 35 °C) for a day or two before evaluating the results. Antimicrobial activity is evaluated by measuring the diameter of the zones of inhibition. *In vivo* testing includes human volunteers who use toothpaste or mouthwash formulations as instructed for a predetermined period of time. Before, during, and after the study, saliva samples are taken from their oral cavity, the samples are inoculated on various plates, either aerobically or anaerobically, and the microbial colonies are counted and compared.[66]
- **Whitening activity** can also be tested *in vitro*, on the teeth removed from animals, for example. The sample teeth can be stained with colored drinks known for their staining power, for example, coffee. The whitening activity of different whitening products can be easily tracked and compared this way. Color is evaluated using a spectrophotometer, colorimeter, or by image analysis before, during, and after the test.[67] Such tests can be performed on human volunteers (*in vivo*) divided into various groups. Selection of subjects, standards on using whitening products as well as other dental and oral care products, and data evaluation must be carefully planned and performed for reliable results.
- There are certain claims, such as **hypersensitivity reduction**, that are better tested on human subjects. Hypersensitivity is uncomfortable due to the pain caused when exposed to hot, cold, spicy, sweet, and other types of triggering factors. If a hypersensitive tooth was removed, it would not have a direct contact with the nerves; therefore, it would not behave the same way as a living tooth. The evaluation of these products is generally conducted by dental professionals in a clinical study that would determine the efficacy of a desensitizing product compared to a placebo, negative, or positive control.[68]

Ingredients Causing Safety Concern

Toothpaste and mouthwash are not intended for ingestion; however, they may be accidentally swallowed by the consumers. Children are of particular concern because toothpaste for children is colored and flavored to enhance their brushing activity. They often swallow the paste during and after brushing their teeth. ⑩ **Ingredients causing safety concerns regarding oral and dental care products include fluoride and hydrogen peroxide.** The findings and current statements on their safety are summarized in the following section.

Fluoride The biggest issue with accidentally or intentionally swallowed toothpaste is its fluoride content. Fluoride is an effective active ingredient when used topically; however, even relatively small doses can induce symptoms of acute fluoride toxicity if consumed. The probably toxic dose of fluoride is generally accepted to be 5 mg fluoride ion per kg body weight.[69] Probably, toxic dose means the dose that could cause toxic signs and symptoms, including death, and that should trigger immediate therapeutic intervention and hospitalization.[70] Early symptoms include gastrointestinal pain, nausea, vomiting, and headaches. It should be noted, however, that toxicity symptoms also occurred at much lower doses (0.3 mg).[71]

DID YOU KNOW?

According to the FDA's recommendations, children under the age of 6 should use only a pea-sized amount of toothpaste. The ratio of a pea-sized amount of toothpaste related to a toothbrush is shown in Figure 6.7. A pea-sized toothpaste typically weighs about 0.25 g (0.0088 oz.).[72] The minimum dose of fluoride at which a toddler weighing 10 kg (22 lbs) would experience toxicity symptoms is 50 mg. It means that a 10 kg toddler, when using toothpaste containing an average concentration of fluoride (0.15% of fluoride), would need to ingest approximately 32 g of the toothpaste (approximately 130 pea-sized amount) to experience one or more of the symptoms. It sounds a large amount of toothpaste; however, it may be an amount as small as one-third of a tube.[73]

Figure 6.7 Toothbrush with pea-sized drop of toothpaste. .Adapted from the National Institute of Dental and Craniofacial Research.

As discussed previously, fluoride intake may increase the risk of fluorosis in young children whose permanent teeth are developing in the gum. To prevent accidental fluoride consumption from toothpaste, the FDA requires the following warning label to be placed on all fluoride toothpaste stating the following: "Keep out of reach of children under 6 years of age. If more than used for brushing is accidentally swallowed, get medical help or contact a Poison Control Center right away."

Hydrogen Peroxide Another ingredient causing safety concerns among patients and professionals is hydrogen peroxide. As discussed earlier, hydrogen peroxide is used in tooth whitening (bleaching) products. Overuse may result in severe damage to the teeth and gum; however, more severe damage is caused by accidental consumption (swallowing of a whitening solution, for example). As discussed, hydrogen peroxide may dehydrate the enamel, which can be accompanied by tooth sensitivity. It can also damage the gumline if it is not shielded, although these effects are temporary and resolve within a couple of days. Ingestion of hydrogen peroxide may cause irritation of the gastrointestinal tract with nausea, vomiting, blood vomiting, and foaming in the mouth; the foam may obstruct the respiratory tract or result in pulmonary aspiration. Painful gastric distension and belching may be caused by the liberation of large volumes of oxygen in the stomach. It is very important, therefore, to make sure that children do not use such formulations, and even adults have to pay attention not to swallow gels or solutions after a whitening treatment.[74]

PACKAGING OF DENTAL AND ORAL CARE PRODUCTS

According to the FDA, liquid oral hygiene products (e.g., mouthwashes and fresheners) must be packaged in **tamper-resistant** packages when sold at retail. A package is considered tamper resistant if it has an indicator or barrier to entry (e.g., shrink or tape seal, sealed carton, tube or pouch, aerosol container), which, if breached or missing, alerts a consumer that tampering has occurred. The indicator must be distinctive by design (breakable cap, blister) or appearance (logo, vignette, other illustration) to preclude substitution. The tamper-resistant feature may involve the immediate or outer container or both. The package must also bear a prominently placed statement alerting the consumer to the tamper-resistant feature. This statement must remain unaffected if the tamper-resistant feature is breached or missing.[75]

The most commonly used packaging materials for oral and dental products include the following:

- **Plastic Tube and Dispenser:** Most toothpastes are marketed in soft, plastic, so-called collapsible tubes supplied with a removable or a nonremovable (flip-flop) cap. This is the classic container that has been used for many years. The volume of the tubes may vary; however, their shape is the same. Some toothpastes are available in standing plastic dispensers where a piston is inside the dispenser. It lifts the paste during application and hermetically seals the container, preventing drying up and loss of the paste. The movement (open–close) of the cap controls the dosing, and it prevents drying out at the same time as it is closed. It is a special and useful dispenser; however, only 1–2% of the toothpastes are available in such a container.

- **Dental Floss Plastic Dispenser:** The standard case for dental flosses is an angled one- or two-piece construction, with an insert that holds the floss spool. Usually, a tiny blade is part of the plastic housing, which helps cut a piece of floss with a desired length during application.
- **Plastic Bottle:** Mouthwashes are typically marketed in plastic bottles that have a twisting cap. The cap also serves as the measuring device; it has a scale on its inner side and helps dose the product. Some toothpaste gels are also supplied in plastic bottles with flip-flop caps.

GLOSSARY OF TERMS FOR CHAPTER 6

ADA seal of acceptance: A designation awarded to products that have met ADA's criteria for safety and effectiveness and whose packaging and advertising claims are scientifically supported.

ADA: American Dental Association.

Bingham plastic: A non-Newtonian rheological behavior characteristic to toothpaste. A Bingham plastic behaves as a solid at rest, but flows as a liquid when force is applied to it.

Cariogenic: An ingredient that can cause tooth decay.

Cementum: A special connective tissue covering the dentin at the root of the teeth. It binds the roots of the teeth firmly to the gums and jawbone.

Demineralization: The process of losing calcium and phosphate from the enamel.

Dental caries: Tooth decay, a common dental problem resulting in the breakdown and destruction of the enamel and appears as tiny brown holes on the surface of the teeth.

Dental floss: A personal care product designed to clean the areas in the oral cavity that are difficult to reach with a brush.

Dentin: A yellowish bone-like layer surrounding the dental cavity. At the crown, it is covered by enamel. At the root, dentin is covered by the cementum.

Enamel: The white outer visible part of the tooth, the hardest substance in the human body. It is mostly composed of hydroxyapatite.

Fluorosis: A condition caused by excessive intake of fluoride in children (under the age of 8 years). It changes the appearance of the enamel.

Gingivitis: Inflammation of the gums, the early stage of gum disease.

Gum disease: Periodontal disease, infection of the gums that can progress to affect the bone that surrounds and supports the teeth.

Halitosis: Bad breath.

Mouthwash: A personal care product designed to refresh the breath and enhance oral hygiene.

Plaque: A sticky, colorless film of bacteria and sugars that constantly forms on our teeth.

Pseudoplastic: A non-Newtonian rheological behavior characteristic of toothpaste gel (and most cosmetic products). A pseudoplastic (shear thinning) ingredient is viscous until a forces is applied, which makes it less viscous and flowable.

Remineralization: The process of replacing the lost calcium and phosphate to the enamel.

Tartar: Dental calculus, a common dental problem caused by hardening of plaques on the teeth, which leads to a yellowish-brownish discoloration and can irritate the gums. It can be removed only by professionals.

Tooth sensitivity: A dental condition characterized by increased sensitivity to various stimuli, including hot and cold, brushing, and air currents.

Toothpaste: A personal care product designed to help remove plaque and stain from the teeth and keep the breath fresh.

Yield stress: The minimum force needed to start a Bingham plastic ingredient flow.

REVIEW QUESTIONS FOR CHAPTER 6

Multiple Choice Questions

1. Which of the following is TRUE about the enamel?
   a) It has a blood and nerve supply
   b) It is the largest part of the human teeth
   c) It is comprised mostly of hydroxyapatite
   d) It is a soft tissue

2. What does the ADA seal of acceptance mean?
   a) The product is approved by the FDA
   b) The safety and efficacy have been evaluated in laboratory and clinical studies
   c) The packaging is tamper resistant
   d) The product is widely accepted by consumers

3. What is the function of abrasives in toothpaste?
   a) Exfoliate dead skin cells
   b) Remove debris and stain
   c) Remove tartar
   d) All of the above

4. Which of the following ingredients CANNOT be used as thickeners in toothpaste?
   a) Waxes
   b) Gums
   c) Cellulose derivatives
   d) All of the above

5. Which of the following is NOT true for sodium monofluorophosphate?
   a) Can be used between the ages of 2 and 6
   b) Forms an insoluble precipitate if used with abrasives-containing calcium
   c) May lead to fluorosis if too much is consumed around the age of 6–8
   d) Anticavities active ingredient

6. Which of the following is used to determine the easiness of squeezing toothpaste out of a tube?
   a) Drop test
   b) Scratch test
   c) Abrasion test
   d) Extrusion test

7. Toothpaste has a special rheology called ___.
   a) Shear-thickening
   b) Thixotropy
   c) Bingham plastic behavior
   d) Newtonian behavior

8. The main advantage of using artificial sweeteners in toothpaste instead of sugars is that ___.
   a) They do not contribute to the development of cavities
   b) They are sweet
   c) They provide protection against drying out of the formula
   d) They help maintain the integral structure

9. Which of the following is a cause of dental caries?
   a) Low pH in the mouth (below 5.5)
   b) Sugar
   c) *Streptococcus mutans*
   d) All of the above

10. What is the function of humectants in toothpaste and mouthwash?
    a) To provide protection against microbial contamination
    b) To moisturize the oral cavity
    c) To provide protection against water loss and drying
    d) To decrease the water content of the formulas

11. What is the typical cause of grittiness in toothpaste?
    a) Water loss from the product
    b) Inappropriate amount and/or type(s) of thickener
    c) Particle size and/or shape of abrasives
    d) Microbiological contamination

12. What is the main safety concern with regard to the use of fluoride in toothpaste?
   a) Consumption can be toxic in a larger dose
   b) Inhalation can be unsafe in a larger dose
   c) Discolors teeth
   d) Causes a bad breath

13. Sodium monofluorophosphate can prevent tooth decay via the following mechanism:
   a) Slowing down demineralization
   b) Eliminating the bacteria in the oral cavity
   c) Forming fluorohydroxyapatite in the enamel
   d) a and b

14. Which of the following is the main application surface for toothpaste in the case of healthy teeth?
   a) Pulp
   b) Enamel
   c) Dentin
   d) Tongue

15. The term "yield stress" refers to ____.
   a) The minimum force needed to be applied to the teeth with the toothbrush during brushing
   b) The minimum force needed to be applied in order to get the toothpaste out of its tube
   c) The stress needed to be applied during the manufacturing process of toothpaste
   d) The maximum force needed to be applied in order to prevent clumping of toothpaste gels

Drug or Cosmetic in the US?

a) Anticaries toothpaste with sodium fluoride: ____________________
b) Mint-flavored dental floss: ____________________
c) Teeth whitening strip: ____________________
d) Mouthwash for masking bad breath: ____________________

Short Answers

1. Part of the tooth that contains nerve endings is called the ______.
2. Fluorosis can happen if too much fluoride is consumed under the age of ______
3. The pH under which the teeth start losing calcium and phosphate ions is ______

4. Hardened plaque on the tooth surface and/or under the gum is called _______
5. Most frequently used teeth whitening ingredient in the US is called _______
6. Enamel is mostly made of a complex chemical compound called _______
7. The process where the teeth regain their lost calcium and phosphate ions is called _______
8. The amount of toothpaste that should be given to 2–6-year-old children is _______

Matching

Match the ingredients in column A with their appropriate ingredient category in column B.

Column A	Column B
_____ A. Alcohol	1. Abrasive
_____ B. Carbomer	2. Anticaries ingredient
_____ C. Chlorhexidine	3. Antihypersensitivity ingredient
_____ D. FD&C Blue #1	4. Antiinflammatory ingredient
_____ E. Hydrated silica	5. Antiplaque/antigingivitis ingredient
_____ F. Hydrogen peroxide	6. Antitartar ingredient
_____ G. Potassium nitrate	7. Artificial sweetener
_____ H. Propolis	8. Astringent
_____ I. Sodium benzoate	9. Bleaching agent
_____ J. Sodium lauryl sulfate	10. Cleansing agent, anionic surfactant
_____ K. Sodium monofluorophosphate	11. Color additive
_____ L. Sorbitol	12. Humectant
_____ M. Tetrasodium pyrophosphate	13. Preservative
_____ N. Witch hazel	14. Solvent
_____ O. Xylitol	15. Thickener

REFERENCES

1. Christopoulos, A.: Mouth Anatomy, Last update: 5/22/2013, Accessed 9/30/2013 at http://emedicine.medscape.com/article/1899122-overview
2. Faller, A., Schünke, G.: *The Human Body: An Introduction to Structure and Function*, New York: Thieme, 2004: 394.
3. Lamont, R. J., Jenkinson, H. F.: *Oral Microbiology at a Glance*, New Jersey: John Wiley & Sons, 2010: 3.

4. Marsh, P. D.: Dental plaque as a microbial biofilm. *Caries Res.* 2004;38(3):204–211.
5. WHO: What is the Burden of Oral Disease?, Accessed 10/2/2013 at http://www.who.int/oral_health/disease_burden/global/en/
6. CDC: Hygiene-Related Diseases, Dental Caries, Last update: 9/22/2009, Accessed 10/2/2013 at http://www.cdc.gov/oralhealth/publications/factsheets/dental_caries.htm
7. Johansson, A. K., Lingström, P., Imfeld, T., et al.: Influence of drinking method on tooth-surface pH in relation to dental erosion. *Eur J Oral Sci.* 2004;112(6):484–489.
8. CDC, Ten great public health achievements - United States, 1900–1999, *CDC Morbid Mortal Wkly Rep* 1999;48(12):241–243.
9. Bratthall, D., Hänsel-Petersson, G., Sundberg, H.: Reasons for the caries decline: what do the experts believe? *Eur J Oral Sci.* 1996;104(4(Pt 2)):416–422.
10. Topping, G., Assaf, A.: Strong evidence that daily use of fluoride toothpaste prevents caries. *Evid Based Dent.* 2005;6(2):32.
11. Rasines, G.: Fluoride toothpaste prevents caries in children and adolescents at fluoride concentrations of 1000 ppm and above. *Evid Based Dent.* 2010;11(1):6–7.
12. Dye, B. A., Tan, S., Smith, V., et al.: Trends in oral health status: United States, 1988–1994 and 1999–2004. *Vital Health Stat* 2007;248:1–92.
13. Bagramian, R. A., Garcia-Godoy, F., Volpe, A. R.: The global increase in dental caries. A pending public health crisis. *Am J Dent.* 2009;22:3–8.
14. National Institute of Dental and Craniofacial Research: Dental Caries, Accessed 10/2/2013 at http://www.nidcr.nih.gov/DataStatistics/FindDataByTopic/DentalCaries/
15. White, D. J.: Dental calculus: recent insights into occurrence, formation, prevention, removal and oral health effects of supragingival and subgingival deposits. *Eur J Oral Sci.* 1997;105(5 Pt 2):508–522.
16. National Institute of Dental and Craniofacial Research: Data and Statistics, Accessed 10/2/2013 at http://www.nidcr.nih.gov/DataStatistics/ByPopulation/Adults/
17. ADA: What is Gum Disease?, Accessed 9/25/2014 at http://www.ada.org/sections/scienceAndResearch/pdfs/forthedentalpatient_jan_2011.pdf
18. Irwin, C. R., McCusker, P.: Prevalence of dentine hypersensitivity in a general dental population. *J Ir Dent Assoc.* 1997;43(1):7–9.
19. Aquafresh Science Academy, Accessed 10/2/2013 at http://www.aquafreshscienceacademy.com/oral-health/tooth-whitening/intrinsic-staining.html
20. Addy, M., Moran, J.: Mechanism of stain formation on teeth, in particular associated with metal ions and antiseptics. *Adv Dent Res* 1995;9:450–456.
21. Nathoo, S. A.: The chemistry and mechanisms of extrinsic and intrinsic discolouration. *J Am Dent Assoc.* 1997;128:6S–10S.
22. Spielman, A. I., Bivona, P., Rifkin, B. R.: Halitosis. A common oral problem. *NY State Dent J.* 1996;62:36–42.
23. Moss, S.: Halitosis and oral malodor. *FDI World.* 1998;5:14–20.
24. Ratcliff, P. A., Johnson, P. W.: The relationship between oral malodor, gingivitis and periodontitis. A review. *J Periodontol* 1999;70:485–489.
25. Delanghe, G., Ghyselen, J., Feenstra, L., et al.: Experiences of a Belgian multidisciplinary breath odour clinic. *Acta Otorhinolaryngol Belg.* 1997;51(1):43–48.
26. Rosenberg, M.: The science of bad breath. *Sci Am.* 2002;72–79.

27. Rosenberg, M.: Clinical assessment of bad breath: current concepts. *J Am Dental Assoc.* 1996;127:475–482.
28. Patrick, B. K., Thompson, J. M.: *An Uncommon History of Common Things*, Washington: National Geographic Books, 2009: 195.
29. Panati, C.: *Extraordinary Origins of Everyday Things*, New York: HarperCollins, 2013: 210–213.
30. Procter and Gamble: Crest, Accessed 10/2/2013 at: http://www.pg.com/en_US/brands/beauty_grooming/crest.shtml
31. Sheffield Pharmaceuticals: History, Accessed 10/2/2013 at http://www.sheffield-pharmaceuticals.com/?p=02
32. Colgate: History of Toothbrushes and Toothpastes, Accessed 9/25/2013 at http://www.colgate.com/app/CP/US/EN/OC/Information/Articles/Oral-and-Dental-Health-Basics/Oral-Hygiene/Brushing-and-Flossing/article/History-of-Toothbrushes-and-Toothpastes.cvsp
33. OralB: The History of Dental Floss, Accessed 10/2/2013 at http://www.oralb.com/topics/history-of-dental-floss.aspx
34. Tandel, K. R.: Sugar substitutes: Health controversy over perceived benefits, *J Pharmacol Pharmacother.* 2011;2(4):236–243.
35. CDC: Community Water Fluoridation, FAQs for Dental Fluorosis, Last update: 7/10/13, Accessed 10/2/2013 at http://www.cdc.gov/fluoridation/safety/dental_fluorosis.htm
36. Colgate: Fluorosis by the Columbia University College of Dental Medicine, Last update: 12/20/2010, Accessed 10/2/2013 at http://www.colgate.com/app/CP/US/EN/OC/Information/Articles/Oral-and-Dental-Health-Basics/Checkups-and-Dental-Procedures/Fluoride/article/Fluorosis.cvsp
37. Cosmetics and Toiletries: Dentists Concerned Over At-home Tooth Bleaching Misuse, Updated 8/8/2011, Accessed 10/2/2013 at C&T website http://www.cosmeticsandtoiletries.com/formulating/category/oralcare/127259973.html?page=1
38. ADA: ADA Seal of Acceptance Program and Products, Accessed 10/4/2013 at http://www.ada.org/sealprogramproducts.aspx
39. Anusavice, K. J., Phillips, R. W., Shen, C., et al.: *Phillips' Science of Dental Materials*, Philadelphia: Elsevier Health Sciences, 2012: 239.
40. Hattab, F. N.: The state of fluorides in toothpastes, *J Dent.* 1989;17(2):47–54.
41. Söderling, E., Hirvonen, A., Karjalainen, S., et al.: The effect of xylitol on the composition of the oral flora: a pilot study, *Eur J Dent.* 2011;5(1):24–31.
42. Cury, J. A., Tenuta, L. M. A.: Enamel remineralization: controlling the caries disease or treating early caries lesions? *Braz Oral Res.* 2009;23(S1):23–30.
43. ten Cate, J. M., Featherstone, J. D.: Mechanistic aspects of the interactions between fluoride and dental enamel. *CRC Crit Rev Oral Biol Med.* 1991;2:283–296.
44. ten Cate, J. M.: Current concepts on the theories of the mechanism of action of fluoride. *Acta Odontol Scand.* 1999;57:325–329.
45. ADA: Toothpaste, Accessed 10/17/2013 at http://www.ada.org/1322.aspx
46. Code of Federal Regulations Title 21 Part 355.50
47. Stoker, T. E., Gibson, E. K., Zorrilla, L. M.: Triclosan exposure modulates estrogen-dependent responses in the female Wistar rat. *Toxicol Sci.* 2010;117(1):45–53.

48. Raut, S. A., Angus, R. A.: Triclosan has endocrine-disrupting effects in male western mosquitofish, *Gambusia Affinis Environ Toxicol Chem.* 2010;29(6):1287–1291.
49. Yazdankhah, S. P., Scheie, A. A., Høiby, E. A., et al.: Triclosan and antimicrobial resistance in bacteria: an overview. *Microb Drug Resist.* 2006;12(2):83–90.
50. Colgate, Colgate Total® toothpaste With Triclosan. Accessed 8/12/2014 at http://www.colgatetotal.com/triclosan-faq
51. FDA: Transcript of FDA Media Briefing on the FDA's Proposed Rule Amending the Tentative Final Monograph for Over-the-Counter Consumer Antiseptics, Last update: 12/16/2013, Accessed 8/12/2014 at http://www.fda.gov/downloads/NewsEvents/Newsroom/MediaTranscripts/UCM378989.pdf
52. Fundamentals of Dentrifice: Oral Health Benefits in a Tube, Last Update: 1/16/2013, Accessed 10/6/2013 at http://media.dentalcare.com/media/en-US/education/ce410/ce410.pdf
53. H. A. Barnes: The yield stress – a review – everything flows? *J Non-Newtonian Fluid Mech.* 1999;81(1–2):133–178.
54. Laba, D.: *Rheological Properties of Cosmetics and Toiletries*, Boca Raton: CRC Press, 1993.
55. Silverson: Manufacture of Toothpaste, Accessed 10/4/2013 at http://www.silverson.com/us/toothpaste.html
56. Butler, H.: *Poucher's Perfumes, Cosmetics and Soaps*, London: Kluwer Academic Publishers, 2000.
57. Rigano, L.: Toothpastes. *Cosmet Toil.* 2012;127(5):324–334.
58. Fitch C P.: Etiology of the discoloration of teeth. *Dent Cosmos.* 1861;3:133–136.
59. Li, Y., Greenwall, L.: Safety issues of tooth whitening using peroxide-based materials. *Brit Dent J.* 2013;15:29–34.
60. Tavares, M., Stultz, J., Newman, M., et al.: Light augments tooth whitening with peroxide. *J Am Dent Assoc.* 2003;134:167–175.
61. Europa: Opinion on Hydrogen Peroxide, in Its Free Form or When Released, in Oral Hygiene Products 2007, Accessed 10/2/2013 at http://ec.europa.eu/health/ph_risk/committees/04_sccp/docs/sccp_o_122.pdf
62. Europa: Public Health, Tooth Whiteners and Oral Hygiene Products, Accessed 10/20/2013 at http://ec.europa.eu/health/opinions/en/tooth-whiteners/#3
63. Attin, T., Hannig, C., Wiegand, A., et al.: Effect of bleaching on restorative materials and restorations—a systematic review. *Dent Mater.* 2004;20:852–861.
64. Goldberg, M., Grootveld, M., Lynch, E.: Undesirable and adverse effects of tooth-whitening products: a review. *Clin Oral Invest.* 2010;14:1–10.
65. ADA Council on Scientific Affairs: Tooth Whitening/Bleaching: Treatment Considerations for Dentists and Their Patients 2009, Accessed 10/2/2013 at http://www.ada.org/sections/about/pdfs/HOD_whitening_rpt.pdf
66. Ciuffreda, L., Boylan, R., Scherer, W., et al.: An in vivo comparison of antimicrobial activities of four commercial mouthwashes. *J Clin Dent.* 1994;5(4):103–105.
67. Torres, C. R., Perote, L. C., Gutierrez, N. C., et al.: Efficacy of mouth rinses and toothpaste on tooth whitening. *Oper Dent.* 2013;38(1):57–62.
68. Holland, G. R.: Guidelines for the design and conduct of clinical trials on dentine hypersensitivity. *J Clin Periodontol.* 1997;24(11):808–813.

69. Shulman, J. D., Wells, L. M.: Acute fluoride toxicity from ingesting home-use dental products in children, birth to 6 years of age. *J Public Health Dent.* 1997;57(3):150–158.
70. Whitford, G. M.: Fluoride metabolism and excretion in children. *J Public Health Dent.* 1999;59(4):224–228.
71. Gessner, B. D., Beller, M., Middaugh, J. P., et al.: Acute fluoride poisoning from a public water system. *N Engl J Med.* 1994;330(2):95–99.
72. DenBesten, P., Ko, H. S.. Fluoride levels in whole saliva of preschool children after brushing with 0.25 g (pea-sized) as compared to 1.0 g (full-brush) of a fluoride dentifrice. *Pediatr Dent.* 1996;18(4):277–280.
73. Akiniwa, K.: Re-examination of acute toxicity of fluoride. *Fluoride.* 1997;30:89–104.
74. Watt, B. E., Proudfoot, A. T., Vale, J. A.: Hydrogen peroxide poisoning. *Toxicol Rev.* 2004;23(1):51–57.
75. CFR Title 21 Part 700.25

7

OTHER PRODUCTS

INTRODUCTION

In addition to the previously discussed cosmetic and personal care products, there are other product categories that should be discussed. Some products are formulated specifically for groups with special needs, such as baby care products and feminine hygiene products, while others are used by a larger segment of the population, such as hair removal products. Individually, these product categories typically have a lower market share when compared to skin care and hair care products or color cosmetics. However, they also play an important role in our everyday lives.

This chapter provides an overview of four product categories, including hair removal products, baby care products, sunless tanners, and feminine hygiene products. The various sections provide an overview of the anatomy and physiological function of the specific body areas and skin types. They also summarize the various product types available in each category, their major ingredients, characteristics, and formulation technology. The potential advantages and disadvantages of using these personal care products are also included in this chapter. Finally, quality control problems, the main testing methods, and consumer needs are discussed here.

Introduction to Cosmetic Formulation and Technology, First Edition. Gabriella Baki and Kenneth S. Alexander.
© 2015 John Wiley & Sons, Inc. Published 2015 by John Wiley & Sons, Inc.

SECTION 1: HAIR REMOVAL PRODUCTS

LEARNING OBJECTIVES

Upon completion of this section, the reader will be able to

1. define the following terms:

Aftertreatment product	Bleaching	BOV	Chemical depilatory
Depilation	Epilation	Epilator	Folliculitis
Ingrown follicle	Pretreatment product	Rosin	Shaving
Shaving product	Threading	Trimming	Tweezing
Waxing			

2. differentiate between vellus and terminal hair;
3. briefly explain how the shape and diameter of hair shafts influence shaving;
4. briefly explain why the hair growth cycle should be taken into consideration when planning permanent hair removal;
5. differentiate between epilation and depilation;
6. name some negative effects that hair removal methods and products may have on the skin and/or hair follicles;
7. list the various required cosmetic qualities and characteristics that an ideal hair removal product should possess;
8. list the various required technical qualities and characteristics that an ideal hair removal product should possess;
9. explain the functions of pre-treatment formulations;
10. list the major ingredient types found in pre-shave formulations, and provide some examples for each type;
11. differentiate between lathering and brushless shaving creams;
12. list the major ingredient types found in shaving creams, and provide some examples for each type;
13. explain how shaving creams should be applied to the skin;
14. explain the difference between shaving creams and aerosol shaving foams;
15. explain how foam and foaming gel are generated from an aerosol shaving can, respectively;
16. list the factors that influence foam stability of aerosol shaving foams;
17. explain how chemical depilatories work and remove hair;
18. name the major ingredient types found in chemical depilatories;
19. differentiate between cold and hot waxes;

20. explain why rosins and rosinates are mixed with other ingredients to formulate hair removal waxes;
21. list the major ingredient types found in hair removal waxes;
22. explain the function of after-treatment products;
23. list the major ingredients in aftershave formulations;
24. list some typical quality issues that may occur during the formulation and/or use of hair removal products, and explain why they may occur;
25. list the typical quality parameters that are regularly tested for hair removal products, and briefly describe their method of evaluation;
26. list the typical containers available for hair removal products.

KEY CONCEPTS

1. Today, there are a number of hair removal techniques and products available. Temporary methods provide hairless skin for a shorter (1–3 days) or longer time (1–3 weeks), depending on the technique and the individual's physiological characteristics. Permanent methods, however, can prolong the duration of hair loss up to years.
2. Pre-treatment products may have various functions depending on the type of hair removal technique used. They are designed to make hair removal easier and more comfortable and reduce the potential for skin irritation.
3. Shaving is still the most common method used for removing unwanted hair by men and women. There are various types of products available, including shaving soap, shaving cream as well as aerosol shaving foam and gel, which aim at making shaving easier and more pleasant.
4. Chemical depilatories work by hydrolyzing and disrupting the disulfide bonds in hair keratin, causing the hair to break in half and allowing the hair to separate from the skin.
5. Waxing is an epilation method that involves applying warm or cold wax onto the hair-bearing skin and quickly stripping off the hardened wax and embedded hairs.
6. After-treatment preparations are intended to reduce redness and alleviate pain and burning sensation experienced after hair removal.
7. The typical quality-related issues of hair removal products include poor foaming of shaving gels and creams, unstable foam, brittle wax, separation of emulsions, microbiological contamination, clumping, and rancidification.
8. Parameters commonly tested to evaluate the quality of hair removal products include foamability, foam stability, foam viscosity, foam density, and foam structure; spray characteristics, aerosol can leakage, pressure test, and actuation force; preservative efficacy; pH; viscosity; and color.
9. The most common ingredients causing safety and environmental concerns include thioglycolates and propellants.

Introduction

Hair removal is an increasingly important sector of the cosmetic and personal care industry. Both men and women are becoming more concerned about the aesthetic aspect of their appearance. Many men and women choose to remove unwanted body hair for cosmetic, social, cultural, or medical reason.[1] A number of hair removal techniques have been developed over the years, including methods for temporary and permanent hair removal. The availability of the current methods and products may be different; most of them can be used at home; however, there are some that can be used only in professional salons and dermatological offices.

This section reviews the various techniques and products for temporary and permanent hair removal. The most popular temporary hair removal methods and products, including their ingredients, properties, formulation technology, quality issues, and packaging materials, are discussed in detail. An overview is provided on how these products may affect the human skin and hair and what the consumers' general requirements are.

Brief Review of the Structure and Function of Human Hair

As discussed in Chapter 5, almost the entire body surface is covered with hair follicles; however, their visibility and color are different. Vellus hairs are fine, thin hairs with light color (i.e., less melanin content); therefore, they are almost invisible. In contrast, terminal hairs are long, thick, dark hairs that are more visible to the eyes. Today, men and women remove their terminal facial and body hair as part of their grooming routine, without having any medical problems.

DID YOU KNOW?

There are certain medical conditions that may trigger the growth of visible hair in places where the vellus hair should grow. An example is abnormal male hormone, i.e., the testosterone level in women. Women normally have a low level of testosterone; however, certain hormonal disturbances may lead to a rise in the testosterone level. In such cases, excess terminal hair can appear in the abdominal region or on the face.[2] This phenomenon is often referred to as excessive or unwanted hair.[3] In such cases, the hair should not be removed without seeking medical help (including dermatologist, gynecologist, and endocrinologist) and finding out the main cause of the condition.

As discussed previously, hair shafts are part of the pilosebaceous unit, which also includes the sebaceous gland and arrector pili muscle (refer to Figure 5.1). These are located in the dermis, except for the hair shaft that goes through the epidermis to the skin surface. It has been demonstrated that facial hair fibers (i.e., beard) has a larger diameter and often an asymmetrical shape.[4] It is important since the structural properties of the hair impact shaving. The force required to cut a hair is proportional to its diameter: the larger the diameter, the larger the force needed. It requires almost three times the force to cut the beard hair than the scalp or leg hair.[5] It is also known that hydration softens the hair shafts, which decreases the resistance to the blade and makes shaving easier. Short-term hydration also improves the skin's elasticity, making it better able to deform and recover as the blade is drawn over its surface.[6]

It was also discussed in Chapter 5 that hair grows in cycles (refer to Figure 5.4). The first phase is the anagen phase, the active growth phase. It governs the length of the hair at different body sites. The second phase is the catagen, the transition phase. During the catagen phase, cell division ceases and the hair stops growing. The third stage is the telogen phase, the resting phase. In this phase, cell replication and subsequent hair growth are inactive for several months. By the end of the telogen phase, the hair falls out. The telogen phase varies in duration from one body area to another. After the resting phase, the cells start to divide again, a new hair shaft is formed, and it starts growing. It is important to take the growth cycle into account if one would like to have permanent hair removal. It has been shown that the hair follicles in the anagen stage are sensitive to laser light and electrolysis since they are still alive; however, those in the second or third stage are irresponsive and will grow back after the treatment.[7]

History of Using Hair Removal Products

Hair removal from various parts of the body has been an important part of beauty for thousands of years. Prehistoric cave paintings and archaeological evidence indicate that people were shaving around 30,000 BC, using different tools, such as shells, flint ax, volcanic glass, and even sharpened animal teeth.[8] Facial hair was in and out of fashion over the course of history. In ancient Egypt, a smooth face was a status symbol. Abrasives, or the friction method, have long been used to remove unwanted hair. Ancient Indian, Roman, and Egyptian women and men also used pumice stones for this purpose. In addition, these cultures used the first shaving razors made of iron, bronze, copper, or gold. It was a straight razor having only a single blade. Shaving at that time was often performed at barbershops by trained barbers. They used soap and water for making a lather that could be applied to the skin before shaving the hair. Sugaring was also a frequently used hair removal method. In ancient cultures, women used beeswax or cream depilatories composed of an alkali to remove the hair from their legs. Women also used razors as well as tweezers and pumice stones.

Ancient Egyptians, early Middle Easterners, and Asian cultures used threading for hair removal. This is a process that removes unwanted hair by the root using a looped thread by twisting the hair.[9] This process is regaining popularity in some cultures today. Shaving was more common during the Middle Ages than before, especially in Europe. It was very fashionable for women to be completely hairless, even on their heads, which allowed them to wear large, ostentatious wigs and headpieces that were in style. To remove the hair on their eyebrows, heads, and necks, women plucked and shaved almost every day. The Aztecs in Central and North America used shaving razors fashioned from volcanic obsidian glass, which was sharp and effective but sometimes fragile.[8]

Early types of razor caused many injuries during shaving and were potentially dangerous; therefore, many attempts were made to improve the razor design. The first L-shaped razor blade with a wooden guard was created in 1760s by a French barber.[10] It was named the Perret razor. This model prevented deep skin cuts, but was still impractical because the guard had to be removed so the blade could be sharpened. The solution to this problem was provided by the Kampfe brothers in the late 19th century late in the US. They created the first safety razor.[8] It was a T-shaped device, which was a great success at that time; however, it still required stropping before each use. With the improvement of the shaving tools, shaving aids were also continuously developed. A special soap was marketed in the early 1800s, which was used by placing it in a cup to which water was added. A brush was used to stir up a lather and apply it to the face.[8]

King C. Gillette created the first double-bladed safety razor in the early 20th century that featured replaceable, disposable blades. This revolutionary invention changed the hair removal industry. By the time of First World War, this invention had caught on; the US government contracted him to supply razors for the army.[8] Shaving was popular between both genders. Later on, improvements of shaving razors included new blade materials, for example, stainless steel, which extended the blade life by preventing corrosion. Additional great improvements in hair removal in the 20th century included the invention of the first electric razor, thioglycolate depilatories, and the first aerosol shaving cream supplied in a can. It soon became the most popular shaving cream, because brush was not needed to apply it. However, the first concerns over aerosols as pollutants started in the 1980s, and as a result, shaving soaps and hand-applied creams made a modest comeback.

Since then and until now, blade technology has improved tremendously. Gillette introduced its five-blade razor for a pleasant and safe shave, and a great variety of shaving aids, pre- and aftertreatment preparations, are now available on the market.

Possible Methods for Removing Hair

❶ **Today, there are a number of hair removal techniques and products available. Temporary methods provide hairless skin for a shorter (1–3 days) or a longer time (1–3 weeks), depending on the technique and the individual's physiological characteristics. Permanent methods, however, can prolong the duration of hair loss up to years.** Figure 7.1 summarizes the currently available hair removal methods. The techniques outlined and shaded refer to those that include the use of cosmetic products. They are discussed in detail in the following parts of this section.

There are two distinct types of temporary hair removal, known as depilation and epilation.

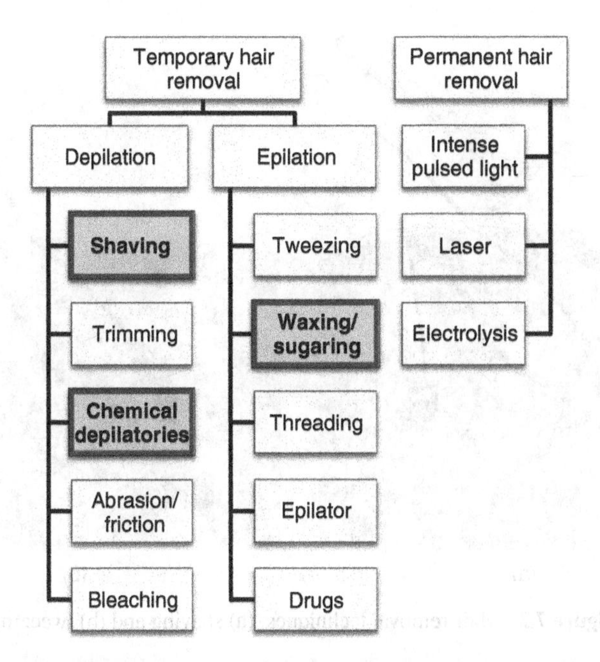

Figure 7.1 Summary of currently available hair removal methods.

- Depilatory techniques and products remove only part of the hair shaft, which protrudes above the skin surface (as illustrated in Figure 7.2a). These techniques are generally pain-free, if there is no skin damage. Depilatory techniques include shaving, trimming, using abrasives, using chemical depilatories, and bleaching. Since only shaving, chemical depilatories, and bleaching require the use of cosmetic products, only these methods will be discussed in detail later on.
    - **Shaving** is a frequently used method since it is fast, easy, painless, effective, and relatively inexpensive. It requires a shaving tool, i.e., razor.
    - **Trimming** is performed using scissors developed for this purpose. This method can be used to trim long eyebrow hair.
    - **Abrasives**, such as pumice stones or gloves made of fine sandpaper, work by physically removing the hair from the skin's surface. This method is very irritating and is rarely used today for hair removal.
    - **Chemical depilatories** dissolve hair fibers causing the hair to break, which can be easily washed away from the skin. Products available include gels, creams, lotions, and aerosols.
    - **Bleaching** is often referred to as a hair removal technique; however, it does not remove any part of the hair shaft, but changes its color.
- Epilatory techniques and products remove the entire hair shaft with its root in the dermis (as illustrated in Figure 7.2b). These techniques are more effective and

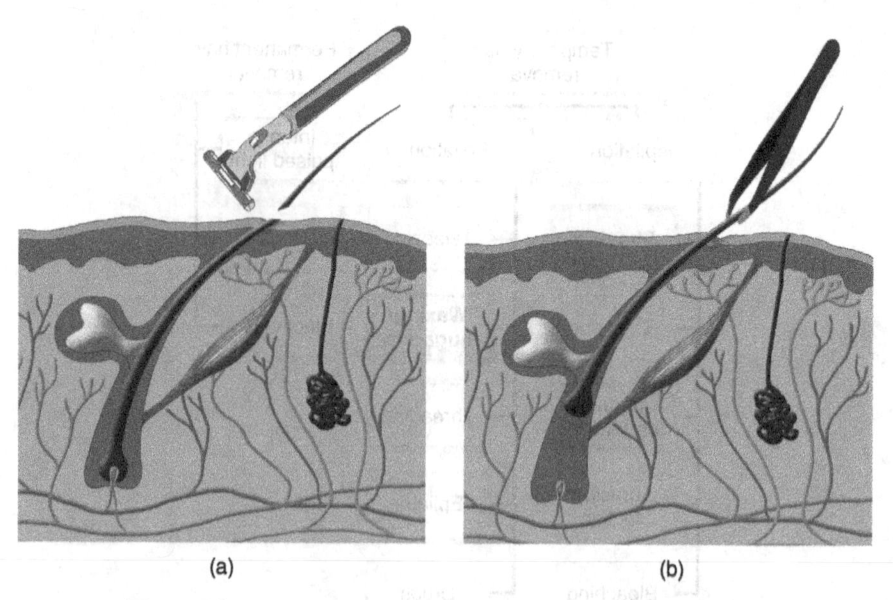

Figure 7.2 Hair removal techniques: (a) shaving and (b) tweezing.

have a longer effect since it takes a longer time for the hair shaft to completely regrow than just regrowing from under the skin as in the case of shaving. Epilation techniques include using epilators, tweezing, waxing and sugaring, and threading and using medications.

- **Tweezing**, or otherwise known as plucking, is best performed using tweezers. It is a beneficial method for removing a small group of hairs found on the eyebrows, for example. This method can be used for small surfaces that are not readily accessible to epilators. However, it is a time-consuming and tedious method.
- **Waxing** involves applying warm or cold wax onto the hairy skin and quickly stripping off the wax and hairs against the direction of hair growth. It is the most effective epilation method since hair is removed in large quantities from a large surface area. Sugaring is similar to waxing, but instead of a wax, a mixture of sugar, lemon juice, and water is used to form a syrup.
- **Threading** is an ancient manual technique that involves the use of a long twisted loop of thread rotated rapidly across the skin. Hairs are trapped within the tight entwined coils and are pulled off. This technique is very popular in many Arabic countries.
- **Epilator** is an electrical device that removes hair by mechanically grasping them and pulling them out.
- **Eflornithine**, a novel method for temporary hair reduction in women, is a topical cream available by prescription only and has been approved by the

FDA for the reduction of unwanted facial hair in women.[11] It is not a hair removal or depilatory product but a topical cream that decreases the rate of hair growth.[12]

DID YOU KNOW?

It is a common belief that shaving causes hair to grow back faster and coarser. However, it is not true. Clinical studies show that shaving has no effect on hair growth, its thickness, or the rate of hair regrowth. Shaving does, however, alter the tip shape of the regrown hair, which may initially change the way hair feels as it grows back. Shaved hair lacks the finer taper seen at the ends of the unshaven hair, giving an impression of coarseness. Shaving removes only the upper part of the hair, not the living section below the skin's surface, therefore, it is unlikely to affect the rate or type of growth.[13]

Permanent hair reduction refers to a significant reduction in the number of terminal hairs after a given treatment, which is stable for a period of time longer than the complete growth cycle of hair follicles at the given body site. Permanent hair removal techniques include the use of laser, intense pulse light source, and electrolysis. Permanent hair removal may sound interesting and a promising attempt to get rid of unwanted hair; however, it should be kept in mind that these procedures can have side effects. It is important that, before starting any permanent hair removal treatment, one should consult a dermatologist to examine his/her skin and see whether other alternatives are available.

- **Laser treatment** stops the active growth of the hair for long periods. Its principle is based on the fact that there is a difference in color between the hair follicle and the skin. The light energy absorbed by the dark pigment in the hair follicle (i.e., melanin) causes damage to the follicle, thereby reducing hair growth. The lighter the skin and the darker the hair, the more selectively the laser will affect the follicle and not the surrounding tissue. A laser instrument produces a powerful, focused light beam, composed of a constant wavelength.[14]
- **Intense pulsed light** devices also use the principle of thermal destruction (i.e., photothermolysis), similar to lasers, to target the melanin in the hair follicle. However, here a range of wavelengths is used, typically between 500 and 1200 nm, instead of one wavelength.[15]
- **Electrolysis**, also termed electrology, involves the insertion of a small, fine needle into the hair follicle, followed by firing of a pulse of electric current that damages and eventually destroys the hair follicle.[16] As a result, hair breaks off at the root and can then be easily pulled off with fine tweezers. It is effective especially in small areas.

DID YOU KNOW?

The efficiency of laser treatments and electrolysis for hair removal partly depends on the growth phase in which the hair is treated. Both methods are truly effective only in the anagen phase. To detect actively growing hair follicles, shaving can be performed before this procedure. Those hairs that grow back after shaving are in their anagen phase, and should be targeted with the hair removal methods. Since not all the hairs are in the anagen phase, even permanent treatments have to be repeated to capture the new cells.[17]

Types and Definition of Hair Removal Products

Products that will be reviewed in detail are all considered cosmetics in the US. They include the following:

- **Pre-treatment products** are designed to be used before hair removal. Hair removal methods that necessitate the use of such products include waxing and dry shaving with electric razors. Preshave products are designed to clean and moisturize the skin surface. Prewax preparations are designed to remove perspiration from the skin, lubricate and moisturize the skin, as well as ensure that the wax sticks only to the hair follicles and not to the skin. Skin-numbing products are also available; these are recommended to be used before waxing and epilation.
- **Shaving products** are applied to the skin to facilitate hair removal. They soften the hair and lubricate the skin in order to provide a pleasant hair removal experience through shaving, which is still the most common method for removing unwanted hair by men and women. This category includes shaving soaps, creams, aerosol foams, and gels.
- **Chemical depilatory** preparations are often referred to as "depilatories." They can be liquids, gels, lotions, foams, and creams that usually contain a thioglycolate salt as the main ingredient. They act by destroying the hair fibers and removing unwanted hair from certain body parts.
- **Waxes** are based on mixtures of rosins and rosinates. They also usually contain skin conditioning agents, colorants, and fragrance. Waxes can be applied hot or cold, and their removal, along with the hair, may be performed by hand or using strips, depending on the type used.
- **After-treatment products** are designed to reduce redness; alleviate the pain and burning sensation after hair removal; as well as cool, refresh the skin, and exert a mild astringent effect. They can contain antiseptics to prevent infections

through any cuts. They can also contain moisturizing ingredients to soothe and hydrate the skin. The available dosage forms include solutions, gels, lotions, creams, balms, and powders. After-wax oils are used to remove the remaining wax from the skin.

How Hair Removal Products May Affect the Skin and Hair?

Removing unwanted hair has become a part of many people's everyday life. Vellus hair turns into terminal hair during puberty, in the beard area, for example; this is one of the reasons why young men start shaving. In some cultures, the hairless face is recognized as more hygienic. Sometimes, hair removal may be mandatory, such as a tradition when reaching a certain age, or on the contrary, it may be forbidden in other cultures for religious reasons. Very frequent and/or inappropriate hair removal can lead to adverse effects and skin reactions. The common side effects of various hair removal techniques are summarized as follows:

- Alcohols found in many pretreatment products and aftershave preparations have an **antibacterial effect** and provide a **cooling sensation** to the skin. It may, however, dry out the skin and lead to a burning sensation when used on freshly shaved skin.
- When using a razor, blades may **cut** the skin, leading to bleeding. If not treated appropriately, infections and inflammation may occur at these skin sites. If the skin is cut severely, it may cause permanent scarring as well.
- Hair removal may lead to temporary **pain, itching, and/or swelling** of the treated area, skin sensitivity, and skin irritation. It is hypothesized that the discomfort associated with shaving is a result of localized skin displacement and/or the rotation and extension of the beard fiber in its follicle. Shaving can remove irregular skin elevations, which may also lead to irritation.[18] The sensory nerves located in the dermis all around the hair follicles' root will transmit a pain signal to the brain. This is why epilation methods are painful. However, the pain is temporary and usually resolves by itself.
- Certain aftershave formulations contain fragrances, which can be **irritating** and lead to contact allergy.[19]
- One of the most common complaints in shaving is **folliculitis**, i.e., inflammation of the hair follicles. Folliculitis starts when the hair follicles are damaged by shaving, for example, or affected by a condition called pseudofolliculitis barbae. Most often, the damaged follicles become infected with *Staphylococcus* bacteria. This condition is often referred to as barber's itch, which occurs more frequently in the face, groin, and underarm than in the legs or arms.[20,21] Tinea barbae is similar to barber's itch, but the infection is caused by a fungus.

DID YOU KNOW?

Pseudofolliculitis barbae is a chronic inflammatory condition most often seen in men and women of African American and Hispanic origin who have tightly curled hair.[22] When they shave, these curly hair follicles may curve back into the skin and cause inflammation.

- Another frequent problem that may occur after hair removal is the **ingrown hair**. An ingrown hair occurs when a shaved or tweezed hair grows back into the skin without entering the skin surface. The result of ingrown hairs is localized pain, irritation, and the appearance of bumps in the hair removal area. In most cases, it can get inflamed as well. It may be mild, which affects only a small number of hair follicles. However, when large areas are affected every time, the users should visit a dermatologist and see if other options are available.
- Wax is designed to adhere to the hair as close to the skin as possible. When the wax is removed, it should adhere to the hair and remove the hair from the follicle. If the wax is not applied appropriately, or if applied at the wrong temperature, or if the skin is not cleaned well, the hair will not be removed. If the wax is too hot, it can cause skin irritation and **burn** the skin causing blistering. Even surface layers of the skin can be removed along with the hair if waxing is not performed correctly.
- Chemical depilatories are highly alkaline products that can possibly damage and degrade the dermal proteins, causing **skin irritation**, excess exfoliation, or contact dermatitis if left on the skin for a longer period of time than recommended. Fortunately, this dermal effect is usually temporary, lasting only a few hours or a day. The FDA warns consumers that they should read the product labels carefully and test the products on a smaller area before use.[23] Evaluation of the test area should be done for 1 or 2 days. It is also recommended that consumers select the formulation appropriate for the intended use, as skin sensitivity varies throughout potential application areas.
- Certain medications may influence the **healing** capabilities of the skin; therefore, certain hair removal techniques should be avoided in patients taking such medications. An example is retinoids, which are frequently used in acne patients. Retinoids may cause permanent scarring in these patients; therefore, they should avoid using electrolysis, chemical depilatories, and waxing during acne treatment.[24,25]
- As discussed earlier, there is no data available on the **long-term effect** of epilation, either temporary or permanent. It is hypothesized that repetitive epilation over several years may result in permanent damage of the hair follicles and production of finer or thinner hairs. However, long-term clinical trials demonstrating the effects of repetitive epilation are lacking.[26]

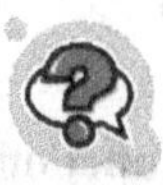

DID YOU KNOW?

It has been a well-established practice among neurosurgeons to shave part of the scalp before surgery based on the belief that hair removal prevents postoperative infections. Today, however, there are studies clearly demonstrating no added benefit of shaving in such situations. Since hair removal neither contributes benefits to the surgery nor decreases the risk of wound infection, but has considerable cosmetic value for the patient, many of the authors recommended that cranial surgeries should be done without hair shaving.[27]

Required Qualities and Characteristics and Consumer Needs

From a consumer perspective, a quality hair removal product should possess the following characteristics:

- Good lubricant properties for skin protection against razor
- Hydrating properties to soften the skin and hair
- Well-tolerated, non-irritating
- Pleasant odor
- Easy application with a tendency to spread easily
- Easy removal from the razor and skin
- Creaminess
- Relatively pain-free
- Pre-treatment products: make hair removal easier and remove oil and sweat from the skin
- After-treatment products: hydrate and cool the skin, alleviate pain and redness, and prevent infections.

The technical qualities of hair removal products can be summarized as follows:

- Long-term stability
- Appropriate texture
- Foaming products: appropriate foam structure, foam density, foam viscosity, and foam stability
- No leakage from aerosol cans
- Appropriate pH
- Dermatological safety.

Product Types, Typical Ingredients, and Formulation of Hair Removal Products

Pre-treatment Products **● Pretreatment products may have various functions depending on the type of hair removal technique used. They are designed to make hair removal easier and more comfortable and reduce the potential for skin irritation.**

Pre-wax Preparations Cleansing products can be used to remove dirt, makeup, sweat, and other chemicals from the skin. They may include antiseptic and astringent compounds, such as witch hazel or alcohol, to reduce the chance of infections. Prewax treatment products, for example, oils and lotions, help hard wax to stick only to the hair shaft but not to the skin. It is also helpful if the patient has very dry skin. However, applying too much of such products can prevent the wax from sticking to the hair.[28] Prewax products primarily contain emollients and botanical extracts, such as soy oil, tea tree oil, grape-seed oil, and aloe vera extract, which may have added calming and soothing properties. The aim of pre-treatment can also be to achieve moisture reduction on the skin using powders, which can also enhance the effectiveness of the waxes. They are more often used for soft waxes. Similar to oils, overuse of powders can decrease the adhesive power of waxes to the hair. Numbing creams are also available on the market, which are intended to be used before waxing. These formulations contain local anesthetics, such as benzocaine, tetracaine, and lidocaine, which can reversibly block nerve conduction near the site of administration, thereby producing temporary loss of sensation in a limited area.[29]

Pre-shave Preparations Pre-shave products are used primarily for dry (electric) shaving, however, products for razor shaving are also available. Pre-shave products of both kinds are designed to make shaving easier, faster, more comfortable, and less irritating to the skin.

Dry shaving is believed to be more efficient if the hair is dry and stiff. Therefore, preelectric shave products are designed to dry and stiffen the hair and skin and make the hair stand up. The products used may also have antiseptic effect as well as cleansing activity to remove perspiration from the face and tighten the pores in the skin. Product types include powders and compressed powder sticks based on lubricating solids, such as talc and lotions primarily with astringents, such as alcohol, witch hazel, zinc oxide, zinc phenolsulfonate, and aluminum chlorohydrate. Emollients, such as diisopropyl adipate or isopropyl palmitate, silicone oils, and lubricating polymers, such as polyvinyl pyrrolidone, may also be used in small amounts to help reduce the friction between the skin and the blade.

As discussed earlier, beard hair requires more force to be removed. The more the pressure needed to shave, the more likely the skin will become nicked, scratched, and irritated. Products to be used before wet shaving lubricate and moisturize the skin to prevent it from damage during hair removal. It is recommended to wash the skin and hair with soap and warm water before wet shaving. It removes dirt and bacteria and degreases the hairs; therefore, they can more easily absorb water, swell, and soften. This way they offer less resistance to cutting, thus avoiding trauma to the skin. Scrubs may also be used to remove dirt and corneocytes from the skin, which may also make the hair to stand up above the skin.[30] Pre-shave products for wet shaving are available as lotions, scrubs, and alcoholic solutions.

Shaving Products ❸ **Shaving is still the most common method used for removing unwanted hair by men and women. There are various types of products, including shaving soap, shaving cream, as well as aerosol shaving foam and gel, which aim at making shaving easier and more pleasant.**

Shaving Soap Shaving soap is a hard soap bar or stick that is applied to the face as foam with a shaving brush. It is mentioned here because of its historical importance, and it was the original product applied to the skin to aid in shaving. Its use has been diminished over the past several decades as more pleasant and convenient products have appeared on the market.

A shaving soap is produced in the same way as any other traditional soap. However, in contrast to a traditional soap, which is generally made of sodium hydroxide, a shaving soap contains a considerable amount of potassium hydroxide. This component with the proper choice of fatty acids and glycerin provides a softer soap base with enhanced lathering characteristics.

Shaving Cream Shaving cream is available in three major types: lathering creams, brushless creams (which do not foam), and aerosols. They are still marketed today; however, they are much less popular than the aerosol-type systems.

LATHERING CREAMS **Lathering creams** are named lathering since they form lather (i.e., layers of bubbles) on the skin. The formation of bubbles ensures a ready supply of water to the beard hairs keeping them hydrated, and lubricates the skin. This way the razor can work more effectively, and the possibility of friction damage is reduced as well. They can be applied with a shaving brush or simply by hand. As the shaving experience is more pleasant with foams and gels, the use of creams has also reduced over the past years.

The main ingredients of lathering creams are summarized as follows:

- **Emulsifiers** create foam with the desired consistency. They are usually a blend of potassium, amine, and sodium salts of fatty acids, but can also be synthetic surfactants.
  - Examples include potassium, sodium, and triethanolamine salts of fatty acids, such as stearic acid, coconut oil, and palmitic acid. Synthetic surfactants include polysorbates, sodium lauryl sulfate, glyceryl stearate, and polyoxyethylene, among others.
- **Superfatting agents** are typically added to all lathering creams to neutralize any free alkali that may be present and to help stabilize the cream and the lather.

- Examples include free fatty acids, such as stearic acid, coconut oil, vegetable oil, mineral oil, and lanolin.

- **Humectants** are added to minimize drying out of the cream and to make the cream slightly softer as well as to hydrate the skin.
  - Examples include glycerin, propylene glycol, and sorbitol.
- **Water** acts as a vehicle for the formulation.
- **Thickeners** are consistency agents that provide appropriate viscosity for the product that is able to "stand up" on the palms before application and on the facial skin during shaving. They also contribute to foam stabilization.
  - Examples include gums, such as xanthan gum; celluloses, such as hydroxyethyl cellulose, polyvinyl pyrrolidone, sodium carboxymethyl cellulose; and other types of polymers.
- **Neutralizing agents** regulate the pH of the product as the residual film should correspond to the skin's pH level.
  - Examples include triethanolamine and potassium hydroxide.
- Since these are soap–glycerin–water systems, the use of **preservatives** is recommended to prevent microbiological growth.
  - Examples include phenoxyethanol and parabens.
- **Emollients** act as skin conditioning agents, providing smoothness and calmness to the skin as well as hydrating the skin.
  - Examples include butters, such as coconut butter and shea butter; oils, such as sunflower oil, mineral oil, and safflower oil; and waxes, such as beeswax.
- Additionally, **natural components** may also be incorporated in shaving creams, which may provide anti-inflammatory, soothing, freshening, and antiseptic effects. Perfume can also be added if needed.
  - Examples include aloe vera extract, chamomile extract, marshmallow extract, calendula extract, peppermint oil, and cinnamon oil.

BRUSHLESS CREAMS Brushless creams are thought to be more acceptable for consumers with dry and sensitive skin than lathering creams. They resemble lather shaving creams in appearance; however, they do not foam. Their advantage is that they do not require a brush for application, which makes shaving faster and easier, and they can give a more comfortable feel due to their greater lubricating ability. The main disadvantages of these creams include that they are difficulty to rinse off the razor, they have less effective softening action due to their higher oil content, and the formulation may leave a greasy feeling on the skin.[31] This type of shaving cream is an emulsion with high concentrations of oils and emulsifying agents. They contain humectants, emollients, thickeners, emulsifiers, water, and preservatives. Additional ingredients, such as fragrance, botanical extracts, and stabilizers, may also be added for additional benefits.

Shaving creams are typically O/W emulsions; thus, the formulation is performed using an emulsification process.

AEROSOL SHAVING FOAMS AND GELS Traditional shaving creams have been largely replaced by aerosol products, such as shaving foams and gels. Aerosol shaving foams are basically diluted shaving creams that are dispensed from the container with the aid of hydrocarbon propellants. Aerosol shaving creams may also include emulsifiers to ensure uniform emulsification of the propellant when shaking the product before dispensing. In general, lathering creams and shaving foams and gels provide similar shaving experience. However, the added benefit of aerosols is their ease of application.

Aerosol shaving foams are complex systems, which can be described as an O/W emulsion where the propellant is in the internal phase and the cream is in the continuous phase. This structure provides a stable foam that is able to stand on the palm and face and is sufficiently stable to last throughout the shaving process. Most propellants are relatively insoluble in water and, therefore, in water-based formulations as well. However, they can be emulsified into such systems using surfactants.

Foam generation starts when the valve is actuated. On shaking, part of the propellant becomes temporarily emulsified in the cream. When the valve is actuated, the propellant pushes the concentrate up the dip tube and out of the valve. On reaching atmospheric pressure, the emulsified propellant expands to form an instant stable foam. Foam stability and skin after-feel should be taken into consideration when formulating foams. Foam stability greatly depends on the type and amount of soaps and emulsifiers used, while the after-feel is mainly influenced by the amount of propellant used. If the propellant level is too low, the foam will be watery, while a too high concentration can lead to a rubbery dry foam. Soaps, surfactants, and emollients also play an important role in the after-feel provided by the product. Figure 7.3 depicts a regular aerosol can and a can used for shaving gels.

Since aerosol shaving creams are basically diluted lathering shaving creams, the ingredients are very similar (refer to the previous section). Additional ingredients incorporated into shaving foams and gels include the following:

- **Foam stabilizers** provide stability to the foam after it has been released from the can. Foam stabilizers orient themselves on the cell walls in a precise manner and cause them to thin to a certain extent.
    - Examples include mainly surfactants, such as those listed under shaving creams. In addition, cosurfactants as foam boosters, such as cocamide DEA, can also be used.
- **Propellants** are either compressed or liquefied hydrocarbon gases that are used to expel the products from aerosol cans.
    - Examples include isopentane (liquefied gas), butane, isobutane, and propane (compressed gases).

Over the past few years, shaving gels, often referred to as postfoaming gels, have become increasingly important in terms of market volume within the shaving product market. The product is not filled in a typical aerosol can, but into a bicompartmental system, often referred to as a BOV, i.e., bag-on-valve system.[32] The bicompartmental

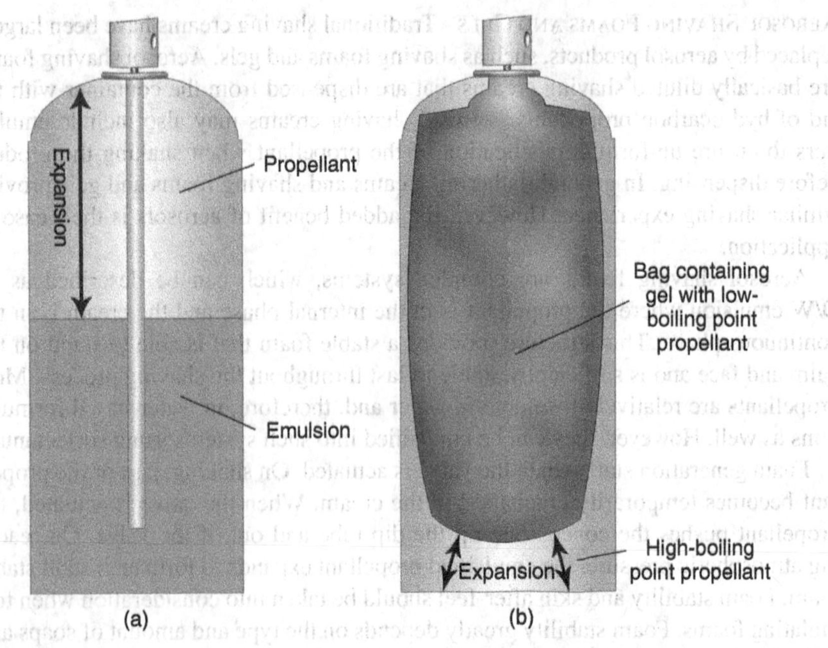

Figure 7.3 Aerosol cans: (a) aerosol can for shaving creams and (b) aerosol can for shaving gels.

system generally consists of an aerosol can, including a valve and a flexible internal bag. The bag contains the product concentrate, i.e., a gel stabilized with water-soluble polymers, and it also contains a low-boiling-point propellant. A higher-boiling-point propellant occupies the volume around the outside of the bag. It operates by causing a pressure on the outside of the bag, which forces the product out when the valve is opened. When the gel is expelled and manipulated between the wet hands before application, the heat from the skin and the mechanical action used vaporize the low-boiling-point hydrocarbon component, transforming the gel into a dense creamy foam.

These products are basically the same as shaving creams; thus, the formulation process is also the same; it is an O/W emulsification process. Similar to other aerosols, aerosol shaving creams and gels are also formulated in special buildings using special equipment. The propellants are highly flammable; thus, specific safety instructions should be followed during filling.

Chemical Depilatories ❶ **Chemical depilatories work by hydrolyzing and disrupting the disulfide bonds in hair keratin, causing the hair to break in half and allowing the hair to separate from the skin** (see Figure 7.4).[33] It can then be removed from the skin surface using a plastic spatula. Depilatories are good for use on the legs, bikini line, face, and underarms, and they perform best when the hair is at a reasonable length. A small site should be tested before a larger area is exposed to the

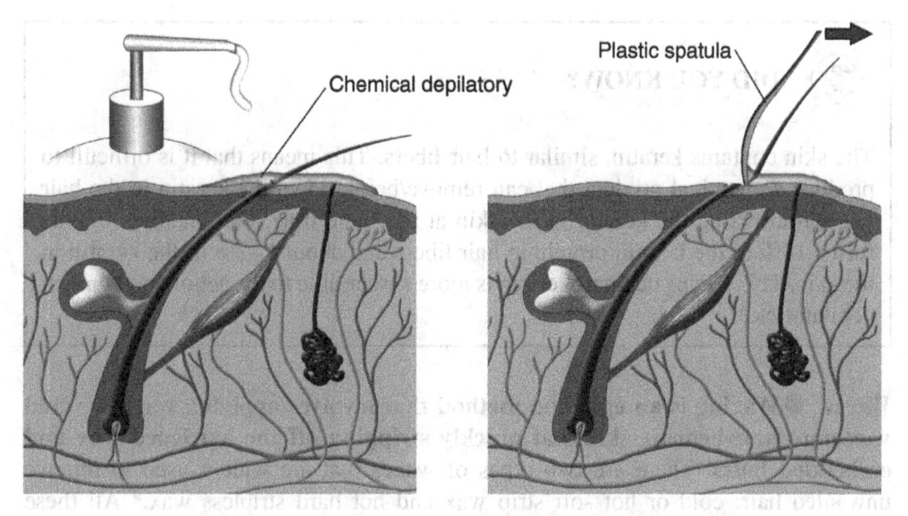

Figure 7.4 Working principle of chemical depilatories.

skin. The products should not be used on the eyebrows, near the mucous membranes, and on broken skin because they may be highly irritative even on healthy skin due to their alkaline pH. Chemical depilatories are generally available as gels, creams, lotions, and foams and in a roll-on form.

Depilatory formulations are able to produce an optimal effect in 5–15 min, depending on the pH of the preparation. The pH must be at least 10, with the quickest depilation occurring at a pH of about 12.5. Preparations for the face generally contain less thioglycolate and a lower pH than those intended for the limbs. Thioglycolates are claimed to be safe at concentrations up to 15%, if used infrequently. The most common hair removal ingredients in chemical depilatories include thioglycolates and sulfides.

- Barium and strontium sulfide (used in a concentration up to 35%)[34] act rapidly; however, hydrogen sulfide forms when using these compounds, which has a repulsive rotten egg smell and irritates the skin.
- Thioglycolates (usually applied in 2–5%), such as sodium or calcium thioglycolate or thioglycolic acid, tend to cause less irritation, and their odor is less offensive compared to sulfides; however, it takes a longer time for the hair to dissolve.

The pH of chemical depilatories is maintained with calcium, strontium, or potassium hydroxide (in a concentration of 5–10%).[34] After removing the hair, the body part is washed with water and various conditioning products are applied to restore the skin's pH to normal. This type of treatment does not destroy the dermal papilla; therefore, the hair grows back relatively soon.[34]

DID YOU KNOW?

The skin contains keratin, similar to hair fibers. This means that it is difficult to produce a chemical product that can remove/break down the keratin in the hair without affecting the keratin in the skin at the same time. Fortunately, cysteine forms 15% of the keratin protein in hair fibers and about 1–2% of the keratin in the skin. This means that hair keratin is more susceptible to the action of chemical depilatories.[35]

Waxes ● **Waxing is an epilation method that involves applying warm or cold wax onto hair-bearing skin and quickly stripping off the hardened wax and embedded hairs.** There are two types of waxes that are widely used to remove unwanted hair: cold or hot soft strip wax and hot hard stripless wax.[9] All these products need some sort of training for proper use.

- **Hot soft wax** has more of a liquid consistency on application. As it is applied to the skin in a hot form, it more readily runs to the base of the hair shaft. It spreads as a thin film and is removed with a nonwoven or muslin strip. It is applied in the same direction as the hair grows and is removed in the opposite direction.
- **Cold strip waxes** work similar to hot soft waxes. Strips precoated with wax or a cool, sugar-based substance are pressed onto the skin in the direction of the hair growth and pulled off in the opposite direction.
- In **hot hard waxing**, a thicker layer of heated wax is applied to the skin. The hair becomes embedded in the wax as it cools and hardens. It is usually much thicker and sets much faster; generally, it is recommended for stronger hair on smaller areas. The wax is pulled off quickly by hand, taking the uprooted hair with it.

The consistency of wax products varies as do their melting points. The melting point of a wax must be greater than the body temperature in order to solidify on the skin. However, its melting point should be low enough to spread over the skin without burning it. In addition, the wax must be sufficiently firm to grip the hair. A proper working temperature for hard wax is between 125 and 140 °F for application.[9] As none of the simple waxes used to make lipstick, for example, meets these criteria, hair removal waxes are based on rosins, often combined with simple waxes, such as beeswax, to modify their melting point and increase their strength.

The main ingredients of waxes are summarized as follows:

- **Hair-binding agents** provide stickiness and enable the wax to "catch" the hair follicles.
  - Rosins are hard and translucent components, derived from pine trees. Examples include wood rosin, gum rosin, dimerized rosin, and esters of rosins.

- Rosinates are glyceryl monoesters made of glycerin and rosins. They contribute to the adhesion of waxes. Examples include methyl hydrogenated rosinate, glyceryl hydrogenated rosinate, polyethylene glycol hydrogenated rosinate, and triethylene glycol rosinate.

- **Skin conditioning and skin protective agents** protect the skin from damage during the waxing process as well as modify the melting point of the rosins and rosinates and make them more flexible.
  - Examples include waxes, such as beeswax, candelilla wax, and carnauba wax; oils, such as mineral oil, almond oil, linseed oil, soybean oil, and safflower oil; butters, such as cocoa butter and shea butter; and silicone oils.
  - Natural components may have additional benefits, such as an anti-inflammatory effect. Examples include honey extract, calendula officinalis flower extract, and aloe extract.
- **Additional ingredients** can include preservatives, such as phenoxyethanol, benzoates, and parabens and antioxidants, such as butylated hydroxytoluene (BHT) and butylated hydroxyanisole (BHA); if water is used in the formulation, water-soluble thickeners may also be added to provide an optimal texture for the formulations. Furthermore, colorants, such as iron oxide, alumina hydroxide, titanium dioxide, and organic pigments and fragrances can also be incorporated into the formulations.

Sugaring is a hair removal method similar to waxing. The sugar mixture is prepared by heating sugar, lemon juice, and water to form a syrup. Examples for sugar components include corn syrup, honey extract, and maltodextrin. The syrup is formed into a ball, flattened onto the skin, and then quickly stripped away. Similar to waxing, the hair is removed entirely from the hair shaft, and sugaring is an alternative for waxing for people sensitive to wax. Sugar-based hair removal products may be applied with the hand or with a spatula. Neither method carries the risk of burning the skin since they are used at room temperature. In addition, the original formulas without resins do not adhere tightly to the skin.[28] Today, manufacturers sometimes incorporate resins in the formulas, making them similar to waxes with the disadvantages of waxes.

The main disadvantage of this technique is that sugar and sugar derivative–based compositions contain substantial amounts of water or alcohol, or mixtures of water and alcohol, which tend to evaporate each time the bottle is opened for product application. This results in the crystallization of sugar or sugar derivatives from such compositions, which causes a loss of hair removing power of the product. It also makes it more difficult to open the bottle before use.

After-Treatment Preparations ❶ **After-treatment preparations are intended to reduce redness and alleviate pain and burning sensation experienced after hair removal.** In addition, they cool and refresh the skin and exert a mild astringent effect. They can also have antibacterial activity to prevent infections while the skin recovers from the mild trauma caused by hair removal. They usually contain moisturizing ingredients to soothe and hydrate the skin. Afterwax products are also used to remove the remaining wax from the skin surface.

Aftershave preparations as well as preparations to be used after the application of chemical depilatories are generally available in the following dosage forms:

- The most popular types of products have been, and continue to be, clear **solutions**, also known as toners, containing about 40–50% of ethanol and an appropriate amount of water.
- **Gels** are also popular as they have a cooling effect resulting from the high water content.
- **Lotions**, also known as balms, and **creams** also tend to be popular, especially among users with sensitive skin who experience irritation to alcoholic products. Balms are easy to apply and provide a soothing and moisturizing effect.

Afterwax preparations are mainly available as **oils** and **wipes** soaked into oily formulations. They may also contain antiseptics and anti-inflammatory ingredients to prevent infections and inflammation of the hair bulbs.

Aftertreatment products typically contain the following types of ingredients:

- **Astringents** contract the pores and help stop bleeding from minor cuts.
  - Examples include ethanol; witch hazel; tea tree oil; boric acid; and aluminum and zinc salts, such as zinc sulfate, zinc chloride, and aluminum sulfate.
- **Emollients** and **humectants** soothe and moisturize the skin.
  - Examples for emollients include olive oil, sweet almond oil, vitamin E, panthenol, decyl oleate or diisopropyl adipate, meadowfoam seed oil, and niacinamide, while those for humectants include glycerin, sorbitol, and propylene glycol.
- **Cooling agents** contribute to the refreshing effect of the formulations.
  - An example is menthol.
- **Herbal extracts** and **essential oils** usually have combined effects, including anti-inflammatory, antiseptic, healing, and soothing.
  - Examples include aloe vera extract, orange peel oil, chamomile extract, calendula extract, lavender oil, and allantoin.
- **Fragrances** are usually also part of the formulations, which, however, may be irritating, especially for sensitive skin.
- **Water** is a basic part of solutions, gels, lotions, and creams.
- **Other ingredients** can include preservatives, emulsifiers in emulsion-based formulations, as well as coloring agents and thickeners.
- Afterwax products used to remove waxes from the skin mainly include **oils**, such as mineral oil, almond oil, and hexyldecyl stearate.

The formulation of aftershave solutions is a simple solution preparation process. Alcohol may act as a solubilizer for the fragrance; however, the incorporation of additional solubilizers may be necessary to obtain a clear solution. The formulation of aftershave gels should be started with hydration of the structuring agent. It should be kept in mind that certain types of thickeners need neutralizers to reach their optimal thickness. Then, all the additional ingredients can be added. The formulation of aftershave balms and creams is an O/W emulsification process.

Typical Quality Problems of Hair Removal Products

❼ **The typical quality-related issues of hair removal products include poor foaming of shaving gels and creams, unstable foam, brittle wax, separation of emulsions, microbiological contamination, clumping, and rancidification.** Issues that were discussed in the previous sections are not reviewed in detail here.

Brittle Wax Hot waxes are used in a warm state after heating them in a microwave oven, for example. If overheated or overcooled, they may become brittle and cannot be easily removed from the skin, leading to potential skin and hair damage and irritation. In the natural state, rosins are hard, brittle materials. If the amount and types of additives added to rosins to form waxes are inappropriate, it may also lead to wax brittleness.

Evaluation of Hair Removal Products

Quality Parameters Generally Tested ❽ **Parameters commonly tested to evaluate the quality of hair removal products include foamability, foam stability, foam viscosity, foam density, and foam structure; spray characteristics, aerosol can leakage, pressure test, and actuation force; preservative efficacy; pH; viscosity; and color.** The range of acceptance and other limiting factors are usually determined by the individual manufacturers. Since these quality parameters were discussed in detail in the previous chapters, they are not reviewed here.

Ingredients Causing Safety Concerns ❾ **The most common ingredients causing safety and environmental concerns include thioglycolates and propellants.** The findings and current statements on their safety are summarized here:

Thioglycolates As discussed previously, thioglycolates are the main ingredients in chemical depilatories. They are also used in permanent hair waving and straightening products in a different chemical environment (refer to Section 3 of Chapter 5). Thioglycolates in chemical depilatories work by dissolving the keratin into a jellylike mass that can be easily wiped off with a plastic spatula. The most common side effects reported for chemical depilatories include skin burns, blistering, stinging, itchy rashes, and skin peeling. Thioglycolates, including calcium, potassium, sodium thioglycolate, and thioglycolic acid, were reviewed by the CIR Panel and are considered to be safe in chemical depilatories when formulated to be nonirritating under conditions of recommended use.[36] The FDA recommends that consumers follow the manufacturer's directions found on the product labels, including performing a preliminary skin test for allergic reactions and irritation before using the product.[23]

Propellants Propellants are the main ingredients in aerosol formulations, including shaving creams and shaving gels. They may be volatile organic compounds (VOCs), which can have negative health effects through the formation of urban smog. Therefore, they have been a target of regulatory restrictions in hair removal products, similar to other aerosol products. The current propellant limit allowed to be used in shaving creams by the Environmental Protection Agency (EPA) is 5%, which is the same in California. As for shaving gels, the federal regulation does allow 7% VOCs, while California's limit is 4%.[37,38]

Packaging of Hair Removal Products

The most commonly used packaging materials for hair removal products include the following:

- **Plastic Tubes and Jars:** Hot stripless waxes are usually supplied in plastic jars that can be placed in a microwave to heat the products before use. Some manufacturers supply their products in jars that have a holder on one side. It allows one to handle the hot wax without burning. They may also supply wood spatulas to apply the product. Depilatory creams, pre-treatment gels, and after-treatment gels are packed into soft tubes (similar to toothpaste) or plastic jars. Some depilatory creams are supplied in plastic containers with a pump head. The package of chemical depilatories usually contains a plastic spatula for the removal of the product and the dissolved hair follicles.
- **Wax Strips:** Cold soft waxes are supplied in a ready-to-use form. Usually, muslin strips are coated with the wax that can be directly applied to the skin.
- **Wax Containers with Roll-On Head:** Hot soft waxes, which are removed with a piece of muslin strip, are usually supplied in plastic containers that have a roll-on head. This package makes their use, starting from heating to application to the skin, comfortable and easy.
- **Glass and Plastic Bottles:** Many pretreatment products and aftershave formulations are supplied in glass and plastic bottles.
- **Aerosol Cans:** Aerosol shaving foams are supplied in one-compartment cans, where the formulation and the propellant are mixed together. Aerosol shaving gels are filled into bicompartment cans, as discussed earlier.

GLOSSARY OF TERMS FOR SECTION 1

After-treatment product: A personal care product designed to reduce redness, to alleviate the pain and burning sensation after hair removal, as well as to cool, refresh the skin, and exert a mild astringent effect.

Bleaching: Removing the hair color.

BOV: Bag-on-valve is a bicompartmental system consisting of an aerosol can, a valve, and a flexible bag. The product concentrate is placed into the bag, while the propellant is filled in the can outside the bag. This technique is used for aerosol shaving gels.

Chemical depilatory: A personal care product designed to remove the hair fibers by hydrolyzing the disulfide bonds in the hair keratin.

Depilation: Removal of the hair that protrudes above the skin.

Epilation: Removal of the entire hair fibers with the root.

Epilator: An electric device designed to remove the entire hair fibers.

Folliculitis: Inflammation of the hair follicles.

Ingrown follicle: A hair fiber that curls back or grows sideways into the skin, which can lead to inflammation and the formation of tender bumps.

Pre-treatment product: A personal care product designed to be used before hair removal, mainly before dry shaving and waxing.

Rosin: A hard and translucent component derived from pine trees.

Shaving product: A personal care product designed to soften the hair and lubricate the skin in order to provide a pleasant hair removal experience.

Shaving: A depilatory technique, removal of hair using a razor.

Threading: An epilatory technique, removal of hair using a thread.

Trimming: A depilatory technique, removal of hair using scissors.

Tweezing: An epilatory technique, removal of hair using tweezers.

Waxing: An epilatory technique, removal of hair using a hot or cold wax and quickly stripping it off the skin.

REVIEW QUESTIONS FOR SECTION 1

Multiple Choice Questions

1. Which of the following is TRUE for brushless shaving creams?
   a) Are more acceptable for dry skin
   b) Have an alkaline pH
   c) Contain a higher amount of oils compared to lathering formulations
   d) A and C are true

2. Which of the following is an example for depilation?
   a) Waxing
   b) Shaving
   c) Laser hair removal
   d) Threading

3. Which of the following is an example for epilation?
   a) Chemical depilatories
   b) Shaving
   c) Threading
   d) Trimming

4. What is the general pH of chemical depilatories?
   a) 2
   b) 5.5
   c) 7
   d) 12

5. The usual working temperature of hard hair removal waxes is ____.
   a) 60–80 °F
   b) 100–120 °F
   c) 125–140 °F
   d) 150–170 °F

6. Which of the following is an advantage of waxing over shaving?
   a) Hairs grow back slower
   b) No pain
   c) Fast
   d) Acidic pH of waxes does not damage the skin

7. What is the main safety issue with the use of thioglycolates as hair removal products?
   a) Allergy
   b) Inhalation toxicity
   c) Skin irritation
   d) Environmental pollution

Matching

Match the ingredients in column A with their appropriate ingredient category in column B.

	Column A	Column B
_____	A. Alcohol	1. Astringent
_____	B. Cocamide DEA	2. Chemical depilatory
_____	C. Glycerin	3. Cooling agent
_____	D. Isopentane	4. Foam booster
_____	E. Menthol	5. Hair-binding agent
_____	F. Rosin	6. Humectant
_____	G. Sodium thioglycolate	7. pH buffer
_____	H. Stearic acid	8. Propellant
_____	I. Tea tree oil	9. Soothing ingredient in prewax products
_____	J. Triethanolamine	10. Superfatting agent

Fill in the Blanks

1. Fill in the blanks using the words in the following box. There are extra words that do not have to be used. Use each word only once.

Oils	Acidic	Viscosity	Epilation
Shaving foam	Depilation	Emulsify	Alkaline
Cooling	Antiseptic	Density	Foam stabilization
Astringent	Expel	Shaving gel	Bleaching

1. ___________ are often used in afterwaxing products to remove the remaining wax from the skin.

2. Alcohol is often incorporated into aftershave lotions because it has a(n) ___________ and _____________ effect.

3. ____________ refers to a type of hair removal product, which is usually packaged into a special container, called BOV.

4. Aftershave preparations often contain menthol for its _________ effect.

5. Depilatory formulations usually have an ___________ pH.

6. _______________ is a hair removal method where the whole hair shaft is removed.

7. A propellant is a liquid or compressed gas that is used to __________ the product.

8. A thickener, such as hydroxyethyl cellulose, is used in the formulation of a hair removal product to provide an appropriate ___________ and contribute to ___________.

9. ___________ refers to a hair removal method where a cosmetic formulation is used to "whiten" the hair follicles.

REFERENCES

1. Bercaw-Pratt, J. L., Santos, X. M., Sanchez, J., et al.: The incidence, attitudes and practices of the removal of pubic hair as a body modification. *J Pediatr Adolesc Gynecol.* 2011;25(1):12–14.
2. Castelo-Branco, C., Cancelo, M. J.: Comprehensive clinical management of hirsutism. *Gynecol Endocrinol.* 2010;26(7):484–493.
3. Bulun, S. E.: Physiology and Pathology of the Female Reproductive Axis, In: Kronenberg, H. M., Melmed, S., Polonsky, K. S., et al., eds: *Williams Textbook of Endocrinology*, 12th Edition, Philadelphia: Saunders Elsevier, 2011.
4. Tolgyesi, E., Coble, D. W., Fang, F. S., et al.: A comparative study of beard and scalp hair. *J Soc Cosmet Chem.* 1983;34:361–382.
5. Deem, D., Rieger, M. M.: Observations on the cutting of beard hair. *J Soc Cosmet Chem.* 1976;27:579–592.
6. Auriol, F., Vaillant, L., Machet, L., et al.: Effects of short-term hydration on skin extensibility. *Acta Derm Venereol.* 1993;73:344–347.
7. Lin, T. Y., Manuskiatti, W., Dierickx, C. C., et al.: Hair growth cycle affects hair follicle destruction by ruby laser pulses. *J Invest Dermatol.* 1998;111(1):107–113.

8. Partick, B., Thompson, J.: *An Uncommon History of Common Things*, Washington: National Geographic, 2011.
9. Bickmore, H.: *Milady's Hair Removal Techniques: A Comprehensive Manual*, Stamford: Cengage Learning, 2003.
10. Sherrow, V.: *For Appearance' Sake: The Historical Encyclopedia of Good Looks, Beauty, and Grooming*, Westport: Greenwood Publishing Group, 2001.
11. Physician and Patient Information Leaflet for VANIQA [package insert]. New York Bristol-Myers Squibb Company; 2000.
12. Balfour, J. A., McClellan, K.: Topical eflornithine. *Am J Clin Dermatol.* 2001;2(3):197–201.
13. Vreeman, R. C., Carroll, A. E.: Medical myths, *BMJ.* 2007;335(7633):1288–1289.
14. Alai, N. N.: Laser-Assisted Hair Removal, Last update: 10/22/2013, Accessed 11/2/2013 at http://emedicine.medscape.com/article/1831567-overview
15. Goldberg, D. J.: Current trends in intense pulsed light. *J Clin Aesthet Dermatol.* 2012;5(6):45–53.
16. Richards, R. N., Meharg, G. E.: Electrolysis: observations from 13 years and 140,000 hours of experience. *J Am Acad Dermatol.* 1995;33(4):662–666.
17. Bashour, M.: Laser Hair Removal, Last update: 4/13/2013, Accessed 11/2/2013 at http://emedicine.medscape.com/article/843831-overview#showall
18. Ertel, K., McFeatm G.: Blade Shaving, In: Draelos, Z. D., ed.: *Cosmetic Dermatology: Products and Procedures*, Oxford: Blackwell Publishing, 2010.
19. Draelos, Z. D.: *Cosmetics in Dermatology*, 2nd Edition, New York: Churchill Livingstone, 1995.
20. Khanna, N., Chandramohan, K., Khaitan, B. K., et al.: Post waxing folliculitis: a clinicopathological evaluation. *Int J Dermatol.* 2013, doi: 10.1111/ijd.12056.
21. Dawber, R. P. R.: *Diseases of the Hair and Scalp*, 3rd Edition, Oxford: Blackwell, 1997.
22. Perry, P. K., Cook-Bolden, F. E., Rahman, Z., et al.: Defining pseudofolliculitis barbae in 2001: a review of the literature and current trends. *J Am Acad Dermatol.* 2002;46(2):S113–S119.
23. FDA: Removing Hair Safely, Last update: 8/22/2013, Accessed 11/7/2013 at http://www.fda.gov/ForConsumers/ConsumerUpdates/ucm048995.htm
24. AAD: Waxing and Isotretinoin FAQs, Accessed 11/7/2013 at http://www.aad.org/dermatology-a-to-z/diseases-and-treatments/i—l/isotretinoin/waxing-and-isotretinoin-faqs
25. AVAGE - Tazarotene Cream, Allergan, Inc. Package Insert, Accessed 11/7/2013 at http://dailymed.nlm.nih.gov/dailymed/lookup.cfm?setid=cd8a3be3-8f83-42cf-8752-6fd16da65150
26. Barba, A.: Nonlaser Hair Removal Techniques, Last Update: 11/11/2013, Accessed 11/7/2013 at http://emedicine.medscape.com/article/1067139-overview#aw2aab6b3
27. Sebastian, S.: Does preoperative scalp shaving result in fewer postoperative wound infections when compared with no scalp shaving? A systematic review. *J Neurosci Nurs.* 2012;44(3):149–156.
28. Gerson, J., D'Angelo, J., Deitz, S., et al.: *Milady's Standard Esthetics: Fundamentals*, 11th Edition, Stamford: Cengage Learning, 2009.
29. Kundu, S., Achar, S.: Principles of office anesthesia: part II. Topical anesthesia. *Am Fam Physician.* 2002;66(1):99–102.

30. Draelos, Z. D.: Male skin and ingredients relevant to male skin care. *Br J Dermatol.* 2012;166(1):13–16.
31. Tadros, T. F.: *Applied Surfactants: Principles and Applications*, Hoboken: John Wiley & Sons, 2006.
32. Bag-on-Valve Systems, Accessed 11/7/2013 at http://www.colep.com/bag-on-valve-systems
33. Olsen, E. A.: Methods of hair removal. *J Am Acad Dermatol.* 1999;40(2):143–155.
34. FDA: Guide To Inspections of Cosmetic Product Manufacturers, 3/13/2009, Accessed 11/7/2013 at http://www.fda.gov/iceci/inspections/inspectionguides/ucm074952.htm
35. Ahluwalia, G.: Management of unwanted hair, In: Ahluwalia, G., ed.: *Cosmetics Applications of Laser & Light-Based Systems*, Norwich: William Andrew, 2008.
36. Burnett, C. L., Bergfeld, W. F., Belsito, D. V., et al.: Cosmetic ingredient review expert panel. Final amended report on safety assessment. *Int J Toxicol.* 2009;68–133.
37. Environmental Protection Agency, Chapter 152: Control of Emissions of Volatile Organic Compounds From Consumer Products, Accessed 4/13/2014 at http://www.epa.gov/region1/topics/air/sips/me/ME_ch152.pdf
38. Summary of State and Federal VOC Limitations for Institutional and Consumer Products, 2013, Accessed 4/13/2014 at http://www.issa.com//data/File/regulatory/VOC%20Limits%20Summary%2010-25-13.pdf

SECTION 2: BABY CARE PRODUCTS

LEARNING OBJECTIVES

Upon completion of this section, the reader will be able to

1. define the following terms:

Acne neonatorum	Atopic dermatitis	Babies	Baby cleansing product
Baby powder	Baby protecting product	Baby wipe	Diaper rash
Eye irritation	Mild surfactant	Nonwoven fabric	Penetration enhancer
Skin protectant			

2. define the age group for whom baby care products are generally formulated;
3. briefly discuss the major structural, functional, and compositional differences between baby skin and adult skin;
4. briefly discuss the major differences between baby hair and adult hair;
5. differentiate between baby cleansing and protecting products;

6. briefly discuss whether baby protecting products are considered drugs or cosmetics in the US;
7. briefly discuss the potential positive effects of baby care products on baby skin;
8. briefly discuss the potential negative effects of baby care products on baby skin;
9. list the various required cosmetic qualities and characteristics that an ideal baby care product should possess;
10. list the various required technical qualities and characteristics that an ideal baby care product should possess;
11. briefly discuss the general considerations that should be taken into account during the formulation of baby care products;
12. list the common product types available for baby cleansing products;
13. name some mild surfactants found in baby shampoos and baby bath products;
14. explain why baby cleansing products have to be milder than adult cleansing products;
15. briefly discuss the concept of a baby wipe and what components it contains;
16. name several ingredients that can be incorporated into baby protecting products as waterproofing agents;
17. briefly discuss the possible causes and treatment options of diaper rash;
18. list some ingredients that can be used as skin protectants in diaper rash formulations;
19. name some factors that may lead to a reduced efficacy of baby sunscreens;
20. list some typical quality issues that may occur during the formulation and/or use of baby care products, and explain why they may occur;
21. list the typical quality parameters that are regularly tested for baby care products, and briefly describe their method of evaluation;
22. name the performance parameters generally tested for baby care products, and describe the method of evaluation;
23. briefly discuss the main safety concern regarding the use of baby cleansing products;
24. list the typical containers available for baby care products.

KEY CONCEPTS

1. Baby care products represent a very special category of products specifically developed for delicate baby skin.
2. There are distinct differences between baby skin and adult skin as well as baby hair and adult hair in structure, function, and composition, which have to be understood and taken into account when formulating products for babies.
3. Based on the function of baby care products, they can be classified as cleansing products and protecting products.

4. Most products are considered cosmetics in the US. However, sunscreens and diaper rash creams and pastes fall under the category of drugs.
5. Cleansing products for babies are generally mild, surfactant-based systems formulated as shampoos, bath products, and wipes.
6. Protecting products for babies are usually emulsions formulated as moisturizers, diaper rash products, powders, and sunscreens.
7. The typical quality-related issues of baby care products include valve clogging for aerosol sunscreens, poor foaming of cleansing products, separation of emulsions, microbiological contamination, clumping, and rancidification.
8. Parameters commonly tested to evaluate the quality of baby care products include spray characteristics; aerosol can leakage and pressure test for aerosol products; spreadability, extrudability, texture, and firmness of lotions, creams, gels, and ointments; foamability, foam stability, foam density, foam viscosity, and foam structure for cleansing products; actuation force; color; preservative efficacy; viscosity; and pH.
9. The most frequently tested parameters referring to the performance (efficacy) of baby care products include the protective effect of diaper rash products, hydrating effect of moisturizing products, and protective effect of sunscreens.
10. Some concerns arose in the past regarding the safe use of baby care products containing talc.

Introduction

❶ **Baby care products represent a very special category of products specifically developed for delicate baby skin.** Baby care products are primarily functional products, rather than decorative products, including baby skin care and hair care cleansing formulations as well as skin protecting products. Due to the sensitivity and vulnerability of the baby skin, special considerations have to be taken into account.

This section reviews the major anatomical and physiological differences between baby skin and adult skin. It also provides an overview of the variety of baby care products available on the market today focusing on their ingredients, characteristics, and testing methods. In addition, the chapter reviews the major consumer needs and the effects that baby care products may have on baby skin and hair during use.

Anatomical and Physiological Differences between Baby and Adult Skin and Hair

The pediatric population represents a spectrum of different physiologies, starting from newborns to adolescents. The needs of a child's skin change with age; therefore, products for children are generally classified into the following groups:

- Products for **infants and toddlers** (0–2 years), which are usually referred to as "baby products." The skin of this group is the most vulnerable; therefore, its primary needs include proper skin cleansing and protection against skin irritants, especially in the diaper zone, as well as the sun.
- Products for **children** (2–11 years). The primary needs of this group include sun protection, proper skin and hair cleansing, and skin protection against dryness that may be caused by water spots.
- Products for **adolescents** (12–16/18 years, depending on the region). The major concerns for this group constitute the skin conditions resulting from changes in their hormone system, such as acne.[1,2]

This section focuses on products formulated for the first group, and the products will be referred to as "baby care products." ● **There are distinct differences between baby skin and adult skin as well as baby hair and adult hair in structure, function, and composition,[3] which have to be understood and taken into account when formulating products for babies.** Therefore, first, a brief summary is provided here.

Skin In general, babies have an anatomically developed skin, which, however, undergoes dramatic physiological and structural changes in the first year of life. About 20 weeks after birth, the skin becomes functional and develops a protective barrier,[4] which is still "unripe" and develops through the first year of life.[5] The major differences between the hair of babies and that of adults are summarized here:

- *Structural Differences in the Skin*: Baby stratum corneum (SC) is approximately 30% thinner than adult SC,[6,7], and the entire epidermis in baby skin is about 20–30% thinner than in adult skin. [8] The cells in the SC are smaller in baby skin, which refers to a faster cell turnover in the epidermis. This serves as an explanation for the improved wound healing properties of babies when compared to adults.[9] Melanocytes are less pigmented, which explains the pale color of newborn skin. It also compromises the sun protective function of the skin. Sun protection is, therefore, even more important for babies than for older children. Additionally, immune protection of the skin and its receptor function also develops dramatically during the early months of life.[3] Structural differences exist also in the dermis as well. Elastic and collagen fibers are present; however, their density is lower.
- *Compositional Differences in the Skin*: Initially, after birth, the water content of the skin of babies is considerably drier than that of adults.[10] However, the water content usually increases in the first month of life. After the first few months, baby skin is considered more hydrated than young adult skin.[10] The sebaceous gland activity and, therefore, sebum production are decreased in babies.[11] Studies show that the concentration of the NMF, which serves as the skin's own humectant, is lower in baby skin than in adult skin.[12] At birth, the skin pH of babies varies between 6.34 and 7.5. Shortly after birth, within the first 2 weeks,

this neutral pH becomes more acidic (approximately 5), which is similar to that of adult skin.[13]

- *Functional Differences in the Skin*: Although the amount of sebum in the skin is significantly lower in babies than in healthy adults, the skin barrier in babies is still competent even at the time of birth.[5] Additional functions of the skin, such as water-retaining ability, still have to improve after birth. Although baby skin is known to be more hydrated than adult skin, it loses water faster than adult skin (has a higher transepidermal water loss, TEWL value). Its water-absorbing ability is superior, meaning that it can absorb more water than adult skin; however, it cannot efficiently keep it.[12] This explains the importance of using moisturizing products to protect baby skin. The surface area in the newborn is relatively higher compared to body weight, which can lead to a more ready absorption through the skin. This high ratio declines progressively during the first year. Eccrine sweat glands exist and function in baby skin; however, their activity is lower than that in adult skin. It can lead to compromised thermoregulation. Apocrine glands become functional only during puberty.

Hair The hair of newborns is well developed; however, it is only faintly pigmented. Their body is mainly covered with vellus hair, which is much lighter in color and thinner than adult hair.[14] After birth, the hairs pass from anagen to the telogen phase, which leads to baby hair fall out about 8 weeks after birth. Afterwards, the hair cycle is similar to that observed in adults, and the hair follicles will be in different stages of the cycle.[15] The number of hair follicles per square centimeter is higher than that in adult skin, which increases the risk of deeper penetration of topical substances applied to the skin.[16] Due to the lower activity of the sebaceous glands, there is not much oil on the hair shafts' surface. Therefore, strong surfactants are not needed to clean the hair.

Other

- **Acne Neonatorum:** Although the sebaceous gland activity in babies is usually lower than in adults, a skin condition called acne neonatorum (i.e., neonatal acne) may develop in babies. It is reported to occur in up to 20% of newborns,[17], usually 2–3 weeks after birth and may persist up to a few months of life. The primary cause is stimulation of the sebaceous glands by maternal or infant androgens. Acne neonatorum generally affects the face, but may also affect the neck, chest, and back. It is more common in males.[17] Neonatal acne usually resolves itself spontaneously within a few months, with treatment required only for cosmetic improvement. The sebaceous gland activity then decreases to a level lower than that in adults, and reactivation typically occurs during puberty.
- **Eyes:** In addition to the differences between baby and adult skin, babies have a greater vulnerability to ocular exposure as well. Babies have neither a fully developed blink reflex nor a behavioral blink response in anticipation of accidental exposure, such as during bathing.[18] Therefore, it is particularly important

that products for babies be extra mild with a minimum number and an extremely low amount of irritating ingredients.

As a result of all these factors, baby skin is more vulnerable and sensitive to the environmental threats than adult skin. Additionally, since the skin keeps developing after birth, the use of proper skin care products is essential to help maintain the natural adaptation and development of the skin barrier.[19]

Types and Definition of Baby Care Products

● **Based on the function of baby care products, they can be classified as cleansing products and protecting products.** Baby care products, in general, provide appropriate cleansing, skin conditioning, as well as protection against water loss, which are cosmetic functions. Therefore, ● **most products are considered cosmetics in the US. However, sunscreens,** which prevent the skin from ultraviolet (UV) radiation, **and diaper rash creams and pastes,** which prevent and/or treat topical skin irritation and inflammation, **fall under the category of drugs.**

- **Cleansing products** for babies are designed to clean the baby skin and hair and remove saliva, nasal secretions, sweat, urine, feces, bacteria, residues of milk, and other dirt and irritants. In general, baby cleansing products are similar to adult cleansing products; however, the surfactants used in these products are much milder. Additionally, many baby products are fragrance-free or contain only a low level of these components. Cleansing products include baby bath products, shampoos, and cleansing wipes.
- **Protecting products**, as their name implies, are designed to protect the baby skin, help maintain its hydration, and, thus, prevent drying. These products are similar to the moisturizing products for adults. Most moisturizing products for babies are O/W emulsions. Sunscreens and formulations for the diaper zone, however, are mainly W/O emulsions, which provide a waterproofing effect. Protecting products include moisturizing lotions, creams, diaper rash creams and pastes, sunscreens, and baby powders.

History of Using Baby Care Products

The history of baby care products is not as well described and detailed as that of adult cleansing and protecting products. The main reason for this may be the fact that the importance of taking special care of babies' skin was not recognized and understood for a long time. Until the late 19th century, the main cleaning aids were handmade soaps made of fats and lye. Later on, syndet bars, based on synthetic detergents, were introduced to the market. Procter & Gamble started to sell pure bar soaps for babies in the late 19th century; however, the same products were used to clean dishes and clothes as well.[20] Johnson & Johnson launched maternity kits in the 1890s. These kits included Johnson's baby powder, which was the first and very successful item of the company's baby care line.[21] Baby powders were popular for a long time for providing lubricity and a dry environment to the diaper zone. Today, although still available, their use has decreased. Introduced in the 1950s, Johnson's® Baby Shampoo was the first specially formulated product to be as mild to a baby's eyes as pure water. The product, known by its famous claim "No More Tears," is still available on the market today.[21,22] Since then, many products have appeared on the market, specifically formulated for babies' delicate skin and sensitive eyes.

How Baby Care Products May Affect Baby Skin and Hair?

Good skin hygiene is essential to the overall health of babies. Baby care products are essential in preserving the skin's barrier function and preventing the entry of bacteria, allergens, particulates, pollution, and other exogenous substances into the body.

- Bacteria, bacterial enzymes, and other soilage on baby skin can lead to irritation, redness, itching, inflammation, and infections. Baby **cleansers** help remove unwanted material and are, therefore, essential in preventing infections and the transmission of microbes from the baby skin surface. Keeping the hands clean in the case of babies with their hand-to-mouth behavior is also essential and can help reduce or prevent oral transmission of microbial contaminants. Water is insufficient for the removal of oil-soluble dirt and soilage and also has poor pH-buffering action[23]. Studies show that cleansing with mild cleansers do not significantly change the TEWL, SC hydration, skin surface pH, and sebum production in babies if care is taken when using such products.[24] The current recommendation of pediatrics and dermatologists is to use liquid, mild, pH-neutral, or slightly acidic cleansers instead of traditional alkaline soaps on baby skin.[25]
- As discussed earlier, baby skin cannot effectively retain water in the SC and deeper layers of epidermis.[26] Therefore, using adequately formulated **emollients** is essential in keeping the skin hydrated. Emollients are most often incorporated into moisturizing lotions and creams that can preserve, protect, and enhance babies' skin barrier function by supplying the skin with water and lipids and helping inhibit water loss.
- Environmental UV **exposure** is another concern for babies. Studies show that UV radiation–induced changes in the skin may begin as early as the first summer of life.[27] It emphasizes the importance of using sun protective aids, which, to most parents, mean sunscreens. However, the significant structural and functional differences between adult and baby skin suggest a greater susceptibility of babies to both absorption of sunscreens through the skin and penetration of UV light. Additionally, there is lack of testing to prove the safety of topical sunscreens for newborns and young babies. Therefore, many US public health organizations, including the Centers for Disease Control and Prevention (CDC), and the National Council on Skin Cancer Prevention, as well as professional organizations, including the American Academy of Dermatology (AAD) and the American Academy of Pediatricians (AAP), actively promote public awareness about comprehensive sun protection. In the first 6 months of life,

sun avoidance is the first-line strategy and not the use of sunscreens. In addition to sun avoidance, especially during peak intensity hours, protective measures include wearing protective clothing and hats, sunglasses, and umbrellas.[28,29]

- **Diaper rash** is one of the most common inflammatory skin reactions in babies. Its overall prevalence is approximately 4–15% in the first month of life,[30], and typically, more than 50% of infants will have at least one episode of diaper rash during the diaper-wearing phase.[31] General emollients are less effective in preventing and/or treating diaper rash; therefore, special products in the form of creams and pastes containing skin protecting ingredients are formulated to specifically aim at this condition.

Although the benefits and need for good hygiene are known, the use of baby care products is often controversial. The most common potential negative effects of baby care products are summarized as follows:

- Skin care products with an inappropriate pH and/or ingredients can significantly alter the skin. Alkaline soaps are effective skin cleansers; however, they can disrupt the skin surface pH, decrease the SC thickness alter the skin lipids, and lead to **dryness and irritation.**[32] All these reasons may make alkaline soaps less preferable for baby skin. Water, in addition to not being an effective cleanser, can also have a drying effect on baby skin.[25] Additionally, cleansing products containing antimicrobial ingredients are not recommended for use in babies because of their harshness and potentially negative effect on skin colonization. Rubbing the skin with a cloth is also discouraged since it may lead to epidermal injury.[33]
- Baby skin has a greater tendency to develop irritant/allergic **contact dermatitis.**[34] Fragrances, certain preservatives, and also some ingredients used in moisturizer products, such as propylene glycol and lanolin, are known irritants and sensitizers for adult skin. Therefore, they can definitely irritate babies' delicate skin, and their use should be minimized in baby care products.
- It is also known that overaggressive cleansing could lead to **atopic dermatitis** (AD).[34] AD, also known as eczema, is a severe inflammatory skin condition that usually develops as a result of hypersensitivity of the immune system. AD occurs in 15–20% of children,[35] typically early in life: 49–70% of childhood AD appears before 6 months of age and 80–90% of babies suffer from AD before 5 years of age.[36] Compromised skin barrier function is believed to be crucial to the development and severity of AD, which is often accompanied by dry, scaly skin and itching.[35] Special care and the use of adequately formulated, mild cleansers, and skin moisturizers are especially important for these children.
- **Eye stinging** and burning sensation is one of the most common drawbacks of many shampoos, for adults, available on the market today. As discussed in Chapter 5, the main ingredients responsible for this negative effect are the primary (anionic) surfactants in the formulations. Anionics are effective cleaning

agents, which are, however, not needed for babies for several reasons: baby hair is weaker and softer than adult hair; it is not as greasy as adult hair can be; and babies do not use hairstyling aids, which may need stronger chemicals to be removed. Baby shampoos generally contain mild, amphoteric, and nonionic surfactants that are nonirritating.

Required Qualities and Characteristics and Consumer Needs

From a consumer perspective, a quality baby care product should possess the following characteristics:

- Cleansing products:
  - Good foamabilty
  - Hydrate the skin
  - Do not alter the skin structure
  - Non-irritant for the eyes and do not cause eye stinging
- Diaper rash products: provide protection against diaper rash
- Sun protection products: provide efficient protection against UVB and UVA radiations
- Pleasant feeling during and after application
- Easy to spread
- Easy to remove
- Contains non-toxic and non-sensitizing materials
- Does not change the surface pH
- Slightly fragranced or not fragranced.

The technical qualities of baby care products can be summarized as follows:

- Sun protection products and diaper rash products: proven efficacy
- Long-term stability
- Appropriate texture
- Appropriate rheological properties
- Mild
- Dermatological safety.

Types, Typical Ingredients, and Formulation of Baby Care Products

The baby care market offers a limited variety of products, usually focused on cleaning and moisturizing. There are some major factors that should be taken into consideration when formulating any type of product for babies. Let us review these factors first before discussing the various baby care products.

- In general, formulators should only use ingredients that have an established **safety profile** and are deemed to be safe on baby skin.
- Ingredients known to be **irritants, sensitizing agents, and/or allergens** should be avoided when formulating products for highly sensitive and immature baby skin.[37] Examples include sodium lauryl sulfate, ethanol, strong or excessive amount of preservatives, and fragrances.
- As most formulations for babies are water based, the use of **preservatives** is a must in these products. Preservative-free formulations are more dangerous to baby skin due to the potential risk of microbiological contamination than formulations with properly selected preservatives. The level of preservatives should be kept as low as possible in formulations. A good solution is to apply a combination of preservatives. This way they may have a broader spectrum, their overall efficacy can be increased, and the concentration of the individual ingredients can be kept low.
- Care should also be taken when using "natural" preservatives or other natural ingredients since they may have not been tested on baby skin; therefore, there is no safety information about their use.
- Formulations should be **mild** and their **pH** should be close to the skin's natural pH in order not to disturb the natural skin flora.
- Ingredients known to be **penetration enhancers**, such as urea or propylene glycol, should not be included in formulations. The main reason for this is that penetration through baby skin is already easier for many ingredients than through adult skin.[27] Enhancing the penetration could easily lead to irritation and even inflammation.
- **Emollients** should be carefully selected for baby formulations in order not to cause allergic reactions, irritation, or other types of problems.
- Formulations should contain ingredients that do not cause **eye irritation** in babies. Cleansing products can often run into the eyes during rinsing. Even leave-on products, such as sunscreens, can cause eye irritation because babies often rub the topical products into or near their eyes. As blinking, which serves as a protective mechanism for the eyes, continues with the eyes maturing during the first year, [38] the babies' eyes are more sensitive and vulnerable to irritants.

Cleansing Products for Babies ❺ **Cleansing products for babies are generally mild, surfactant-based systems formulated as shampoos, bath products, and wipes.** The main types of ingredients, except for surfactants, are similar to adult cleansing products; therefore, they are not discussed in detail here (see Section 2 of Chapter 4 for more detail). This section discusses the main characteristics of these products and ingredients different from those of adult products.

Baby Bath Products For baby bathing, the ideal water temperature is 38–40 °C, and bath time should be limited to 5–10 min.[33,39] Baby bath products do not need to possess high cleansing activity, because the extent of soiling of a baby's skin is generally low when compared with that of adult skin. For the same reason, adult skin cleansers should not be used on baby skin. As discussed earlier, strong and effective anionic surfactants can remove components of SC lipids, NMF, and can even insert themselves into the skin.[40] All these can quickly lead to alteration and/or disruption of the skin barrier and development of symptoms, such as irritation, dryness, redness, and itching.[40–42] Primary requirements of a baby bath are that it should be functional, yet mild on the skin, and nondrying. When diluted in bath water, baby bath products are often used to shampoo baby hair; therefore, low eye irritancy is another essential factor. For this reason, baby bath formulations generally contain a low level of secondary surfactants, preferably amphoterics (such as cocamidopropyl betaine) and nonionics (such as polyethylene glycol (PEG)-80 sorbitan laurate). In addition, they may also contain mild anionic surfactants (such as sulfosuccinates, e.g., disodium laureth sulfosuccinate, and isothianates, e.g., sodium cocoyl isethionate) in lower levels. When using a combination of such surfactants, the micelles forming above the critical micelle concentration (CMC) are generally larger with a lower micellar charge. These result in reduced aggressiveness of the surfactant system to the skin and better tolerance.[43] Viscosity is generally a major factor to consider since it can prevent running of the formulation into the eyes. Typically, hydrophilic thickeners, such as cellulose derivatives, gums, or acrylate polymers are used. Additional ingredients include solvents, typically water; preservatives; antioxidants; and chelating agents. Moisturizers, such as humectants and emollients as well as cationic polymers, may also be included in the formulations to minimize any drying effect and enhance the softness and feel of the skin after use. Colorants and fragrances may also be included.

The formulation of baby bath products is similar to that of adult products. It usually starts with dispersion and hydration of the thickeners, while surfactants are added in the final step with gentle mixing to avoid excessive foam formation.

Bar soaps have an alkaline pH and are, therefore, harsher to the skin. They can shift the skin's normal pH to the alkaline range, which can damage the skin barrier and lead to irritation and dryness.[44] Soaps were popular in the past; however, today, their use has diminished.

DID YOU KNOW?

In addition to using nonionic surfactants in baby cleansers, there are other ways of lowering the harshness of cleansing products as well. More commonly, hydrophobically modified polymers (HMPs) are added to surfactant-based cleansers, which can help create large polymer–surfactant complexes. These complexes are less irritating to the SC lipid barrier. Additionally, HMPs can reduce the effective concentration of free surfactant micelles in solution and facilitate foam formation.[45]

Baby Shampoos Babies do not have as much lipids on their hair as adults, and they do not use any hair fixatives either. Therefore, their hair is easy to clean. Baby shampoos should be mild, and they do not have to be applied every day. Similar to baby bath products, baby shampoos are usually based on mild secondary surfactants, such as nonionics (e.g., PEG-80 sorbitan laurate) and amphoterics (e.g., cocamidopropyl hydroxysultaine).[14] These surfactants are mild to the eyes. Irritation can be further reduced by increasing the product's viscosity since it can prevent the product from running into the eyes. These surfactants as well as anionic surfactants do not foam; therefore, baby shampoos may be considered low-performing products by some parents. However, as discussed previously, foaming in not related to the cleansing efficacy of shampoos; a low-foaming formulation can be an effective cleanser just as a highly foaming formulation.

The formulation technology of baby shampoos is similar to that of adult shampoos as well as baby bath products.

Baby Wipes Disposable baby wipes represent one of the fastest growing sectors of the baby products market. They have become a very popular alternative liquid cleansing products, particularly for the diaper area. Baby wipes usually consist of a nonwoven disposable cloth soaked with an aqueous surfactant solution or an O/W emulsion enriched with emollients.[30] Wipes are usually non-lathering formulations, which are used to wipe off the baby skin without rinsing it afterwards. Typical cleansing ingredients used in such solutions and emulsions include mild nonionic surfactants, such as coco glucoside and lauryl glycoside, as well as amphoteric surfactants, such as disodium cocamphodiacetate. Humectants can also be incorporated into the solutions or emulsions, which provide moisturizing benefits for the products. Thickeners, usually hydrophilic ones, such as xanthan gum, are often also formulated into these products to provide optimal viscosity for the soaking solution or emulsion. Due to the water content of the emulsions, preservatives must be used to ensure that the product will not be contaminated during its normal lifetime. Preservatives and fragrances, which may be used to give a nice scent to products, can be irritants though.[29] Formulators can also incorporate soothing and anti-inflammatory ingredients in the solutions or emulsions that can reduce the signs of any irritation. Examples for such ingredients include allantoin, and plant extracts, such as chamomile extract, pot marigold extract, and aloe extract. In general, baby wipes are shown to be mild and well-tolerated by baby skin, even when used on sensitive skin or dermatitis skin.[46–48] Important factors that should be considered are the pH of wipes, fragrance selection, and the types and amount of surfactants included. Strong irritants, such as alcohol, fragrances not tested for their allergic potential, and pH significantly different from that of the skin should be avoided in order not to damage the baby skin or worsen diaper rash.

The formulation of baby wipes is similar to that of facial cleansing wipes. The surfactant solution and wipes are prepared separately. In the final step, wipe cloths are soaked in the cleansing solution and packaged either individually into sachets or together into plastic bags or other types of containers.

Moisturizing and Protecting Products for Babies **Protecting products for babies are usually emulsions formulated as moisturizers, diaper rash products, powders, and sunscreens.** Baby skin can quickly dry out, which, however, can be prevented using appropriate moisturizing products containing humectants and emollients. In order to prevent diaper dermatitis and other irritation reactions in the diaper zone, formulations that can make the skin more waterproof, usually based on emollients and occlusives, are needed to be used. Sun protection is another important aspect of skin protection, which is also discussed in this section. Similar to cleansing products, only the major characteristics and differences from adult products are detailed.

Baby Lotions and Creams Simple moisturizing lotions and creams for babies are emulsions, generally, light O/W formulations that are easy to spread and quick to absorb. The oil phase generally consists of various emollients and also occlusive ingredients, which can significantly lower TEWL in babies if used after skin cleansing.[24] Most frequently, mineral oil and various vegetable oils, such as sweet almond oil, sunflower oil, shea butter, meadowfoam seed oil, and palm oil, are used. A frequent problem with many vegetable oils is their sensitivity to oxidation, which can lead to rancidity. Using unstable emollients or those that degrade quickly may lead to undesirable effects, especially on baby skin that is undergoing maturation. Mineral oil is a stable ingredient that is preferable for use on baby skin. It is an occlusive ingredient that forms a protective waterproof layer over the skin, reducing water loss. It may also act as an emollient. Another beneficial property of mineral oil is its long record of safe use.[49] It is unlikely to go rancid even in hot, humid climates. Silicone oils have also received more attention recently due to their beneficial effects on the skin. Similar to cleansing wipes, soothing and anti-inflammatory ingredients, such as allantoin, panthenol, and natural extracts, such as calendula extract, can also be added to the products. The water phase often contains humectants; most commonly, glycerin for its beneficial moisturizing effect. Antioxidants, such as tocopherol, and chelating agents, such as tetrasodium glutamate diacetate, are usually also incorporated into the products to prevent deterioration accelerated by metal ions and free radicals. The formulation steps for baby lotions and creams follow the general emulsification process.

Diaper Rash Creams and Ointments Diaper rash, also known as diaper dermatitis, is one of the most common dermatologic conditions encountered in babies while using diapers.[50] It usually occurs on the groin, thighs, buttocks, and perianal area of babies. Its development is multifactorial (see Figure 7.5). Factors contributing to its development include skin wetness due to incontinence in babies along with infrequent diaper changes, which lead to a longer contact time between the skin and potential skin irritants.[29] Moisture also makes the skin more fragile and increases its susceptibility to frictional damage caused by the diaper, pressure, and chafing. Diaper rash usually occurs as a primary reaction to irritants in urine and feces, as well as moisture and friction.[51] If urine and feces are present in the diaper zone, the pH shifts to a more alkaline value. It irritates the skin and leads to degradation of the skin proteins and lipids and an impaired barrier function.[30] Additionally, alteration of the skin pH allows for the growth of microorganisms, including *Candida*, *Staphylococcus*, and *Streptococcus* species.[29] Signs and symptoms of diaper rash include uncomfortable erythema and mild scaling.[52] If left untreated, papules and edema can occur, or, in severe cases, ulcerated lesions may develop.[33]

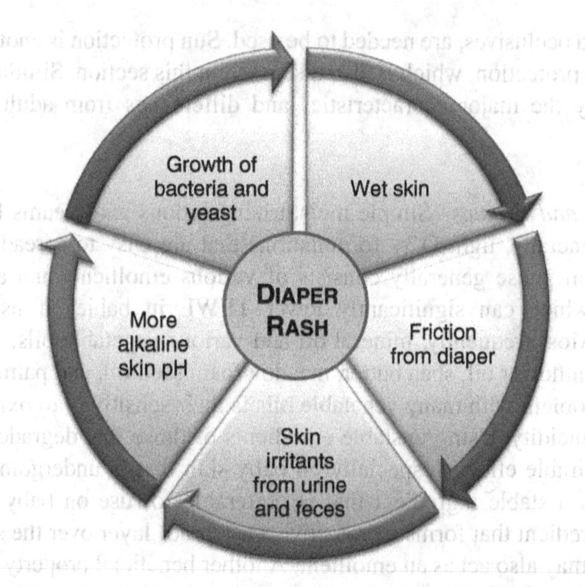

Figure 7.5 Major factors contributing to the development of diaper rash.

In order to avoid worsening of the condition, more frequent diaper change should be performed with proper cleansing of the diaper zone. It limits the amount of time the skin is exposed to urine and feces.[53] Additionally, formulations aimed to treat diaper rash should be applied on the skin surface.

Products specifically formulated for the diaper zone are usually W/O emulsions or anhydrous pastes, which are thicker, longer-lasting, and more waterproof than water-based formulations. These products serve as protective formulations to protect the skin from urine and feces, and most of them are considered drugs in the US. The OTC monograph for skin protectants[54] includes a list of ingredients and their concentration that, if used in the concentration stated, are considered drugs. In such cases, the ingredients are listed as active ingredients. Below this concentration, the products are considered cosmetics, even if the same ingredients are used. Examples for such ingredients frequently used in diaper rash treatment products with their concentration as listed in the OTC monograph include the following:[54]

- Allantoin: 0.5–2%
- Cocoa Butter: 50–100%
- Dimethicone: 1–30%
- Glycerin: 20–45%
- Kaolin: 4–20%
- Lanolin: 12.5–50%
- Petrolatum: 30–100%
- Zinc Oxide: 1–25%.

Natural extracts, such as aloe vera extract, can be included for additional soothing and anti-inflammatory benefits, and silicones are also often added to improve water resistance. These products often leave a protective film on the baby skin, thus reducing chaffing or soreness, and leaving the skin soft and smooth. If the diaper rash shows evidence of *Candida* infection – diagnosed by a dermatologist – antifungal therapy may be indicated.[55]

During the formulation of diaper rash creams and pastes, it is important to provide homogenous distribution of the solid ingredients in the semisolid phase. Typically, first the emulsion or anhydrous base is prepared. In the case of emulsions, hydrophilic thickeners should be thoroughly mixed and hydrated in the water before mixing it with the oil phase. After preparing the base, the powder ingredients, such as kaolin and zinc oxide, are added to the base with intensive mixing and thoroughly worked into the base using a mortar and pestle in laboratory settings or a grinding mill or triple roll mill for larger batches. These tools and equipment ensure a smooth, clump-free appearance for the products.

DID YOU KNOW?

Zinc oxide can be used as both a skin protectant and a sunscreen, but not in the same particle size. Diaper rash formulations typically contain zinc oxide with a larger particle size. These products leave a visible white layer on the skin after application. On the other hand, sunscreens contain micronized zinc oxide, which means that its particle size is decreased to the micro- or nanorange (1 mm = 1000 μm = 1,000,000 nm). Due to the smaller particle size, they do not provide a visible white appearance, but are completely transparent.

DID YOU KNOW?

Diaper technology has significantly improved. There are diapers that can deliver petroleum jelly and zinc oxide to the skin surface, providing a waterproof layer on the skin. These have been shown to significantly reduce the prevalence and severity of diaper rash.[56] Another innovation is diapers with an absorbent gelling material that wicks moisture away. Studies suggest that these diapers are associated with less severe diaper rash.[57]

Baby Powder Powder is among the earliest and best known of all baby care products. The primary function of a baby powder is to absorb the residual moisture on the skin, provide a certain degree of mild lubrication, and prevent irritation.[39] The main ingredients providing lubricity in baby powders are talc and starch, primarily corn starch. Additional ingredients may include other absorbents, such as kaolin; fragrances; natural extracts, such as aloe extract; and antioxidants. Zinc oxide may also be included due to its mild antibacterial properties.

Today, corn starch–based baby powders are more popular due to safety issues, which arose in the past for talc-based products. (The safety issues are discussed later under evaluation of products.)

DID YOU KNOW?

Particles considered inhalable (i.e., those with a particle size >10 μm) are regulated by the Environmental Protection Agency (EPA).[58] These particles can significantly affect the heart and lungs and can cause serious health effects. Such particles can be emitted from a variety of sources, including construction sites, unpaved roads, fields, smokestacks, power plants, industries, and automobiles.

DID YOU KNOW?

Talc is a very frequently used white powder ingredient in formulations due to its versatile properties: it has abrasive, absorbent, anticaking, bulking, opacifying, skin protective, and slip modifier properties.[59] It is used in all types of cosmetics and personal care products, including face powders and eye shadows, aerosol makeup bases, aerosol deodorants, and baby powders, among others.

Sunscreens for Babies Due to the significant structural and functional differences between adult and baby skin, penetration of UV light into baby skin is easier than into adult skin, which makes babies more vulnerable to sunlight. Newborns today have a projected 1:33 risk of melanoma versus the 1:1500 chance of children born in 1935.[60] Therefore, sun protection is especially important in their case. According to US public health organizations, and professional organizations, the best way to protect babies younger than 6 months from the sun is to avoid the sun and cover their body and head with hats and clothing. In the cases where sun exposure cannot be avoided, sunscreens should be applied to the uncovered body surfaces.[61,62] When sunscreens are used, preferably, products should provide protection against UVA and UVB radiations and offer a high sun protective factor.[63]

For appropriate protection, it should be ensured that a reasonable amount of product is applied to the skin. Most people apply considerably less amount of sunscreen to the skin than recommended (2 mg/cm^2 of skin for adults). In such cases, a sunscreen of a given SPF may not be as protective as claimed.[64] Another typical problem that may reduce the efficacy of sunscreens is that certain body parts may be adequately covered, while some parts are missed. Commonly missed skin areas include the ears, hairline, neck, and feet.[65,66] Reapplication is also necessary during the day since sweating and swimming can wash some product off the skin, which can decrease its protective function (especially if the originally applied amount was lesser than recommended).[67]

Due to the weaker defense system of the baby skin, sunscreen ingredients can penetrate into the baby skin much easier than into adult skin. For babies, often, physical UV filters, including micronized and nanonized forms of zinc oxide and titanium dioxide, are used as an alternative for chemical sunscreens. Physical filters penetrate only a few layers into the SC, which reduces the potential irritation or sensitization.[68] Additionally, they are effective and safe for babies.[69,70]

Water-resistance is an important aspect of sunscreens; therefore, most formulations include waterproofing ingredients, similar to adult products (see Chapter 3). The formulation technology of sunscreens is similar to that of adult sunscreen products. Sunscreen products for babies are available in a variety of dosage forms, including aerosols, creams and lotions, pump sprays, and sticks.

Typical Quality Problems of Baby Care Products

●**The typical quality-related issues of baby care products include valve clogging for aerosol sunscreens, poor foaming of cleansing products, separation of emulsions, microbiological contamination, clumping, and rancidification.** As these problems were discussed in detail earlier, they are not reviewed here.

Evaluation of Baby Care Products

Quality Parameters Generally Tested ●**Parameters commonly tested to evaluate the quality of baby care products include spray characteristics; aerosol can leakage and pressure test for aerosol products; spreadability, extrudability, texture, and firmness of lotions, creams, gels, and ointments; foamability, foam stability, foam density, foam viscosity, and foam structure for cleansing products; actuation force; color; preservative efficacy; viscosity; and pH.** As these evaluations were discussed in the previous sections, they are not reviewed in detail here.

Performance (Efficacy) Parameters Generally Tested ●**The most frequently tested parameters referring to the performance (efficacy) of baby care products include the protective effect of diaper rash products, hydrating effect of moisturizing products, and protective effect of sunscreens.** Efficacy testing for baby sunscreens is identical to that of adult products. Since the testing conditions were detailed in Section 5 of Chapter 3, and the performance testing of moisturizing products was discussed in Section 2 of Chapter 3, these tests are not discussed here again. For more detail, refer to Chapter 3.

Efficacy of Diaper Rash Products As discussed previously, the main function of diaper rash products is to form a waterproof layer over the skin and protect the skin from irritating ingredients in urine and feces. The efficacy of the products is generally tested *in vivo*, on babies, and the results are evaluated by experts. A general study procedure can be described as follows. The study subjects are randomized into two groups: one of the groups receives the diaper rash product, while the other group receives a control (placebo) product or just usual care. The diaper rash product and the placebo product are applied to the diaper zone after gentle cleansing at every diaper change and following bathing of the baby once daily for a few weeks. Clinical evaluation of the diaper rash is performed by experts before the study (Day 0) during the treatment and on the last day of the treatment. The severity of diaper rash is assessed for specific sites (such as the buttock, genital region, perianal region, back, and legs, among others) on a predetermined scale. In addition, the overall condition of the diaper zone is also evaluated on a similar scale.[71] Microbiological determination can also be performed for the presence of bacteria (*Streptococcus* and *Staphylococcus* species) and yeast (primarily *Candida albicans*).[72] The results are evaluated at the end of the study, and statistical analysis is performed to see whether the differences are statistically significant or not.

Safety Concerns Due to the increased vulnerability of baby skin, ⑩ **some concerns arose in the past regarding the safe use of baby care products containing talc.** The main scientific findings are summarized here. Additionally, as eye irritation is a great concern for baby care products, the major testing methods for such irritation are also discussed here.

Talc A variety of safety concerns arose in the past with regard to the use of talc in baby powders. The issues were focused on the potential asbestos contamination of talc, its potential inhalation during use, and potential microbiological contamination over time. Since the safety issues and scientific findings regarding the asbestos contamination of talc were discussed under color cosmetics (Chapter 2), they are not discussed again here.

Products containing talc when applied to baby skin (or used in cosmetic sprays and face powders) may potentially be inhaled. Particles that can be potentially inhaled and get deep into the lungs are referred to as "respirable" or "inhalable" particles. In order for a particle to be potentially inhaled, its diameter should be smaller than 10 μm. Larger particles are generally deposited in the upper respiratory tract.[58] The particle size (diameter) of talc varies widely depending on the type of product it is used in and also on the manufacturer selling talc. However, the particle size of talc in loose powders, such as baby powders, is generally larger than that of inhalable particles. Therefore, talc from baby powders will not get into the lungs. Additionally, in loose powders, agglomeration of particles may also occur due to various factors, leading to an increase in the particle size. Based on this information, talc is found to be non-inhalable and safe for use in cosmetics by the CIR Panel.[73]

An additional concern that arose with regard to talc was its potential microbiological contamination over time. Microbial contamination could lead to irritation on the applied skin surface and skin infections. There were issues in the 1940s in New Zealand, when baby powder was infected with bacteria.[74] Since talc used in cosmetics today (referred to as "cosmetic grade talc") is sterilized by heat treatment, it has a low chance to being contaminated.[75] Additionally, manufacturers usually check the microbial contamination of the final products as part of the quality evaluation. The so-called anaerobic plate count test is recommended by the FDA to test whether talc and powders are contaminated with microorganisms (especially *Clostridium tetani*).[76]

Eye Irritation of Baby Care Products As discussed in Section 2 of Chapter 4, the traditional method for testing eye irritation for color cosmetics and personal care products is the Draize eye irritation test, which is performed using rabbits. Since animal testing of cosmetic products is banned in the EU, companies are looking for new *in vitro* and *ex vivo* tests as alternatives for the Draize eye test. The alternative methods for testing eye irritation involve organotypic models such as isolated eyes (such as isolated chicken eye) or components, tissue and cell culture systems, and physicochemical tests.[77] The fluorescein leakage test is a new, widely used *in vitro* test to predict the aggressiveness of surfactants on the eyes as well as to evaluate their compatibility with the cells and tissues.[78] The test method models ocular irritation caused by damage to the intercellular tight junctions and the cell membranes of the corneal and conjunctival epithelia. This test enables effective differentiation of mild and harsh cleansers. In the study, the cells are exposed to the surfactant solution for a predetermined period of time. If disruption of the cell layer occurs, the dye can diffuse through the cells and its amount can be measured. Therefore, in the test, the integrity of the cell monolayer and its tight junctions is assessed via the amount of fluorescein dye leaking through the cell layer. Using this test, various cleanser solutions can be quantitatively evaluated and their irritation potential can be numerically compared. The results are expressed in comparison to the control values. Usually, the concentrations of the test substance causing a leakage of 20% or 50% are quoted as the toxicological endpoint.[79,80]

Packaging of Baby Care Products

The most commonly used packaging materials for baby care products include the following:

- **Plastic Bottles** The majority of cleansing and protective products as well as baby sunscreens are supplied in plastic bottles, usually with a flip-top cap, screw head, tilted head, or pump head. Baby powder containers are made of plastic and have a special dispenser cap. There are several small holes in the cap, and the user has to twist the cap to open/close the openings in the lid. It is very practical and ensures that the product lasts as long as possible, and none is wasted or accidentally poured out in large amounts.
- **Soft Tubes and Jars** Some protective products and baby sunscreens are packed into soft tubes (similar to hand creams), while others are packaged into jars.
- **Aerosol Cans** Sunscreens may also be available in aerosol containers, similar to adult products.
- **Containers for Baby Wipes** Baby wipes are designed to be durable enough for cleaning baby skin, yet still be disposable. Several packaging materials are available today, including sachets for individually packaged wipes as well as soft or hard plastic containers for larger quantities. The majority of baby cleansing wipes are folded together and packaged into such plastic containers (tubs and canisters) with a resealable label or flip-flop lid. These containers are

designed to easily dispense single sheets while keeping the wipes moist until ready for use.

GLOSSARY OF TERMS FOR SECTION 2

Acne neonatorum: Neonatal acne, a skin condition that may occur in babies.

Atopic dermatitis: A severe inflammatory skin condition, which usually develops as a result of hypersensitivity of the immune system.

Babies: Very young children between the ages of 0 and 2.

Baby cleansing product: A personal care product designed to clean the baby skin and hair and remove saliva, nasal secretions, sweat, urine, feces, bacteria, residues of milk, and other dirt and irritants.

Baby powder: A personal care product designed to absorb moisture from the skin, lubricate the skin, and prevent irritation.

Baby protecting product: A personal care product designed to protect baby skin, help maintain its hydration, and, thus, prevent drying.

Baby wipe: A personal care product designed to clean the skin, particularly the diaper zone.

Diaper rash: A common inflammatory skin reaction in babies in the diaper area.

Eye irritation: A common negative effect of baby shampoos.

Mild surfactant: A surfactant molecule that is not as harsh to the skin as others due to its chemical structure and size.

Nonwoven fabric: A flat, porous sheet that is made directly from separate fibers or from molten plastic or plastic film. It is not made by weaving or knitting.

Penetration enhancer: An ingredient that can increase the skin penetration of other ingredients.

Skin protectant: An OTC category of active ingredients used to provide protection for the skin. Many diaper rash ingredients belong to this category.

REVIEW QUESTIONS FOR SECTION 2

Multiple Choice Questions

1. Which of the following is NOT true for baby skin?
   a) Thinner than adult skin
   b) Sebaceous glands are less active than in adult skin
   c) None of the sweat glands are working
   d) Loses water more easily than adult skin

2. Which of the following is NOT true for baby wipes?
   a) They are nonlathering cleansing formulations
   b) They are disposable
   c) They are anhydrous
   d) They can contain natural extracts
3. The critical age under which sunscreens should not be used on babies is:
   a) 6 weeks
   b) 6 months
   c) 1 year
   d) 6 years
4. Which of the following is the most important requirement for baby products?
   a) Mildness
   b) Good smell
   c) Attractive color
   d) High foaming power
5. Which of the following is TRUE for baby hair?
   a) Less developed than adult hair
   b) Pigmented as much as adult hair
   c) The number of hair follicles per square centimeter is higher than that in adult skin
   d) Sebaceous glands are highly active, making the hair greasy
6. Diaper zone products are typically available in the following dosage form:
   a) O/W emulsion
   b) W/O emulsion
   c) Stick
   d) Aerosol spray
7. Micelles in baby care products tend to be ___ than those in adult cleansing products.
   a) Larger with a lower micellar charge
   b) Smaller with a lower micellar charge
   c) Larger with a higher micellar charge
   d) Smaller with a higher micellar charge
8. Which of the following is NOT true for baby sun protection?
   a) Typically, inorganic sunscreens are used
   b) Baby skin is more vulnerable to sunlight than adult skin
   c) Water resistance is not important for baby sunscreen products
   d) The best sun protection for very young babies is to avoid the sun

9. Which of the following is a main safety issue for baby care products?
  a) Eye irritation
  b) Nail infection
  c) Hair loss
  d) Environmental pollution

Drug or Cosmetic in the United States?

a) Diaper rash cream with 5% zinc oxide ____________________
b) Baby sunscreen SPF 30 ____________________
c) Baby powder with aloe vera ____________________
d) Baby shampoo ____________________

Matching

Match the ingredients in column A with their appropriate ingredient category in column B.

Column A	Column B
_____ A. Aloe vera	1. Absorbent
_____ B. Cocamidopropyl betaine	2. Amphoteric surfactant
_____ C. Corn starch	3. Anti-inflammatory, soothing ingredient
_____ D. Disodium laureth sulfosuccinate	4. Antioxidant
_____ E. PEG-80 sorbitan laurate	5. Emollient and occlusive ingredient
_____ F. Petrolatum	6. Hydrophilic thickener
_____ G. Titanium dioxide	7. Mild anionic surfactant
_____ H. Tocopherol	8. Nonionic surfactant
_____ I. Water	9. Physical sunscreen
_____ J. Xanthan gum	10. Solvent

REFERENCES

1. Zàhejsky, J.: Cosmetic for children from the point of view of dermatologie, *Eurocosmetics*. 1994;34:3–4
2. WHO, Promoting Safety of Medicines For Children, 2007, Accessed 4/10/2014 at http://www.who.int/medicines/publications/essentialmedicines/Promotion_safe_med_childrens.pdf

3. Stamatas, G. N., Nikolovski, J., Mack, M. C., et al.: Infant skin physiology and development during the first years of life: a review of recent findings based on *in vivo* studies *Int J Cosmet Sci.* 2011;33(1):17–24.
4. Hoath, S. B., Maibach, H. I.: *Neonatal Skin: Structure and Function*, 2nd Edition, New York: Marcel Dekker, 2003.
5. Fluhr, J. W., Darlenski, R., Taieb, A., et al.: Functional skin adaptation in infancy—almost complete but not fully competent. *Exp Dermatol.* 2010;19(6):483–492.
6. Stamatas, G. N., Nikolovski, J., Luedtke, M. A., et al.: Infant skin microstructure assessed *in vivo* differs from adult skin in organization and at the cellular level. *Pediatr Dermatol.* 2010;27(2):125–131.
7. Stamatas, G., Martin, K.: Baby skin vs. adult skin structure, functions and compositions, *Cosmet Toilet.* 2009;124(4):50–53.
8. Nikolovski, J., Luedtke, M., Chu, M., et al.: Visualization of infant skin structure and morphology using in vivo confocal microscopy. *J Am Acad Dermatol.* 2007;56:AB4.
9. Stamatas, G., Estanislao, R. B., Suero, M., et al.: Facial skin fluorescence as a marker of the skin's response to chronic environmental insults and its dependence on age, *Br J Dermatol.* 2006;154(1):125–132.
10. Hoeger, P. H., Enzmann, C. C.: Skin physiology of the neonate and young infant: a prospective study of functional skin parameters during early infancy. *Pediatr Dermatol.* 2002;19(3):256–262.
11. Agache, P., Blanc, D., Barrand, C., Laurent, R.: Sebum levels during the first year of life. *Br J Dermatol.* 1980;103(6):643–649.
12. Nikolovski, J., Stamatas, G. N., Kollias, N., et al.: Barrier function and water-holding and transport properties of infant stratum corneum are different from adult and continue to develop through the first year of life. *J Invest Dermatol.* 2008;128(7):1728–1736.
13. Lund, C., Kuller, J., Lane, A., et al.: Neonatal skin care: the scientific basis for practice. *Neonatal Network.* 1999;18(4):15–27.
14. Trüeb, R. M.: Shampoos: composition and clinical applications. *Hautarzt.* 1998;49:895–901.
15. Sams, W. M.: Structure and Function of the Skin. In: Sams, W. M., Lynch, P. J., eds: *Principles and Practice of Dermatology*, New York: Churchill-Livingstone, 1990.
16. Rigano, L.: Baby Creams, *Cosmetic Toilet.* 2013;128(7):454–461.
17. Katsambas, A. D., Katoulis, A. C., Stavropoulos, P.: Acne neonatorum: a study of 22 cases. *Int J Dermatol.* 1999;38(2):128–130.
18. Walters, R. M., Fevola, M. J., LiBrizzi, J. J., et al.: Designing cleansers for the unique needs of baby skin. *Cosmetic Toilet.* 2008;123(12):53–60.
19. Darmstadt, G. L., Dinulos, J. G.: Neonatal skin care. *Pediatr Clin North Am.* 2000;47:757–782.
20. *Scribner's Magazine* advertisement, author unknown, 1990;136.
21. Johnson & Johnson History, Accessed 4/2/2014 at http://www.jnj.com/about-jnj/company-history
22. US 791,169; 1965.
23. Gelmetti, C.: Skin cleansing in children. *J Eur Acad Dermatol Venereol.* 2001; 15(1):12–15.
24. Bartels N. G., Scheufele, R., Prosch, F., et al.: Effect of standardized skin care regimens on neonatal skin barrier function in different body areas. *Pediatr Dermatol.* 2010;27(1):1–8.

25. Blume-Peytavi, U., Cork, M. J., Faergemann, J., et al.: Bathing and cleansing in newborns from day 1 to first year of life: recommendations from a European round table meeting. *J Eur Acad Dermatol Venereol.* 2009;23(7):751–759.
26. Saijo, S., Tagami, H.: Dry skin of newborn infants: functional analysis of the stratum corneum. *Pediatr Dermatol.* 1991;8(2):155–159.
27. Paller, A. S., Hawk, J. L. M., Honig, P., et al.: New insights about infant and toddler skin implications for sun protection; *Pediatrics.* 2011;128(1):92–102.
28. FDA: Should you Put Sunscreen on Infants? No Usually, Last update: 2/10/2014, Accessed 3/10/2014 at http://www.fda.gov/downloads/ForConsumers/ConsumerUpdates/UCM309579.pdf
29. Eichenfield, L. F., Frieden, I. J., Esterly, N. B.: *Neonatal Dermatology*, 2nd Edition, Philadelphia: Saunders Elsevier, 2008.
30. Adam, R.: Skin care of the diaper area. *Pediatr Dermatol.* 2008;25:427–433.
31. Adalat, S., Wall, D., Goodyear, H.: Diaper dermatitis frequency and contributory factors in hospital attending children. *Pediatr Dermatol.* 2007;24(5):483–488.
32. Ananthapadmanabhan, K. P., Moore, D. J., Subramanyan, K., et al.: Cleansing without compromise: the impact of cleansers on the skin barrier and the technology of mild cleansing. *Dermatol Ther.* 2004;17(1):16–25.
33. Association of Women's Health, Obstetric and Neonatal Nurses: *Neonatal Skin Care*, Washington, DC: Association of Women's Health, Obstetric and Neonatal Nurses, 2007.
34. Callard, R. E., Harper, J. I.: The skin barrier, atopic dermatitis and allergy: a role for Langerhans cells? *Trends Immunol.* 2007;28(7):294–298.
35. Laughter, D., Istvan, J. A., Tofte, S. J., et al.: The prevalence of atopic dermatitis in Oregon schoolchildren. *J Am Acad Dermatol.* 2000;43(4):649–655.
36. Niwa, Y., Sumi, H., Kawahira K., et al.: Protein oxidative damage in the stratum corneum: evidence for a link between environmental oxidants and the changing prevalence and nature of atopic dermatitis in Japan. *Br J Dermatol.* 2003;49:248–254.
37. Abrutyn, E. S.: Baby care. *Cosmetic Toilet.* 2011;126(7):486-491.
38. Lawrenson, J. G., Birhah, R., Murphy, P. J.: Tearfilm lipid layer morphology and corneal sensation in the development of blinking in neonates and infants. *J Anat.* 2005;206(3):265–270.
39. Dhar, S.: Newborn skin care revisited. *Indian J Dermatol.* 2007;52:1–4.
40. Imokawa, G., Akasaki, S., Minematsu, Y., et al.: Importance of intercellular lipids in water-retention properties of the stratum corneum: induction and recovery study of surfactant dry skin. *Arch Dermatol Res.* 1989;281:45–51.
41. Lopez-Castellano, A., Cortell-Ivars, C., Lopez-Carballo, G., et al.: The influence of Span 20 on the stratum corneum lipids in Langmuir monolayers: comparison with azone. *Int J Pharma.* 2000;203:245–253.
42. Saad, P., Flach, C. R., Walters, R. M., Mendelsohn, R.: Infrared spectroscopic studies of sodium dodecyl sulphate permeation and interaction with stratum corneum lipids in skin. *Int J Cosmet Sci.* 2012;34:36–43.
43. Löffler, H., Happle, R.: Profile of irritant patch testing with detergents: sodium lauryl sulfate, sodium laureth sulfate and alkyl polyglucoside. *Contact Dermatitis.* 2003;48(1):26–32.

44. Cork, M. J., Danby, S., Vasilopoulos, Y., et al.: Epidermal Barrier Dysfunction. In: Reitamo, S., Luger, T. A., Steinhoff, M., eds: *Textbook of Atopic Dermatitis*, London: Informa Healthcare; 2008.
45. Fevola, M. J., LiBrizzi, J. J., Walters, R. M.: Reducing irritation potential of surfactant-based cleansers with hydrophobically-modified polymers. *Polymer Preprints.* 2008;49(2):671–672.
46. Ehretsmann, C., Schaefer, P., Adam, R.: Cutaneous tolerance of baby wipes by infants with atopic dermatitis, and comparison of the mildness of baby wipe and water in infant skin. *J Eur Acad Dermatol Venereol.* 2001;15(1):16–21.
47. Odio, M., Streicher-Scott, J., Hansen, R. C.: Disposable baby wipes: efficacy and skin mildness. *Dermatol Nurs.* 2001;13(2):107–112.
48. Lavender, T., Furber, C., Campbell, M., et al.: Effect on skin hydration of using baby wipes to clean the napkin area of newborn babies: assessor-blinded randomised controlled equivalence trial, *BMC Pediatr*. 2012;12:59.
49. Nash, J. F., Gettings, S. D., Diembeck, W., et al.: A toxicological review of topical exposure to white mineral oils. *Food Chem Toxicol.* 1996;34(2):213–225.
50. Liptak, G.S.: Diaper Rash. In: Hoekelman, R., Adam, H. M., Nelson, N. M., et al., eds: *Pediatric Primary Care*, Maryland Heights: Mosby, 2001.
51. Ravanfar, P., Wallace, J. S., Pace, N. C.: Diaper dermatitis: a review and update. *Curr Opin Pediatr*. 2012;24(4):472–479.
52. Atherton, D. J.: The aetiology and management of irritant diaper dermatitis. *J Eur Acad Dermatol Venereol.* 2001;15(1):1–4.
53. Nield, L. S., Kamat, D.: Prevention, diagnosis, and management of diaper dermatitis. *Clin Pediatr*. 2007;46(6):480–486.
54. CFR Title 21 Part 347 Skin Protectant Drug Products for Over-The-Counter Human Use
55. Schönrock, U.: Baby Care. In: Barel, A. O., Paye, M., Maibach, H., eds: *Handbook of Cosmetic Science and Technology*, New York: Marcel Dekker Inc., 2001.
56. Odio, M. R., O'Connor, R. J., Sarbaugh, F., et al.: Continuous topical administration of a petrolatum formulation by a novel disposable diaper. 2. Effect on skin condition. *Dermatology*. 2000;200(3):238–243.
57. Farrington, E.: Diaper dermatitis. *Pediatric Nursing*. 1992;18(8):81–82.
58. EPA: Particulate Matter (PM): Basic Information, Last updated: 3/18/2013, Accessed 3/12/2014 at http://www.epa.gov/airquality/particulatematter/basic.html
59. Gottschalck, T. E., Breslawec, H. P.: *International Cosmetic Ingredient Dictionary and Handbook*, 14th Edition, Washington, DC: Personal Care Products Council, 2012.
60. Rigel, D.S.: Cutaneous ultraviolet exposure and its relationship to the development of skin cancer. *J Am Acad Dermatol*. 2008;58(5):S129–S132.
61. CDC: Sun-protection behaviors used by adults for their children: United States, 1997. *MMWR Morb Mortal Wkly Rep*. 1998;47(23):480–482.
62. Etzel, R., Balk, S.: *Pediatric Environmental Health*. 2nd Edition, Elk Grove Village: American Academy of Pediatrics, 2003.
63. Diffey, B. L. Sunscreens: Use and Misuse, In: Giacomoni, P. U., ed.: *Sun Protection in Man*, Amsterdam: Elsevier Science, 2001.
64. Azurdia, R. M., Pagliaro, J. A., Rhodes, L. E.: Sunscreen application technique in photosensitive patients: a quantitative assessment of the effect of education. *Photodermatol Photoimmunol Photomed*. 2000;16(2):53–56.

65. Diffey, B. L.: Has the sun protection factor had its day? *BMJ*. 2000;320:176–177.
66. Bandi, P., Cokkinides, V. E., Weinstock, M. A., et al.: Sunburns, sun protection and indoor tanning behaviors, and attitudes regarding sun protection benefits and tan appeal among parents of U.S. adolescents-1998 compared to 2004. *Pediatr Dermatol*. 2010;27(1):9–18.
67. Diffey, B. L.: When should sunscreen be reapplied? *J Am Acad Dermatol*. 2001;45:882–885.
68. Cross, S. E., Innes, B., Roberts, M. S., et al.: Human skin penetration of sunscreen nanoparticles: in-vitro assessment of a novel micronized zinc oxide formulation. *Skin Pharmacol Physiol*. 2007;20(3):148–154.
69. Schlossman, D., Shao, Y.: Inorganic Ultraviolet Filters, In: Shaath, N. A., ed.: *Sunscreens, Regulations and Commercial Development*, 3rd Edition, Boca Raton: Taylor and Francis, 2005.
70. van der Molen, R. G., Hurks, H. M. H., Out-Luiting, C.: Efficacy of micronized titanium dioxide containing compounds in protection against UVB-induced immunosuppression in humans *in vivo*. *J Photochem Photobiol B*. 1998;44:143–150.
71. Alonso, C., Larburu, I., Bon, E., et al.: Efficacy of petrolatum jelly for the prevention of diaper rash: a randomized clinical trial *J Special Pediatr Nurs*. 2013;18:123–132.
72. Johnson & Johnson Consumer Companies, Inc. PEDIASTAT™ (miconazole nitrate, USP 0.25%) Diaper Rash Ointment NDA 21-026, Accessed 4/4/2014 at http://www.fda.gov/ohrms/dockets/ac/00/backgrd/3626b2a_Efficacy.pdf
73. Safety Assessment of Talc as Used in Cosmetics, Final Report, Release Date: 4/12/2013
74. Tremewan, H. C.: Tetanus neonatorum in New Zealand. *N Z Med J*. 1946;45:312.
75. Industrial Minerals Association-North America (IMA-NA) and EUROTALC. RE: Scientific Literature Review: Talc as Used in Cosmetics. 10/19/2012.
76. FDA: BAM: Microbiological Methods for Cosmetics, Last Update: 4/8/2013, Accessed 3/12/2014 at http://www.fda.gov/food/foodscienceresearch/laboratorymethods/ucm073598.htm
77. EURL ECVAM Progress Report on the Development, Validation and Regulatory Acceptance of Alternative Methods, 2013, Accessed 3/15/2014 at http://ec.europa.eu/environment/chemicals/lab_animals/pdf/EURL_ECVAM_progress_report_cosmetics_2013.pdf
78. Eye Irritation, Accessed 3/15/2014 at http://ec.europa.eu/consumers/sectors/cosmetics/files/doc/antest/%285%29_chapter_3/3_eye_irritation_en.pdf
79. Cottin, M. Zanvit, A.: Fluorescein leakage test: a useful tool in ocular safety assessment. *Toxicology In Vitro*. 1997;11(4):399–405.
80. Fluorescein Leakage (FL) Test, DB-ALM Protocol n° 71, Accessed 3/15/2014 at http://ecvam-dbalm.jrc.ec.europa.eu/public_view_doc2.cfm?id=C06709AA034D363C22C6F6AB5BAFADFF7180BB0BC12CB10496CDA74B54630A05A3291B895581F634

SECTION 3: SUNLESS TANNING PRODUCTS

LEARNING OBJECTIVES

Upon completion of this section, the reader will be able to

1. define the following terms

Airbrush technique	Bronzer	DHA	Fading
Melanoid	Nail discoloration	Spray tan	Sunless tanner

2. define sunless tanning products;
3. list the various required cosmetic qualities and characteristics that an ideal sunless tanning product should possess;
4. list the various required technical qualities and characteristics that an ideal sunless tanning product should possess;
5. briefly summarize the major advantages of using sunless tanners;
6. list some negative effects that sunless tanners may have on the skin;
7. name the most commonly used sunless tanning agent in the US;
8. briefly explain how sunless tanning products provide a tanned color upon application;
9. list the major product forms of sunless tanners available today;
10. briefly discuss why the following factors are important to consider during the formulation of sunless tanning products: pH, heat, and interactions;
11. explain from where the characteristic odor of sunless tanning products derives;
12. list some typical quality issues that may occur during the formulation and/or use of sunless tanning products, and explain why they may occur;
13. list the typical quality parameters that are regularly tested for sunless tanner products, and briefly describe their method of evaluation;
14. briefly explain why spray tanning causes safety concerns;
15. list the typical containers available for sunless tanning products.

KEY CONCEPTS

1. Sunless tanning products are designed to produce a skin color similar in appearance to a traditional suntan without exposure to UV light.
2. The most widely used and most effective ingredient for providing sunless tan is dihydroxyacetone (DHA).

3. DHA is primarily formulated as lotions or creams. The latest technologies and products launched include self-tanning wipes as well as no-rub mists.
4. Formulations containing DHA are sensitive to a variety of formulation factors, including pH, heat, polymeric thickeners, fragrance, nitrogen-containing ingredients, and metal oxides limiting the shelf-life of the finished products.
5. The typical quality-related issues of sunless tanning products include discoloration and malodor formation, separation of emulsions, microbiological contamination, clumping, and rancidification.
6. Parameters commonly tested to evaluate the quality of self-tanning products include spray characteristics; aerosol can leakage and pressure test for aerosol products; spreadability, extrudability, texture, and firmness of lotions and creams; actuation force; color; preservative efficacy; viscosity; and pH.
7. The most frequently tested performance parameters include the onset of color, color intensity, and color longevity.
8. Sunless tanning products containing DHA have been used for almost 30 years and are generally recognized as safe.

Introduction

Today, tanned skin is considered a sign of healthiness and even affluence in many cultures. Sunless tanners are gaining popularity as we learn more about the potential damaging effect of sunlight (for more details, refer to Section 5 of Chapter 3). Self-tanning products, therefore, offer an alternative approach to those desiring the tanned look but not wishing to expose themselves to UV light. When applied topically, sunless tanning products impart a temporary coloration to the skin, mimicking the color of naturally suntanned skin.

This section reviews the main ingredients used in sunless tanning products, their mechanism of action, and other major characteristics. It also summarizes the consumer needs as well as the positive and potential negative effects of self-tanners on the human body. The main tests used to evaluate the quality and performance of self-tanning products are also included in this section.

Types and Definition of Sunless Tanning Products

❶ **Sunless tanning products**, otherwise known as self-tanning products, artificial tanning products, or UV-free tanners, **are designed to produce a skin color similar in appearance to traditional suntan without exposure to UV light.** Depending on the product and active ingredients used, the onset of color formation can be immediate up to 1 week. Sunless tanning products available today include gels, creams, lotions, wipes, spray-on tans, and bronzers. All these products are considered cosmetics in the US.

History of Using Sunless Tanning Products

Throughout the history, skin color has played an important role in identity, self-esteem, and social class. Although tanned skinned is currently considered to be attractive in many parts of the world, it has not always been the case. During the 19th century, higher class women were always white skinned since tanned skin referred to outdoor work, which was characteristic of the lower class. At those times, wealthy women would carry parasols and protect their skin from the sun to prevent tanning and convey the image of upper-class status. Even today, there are some cultures in which tanned skin is not considered attractive, and therefore, skin exposure is avoided as much as possible.[1] Later on, in the 20th century, tanned skin became a sign of financial stability, showing that one could afford to have a longer vacation in the sun. Coco Chanel, a fashion icon from Europe, had a huge influence on this in the 1920s when she tanned her skin during a vacation on the French Riviera. Women liked her appearance and started to copy her.[2,3] DHA, the most commonly used self-tanning ingredient today, was recognized as a self-tanner in the 1950s when it was administered orally for a glycogen storage disease.[4] The children received large doses of DHA by mouth and sometimes spat or spilled the substance onto their skin. Eva Wittgenstein at the University of Cincinnati was leading the experiments, and she noticed that the skin turned brown after a few hours of DHA exposure.[5] After experimenting with DHA solutions, she was able to reproduce the skin pigmentation repeatedly. The first self-tanning products were introduced into the market in the late 1950s. These products, however, led to unattractive results such as unnatural orange tone, coloration of hands and fingernails, streaking, and poor, uneven coloration of the body.[6] Therefore, they were not very popular those days. Since then, formulations have significantly improved and are available in a more refined form, which provides even coverage in a more natural shade and is more resistant to fading. Their sales, therefore, grew exponentially in the past decades. Today, operator-assisted spray tans are also available in spas and salons, which are quite popular, especially in the US and Europe. In the US, self-tanning products are among the 10 fastest growing industries.[7]

Required Qualities and Characteristics and Consumer Needs

From a consumer perspective, a quality sunless tanning product should possess the following characteristics:

- Easy application with a tendency to spread easily
- Provides even application and uniform color
- Provides naturally tanned color
- Does not color the hands, nails, and clothing
- Well-tolerated, nonirritating
- Pleasant odor.

The technical qualities of sunless tanning products can be summarized as follows:

- Long-term stability
- Appropriate rheological properties
- No leakage from aerosol cans
- Appropriate pH
- Dermatological safety.

How Sunless Tanners May Affect the Human Body?

Sunless tanning products offer an alternative to those who wish to tan without any exposure to the sun. The advantages of using sunless tanners are the following:

- The primary advantage of sunless tanning products is that they can be used without **UV exposure**. Therefore, they do not carry the risk of premature aging or skin cancer.
- Additionally, sunless tanning products provide a tanned color relatively **quickly**. Tanning with certain products can be seen within hours after application.
- Unlike sunbathing, where first a reddish color appears, which, hours or days later, develops into a tanned color, sunless tanning products provide a **tanned color** immediately.

In addition to the advantages, some negatives effects can also be identified when using sunless tanning products. A short summary is provided here.

- Sunless tanning products are often recommended as a safer alternative for indoor and outdoor tanning, since they do not require UV exposure to provide a tan color. However, the use of sunless tanners does not necessarily lead to a decrease in the **use of tanning booths** or incidence of **outdoor tanning**.[8,9]
- It is important to understand, for users of sunless tanning agents, that these products do not provide appropriate **sun protection**. Studies have shown that tanning from sunless tanners may provide a minimal sun protection at the low end of the visible spectrum, limited UVA protection,[10] and modest UVB protection with an SPF of approximately 3–4. The SPF value depends on the concentration of the ingredient used and also on the number of applications as well.[11] Such a high (low) protection can be expected from products containing 20% DHA, the most commonly used tanning ingredient.[12] However, most commercial formulations contain only about 5% DHA. Products with 5% DHA provide much less protection.[11] In addition, the SPF is present only for several hours after application and does not last for the duration of the tan.[3] Referring to Chapter 3, it was discussed that sunscreens with an SPF of at least 15 and broad-spectrum protection provide adequate protection against skin burning, aging, and skin cancer. Based on this information, it is clear that only sunless tanners that contain

sunscreen ingredients and are labeled with SPF numbers provide appropriate sun protection. All suntanning preparations that do not contain sunscreen ingredients are required, in the US to carry the following warning statement on the label: "Warning – This product does not contain a sunscreen and does not protect against sunburn. Repeated exposure of unprotected skin while tanning may increase the risk of skin aging, skin cancer, and other harmful effects to the skin even if you do not burn."[13]

There are some products available that contain both DHA and sunscreen agents. They are regulated as OTC drugs in the US. Such products require very clear and careful labeling to avoid confusion and are useful where a consumer requires a high level of protection from the sun and yet still desires a tan.

- When using DHA-containing products, **discoloration** of the nails, hands, and even clothes may happen. The hands need to be washed immediately after use to avoid darkening of the palms, fingers, and nails. There are products available that are supplied with gloves. These are recommended to be used when applying the product to the skin to avoid any discoloration. When undergoing sunless booth tanning, often the use of barrier creams is recommended to avoid discoloration of the fingernails, hands, toenails, and feet. Clothing may rub off some of the product from the skin, which will be visible on the clothing as an orange-brownish stain. This is usually easy to remove with a regular laundry cycle.
- One of the main disadvantages of sunless tanning products from a consumer perspective is that it is not easy to obtain a uniform coverage by self-application. **Uneven application** can result in streaking and lighter or darker spots, especially on the back.[14] There are certain skin conditions that may also result in an uneven, unattractive tan. These include the skin that has been previously damaged by the sun, aging skin, mottled or freckled skin, and scarred as well as thickened skin on the elbows, for example. Exfoliating products used before sunless tanning may help reduce the thickness in these body areas.
- An additional disadvantage of sunless tanning products containing DHA is the formation of an **odor** perceived by most users as unpleasant. As odor formation is the result of the tanning process, it occurs every time sunless tanning products are used.
- The early DHA formulations provided a quite **unnatural** orange **color** to the skin, which was unappealing for most users. Since then, formulations have improved significantly. However, the development of a color toward orange by the dominance of the yellow component is still a frequent problem when using self-tanning substances, especially DHA. This is frequently regarded by many users as a disadvantage and perceived as unnatural.[14]
- In order to maintain the tanned color for more than a few days, **continuous application** is needed as the color fades over time. This may also be considered a negative effect by users, especially if we take into consideration the above-mentioned disadvantages.

Sunless Tanning Products

Tanning Ingredients ● **The most widely used and most effective ingredient for providing sunless tan is DHA.** It is the only ingredient currently recognized as a self-tanning agent by the FDA.[15] It can only be used externally to impart a color to the human body, i.e., as a sunless tanner. DHA is a sugar derivative that has been used for decades. It initiates color change in the epidermis by a chemical reaction known as the Maillard reaction, also known as nonenzymatic browning. This reaction usually can be seen 1–2 h after application of DHA and can take up to 24–48 h to fully develop.[16] The color has a similar hue as that of natural suntan. DHA can penetrate into the epidermis, where it interacts with certain amino acids of epidermal proteins in the dead cells and darkens the skin. This color is permanent and disappears gradually as the cells of the outer layers of the epidermis proceed toward the surface of the skin and are shed naturally. Until the outer cells are shed from the skin, the color resulting from the use of this substance cannot be removed. DHA content in self-tanning products depends on the desired browning intensity on the skin; it can range from 2.5% to 10%, with the usual concentration of 5%.[17] Products are usually labeled as light, medium, and dark, which can be very helpful for consumers. Antioxidants such as caffeic acid phenethyl ester may be added to the self-tanner[18] to mitigate the artificial-looking orange or yellow color DHA can produce on the skin and achieve a more natural tone.[19]

DID YOU KNOW?

DHA is a colorless dye, i.e., it is a white powder. It imparts temporary color by a chemical reaction, not simply by coloring the skin.

DID YOU KNOW?

While the FDA allows DHA to be "externally applied" on the skin, there are restrictions on its use. DHA should not be inhaled, ingested, or exposed to areas covered by mucous membranes, including the lips, nose, and areas in and around the eyes.[20]

Additional ingredients may also be used for coloring the skin. The chemical reaction initiated by DHA can also be initiated by other ingredients similar in chemical structure. Erythrulose, another sugar, has also been used for years now. It produces a more gradual tan than DHA. Some formulations combine DHA and erythrulose for a longer-term gradual tanning.[21,22] **Bronzers** are also available on the market today. These are products that contain water-soluble dyes, such as carmine and caramel, which color the surface of the skin, making it look like tanned. A bronzer works like a color cosmetic that instantly provides a tan color. Since these products do not chemically react with the skin, they can be simply washed off with water and soap. Products are available as gels, lotions, and creams. They may be advantageous for those needed an instant tanning effect without any permanency.

As mentioned earlier, DHA is a white, water-soluble powder; therefore, when it is formulated into emulsions, the product will be white. Many sunless tanners, however, are tinted today, which provide an instant color upon application. These are two-in-one products, containing both sunless tanners and bronzers. Using colored products helps ensure even application and provides instant color in the short term, while DHA generates a color that appear more slowly but lasts substantially longer.

DID YOU KNOW?

As discussed in Chapter 3, UVA and UVB radiations lead to increased melanin production, which results in tanning. Tanning by the sun, however, is a sign of DNA damage, which on the long term can lead to the development of skin cancer. When using sunless tanning agents, no such damage occurs. The tan from most sunless tanners comes from DHA. DHA binds to the proteins on the skin surface, forming brownish melanin-like molecules called "melanoids," which produce a tan-like appearance. Unlike UV, DHA does not penetrate deep into the skin; it is active only in the SC.

Main Product Forms ● **DHA is primarily formulated as lotions or creams. The latest technologies and products launched include self-tanning wipes as well as no-rub mists.** Gels are not as popular as other forms due to the skin feel they provide, and their formulation is quite a challenge from a stability perspective as well.

Sunless Tanning Creams and Lotions Similar to other skin care products, O/W lotions and creams are generally the most popular and most commonly accepted by consumers due to their ease of spreadability and aesthetics. Since creams have a higher viscosity, they are usually applied as a thicker film and, therefore, can produce a more intense tan. Ingredient selection plays an important role in the stability of the products, since one of the biggest challenges of formulating sunless tanning products with DHA is the instability of this chemical ingredient. Products usually contain a variety of emollients, including oils, butters, waxes, and silicones; water; and thickeners, including both hydrophilic molecules, such as xanthan gum, and lipophilic thickeners, such as waxes or even certain nonionic emulsifiers. Nonionic emulsifiers may also increase the overall stability of the product.[23] Lower viscosity formulations are also available as spray-on formulations. Today, sunless tanning products formulated as W/Si emulsions are also available on the market. In appearance, they are identical to gels but provide an improved skin feel without any cooling sensation. These products are usually referred to as "gelées" on the labels.

Aerosols Aerosols are also popular products due to their easy and conformable application. Spray formulations are usually aqueous or hydroalcoholic solutions, which along with the valve system provide a very small particle size and offer quick evaporation. One of the latest forms of aerosol products is the no-rub mists or spray tanners. These are usually offered at salons. As their name states, no rubbing action is needed to spread the product on the skin after application. The product is usually applied to the entire body surface in a booth from a multiangle applicator system. Spray booths are shower-sized stalls equipped with nozzles that spray the DHA-containing product onto the users' skin. This spray system allows for simple, even, and often hands-free application.[24] Thus, all the inconveniences attributed to the use of self-tanning preparations, including discoloration of the nails and palms, and uneven application resulting in darker shade at certain body parts can be eliminated. More recently, the airbrush tanning technique was introduced. In this tanning technique, a trained technician (cosmetician) sprays the tanning product onto the client's skin from a handheld airbrush device. These techniques are generally more expensive than regular, self-applied tanning products.[24] The main disadvantage of full-body aerosol products is the risk of inhalation, which was discussed earlier.

Wipes Sunless tanning wipes are also a new addition of products to the arena of artificial tanners. These are similar to facial cleanser wipes in terms of the ingredients and formulation. These products are ready-to-use, pre-wetted wipes that can be used to provide a skin tone darker than the original tone without exposure to the sun. They usually contain additional emollients as well.

Formulation Concerns

As discussed previously, currently, DHA is the most effective sunless tanning agent and the only ingredient approved by the FDA for this purpose. Most products, therefore, contain it as the main "tanning" ingredient. There are, however, some important factors that should be taken into consideration when formulating products containing DHA, not to compromise its stability and, thus, efficacy and safety. Stable as a crystalline powder, when dissolved in water, DHA gradually degrades. ● **Formulations containing DHA are sensitive to a variety of formulation factors, including pH, heat, polymeric thickeners, fragrance, nitrogen-containing ingredients, and metal oxides limiting the shelf-life of the finished products.**[23,25] Since many of these factors may be present at the time of formulation, stabilization can be quite challenging. Upon degradation, a brownish color and odor develop referring to chemical changes in the molecule. Furthermore, degradation of DHA can lead to the production of by-products, such as formaldehyde, formic acid, and acetic acid.[26] These by-products are responsible for reduced pH values, which can contribute to an acidic smell. In addition, they may be potential irritants. Therefore, their development should be avoided using a proper formulation technique. If DHA is degraded, it cannot provide the proper reaction to the skin.

- **pH:** During product storage, the pH of a DHA-containing formulation drifts over time to 3–4. At this pH, DHA is stable; however, a higher pH enhances the degradation of DHA.[27] In order to ensure end-product stability, the formulation's pH is recommended to be adjusted to 3–4. The adoption of this rule, however, is a challenge due to the risk of skin irritation, as well as strong limitations in the choice of the acidifying agent. Acetic acid/sodium acetate buffer, and perfluoropolyether phosphate,[25] tested recently, can be used to adjust the pH of the formulations. Hydroxy acids, such as lactic acid and citric acid, as well as phosphoric acid/phosphate buffers, may increase its instability; therefore, they should be avoided.[25]
- **Heat:** High temperature (>40 °C) causes a rapid degradation of DHA; therefore, it should be avoided during storage and processing. During formulation processes that require heating (such as the formulation of certain emulsions), DHA should be handled as any other sensitive ingredients and should be added at the end of the formulation process. Additionally, finished products containing DHA should be packaged into an opaque container that can provide light protection.
- **Interactions:** Ionic emulsifiers should be avoided as they may degrade DHA. However, nonionic emulsifiers can be used. Thickener selection is also important. Thickeners with proper efficacy at lower pH should be selected. Additionally, thickeners that do not require neutralization to a higher pH should be chosen, since increasing the pH degrades DHA. Using cellulose-derived ingredients, such as methylcellulose, as well as silica, polyquaternium-10, waxes, and fatty alcohols, is recommended. Acid-stable acrylate-based polymers are also available today.[28] DHA can react with oxygen- and nitrogen-containing compounds, for example, collagen, amino acids, and proteins. The reactivity of DHA toward these compounds can lead to its degradation, therefore resulting in the loss of efficacy, color, and odor formation. Natural extracts should be checked for compatibility with DHA before adding them to the formulation. Metal oxides, such as iron oxides and titanium dioxide, may also lead to instability. Therefore, the use of chelating agents is recommended.[16]
- **Odor:** As mentioned earlier, odor formation is a characteristic sign of degradation; however, odor formation can also be experienced during the tanning process. The skin has a characteristic odor when using a self-tanning agent, which is perceived as an unpleasant experience by most consumers.[29] Numerous attempts have been made to conceal this odor by masking with fragrances and sequestering using cyclodextrins. Since fragrances are potential irritants, increasing their amount is not recommended. As for cyclodextrins, it is not clear whether they can conceal the odor continuously developing during tanning.[30] Currently, malodor formation is one of the major challenges for sunless tanning product formulators.
- **Other:** DHA is slightly hygroscopic; therefore, it should be properly stored, in a moisture-poor environment. Additionally, its efficacy should be tested periodically to ensure it has not degraded.

Typical Quality Problems of Sunless Tanning Products

● **The typical quality-related issues of sunless tanning products include discoloration and malodor formation, separation of emulsions, microbiological contamination, clumping, and rancidification.** Quality issues that were discussed in the previous sections are not reviewed here again.

Discoloration and Malodor Formation As discussed previously, these problems may occur upon degradation of the product. Formulation steps and techniques to avoid these phenomena were discussed earlier.

Evaluation of Sunless Tanning Products

Quality Parameters Generally Tested ● **Parameters commonly tested to evaluate the quality of self-tanning products include spray characteristics; aerosol can leakage and pressure test for aerosol products; spreadability, extrudability, texture, and firmness of lotions and creams; actuation force; color; preservative efficacy; viscosity; and pH.** As these tests were discussed in the previous sections, they are not reviewed here.

Performance Evaluation The main function of sunless tanners is to color the skin. ● **The most frequently tested performance parameters include the onset of color, color intensity, and color longevity.** For products containing sunscreens, additional efficacy testing needs to be performed. Sunscreen efficacy testing was discussed in Section 5 of Chapter 3.

Onset of Color, Color Intensity, and Color Longevity Various spectrophotometers can be used to evaluate the coloring performance of sunless tanners. Sunless tanners are generally offered in multiple shades with each shade recommended for a particular base color. Spectrophotometers can be used to detect the base color and the "new" color developed after application of the sunless tanning product. These instruments are able to distinguish between colors that look identical to the human eyes. In addition, colorimeters can also be applied to test the color of products.

Safety Concerns ● **Sunless tanning products containing DHA has been used for almost 30 years and are generally recognized as safe.** Only a few cases of allergy and contact dermatitis were reported.[31] These are mainly related to the additional ingredients in the formulation, such as fragrances, emollients, preservatives, but not the tanning agent, DHA.[32] Adverse reactions are more likely to occur on the basis of irritation and not true allergy.

The major concern regarding DHA is its potential inhalation when applied to the skin in tanning booths or by the airbrush technique. In tanning booths, customers stand in the aerosol mist for a few seconds. They are generally advised to keep their eyes and mouth closed or use eye and nose protection during the treatment. However, accidental inhalation may occur. The exposure may be single (in customers using tanning booths less than once per month) or repeated (in the case of multiple uses in a month, e.g., every week). The Scientific Committee on Consumer Safety (SCCS, see Section 3 of Chapter 2) considers the use of DHA as a self-tanning agent in spray cabins up to 14% of concentration not to cause any health risks to consumers.[33] In the US, this ingredient is approved only as an external color additive, excluding its application close to the lips, eyes, and other body surfaces covered by mucous membrane. The industry has not provided safety data to FDA in order for the agency to consider approving it for use on these exposure routes, including "misting" from tanning booths. Additionally, the industry has not provided safety data to FDA in order for the agency to consider approving it for use in the area of the eye.[34]

Packaging of Sunless Tanning Products

The most commonly used packaging materials for sunless tanning products include the following:

- **Plastic Bottles:** Most self-tanning products are supplied in plastic bottles with a flip-flip cap, pump head, or scroll-on cap. Often, rubber gloves are also provided in the package to prevent discoloration of the nails and hands. Some packages also contain a mitt, which helps application and even spreading and also provides protection against discoloration.
- **Aerosol Cans:** Aerosol products are supplied in aluminum cans.
- **Wipes:** Wipes are available in containers similar to those of sunscreen wipes. Most of them are packaged individually.

GLOSSARY OF TERMS FOR SECTION 3

Airbrush technique: A technique used to produce an artificial suntan without exposure to UV light by spraying the tanning product from a handheld device.

Bronzer: A color cosmetic containing water-soluble dyes used to give the appearance of a suntan. It colors only the surface without any chemical reaction and washes off with water.

DHA: Dihydroxyacetone, the most commonly used self-tanning ingredient.

Fading: Gradually fainting.

Melanoid: A brownish melanin-like molecule.

Nail discoloration: An orange-brownish color under the nails, a common side effect of using sunless tanners without gloves.

Spray tan: A personal care product designed to be sprayed on the body in a booth to produce an artificial suntan without exposure to UV light.

Sunless tanner: A personal care product designed to produce a skin color similar in appearance to a traditional suntan without exposure to UV light. It colors the skin via a chemical reaction.

REVIEW QUESTIONS FOR SECTION 3

Multiple Choice Questions

1. Which of the following is NOT true for sunless tanners?
   a) They provide sun protection
   b) They can be used without UV exposure
   c) They may discolor clothes
   d) They have to be applied every day in order to provide an even color

2. Which of the following is the most commonly used sunless tanner ingredient in the United States?
   a) Melanin
   b) Dihydroxyacetone
   c) Acetone
   d) Erythrulose

3. At which pH is dihydroxyacetone the most stable?
   a) 1–2
   b) 3–4
   c) 7
   d) 10–12

4. Where does the characteristic odor of sunless tanning products often considered unpleasant come from?
   a) Can be an organoleptic sign of degradation
   b) It is a result of the tanning process
   c) It is provided by the fragrances used in the product
   d) Both a and b are true

5. Sunless tanning products provide tanned skin via (a) ____.
   a) Increased melanin production
   b) Increased melanocyte production
   c) Chemical reaction with the proteins on the skin's surface
   d) Physical coloring process

6. What is the main safety concern regarding the use of spray tanning products?
   a) Dihydroxyacetone can cause skin irritation
   b) Dihydroxyacetone should not be inhaled according to the FDA
   c) Dihydroxyacetone can cause hair breakage
   d) Dihydroxyacetone damages the ozone layer

7. Which of the following can cause degradation of dihydroxyacetone?
   a) pH
   b) High temperature
   c) Metal oxides
   d) All of the above

8. Sunless tanners are considered ___ in the United States.
   a) Cosmetics
   b) OTC drugs
   c) Prescription-only drugs
   d) Cosmeceuticals

Fact or Fiction?

______ a) Sunless tanners can discolor the fingernails.
______ b) Sunless tanners provide UV protection.
______ c) Sunless tanning products provide an instant tan in hours after application.

REFERENCES

1. Butler, H.: *Poucher's Perfumes, Cosmetics and Soaps*, New York: Springer, 2000:117–123.
2. Campo, R.: *Best in Beauty: An Ultimate Guide to Makeup and Skincare Techniques, Tools, and Products*, New York: Atria Books, 2010.
3. Draelos, Z. D.: Self-tanning lotions: are they a healthy way to achieve a tan? *Am J Clin Dermatol.* 2002;3:317–318.
4. Goldman, L., Barkoff, J., Blaney, D.: Investigative studies with the skin coloring agents dihydroxyacetone and glyoxal: preliminary report. *J Invest Dermatol.* 1960;35:161–164.
5. Guest, G. M., Cochrane, W., Wittgenstein, E.: Dihydroxyacetone tolerance test for glycogen storage disease. *Mod Prob Paediat.* 1959;4:169–178.
6. Wittgenstein, E., Berry, H. K.: *Staining of skin with dihydroxyacetone Science* 1960;132(3431):894-895.
7. Statista: Industry Revenue Forecast for the 10 Fastest-Growing Industries in the United States 2012 and 2017 (in million U.S. dollars), Accessed 4/12/2014 at http://www.statista.com/statistics/224657/revenue-forecast-for-10-fastest-growing-us-industries/
8. K Brooks: Use of artificial tanning products among young adults. *J Am Acad Dermatol.* 2006;54(6):1060–1066.
9. Sheehan, D. J., Lesher, J. L.: The effect of sunless tanning on behavior in the sun: a pilot study. *South Med J.* 2005;98(12):1192–1195.

10. Fusaro, R. M., Johnson, J. A.: Protection against long ultraviolet and/or visible light with topical dihydroxyacetone. *Dermatologica.* 1975;150:346–351.
11. Faurschou, A., Janjua, N. R., Wulf, H. C.: Sun protection effect of dihydroxyacetone. *Arch Dermatol.* 2004;140:886–887.
12. Faurschou, A., Wulf, H. C.: Durability of the sun protection factor provided by dihydroxyacetone. *Photodermatol Photoimmunol Photomed.* 2004;20(5):239–242.
13. CFR Title 21 Part 740.19
14. Sunless Tanning Stats Show Women Seeking a Faux Glow, Posted: 9/12/2013, Accessed 5/1/2014 at http://www.gcimagazine.com/marketstrends/segments/suncare/Sunless-Tanning-Stats-from-The-Beauty-Company-Show-Women-Seeking-a-Faux-Glow-223466621.html
15. CFR Title 21 Part 73.1150
16. Tsolis, P., Lahanas, K. M.: DHA-based self-tan products. *Cosmetic Toilet.* 2013;128(4):230–233.
17. Maes, D. H., Marenus, K. D.: Self-Tanning Products. In: Baran, R., Maibach, H. I., eds: *Textbook of Cosmetic Dermatology*, 3rd Edition, London: Taylor and Francis, 2005:225–227.
18. Sud'ina, G. F., Mirzoeva, O. K., Pushkareva, M. A.: Caffeic acid phenethyl ester as a lipoxygenase inhibitor with antioxidant properties. *FEBS Lett.* 1993;329(1–2):21–24.
19. Muizzuddin, N., Marenus, K. D., Maes, D. H.: Tonality of suntan vs. sunless tanning. *Skin Res Technol.* 2000;6(4):199–204.
20. FDA: Tanning Products, Last update: 9/4/2013, Accessed 5/1/2014 at http://www.fda.gov/radiation-EmittingProducts/RadiationEmittingProductsandProcedures/tanning/ucm116434.htm
21. Too Faced: Tanning Bed in a Tube, Accessed 5/2/14 at http://www.toofaced.com/p/bronzers/tanning-bed-in-a-tube/
22. U.S. Pat 6451293, Schreier, T., Jermann, R., Combination of erythrulose and a reducing sugar with self-tanning properties, 2002.
23. Chaudhuri, R. K.: Formulating with Dihydroxyacetone (DHA), In: Rhein, L. D., Schlossman, M., O'Lenick, A., eds: *Surfactants in Personal Care and Decorative Cosmetics*, 2nd Edition, New York: Marcel Dekker, 2006.
24. Heckman, C. J., Manne, S. L.: *Shedding Light on Indoor Tanning*, Berlin: Springer, 2011.
25. Pantini, G., Ingoglia, R., Brunetta, F., et al.: Sunless tanning products containing dihydroxyacetone in combination with a perfluoropolyether phosphate. *Int J Cosmet Sci.* 2007;29(3):201–219.
26. Ostrovskaya, A., Rosalia, A. D., Landa, P. A., et al.: Stability of dihydroxyacetone in self-tanning cosmetic products, *J Cos Sci.* 1996;47:275–278.
27. Chaudhuri, R. K.: Dihydroxyactone: Chemistry and Applications in Self-Tanning Products, In: Schlossman, M. L., ed.: *The Chemistry and Manufacture of Cosmetics*, Volume III, Carol Stream: Allured Publishing, 2002.
28. Gel-Creams, Seppic website, Accessed 4/20/2014 at www.seppic.com/cosmetics/gel-cream-@/1019/view-399-searticle.html?lang=en
29. Caswell, M.: Sunscreen Formulations and Tanning Formulations, In: Schlossman, M. L., ed.: *The Chemistry and Manufacture of Cosmetics.* Volume II, Carol Stream: Allured Publishing, 2000.

30. Reineccius, T. A., Reineccius, G. A., Peppard, T. L.: Encapsulation of flavors using cyclodextrins: comparison of flavor retention in alpha, beta and gamma types, *J Food Sci.* 2002;67(9):3271–3279.
31. Morren, M., Dooms-Goossens, A., Heidbuchel, M., et al.: Contact allergy to dihydroxyacetone. *Contact Dermatitis.* 1991;25:326–327.
32. Foley, P., Nixon, R., Marks, R., et al.: The frequency of reaction to sunscreens: results of a longitudinal population-based study on the regular use of sunscreens in Australia. *Br J Dermatol.* 1993;128:512–518.
33. Scientific Committee on Consumer Safety, Opinion on Dihydroxyacetone, Latest update: 12/10/2010, Accessed 4/23/2014 at http://ec.europa.eu/health/scientific_committees/consumer_safety/docs/sccs_o_048.pdf
34. CFR Title 21 Part 70

SECTION 4: FEMININE HYGIENE PRODUCTS

LEARNING OBJECTIVES

Upon completion of this section, the reader will be able to

1. define the following terms

Cleansing and deodorizing product	Condom compatibility	Douching
External feminine genitalia	Glycogen	Internal feminine genitalia
Intimate moisturizer and lubricant	Lactic acid	*Lactobacillus*
Osmolality Spermicide	Topical anti-itch product	Topical antifungal product

2. provide examples for female hygiene products that are applied to the external and internal female genital areas, respectively;
3. explain how lactic acid is produced in the vagina;
4. explain why lactic acid is important to be present in the vagina;
5. name the major types of feminine hygiene products;
6. explain the main function of feminine cleansing and deodorizing products;
7. explain what spermicides are and how they are regulated in the US;
8. explain how lubricants are regulated in the US;
9. summarize the positive effects of using feminine hygiene products;

10. summarize the potential negative effects of using feminine hygiene products;
11. list the various required cosmetic qualities and characteristics that an ideal feminine hygiene product should possess;
12. list the various required technical qualities and characteristics that an ideal feminine hygiene care product should possess;
13. name the major types of feminine cleansing products, and explain their characteristics;
14. briefly explain the function of vaginal lubricants, and name the main ingredients that are generally used for their formulation;
15. explain why the pH of feminine care products is important to consider;
16. explain why osmolality of lubricants is important to consider;
17. briefly explain the difference between the three types of lubricants and their advantages/disadvantages;
18. briefly explain the function of feminine anti-itch and antifungal medications;
19. list a few examples for active ingredients commonly used in feminine anti-itch and antifungal medications;
20. list some typical quality issues that may occur during the formulation and/or use of feminine hygiene products, and explain why they may occur;
21. list the typical quality parameters that are regularly tested for feminine hygiene products, and briefly describe their method of evaluation;
22. briefly explain how the performance of feminine hygiene products is evaluated;
23. list the typical containers available for feminine hygiene products.

KEY CONCEPTS

1. The female genital area has a complex anatomy. It can be subdivided into the internal and external genitalia.
2. The presence of bacteria along with glycogen and the acidic pH is a prerequisite for a healthy vaginal condition.
3. Cleansing and deodorizing feminine products are applied to either the external genital area or the vagina, depending on the type of the cleansing product. These products are available in a variety of forms, including feminine washes, wipes, powders, solutions, and sprays.
4. Lubricants and feminine moisturizers help relieve dryness and are commonly used to enhance pleasure with possibly reducing trauma during sexual intercourse.
5. Spermicides are topical contraceptives that contain spermicidal (i.e., sperm killing) ingredients.
6. Medicated anti-itch and antifungal creams are marketed for relief of the external genital and vaginal feminine itching, irritation, burning sensation, discharge, and infection.

7. The typical quality-related issues of feminine hygiene products include valve clogging for aerosol products, poor foaming for feminine washes, separation of emulsions, abnormal shape for vaginal suppositories, microbiological contamination, clumping, and rancidification.
8. Parameters commonly tested to evaluate the quality of feminine hygiene products include osmolality; spray characteristics; aerosol can leakage and pressure test for aerosol products; spreadability, extrudability, texture, and firmness of lotions and creams; foamability, foam stability, foam density, foam viscosity, and foam structure for feminine cleansing products; actuation force; color; viscosity; osmolality; preservative efficacy; and pH.
9. The most frequently tested parameters referring to the performance (efficacy) of feminine hygiene products include condom compatibility testing with vaginal lubricants, spermicidal efficacy, and antifungal efficacy.
10. Some safety concerns arose in the past regarding the safe use of feminine hygiene products. The main concerns were related to talc in dusting powders and the negative effect of lubricants on the normal *Lactobacillus* flora.

Introduction

The term "hygiene" refers to procedures and principles designed to keep conditions healthy.[1] Each woman has her own unique feminine care needs based on the hygiene concerns, personal preferences, and symptoms she may experience during each cycle. Healthiness and well-being of the female genital area are of big concern. Numerous women use feminine hygiene products every day. Feminine hygiene products include cleansing and deodorant products, topical creams and lotions for managing bacterial and fungal infections, spermicides for local contraception, lubricants and moisturizers for relieving dryness symptoms, as well as sanitary pads, panty liner, tampons, and incontinence pads. A wide variety of products are available today in various dosage forms.

This section provides an overview of the female genital area, the advantages and disadvantages of using hygiene products, as well as the most common product types, their ingredients, and formulation considerations. It also reviews the major quality control issues and tests generally used to evaluate the performance and quality of these products. The most common safety issues are also discussed.

Anatomy and Physiology of the Female Genital Area

● **The female genital area has a complex anatomy. It can be subdivided into the internal and external genitalia.** The internal genitalia include the vagina, cervix, uterus, fallopian tubes, and ovaries. The external genitalia, often referred to as the vulva as a collective term, include the mons pubis (the rounded area in front of the pubic bones that becomes covered with hair at puberty), inner and outer lips of the vagina, clitoris, urethral opening, vaginal opening and its glands, and the perineum (the area between the vaginal and the anal opening).[2,3] The mons pubis and outer lip are keratinized and stratified, similar to the skin in other areas.[4] The inner lips are thinner, not keratinized, and devoid of hair follicles and sweat glands.[5]

Vulva Cleansing agents, wipes, and certain moisturizers and medicated creams, as well as deodorants, are applied to the vulva. Additionally, sanitary pads and panty liners are in continuous contact with the vulvar skin. The vulva differs functionally from the other skin surfaces in numerous characteristics. The genital areas have the thinnest SC in the body.[6] The vulva is more hydrated than other body parts due to the occluded environment. Additionally, the density of hair follicles and sweat glands is higher in the vulva per surface area than in other body surfaces; the friction is higher in the vulvar area, which may make it more vulnerable to mechanical damage; the vulvar area has a higher rate of water loss than the forearm, and it is more penetrable for chemical substances.[7] Based on all these data, it can be seen that the barrier function of the vulva is imperfect when compared to the other skin surfaces. It may lead to the assumption that the vulvar area is more sensitive to irritants and allergic reactions may have a higher incidence in this region. However, studies show that the vulvar irritant reactivity is not higher than that of the forearm for all irritants.[4] In many cases, consumers who have allergic reactions to female hygiene products are generally more sensitive to chemicals on other body surfaces as well.[8] The vulvar pH is higher than that of the forearm (it is around pH 6), which has an important role in bacterial colonization.

Vagina Vaginal lubricants, douches, certain topical medicated creams and suppositories, spermicides, and tampons are applied to the vagina. The vaginal channel, which is open to the environment through the external genital organs, is characterized by a unique and dynamic ecosystem continuously changing throughout a woman's life. Changes in the ecosystem are generated by variations in the hormone levels, diseases, sexual activity, medications, and personal hygiene practices, among others.[9] During the reproductive years, the fluctuating hormone levels, which regulate the menstrual cycle, are important influencing factors of the vaginal microbiota.[10]

The vaginal vault is colonized within 24 h after birth, and it remains colonized until death.[11] The vagina of healthy fertile women harbors an extensive number of bacteria, including *Lactobacillus*, *Staphylococcus*, *Ureaplasma*, *Corynebacterium*, *Streptococcus*, *Peptostreptococcus*, *Gardnerella*, *Bacteroides*, *Mycoplasma*, *Enterococcus*, *Escherichia*, *Veillonella*, *Bifidobacterium* species, and *Candida*,[12,13] of which *Lactobacillus* spp. predominate.[14]

In a fertile woman, the desquamated vaginal epithelial cells release glycogen, a carbohydrate, which supplies *Lactobacillus* spp. with nutrients. These bacteria degrade glycogen and convert it to lactic acid to create an acidic environment, which restricts the growth of pathogenic microorganisms.[15] This cycle (shown in Figure 7.6) is extremely important in order to maintain vaginal health. In healthy fertile women, the normal pH in the vagina ranges from 3.5 to 4.5, with a typical value of 4.2.[16] ● **The presence of bacteria along with glycogen and the acidic pH is a prerequisite for a healthy vaginal condition.**[17]

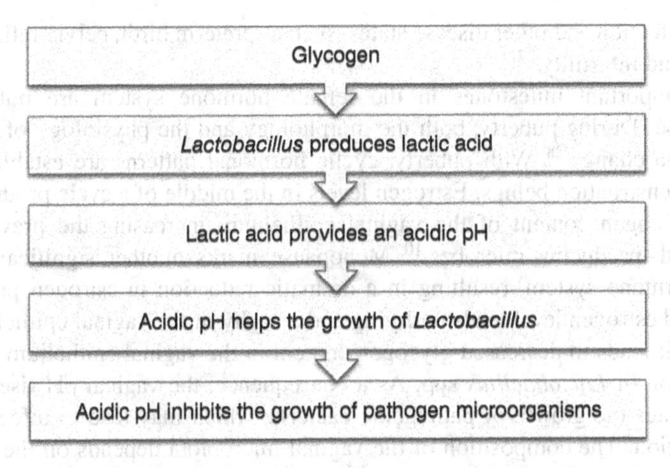

Figure 7.6 Protective effect of lactic acid and *Lactobacillus* bacteria in the vagina.

Pathogenic microorganisms can enter the vagina from the environment and cause discomfort and inflammation. *Lactobacilli* are beneficial for vaginal health because they compete with exogenous microbes for nutrients and block the attachment of other microorganisms to the vaginal epithelium.[18,19] The protective role is facilitated by the production of lactic acid and hydrogen peroxide.[20] Hydrogen peroxide is toxic to other microorganisms that produce little or no hydrogen-peroxide-scavenging enzymes. Thus, hydrogen-peroxide-producing *Lactobacilli* regulate the growth of other vaginal microbes, making the environment less hospitable to others. *Lactobacillus* spp. prevent the colonization and growth of pathogenic exogenous microorganisms via several mechanisms, including keeping the pH below 5 and blocking attachment to the vaginal epithelium.[14,21] A functional equilibrium can also inhibit the transmission of sexually transmitted diseases (STDs).[22]

Despite the paucity of glands, the vaginal epithelium is usually kept moist by a viscous surface film that is generally referred to as the vaginal fluid or mucus. This film is mostly transudate from the vaginal and cervical cells and also contains secretions from the sebaceous glands, sweat glands, and other glands located at the vaginal opening as well as the cervical mucus, with additional fluids from the endometrium and fallopian tubes.[23] The vaginal secretion is a mixture of several components, including ions (sodium, calcium, and chloride), proteins/peptides, glycoproteins, lactic acid, acetic acid, glycerol, urea, and glycogen, which vary depending on the level and ratio of estrogen and progesterone, sexual stimulation, and the status of microbiota.[23,24]

There are a number of factors that can modify the normal pH balance of the vagina, including lack of hygiene, frequent cleansing, excessive fragrances in cleaning products and deodorants, inadequate rinsing of cleansing products, carbohydrate intake, and medications, such as antibiotics and oral contraceptives. These factors can disturb the normal bacterial flora and also the physiology of the vagina.[25,26] It is known that disturbances in the vaginal microbiota can be the first step in the development of vaginal infection and other disease states, such as preterm birth, pelvic inflammatory disease, and infertility.[27]

Two important milestones in the female hormone system are puberty and menopause. During puberty, both the morphology and the physiology of the vulva and vagina change.[28] With puberty, cyclic hormonal patterns are established and regular menstruation beings. Estrogen levels in the middle of a cycle produce peaks in the glycogen content of the vaginal epithelium, increasing the prevalence of lactic acid–producing microbes.[10] Menopause marks another significant change in the hormone system, resulting in a dramatic reduction in estrogen production. Decreased estrogen level leads to drying and atrophy of the vaginal epithelium.[29] In addition, it leads to decreased glycogen content in the vaginal epithelium, resulting in depletion of *Lactobacillus* spp. As a consequence, the vaginal pH rises.[30] High pH promotes the growth of pathogenic bacteria, which may lead to infections and inflammation. The composition of the vaginal microbiota depends on the duration, rate, and severity of estrogen deficiency.[31]

Types and Definition of Feminine Hygiene Products

A wide variety of feminine hygiene products are available today for a variety of uses on the internal or external genital area.

- **Cleansing and deodorizing products** form the largest category of intimate hygiene products. Cleansers are designed to remove debris, including dead cells, sweat, urine residue, vaginal secretion, and blood during menstruation from the genital area. Additionally, they refresh the genital area and help maintain its normal pH. Deodorants are designed to refresh the genital area and mask the characteristic odor developing during the day due to the tightness of the genital area, bacterial activity, and various residues. Products in this category include intimate deodorant sprays, pre-moistened cleansing wipes, non-medicated douches, dusting powders, and cleansing body washes. These products are considered cosmetics in the US.
- **Absorbents** are a special category of products that are designed to absorb fluids during the menstrual cycle and in the case of incontinence. These products include tampons, sanitary pads, panty liners, and incontinence control pads. They are considered and regulated as medical devices in the US. Their ingredients and manufacturing technology greatly differ from those of other products, which does not belong to the scope of this section.
- **Intimate moisturizers and lubricants** can provide relief for women struggling with symptoms of vaginal dryness. Lubricants may be regulated as medical devices in the US if they are used to insert medical tools, douches, tampons, and condoms. These products are also widely used in gynecological and hospital procedures. Other products claiming to relieve discomfort caused by dryness or enhance sexual pleasure are considered OTC drugs according to the FDA's current statement.[32] Moisturizers usually claim long-term relief of vaginal dryness

rather than just being sexual lubricants providing temporary effect. The FDA considers these claims to be drug claims as well.

- **Topical anti-itch and antifungal products** are a special category of products that include active ingredients with the purpose of preventing and/or treating a specific disease. These products are, therefore, considered either OTC drugs or prescription-only drugs in the US depending on the type of the active ingredient used. There are a number of anti-itch and antifungal creams and vaginal suppositories available for the external genital areas and vagina. These are generally regulated as OTC drugs. Medicated douches are also available, which are also considered OTC drugs.
- **Spermicides** are products designed for topical vaginal application to prevent unwanted pregnancy. These products are considered OTC drugs in the US. Available product forms include sponges, foams, jellies, creams, vaginal suppositories, and films.

DID YOU KNOW?

According to the FDA, a medical device is "an instrument, apparatus, implement, machine, contrivance, implant, in vitro reagent, or other similar or related article, including a component part, or accessory which is: recognized in the official National Formulary, or the United States Pharmacopoeia, or any supplement to them; intended for use in the diagnosis of disease or other conditions, or in the cure, mitigation, treatment, or prevention of disease, in man or other animals; or intended to affect the structure or any function of the body of man or other animals, and which does not achieve any of its primary intended purposes through chemical action within or on the body of man or other animals and which is not dependent upon being metabolized for the achievement of any of its primary intended purposes."[33]

DID YOU KNOW?

Products classified as medical devices need not disclose the ingredients on packaging.

History of Using Feminine Hygiene Products

The history of various feminine hygiene products dates back to the ancient ages. The first cleansing agents included soaps made at home; these were used for cleaning the body and external genital parts as well. As discussed in the previous chapters, synthetic surfactants were developed only in the early 20th century. Since then, more sophisticated and milder cleansing agents have been made available as an alternative for soaps.

One of the earliest records from Egypt dating back to the 19th century BC described various methods of inhibiting contraception with simple mixtures of natural ingredients, such as acacia, sodium carbonate, and honey. These mixtures acted as a physical barrier against sperms, and acacia also helped maintain the acidic pH of the vagina.[34] Other cultures used herbs and herbal extracts, such as flowers and roots as well as rice water, to prevent conception. Olive oil was also a frequently used contraceptive agent in various cultures. Another method used as a contraceptive was rock salt mixed with oil and placed in the vagina. This could act as a sperm-killing agent; however, it could have been painful if women had vaginal lesions.[35] Douching has also been used since ancient times to clean the vagina and prevent unwanted pregnancy. Early douching solutions were made of mixtures of vinegar and water, lemon juice and water, and other acidic fruit extracts. In the early 20th century, homemade products, mainly douches and suppositories, contained ingredients such as lint or cotton mixed with olive oil or acidic fruit juices.[36] Since the 1920s, Lysol®, a famous disinfectant, began to be sold as a feminine hygiene product as well. Product advertisements claimed that the cleaner can be used as a vaginal douche, which is effective and safe in combating both germs and odors.[37] Today, scientific data show that douching can have severe negative health effects; therefore, douching is not recommended as a daily routine for women. As for vaginal lubricants, in the early ages, vegetable oils, most commonly olive oil, were used. The first water-based lubricant was created in the early 20th century. It was named K-Y Jelly®, which is still on the market today.[38] Today, in addition to the water- and oil-based products, silicone-based products are also available.

How Feminine Hygiene Products May Affect the Human Body and Genital Area?

A wide variety of intimate hygiene products are used daily by many women due to the beneficial effects of such formulations. The main advantages of using feminine hygiene products are summarized here.

- Cleansing products help **remove** the residue of dead cells, residue of urine and vaginal secretions, sweat, and other debris and help keep the external genital area clean. Products formulated specifically for the sensitive and unique genital environment help maintain the normal acidic pH and bacterial flora, which in turn can prevent the colonization and overgrowth of pathogenic bacteria.
- The use of disposable sanitary pads, panty liners, and tampons is ubiquitous in the industrialized countries. They provide a comfortable way for **menstrual protection**.
- Lubricants and moisturizers offer a relief for **vaginal dryness** symptoms and can enhance sexual intercourse. These are definite advantages for women temporarily or permanently suffering from vaginal dryness. In addition, medical

and gynecological procedures are more comfortable for patients when using such products.

- **Spermicides** provide an easy and comfortable contraceptive option for women not desiring to get pregnant. One of the biggest advantages of spermicide products is that their use does not need cooperation from the partner. These products are beneficial for those who do not want to take pills every day or use another systemic contraceptive method, such as intrauterine devices.

Although feminine hygiene products are advantageous in most cases, potential adverse effects may also occur. These are summarized here.

- The principal health concern related to **tampon** use is its association with menstrual **toxic shock syndrome** (TSS). TSS is a rare but serious disease that may even cause death. It is a recognizable and treatable disease caused by toxin-producing strains of *Staphylococcus aureus* and *Streptococcus pyogenes*. The disease was associated with using highly absorbent tampons; however, the exact connection remains unclear.[39] These materials are no longer used in the US. In the US, statements on the packages suggest that women should use the lowest absorbency product in order to reduce the risk of contacting TSS.[40] In addition, the FDA does not recommend using tampons 24 h a day, 7 days a week, but alternated with pad use. Due to all these factors, the number of reported TSS cases has decreased significantly in recent years.
- **Douching**, the insertion of a device into the vagina for flushing liquid into the vaginal vault, is a common practice worldwide. Women have a number of reasons for douching, including killing germs, preventing pregnancy, preventing STDs, help relieving vaginal itching and discharge, odor control, and improve cleanliness.[41,42] Theoretically, douching is advantageous to help normalize the vaginal pH and microflora; however, some concerns arose in the past decade. Due to the large volume, douches can wash away a variety of vaginal defenses and can promote colonization of bacteria or alter the vaginal pH, allowing pathogenic bacteria and yeast to proliferate. Douching has been associated with many adverse outcomes, including pelvic inflammatory disease, bacterial vaginosis, cervical cancer, low birth weight, preterm birth, HIV transmission, STDs, ectopic pregnancy, recurrent vulvovaginal candidiasis, and infertility.[43,44] The current notion is that douching is not needed to be practiced for normal vaginal hygiene since the vagina is naturally self-cleaning.[45] Women with vaginal symptoms such as odor and yellow vaginal discharge should consult with a pharmacist or other healthcare professionals in order to identify an underlying infection requiring treatment and a visit to their physician.[43]
- **Lubricants** can provide relief for dryness symptoms and can enhance sexual experience. Among couples trying to conceive, incidence of vaginal dryness appears to be increased, and therefore, couples often use lubricants while trying

to conceive.[46] However, studies show that some vaginal lubricants can negatively affect *in vitro* sperm motility and, therefore, can negatively affect **fertility** by inhibiting fertilization. For this reason, couples trying to conceive are advised not to use vaginal lubricants during intercourse.[47]

- **Oil-based lubricants** can cause latex condoms to lose strength rapidly, making them ineffective against unplanned pregnancy and STDs. Therefore, they are not recommended to be used with latex condoms.[48,49]
- Vaginal yeast and bacterial infections are common among women. These conditions are often **self-treated** using OTC medications without visiting a physician. However, without a proper medical diagnosis, these infections often cannot be effectively treated. Additionally, the number of resistant strains can be increased, which can make future treatment more challenging.[50] Therefore, women are recommended to visit their physician in case they experience signs and symptoms of vaginal infections.[51]
- Common issues from the use of various feminine hygiene products are **allergic reactions** and irritation. As discussed earlier, however, the vulvar area is not more sensitive to chemicals than other body parts. Patients experiencing irritation reactions may be hypersensitive to other topical agents as well.[52,53]
- **Overuse** of cleansing products may damage the already thin SC and can also alter the vulvar and vaginal microbiota, leading to irritation. Therefore, it is important not to use these products too often in order to maintain the healthy state of the genital area.
- Spermicide creams, gels, films, sponges, and other product forms containing **nonoxynol-9** as a spermicide may disrupt the vaginal microbiota on the long term. Since this active ingredient kills sperms, damage of the normal vaginal bioflora may also occur.[54,55]

Required Qualities and Characteristics and Consumer Needs

From a consumer perspective, a quality feminine hygiene product should possess the following characteristics:

- **Absorbents:**
  - Absorb a large amount of fluids
  - Provide a fresh smell or be fragrance-free
  - Good attachment to the underwear
  - Odorless fluid-absorbing property
  - Do not irritate the genital area
- **Cleansing Agents and Deodorants:**
  - Provide a fresh smell
  - Be gentle to the sensitive genital area
  - Provide a pH close to that of the genital area
  - Remove debris

- **Moisturizers and Lubricants:**
  - Good lubricating properties for the genital area
  - Provide dryness relief
  - Enhance pleasure during sexual intercourse
  - Do not interfere with latex condoms
  - Do not negatively affect the vaginal wall.
- **Medications:**
  - Provide relief from infection and irritation
  - Spermicide: provide contraceptive effect.
- Well-tolerated, nonirritating
- Easy application
- Comfortable feeling
- Easy removal.

The technical qualities of feminine intimate care products can be summarized as follows:

- Long-term stability
- Foaming Products: appropriate foam structure, density, viscosity, and stability
- Appropriate pH
- Appropriate rheological properties
- Dermatological safety.

Types, Typical Ingredients, and Formulation of Feminine Hygiene Products

Cleansing and Deodorizing Products As discussed earlier, ● **cleansing and deodorizing feminine products are applied to either the external genital area or the vagina, depending on the type of the cleansing product. Products are available in a variety of forms, including feminine washes, wipes, powders, solutions, and sprays.** Cleansing products for the genital area must be mild; surfactant selection should be based on the needs of this delicate area. These cleansers are generally nonfoaming or low-foaming formulations with a refreshing effect and appropriate pH. Antibacterial agents may also be included in these products.

Feminine Washes Feminine washes are similar to body washes in terms of texture and composition; however, there are some important considerations that should be kept in mind. These include the mildness of the surfactants and the pH of the products, which should be close to that of the genital area. Feminine washes are usually water-based acidic surfactant solutions with added beneficial ingredients.

- **Surfactants** generally include a mixture of anionics, such as sodium or ammonium laureth sulfate, or milder versions, such as sulfosuccinates or isethionates, usually in a lower concentration, and nonionics, e.g., decyl glucoside, and amphoterics, e.g., cocamidopropyl betaine and disodium cocoamphodiacetate.

Rinsability is an important consideration of the products, taking into account the short period of time generally taken to rinse these products. Additionally, the genital hair may also make rinsing more difficult.

- **Skin conditioners** can be employed to decrease the aggressiveness of the cleansers.
  - Examples include quats and hydrolyzed proteins.
- **Natural extracts** can be used due to their beneficial, soothing, anti-inflammatory, and anti-itching properties.
  - Examples include aloe, chamomile, green tea, and pomegranate extract.
- **Thickeners** play an important role in controlling the product texture and its flowing properties. The pH sensitivity of some thickeners should be kept in mind when selecting thickeners for feminine cleansing products.
  - Examples for thickeners used in feminine care products include gums, such as xanthan gum, and cellulose derivatives, such as hydroxyethyl cellulose and aluminum silicates. Sodium chloride can also be used as a thickener as these systems contain anionic surfactants.
- **pH buffers** are very important in feminine care products to provide the with a pH same as that of the genital area. Products are typically formulated to be acidic, which may be shifted to the natural range when used and diluted with water.
  - Examples for pH buffer commonly used include lactic acid, citric acid, and sodium hydroxide.
- **Preservatives** are also essential in these products as they are water based.
  - Examples include phenoxyethanol, methylchloroisothiazolinone, methylisothiazolinone, and potassium sorbate.
- **Additional ingredients** of cleansing products are similar to those of body cleansing products and can include chelating agents, antioxidants, humectants, colorants, and fragrances. Fragrances should be mild and refreshing and have the ability to mask characteristic odor.

The formulation of feminine washes is similar to that of regular body washes.

Douches Douches are acidic solutions used to irrigate the vagina for reasons such as cleansing, preventing infections,[56] preventing pregnancy, and odor control, among others.[45] Users may make their own nonmedicated douches at home using salt, vinegar, and water or can use douches marketed in stores. However, it has been shown that douching may have negative health benefits. The general consensus is that douching is not needed to maintain vaginal hygiene as the vagina is self-cleaning.

Feminine Cleansing Wipes Premoistened feminine cleansing wipes are similar to cleansing wipes used for the diaper zone. These are disposable and usually flushable (biodegradable) cloths made of nonwoven fabric. They are soaked into a mild surfactant solution and packaged either individually or into larger plastic resealable bags. Preservatives are essential ingredients for these products as well. Skin conditioners may also be incorporated and deposited on the skin surface as the products do not require rinsing after application. Women who use sanitary pads to manage their monthly flow can also benefit from wipes. The only disadvantage of feminine wipes is that they do leave the skin wet and require a few seconds to dry.

Feminine Dusting Powders Feminine dusting powders can be used for odor control with the same concept as baby powders. Ingredients with good absorbing capacity are used, which can absorb sweat, urine residue, and vaginal secretion, while the added fragrance may mask the characteristic odor, leaving the skin drier, cleaner, and fresher. Additional odor neutralizing ingredients, such as sodium bicarbonate, can also be added. Application may occur directly to the vulvar area or indirectly to the sanitary pads. Today's dusting powders usually contain corn starch as the main absorbent. The main disadvantage of dusting powders is that they can be complicated and messy to apply.

Feminine Deodorant Sprays Feminine deodorant sprays are either aerosol products or nonaerosol pump sprays applied to the external genital area. They are designed to eliminate odor and absorb moisture. The concept is similar to that of underarm deodorants; however, feminine products are usually nonalcohol based to avoid irritation. Odor masking ingredients, similar to underarm deodorants, can include fragrances, zinc ricinoleate, sodium bicarbonate, and corn starch as an absorbent. Deodorant sprays may contain antimicrobial agents and astringents, such as witch hazel.

Feminine Deodorant Suppositories Feminine deodorant suppositories are also available on the market. These are nonmedicated suppositories that melt at body temperature when inserted into the vagina and provide a clean, fresh feeling and scent. Typical ingredients include suppository bases, such as PEG 20, PEG 32, and PEG 20 stearate. The main disadvantage of such products is that users may have to wear panty liners since slight leakage may occur.

DID YOU KNOW?

Antiperspirants are not recommended to be used in the genital region since their acidic pH may be irritating to this sensitive area.[57]

Feminine Moisturizers and Lubricants Vaginal dryness can occur as the hormone levels (especially estrogen) fluctuate every month, and it is a common symptom of aging. ● **Lubricants and feminine moisturizers help relieve dryness and are commonly used to enhance pleasure with possibly reducing trauma during sexual intercourse.**

Vaginal moisturizers are usually used by women suffering from chronic dryness, which can be characterized as a dry feel, itching, and irritation during walking or even sitting.[58] These women can use vaginal moisturizers, which may be creams or suppositories that hydrate the vaginal tissue. The products are typically used every 2–3 days to improve dryness, itching, and irritation. The main ingredients providing prolonged relief include polycarbophil and pectin. Polycarbophil has bioadhesive properties (i.e., adheres to the cells in the vaginal wall) and only detaches upon desquamation of the outer layer of cells.

Women with dryness can also use lubricants, which, however, provide temporary relief. Therefore, these products are generally used before and during sexual intercourse on the female or both genders' genitals to reduce friction and pain. Lubricants are available as water-, oil-, and silicone-based products.

- **Water-based lubricants** contain water; thickeners, such as gums, celluloses, and carbomer; humectants, such as glycerin and propylene glycol; pH buffers; and preservatives, such as parabens and sodium benzoate, to prevent microbial contamination. These are safe to use with latex condoms; however, they tend to dry quickly.
- **Oil-based lubricants** are based on petroleum jelly, vitamin E oil, and vegetable oils. These products, however, are contraindicated with latex condoms. They may damage the latex, making them ineffective against unplanned pregnancy and STDs.
- **Silicone-based products** provide longer-lasting lubrication than water-based lubricants. They are also safe with latex condoms. They are based on various types of silicones. Some examples include dimethicone, dimethiconol, caprylyl methicone, and cyclopentasiloxane.

Additional ingredients of lubricant include preservatives, surfactants, antioxidants, and pH buffers. Natural extracts, such as aloe vera, may also be incorporated.

An important fact that should be mentioned under lubricants is that the preservatives added to ensure proper shelf life can have a negative impact on the genital microflora.[59] Some products contain chlorhexidine as the preservative, which is a bactericidal compound frequently used in preoperative skin preparations and antigingivitis mouthwash.[60] This ingredient can completely eliminate the *Lactobacillus* species in the vagina. Parabens seems to be safer from the perspective of the vaginal flora.

DID YOU KNOW?

Osmolarity is a measure of the osmoles of solute per liter of solution (mol/l, molar, or M). As the volume of solution changes with the amount of solute added as well as with changes in the temperature and pressure, osmolarity is difficult to determine. **Osmolality** is a measure of the moles (or osmoles) of solute per kilogram of solvent (mol/kg, molal, or m). As the amount of solvent remains constant regardless of the changes in temperature and pressure, osmolality is easier to evaluate and is more commonly used in practical osmometry.

Another important fact that should be considered when formulating lubricants is their osmotic concentration. Some lubricants are hyperosmotic, which extract water from the vaginal tissues, making the vaginal barrier weaker. This can lead to vulnerability to damage and a higher rate of transmission of diseases.[61] Therefore, osmolality of the product should be checked as part of product testing. The primary factor determining the osmolality of the majority of lubricants is the concentration of glycols. Examples for glycols include glycerin and propylene glycol, which are added to lubricants as humectants. Most commercial personal lubricants have high osmolalities (2000–6000 mOsm/kg).[62] By comparison, the normal osmolality of female vaginal secretions is 260–290 mOsm/kg.[16] Ideally, the osmolality of a personal lubricant should not exceed 380 mOsm/kg to minimize any risk of epithelial damage. However, osmolalities up to 1200 mOsm/kg are still considered fine according to the WHO. The final osmolality will depend on the amount and/or ration of glycols used.[49]

DID YOU KNOW?

Osmolarity, otherwise known as osmotic concentration, is the concentration (measure) of osmotically active particles in a solution. Solutions that contain salts in equal amounts to the normal tissues are referred to as isosmotic solutions. Hyperosmotic solutions contain a higher concentration of salts and other dissolved materials than normal tissues, while hypoosmotic solutions contain a lower concentration of salts and other osmotically active particles than normal tissues.

DID YOU KNOW?

If a hyperosmotic solution is applied to the skin, water leaves the skin to dilute the hyperosmotic solution. In this cases, cells shrink and can fracture.

Spermicides ● **Spermicides are topical contraceptives that contain spermicidal (i.e., sperm killing) ingredients.** Currently, the most commonly used spermicidal product is nonoxynol-9. Spermicides are available in a number of products forms, such as gels, which are similar to non-spermicidal water-based lubricants; sponges that are inserted into the vagina after wetting them; as well as creams, films, and vaginal suppositories. Many spermicides are supplied with applicators, which help the insertion of the products into the vagina. These products are effective in killing sperms and preventing pregnancy, however, are not effective against other diseases. Current labeling of these products state that the products do not provide protection against the human immunodeficiency virus (HIV) or against getting other STDs.[32]

Topical Anti-Itch and Antifungal Medications ● **Medicated anti-itch and antifungal creams are marketed for relief of the external genital and vaginal feminine itching, irritation, burning sensation, discharge, and infection.** Itching may be caused by a variety of factors, including bacterial and yeast infections,[63] changing hormone levels during menopause,[64] chemicals in detergents, fabric softeners, vaginal deodorants, douches, and other topically applied products, among other reasons.[65] Anti-itch and antifungal products are available as creams, lotions, and vaginal suppositories with active ingredients. Commonly used anti-itch active ingredients include benzocaine and pramoxine hydrochloride, which are local anesthetics (i.e., numbing agents), as well as resorcinol, a topical antiseptic. Commonly used antifungal active ingredients include clotrimazole, miconazole nitrate, and butoconazole nitrate.

As mentioned earlier, the main concern with self-treatment is that, without a proper diagnosis, patients may not use the proper active ingredient and the potential for developing drug resistance.

The formulation of these products is identical to that of the traditional nonmedicated dosage forms.

Typical Quality Problems of Feminine Hygiene Products

● **The typical quality-related issues of feminine hygiene products include valve clogging for aerosol products, poor foaming for feminine washes, separation of emulsions, abnormal shape for vaginal suppositories, microbiological contamination, clumping, and rancidification.** Quality issues that were discussed in the previous sections are not reviewed here.

Abnormal Shape for Vaginal Suppositories Vaginal suppositories usually have an oblong shape for easier application. Abnormal shape may be caused by a variety of factors, such as storage at high temperatures, faulty molding technique, and inappropriate composition that leads to softening even at room temperature. During molding, it is important to ensure not to include air bubbles in the melted mixture since their presence can lead to holes if they cannot escape before the suppositories solidify.

Evaluation of Feminine Hygiene Products

Quality Parameters Generally Tested ● **Parameters commonly tested to evaluate the quality of feminine hygiene products include osmolality; spray characteristics; aerosol can leakage and pressure test for aerosol products; spreadability, extrudability, texture, and firmness of lotions and creams; foamability, foam**

stability, foam density, foam viscosity, and foam structure for feminine cleansing products; actuation force; color; viscosity; osmolality, preservative efficacy; and pH. Tests that were discussed in the previous sections are not reviewed here in detail. Quality control testing of vaginal suppositories includes a number of physical and chemical parameters. Physical analysis includes physical appearance, weight uniformity, uniformity of texture, melting point, liquefaction time, melting and solidification time, and mechanical strength. Chemical tests include analysis of the activity and dissolution testing. All these tests are described in detail in the United States Pharmacopoeia (USP).

Osmolality Osmolality is measured with an osmometer. There are three major types of osmometers available commercially, including **freezing point osmometers**, which determine the osmotic strength of solution by using freezing point depression; **vapor pressure osmometers**, which determine the concentration of osmotically active particles that reduce the vapor pressure of the solution; and **membrane osmometers**, which measure the osmotic pressure of a solution separated by a semipermeable membrane.[66]

Performance (Efficacy) Parameters Generally Tested **● The most frequently tested parameters referring to the performance (efficacy) of feminine hygiene products include condom compatibility testing with vaginal lubricants, spermicidal efficacy, and antifungal efficacy.**

Condom Compatibility Testing with Vaginal Lubricants Weakening of natural rubber latex is known to occur after contact with certain lubricants, particularly petroleum-based products. Therefore, biocompatibility testing of vaginal lubricants is essential to confirm the safety for intended use. The current method used is an ASTM D method [67], which determines whether a vaginal lubricant has a significant effect on the tensile strength and airburst properties of a natural rubber latex condom. In the tensile test (stretch test), a band from the shaft of a condom is tested for its stretchability. In the airburst test, condoms are inflated with air until they burst; the maximum volume of air tolerated is used as the measure of strength.

Spermicidal Efficacy The spermicidal effect of gels, films, sponges, and other products containing nonoxynol-9 is essential to test in order to ensure proper product performance. The test can be performed on animals *in vitro* and *in vivo* on human subjects in actual use studies. In the actual use studies, subjects are randomized into groups. Generally, the spermicidal effect of the product is compared to either no contraceptives or a contraceptive with known efficacy. According to the FDA and WHO, spermicides are the least effective contraceptive methods in general.[68,69] Therefore, subjects should be informed that they may become pregnant during the study.

Antifungal Efficacy The efficacy of antifungal products can be tested *in vitro* using agar plates and fungi generally causing fungal vaginal infection or via *in vivo* actual use studies with human subjects. In the latter case, the subjects are randomized into groups using different products or the study product and a placebo. The efficacy of the products is compared at predetermined time periods and the statistical difference is calculated.

Safety Concerns **⓾ Some safety concerns arose in the past regarding the safe use of feminine hygiene products. The main concerns were related to talc in dusting powders and the negative effect of lubricants on the normal *Lactobacillus* flora.**

Talc Similar to face powders and baby powders, talc caused safety concerns in the case of feminine powders as well. First, asbestos contamination was one of the biggest concerns. However, cosmetic grade talc was evaluated by many scientific organizations and was found not to be contaminated with asbestos. For more detail, refer to Section 3 of Chapter 4.

An additional concern with talc was whether it can cause ovarian cancer if it is used on the external genital area. It has been suggested that talcum powder might cause ovarian cancer if applied to the genital area (or sanitary napkins, diaphragms, or condoms). Studies have been conducted to establish a link between talc application and ovarian cancer. The findings are mixed, with some studies reporting a slightly increased risk and some reporting no increase.[70–72] Based on the limited evidence from human studies, the International Agency for Research on Cancer (IARC) has classified genital exposure to talc as group 2B (possibly carcinogenic to humans).[70,73] Therefore, today's feminine powders (similar to baby powders) are based on corn starch and not on talc.

Lubricants' Effect on the Lactobacillus *Species* As discussed earlier, non-spermicidal lubricants may also have a negative effect on the normal vaginal *Lactobacillus* flora. *Lactobacillus* species are important for the maintenance of the female genital tract as they produce lactic acid and hydrogen peroxide. Loss of *Lactobacillus* spp. can lead to infections and easier transmission of diseases. Therefore, it is important to test the viability of the *Lactobacillus* species after applying a lubricant. The test can be performed *in vitro* culturing *Lactobacillus* species on agar plates, applying the lubricant and evaluating the zone of inhibition and the number of colony-forming units (CFUs). Additionally, *ex vivo* studies can be performed using cell lines.[59]

Packaging of Feminine Hygiene Products

Similar to liquid oral care products, vaginal products must also be packaged into tamper-resistant packaging under the US law. It is the responsibility of the manufacturers and packagers to ensure tamper-resistant packaging if a product is accessible to the public while held for sale[74] (for more detail on what a tamper-resistant package is, refer to Chapter 6). Additionally, a warning statement should be placed on the label that informs customers about the tamper-resistant nature of the packaging and advises them not to use it if the package is broken or missing.

The most commonly used packaging materials for hair removal products include the following:

- **Plastic Tubes and Bottles:** Most medicated creams and gels are packaged into plastic tubes with a screw-top cap or flip-flop cap. Lubricants are usually packaged into plastic tubes or plastic bottles with a flip-flip cap or pump head. Feminine wash products are generally packaged into plastic bottles, similar to body washes that may be supplied with a flip-flop cap or pump head. Feminine powders are marketed in plastic dusting powder containers, similar to baby powders.
- **Foil Packet:** Medicated suppositories are generally available in individually wrapped plastic or aluminum sachets. Since medicated products are inserted into the vagina, most products are supplied with a special disposable applicator that aids in intravaginal application.
- **Sachets:** Spermicide films, sponges, and suppositories are generally packaged individually into plastic or aluminum sachets. Sanitary pads and tampons are usually packaged individually into plastic sachets. Panty liners and incontinence pads may be packaged in bulk into paper boxes or sealed plastic bags wrapped individually, similar to other absorbents.
- **Aerosol Cans:** Feminine deodorants are usually supplied in aerosol cans.

GLOSSARY OF TERMS FOR SECTION 4

Cleansing and deodorizing product: A personal care product for women designed to remove debris, including dead cells, sweat, urine residue, vaginal secretion, blood during menstruation, and other debris from the genital area.

Condom compatibility: A performance test performed to evaluate the compatibility of personal lubricants with latex condoms.

Douching: Inserting a device into the vagina for flushing liquid into the vaginal vault to kill germs, prevent pregnancy and STDs, help relieve vaginal itching and discharge, control vaginal odor, and improve cleanliness.

External feminine genitalia: Feminine reproductive organs outside the body, including the mons pubis, inner and outer lips of the vagina, clitoris, urethral opening, vaginal opening and its glands, and perineum.

Glycogen: A carbohydrate released by the desquamated vaginal epithelial cells, which supplies *Lactobacillus* spp. with nutrients.

Internal feminine genitalia: Feminine reproductive organs within the body, including the vagina, cervix, uterus, fallopian tubes, and ovaries.

Intimate moisturizer and lubricant: A personal care product designed to provide relief for women struggling with symptoms of vaginal dryness.

Lactic acid: The main ingredient controlling the acidic pH in the vagina and the growth of pathogenic microorganisms.

***Lactobacillus*:** The dominant vaginal bacteria under normal conditions.

Osmolality: The measure of osmotically active particles in a solution.

Spermicide: A medicated product designed for topical vaginal application to prevent unwanted pregnancy.

Topical antifungal product: A medicated product designed to prevent and/or treat fungal diseases of the external or internal genitalia.

Topical anti-itch product: A medicated product that provides relief from itching of the external and internal genitalia.

REVIEW QUESTIONS FOR SECTION 4

Multiple Choice Questions

1. What is the normal vaginal pH?
   a) 1.5–2.5
   b) 3.5–4.5
   c) 5.5
   d) 7

2. What is the main function of *Lactobacillus* in the vagina?
   a) To maintain an acidic pH
   b) To provide protection against pathogenic microorganisms
   c) To produce vaginal mucus
   d) a and b

3. What is the purpose of using spermicides?
   a) To provide a contraceptive effect
   b) To provide protection against STDs
   c) To lubricate the vagina
   d) All of the above

4. What is the purpose of using lubricants?
   a) To provide a contraceptive effect
   b) To enhance sexual intercourse by reducing friction
   c) To relieve vaginal dryness via lubrication
   d) B and C

5. What is the main disadvantage of oil-based lubricants?
   a) They tend to dry quickly
   b) They may damage latex condoms
   c) They may discolor the skin
   d) They may dry out the vagina

6. What is the main safety concern with regard to the use of vaginal douches?
a) Their alkaline pH damages the vaginal wall
b) They enhance lactic acid production, leading to a highly acidic pH
c) They can wash away beneficial bacteria, allowing pathogenic microorganisms to grow
d) They often lead to allergic reactions

7. Which of the following can lead to depletion of the *Lactobacillus* flora in the vagina?
a) Preservatives in lubricants
b) Spermicidal ingredients
c) Lower level of lactic acid in the vagina
d) All of the above

8. Which of the following contains surfactants?
a) Feminine dusting powder
b) Spermicide gel
c) Lubricant gel
d) Feminine liquid wash

Drug or Cosmetic in the United States?

a) Spermicide cream: ______________
b) Vaginal antifungal cream with miconazole: ______________
c) Vaginal douche: ______________
d) Feminine dusting powder: ______________

Matching

Match the ingredients in column A with their appropriate ingredient category in column B.

Column A	Column B
_____ A. Benzocaine	1. Antifungal ingredient
_____ B. Dimethicone	2. Anti-itch ingredient
_____ C. Hydrolyzed protein	3. Hydrophilic thickener
_____ D. Hydroxyethyl cellulose	4. Oily lubricant
_____ E. Lactic acid	5. pH buffer
_____ F. Miconazole	6. Preservative
_____ G. Nonoxynol-9	7. Silicone lubricant
_____ H. Petroleum jelly	8. Skin conditioner
_____ I. Phenoxyethanol	9. Spermicide
_____ J. Polycarbophil	10. Vaginal moisturizer

REFERENCES

1. *Dictionary of Medical Terms*, London: A&C Black, 2007.
2. Chung, K. W.: *Gross Anatomy*, 4th Edition, Philadelphia: Lippincott Williams & Wilkins, 2000.
3. Fritsch, H., Kuehnel, W., Fritsch, H.: *Color Atlas of Human Anatomy*, Volume 2, Stuttgart: Thieme, 2007.
4. Farage, M. A.: Vulvar susceptibility to contact irritants and allergens: a review. *Arch Gynecol Obstet.* 2005;272:167–172.
5. Farage, M. A., Maibach, H. I.: The vulvar epithelium differs from the skin: implications for cutaneous testing to address topical vulvar exposures. *Contact Dermatitis.* 2004;51:201–209.
6. Tagami, H.: Racial differences on skin barrier function. *Cutis.* 2002;70:6–7.
7. Elsner, P., Wilhelm, D., Maibach, H. I.: Frictional properties of human forearm and vulvar skin: influence of age and correlation with transepidermal water loss and capacitance. *Dermatologica.* 1990;181:88–91.
8. Nardelli, A., Degreef, H., Goossens, A.: Contact allergic reactions of the vulva: a 14-year review. *Dermatitis.* 2004;15:131–136.
9. Eschenbach, D. A., Thwin, S. S., Patton, D. L.: Influence of the normal menstrual cycle on vaginal tissue, discharge, and microflora. *Clin Infect Dis.* 2000;30:901–907.
10. Farage, M. A., Maibach, H.: Lifetime changes in the vulva and vagina. *Arch Gynecol Obstet.* 2006;273:195–202.
11. Marshall, W., Tanner, J.: Puberty, In: Davis, J., Dobbing, J., eds: *Scientific Foundations of Paediatrics*, London: Heinemann, 1981.
12. Larsen, B., Monif, G. R.: Understanding the bacterial flora of the female genital tract. *Clin Infect Dis.* 2001;32:e69–e77.
13. Redondo-Lopez, V., Cook, R. L., Sobel, J. D.: Emerging role of lactobacilli in the control and maintenance of the vaginal bacterial microflora. *Rev Infect Dis.* 1990;12:856–872.
14. Witkin, S. S., Linhares, I. M., Giraldo, P.: Bacterial flora of the female genital tract: function and immune regulation. *Best Pract Res Clin Obstet Gynaecol.* 2007;21(3):347–354.
15. Boskey, E. R., Cone, R. A., Whaley, K. J., et al.: Origins of vaginal acidity: high D/L lactate ratio is consistent with bacteria being the primary source. *Hum Reprod.* 2001;16(9):1809-1813.
16. Owen, D. H., Katz, D. F.: A vaginal fluid simulant. *Contraception.* 1999;59(2):91-95.
17. Boskey, E. R., Telsch, K. M., Whaley, K. J., et al.: Acid production by vaginal flora in vitro is consistent with the rate and extent of vaginal acidification, *Infect Immun.* 1999;67(10):5170–5175.
18. Osset, J., Bartolome, R. M., Garcia, E., et al.: Assessment of the capacity of Lactobacillus to inhibit the growth of uropathogens and block their adhesion to vaginal epithelial cells. *J Infect Dis.* 2001;183(3):485-491.
19. Boris, S., Suarez, J. E., Vazquez, F., et al.: Adherence of human vaginal lactobacilli to vaginal epithelial cells and interaction with uropathogens. *Infect Immun.* 1998;66(5):1985-1989.
20. Larsen, B., Monif, G. R.: Understanding the bacterial flora of the female genital tract. *Clin Infect Dis.* 2001;32:69-77.

21. Boris, S., Barbes, C.: Role played by lactobacilli in controlling the population of vaginal pathogens. *Microbes Infect.* 2000;2(5):543-546.
22. Newton, E. R., Piper, J. M., Shain, R. N., et al.: Predictors of the vaginal microflora. *Am J Obstet Gynecol.* 2001;184:845-853.
23. Paavonen, J.: Physiology and ecology of the vagina. *Scand J Infect Dis Suppl.* 1983;40:31-35.
24. Huggins, G. R., Preti, G.: Vaginal odors and secretions. *Clin Obstet Gynecol.* 1981; 24(2):355-377.
25. Ravel, J: Vaginal microbiome of reproductive- age women, *Proc Natl Acad Sci.* 2011;108(1):4680-4687.
26. Delucchi, L., Fraga, M., Perelmuter, K., et al.: Vaginal lactic acid bacteria in healthy and ill bitches and evaluation of in vitro probiotic activity of selected isolates. *Can Vet J.* 2008;49(10):991–994.
27. Hillier, S. L., Krohn, M. A., Cassen, E., et al.: The role of bacterial vaginosis and vaginal bacteria in amniotic fluid infection in women in preterm labor with intact fetal membranes. *Clin Infect Dis.* 1995;20(2):276–278.
28. Jones, I. S.: A histological assessment of normal vulval skin. *Clin Exp Dermatol.* 1983;8:513–521.
29. Galhardo, C. L., Soares, J. M. J., Simões, R. S., et al.: Estrogen effects on the vaginal pH, flora and cytology in late postmenopause after a long period without hormone therapy. *Clin Exp Obstet Gynecol.* 2006;33:85–89.
30. Gupta, S., Kumar, N., Singhal, N., et al.: Vaginal microflora in postmenopausal women on hormone replacement therapy. *Indian J Pathol Microbiol.* 2006;49:457–461.
31. Farage, M. A., Miller, K. W., Sobel, J. D.: Dynamics of the vaginal ecosystem—hormonal influences *Infect Dis.* 2010:3:1–15.
32. CFR Title 21 Part 201, Over-the-Counter Vaginal Contraceptive and Spermicide Drug Products Containing Nonoxynol 9; Required Labeling 2007;72(243):71769–85. Final rule.
33. FDA: What is a Medical Device?, Last update: 04/22/2014, Accessed 5/13/2014 at http://www.fda.gov/aboutfda/transparency/basics/ucm211822.htm
34. Touitou, E., Barry, B. W.: *Enhancement in Drug Delivery*, Boca Raton: CRC Press, 2010.
35. Stromquist, N. P.: *Women in the Third World: An Encyclopedia of Contemporary Issues*, London: Routledge, 2014.
36. Michael, A.: *Bellesiles: Lethal Imagination: Violence and Brutality in American History*, New York: NYU Press, 1999
37. Advertisement, Accessed 5/12/2014 at http://torontoist.com/attachments/toronto_jamieb/20100216lysol48.jpg
38. K-Y®: FAQ, Accessed 5/13/2014 at http://www.k-y.ca/faq
39. Berkley, S. F.: The relationship of tampon characteristics to menstrual toxic shock syndrome. *JAMA.* 1987;258:917.
40. CFR Title 21 Part 801.430 – User labeling for menstrual tampons, Last update: 1/1/2013, Accessed 5/14/2014 at http://www.accessdata.fda.gov/scripts/cdrh/cfdocs/cfcfr/cfrsearch.cfm?fr=801.430
41. Oh, M. K., Funkhouser, E., Simpson, T., et al.: Early onset of vaginal douching is associated with false beliefs and high-risk behavior. *Sex Transm Dis.* 2003;30(9):689-693.

42. Wilson, T. E., Uusküla, A., Feldman, J., et al.: A case-control study of beliefs and behaviors associated with sexually transmitted disease occurrence in Estonia. *Sex Transm Dis.* 2001;28(11):624-649.
43. Cottrell, B. H.: An updated review of evidence to discourage douching. *MCN Am J Mat Child Nurs.* 2010;35(2):102-107.
44. Misra, D. P., Trabert, B.: Vaginal douching and risk of preterm birth among African American women. *Am J Obstet Gynecol.* 2007;196(2):140:1–8.
45. Martino, J. L., Vermund, S. H.: Vaginal douching: evidence for risks or benefits to women's health. *Epidemiol Rev.* 2002;24:109–124.
46. Ellington, J., Daugherty, S.: Prevalence of vaginal dryness in trying-to-conceive couples. *Fertil Steril.* 2003;79(2):21–22.
47. Practice Committee of American Society for Reproductive Medicine in collaboration with Society for Reproductive Endocrinology and Infertility: Optimizing natural fertility. *Fertil Steril.* 2008;90(5):Supplement 1:S1–S6.
48. Steiner, M., Piedrahita, C., Glover, L., et al.: The impact of lubricants on latex condoms during vaginal intercourse. *Int J STD AIDS.* 1994;5(1):29-36.
49. WHO: Use and Procurement of Additional Lubricants for Male and Female Condoms: WHO/UNFPA/FHI, Accessed 5/3/2014 at http://apps.who.int/iris/bitstream/10665/76581/1/WHO_RHR_12.34_eng.pdf
50. Sihvo, S.: Self-medication with vaginal antifungal drugs: physicians' experiences and women's utilization patterns. *Fam Pract.* 2000;17(2):145–149.
51. McCaig, L. F., McNeil, M. M.: Trends in prescribing for vulvovaginal candidiasis in the United States. *Pharmacoepidemiol Drug Saf.* 2005;14(2):113–120.
52. Wujanto, L., Wakelin, S.: Allergic contact dermatitis to colophonium in a sanitary pad—an overlooked allergen? *Contact Dermatitis.* 2012;66(3):161–162.
53. Williams, J. D., Frowen, K. E., Nixon, R. L.: Allergic contact dermatitis from methyldibromo glutaronitrile in a sanitary pad and review of Australian clinic data. *Contact Dermatitis.* 2007;56(3):164–167.
54. Schreiber, C. A., Meyn, L. A., Creinin, M. D., et al.: Effects of long-term use of nonoxynol-9 on vaginal flora. *Obstet Gynecol.* 2006;107(1):136–143.
55. Fashemi, B., Delaney, M. L., Onderdonk, A. B., et al.: Effects of feminine hygiene products on the vaginal mucosal biome. *Microb Ecol Health D.* 2013;24:19703.
56. Aral, S. O., Mosher, W. D., Cates, W.: Vaginal douching among women of reproductive age in the United States: 1988. *Am J Public Health.* 1992;82(2):210–214.
57. Rigano, L.: Feminine hygiene products, *Cosmetic Toilet.* 2012;127(12):838-845.
58. Lobo, R. A.: Menopause and Care of the Mature Woman, In: Lentz, G. M., Lobo, R. A., Gershenson, D. M., Katz, V. L., eds: *Comprehensive Gynecology*, 6th Edition, Philadelphia: Mosby Elsevier, 2012.
59. Dezzutti, C. S. Brown, E. R., Moncla, B., et al.: Is wetter better? An evaluation of over-the-counter personal lubricants for safety and anti-HIV-1 activity. *PLoS ONE.* 2012;7(11):e48328.
60. Basrani, B., Lemonie, C. Chlorhexidine gluconate. *Aust Endod J.* 2005,31:48–52.
61. Blaskewicz, C. D., Pudney, J., Anderson, D. J.: Structure and function of intercellular junctions in human cervical and vaginal mucosal epithelia. *Biol Reprod.* 2011;85:97–104.

62. Begay, O., Jean-Pierre, N., Abraham, C. J.: Identification of personal lubricants that can cause rectal epithelial cell damage and enhance HIV Type 1 replication in vitro. *AIDS Res Hum Retrov.* 2011;27(9):1019-1024.
63. Wilson, C.: Recurrent vulvovaginitis candidiasis: an overview of traditional and alternative therapies. *Adv Nurse Pract.* 2005;13(5):24–29.
64. Perrotta, C., Aznar, M., Mejia, R.: Oestrogens for preventing recurrent urinary tract infection in post-menopausal women. *Cochrane Database Syst Rev.* 2008:CD005131.
65. Rietschel, R. L., Fowler, J. F.: *Fisher's Contact Dermatitis.* 5th Edition, Philadelphia: Lippincott Williams and Wilkins, 2001.
66. Burkitt Creedon, J. M., Davis, H.: *Advanced Monitoring and Procedures for Small Animal Emergency and Critical Care*, Hoboken: Wiley and Sons, 2012.
67. ASTM D7661 – 10, Standard Test Method for Determining Compatibility of Personal Lubricants with Natural Rubber Latex Condoms.
68. NATAZIA: Prescribing Information, 2010, Accessed 5/5/2014 at http://www.accessdata.fda.gov/drugsatfda_docs/label/2012/022252s001lbl.pdf
69. Shears, K. H., Aradhya, K. W.: Helping women understand contraceptive effectiveness. *Fam Health Int.* 2008:3–6.
70. Langseth, H., Hankinson, S. E., Siemiatycki, J., et al.: Review Perineal use of talc and risk of ovarian cancer. *J Epidemiol Community Health.* 2008;62(4):358-360.
71. Wu, A. H., Pearce, C. L., Tseng, C. C., et al.: Markers of inflammation and risk of ovarian cancer in Los Angeles County. *Int J Cancer.* 2009;124(6):1409-1415.
72. Rosenblatt, K. A., Weiss, N. S., Cushing-Haugen, K. L., et al.: Genital powder exposure and the risk of epithelial ovarian cancer. *Cancer Causes Control.* 2011;22(5):737–742.
73. Baan, R., Straif, K., Grosse, Y., et al.: Carcinogenicity of carbon black, titanium dioxide, and talc. *Lancet Oncol.* 2006;7(4):295-296.
74. CFR Title 21 Part 700.25

KEY FOR REVIEW QUESTIONS

CHAPTER 1: GENERAL CONCEPTS

Section 1: Basic Definitions

Multiple Choice Questions: 1-d, 2-c, 3-b, 4-a, 5-d, 6-c, 7-c, 8-a, 9-b, 10-b, 11-d, 12-a, 13-d, 14-b, 15-c, 16-b

Fact or Fiction? 1-fiction, 2-fact, 3-fiction, 4-fact, 5-fiction

Figure: A-cosmetic, B-OTC drug–cosmetic product, Justification: Hand soap removes dirt from the hands. Antibacterial hand soap also cleans the hands, and additionally, it kills germs on the hands and helps prevent infections. This latter function makes it to be considered a drug, since prevention of disease meets the definition of drugs.

Section 2: Classification of Cosmetics and OTC Drug–Cosmetic Products. Cosmetic Ingredients and Active Ingredients Used in Cosmetics and OTC Drug–Cosmetic Products

Multiple Choice Questions: 1-c, 2-b, 3-a, 4-a, 5-c, 6-c, 7-d, 8-c

Matching 1: 1-P, 2-L, 3-O, 4-I, 5-B, 6-A, 7-G, 8-C, 9-F, 10-H, 11-D, 12-K, 13-M, 14-E, 15-J, 16-N

Matching 2: 1-F, 2-D, 3-J, 4-E, 5-H, 6-L, 7-K, 8-I, 9-C, 10-B, 11-G, 12-A

Introduction to Cosmetic Formulation and Technology, First Edition. Gabriella Baki and Kenneth S. Alexander.
© 2015 John Wiley & Sons, Inc. Published 2015 by John Wiley & Sons, Inc.

Section 3: Dosage Forms for Cosmetics and OTC Drug–Cosmetic Products

Multiple Choice Questions: 1-c, 2-a, 3-a, 4-d, 5-a, 6-c, 7-b, 8-d, 9-c, 10-c

Fact or Fiction? 1-fact, 2-fiction, 3-fact, 4-fact

Matching 1: 1-D, 2-H, 3-F, 4-J, 5-I, 6-G, 7-L, 8-K, 9-C, 10-B, 11-A, 12-E

Matching 2: 1-F, 2-E, 3-B, 4-H, 5-G, 6-A, 7-D, 8-C, 9-J, 10-I, 11-K

CHAPTER 2: LEGISLATION FOR COSMETICS AND OTC DRUG–COSMETIC PRODUCTS

Section 1: Current Rules and Regulations for Cosmetics and OTC drug–cosmetic products in the United States and in the European Union

Multiple Choice Questions: 1-a, 2-d, c-d, 4-b, 5-b, 6-a, 7-d, 8-d, 9-c, 10-a, 11-d, 12-c

Fact or Fiction? 1-fact, 2-fact, 3-fiction, 4-fiction, 5-fact

Misbranded or Adulterated? 1-A, 2-M, 3-A, 4-M, 5-A, 6-M, 7-A, 8-M, 9-A, 10-A, 11-A, 12-A

Section 2: Labeling Tutorial for Cosmetics and OTC Drug–Cosmetic Products Marketed in the United States

Multiple Choice Questions: 1-a, 2-b, 3-a, 4-b, 5-c, 6-a, 7-d, 8-b

Matching 1

a) 1-principal display panel, 2-information panel, 3-identity statement, 4-product claims, 5-net weight, 6-directions for safe use, 7-general warning and caution statement(s), 8-list of ingredients, 9-expiration date, 10-manufacturer's name and address, 11-country of origin, 12-bar code

b) 1-yes, 2-yes, 3-yes, 4-no, 5-yes, 6-yes, 7-yes, 8-yes, 9-no, 10-yes, 11-only if imported, 12-no

Matching 2: 1-O, 2-I, 3-M, 4-N, 5-J, 6-F, 7-E, 8-A, 9-G, 10-K, 11-C, 12-D, 13-L, 14-H, 15-B

Section 3: Government and Independent Organizations in the Cosmetic Industry

Multiple Choice Questions: 1-c, 2-a, 3-a, 4-b, 5-a, 6-d

Matching: 1-C, 2-D, 3-B, 4-H, 5-F, 6-D, 7-G, 8-I, 9-A

Section 4: Cosmetic Good Manufacturing Practices

Multiple Choice Questions: 1-c, 2-b, 3-a, 4-b, 5-a, 6-d

Fact or Fiction? 1-fiction, 2-fact, 3-fact, 4-fact, 5-fiction

CHAPTER 3: SKIN CARE PRODUCTS

Section 1: Skin Anatomy and Physiology

Multiple Choice Questions: 1-b, 2-c, 3-a, 4-b, 5-a, 6-c, 7-d, 8-c, 9-d, 10-b, 11-a, 12-d, 13-d, 14-b, 15-a

Fact or Fiction? 1-fact, 2-fact, 3-fiction, 4-fiction, 5-fact

Section 2: Skin Cleansing Products

Multiple Choice Questions: 1-d, 2-b, 3-b, 4-a, 5-d, 6-b, 7-c, 8-b, 9-a, 10-d, 11-c, 12-c, 13-a, 14-c, 15-d

Fact or Fiction? 1-fiction, 2-fact, 3-fact, 4-fact

Matching: A-6, B-9, C-11, D-3, E-8, F-2, G-1, H-4, I-5, J-12, K-13, L-7, M-10

Section 3: Skin Moisturizing Products

Multiple Choice Questions: 1-a, 2-b, 3-c, 4-c, 5-d, 6-b, 7-d, 8-a, 9-d, 10-a

Fact or Fiction? 1-fiction, 2-fact, 3-fiction, 4-fiction

Matching: A-10, B-9, C-1, D-4, E-7, F-8, G-3, H-6, I-2, J-5

Section 4: Products for Special Skin Concerns – Aging and Acne

Multiple Choice Questions: 1-a, 2-c, 3-c, 4-d, 5-b, 6-a, 7-c, 8-d, 9-a, 10-b

Fact or Fiction? 1-fiction, 2-fact, 3-fact, 4-fact, 5-fiction

Matching: A-6, B-3, C-4, D-1, E-5, F-7, G-2

Section 5: Sun Care Products

Multiple Choice Questions: 1-b, 2-a, 3-b, 4-c, 5-d, 6-c, 7-d, 8-c, 9-b, 10-d, 11-c, 12-b, 13-c, 14-c, 15-c

Short Answers: a-6 months, b-370, c-15, d-80, e-100

Matching: A-9, B-1, C-5, D-10, E-4, F-8, G-3, H-2, I-6, J-7

Section 6: Deodorants and Antiperspirants

Multiple Choice Questions: 1-b, 2-d, 3-a, 4-b, 5-b, 6-c, 7-b, 8-a, 9-c, 10-b

Fact or Fiction? 1-fact, 2-fact, 3-fiction, 4-fiction

Matching: A-2, B-10, C-8, D-1, E-3, F-5, G-4, H-9, I-7, J-6

CHAPTER 4: COLOR COSMETICS

Section 1: Lip Makeup Products

Multiple Choice Questions: 1-d, 2-a, 3-c, 4-b, 5-c

Matching 1: A-1, B-2, C-5, D-4, E-3, F-8, G-9, H-7, I-6

Fill in the Blank: 1-beeswax, 2-antimicrobial, 3-titanium dioxide, 4-flaming, 5-lip balms, 6-lead, 7-melting point, 8-menthol

Matching 2: A-6, B-4, C-3, D-5, E-8, F-2, G-10, H-9, I-1, J-7

Section 2: Eye Makeup Products

Multiple Choice Questions: 1-d, 2-b, 3-c, 4-c, 5-a, 6-d, 7-d, 8-c, 9-b, 10-c

Fact or Fiction? 1-fiction, 2-fact, 3-fact, 4-fiction

Matching: A-10, B-6, C-1, D-2, E-9, F-3, G-5, H-7, I-8, J-4

Section 3: Facial Makeup Products

Multiple Choice Questions: 1-c, 2-b, 3-a, 4-b, 5-a, 6-d, 7-d, 8-a

Fact or Fiction? 1-fiction, 2-fiction, 3-fact, 4-fact

Matching: A-8, B-2, C-10, D-6, E-1, F-7, G-4, H-5, I-3, J-9

Section 4: Nail Care Products

Multiple Choice Questions: 1-a, 2-b, 3-b, 4-b, 5-a, 6-c, 7-b, 8-a, 9-d, 10-c

Fact or Fiction? 1-fact, 2-fact, 3-fiction, 4-fact

Matching: A-6, B-5, C-9 D-2, E-7, F-8, G-3, H-10, I-4, J-1

CHAPTER 5: HAIR CARE PRODUCTS

Section 1: Hair Anatomy and Physiology

Multiple Choice Questions: 1-c, 2-a, 3-b, 4-d, 5-c, 6-b, 7-a, 8-a, 9-c, 10-b

Fact or Fiction? 1-fiction, 2-fact, 3-fact, 4-fiction

Section 2: Hair Cleansing and Conditioning Products

Multiple Choice Questions: 1-a, 2-b, 3-c, 4-b, 5-d, 6-a, 7-b, 8-b, 9-c, 10-a

Fact or Fiction? 1-fact, 2-fiction, 3-fact, 4-fact

Matching: A-7, B-1, C-9, D-5, E-8, F-4, G-6, H-10, I-2, J-3

Section 3: Hair Styling Products, Hair Straightening Products, and Hair Waving Products

Multiple Choice Questions: 1-c, 2-a, 3-d, 4-d, 5-c, 6-a, 7-d, 8-a, 9-b, 10-c

Fact or Fiction? 1-fiction, 2-fact, 3-fiction, 4-fiction

Matching: A-9, B-5, C-10, D-6, E-8, F-7, G-2, H-3, I-4, J-1

Section 4: Hair Coloring Products

Multiple Choice Questions: 1-a, 2-d, 3-a, 4-c, 5-b, 6-d, 7-d, 8-b

Fact or Fiction? 1-fact, 2-fiction, 3-fact, 4-fiction

Matching: A-2, B-7, C-4, D-6, E-1, F-3, G-5

CHAPTER 6: ORAL AND DENTAL CARE PRODUCTS

Multiple Choice Questions: 1-c, 2-b, 3-b, 4-a, 5-b, 6-d, 7-c, 8-a, 9-d, 10-c, 11-c, 12-a, 13-d, 14-b, 15-b

Drug or Cosmetic in the United States? a-drug, b-cosmetic, c-cosmetic, d-cosmetic

Short Answers: a-pulp, b-eight, c-5.5, d-tartar, e-hydrogen peroxide, f-hydroxyapatite, g-remineralization, h-pea size

Matching: A-14, B-15, C-5, D-11, E-1, F-9, G-3, H-4, I-13, J-10, K-2, L-12, M-6, N-8, O-7

CHAPTER 7: OTHER PRODUCTS

Section 1: Hair Removal Products

Multiple Choice Questions: 1-d, 2-b, 3-c, 4-d, 5-c, 6-a, 7-c

Matching: A-1, B-4, C-6, D-8, E-3, F-5, G-2, H-10, I-9, J-7

Fill in the Blank: 1-oils, 2-astringent, antiseptic, 3-shaving gel, 4-cooling, 5-alkaline, 6-epilation, 7-expel, 8-viscosity, foam stabilization, 9-bleaching

Section 2: Baby Care Products

Multiple Choice Questions: 1-c, 2-c, 3-b, 4-a, 5-c, 6-b, 7-a, 8-c, 9-a

Drug or Cosmetic in the United States? a-drug, b-drug, c-cosmetic, d-cosmetic

Matching: A-3, B-2, C-1, D-7, E-8, F-5, G-9, H-4, I-10, J-6

Section 3: Sunless Tanning Products

Multiple Choice Questions: 1-a, 2-b, 3-b, 4-d, 5-c, 6-b, 7-d, 8-a

Fact or Fiction?: 1-fact, 2-fiction, 3-fact

Section 4: Feminine Hygiene Products

Multiple Choice Questions: 1-b, 2-d, 3-a, 4-d, 5-b, 6-c, 7-d, 8-d

Drug or Cosmetic in the United States? a-drug, b-drug, c-cosmetic, d-cosmetic

Matching: A-2, B-7, C-8, D-3, E-5, F-1, G-9, H-4, I-6, J-10

INDEX

Note: Page numbers followed by *t* refer to *tables*, and those *f* refer to *figures*.

Introduction to Cosmetic Formulation and Technology, First Edition. Gabriella Baki and Kenneth S. Alexander.
© 2015 John Wiley & Sons, Inc. Published 2015 by John Wiley & Sons, Inc.